Case Studies in Forensic Anthropology
Bonified Skeletons

Case Studies in Forensic Anthropology

Bonified Skeletons

Edited by

Heather M. Garvin, PhD, D-ABFA
&
Natalie R. Langley, PhD, D-ABFA

CRC Press
Taylor & Francis Group
Boca Raton London New York

CRC Press is an imprint of the
Taylor & Francis Group, an **informa** business

CRC Press
Taylor & Francis Group
6000 Broken Sound Parkway NW, Suite 300
Boca Raton, FL 33487-2742

© 2020 by Taylor & Francis Group, LLC

CRC Press is an imprint of Taylor & Francis Group, an Informa business
No claim to original U.S. Government works

International Standard Book Number-13: 978-1-138-34765-6 (Hardback)

Library of Congress Cataloging-in-Publication Data

Names: Garvin, Heather M., author. | Langley, Natalie R., author.
Title: Case studies in forensic anthropology : bonified skeletons / Edited
by Heather M. Garvin and Natalie R. Langley.
Description: Boca Raton, FL : CRC Press, Taylor & Francis Group, [2020] |
Includes bibliographical references and index. | Summary: "Through a set
of unique cases presented by a diverse international group of practicing
forensic anthropologists, Case Studies in Forensic Anthropology seeks to
prepare students and professionals for future cases they may confront"
Provided by publisher.
Identifiers: LCCN 2019020891 (print) | LCCN 2019022358 (ebook) | ISBN
9781138347656 (hardback : alk. paper)
Subjects: LCSH: Forensic anthropology--Case studies. | Forensic
osteology--Case studies. | Human remains (Archaeology)--Case studies.
Classification: LCC GN69.8 .C37 2020 (print) | LCC GN69.8 (ebook) | DDC
614/.17--dc23
LC record available at https://lccn.loc.gov/2019020891
LC ebook record available at https://lccn.loc.gov/2019022358

Visit the Taylor & Francis Web site at
http://www.taylorandfrancis.com

and the CRC Press Web site at
http://www.crcpress.com

Contents

SECTION V   OTHER CONSIDERATIONS

Acknowledgments

The editors would like to thank the contributing authors for their efforts and incredible cooperation in putting this book together. We were overwhelmed by how many colleagues were eager to contribute a case study, and appreciate their diligence in submitting thoughtful and timely contributions. We enjoyed reading and editing each chapter and learned more about our field in the process. We are confident that students and practitioners will take something away from each case study, as well. Thank you for making this book a success.

This text includes actual forensic anthropological casework for educational purposes. All efforts were made to ensure materials do not contain identifiable information and publication will not impede any investigations or legal pursuits. Contributors acknowledged that they obtained all necessary permissions to publish the case materials.

In a sense, the true educators in these case studies are the decedents. We express sincere gratitude to them, and our thoughts go out to their friends and loved ones. We hope publishing these case studies not only helps current and future forensic anthropologists grow professionally as practitioners, but also helps the discipline grow by increasing awareness of the field and the diverse challenges it faces. The cases in this text illustrate the importance of continuing scientific advancements, rigorous and innovative methods, and multidisciplinary approaches in bringing justice and closure to those who have lost loved ones.

Editors

Heather M. Garvin, PhD, D-ABFA, began her journey in forensic anthropology as an undergraduate volunteer at the C.A. Pound Human Identification Laboratory at the University of Florida. Graduating with a BA in Anthropology and a BS in Zoology, she then earned an MS in Forensic and Biological Anthropology from Mercyhurst College and completed a PhD in Functional Anatomy and Evolution from Johns Hopkins University School of Medicine. From 2012–2017, she taught undergraduate and graduate students in Forensic Anthropology at Mercyhurst University, and was heavily involved in casework and research. Dr. Garvin became a Diplomate of the American Board of Forensic Anthropology in 2017 and has served on the Editorial Board for the *Journal of Forensic Sciences* since 2015. Her research interests include forensic anthropological methods, human skeletal variation, functional morphology, 3D scanning, and geometric morphometrics. She has more than 30 publications and 50 national presentations related to this research and is a Fellow in the American Academy of Forensic Sciences and a member of the American Association of Physical Anthropology. Dr. Garvin is currently an Associate Professor of Anatomy at Des Moines University, where she consults on forensic anthropology cases for the State of Iowa and continues human skeletal research.

Natalie R. Langley, PhD, D-ABFA, began her training in forensic anthropology as a master's student at Louisiana State University. She earned a BA (1998) and an MA (2001) in anthropology from LSU and PhD (2009) from the University of Tennessee in Knoxville (previous names: Natalie Shirley, Natalie Langley-Shirley). In 2007 the American Academy of Forensic Sciences Forensic Science Foundation awarded her the Emerging Forensic Scientist Award for her doctoral research in skeletal maturation. She also starred as the forensic anthropologist on the 2012 National Geographic Channel television series *The Great American Manhunt*. Dr. Langley became a Diplomate of the American Board of Forensic Anthropology in 2015 and has consulted for the Georgia Bureau of Investigation and the US Department of Justice's International Training Assistance Program in Bogotá, Colombia. She is a Fellow of the American Academy of Forensic Sciences and a member of the American

Association of Anatomists and American Association of Clinical Anatomists. Dr. Langley has authored numerous publications, including two forensic anthropology textbooks and the Data Collection Procedures 2.0 laboratory manual. Her research interests include modern skeletal biology, forensic anthropology methods, clinical anatomy, and medical education. Dr. Langley is currently Associate Professor at Mayo Clinic Arizona in the Department of Laboratory Medicine and Pathology, Division of Anatomic Pathology.

Contributors

*Eric J. Bartelink, PhD, D-ABFA
Department of Anthropology
California State University–Chico
Chico, California

*Angela Berg, MA
Oklahoma Office of the Chief Medical
    Examiner
Eastern Office
Tulsa, Oklahoma

*Gregory E. Berg, PhD, D-ABFA
Defense POW/MIA Accounting Agency,
    Laboratory
Joint Base Pearl Harbor–Hickam
Hawaii

Carlos Martin-Beristain, MD
Department of Forensic Anthropology
University of Basque Country
Donostia-San Sebastián, Spain

Jonathan D. Bethard, PhD, D-ABFA
Department of Anthropology
University of South Florida
Tampa, Florida

Cate E. Bird, PhD
Forensic Anthropology Consultant
Washington, D.C.

*Soren Blau, PhD
Victorian Institute of Forensic Medicine/
    Department of Forensic Medicine
Monash University
Melbourne, Australia

Katelyn Bolhofner, PhD
School of Mathematical and Natural
    Sciences
Arizona State University
Glendale, Arizona

Kent J. Buehler
Oklahoma Archaeological Survey
    (Emeritus)
Norman, Oklahoma

Jennifer F. Byrnes, PhD
Department of Anthropology
University of Nevada, Las Vegas
Las Vegas, Nevada

Francisca Cardona
Department of Physical Anthropology
Aranzadi Society of Sciences
Gipuzkoa, Spain

Juanjo Marí Casanova
Societat Arqueològica d'Eivissa i
    Formentera
Eivissa, Spain

*Erin N. Chapman, PhD
Erie County Medical Examiner's Office
Buffalo, New York

Luz Elena Cifuentes, MSc (candidate)
Forensic Biology Group
National Institute of Legal Medicine and
    Forensic Sciences
Bogotá D.C., Colombia

---

* Denotes first authors

Cristina Cordeiro, MSc, MD
National Institute of Legal Medicine and
    Forensic Sciences
Coimbra, Portugal

and

Faculty of Medicine
University of Coimbra
Coimbra, Portugal

*Eugénia Cunha, PhD
National Institute of Legal Medicine and
    Forensic Sciences
Lisbon, Portugal

and

Laboratory of Forensic Anthropology,
    Department of Life Sciences, Center for
    Functional Ecology
University of Coimbra
Coimbra, Portugal

Detective Frank Di Modica (ret)
Homicide Unit/Crimes Against Children
    Unit
Phoenix Police Department
Phoenix, Arizona

Dennis C. Dirkmaat, PhD, D-ABFA
The Department of Applied Forensic
    Sciences
Mercyhurst University
Erie, Pennsylvania

Rhian R. Dunn, MS
The Department of Applied Forensic
    Sciences
Mercyhurst University
Erie, Pennsylvania

Kelley S. Esh, MA
Defense POW/MIA Accounting Agency,
    Laboratory
Joint Base Pearl Harbor–Hickam
Hawaii

*Francisco Etxeberria, MD, PhD
Department of Forensic Anthropology
University of Basque Country
Donostia–San Sebastián, Spain

Maria Teresa Ferreira, PhD
Laboratory of Forensic Anthropology
Department of Life Sciences, Center for
    Functional Ecology
University of Coimbra
Coimbra, Portugal

*Laura C. Fulginiti, PhD, D-ABFA
Maricopa County Office of the Medical
    Examiner
Phoenix, Arizona

Jan C. Garavaglia, MD
Forensic Pathology Consultant

*Almudena García-Rubio, PhD
Department of Physical Anthropology
Aranzadi Society of Sciences
Donostia–San Sebastián, Spain

and

Societat Arqueològica d'Eivissa i
    Formentera
Eivissa, Spain

Glenda Graziani
Societat Arqueològica d'Eivissa i
    Formentera
Eivissa, Spain

and

Departament de Ciències de l'Antiguitat i
    de l'Edat Mitjana
Universitat Autònoma de Barcelona
Barcelona, Spain

Ashley Green, MA, BSN, RN
Department of Anthropology
University of Central Florida
Orlando, Florida

---

* Denotes first authors

Marie H. Hansen, MD
Districts 9 and 25 Medical Examiner's
    Office
Orlando, Florida

*Kristen Hartnett-McCann, PhD, D-ABFA
Office of the Chief Medical Examiner
Farmington, Connecticut

*Joseph T. Hefner, PhD, D-ABFA
Department of Anthropology
Michigan State University
East Lansing, Michigan

Lourdes Herrasti, MA
Department of Physical Anthropology
Aranzadi Society of Sciences
Donostia–San Sebastián, Spain

Rimantas Jankauskas, MD, PhD
Faculty of Medicine
Vilnius University
Vilnius, Lithuania

*Ashley E. Kendell, PhD
Department of Anthropology
California State University–Chico
Chico, California

*Alexandra R. Klales, PhD, D-ABFA
Forensic Anthropology Program
Washburn University
Topeka, Kansas

Ruth E. Kohlmeier, MD
Suffolk County Office of the Medical
    Examiner
Hauppauge, New York

Ericka N. L' Abbé, PhD, D-ABFA
Department of Anatomy, Forensic
    Anthropology Research Center
University of Pretoria
Pretoria, South Africa

Mark M. LeVaughn, MD
Office of the State Medical Examiner
Pearl, Mississippi

Maria Alexandra Lopez-Cerquera, PhD
Department of Natural and Behavioral
    Sciences
Pellissippi State Community College
Knoxville, Tennessee

*Murray K. Marks, PhD, D-ABFA
Department of General Dentistry
University of Tennessee Health Science
    Center
Graduate School of Medicine
Knoxville, Tennessee

and

Regional Forensic Center
Knoxville, Tennessee

Nicholas Márquez-Grant, PhD
Cranfield Forensic Institute
Cranfield University
Defense Academy of the United Kingdom
Cranfield, United Kingdom

and

Societat Arqueològica d'Eivissa i
    Formentera
Eivissa, Spain

*Diana L. Messer, MS
Department of Applied Forensic Sciences
Mercyhurst University
Erie, Pennsylvania

Darinka Mileusnic-Polchan, MD, PhD
Knox County Regional Forensic Center
Knoxville, Tennessee

and

Department of Pathology
University of Tennessee Graduate School
    of Medicine
Knoxville, Tennessee

---

* Denotes first authors

Sergi Moreno
Societat Arqueològica d'Eivissa i
    Formentera
Eivissa, Spain

Turhon A. Murad, PhD, D-ABFA
(Deceased)
Department of Anthropology
California State University–Chico
Chico, California

Ronald A. Murdock II, MS
Forensics Unit, Orange County Sheriff's
    Office
Orlando, Florida

Owen L. O'Leary, PhD
Defense POW/MIA Accounting Agency,
    Laboratory
Joint Base Pearl Harbor–Hickam
Hawaii

*Andrea M. Ost, MS
The Department of Applied Forensic
    Sciences
Mercyhurst University
Erie, Pennsylvania

*James T. Pokines, PhD, D-ABFA
Office of the Chief Medical Examiner
Boston, Massachusetts

and

Forensic Anthropology Program,
    Department of Anatomy and
    Neurobiology
Boston University School of Medicine
Boston, Massachusetts

David Ranson, DMJ(Path), LLB
Victorian Institute of Forensic Medicine/
    Department of Forensic Medicine
Monash University
Melbourne, Australia

Katie M. Rubin, PhD
Department of Anthropology
University of Florida
Gainesville, Florida

*César Sanabria-Medina, PhD,
DLAF-006
Biomedical Sciences Research Group,
    School of Medicine
Universidad Antonio Nariño
Bogotá D.C., Colombia

and

Branch of Scientific Research
National Institute of Legal Medicine and
    Forensic Sciences
Bogotá D.C., Colombia

Andrew Seidel, PhD
Center for Bioarchaeological Research,
    School of Human Evolution and Social
    Change
Arizona State University
Tempe, Arizona

*John J. Schultz, PhD
Department of Anthropology
University of Central Florida
Orlando, Florida

and

National Center for Forensic Science
Orlando, Florida

Joshua D. Stephany, MD
Districts 9 and 25 Medical Examiner's
    Office
Orlando, Florida

*Michala K. Stock, PhD
Department of Exercise Science
High Point University
High Point, North Carolina

* Denotes first authors

Pau Sureda, PhD
Instituto de Ciencias del Patrimonio
(Incipit)–Consejo Superior de
Investigaciones Científicas (CSIC)
Santiago de Compostela, Spain

and

McDonald Institute for Archaeological
Research
Cambridge, United Kingdom

and

Homerton College
University of Cambridge
Cambridge, United Kingdom

and

Societat Arqueològica d'Eivissa i
Formentera
Eivissa, Spain

*Steven A. Symes, PhD, D-ABFA
Office of the State Medical Examiner
Pearl, Mississippi

and

Department of Anatomy,
Forensic Anthropology
Research Center
University of Pretoria
Pretoria, South Africa

***Lindsay H. Trammell, PhD, D-ABMDI,
D-ABFA**
St. Louis County Office of the Medical
Examiner
St. Louis, Missouri

Duarte Nuno Vieira, PhD, MD
Department of Forensic Medicine, Ethics
and Medical Law
Faculty of Medicine, University of
Coimbra
Coimbra, Portugal

*Jennifer Vollner, PhD
Pima County Office of the Medical
Examiner
Tucson, Arizona

*Daniel J. Wescott, PhD
Department of Anthropology, Forensic
Anthropology Center at Texas State
Texas State University
San Marcos, Texas

Carlos Zambrano, PhD
Oklahoma Office of the Chief Medical
Examiner, Central Office
Oklahoma City, Oklahoma

Jorge Andrés Franco Zuluaga, MSc
Faculty of Medicine
Pontificia Universidad Javeriana
Bogotá D.C., Colombia

---

* Denotes first authors

Introduction

Case studies are becoming overshadowed in the forensic literature by an emphasis on research-based studies, but the educational value of casework and case reports endures as an authentic learning opportunity for practitioners and a means of informing our research. As professionals, we are obliged to share these exceptional learning experiences to foster continuing professional development and the quest for knowledge. Thus was born *Case Studies in Forensic Anthropology: Bonified Skeletons*, an edited volume of previously unpublished forensic anthropology case studies contributed by a diverse set of US and international practitioners. These unique cases provide a valuable opportunity to learn from our experiences and better prepare students and practitioners for future encounters. As a side note, "bonified" is not a misspelling. "Bonify" is an archaic word that means "to convert into good." We felt this pun was appropriate given that forensic anthropologists take something bad (the death of an individual) and do something good with the information by helping to identify them, unravel the circumstances around their death, and bring closure to loved ones.

Every forensic anthropology case is unique, and practitioners are routinely faced with new challenges and unexpected outcomes. Introductory forensic anthropology courses and texts generally present best practice for handling conventional cases. In practice, however, forensic anthropologists frequently must improvise, and all practitioners have encountered cases where they obtained surprising results from laboratory analyses, or the results did not conform to known case information. By the same token, our knowledge is influenced and often limited by our experiences. For example, a forensic anthropologist performing casework at a medical examiner's office in the United States may not understand fully the challenges and resource limitations faced by an international team investigating a potential human rights violation. Likewise, would a forensic anthropologist know how to interpret the taphonomic effects of acid on bones if they had never encountered such a scenario? Most forensic anthropologists would agree that their didactic education provided the foundations of forensic anthropology, but they learned the most from actual case experiences. Indeed, authentic learning from case-based instruction is a teaching method frequently used in forensic anthropology classrooms. The knowledge gained by those students, however, may be limited by the casework received at that specific lab and by the experiences of their mentor(s). This edited volume combines the experiences of a diverse array of forensic anthropologists to expose students and practitioners to cases encountered, methods employed, and conclusions drawn by others in our field. We aim to provide a much-needed awareness about our field at large and allow us to learn from one another.

Case Studies in Forensic Anthropology: Bonified Skeletons is organized broadly into five sections: I. Biological Profile and Positive Identification; II. Forensic Taphonomy; III. Skeletal Trauma; IV. Human Rights and Mass Disasters; V. Other Considerations.

Cases were selected for their application of new resources and methods, the experiential knowledge they offer to readers, and their link to pertinent issues in the field. Section I (Biological Profile and Positive Identification) presents six cases dealing with the challenges of identifying unknown remains. Several cases demonstrate innovative or resourceful means for arriving at an identification (e.g., social media, medical implants, biomechanical analyses of long bones). Others discuss conflicting results from metric and macromorphoscopic data, the challenges of ancestry estimation in an increasingly globalized society, and the value of multidisciplinary collaborations to produce information leading to the identity of the decedent.

The cases in Section II (Forensic Taphonomy) showcase the broad range of taphonomic agents that may complicate forensic analyses and recovery efforts, including fire, water, corrosive agents, and unusual scavengers. In some instances, the remains displayed alterations characteristic of multiple taphonomic agents that had to be sorted to decipher the circumstances leading to the observed condition. These case studies also illustrate failed attempts to obscure and dispose of bodies, the ability of forensic anthropologists to identify signs of trauma despite taphonomic modifications, and how with proper documentation the circumstances around the death can be unraveled.

Section III (Trauma) illustrates the role of the forensic anthropologist in assessing skeletal trauma and contributing information toward the cause and manner of death. The cases emphasize the use of biomechanics and anatomical knowledge for interpreting skeletal trauma (e.g., weapon class, sequencing of injuries, and direction of impact). Several cases also discuss the information forensic anthropologists can contribute to the hypothesized circumstances around the death and testimony corroboration.

Section IV (Human Rights/Mass Disasters) cases were contributed primarily by international authors. They reflect the global scope of forensic anthropology and its unique capacity to address mass burials and mass disasters. Challenges faced during recovery efforts and analyses are highlighted, especially those due to extreme environments, international logistics, and commingled scenarios. The cases also draw attention to forensic anthropology's service to humanity in sorting out the past and bringing justice and closure after many years.

Section V (Other Considerations) further illustrates the diverse competencies required to execute casework successfully, including knowledge of the scientific method; communication, teamwork, and leadership skills; flexibility and innovation; and tacit knowledge of various subject areas (e.g., taphonomy, archaeology, skeletal biology, human variation, and cultural anthropology). One chapter describes the ritualistic use of human remains, another focuses on the need to contextualize injury patterns, and a third highlights the importance of inter-agency collaboration in a case involving pack rat nests and multiple aliases. The final chapter describes the typical distribution of forensic cases at one medical examiner's offices, including the commonality of non-forensically relevant remains.

Although each case was attributed to a specific section of the book, readers will find that many cases have relevance to multiple sections, reflective of the complex nature of forensic casework. Despite the diversity of practitioners and case scenarios, several common threads are present throughout the book: the benefits of multidisciplinary collaborations, the value of hypothesis testing, and the importance of avoiding confirmation bias in our casework. At the end of each chapter, the authors summarize the main lessons they took away from the case in a Lessons Learned section. Several Discussion Questions that encourage critical thinking are also included for each chapter, facilitating further conversations and debates.

We recognize differing professional opinions about preferred methodologies and interpretations and hope these case studies spur rich discussions among colleagues, enabling our collective growth as a discipline. As these cases illustrate, forensic anthropological casework is complex, with each case presenting its own challenges. The competencies required to perform forensic anthropology casework entail multifaceted education, training, and experience that combines classroom and laboratory learning with authentic case-based learning experiences. Sharing case experiences among the forensic anthropology community guarantees the availability of these types of learning experiences, and this shared knowledge is vital to the continued success of our field.

Estimation of the Biological Profile and Positive Identification

Death Along the Tracks: The Role of Forensic Anthropology and Social Media in a Homicide Investigation

Eric J. Bartelink

CONTENTS

INTRODUCTION

Forensic anthropology casework conducted through the California State University, Chico Human Identification Laboratory routinely involves postmortem examination of skeletonized remains from rural outdoor scenes. Over the past decade, decomposed remains cases have comprised a larger proportion of the caseload. These cases are usually examined by a forensic pathologist prior to the anthropological analysis. This case study highlights the importance of using online searches of antemortem decedent information in the identification process and the role of forensic anthropology in detailing perimortem trauma patterns to corroborate witness statements.

CASE BACKGROUND

In 2013, law enforcement responded to a report of human remains discovered near a California railroad yard frequented by transients. The body was located along a creek bank near the railroad yard and represented an unclothed individual in an advanced stage of decomposition. Following the recovery of the body by sheriff's office personnel, the remains were transferred to the county medical examiner's office for autopsy. Due to the mummified nature of the remains and lack of internal organs, the case was delivered to the Human Identification Laboratory at California State University, Chico for anthropological analysis. The remains consisted of an intact, articulated skeleton encased within mummified skin. The skin was removed following the documentation of a tattoo on the right leg, and final processing was completed using a dermestid beetle colony. Once skeletonized, the bones were soaked in water, dried, and laid out on a table in anatomical position for analysis.

BIOLOGICAL PROFILE

The biological profile assessment was conducted using all available skeletal elements. Sex was determined to be male based on morphological features of the pelvis and skull and postcranial osteometrics. Age-at-death was estimated based on the assessment of the pubic symphysis, auricular surface, right 4th sternal rib, and epiphyseal union. Of note, several epiphyses were in the late stages of fusion (e.g., iliac crests, vertebral centra, distal radii and ulnae, proximal humeri and tibiae), whereas the medial clavicle was unfused. The permanent teeth were also fully erupted with two exceptions: the left central mandibular incisor was represented by a partial deciduous tooth and all third molars appeared to have been recently extracted. Collectively, these indicators suggested an age interval of approximately 17–21 years. Ancestry was assessed from metric and morphoscopic skull traits. Fordisc 3.0 classified the skull as a white male (posterior probability = 0.940; typicality probability = 0.639) based on osteometric data (Jantz & Ousley, 2005). The morphoscopic results also classified the decedent as white (European ancestry). However, the presence of shovel-shaped maxillary incisors suggested that Asian ancestry should not be excluded as a possibility. Stature was estimated at 5 feet, 6 ± 3.9 inches (95% prediction interval) using a 20th-century formula for white males. In sum, the decedent appeared to be a male of either European or Asian ancestry, between 17 and 21 years of age, with a stature of 5 feet, 6 ± 3.9 inches.

ANTEMORTEM CHARACTERISTICS

Prior to processing the remains to remove the mummified skin, identifying features were assessed visually. The examination revealed the presence of a tattoo consisting of two parallel black lines running transversely along the upper right thigh (see Figure 1.1). Because no other identifying soft tissue features were noted, the remaining skin was removed, and the processing was completed using a dermestid beetle colony. No evidence of antemortem trauma was noted on the skeleton. However, evidence of remodeling of the tooth sockets of the third molars suggests that these teeth were recently extracted prior to death. Occlusal surface restorations were noted on ten teeth, and buccal restorations were noted on five teeth. In addition, the left central mandibular incisor was represented by a partial deciduous tooth with evidence of an antemortem fracture of the occlusal surface (see Figure 1.2).

FIGURE 1.1 Photograph of the mummified skin of the right leg showing black linear tattoos.

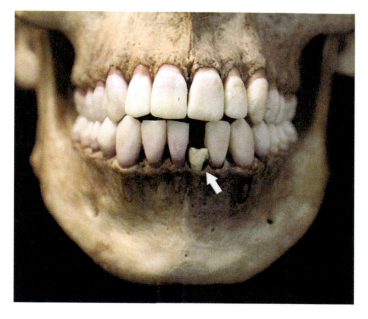

FIGURE 1.2 Facial view of the dentition showing the retained deciduous left mandibular central incisor.

PERIMORTEM TRAUMA

All skeletal elements were examined for evidence of perimortem trauma and taphonomic alterations. All taphonomic alterations were related to autopsy damage (removal of the cranial vault, cuts through several sternal rib ends, and cuts through the lunate, scaphoid, a proximal phalanx, and an intermediate hand phalanx to remove part of the right hand for fingerprinting). No other postmortem alterations (e.g., carnivore scavenging) were noted on the remains.

Perimortem trauma is defined here as fractures of fresh bone that may be associated with the death event. Perimortem blunt force trauma was observed on the left maxilla and nasal bones, the sternum, the left thorax, and the lumbar spine. These fractures show evidence of displacement, as well as a wet bone response, consistent with perimortem trauma (Galloway, 1999; Sauer, 1998).

Skull

The facial skeleton showed probable evidence of perimortem trauma, including fractures of the inferior margin of the left and right nasal bones and the medial margin of the frontal process of the left maxilla (see Figure 1.3). Although these bones are thin and fragile and often show postmortem damage, mummified skin protected the midface region.

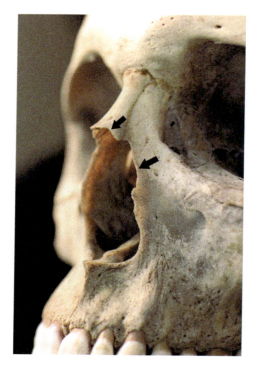

FIGURE 1.3 Oblique view of the left side of the skull showing probable perimortem fractures to the inferior margin of the left and right nasal bones and the medial aspect of the frontal process of the left maxilla.

Thorax

The sternal body shows evidence of two perimortem transverse fractures (see Figure 1.4). The first fracture is located on the superior end and traverses the sternal body. This fracture only affects the anterior surface. The second fracture is located on the inferior end of the sternum near the junction with the xiphoid process and affects both the anterior and posterior surfaces. One fracture line radiates superolaterally toward the sixth costal notch on the left side of the sternal body, and a second fracture line radiates toward the seventh costal notch on the right side. Sternal fractures often result from direct violence to the chest area (Knight, 1991), but can occur in automobile accidents involving unrestrained drivers due to an impact of the chest with the steering wheel (Galloway, 1999).

Seven perimortem rib fractures were identified on four ribs from the left thorax. Two incomplete transverse fractures were identified on the shafts of left ribs 8–10 (see Figure 1.5). In addition, left rib 11 showed evidence of a single incomplete transverse shaft fracture. These fractures all show evidence of failure under tension on their internal surfaces, consistent with blunt force directed at the left side of the thorax. When these ribs are articulated, the fractures align between consecutive ribs. These fractures may have resulted from a single traumatic impact distributed over a broad area of the left thorax or may have resulted from multiple impacts. Rib fractures can result from accidents, falls, and direct blows to the thorax, with transverse fractures commonly resulting from direct blows to the chest (Galloway, 1999). When considered together, the fractures of the sternum and left thorax may be consistent with a "flail chest," where multiple ribs are fractured in two or more places, resulting in movement of the flail (i.e., fractured) section

FIGURE 1.4 Anterior view of the sternum showing perimortem fractures to the superior and inferior aspects of the sternal body.

FIGURE 1.5   Superior view of left ribs nos. 8–11 showing evidence of perimortem fractures to the rib shafts.

in the opposite direction than the remaining chest wall during respiration. Flail chest is a serious medical condition often associated with bruising of the lungs, cardiac injuries, labored breathing (Knight, 1991), and also puncture of the parietal pleura, which can result in pneumothorax (Ciraulo et al., 1994). These injuries commonly result from a direct blow to the chest, including cases where an individual is kicked and stomped, and in motor vehicle accidents (Galloway, 1999).

Vertebral Column

Perimortem fractures were identified on three lumbar vertebrae (L1–L3). All three vertebrae showed complete fractures through the left transverse processes, consistent with one or more blunt force impacts to the left side of the body (see Figure 1.6). Fractures of the transverse processes are common and often result from direct blows to the lower back and from motor vehicle accidents (Galloway, 1999). One retrospective clinical study found that nearly half of the patients who presented with transverse process fractures of lumbar vertebrae had a significantly higher chance of abdominal organ injuries than patients who presented with fractures to the vertebral body, pedicle, or spinous process (Miller et al., 2000). These findings suggest that transverse process fractures to the lumbar spine may be more serious than previously recognized in the clinical literature.

CASE RESOLUTION

During the early stages of the skeletal analysis, the author conducted online searches to identify missing persons matching the decedent's biological profile. At this point in the

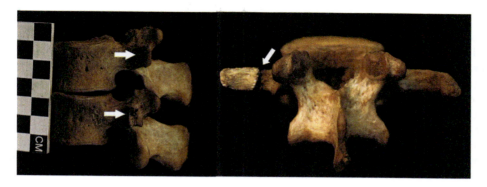

FIGURE 1.6   Left lateral view of L1 and L2 (left image) and posterior view of L3 (right image) showing evidence of perimortem fractures to the transverse processes.

investigation, law enforcement had no leads regarding the identity of the decedent or the cause and manner of death. A search of NamUs and the Doe Network failed to identify any missing persons consistent with the decedent. However, the California Department of Justice's Missing Persons database contained a missing persons report consistent with the decedent's biological profile. An internet search of the missing person's name revealed additional identifying information on Instagram. Two postings relayed that a group of people was searching for their friend, a 19-year-old male from southern California last known to be in the area of the railroad yard two months prior. The postings also mentioned that the missing person had a partial lower incisor and linear tattoos on his thighs. Noting several points of similarity between the missing persons' postings and the decedent, the author contacted the sheriff's office to provide this information as a possible lead. The sheriff's office obtained dental records for the missing person, and a forensic odontologist was able to make a positive identification of the decedent.

The decedent was identified as a 19-year-old male of Asian ancestry, with a reported stature of 5 feet, 6 inches. This was consistent with the biological profile, which suggested a male of either European or Asian ancestry, between 17 and 21 years of age, with a stature of 5 feet, 6 ± 3.9 inches. With the decedent identified, investigators began the lengthy search for the person or persons responsible for the homicide.

The decedent, a college student, reportedly hopped a train to join a group of experienced train riders heading toward the Sacramento Valley. He maintained contact with friends and family every day for approximately two weeks. His last contact was in the vicinity of the railroad yard where his body was found. When a concerned relative called his cell phone, a woman answered and claimed that he had walked away from the group, and she didn't know his whereabouts. Investigators began to piece together the decedent's movements through cell phone records and his social media account. Further digging into social media postings revealed that the decedent had been traveling by train with four experienced train riders. The investigation led to the discovery of the identities of the four individuals (two men and two women), as well as incriminating photographs from their social media posts that placed them at the scene of the homicide. Authorities believed that some photographs posted on social media were taken by the decedent just hours prior to his death. The investigators obtained a court order to place a wiretap on their phones and recorded information that incriminated all four suspects for their involvement in the homicide. Following an intensive manhunt, investigators tracked down the two male suspects out-of-state, where they were arrested during a scheduled train stop and extradited to California. Both female suspects were later also arrested out-of-state and extradited to California.

Details of the homicide emerged after extensive interrogations of the suspects and interviews of other informants. The victim hopped the train with the four suspects with the intention of experiencing "riding the rails" during spring break; he planned on returning home after two weeks. His last communication with friends and family indicated that he was planning to return home shortly. While at the railroad yard, the four suspects and victim began to drink heavily around a bonfire. At some point in the evening, the victim expressed his desire to leave the group and return home. The suspects began to bully the victim, claiming that he was weak and needed to be taught a lesson. A physical altercation ensued between the victim and the first female suspect, who allegedly bit him. All four suspects proceeded to punch and kick the victim repeatedly. Although still breathing, the victim was badly injured. At some point later that evening, one of the male suspects and the second female suspect dragged the victim away from the camp. The second female suspect allegedly stated that her "boots" were "going to end this" and proceeded to kick and stomp the victim to death. The suspects removed the victim's clothing and personal effects and burned them in the bonfire. They then moved the victim's body near the edge of the creek bank and covered it with brush before catching a train out of the area.

During an interrogation with the first female suspect, she attempted to place blame on another train rider who had recently died in a train accident. This angered other members of the train rider community, many of whom agreed to speak with authorities to provide corroborating evidence against the four suspects. As the case began to move toward trial, all four suspects agreed to take a plea deal in exchange for information regarding their involvement in the homicide. Two of the suspects plead no contest to voluntary manslaughter and received 11 years in state prison. Another suspect plead no contest to "assault with a deadly weapon with force likely to cause great bodily injury," and a special allegation of "causing great bodily injury," resulting in a seven-year sentence. The fourth suspect, the female who kicked and stomped the victim to death, plead no contest to second-degree murder and was sentenced to 15 years to life in prison.

LESSONS LEARNED

- This case study highlights the value of online searching of missing persons records using biological data derived from the skeletal analysis. In this case, the decedent's biological profile in conjunction with information on the retained deciduous tooth and the tattoos were instrumental in suggesting a possible match to investigators. This lead was critical in the case, because once the decedent was identified, investigators were able to move forward with the identification of possible suspects.
- The trauma analysis also proved critical in this case. The investigators were able to compare notes between the fractures described in the forensic anthropology report with the information provided by each of the four suspects and other informants. The facial, thoracic, and lumbar fractures documented in the forensic analysis were all consistent with the record of events reported by the suspects as part of their plea deals.

DISCUSSION QUESTIONS

1.1 In the current case, how was social media valuable in not only the identification of the decedent but also in the investigation of the homicide? If it weren't for social media, do you think the case would be closed?

1.2 What are possible limitations of perimortem trauma analysis when human remains are discovered weeks to months following death?

1.3 In this particular case, the forensic anthropologist did not have to testify in court given that the suspects all took plea deals. If there had been a court trial, however, what types of questions might you expect the prosecution and defense to ask the anthropologist?

REFERENCES

Ciraulo, D. L., Elliott, D., Mitchell, K. A., & Rodriguez, A. (1994). Flail chest as a marker for significant injuries. *Journal of the American College of Surgeons, 178*(5), 466–470.

Galloway, A. (Ed.). (1999). *Broken bones: Anthropological analysis of blunt force trauma*. Springfield, IL: Charles C. Thomas.

Jantz, R. L., & Ousley, S. D. (2005). FORDISC 3.0: Personal computer forensic discriminant functions. Knoxville, TN: University of Tennessee.

Knight, B. (1991). *Forensic pathology*. Oxford, UK: Oxford University Press.

Miller, C. D., Blyth, P., & Civil, I. D. S. (2000). Lumbar transverse process fractures—A sentinel marker of abdominal organ injuries. *Injury, 31*(10), 773–776.

Sauer, N. (1998). The timing of injuries and manner of death: Distinguishing among antemortem, perimortem, and postmortem trauma. In K. Reichs (Ed.), *Forensic osteology: Advances in the identification of human remains* (pp. 321–332). Springfield, IL: Charles C. Thomas.

CHAPTER 2

The Skull in Concrete: A Multidisciplinary Approach to Identification

Kristen Hartnett-McCann and Ruth E. Kohlmeier

CONTENTS

INTRODUCTION

In 2016, a bicyclist went to look for a place to relieve himself in the woods off of a highway in New York when he discovered a human skull partially embedded in concrete. He was intrigued by some antique glassware he saw on the ground and then noticed a concrete mound with what appeared to be a human skull inside. The bicyclist contacted the county law enforcement, who responded to the scene. At the scene, law enforcement confirmed a cement mound with an opening at the top, exposing a human skull partially embedded in cement and lying on its right side (see Figure 2.1). A cervical vertebra was located adjacent to the cement mound, and another cervical vertebra was resting on top of the skull. A search of the adjacent wooded area by law enforcement detected no additional bones or evidence. The concrete mound and associated bones were transported to the county medical examiner's office where they were documented, photographed, and x-rayed. The medical examiner assigned to the case recognized a multidisciplinary analytical approach would be best, given the unusual circumstances. Thus, the medical examiner enlisted the expertise of a forensic anthropologist, forensic odontologist, forensic artist, trace evidence scientist, and a DNA specialist to maximize the information that could be gleaned from the remains and concrete.

FIGURE 2.1   Concrete mound with embedded human skull found near a highway in New York.

FORENSIC ANTHROPOLOGICAL ANALYSIS

After the initial examination by the medical examiner, the forensic anthropologist carefully extracted the skull from the cement mound using a hammer and a flat-head screwdriver as a chisel. Small sections of the upper portion of the cement mound were chiseled away until the skull could be lifted out. The remains consisted of one nearly complete human skull (see Figure 2.2), the body and right greater horn of the hyoid, and three cervical vertebrae (C2, C3, and C4). The teeth were in good condition, and several restorations were observed. The remains were skeletonized, dry with no soft tissue, insects,

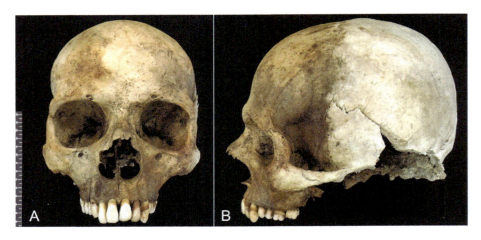

FIGURE 2.2   Anterior (A) and left lateral (B) views of the cranium present for analysis. The mandible is present but not shown here.

or odor of decomposition, and exhibited brown staining as well as sun bleaching. A small amount of scalp hair was found with the cranium, also embedded in the concrete. Overall, the remains exhibited taphonomic changes consistent with internment in concrete (while still fleshed) for an extended period of time – at least two years but likely longer.

The forensic anthropologist had a difficult time estimating ancestry because the cranium exhibited features consistent with European (White), Asian, and Hispanic ancestries (Hefner, 2009; Rhine, 1990), but did not seem to fall clearly into one group. European (White) traits included receding zygomatic bones, projecting anterior nasal spine, and no alveolar prognathism. Asian traits included a moderately wide nasal aperture with a slightly guttered nasal sill and a round cranium (see Figure 2.2). Metric discriminant function analyses were also performed using Fordisc 3.1 (Jantz & Ousley, 2005). When tested against all males in Fordisc 3.1, this individual grouped with both Asian and European (White) individuals but was not very typical of either group.

Forensic stature was not estimated because the postcranial limb elements were absent. Sex was estimated using the morphological features of the cranium (see Figure 2.2): the supraorbital tori were large, the mastoid processes were large, the supraorbital margins were blunt and thick, the nuchal muscle attachments were large, and the mental eminence on the mandible was large and square (Bass, 2005; Buikstra & Ubelaker, 1994). In addition, a logistic regression discriminant analysis of the morphology of the skull yielded a determination of male (Walker, 2008), and a metric analysis of the cranium using Fordisc 3.1 (Jantz & Ousley, 2005) was also consistent with a male individual. The broad age interval was based on the cranial sutures (which appeared mostly open), and the presence of a fully erupted mandibular left third molar.

The forensic anthropologist observed both antemortem and perimortem trauma on the remains present for analysis (see Figure 2.3). Antemortem (healed) fractures were visible on the nasal bones. A combination of perimortem sharp and blunt force trauma was observed on the right parietal, temporal, and frontal bones (see Figure 2.3). The sharp force defect had a sharp superior margin as well as associated concentric, depressed, and radiating fractures. Hair was embedded in the large concentric and depressed fracture on the parietal bone. A fracture was also located on the right maxilla, medial to the zygomaticomaxillary suture.

Sharp force defects were observed on the squamous and basilar portions of the occipital bone, on the right mastoid process, and on the ascending ramus of the right side of the mandible. A portion of the occipital was missing, and concrete was present in the defects. These defects may be perimortem sharp force trauma to the back of the head but could also be associated with postmortem removal of the head at the neck. Furthermore, sharp force defects were noted on the superior and inferior aspects of the body, spinous process, and inferior articular processes of the fourth cervical vertebra. These defects are likely associated with the removal of the head through transection with a sharp instrument between C4 and C5.

IDENTIFICATION

A forensic odontologist performed a dental examination with full mouth digital X-rays and charting. Teeth #1, 16, and 29 were missing. Silver amalgams were identified on the occlusal surfaces of teeth #2, 3, and 30, and on the occlusal and lingual surfaces of tooth #31. Tooth #32 was unerupted and horizontally impacted. Unfortunately, a manual

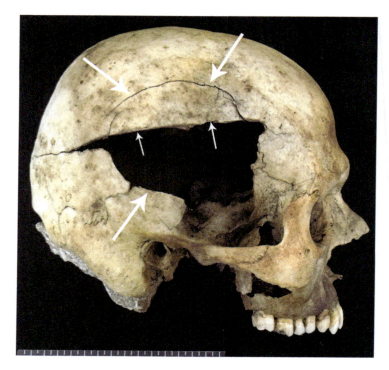

FIGURE 2.3   Right lateral aspect of the cranium. A perimortem sharp force defect with associated radiating, concentric, and depressed fractures is present on the right parietal, temporal, and frontal bones. The concentric fractures are indicated by the large white arrows and the sharp margin of the defect is indicated by the smaller white arrows.

review of possible matching cases in NamUs was unsuccessful; no antemortem dental radiographs or records matched the postmortem radiographs of the unknown decedent.

Hair was attached to the skull and the cement where the skull had been partially embedded. Hair samples were submitted to the Trace Evidence division of the County Crime Laboratory. Microscopic examination of a representative sample revealed light brown, brown, and dark brown human head hair fragments exhibiting European (White) racial characteristics. The hair fragments were unsuitable for microscopic comparison and nuclear DNA analysis but appeared suitable for mitochondrial DNA analysis.

A portion of the mandible and one tooth were submitted to the Forensic Biology division of the County Crime Laboratory for nuclear DNA analysis. DNA from the tooth was extracted using the EZ1 DNA Investigator Kit. Quantitation of the DNA extracts was performed using the Quantifiler Duo Kit. The polymerase chain reaction amplification of short tandem repeats was performed using the Identifiler Plus PCR Amplification Kit, and the amplified DNA was analyzed with capillary electrophoresis. The bone sample was extracted and quantitated but no further analysis was conducted on this sample because positive results were already obtained from the tooth. The DNA profile from the tooth was entered into the local DNA database and submitted to the state and national DNA databases. There was a match between the tooth and another sample in the local CODIS database, which happened to be bone submitted from a 2009 case in the same county involving dismembered human remains discovered in a landfill. The 2009 case had been previously positively identified via DNA from a toothbrush of a 36-year-old male from Pakistan who had been missing since 2007.

The medical examiner reviewed the autopsy report, photographs, investigation notes, and anthropology report from the 2009 case, illustrating how clear and accurate documentation is important because it may be relevant to cases down the road. According to the 2009 documentation, the decomposing and dismembered remains were discovered in a New York landfill by sanitation workers. This particular landfill was very busy and received construction trash from multiple sites, including New York City. The remains included only the mid-cervical neck, torso, and portions of the upper and lower extremities. Clothing was present, and the remains were enclosed in black plastic bags with silver tape. The medical examiner described the remains as an adult European (White) male, based on the soft tissue still present. A forensic anthropologist determined the age was between 30 and 50 years based on the examination of the sternal rib ends, pubic symphyses, and 6th rib histology. Sharp and blunt-force trauma consistent with postmortem dismemberment was observed on two mid-cervical vertebrae, at the distal left humerus and olecranon process of the left ulna, distal right humerus, and midshaft of the right and left femora. The skeletal evidence indicated that the head, forearms, and lower limbs at mid-thigh were removed using a sharp implement, such as a knife or other beveled instrument, in a hacking or chopping motion. The dismemberment trauma observed on the vertebrae in the 2016 case was consistent with the dismemberment trauma described on the vertebrae of the 2009 case.

The medical examiner ruled the cause of death in the 2009 case as "Homicidal Violence, Type Undetermined," and the manner of death a "Homicide." At the time, a segment of femur was submitted for DNA analysis. A local CODIS database (LDIS) search yielded a match between the mitochondrial DNA (mtDNA) sequences and nuclear DNA profiles generated from the femur of the 2009 unidentified male to that of a toothbrush used by a missing person. Coincidentally, the forensic photographer documenting the skull in 2016 vividly remembered the 2009 torso because he was interviewing for his position the very day it came in. In 2016, he still remembered that the 2009 dismemberment was missing the skull and was the first to suggest they might be from the same person! To date, the right and left forearms and hands, as well as the left and right legs from mid-thigh to the feet of this individual are unaccounted for.

After removal of the skull from the concrete, a facial impression was clearly visible in the cement. Under the forensic anthropologist's supervision, a silicone material (*Dragon Skin*® 10 MED) was used to create a mold of the impression inside the concrete (https://www.smooth-on.com/product-line/dragon-skin/). Dragon Skin® is a high-performance silicone rubber that can be used for a variety of applications ranging from movie special effects and skin effects to medical prosthetics, orthopedics, and cushioning applications. It is clear (pigment can be added for a desired color), can stretch and spring back to its shape without distortion, and is relatively inexpensive. The Dragon Skin® comes in two thick liquid parts (parts A and B), which when mixed in equal parts together will cure chemically and solidify at room temperature. No measuring, calculations, or scales are needed. After mixing parts A and B, the forensic anthropologist poured the mixture into the concrete mound where the facial impressions were observed. Unfortunately, during this first attempt of making a mold of the inside of the concrete, small holes in the base of the concrete caused the Dragon Skin® to leak out. Although the first attempt was not successful, the silicone from the first pour sealed all of the holes. The second attempt yielded a detailed 3D Dragon Skin® rendering of the individual's face and injuries after death, when the head was placed into the wet concrete (see Figure 2.4). The structure

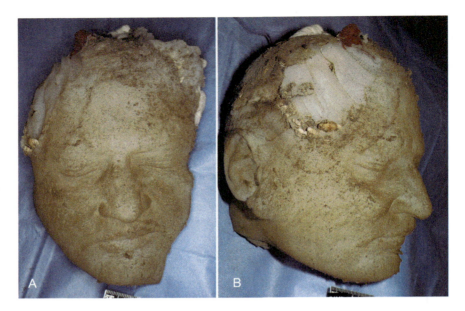

FIGURE 2.4   Anterior (A) and right lateral (B) views of the silicone mold created from the concrete.

of the face was clear, and details such as lack of facial hair were noted (i.e., a beard or mustache, which cannot be deduced from the skeleton). The eyes and lips were closed; no teeth were visible. The trauma to the cranium was also visible in the mold, as well as possible swelling/injuries to the soft tissue of the face. No pigment had been added to the silicone upon mixing, so soil was rubbed onto the final mold to provide a color contrast which made features more visible.

A forensic artist was able to create a two-dimensional forensic skull reconstruction in the blind based on the examination of the skull to corroborate identification (see Figure 2.5). A forensic skull reconstruction is an artistic and scientific approach to helping aid in the identification of human remains. Through the use of tissue depth markers, which are set in 21 anthropological landmarks, the face is built up to resemble what he/ she may have looked like in life. The tissue depth markers are erasers cut by hand and glued onto the skull using Duco Cement. The first ten markers run down the middle of the face from the forehead to the chin. Markers 11–21 contain two sets, for either side of the face. These tissue depth markers represent the thickness of the muscles, tendons, and skin surface and act as a guideline for the topographic drawing. Once the tissue depth markers are glued on and set, a photograph is taken using a 1:1 scale, which is life size. A clear vellum paper is overlaid, and a drawing is done based on the photographed skull with adhering tissue depth markers underneath. This 2D technique was pioneered by Karen T. Taylor (Taylor, 2000).

Overall, the multidisciplinary team of forensic scientists who worked on the skull in concrete were able to obtain as much information as possible from the skeletonized and fragmented remains. Positive identification was made ultimately with DNA technology, but other modalities such as dental evidence, the biological profile, a 3D silicone cast of a facial impression left in concrete, and the forensic skull reconstruction provided additional support to the DNA identification in the event that it was

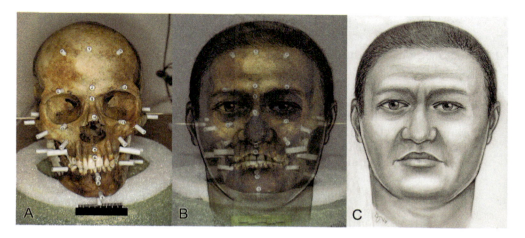

FIGURE 2.5   The soft tissue depth markers on the skull placed by the forensic artist (A), a 2D drawing created by the forensic artist based on the tissue depth markers and the appearance of the face from the mold (B), and an overlay of the drawing and skull with tissue depth markers (C).

questioned in a court of law. The skeletal trauma on the cervical vertebrae was consistent with dismemberment. Sharp and blunt force trauma were caused by a sharp or beveled instrument using a hacking or chopping motion. Unfortunately, although this case is a homicide and the perpetrator is known, he will not be brought to justice unless he returns to the United States.

LESSONS LEARNED

- A multidisciplinary approach to this case produced more information regarding the decedent and cause/manner of death than a single-field approach.
- Ancestry estimation using morphology and metrics can be challenging and sometimes contradictory; understanding the limits of anthropology methods, techniques, and databases for estimating ancestry is imperative. The morphological ancestry estimation was puzzling to the forensic anthropologist in this case because of the mix of morphological features and the low Fordisc typicalities. Low typicality values indicate that the reference sample for the unknown may not be in included in the comparative database. The decedent was from Pakistan, a population which is not represented in the Fordisc reference sample. Furthermore, individuals from Pakistan and other surrounding areas have morphological cranial features that do not always fall neatly into one of the three major ancestral designations (European [White], African [Black], and Asian).
- A mold-making silicone material, such as Dragon Skin®, can be used to create 3D molds from impressions in concrete. The 3D molds can provide additional evidence in a case. A forensic artist also produced a 2D facial reproduction. While this case was identified via DNA, the facial mold and reconstruction would have been helpful in the absence of other available methods of positive identificastion and provided an additional line of support for the DNA identification.

DISCUSSION QUESTIONS

2.1 What methods of identification were discussed in this case study and how were they utilized?

2.2 How does this case study illustrate the challenges with ancestry estimation in forensic anthropology, as well as the concept of race in general?

2.3 Discuss the role that the detailed documentation of the 2009 case played in linking it with the 2016 remains. If DNA had not been taken from the 2009 remains, would conclusive linking of the two cases have been possible? What other information could forensic anthropologists use to illustrate an association between the two cases?

ACKNOWLEDGMENTS

The authors would like to thank the following individuals for their contributions to this interesting case: Dr. Michael Caplan, Detective Alfred Ciccotto, Robert Baumann, Joshua Denenberg, Danielle Gruttadaurio, Dr. David Lynn, Dr. Lillian Nawrocki, and Clyde Wells.

REFERENCES

Bass, W. M. (2005). *Human osteology*. Columbia, MO: Missouri Archaeological Society.

Buikstra, J. E., & Ubelaker, D. H. (1994). *Standards for data collection from human skeletal remains. Proceedings of a seminar at the field Museum of Natural History* (Arkansas Archeological Survey Research Series No. 44). Fayetteville, NC: Arkansas Archeological Survey.

Hefner, J. T. (2009). Cranial nonmetric variation and estimating ancestry. *Journal of Forensic Sciences, 54*(5), 985–995.

Jantz, R. L., & Ousley, S. D. (2005). FORDISC 3: Computerized forensic discriminant functions (Version 3.1). Knoxville, TN: The University of Tennessee.

Rhine, S. (1990). Non-metric skull racing. In G. W. Gill, & S. Rhine (Eds.), *Skeletal attribution of race, methods for forensic anthropology* (pp. 9–20). Albuquerque, NM: Maxwell Museum of Anthropology.

Taylor, K. T. (2000). *Forensic art and illustration*. Boca Raton, FL: CRC Press.

Walker, P. L. (2008). Sexing skulls using discriminant function analysis of visually assessed traits. *American Journal of Physical Anthropology, 136*(1), 39–50.

CHAPTER 3

The Use of Medical Implants to Aid in the Identification Process

Jennifer Vollner

CONTENTS

INTRODUCTION

The Pima County Office of the Medical Examiner (PCOME) located in Tucson, Arizona, serves a great portion of southern Arizona. The combination of the desert environment, with high temperatures and arid conditions, along with the clandestine crossing of the US–Mexico border creates a need for forensic anthropologists at the PCOME. Individuals who perish in the Sonoran Desert quickly become visually unrecognizable. Because the volume of unidentified individuals coming into the PCOME is high, procedures have been developed to increase the rate of successful identifications, currently at ~65%.

Since the year 2000, close to 3000 unidentified individuals have been examined at the PCOME. These unidentified cases range from visually recognizable individuals to severely weathered skeletal elements. As the state of decomposition progresses, there are fewer possible identification methods. The PCOME has attempted to increase these possibilities by developing methods for fingerprint rehydration (Hernandez & Hess, 2014), using infrared camera lenses to document tattoos on mummified skin (Cain, Roper, & Atherton, 2016), and rehydrating areas of skin to determine the coloring of tattoos. However, if an individual is fully skeletonized, as is often the case at the PCOME, these methods are not applicable.

Collaboration has been key for the PCOME's high identification rate for the challenging volume of unidentified individuals examined. Collaboration within the office between the pathologists, forensic autopsy technicians, morgue supervisor, investigators, and anthropologists is crucial. However, collaboration with outside institutions and organizations is equally important in the identification process. This process often relies on other entities for fingerprint comparison and DNA analyses, as well as other agencies (e.g., law enforcement or non-profits) that collect missing persons information which can be used to either exclude or fail to exclude someone from further comparisons. This web of resources is used to its fullest extent in order to identify individuals that enter the PCOME without their name. The following case study highlights the use of medical implants in the process of identification at the PCOME.

CASE STUDY

A dry, sun-bleached skull was discovered in a remote desert region in southwest Arizona. This region was a well-known route for undocumented migrants crossing into the United States.

A standard anthropological examination to construct a biological profile was completed and indicated the decedent was likely a male with indications of Amerindian ancestry. The age was conservatively estimated to be 20–50 years old based on cranial sutures (Meindl & Lovejoy, 1985), maxillary sutures (Mann, Jantz, Bass, & Willey, 1991) and dental wear. Unfortunately, this basic biological profile fit the majority of the more than 1000 unidentified decedents who have been examined at the PCOME since the year 2000. The cortical surface of the skull was exfoliated and sun-bleached, indicating that this individual had been exposed to the desert environment for at least a year and up to several years (Galloway, Birkby, Jones, Henry, & Parks, 1989; Trammel, Soler, Milligan, & Reineke, 2014).

However, this individual had two separate surgical plates implanted in the mandible. One implant was a simple band located on the anterior portion of the body just lateral to the mental eminence held in place by three screws as seen in Figure 3.1. The other implant was a more complex strut piece on the left ramus below the socket for the third molar held in place by two screws as seen in Figure 3.2. Both plates had significant reactive bone growth covering portions of the implant. Additionally, the lower border of the mandible had become misaligned when healing. Both nasal bones were fractured and had not been realigned before healing as seen in Figure 3.3. These fractures would have impacted the profile of this individual's nose significantly. Based on the amount of bony overgrowth on the mandibular plates and the state of the healed nasals it was possible both injuries

FIGURE 3.1   Oblique view of the mandible with the thin band in place. Note the bony growth over the lateral portions of the plate.

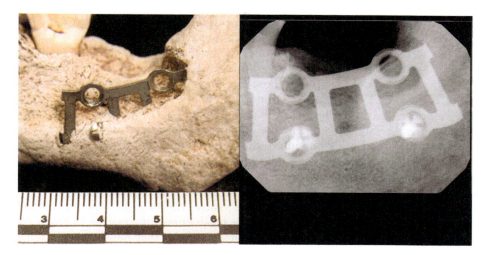

FIGURE 3.2   Lateral photograph of the strut plate on the left and a lateral radiographic image of the plate and two screws on the right. Note the bony overgrowth obscuring the shape of the plate and one of the screws.

FIGURE 3.3   A close-up of the healed fractures in the left and right nasals.

occurred at the same time. All information, including where the skull was originally discovered, was entered into the original NamUs unidentified person profile in hopes that a loved one would recognize the information.

Most surgical supply companies utilize logos to mark their products (Ubelaker & Jacobs, 1995; Wilson, Bethard, & DiGangi, 2011), and these plates were no exception. Unfortunately, most surgical implants do not have a unique identifying number which can be used for personal identifications. Surgical implants instead may have a product number that correlates to the type of plate as well as a lot number, which is a unique number assigned to a set number of plates produced within a specific time frame. PCOME's anthropologists have had positive experiences contacting surgical supply companies for assistance in tracking their products. Based on the combination of numbers most companies can provide a range of years that a product was manufactured and help narrow the window during which that individual may have had surgery. Depending on how many plates were produced these companies might be willing to share the facilities that were sent the particular lot number. Oftentimes, several dozen facilities were sent those implants going back a decade, and in those circumstances, it is often too challenging for medical offices to track down information for those potential patients.

The thinner plate on the body of the mandible had a visible logo, product number, and lot number. The strut plate near the gonial angle had extensive bone growth over much of the strut obscuring a portion of the product number and the entirety of the lot number. Radiographic images were taken to visualize the extent of the strut plate. A Dremel-like tool was used to remove the bony overgrowth which had to be done carefully as the drill could easily scratch the numbers and obscure them permanently.

Once all the numbers on each plate were visible, the surgical supply company was contacted. The company was able to pull the production information on each specific plate's lot numbers and pair it with the medical facilities that had received each plate. The thinner mandibular plate was relatively common, and a high volume of plates were produced within a brief amount of time and shipped across the country to several medical facilities. The more complex strut plate had fewer plates manufactured in that lot and shipped across the country. Separately, attempting to track these plates would be a huge task for both the anthropologist and the medical facilities. However, by cross-checking the plates with the medical facilities that received both plates within the specific shipment periods, only four medical facilities had received both lot numbers for the specific plates across the entire country. The manufacturer was able to provide the general time frame the order was filled, so a basement date for the decedent's surgery could be inferred.

After viewing the catalog entry for the complex strut plate as it would have been shipped in comparison to the shape in the radiographic image, it was apparent that the plate implanted in the decedent had been altered from its original form. This plate alteration and the information on possible surgery locations and date cannot be used to identify someone, but these parameters can be used to exclude individuals. During this investigative time period, our office was contacted by loved ones searching for missing persons, and the PCOME was able to exclude those missing based on the surgery location or general date. NamUs was updated to reflect the four states in which the surgery may have occurred as well as a general timeframe.

Each medical facility was contacted to determine if it would be possible to track a surgery that likely occurred close to a decade ago. As to be expected, this is not a possibility in every facility. Each facility was provided with as much information about the unidentified decedent as possible including the sex estimation, ancestry assessment, age estimation and information about where these plates were located on the mandible. Information regarding when each medical facility received the surgical implants was provided to aid in the records search.

Two of the four medical facilities contacted our office to inform us that they regrettably would be unable to track any information on the plates that were received by their facility. While this kind of notification is disappointing, it is common. These two facilities could not be ruled out and remained possible sites where the decedent may have had surgery.

An oral surgeon contacted the PCOME from a third facility indicating that their staff was able to locate the patient file for an individual that had both plates implanted into his mandible. In order to protect the identity of this former patient, the PCOME called the surgeon and spoke in further detail regarding where the plates had been placed which seemed consistent with the former patient. However, after discussing the alterations to the strut plate found in the decedent and the extraction of the former patient's tooth, it was obvious that this former patient could be excluded as the decedent. Although disappointing, this exclusion narrowed the list of medical facilities that performed the surgery.

A surgery coordinator from the fourth and final facility contacted the authors asking for specific information regarding the placement of the plates and was provided a schematic of the mandible. After several months, this facility contacted the PCOME indicating that they were able to track one of the plates to a particular patient, and the surgeon was willing to discuss the case with our office. This notification was disheartening as there was no mention of the second plate.

The author was fully prepared to thank this surgeon for their time in tracking this patient but was anticipating an exclusion based on the absence of the strut plate. However, the surgeon indicated that their facility was only able to definitively track the thin plate because a post-operative photograph captured the lot number. Nearly a decade ago it was not part of the procedures to record and save lot numbers from plates (Ubelaker & Jacobs, 1995; Wilson, Bethard, & DiGangi, 2011). This surgeon routinely altered plates to fit the patient and had recorded the alteration of the strut plate in the notes. However, the surgeon mentioned a third plate had been used to stabilize the inferior border of the mandible and was attached to the thin band on the anterior portion of the mandible body. This discrepancy could be explained by the removal of this third plate at a later time. This theory was supported by the missing fourth screw in the thin plate and the uneven healing of the inferior border of the mandible. The surgeon indicated that post-operative computed topography (CT) scans were available of this individual that could be rendered to show the cranium's surfaces.

The CT scan depicted displaced nasal fractures in a very similar positioning to the antemortem nasal fractures seen in the decedent. The CT scan and the decedent both had maxillary tori on the alveolar bone above teeth 12, 13, 14, 15, and 16. The positioning of both plates present in the decedent were the same as the plates seen in the patient CT scan. Several other areas of osseous similarity were noted: size and shape of the mastoid processes, size and shape of the styloid processes, and general shape of the supraorbital browridges. However, a significant amount of postmortem damage to both the cranium and the mandible did not allow for comparison of all bony features noted in the CT scans. The CT scans depicted the third plate as well as wiring to close the mouth that was not present in the decedent.

The final comparison which was able to identify the decedent as the patient seen in the CT scans was the anterior-posterior (AP) scout view. A scout view is a preliminary radiographic image taken before the CT scan to ensure that the study area is fully encompassed. This method of scout shot comparison to a postmortem radiograph has been used in the field before as described by Haglund & Fligner (1993). The decedent had a unilateral frontal sinus with a distinct bilobed shape. The AP scout shot of the patient had a unilateral bilobed shape in the same position and orientation as the decedent as seen in Figure 3.4. Frontal sinus morphologies have been shown to be individually unique and useful for positive identifications (Christensen 2005; Quatrehomme, Fronty, Sapanet, Grévin, Bailet, & Ollier 1996); thus this confirmed that the decedent and this former patient were the same person.

Unfortunately, the name and date of birth the patient provided has not been verified by family or friends and could be an alias. Local law enforcement in the city and county

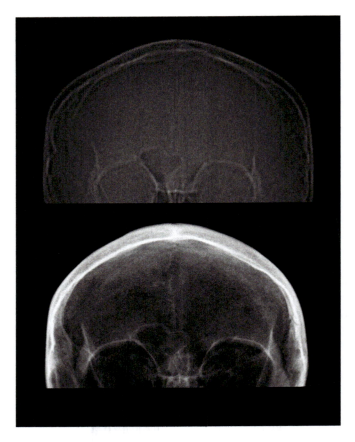

FIGURE 3.4   The antemortem anterior-posterior scout shot of the patient (top) as compared to the AP radiographic image of the unidentified cranium (bottom).

where the surgery took place was notified to inquire if there had been any contact with the decedent. No new information was obtained. The full hospital records for the decedent were requested, but no additional information on next of kin, employment, or nationality was determined. The foreign consulates and non-profit agencies were contacted to ascertain if a missing persons report had been filed with the name the decedent had provided at the hospital, but no such report had been filed. The US Customs and Border Protection was contacted to see if any record of previous apprehension existed for this individual, but there was no record under the name provided. Although the PCOME has a lot of information on this decedent, no next of kin has been discovered who can confirm this individual's true identity.

LESSONS LEARNED

- Surgical implants are helpful in aiding the identification process, but usually do not have a unique number that can be used as a sole source of an identification.
- Record keeping at surgical implant manufacturers and medical facilities varies from place to place as well as throughout time.

- As such, the presence of a surgical implant does not guarantee that an identification can be made, and in this case, even the identification of an individual from the surgical medical records did not result in the true identity of the decedent. Although tracking numbers found on surgical implants may not lead to an identification, they can provide valuable information for investigating agencies.

DISCUSSION QUESTIONS

3.1 Should ALL surgical devices be required to have unique identification numbers engraved on them? What would be some potential downfalls to implementing such protocols?

3.2 Without the presence of the surgical devices, what do you think the chances would have been of obtaining an ID?

3.3 What ways can a CT scan be used to help identify a cranium? What kind of information is required to make a positive identification?

3.4 Why should caution be exercised when attempting to identify someone from records that were self-reported?

REFERENCES

Cain, M. D., Roper, D., & Atherton, D. S. (2016). Use of infrared photography to visualize a tattoo for identification in advanced decomposition. *Academic Forensic Pathology*, 6(2), 338–342.

Christensen, A. (2005). Testing the reliability of frontal sinuses in positive identification. *Journal of Forensic Sciences*, 50(1), 1–5.

Galloway, A., Birkby, W. H., Jones, A. M., Henry, T. E., & Parks, B. O. (1989). Decay rates of human remains in an arid environment. *Journal of Forensic Sciences*, 34(3), 607–616.

Halgund, W. D., & Fligner, C. L. (1993). Confirmation of human identification using computerized tomography (CT). *Journal of Forensic Sciences*, 38(3), 708–712.

Hernandez, G., & Hess, G. L. (2014). Rehydrating mummified hands: The Pima County experience. *Academic Forensic Pathology*, 4(1), 114–117.

Mann, R. W., Jantz, R. L., Bass, W. M., & Willey, P. S. (1991). Maxillary suture obliteration: A visual method for estimating skeletal age. *Journal of Forensic Sciences*, 36(3), 781–791.

Meindl, R. S., & Lovejoy, C. O. (1985). Endocranial suture closure: A revised method for the determination of skeletal age at death based on the lateral-anterior sutures. *American Journal of Physical Anthropology*, 68(1), 57–66.

Quatrehomme, G., Fronty, P., Sapanet, M., Grévin, G., Bailet, P., & Ollier, A. (1996). Identification by frontal sinus pattern in forensic anthropology. *Forensic Science International*, 83(2), 147–153.

Trammel, L., Soler, A., Milligan, C. F., & Reineke, R. C. (2014). The postmortem interval: A retrospective study in desert open-air environments. Paper presented at 66th Annual Scientific Meeting for American Academy of Forensic Sciences.

Ubelaker, D. H., & Jacobs, C. H. (1995). Identification of orthopedic device manufacturer. *Journal of Forensic Sciences*, 40(2), 168–170.

Wilson, R. J., Bethard, J. D., & DiGangi, E. A. (2011). The use of orthopedic surgical devices for forensic identification. *Journal of Forensic Sciences*, 56(2), 460–469.

CHAPTER 4

Biomechanical Analysis of Long Bones Provides the Crucial Break in Decedent Identification

Daniel J. Wescott

CONTENTS

INTRODUCTION

Law enforcement and medical examiners frequently enlist forensic anthropologists to aid in the identification of skeletal or severely decomposed human remains. This process begins by developing a biological profile, which involves the estimation of sex, age, ancestry, stature, and potentially individualistic characteristics (e.g., surgical implants, diseases, healed fractures). These unique characteristics may aid in the identification process by excluding or failing to exclude missing individuals and by providing information that can be compared to antemortem (before death) medical records.

In this case, the partially decomposed and headless corpse of a woman was discovered in a shallow grave by an off-duty sheriff's deputy who was mushroom hunting in the woods near a major river in the Midwest. The remains were brought to the local medical examiner's office for autopsy, and I was contacted by the medical examiner to provide forensic anthropological assistance. The primary purpose of my involvement was to estimate the age of the individual and the postmortem interval (i.e., time since death). However, upon seeing the lower limb long bones of the deceased woman, I realized this would be a different type of case, and my expertise in bone biomechanics might be useful. This case study demonstrates how long bone biomechanical analysis was used to help identify the body of a woman discovered near a small Midwest town.

Autopsy/Anthropological Findings

I arrived at the medical examiner's office with a graduate student. To begin our analysis, we conducted an overall assessment of the body and examined all exposed bones. From the soft tissue, it appeared that the individual was a white female. An age of 30–50 years was estimated using the pubic symphysis and sternal end of the rib. Stature was estimated as 5'2"–5'8" using the 20th-century white female reference sample in Fordisc (Jantz & Ousley, 2005).

We conducted a gross examination of the body to estimate time since death. The torso was in early decomposition with mostly intact internal organs, mild skin discoloration, and no evidence of bloating. The lower limbs, on the other hand, exhibited nearly complete loss of muscle tissue and exposure of the bones. The overlying skin was discolored, and maggots were present on the lower limbs. The hands were in good condition and fingerprints were collected. We provided a broad estimation of time since death of between three and 30 days.

We then examined the long bones. The femora showed evidence of coxa valga, anteversion, and a smaller than expected midshaft diameter, especially in the mediolateral dimension. These characteristics are commonly associated with reduced biomechanical stress of the bone during growth and development. Coxa valga is a higher than normal neck-shaft angle (>135°), which is the angle formed by a line drawn through the center of the femoral head and neck and a line through the center of the long axis of the shaft. Version refers to the torsion or twisting present from the proximal (neck and head) to the distal ends (knee joint) of the bone. Typically, the head and neck are projected forward at an angle of approximately 15°. Anteversion occurs when the torsion is projected more forward than normal. The expected shaft diameter is relative to the bone length, body mass, and sex of the individual. Tall and heavy individuals, for example, typically have relatively larger shaft diameters than shorter and lighter individuals to compensate for the greater forces experienced by the bone. The femoral head diameter is used as a proxy for body mass since it provides a good estimation of lean body mass (Auerbach & Ruff, 2004). In this case, the woman had a femur length of 432 mm and a femoral head dimeter of 42 mm, all within the normal range for a white female. However, her midshaft mediolateral diameter was 18 mm, which is outside the range of the comparative white females and far below the average (average = 24.7 mm with a range of 20 to 33 mm). This would suggest that mechanical stimulus was not sufficient to develop or maintain a normal shaft diameter development, a common characteristic of immobility or reduced mobility. In addition, the medical examiner informed us that a Baclofen pump had been removed during autopsy. The pump is a surgical implant that delivers medication to relieve spasticity associated with spinal cord diseases, which are also commonly associated with immobility (Medical Advisory Secretariat, 2005).

BACKGROUND IN BIOMECHANICAL ANALYSES

Biomechanical analyses examine the size and shape of bones, especially long bones of the upper and lower limbs. The cross-sectional size and shape of long bone shafts are examined using the principles of bone functional adaptation (Ruff et al., 2006). Most biomechanical studies in anthropology use these principles to examine activity patterns and mobility within and among populations (Ruff, 2008). However, the same principles can be used to examine changes expected due to immobility (Schlecht et al., 2012; Sievänen, 2010).

Although people often think of bone as static or unchanging, bone responds to mechanical stimuli caused by dynamic loading during normal activities such as walking, running, and jumping throughout life. Forces created by muscles and forces exerted by the ground on the body (ground reaction force) during normal activities can cause physical deformation or strain on the bones (Wallace, 2014). Under sufficient compression, a long bone deforms by shortening and slightly bulging. Under tension, it lengthens and narrows slightly. Shear forces cause angular deformation. Bone cells known as osteocytes detect strain, and bone responds by adding bone to specific locations to reduce future strain or by removing bone in areas where strain is low (Robling et al., 2014; Wallace, 2014). In other words, bone responds to mechanical forces by forming bone in areas that receive a signal that increased strength is necessary and removing bone where strength is less important. As a result, bone size and shape changes throughout life to reduce mechanical strain caused by the stress of daily activities. Activities such as jumping create greater strain than walking, and therefore evoke greater change.

Bone can be added, removed, or maintained at four "envelopes": periosteal, endosteal, intracortical, and trabecular envelopes (Robling et al., 2014). The periosteal surface is the outermost surface of the cortical bone, while the endosteal surface is the inner surface next to the medullary cavity of long bones. Intracortical refers to the cortical bone between the endosteal and periosteal surfaces. This chapter focuses on the periosteal, endosteal, and intracortical envelopes, which are prominent in the bone shaft.

Bone is added or removed through two processes – modeling and remodeling (Allen & Burr, 2014). During modeling, bone size and shape changes as osteoblasts and osteoclasts sculpt the bone. This occurs primarily during growth and development and less commonly in adulthood. The bone cells that deposit the organic matrix of bone (osteoblasts) and those that remove bone (osteoclasts) are decoupled during modeling. In other words, they act independently on different surfaces. For example, if you imagine a round cross-section of a long bone, adding bone to the anterior and posterior periosteal surfaces but not the medial or lateral surfaces will result in a more anteroposterior-oriented and oval cross-section. Remodeling, on the other hand, is a process that targets specific areas of bone requiring repair due to micro-damage. During remodeling, bone is removed and then replaced in the same area, forming structures referred to as secondary osteons (Allen & Burr, 2014). Remodeling primarily occurs in the cortical bone or intracortical envelope.

The magnitude, rate, and frequency of the load determine how bone responds to mechanical signals (Robling et al., 2014). The conceptual model known as Mechanostat developed by Harold Frost can be used to predict if bone will be added or removed as well as the location where the addition/resorption will occur (Frost, 1987; 2003; Robling et al., 2014). Frost argues that that bone adaptation works as a negative feedback system much like a thermostat in your house with three different set points or minimum effective strains (MES) for bone loss, bone formation, and bone repair (see Figure 4.1). These three set points define four mechanical usage windows – disuse, maintenance, overuse, and overload. If the strain is within normal levels (maintenance window), the size and shape of the bone will be maintained. However, if the strain falls below the remodeling threshold (disuse window) bone will be removed intracortically due to increased remodeling causing porosity or loss of density. If strain exceeds the modeling threshold (overload window), bone will be added in areas via modeling to reduce future deformation and bring the bone back to optimal or maintenance levels of strain.

According to the Mechanostat model, increased strain will result in the deposition of bone on the periosteal surface through modeling, retarded bone loss on the

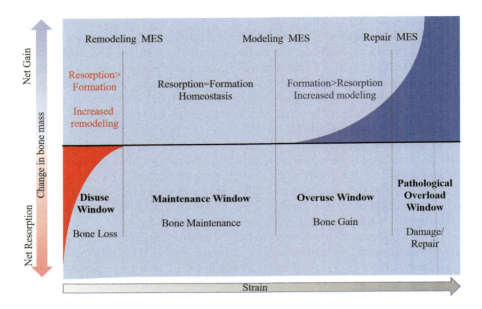

FIGURE 4.1 Illustration of the Mechanostat model and the four mechanical usage windows. According to the Mechanostat model, bone is only added or resorbed when strain is above or below a certain threshold. These thresholds define the four mechanical usage windows. Loads above the modeling minimum effective strain (MES) result in increased modeling and inhibited remodeling, while loads below the remodeling MES inhibit modeling and increase remodeling. Bone formation and resorption are nearly equal in the maintenance window. Adapted from Robling et al. (2014).

endosteal surface during modeling and remodeling, and decreased porosity of the cortical bone during remodeling (Robling et al., 2014). Decreased strain, on the other hand, will inhibit deposition of bone on the periosteal surface, accelerate loss of bone on the endosteal envelope, and increase cortical bone porosity due to remodeling. In other words, when deformation necessitates increasing the strength of the bone, the external size of a long bone cross-section increases, the medullary cavity remains relatively constant and the cortical bone remains dense. When bone strength isn't necessary (e.g., due to disuse), the body removes bone from the endosteal surface, expanding the medullary cavity, and the cortical bone becomes more porous, but the periosteal surface (external bone diameter) generally remains unchanged. This is also true with pathological conditions such as osteoporosis. Bone is removed primarily from the endosteal surface as it affects overall bone strength less than removing bone from the periosteal surface.

Prediction of Bone Strength

Since bone material properties are relatively consistent throughout the skeleton, anthropologists can use the mechanical properties that quantify the cross-sectional size and shape to predict how resistant whole bones are to deformation, or the bone's rigidity (Ruff, 2006; 2008). The area of the bone (i.e., the amount of cortical bone in the cross-section) provides an estimation of the bone's resistance to compressive and tensile strains (see Figure 4.2). Second moments of area (I) help determine the bone's resistance to deformation during bending in a particular plane across the distribution of the mass.

The moments of area (I) are a measure of the amount of bone and its cross-sectional distribution about a given axis (e.g., the anatomical anteroposterior and mediolateral, or the minimum and maximum principle axes). The further away the bone is from the neutral axis or center of the bone, the greater its resistance to deformation (Wallace, 2014). For example, if you tried to bend a wooden "two by four" board, it would be easier to bend the board along the shorter dimension than the longer dimension of the rectangle. When a bone undergoes bending, one side of the shaft is under tension while the other is under compression. That is, one side of the bone lengthens and the other shortens. At some point near the middle of the shaft, the compression and tension cancel each other, and the strain is zero (see Figure 4.3). This is referred to as the neutral axis. Since bone is stronger the further away it is from the neutral axis, a bone with relatively thin cortical bone can be as strong or stronger than a bone with thicker cortical bone under bending loads if the overall diameter of the bone is larger. During growth and development, the long bones increase in size and strength to meet the demands placed on them during daily activities. For more detailed explanations on bone biomechanics, see Ruff (2003).

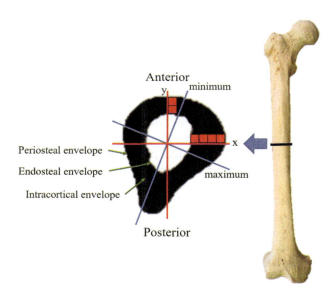

FIGURE 4.2 Femoral midshaft cross-section illustrating the periosteal, endosteal, and intracortcical envelope (green arrows) locations and cross-sectional axes used to calculate second moments of inertia. Second moments of inertia (I) reflect the cross-sectional shape and mass of the bone. They are measured in relationship to particular axes as $\Sigma a_i d_i^2$, where a_i is a unit area (red squares) and d_i is the perpendicular distance from the neutral axis to the center of the unit area (distance from red crosshair to red square). The bending rigidity in the anteroposterior and mediolateral planes are estimated based on I_x and I_y, respectively (red lines). The maximum (I_{max}) and minimum (I_{min}) moments are the greatest and least bending rigidity, respectively. The I_{max} axis is along the narrowest dimension, while the I_{min} axis is along the greatest dimension (blue lines).

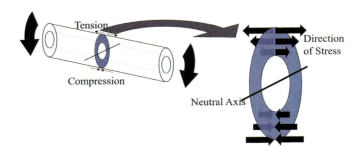

FIGURE 4.3   Cartoon of bone shaft undergoing bending. During bending, tension occurs on one surface while compression occurs on the opposite surface. These two strains cancel each other to form an area of zero strain known as the neutral axis.

IDENTIFICATION

No record of a missing person fit the characteristics of the unknown woman. Although fingerprints were recovered, they provided no leads to the identity of the woman. For fingerprints to be useful for identification, her fingerprints must be in a database such as the Integrated Automated Fingerprint Identification System. DNA analysis also requires access to a premortem source of the decedent's DNA or a family reference sample. Therefore, the next step was to provide the media with a description of the unknown woman in hopes that someone from the public would come forward with information about the unidentified woman.

Based on the cross-sectional morphology of the femur from computed tomography scans, we were able to determine the individual had lower than normal mechanical stress on the lower limbs during growth and development of bones, but we could also tell that she had walked for a while. So, how do we know that this occurred during growth and not as an adult? We examined the mechanical properties of her bones as well as the bone formation/resorption processes (modeling versus remodeling). As shown in Figure 4.4, the second moments of area around the maximum and minimum cross-sectional bone axes (I_{max} and I_{min}, respectively) for the unidentified woman were lower than expected, suggesting that her femora were not able to resist deformation as much as women with normal mobility. If the woman had lost mobility as an adult, we would expect the external cortical dimensions to be relatively normal but porosity or loss of bone in the intracortical envelope due to remodeling, and perhaps a greater medullary cavity size due to endosteal bone loss (Gleiber, 2017). Figure 4.5 shows the femoral cross-sections of three women. The first (A) is a female with normal mobility who was age-matched to the unidentified woman. The middle cross-section (B) is from this case, while the third (C) is a woman who became immobile as an adult. Notice the difference in the size of the cross-sections and the porosity of the cortical bone. The moments of area for the unidentified woman were reduced, suggesting insufficient mechanical stimulus to cause the addition of bone on the periosteal surface through modeling (see Figure 4.4). We hypothesized she had been mobile for a period of her life because, while the femur shaft size was small, it was not as small as would be expected if she had never walked (see Figures 4.5 and 4.6). Figure 4.6 shows anterior views of three femora. Femur "A" is from a female with normal mobility, femur "B" is from a young individual who never walked, and femur "C" is that of the unidentified woman. As you can see, the unidentified woman's femur is smaller

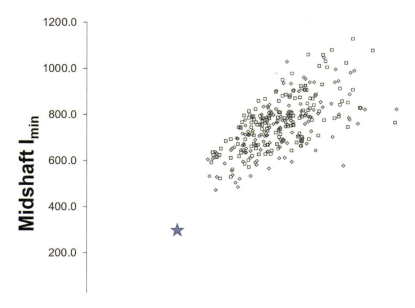

FIGURE 4.4 Bivariate plot comparing the femur midshaft maximum and minimum moments of inertia for the unidentified woman to normal mobility females. The unidentified woman is indicated by the star, while the normal mobility females are represented by black squares (left bones) and diamonds (right bones).

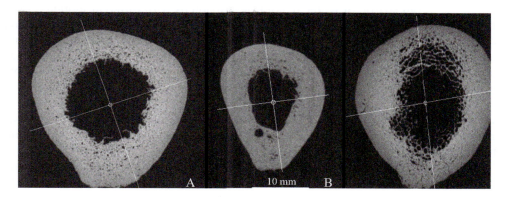

FIGURE 4.5 Comparison of size, shape, and density of femoral midshaft cross-sections for three females. The computed tomography cross-section on the left (A) represents a normal mobility female who was age-matched to the unidentified woman. The cross-section in the middle (B) is from the unidentified woman, while the cross-section on the right (C) is from a female with adult onset mobility impairment. Note the difference in the cross-section size and cortical bone density. The woman who became immobile as an adult has a normal external cortical size but more cortical porosity due to remodeling in the intracortical and endosteal envelopes. The unidentified woman has a smaller cross-sectional size but similar porosity to the age-matched normal mobility female. The lines represent the maximum and minimum axes. Images are courtesy of Devora Gleiber.

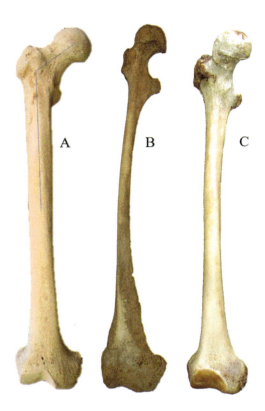

FIGURE 4.6   Comparison of whole femora of individuals with different mobility. From left to right: anterior views of the right femur of a person with normal mobility showing neck-shaft angle (A), an individual who never walked (B), and the unidentified woman (C). Note the differences in the breadth and twist (version) of the shaft and the neck-shaft angle. Both individuals with mobility impairment have higher neck-shaft angles (coxa valga) and anteversion (higher than normal twisting).

in diameter than the normal mobility female but much larger than the person who was immobile throughout life. You can also see the coxa valgus (high neck-shaft angle) and anteversion in Figure 4.6.

   We recommended that the description in the newspaper state she had mobility impairment, possibly used a wheelchair, but not for an extended time, and had walked, at least with assistance, during a portion of her growth period. They followed our advice, and the break in the case came when a local wheelchair salesman contacted law enforcement and provided them with a name. The husband of the woman had contacted the salesman about purchasing an electric wheelchair. The police arrived at her home and after further investigation arrested her husband for murder. According to his later confession, he strangled his wife and discarded her body near an access point of a major Midwest river. He cut off her head because he believed she could not be identified if her head was not discovered. The woman was later positively identified based on DNA using a family reference sample. Records indicate she had cerebral palsy, which is a neurological condition that commonly results in mobility issues. Individuals with cerebral palsy often have difficulty walking because of pain. She had begun using a wheelchair a few years before her death but could walk with a walker after having a Baclofen pump installed a few months before her death.

LESSONS LEARNED

- This case demonstrates how the examination of long bone morphology using the principles of bone functional adaptation provided the clues necessary for the eventual positive identification of the woman. While the biomechanical analysis did not positively identify the woman, describing her condition in the local newspaper initiated the necessary leads for law enforcement to identify her remains and bring her killer to justice.
- Forensic anthropologists should be familiar with the biomechanical properties of bone and the principles of bone biology and remodeling which explain the functional adaptation of bone and predict where bone will be added or removed based on strain.
- A complete osteological analysis goes beyond conducting the typical methods of biological profile analysis. The forensic anthropologist is responsible for documenting all observed skeletal phenomena and explaining the observations whenever possible. At the same time, the anthropologist should make sure authorities understand the level of confidence of such predictions, while ensuring that they do not provide more information than the evidence supports.

DISCUSSION QUESTIONS

4.1 What are the expected differences in bone cross-sectional morphology between individuals who become immobile prior to the completion of growth and development and individuals who become immobile as adults?

4.2 Describe how other conditions or activities might affect bone cross-sectional properties based on the principles of bone functional adaptation. Do you expect all changes in cross-sectional properties to be symmetric, affecting left and right sides equally?

4.3 How does fracture occurrence relate to the Mechanostat model? How do modeling and remodeling contribute to fracture repair?

REFERENCES

Allen, M. R., & Burr, D. B. (2014). Bone modeling and remodeling. In D. B. Burr, & M. R. Allen (Eds.), *Basic and applied bone biology* (pp. 75–89). New York: Academic Press.

Auerbach, B. M., & Ruff, C. B. (2004). Human body mass estimation: A comparison of "morphometric" and "mechanical" methods. *American Journal of Physical Anthropology, 125*, 331–342.

Frost, H. M. (1987). Bone "mass" and the "mechanostat": A proposal. *Anatomical Record, 219*, 1–9.

Frost, H. M. (2003). Bone's mechanostat: A 2003 update. *Anatomical Record, 275A*, 1081–1101.

Gleiber, D. S. (2017). *The effect of mobility impairment on femoral cortical and trabecular structure* (MA thesis).

Jantz, R. L., & Ousley, S. D. (2005). FORDISC 3: Computerized forensic discriminant functions (Version 3.1). Knoxville, TN: The University of Tennessee.

Medical Advisory Secretariat. (2005). Intrathecal baclofen pump for spasticity: An evidence-based analysis. *Ontario Health Technology Assessment Series, 5*(7), 1–93.

Robling, A. G., Fuchs, R. K., & Burr, D. B. (2014). Mechanical adaptation. In D. B. Burr, & M. R. Allen (Eds.), *Basic and applied bone biology* (pp. 175–204). New York: Academic Press.

Ruff, C. B. (2003). Growth in bone strength, body size, and muscle size in a juvenile longitudinal sample. *Bone, 33*, 317–329.

Ruff, C. B. (2006). Gracilization of the modern human skeleton. *American Scientist, 94*, 508–514.

Ruff, C. B. (2008). Biomechanical analyses of archaeological human skeletons. In M. A. Katzenberg, & S. R. Saunders (Eds.), *Biological anthropology of the human skeleton* (2nd ed., pp. 183–206). Hoboken, NJ: Wiley and Sons.

Ruff, C. B., Holt, B., & Trinkaus, E. (2006). Who's afraid of the big bad Wolff? "Wolff's law" and bone functional adaptation. *American Journal of Physical Anthropology, 129*, 484–498.

Schlecht, S. H., Pinto, D. C., Agnew, A. M., & Stout, S. D. (2012). Brief communication: The effects of disuse on the mechanical properties of bone: What unloading tells us about the adaptive nature of skeletal tissue. *American Journal of Physical Anthropology, 149*, 599–605.

Sievänen, H. (2010). Immobilization and bone structure in humans. *Archives of Biochemistry and Biophysics, 503*, 146–152.

Wallace, J. M. (2014). Skeletal hard tissue biomechanics. In D. B. Burr, & M. R. Allen (Eds.), *Basic and applied bone biology* (pp. 115–130). New York: Academic Press.

CHAPTER 5

Race and the Role of Sociocultural Context in Forensic Anthropological Ancestry Assessment

Michala K. Stock and Katie M. Rubin

CONTENTS

INTRODUCTION

The C. A. Pound Human Identification Laboratory (CAPHIL) was asked to perform osteological analyses on unidentified remains by a regional Medical Examiner's Office to aid in decedent identification and determination of cause of death. Analyses at the CAPHIL revealed the remains were those of a young adult male with indications of high-velocity projectile trauma to the thorax. The perimortem trauma findings were fairly straightforward, but the ancestry assessment spurred discussion among the practitioners, and thus constitutes the focus of this chapter.

CASE ANALYSIS

Trauma

At least one trajectory of damage from perimortem, high-velocity projectile trauma to the thorax was noted following maceration and rearticulation (see Figure 5.1). A semi-circular

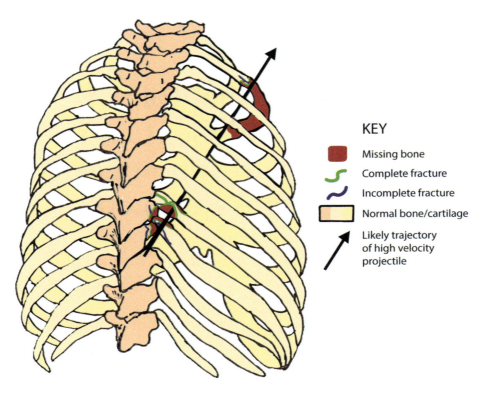

KEY

	Missing bone
	Complete fracture
	Incomplete fracture
	Normal bone/cartilage
	Likely trajectory of high velocity projectile

FIGURE 5.1   Homunculus of the thorax demonstrating the reconstructed trajectory of the projectile based on skeletal analysis of the remains.

defect was present on the inferior aspect of the right 8th rib's neck. This defect was beveled inwardly onto the rib's pleural surface, indicating the projectile entered the decedent's body between adjacent mid-thoracic vertebral transverse processes and vertebral rib ends. The angle of beveling at this entrance defect, in conjunction with fractured and missing bone at the sternal end of the right 2nd rib and damage to the coracoid process of the right scapula, indicated that the projectile traveled through the body along an anterosuperior trajectory and exited through the 2nd rib's shaft. The projectile was neither recovered nor examined by CAPHIL analysts, and no radiopaque foreign material was noted in radiographs or during microscopic examinations of the remains conducted at the CAPHIL.

Biological Profile – Sex and Age

Non-metric analyses performed on the skull (i.e., Buikstra & Ubelaker, 1994) and pelvis (i.e., Phenice, 1969) both indicated male biological sex. Age was assessed in this individual's remains predominantly via stage of epiphyseal fusion at select sites (e.g., medial clavicles) and metamorphic changes to select bone surfaces (e.g., pubic symphyses and sternal rib ends). All analyses indicated young adulthood (22–29 years) at time of death. Neither sex nor age assessment was ambiguous in this case.

Biological Profile – Ancestry

Ancestry was assessed using both metric and non-metric methods. The computer program Fordisc 3.1 (Jantz & Ousley, 2005) uses discriminant function analysis of cranial measurements to classify an individual into one of the reference populations in the Forensic Data Bank (FDB). During case analysis, this individual's cranial measurements were run against cranial measurements from all male groups in the FDB using no transformations; Fordisc indicated this individual was atypical of all groups other than "Black males" and "Hispanic males" (see Table 5.1; Runs 1 and 2). Non-metric traits were also described (i.e., Gill, 1998) and scored (i.e., Hefner, 2009) for the craniofacial skeleton of this individual. The mid-facial skeleton of this individual displayed a medium interorbital breadth, an inverted V-shaped nasal bone contour, a moderate sill on the inferior nasal aperture, a pronounced, markedly projecting anterior nasal spine, and an "M-shaped" transverse palatine suture (see Figure 5.2).

Taken individually, these traits occur most frequently in individuals of primarily European descent (Gill, 1998; Hefner, 2009). However, other features such as marked maxillary prognathism with a hyperbolic dental arcade, a dolichocephalic vault shape, and crenulated molars are most common among individuals of primarily African descent (Gill, 1998; Zinni & Crowley, 2012). A postbregmatic depression is also present in this individual, however Hefner & Ousley (2014) state that other preliminary analyses suggest that this trait may not be very informative in ancestry classification.

TABLE 5.1   Fordisc Results when all Available Cranial Measurements Were Run Against all Male Groups in the FDB Using no Transformations (Original Run, Version 3.1.312: Run 1), as well as a Run from the Most-Updated Version as of December 2018 Employing Stepwise Variable Selection (Version 3.1.315: Run 2)

	Group	Distance From	Posterior Probability	Typicality – F	Typicality – Chi	Typicality – R
Run 1	BM	19.3	0.717	0.223	0.198	0.311
	HM	22.2	0.176	0.119	0.104	0.129
	JM	25.2	0.038	0.059	0.047	0.047
	AM	25.7	0.031	0.055	0.042	0.208
	GTM	26.3	0.022	0.046	0.035	0.057
	CHM	27.6	0.011	0.032	0.024	0.029
	VM	29.8	0.004	0.019	0.013	0.020
	WM	32.6	0.001	0.007	0.005	0.010
Run 2	HM	26.7	0.563	0.168	0.144	0.259
	BM	27.4	0.388	0.149	0.123	0.340
	AM	32.5	0.031	0.055	0.039	0.212
	GTM	34.5	0.011	0.033	0.023	0.044
	VM	37.2	0.003	0.018	0.011	0.041
	JM	38.9	0.001	0.010	0.007	0.005
	WM	39.4	0.001	0.009	0.006	0.021
	CHM	40.4	0.001	0.007	0.004	0.014

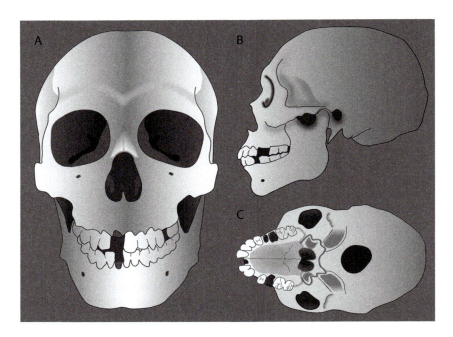

FIGURE 5.2 Views of the skull to illustrate the morphology considered during ancestry assessment. (A) Anterior view of the skull. Note the narrow breadths of mid-facial traits. (B) Left lateral view of the skull. (C) View of the cranial base. Note the pronounced maxillary prognathism, the crenulated molars, and the sharp, protrusive anterior nasal spine. Line drawings are presented here to protect the decedent's identity and the rights of his family; they were created freehand by the authors while consulting case photographs and then compared to these to ensure accurate representation of this individual's morphology.

Studies have suggested that more weight should be placed on the upper mid-facial traits such as inferior nasal aperture and nasal aperture width (Hefner & Ousley, 2014), which in this case supported the European designation. In addition, primarily European descent was corroborated by statistical analysis of traits using the program OSSA (Hefner & Ousley, 2014, v. 2.1); note that these analyses were carried out post hoc and were not employed during the original evaluation. OSSA analyses returned a summed score of 5; in the program's database, 93.75% of individuals with a score of 5 are "White," while only 6.25% of these individuals are "Black" (Hefner & Ousley, 2014). The authors of OSSA also present a distinct decision tree for ancestry assessment; this tree likewise indicates the decedent was "White," with 30 White, 6 Black, and three Hispanic individuals from their database falling in the "leaf" into which the decedent was sorted (Hefner & Ousley, 2014). Additionally, based on the scores of this individual's macromorphoscopic traits and Bayesian statistics, the Osteomics program HefneR gives a 98.5% probability of the individual being European and a 1.2% probability of the individual being African (Navega & Coelho, 2018).

Overall, the preponderance of traits present in the upper mid-facial skeleton was most characteristic of individuals of European ancestry, as was corroborated by the OSSA analyses, but other non-metric traits and the Fordisc analyses were more indicative of individuals of a primarily non-European background, such as individuals of African ancestry or Hispanic ethnicity. Standard practice at the CAPHIL at the time these remains were analyzed was to report ancestry as "primarily European," "primarily African," "primarily Asian," or "primarily American Indian," with elaboration for individuals who may socially classify into specific ethnic groups, such as "Hispanic." Although laboratories and individual anthropologists vary in their language of ancestry reporting, "primary ancestry" bins are used by many forensic practitioners. Because CAPHIL analysts frequently confront the well-known issue of poor representation of Florida Hispanic individuals within the Hispanic reference groups in the current FDB (Tise et al., 2014), they tend to heavily consider non-metric assessments of ancestry when reporting a primary ancestry; a large proportion of decedents analyzed at this laboratory yield low (but not always statistically insignificant) typicality probabilities in Fordisc, and non-metric analyses assist analysts in gauging and interpreting the biological significance of these results. Based on the above enumeration of traits and best practices at the laboratory, should the analysts then have reported this individual as being of "primarily European ancestry" to the Medical Examiner's Office and the investigating authorities?

BROADER CONTEXT

Forensic anthropologists are called upon routinely to assess the biological profile of an unidentified individual, i.e., to give an estimate of demographic characteristics to search and narrow the potential list of missing persons. While debates persist among scholars within the field regarding the best methodology for assessing sex, age, and stature, whether to pursue these assessments and how to communicate findings to a non-academic audience is generally not the focus of debates. However, the practice and communication of ancestry assessment has been more controversial in forensic anthropology.

To delve into the question of how to best interpret and report the ancestry markers for this individual, we must first discuss the relationship of race and ancestry, within both the field of biological anthropology, of which forensic anthropology is a subfield, and the cultural milieu of the United States (US). The biological race concept, through which humans can be discretely divided into four or five races based on each person's biological characteristics, was pervasive among the early founders of physical anthropology and became entrenched through their works. Modern anthropologists have largely rejected the concept of biological race in the aftermath of World War II. As an alternative, race is understood primarily as a social construct and is not necessarily reflective of biological reality (Sauer, 1992). That race is socially defined is illustrated by the fact that racial categories, and criteria for an individual falling within any given category, vary by country and can change over time. Many anthropologists have used the social nature of race to repudiate race as a topic of relevance to human skeletal assessment.

However, social constructs cannot be dismissed summarily by biological anthropologists as unscientific or beyond our purview. As an apparatus of social organization, race is fundamental to how the modern US functions, both at structural/institutional and individual levels, including self-identity. Official governmental documentation, such as the US Census and Missing Persons databases like CODIS, and eyewitness accounts and reports routinely use race-based terms like "White" or "Black" to describe people. As a result, race is also enmeshed in the infrastructure of medicolegal death investigation.

In addition, people categorized into certain groups face individual and structural inequality, which often have tangible, negative biological ramifications, including maternal epigenetic effects that can impact development *in utero* (e.g., Mulligan, 2016).

Therefore, to claim that race is not "real" is to negate and subvert the lived experiences and systematic inequalities faced by individuals whose groups face systemic discrimination; rather, race can be construed as a social reality that produces biological reality, inverting the typical conception of race held by most Americans (Gravlee, 2009). Further, the social construct of race still influences mating decisions for many people in US society (whether explicitly or implicitly), thereby shaping gene flow and the phenotypic traits of subsequent generations.

Furthermore, while race is socially defined, there does exist observable phenotypic, or physically expressed, variation among human groups that is biological in nature. However, patterned variation in human variation does not equate to race; race is typically conceived of as discrete bins, whereas biological variation is clinal – that is, much of human physical variation occurs along gradients (typically geographic in nature) with no well-defined or distinct boundaries. The systematic nature of human variation means that more closely related populations of individuals tend to share greater similarity in their phenotypic expression – including skeletal traits – than do more distantly related groups, even though expression of those traits is not homogenous within any group (i.e., there is significant variation within groups) and any given expression of a trait is not unique to any one group (Hefner, 2009: Tables 3–13). Perhaps the simplest clinal or geographically patterned phenotypic trait to appreciate in humans is skin color (Relethford, 2002). Yet, when it comes to race, an individual's skin color does not obligate them to self-identify by others' external categories; for example, individuals who self-identify as "Black" can have less melanin (i.e., lighter skin tone) than those who self-identify as "White."

Based on both craniometric data (e.g., Howells, 1973, 1989; Roseman & Weaver, 2004) and non-metric analyses (e.g., Hanihara et al., 2003), the systematic patterning of human skeletal variation, particularly in the skull, appears to be tied to geographic origin. Therefore, forensic anthropologists have data that support broadly categorizing the ancestral origin (usually to the level of continent, as was described above in current practices at the CAPHIL) of individuals based on their skeletal remains. However, this is not the parlance lay people use to describe themselves or each other, nor does this terminology aid investigators who are conducting interviews and searching databases. One might argue that "primarily of European ancestry" equates to "White," while "primarily of African ancestry" equates to "Black," and indeed ancestral origin and racial identification are often closely linked. However, even if the substitution of ancestral origin and racial classifications may be applicable in some cases, i.e., there is a match between biological affinity and social race, this is certainly not always the case, and this assumption can be damaging as it belies the social and historical context of race in the US. Furthermore, many individuals may have genetic contributions from multiple ancestries and may self-identify as bi- or multiracial; the 2000 US census took a first step away from treating races as discrete, non-overlapping entities by allowing citizens to identify themselves as more than one race for the first time. In the two censuses following this change (2000 and 2010), between 2–3% of the US population, or somewhere between six and nine million people, self-identified as belonging to two or more races (Jones & Bullock, 2012). Therefore, when assessing ancestry in forensic anthropology it is important to understand when, how, and why an individual's suite of craniofacial traits, primary ancestral origin, and racial affiliation may not all three align.

Despite this budding awareness of the complexity of racial (self-) identification, the vestiges and impact of the one-drop rule remain felt in the US even today (Khanna, 2010). Originally a legal classification, the edict that any person with "one drop" of African blood could not be "White" served to reinforce White supremacy and guard racial purity, especially in the South in the aftermath of the Civil War and into the Jim Crow era. More recently, sociologists have argued that in the mid-20th century the Civil Rights Movement embraced the one-drop rule as a means to engender inclusivity and pride, helping to strengthen the movement (Khanna, 2010). Regardless of the underlying reasons, the concept persists in our society, influencing Americans' perception of race to this day. In practice, this conceptualization of race leads to wider phenotypic variation, including greater variation in skeletal traits or suites of traits, within certain racial group in the US.

This historical and social milieu, especially when issues of self-identification versus external labeling come into play, render a ubiquitous equivalency between ancestry – as estimated via a suite of skeletal traits – and a racial identification untenable. Anthropologists have argued this point (see Goodman & Armelagos, 1996) and gone even further, arguing that any acknowledgment of or articulation with race by forensic anthropologists reifies the biological race concept (Goodman, 1997). Yet, forensic anthropologists routinely incorporate information about the assessment of ancestry, and often a likely racial self-identification, in order to provide the most salient information to investigators.

FORENSIC RELEVANCE

Ancestry is commonly assessed by considering the expression of a list of non-metric cranial traits in a given individual and reporting the analytical conclusion based on the preponderance of traits observed. This practice essentially gives equal weight to all traits while treating craniofacial skeletal features as independent, which is problematic for several reasons. Firstly, cranial traits that are routinely scored for ancestry estimation (e.g., Gill, 1998; Hefner, 2009) are often in close physical proximity to one another (such as the anterior nasal spine and the inferior nasal aperture) and may be both functionally and developmentally linked (Bastir, 2008). The integration between traits means that certain subsets or suites of traits likely covary with one another in terms of their expression, meaning that they cannot be considered independent variables and will create bias in the subsequent analysis if treated as such. This consideration leads to the question: is preponderance of traits actually a reliable way of assessing ancestry?

And if considering the balance of a long list of traits (as preponderance suggests) is not particularly useful or scientifically supported, the logical corollary is to ask whether any specific traits are better at separating groups than others. In other words, are all non-metric traits equally salient when it comes to assessing ancestry? Recent research has hinted at the efficacy of particular traits for ancestry estimation (e.g., Hefner & Ousley, 2014) and tried to capitalize on these efficacies using statistical methods; however, these methods do not always include all potentially pertinent traits (such as facial prognathism) or consider their phenotypic relevance. Further, there have been no published discussions of the saliency of different traits for distinguishing between particular groups.

Given the discussion in the previous section about the potential intersection (or lack thereof) between geographical ancestral origin and race, the question posed in the paragraph above becomes more complicated. Not only do forensic anthropologists need to consider the relevance of particular traits for primary ancestral origin, we also need to ponder if all traits are equally salient when it comes to racial identification. And do the ancestral origin saliences and race saliences of cranial traits align? These questions are being discussed in many institutions and laboratories where casework is conducted in the US, resulting in some changes, including that both geographical ancestral origin and peer-perceived ancestry (essentially, US race) categories are incorporated into the newly released Macromorphoscopic Databank (Hefner, 2018).

That being said, there is certainly opportunity for further research into potential differences between how non-metric trait expression/scores articulate with geographical ancestral origin versus racial identification among US populations. For instance, some skeletal features such as maxillary prognathism, zygomatic flaring, and interorbital breadth are visible when observing another, living person, whereas other traits, like suture shape or the morphology of the inferior nasal aperture, are obscured by overlying soft tissues. Might these readily obvious traits provide greater saliency for peer- or self-identified race than other craniofacial traits? Overall, the need to reconcile the issues plaguing ancestry assessment endures, especially since we live in a society that includes a large and growing population of people for whom primary ancestral geographic origin and racial identification do not "match," particularly when linguistic or ethnic grouping contributes to racial identification.

Although in this case study the individual under analysis at the CAPHIL displayed a far greater number of non-metric traits associated with European ancestry than those associated with African ancestry, the analysts felt that an individual with the observed degree of maxillary prognathism likely would not have identified as "White" in the US, even if the individual had substantial – or even predominately – European ancestral contributions. This assessment was based partially on the analysts' experience in assessing ancestry from the skeletal remains of local Floridians in numerous cases over multiple years, as well as a "gestalt," overall impression of the individual's craniofacial morphology, which was supported by Run 1 in Fordisc (individual sorted into "BM" group; Table 5.1). Although the results of all tests were detailed in the notes and discussed in the final report, as is standard practice at the CAPHIL, in this case study, the analysts reported a final conclusion that the individual likely self-identified as "Black" in the US, thus only giving a likely racial classification and deviating from the routine laboratory practice of reporting a likely primary ancestral origin. While basing ancestral assessment on one or two features is certainly less than ideal, and forensic anthropologists favor discussion of biological ancestry over social race, the analysts felt that this conclusion was justified by the scientific observations, forensic nature of the case, and the predominant social constructs in the US (see Sauer, 1992). A conclusion of "primarily European ancestry" may have negatively impacted identification of the remains, while a conclusion of "primarily African ancestry" could not be robustly, scientifically supported; despite the Fordisc results that did not eliminate "Black males" as the group membership for this individual (see Table 5.1), in the updated version of Fordisc this individual was sorted consistently into the "Hispanic male" group, regardless of any transformations applied to the analysis.

Radiographic dental comparison conducted by one of the CAPHIL analysts led to the official identification of the remains by the Medical Examiner's Office, which corroborated the racial analytical conclusion of "Black."

LESSONS LEARNED

- This case caused the CAPHIL to reflect on how we report ancestry as a laboratory and to adapt our conclusions regarding ancestry to be more reflective of the fact that many individuals who self- and/or socially identify as a given racial category may have genetic contributions from groups with disparate geographic origins. An individual who identifies as "Black" may also have genetic contributions from or express phenotypic traits more common among European or American Indian populations, or vice-versa. In such situations, any one skeletal trait with an extreme expression does not necessarily justify a racial, or even a primary ancestral, designation for a particular individual in all cases. However, in situations – like this case – where remains display a trait that is strongly associated with a phenotypic feature already used by Americans to classify humans socially, such as marked facial prognathism, clear and open discussions of potential racial identity may be most appropriate, especially in light of the practitioners' experience in that geographic region.

- Additionally, this case study highlights the important role that experience still plays for forensic anthropologists during biological profile analyses, where some work has demonstrated that experienced observers' more subjective analyses may be more accurate than standardized methods (Berg & Tersigni-Tarrant, 2014).

DISCUSSION QUESTIONS

5.1 What is meant by "preponderance of traits," and how/why do anthropologists use this idea when assessing ancestry from skeletal remains? Is this the most appropriate method? And if metric and non-metric methods yield conflicting results, how does/should the anthropologist reconcile this discrepancy?

5.2 How do the particular social and historical contexts in the US impact our scientific approach to assessing ancestry from skeletal remains?

5.3 How are ancestral origin and race different, and how does this difference impact communication with the medicolegal community?

5.4 In this particular case, the analysts diverted from their laboratory procedures reported a racial rather than ancestral primary bin. What positive and negative implications could such an action have? If this case went to trial, how could they defend their decision? How would the case have been impacted if they had instead relied on the preponderance of traits?

5.5 Why doesn't a mixture of ancestry trait states necessarily mean an individual is of "mixed" ancestry or race? Can forensic anthropologists accurately estimate that a person is multiracial? How does admixture confound ancestry estimation?

ACKNOWLEDGMENTS

The authors thank Drs. Alexander Bennett and Phoebe R. Stubblefield for reviewing and providing helpful feedback on this chapter.

REFERENCES

Bastir, M. (2008). A systems-model for the morphological analysis of integration and modularity in human craniofacial evolution. *Journal of Anthropological Sciences*, 86(1), 37–58.

Berg, G. E., & Tersigni-Tarrant, M. A. (2014, February 17–22). *Sex and ancestry determination: Assessing the 'gestalt'. Proceedings of the 66th annual meeting of the American Academy of Forensic Sciences, Seattle, WA*. Colorado Springs, CO: American Academy of Forensic Sciences.

Buikstra, J., & Ubelaker, D. (1994). Standards for data collection from human skeletal remains. In *Proceedings of a seminar at the field museum of natural history*, 68(1), Arkansas Archaeological Survey.

Gill, G. (1998). Craniofacial criteria on the skeletal attribution of race. In K. Reichs (Ed.), *Forensic osteology* (2nd ed., pp. 293–317). Springfield, IL: Charles C Thomas.

Goodman, A. H. (1997). Bred in the Bone? *The Sciences*, 37(2), 20–25.

Goodman, A. H., & Armelagos, G. J. (1996). The resurrection of race: The concept of race in physical anthropology in the 1990s. In L. T. Reynolds, & L. Lieberman (Eds.), *Race and other misadventures: Essays in honor of Ashley Montagu in his ninetieth year* (pp. 174–186). Dix Hills, NY: General Hall, Inc.

Gravlee, C. C. (2009). How race becomes biology: Embodiment of social inequality. *American Journal of Physical Anthropology* 139(1), 47–57.

Hanihara, T., Ishida, H., & Dodo, Y. (2003). Characterization of biological diversity through analysis of discrete cranial traits. *American Journal of Physical Anthropology*, 121(3), 241–251.

Hefner, J. T. (2009). Cranial nonmetric variation and estimating ancestry. *Journal of Forensic Sciences*, 54(5), 985–995.

Hefner, J. T. (2018). The macromorphoscopic databank. *American Journal of Physical Anthropology*, 166(4), 994–1004.

Hefner, J. T., & Ousley, S. D. (2014). Statistical classification methods for estimating ancestry using morphoscopic traits. *Journal of Forensic Sciences*, 59(4), 883–890.

Howells, W. W. (1973). Cranial variation in man. Papers of the Peabody Museum of Archaeology and Ethnology, Vol. 67. Cambridge, MA: Harvard University.

Howells, W. W. (1989). Skull shapes and the map. Craniometric analyses in the dispersion of modern Homo. Papers of the Peabody Museum of Archaeology and Ethnology, Vol. 79. Cambridge, MA: Harvard University.

Jantz, R. L., & Ousley, S. D. (2005). FORDISC 3.0: Personal computer forensic discriminant functions. Knoxville, TN: University of Tennessee.

Jones, N. A., & Bullock, J. (2012). The Two or More Races Population: 2010. United States Census Bureau, 2010 Census Briefs. Retrieved from https://www.census.gov/prod/cen2010/briefs/ c2010br-13.pdf

Khanna, N. (2010). "If you're half black, you're just black": Reflected appraisals and the persistence of the one-drop rule. *The Sociological Quarterly*, 51(1), 96–121.

Mulligan, C. J. (2016). Early environments, stress, and the epigenetics of human health. *Annual Review of Anthropology*, 45(1), 233–249.

Navega, D. S., & Coelho, J. d'O. (2018). HefneR [Computer program]. Retrieved from http://osteomics.com/hefneR/

Phenice, T. W. (1969). A newly developed visual method of sexing the os pubis. *American Journal of Physical Anthropology*, 30(2), 297–301.

Relethford, J. H. (2002). Apportionment of global human genetic diversity based on craniometrics and skin color. *American Journal of Physical Anthropology*, 118(4), 393–398.

Roseman, C. C., & Weaver, T. D. (2004). Multivariate apportionment of global human craniometric diversity. *American Journal of Physical Anthropology*, 125(3), 257–263.

Sauer, N. J. (1992). Forensic anthropology and the concept of race: If races don't exist, why are forensic anthropologists so good at identifying them? *Social Science & Medicine*, 34(2), 107–111.

Tise, M. L., Kimmerle, E. H., & Spradley, M. K. (2014). Craniometric variation of diverse populations in Florida: Identification challenges within a border state. *Annals of Anthropological Practice, 38*(1), 111–123.

Zinni, D. P., & Crowley, K. M. (2012). Human odontology and dentition in forensic anthropology. In M. A. Tersigni-Tarrant, & N. R. Shirley (Eds.), *Forensic anthropology: An introduction* (pp. 70–86). Boca Raton, FL: CRC Press.

CHAPTER 6

Globalization, Transnationalism, and the Analytical Feasibility of Ancestry Estimation

Joseph T. Hefner and Jennifer F. Byrnes

CONTENTS

INTRODUCTION

Forensic anthropologists approach the estimation of ancestry in many ways – from traditional cranial and dental metric methods incorporating large reference samples and robust classification statistics (Edgar, 2013; Hughes et al., 2019; Jantz & Ousley, 2005; L'Abbé et al., 2013; Maier et al., 2015; Murphy & Garvin, 2018; Pilloud et al., 2014; Spradley & Jantz, 2016), to more subjective methods of analysis relying on trait lists and the observer's "eye" for human variation (Birkby et al., 2008; Gill, 1995; Gill & Rhine, 1990; Hefner, 2014; Hefner & Ousley 2014; Hurst, 2012; Klales & Kenyhercz, 2015; Maddux et al., 2015; Rhine, 1990). Statistical analyses of macromorphoscopic traits – a suite of cranial nonmetric traits preferred by forensic anthropologists for the estimation of ancestry – have gained popularity in recent years (Hefner & Linde, 2018), in part because they offer more rigorous methods of analysis over the trait list approach, but also because macromorphoscopic traits still involve observer expertise and participant judgment beyond simple majority voting and trait familiarity (Thomas et al., 2017).

One impetus for the revival in macromorphoscopic trait (MMS) analysis is Hefner's (2009) attempt to remove subjectivity from cranial nonmetric trait evaluations. His research culminated in the collection of 17 MMS traits (see Hefner & Linde, 2018), scored using categorical scales for character states representing binary (i.e., presence/absence), nominal (e.g., shape variables), and ordinal (e.g., small, medium, large) expressions (Hefner, 2009). Using these data, Hefner and Ousley (2014) presented a series of statistical classification methods useful for forensic anthropological casework. However, the methods they proposed – such as Optimized Summed Scored Attributes (OSSA), which works for only two groups, and decision tree analysis, which does not permit missing data – are not always appropriate and do not necessarily permit a level of resolution on a scale beyond three or four human groups.

The Macromorphoscopic Databank (MaMD) was established to provide MMS trait data for various populations from around the world (see Table 6.1; Hefner, 2018). The MaMD comprises four data tables that include demographic data and trait scores. The

TABLE 6.1 Correspondence between Three-Group Ancestry and Finer Levels of Resolution in the MaMD

Three-Group (Typological) Ancestry	Ancestry (Geographic)	Peer-Perceived Ancestry (US)	Geographic Origin	
African	African	American Black	American (modern) Black	
			American (19th c.) Black	
			East Africa	
			West Africa	
Asian	Asian	Asian American	Asian American	Thailand
		Chinese	Chinese	
		Japanese	Japanese	
	Pacific Island		Australia	New Zealand
			Borneo	Papua New Guinea
			Easter Island	Philippine Islands
			Fiji	Samoa
			French Polynesia	Solomon Islands
			Guam	Sri Lanka
			Indonesia	Sumatra
			Malaysia	Vanuatu
			Maori	
	Amerindian	Amerindian	Midwest Amerindian	Southeast Amerindian
			Northeast Amerindian	Southwest Amerindian
			Northwest Amerindian	Eskimo
			Pacific Northwest Amerindian	
		Hispanic	Mexico	Honduras
			Colombia	Peru
			El Salvador	Southwest Hispanic
			Guatemala	
European	European	American White	America (modern) White	
			America (19th c.) White	
			Germany	
			Holland	

MaMD data tables contain values for cranial macromorphoscopic traits largely following the definitions provided by Hefner (2009, 2018) and Hefner and Linde (2018). The demographic data tables include the following, when available, for each individual: three-group ancestry (typological), ancestry (i.e., geographic), peer-perceived ancestry (i.e., US-designation), geographic origin (i.e., place of birth), tribe (Amerindian sample), age-at-death, sex, and birth year.

The three-group ancestry level reflects the three-group model approach to ancestry (cf., Barnholtz-Sloan et al., 2005; Caufield et al., 2009; Feldman, Lewontin & King, 2003; Hefner et al., 2015; Li et al., 2008) and should be used as an initial level of classification (*African, Asian, European*). In some ways, the ancestry (geographic) level falls into a geographically based typological category, although more variation is accounted for at this level than in the three-group model. Five categories are included: African, Asian, Amerindian, European, and Pacific Islanders. The Peer-Perceived Ancestry category contains seven US-specific designations. While many of these classifications are redundant to geographic origin, this level of analysis is sufficient in many forensic anthropological investigations. Geographic origin corresponds to a specific geographical location of birth and is the finest level of analysis for the modern dataset in the MaMD. In many instances, this level of resolution may be too exacting for a forensic investigation. The overlap between groups in the expression of macromorphoscopic traits may be too great for this level of analysis to be used effectively (Hefner, 2018). However, as will be demonstrated below, the results of an analysis at this level can help guide medicolegal death investigators and law enforcement officials to the eventual identification of an unknown individual.

Transnationalism, globalization, and the changing demographic structure of the United States (US) necessitate levels of analysis finer than the three-group model can achieve. Transnationalism is the building of "social fields" (e.g., economic, political, and sociocultural activities) between settled immigrants and their country of origin (Schiller et al., 1992). However, the concept is so closely tied to trends in globalization and immigration that forensic anthropologists are faced with an increasingly difficult task of identifying individuals who may not have been encountered during forensic anthropological casework 20 years ago.

The US demographic composition is different today than how it was when compared to the 1950s (Shrestha & Heisler, 2011). US immigration has more than doubled each decade since the 1930s (see Figure 6.1), changing how forensic anthropologists approach the estimation of ancestry. Distinguishing an *American Black* from an *American White* or the occasional *Amerindian* is no longer sufficient (Birkby, 1966; Giles & Elliot, 1962; Hefner & Ousley, 2014). Today's leading source countries for immigration into the US include Mexico, China, the Philippines, India, Cuba, and the Dominican Republic. As well, there are smaller contributors from countries such as El Salvador, Guatemala, Colombia, and various Oceanic regions including Fiji, the Solomon Islands, and numerous sub-regions of Australia and New Zealand that remain important to this changing American-immigrant melting pot (see Figure 6.2; US Census Bureau, 2016). The traditional approach, the three-group model, forced all of human variation into three cohorts (*African, Asian,* or *European*), but when finer levels of geographical origin are compared to this three-group pooling, the numbers tell two different stories (see Figure 6.2, a, b, and c).

Certain assumptions are necessary to pool these data into three groups. To which geographic group should one include the *Hispanic* populations? Do the pooled groups – for example, the *African* sample – represent a single, homogenous unit of analysis? Is a

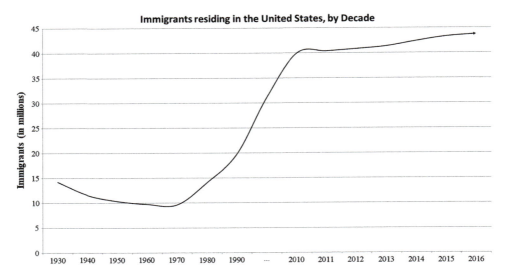

FIGURE 6.1 US immigrants living in the United States from 1930 to 2010. Note: Data derived from US Census Bureau 2016.

single classification (e.g., *Asian* for an individual born in Sumatra) specific enough for law enforcement and medicolegal death investigators to reach an identification? In the sections that follow, we present and discuss two case studies incorporating various levels of resolution. As with most casework, we combine different methods of analysis and levels of resolution.

A SPECIFIC PACIFIC (ISLANDER)?

The skeletal remains of an unknown individual were recovered on the island of Oʻahu, Hawaiʻi (see Figure 6.3). Local authorities from the City and County of Honolulu's Department of the Medical Examiner requested a full forensic anthropology report by a forensic anthropologist (JFB). All analyses were conducted without reference to any known demographic or contextual data (i.e., blindly).

Prior to a positive identification, biological profile data were generated using standard forensic anthropological methods of analyses, including an estimation of ancestry using craniometric and MMS trait data. Skeletal indicators of sex and age were most consistent with a middle-to-old-aged female. Therefore, all subsequent analyses focused on adult female samples.

An initial estimation of ancestry utilized craniometric data and Fordisc 3.1 (FD3) (Jantz & Ousley, 2005). All females were included in this analysis. FD3 used 10 Forward mean %-selected variables (BPL, ZYB, WFB, MAB, BNL, BBH, FOB, AUB, MAL, and XCB[1]) in a cross-validated discriminant function analysis (DFA) that correctly classified 74.1% of the reference sample (see Table 6.2). Using this model, the unknown individual was classified closest to the *Hispanic female* sample (see Table 6.2). However, the Mahalanobis distances ($d = 14.4–32.2$) to the group centroids were moderately high for all groups and visualization of the DFA in multivariate space located the unknown individual outside the 95% confidence ellipse for *Hispanic females* (see Figure 6.4). The

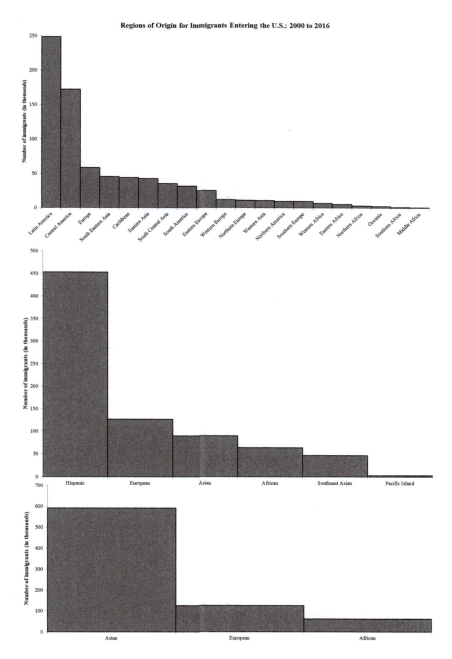

FIGURE 6.2 Number of Immigrants entering the US between 2000 and 2016, by (a) world region, (b) broad country of origin, and (c) three-group pooled ancestry. Note: Data derived from US Census Bureau 2016.

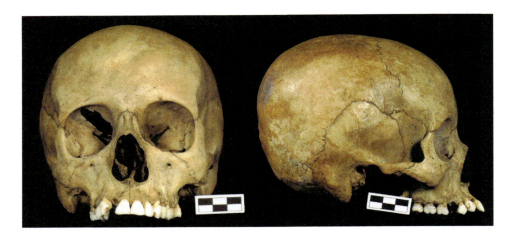

FIGURE 6.3   Anterior and right lateral views of the cranium.

TABLE 6.2   Final Classification Tables from Fordisc 3.1 (Top) and the CAP (Bottom) Analyses

FD3			
Group	Distance	Posterior Probability	Typicality (F) Probability
Hispanic Female*	14.4	0.473	0.192
Black Female	14.9	0.361	0.164
White Female	16.5	0.162	0.104
Japanese Female	24.0	0.004	0.011
Amerindian Female	32.2	0.000	0.001

CAP			
Group	Distance	Posterior Probability	Typicality (F) Probability
French Polynesia*	3.4	0.268	0.909
Papua New Guinea	3.5	0.246	0.946
Philippine	4.1	0.186	0.880
American Black	4.9	0.123	0.778
American White	5.0	0.121	0.778
Asian	6.5	0.056	0.871

Note: Each "*" indicates classification for that method.

rather anomalous nature of these results is generally characteristic of (1) measurement error, (2) atypicality (e.g., an outlier), or (3) potential misclassification due to nonrepresentative reference data. Therefore, all measurements were reassessed to safeguard against error introduced during data collection. The cranium was assessed visually for pathologies and other irregularities that may affect craniometric analysis. No errors or abnormalities were detected.

MMS traits were collected from the cranium using Hefner's (2009) trait descriptions and scores (Hefner & Linde, 2018). Using a canonical analysis of the principal coordinates (CAP with classification; Hefner & Ousley, 2014), these data were compared to

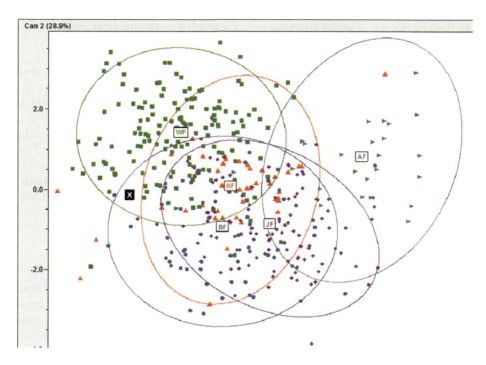

FIGURE 6.4   Visualization of the DFA from Fordisc 3.1 in multivariate space. Note: Unknown individual designated "X."

reference data from multiple populations in the MaMD database. A CAP analysis is simi-lar in scope and function to the DFA in FD3, with the exception that MMS data, rather than craniometric data, are applied to the model. As with FD3, groups and variables are selected for analysis. Future versions of the now-beta version of this MaMD program will also permit the end-user to select from various machine learning models (e.g., artificial neural networks, random forest models, *k*-nearest neighbor).

The CAP analysis (see Table 6.2) began with the three-group model (i.e., *African*, *Asian*, and *European*) using all available variables (see Hefner, 2016 for a more detailed description of the CAP method). The three-group model approach is preferred as the initial assessment to establish the broad-spectrum ancestry of an unknown individual (Hefner, 2018). Those results specified *Asian* ancestry as the most likely origin. Next, each of the ancestry (geographic) levels for the Asian samples was included in the CAP analysis. This included samples representing the following populations: *Asian* (Mainland and Japan); *Pacific Island*; and *Amerindian*. In every analysis (full model, stepwise selec-tion, etc.) the unknown individual classified closest to the *Pacific Island* sample. The final model further reduced the level of resolution to *geographic origin*. To guarantee a robust analysis, the following samples were included: *American Black*, *American White*, *Asian* (mainland and Japan), *French Polynesia*, *Indonesia* (including Sumatra as a sub-sample), *Papua New Guinea*, *Philippine* and *Solomon Islands*, and *Thailand*. The unknown cra-nium classified closest to the *French Polynesia* (d = 3.4), *Papua New Guinea* (d = 3.5), and *Philippine* (d = 4.1) samples. The relative similarity of these three groups in the classification suggested this level of resolution may be too fine to report, so the ancestry (geographic) level – i.e., *Pacific Islander* – was provided to the medical examiner's office in the final forensic anthropology report.

Circumstantial evidence suggested the skeletal remains may be from a person reported missing. Medical records – including antemortem radiographs – were obtained for the missing person and used to obtain a positive identification. Following the positive identification, the individual was confirmed to be a middle-aged adult female originating from the *Philippines*.

In this example, MMS data and reference samples from the MaMD resulted in greater precision in ancestry estimation than has traditionally been possible. The lack of a suitable reference sample in FD3 led to an "incorrect" classification; however, the fail-safes built into FD3 caveated that classification and warranted additional analyses.

In the next example, we highlight a case study where craniometric and MMS trait data could not rule out individual idiosyncrasy and atypicality.

HEARING HOOVES AND THINKING ZEBRAS

In 2017, a contract archaeologist performing a survey on private property discovered human skeletal remains (see Figure 6.5) scattered over an approximately 69m area. The Kauai Police Department (KPD) was contacted by the archaeologist; KPD subsequently requested the assistance of forensic anthropologists to assist in the recovery of the skeletal remains and associated material evidence. A rucksack containing a state identification was also recovered from the scene and presumably belonged to the decedent. All forensic anthropological analyses were conducted blindly.

Prior to the positive identification, biological profile data were generated using standard forensic anthropological methods of analyses, including an estimation of ancestry using craniometric, macromorphoscopic, and mandibular trait data. Skeletal indicators

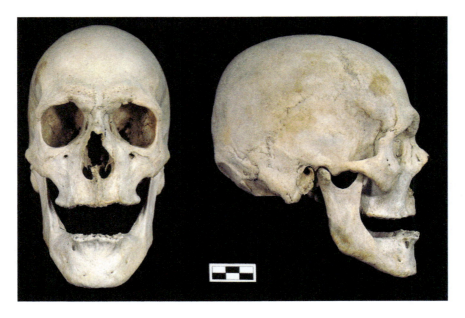

FIGURE 6.5   Anterior and right lateral views of the skull.

of sex and age were most consistent with a middle-to-old-aged male. Therefore, all subsequent analyses focused on adult male samples.

All males were included in this analysis. The FD3 was used in Fordisc 3.1 (Jantz & Ousley, 2005) with an eight-group model using 16 variables (AUB, BBH, BNL, BPL, FRC, GOL, NLB, NLH, OBB, OBH, OCC, PAC, UFHT, WFB, XCB, ZYB[1]) in a cross-validated DFA that correctly classified 61.2% of the reference sample (see Table 6.3). All typicality probabilities were below 0.05 except for *White males* (TypF = 0.099); Mahalanobis distances ranged from 24.1 (*White male*) to 45.9 (*Amerindian male*) (see Table 6.3). Visualization of the DFA was consistent with an atypical individual (see Figure 6.6). As in the previous case study, all measurements were reassessed to ensure no errors were made during data collection, and the cranium was assessed visually for pathologies and other anomalies that may affect craniometric analysis. Again, no irregularities or abnormalities were detected that could not be explained by alveolar resorption (nearly edentulous maxilla and mandible) or general morphological changes associated with aging.

MMS traits were collected (see Table 6.4). As with the first case study, these data were subjected to a CAP using reference data from the MaMD.

The three-group analysis indicated the unknown individual was most similar to the *European* sample. The results were not compelling; moderately low typicality values and high distances from all three group centroids were noted, consistent with the results from Fordisc. Additional analyses at finer levels of resolution (see Table 6.3) indicated the trait scores were closest to the *American White* sample, but *Southwest Hispanic/Guatemalan* samples were closely situated to the unknown individual as well.

Additional analysis of the mandible using (hu)MANid (Berg & Kenyhercz, 2017) did not clarify the already muddled ancestry estimates (see Table 6.3). Analysis of metric and morphoscopic mandibular traits (again) classified the unknown individual closest to the *White male* sample (posterior probability = 0.582; Chi-Square typicality = 0.195), but the *Black male* sample (posterior probability = 0.338) was a strong second and had the highest Chi-Square typicality (C-Stp = 0.680).

The cranium and mandible do not appear morphologically idiosyncratic (see Figure 6.5). The vault is high and relatively short. Although the nasal bones were fractured (antemortem), the steep-sided walls and modest surface plateau with moderate nasal suture pinching is a frequent observation in the *American White* sample in the MaMD (Hefner & Linde, 2018). The sutures – fused or fusing are not pathologically altering the general shape of the cranium. No one of the MMS trait scores (see Table 6.4) is outside of the range expected in any of the samples from the MaMD, let alone the *American White* sample. In fact, there is nothing of anthropological interest with this cranium. Yet, in the craniometric, macromorphoscopic, and mandibular analyses, the results indicated a somewhat uncommon suite of shape features.

Should the results from these analyses be reported even though the models indicate a level of individual idiosyncrasy? And how do forensic anthropologists objectively evaluate their expert-level surety that there is an "indefinable something" (Stewart, 1979, p. 231) *weird* about the analytical results, but not the skull, which looks fairly typical? Consistently, each method quantitatively identified *American White* as the most likely candidate group, but not without sounding alarms (Spradley & Jantz, 2016). Forensic anthropologists routinely use their initial impression of ancestry from the cranial *gestalt* as a starting point for their analysis (Berg & Tersigni-Tarrant, 2014; Hefner & Ousley, 2014). Based on her experience with reference collections and casework, the forensic anthropologist (JFB) subjectively believed the cranium was most consistent with other

TABLE 6.3   Final Classification Tables from Fordisc 3.1 (Top), CAP (Middle), and (hu)MANid (Bottom) Analyses

FD3

Group	Distance from Centroid	Posterior Probability	Typicality (F) Probability
White Male*	24.1	0.948	0.099
Vietnamese Male	32.0	0.019	0.015
Chinese Male	32.0	0.019	0.014
Black Male	33.8	0.007	0.008
Japanese Male	35.1	0.004	0.006
Hispanic Male	35.7	0.003	0.004
Guatemalan Male	45.3	0.000	0.000
Amerindian Male	45.9	0.000	0.000

CAP

Group	Distance from Centroid	Posterior Probability	Typicality (F) Probability
American White*	7.8	0.431	0.373
Guatemalan	8.8	0.262	0.315
SW Hispanic	10.5	0.114	0.214
Colombian	11.7	0.062	0.126
Thailand	12.0	0.054	0.105
Fiji	12.0	0.053	0.100
Japanese	14.1	0.019	0.421
American Black	17.1	0.004	0.022

(hu)MANid

Group	Distance from Centroid	Posterior Probability	Typicality (X^2) Probability
White Male*	——	0.582	0.195
Black Male	——	0.338	0.680
Hispanic Male	——	0.057	0.148
Thai Male	——	0.012	0.002
Korean Male	——	0.004	0.000
Guatemalan Male	——	0.003	0.000
Chinese Male	——	0.002	0.000
Cambodian Male	——	0.001	0.000
Vietnamese Male	——	0.000	0.000

Note: Each "*" indicates classification for that method.

American White male skulls. Unlike classification statistics, there is no way to quantify the *gut*; there is no typicality probability for experience.

As noted earlier, antemortem trauma to the nose prevented a score for anterior nasal spine. Heavy methamphetamine use rendered this individual nearly edentulous, which affected cranial and mandibular morphology. These factors may have

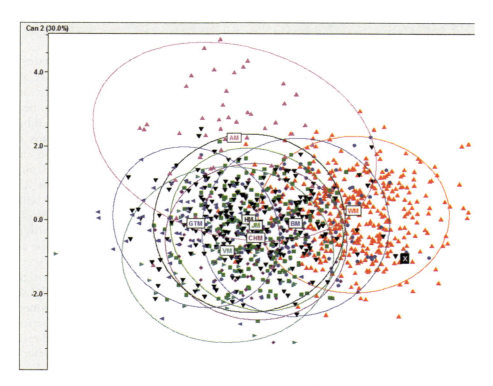

FIGURE 6.6 Visualization of the DFA from Fordisc 3.1 in multivariate space. Note: Unknown individual designated "X."

TABLE 6.4 Macromorphoscopic Trait Scores Taken from the Unknown Cranium and Used in the CAP Analysis

Trait	Score
Anterior Nasal Spine	——
Inferior Nasal Aperture	5
Interorbital Breadth	1
Malar Tubercle	2
Nasal Aperture Width	2
Nasal Bone Contour	3
Nasal Overgrowth	0
Postbregmatic Depression	0
Supranasal Suture	0
Transverse Palatine Suture	1
Zygomaticomaxillary Suture	0

influenced the anomalous nature of the classification statistics. However, the suspected decedent was a *White* male, reported missing by his family. The KPD requested DNA analysis, which subsequently produced the positive identification. There is nothing anomalous about the decedent we could ascertain, and nothing in his family or personal history (with the possible exception of heavy drug use) explained the aberrant results.

LESSONS LEARNED

- Accurately and confidently estimating ancestry from human skeletal remains is challenging. This is especially true when one understands that an accurate and confident estimate should also account for (or at a minimum consider) the total range in human variation. Yes, the typological three-group model takes some of the pressure off of a forensic anthropologist, because it collapses group membership into three choices: *Asian*, *African*, or *European*. But these three groups are not always appropriate for narrowing down antemortem record searches because they do not satisfactorily describe today's demographic population in the United States. Some of the difficulty associated with ancestry estimation in the 21st century (Plemons & Hefner, 2016; Spradley & Hefner, 2016) is tied to the changing US demographic landscape. As many forensic anthropologists have muttered to themselves in their laboratory, "Sure, this skull is pretty consistent with an *American Black*, but all of the results are peculiar – and really it could be any group – so I can't rule out <insert group name> or mixed ancestry."
- In the United States, social identity is closely associated with country of origin; transnationalism necessitates and warrants maintaining cultural and social ties to shared cultural and biological heritage. This means *Asian* is not the same as *Pacific Islander* or *Indonesian*, just like *European* is not the same as *American White*. The exponential growth in the number of immigrants entering and residing in the United States will be concomitantly linked to a rise in the number of individuals from those groups entering the forensic anthropological record (i.e., casework), making the already arduous task of ancestry estimation much more demanding. Nevertheless, forensic anthropologists are "still good at identifying" ancestry from human skeletal remains (Ousley et al., 2009; Sauer, 1992). "How-they-do-it" may be the only part of the entire ancestry estimation process that is undefinable and unquantifiable.
- The preceding case studies highlight two explanations (maybe they define and quantify the process to a certain degree) of how forensic anthropologists estimate ancestry, namely: *data* and *experience*. In the first case study, an appropriate reference sample drawn from the MaMD was necessary to refine the ancestry estimate. Any apparent misclassification by FD3 is not anathema to the tool, but an illustration of what Jantz and Ousley (2005) have repeatedly argued: as with any tool, Fordisc 3.1 is limited by available resources (*contra* Armelagos & Goodman, 1998; Goodman, 1997; Goodman & Armelagos, 1996). The Forensic Anthropology Databank (Ousley & Jantz, 1998) is the primary source for reference data in FD3. As the Databank grows in size – in other words, as more forensic anthropologists furnish identified cases to the FDB – reference data gaps within Fordisc will decrease and facilitate finer levels of resolution.
- At the beginning of this chapter, observer expertise and participant judgment were referenced as an advantage to macromorphoscopic trait analysis. But observer experience and participant judgment does not start or stop with MMS trait analysis. The second case study highlights the importance of experience when interpreting the results of a test method. The initial, *gestalt* impression of the cranium and mandible in the second case study did not indicate any group other than *American White*. That each test method classified the skull as *American White* was not unexpected. What was unexpected, however, was the atypicality of such a seemingly typical skull. Conflicting or paradoxical results

were detected in every analysis. The most parsimonious explanation for these atypical results would be individual idiosyncrasy; yet, the impact of human immigration as a globalizing factor on forensic anthropological estimations of ancestry has swung the pendulum from extreme typology and "cherry-picked data" (Ousley et al., 2018, p. 71) to indecision, or at the very least uncertainty, in ancestry estimation.

- Forensic anthropologists will continue to use a variety of methods and tools to estimate ancestry from skeletal remains. To do so, they will draw on their past experiences: skulls they have seen in the laboratory or at a museum or a particularly problematic Fordisc analysis that was eventually resolved and is now used as an anecdotal case described to colleagues over a whiskey. Or, maybe that experience is an unresolved case, a constant reminder sitting in a labeled box on a shelf in the lab. Experience plays a role in forensic anthropological analysis, including the estimation of ancestry. The ability to look at a skull and gauge geographic origin is not magic; it is a learned skill predicated on the observation of a large number of skulls from all over the world. And while estimating ancestry is not some act of prestidigitation or divination, it is also not quantifiable or capable of empirical testing. In the end, ancestry estimations must be based on quantified methods of analysis. But the interpretation of those results and their meanings will always be based on the expert's appreciation of human variation.

DISCUSSION QUESTIONS

6.1 Classification statistics have built in measures for error and atypicality, but they may not always alert end-users to such issues because of the nature of human variation. What are some of those measures, how can you use experience to interpret them, and how would you caveat an estimation of ancestry if your "gut" told you some aspect of that estimation was atypical?

6.2 Globalization has implications for many aspects of human culture and biology. How will population movement and increased gene flow between once geographically or politically separated human populations affect craniofacial morphology and ancestry estimation?

6.3 How is craniometric data different from macromorphoscopic data? Contrast how macromorphoscopic data and craniometric measurements have been used to estimate ancestry in the past and how they are used today.

6.4 Ancestry is occasionally described as one of the most difficult estimations a forensic anthropologist faces in the process of creating the biological profile from unknown human remains. Why would that be the case? Are there aspects of ancestry estimation requiring improvement? Given unlimited resources, how would you improve ancestry estimation from the skeleton?

ACKNOWLEDGMENTS

We are grateful to the editors for inviting us to contribute to these case studies. We would like to thank the Kauai Police Department and the City & County of Honolulu's Department of the Medical Examiner for permitting us to use these cases as an educational tool. R. Kalani Carreira and Stephanie Medrano volunteered their time to assist with the field and/or lab processing of these cases. Finally, we would like to thank the Macromorphoscopic Databank for access to the reference data. Kelly Kamnikar, Amber Plemons, and Micayla Spiros provided useful insight and assistance with earlier versions of this manuscript. All errors and/or omissions are ours.

NOTE

1. For a definition of the abbreviations used in Fordisc, please see the Fordisc 3.1 Help File or Langley et al. 2016

REFERENCES

Armelagos, G. J., & Goodman, A. H. (1998). Race, racism, and anthropology. In A. H. Goodman & T. L. Leatherman (Eds.), *Building a new biocultural synthesis: Political-economic perspectives on human biology* (pp. 359–378). Ann Arbor, MI: University of Michigan Press.

Barnholtz-Sloan, J. S., Chakraborty, R., Sellers, T. A., & Schwartz, A. G. (2005). Examining population stratification via individual ancestry estimates versus self-reported race. *Cancer Epidemiology, Biomarkers & Prevention, 14*(6), 1545–1551.

Berg, G. E., & Kenyhercz, M. W. (2017). Introducing human mandible identification [(hu) MANid]: A free, web-based GUI to classify human mandibles. *Journal of Forensic Sciences, 62*(6), 1592–1598.

Berg, G. E., & Tersigni-Tarrant, M. A. (2014). Sex and ancestry determination: Assessing the "Gestalt." *Proceedings of the 66th annual meeting of American Academy of Forensic Sciences Annual Meeting*, Seattle, WA (Vol. 20, pp. 414–415).

Birkby, W. H. (1966). An evaluation of race and sex identification from cranial measurements. *American Journal of Physical Anthropology, 24*, 21–28.

Birkby, W. H., Fenton, T. W., & Anderson, B. E. (2008). Identifying Southwest Hispanics using nonmetric traits and the cultural profile. *Journal of Forensic Sciences, 53*(1), 29–33.

Caufield, T., Fullerton, S. M., Ali-Khan, S. E., Arbour, L., Burchard, E. G., Cooper, R. S., …, & Daar, A. S. (2009). Race and ancestry in biomedical research: Exploring the challenges. *Genome Medicine, 1*(8), 8.1–8.8.

Edgar, H. J. H. (2013). Estimation of ancestry using dental morphological characteristics. *Journal of Forensic Sciences, 58*(S1), S3–S8.

Feldman, M. W., Lewontin, R. C., & King, M-C. (2003). Race: A genetic melting pot. *Nature, 424*, 374.

Giles, E., & Elliot, O. (1962). Race identification from cranial measurements. *Journal of Forensic Sciences, 7*(2), 147–157.

Gill, G. W. (1995). Challenge on the frontier: Discerning American Indians from whites osteologically. *Journal of Forensic Sciences, 40*, 783–788.

Gill, G. W., & Rhine, S. (1990). *Skeletal attribution of race: Methods for forensic anthropology*. Albuquerque, NM: Maxwell Museum of Anthropology.

Goodman, A. H., & Armelagos, G. J. (1996). The resurrection of race: The concept of race in physical anthropology in the 1990s. In L. Reynolds & L. Lieberman (Eds.), *Race and other misadventures: Essays in honor of Ashley Montagu in his ninetieth year* (pp. 174–186). Dix Hills, NY: General Hall.

Goodman, A. H. (1997). Bred in the bone? *The Sciences, 37*(2), 20–25.

Hefner, J. T. (2009). Cranial nonmetric variation and estimating ancestry. *Journal of Forensic Sciences, 54*(5), 985–995.

Hefner, J. T. (2014). Cranial morphoscopic traits and the assessment of American Black, American White, and Hispanic Ancestry. In S. Ta'ala & G. Berg (Eds.), *Biological affinity in forensic identification of human skeletal remains: Beyond black and white* (pp. 27–42). Boca Raton, FL: Taylor & Francis Group, LLC.

Hefner, J. T., & Ousley, S. D. (2014). Statistical classification methods for estimating ancestry using morphoscopic traits. *Journal of Forensic Sciences, 59*(4), 883–889.

Hefner, J. T. (2016). Biological distance analysis, cranial morphoscopic traits, and ancestry assessment in forensic anthropology. In M. A. Pilloud & J. T. Hefner (Eds.), *Biological distance analysis: Forensic and bioarchaeological perspectives* (pp. 301–315). San Diego, CA: Elsevier, Academic Press.

Hefner, J. T. (2018). The macromorphoscopic databank. *American Journal of Physical Anthropology, 166*(4), 994–1004.

Hefner, J. T., & Linde, K. (2018). *Atlas of human cranial macromorphoscopic traits*. San Diego, CA: Elsevier, Academic Press.

Hefner, J. T., Pilloud, M. A., Black, C. J., & Anderson, B. E. (2015). Morphoscopic trait expression in "Hispanic" populations. *Journal of Forensic Sciences, 60*(5), 1135–1139.

Hughes, C. E., Dudzik, B., Algee-Hewitt, B. F. B., Jones, A., & Anderson, B. E. (2019). Understanding (mis)classification trends of Hispanics in Fordisc 3.1: Incorporating cranial morphology, microgeographic origin, & admixture proportions for interpretation. *Journal of Forensic Sciences, 64*(2), 353–366.

Hurst, C. V. (2012). Morphoscopic trait expressions used to identify Southwest Hispanics. *Journal of Forensic Sciences, 57*(4), 859–865.

Jantz, R. L., & Ousley, S. D. (2005). Fordisc (Version 3.0). Knoxville, TN: University of Tennessee.

Klales, A. R., & Kenyhercz, M. W. (2015). Morphological assessment of ancestry using cranial macromorphoscopics. *Journal of Forensic Sciences, 60*(1), 13–20.

L'Abbé, E. N., Kenyhercz, M. W., Stull, K. E., Keough, N., & Nawrocki, S. (2013). Application of Fordisc 3.0 to explore differences among crania of North American and South African Blacks and Whites. *Journal of Forensic Sciences, 58*(6), 1579–1583.

Langley, N. R., Jantz, L. M., Ousley, S. D., Jantz, R. L., & Milner, G. (2016). *Data collection procedures for forensic skeletal material 2.0*. University of Tennessee and Lincoln Memorial University.

Li, J. Z., Absher, D. M., Tang, H., Southwick, A. M., Casto, A. M., Ramachandran, S., ... & Myers, R. M. (2008). Worldwide human relationships inferred from genome-wide patterns of variation. *Science, 319*, 1100–1104.

Maddux, S. D., Sporleder, A. N., & Burns, C. E. (2015). Geographic variation in zygomaxillary suture morphology and its use in ancestry estimation. *Journal of Forensic Sciences, 60*(4), 966–972.

Maier, C. A., Zhang, K., Manhein, M. H., & Li, X. (2015). Palate shape and depth: A shape-matching and machine learning method for estimating ancestry from human skeletal remains. *Journal of Forensic Sciences, 60*(5), 1129–1134.

Murphy, R. E., & Garvin, H. M. (2018). A morphometric outline analysis of ancestry and sex differences in cranial shape. *Journal of Forensic Sciences, 63*(4), 1001–1009.

Ousley, S. D., & Jantz, R. (1998). The forensic data bank: Documenting skeletal trends in the United States. In K. J. Reichs (Ed.), *Forensic osteology: Advances in the identification of human remains* (2nd ed., pp. 441–458). Springfield, IL: Charles C. Thomas Publisher, Ltd.

Ousley, S., Jantz, R., & Freid, D. (2009). Understanding race and human variation: Why forensic anthropologists are good at identifying race. *American Journal of Physical Anthropology, 139*, 68–76.

Ousley, S., Jantz, R. L., & Hefner, J. T. (2018). From Blumenbach to Howells: The slow, painful emergence of theory through forensic race estimation. In C. C. Boyd & D. C. Boyd (Eds.), *Forensic Anthropology: Theoretical framework and scientific basis* (pp. 67–97). Hoboken, NJ: Wiley.

Pilloud, M. A., Hefner, J. T., Hanihara, T., & Hayashi, A. (2014). The use of tooth crown measurements in the assessment of ancestry. *Journal of Forensic Sciences, 59*(6), 1494–1500.

Plemons, A., & Hefner, J. T. (2016). Ancestry estimation using macromorphoscopic traits. *Academic Forensic Pathology, 6*(3): 400–412.

Rhine, S. (1990). Nonmetric skull racing. In G. Gill & S. Rhine (Eds.), *Skeletal attribution of race: Methods for forensic anthropology* (pp. 9–20). Maxwell Museum of Anthropological Papers No. 4. Albuquerque, NM: University of New Mexico.

Schiller, N. G., Basch, L., & Blanc-Szanton, C. (1992). Transnationalism: A new analytic framework for understanding migration. *Annals of the New York Academy of Sciences, 645*(1), 1–24.

Shrestha, L. B., & Heisler, E. J. (2011). *The changing demographic profile of the United States.* Washington, DC: Congressional Research Service.

Spradley, M. K., & Hefner, J. T. (2016). Using non-metric traits to estimate ancestry in the 21st century. *Proceedings of the 85th annual meeting of American Association of Physical Anthropologists*, Atlanta, GA.

Spradley, M. K., & Jantz, R. L. (2016). Ancestry estimation in forensic anthropology: Geometric morphometric versus standard and nonstandard interlandmark distances. *Journal of Forensic Sciences, 61*(4), 892–897.

Stewart, T. D. (1979). *Essentials of forensic anthropology – Especially as developed in the United States.* Springfield, IL: Charles C Thomas.

Sauer, N. J. (1992). Forensic anthropology and the concept of race: If races don't exist, why are forensic anthropologists so good at identifying them? *Social Science & Medicine, 34*(2), 107–111.

Thomas, R. M., Parks, C. L., & Richard, A. H. (2017). Accuracy rates of ancestry estimation by forensic anthropologists using identified forensic cases. *Journal of Forensic Sciences, 62*(4), 971–974.

US Census Bureau. (2016). Foreign-born populations, 1930–2016. Retrieved from https://www.census.gov/topics/population/foreign-born/guidance/acs-by-table-number.html. Accesssed on September 14, 2018.

Forensic Taphonomy

CHAPTER 7

What Forensic Taphonomy Can Do for You: A Case Study in Rural Pennsylvania

Andrea M. Ost, Rhian R. Dunn, and Dennis C. Dirkmaat

CONTENTS

INTRODUCTION

It was a hot summer day when Dr. Dennis Dirkmaat received a call in his office at Mercyhurst University in Erie, Pennsylvania, regarding human remains found at a rural residence. Dirkmaat was no stranger to calls of this nature, as he and his department average over one hundred forensic cases a year. These range from the determination of forensic significance via email and text messages (essentially human versus non-human) to assisting with the aftermath of mass disasters, such as United Flight 93 in Shanksville, Pennsylvania, in 2001. The call on this particular day, however, concerned a potential double homicide.

The initial contact came from a criminal investigator with the Pennsylvania State Police (PSP) regarding a married couple who was presumed dead. The couple's death was reported initially by the surviving son as a fiery motor vehicle accident, but upon further inspection of the circumstances surrounding the couple's death, some factors of this story lacked consistency. Grieving relatives of the couple probed the son for more details regarding the circumstances of the accident. When he was unable to produce a death certificate, relatives began to suspect that foul play had been involved and contacted the authorities. Law enforcement took notice of their concerns and decided to search the late couple's expansive property. During this search, investigators discovered what appeared to be a partial human cranium along with a number of other skeletal elements scattered along and in a pond in the backyard. A PSP trooper sent a photo to Dr. Dirkmaat to verify that the remains were human and, with that confirmation, requested his aid with the search and recovery. The photographs showed a partial human cranium and mandible exhibiting signs of burning lying in the grass next to the pond, as well as human postcranial remains in the shallow waters of the pond.

SCENE RECOVERY

Dirkmaat and the Mercyhurst Forensic Scene Recovery Team (M-FSRT), composed of other faculty members and graduate students in the Department of Applied Forensic Sciences, immediately mobilized and arrived at the scene within a few hours of the call. The property included a large house on sprawling grounds and a pond, bordered by sparse woods. Several additional large buildings containing farm equipment, vehicles, and tools were located on the property. Upon arrival, the M-FSRT van was directed to park on a gravel driveway on the property next to a burn pit and, upon stepping out of the Mercyhurst vehicle, one of the team members noticed a rounded, white object lying on the gravel surface – a human patella. Upon inspection, a chalky white appearance indicated prolonged exposure to thermal alteration. A quick search of the roughly four-foot-tall mound of dirt next to the garbage-burning pit revealed a distinctive ash lens overlaying a different colored dirt layer, within which a number of fragmented and burned (primarily calcined) human remains were located. At this point, two distinct forensic scenes required processing: one adjacent to the pond, where remains were initially discovered by law enforcement (labeled as S for mapping purposes), and another near the household garbage burn pit, designated as Site PL (see Figure 7.1). This discovery brought a new hypothesis to light: that the individual(s) likely were purposefully and extensively burned.

Forensic anthropologists must take measures to reduce the effects of bias in their analyses and to resist tailoring results to the expectations of the police. It may be easy to fit subjective analyses to match a potential ID, especially when little can be gained from the remains themselves (e.g., when only one or two bones are found, etc.). Due to this, some believe that a forensic investigator should not hear details about a case beyond their scope of the investigation because outside information provided by law enforcement may contribute to analysis bias. While bias is an important concern, remaining ignorant of law enforcement's information when entering a crime scene may limit the capabilities of a forensic scientist. Background information enables the formation of hypotheses, which can be tested throughout the recovery and analysis processes. Just as with any other scientific discipline, a forensic analysis should adhere to the scientific method, beginning with the formation of a hypothesis to test the initial question of what events occurred at the crime scene. Background information, or information given by law enforcement, facilitates the creation and testing of hypotheses.

The inclusion of a forensic anthropologist during the early stages of the outdoor scene recovery is essential for outdoor crime scene investigation for several reasons. First, forensic anthropologists are highly skilled in recognizing and properly handling biological tissues, especially decomposing and skeletonized remains, whether commingled, fragmented, or thermally altered. These skills make forensic anthropologists an asset to law enforcement who have not had the same level of osteological training. Second, while forensic anthropologists are trained primarily to assist with modern cases through the analysis of skeletal tissues with respect to biological profile, trauma, and taphonomic analyses, they are not far removed from their roots, namely those in the anthropological subdiscipline of archaeology. While archaeology is often sensationalized (one has only to watch an Indiana Jones film to see this), at its core, archaeology aims to understand the context and association of artifacts to accurately reconstruct past events. This means

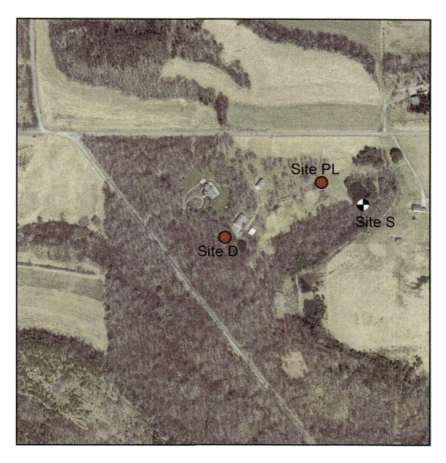

FIGURE 7.1 Aerial photograph of the property; sites where human remains were found are indicated.

attempting to account for any and all taphonomic influences, such as plants, animals, soils, climate, topography, or even (especially) subsequent human activity, in the modification of the remains and artifacts from the time of initial emplacement on the scene. Thus, this process plays a key role in the interpretation of past events at a forensic scene. When dealing with an outdoor crime scene that is "exposed to the elements," unlike indoor scenes, there is seemingly an infinite number of factors to deal with, and often the sense is that the scene is an acutely disturbed one that may be impossible to analyze properly. However, forensic anthropologists are the best-equipped investigators to reconstruct past events with scientific validity at the outdoor forensic scene. They are armed with an understanding of what can be gained by employing archaeological methods, practices, and principles to properly locate, document, and collect the maximum amount of evidence found at the outdoor scene, as well as formulate hypotheses concerning taphonomic agents (Dirkmaat & Cabo, 2016).

Creating hypotheses takes scene analysis one step further and improves efficiency and strategy. Background information from the police can be used to formulate hypotheses and, therefore, a plan of attack for the investigation (Dirkmaat & Cabo, 2016). For example, if the police have knowledge that an individual may have been buried behind a potted plant in the backyard, a forensic anthropologist can "test" this hypothesis by excavating that location. If remains are discovered, proper excavation and documentation techniques have already commenced and can continue. If this hypothesis is false, a new hypothesis can be generated and tested.

Thus, at the scene of the rural Pennsylvania residence, the formation of a hypothesis involving the intentional burning of human remains triggered a more refined comprehensive search of the property. This hypothesis suggested that the search may include bones and tissues that were thermally altered, and therefore more difficult to identify than unburned bones. Team members were notified to search areas of interest for similar thermally altered, fragmented remains, as the original elements found by PSP did not display the same level of alteration.

The commencement of the search necessitated the initiation of proper documentation in the form of written, photographic, and eventually, cartographic notation. Without proper documentation, the context and association of evidence are lost (Dirkmaat et al., 2008). Once an object is moved, it can never be put back in exactly the same place. By documenting the conditions thoroughly, it is possible to preserve (via notes and photographs) the original crime scene at the level of detail required for later interpretation in the laboratory. With laboratory evidence alone and a very limited understanding of taphonomic influences on the remains, a proper scientific reconstruction of past events at the scene is not possible. Additionally, with proper documentation and search techniques, team members are able to search 100% of an area and know where they have been, preventing the loss of context and the chance of missing any evidence.

In this case, additional support from a cadaver dog and handler was requested due to the property's expansive grounds. Although cadaver dogs are not usually 100% accurate depending on a myriad of factors such as the postmortem interval, the presence of strong scents such as accelerants, and the condition of the remains, additional expertise can be valuable when covering a substantial area (Dirkmaat el al., 2008). The cadaver dog searched the property in the vicinity of the pond where the human skull and postcranial elements were discovered but did not alert to any additional remains.

Following the discovery of burned and calcined remains near the residence's trash-burning pile, the forensic archaeology team faced two distinct scenes to process. As the time of day was nearing dusk, it was decided that the scenes would be preserved and protected overnight by law enforcement, and processing would commence the next morning. However, before leaving for the day, the two burn barrels in close proximity to the trash-burning area were excavated, and the contents hand-sorted on tarps. This method prevents damage to brittle remains that would be incurred with screening and enables more rapid sorting of the barrel contents (as screens can only sort through so much material at once), resulting in the most efficient and effective search and sorting strategy (Dirkmaat et al., 2012). No human remains were found. Thus, one potential hypothesis – that the remains may have been burned in the barrels – was falsified.

The M-FSRT arrived the next morning with a new strategy. Two teams were created to address both areas of interest so far identified: the pond and the conical mound of dirt and ash next to the trash-burning pit. The locations of these two sites were noted with a survey-grade GPS unit. This geospatial data, combined with other points such as the perimeters of the property, enabled incorporation into Geographic Information Systems (GIS) and permitted further detailed analysis of the two sites in relation to the overall scene.

At the burn pile (Site PL), excavation of the ash lens began and led to the collection of calcined and fragmented human remains. It was determined that the remains had not been burned on that dirt pile and were secondarily deposited there. The location of the ash and burned bone feature was noted via GPS, but the specific location, position, and orientation of each individual piece of bone was not noted as the provenience data was deemed irrelevant (secondarily deposited remains cannot provide information about the original orientation of the remains to aid in the reconstruction of the death event). The condition of the remains found in this ash lens, consistent with that of the patella found on the driveway, indicated a long burning episode.

While the first team completed the recovery of the calcined remains at the house's trash pile location, another recovery team processed the human remains scattered at the pond's edge and in the water. This began by determining the extent of the immediate scene via a pedestrian search of the area, followed by the clearing of all vegetation and brush (denuding/defoliating) around and over the remains to expose any underlying evidence and understand the underlying microtopography. This defoliation and hands-and-knees search effort began at the outer perimeter of the immediate scene and moved slowly inward toward the central concentration of evidence. Denuding gives a better idea of the overall distribution pattern of the immediate scene (Dirkmaat & Cabo, 2016). Small items that are indicative of human activity (and potentially intervention) at the site, such as footprints or cigarette butts, will also be exposed through this denuding effort. The scattered position of the elements at the pond location, the lack of a decomposition stain, and the fact that the remains were partially burned, but not the soil in the area, indicated a secondary deposition. This led to a question that needed to be addressed: where were the remains burned?

The pattern of the dispersal of human elements in the pond and along the pond's edge was rather random, although initial impressions were as if the remains had been dumped from a wheelbarrow. Two small ash features were arranged at the pond's edge in a longitudinal orientation perpendicular to the water's edge, along with the cranium and mandible. Other larger elements were scattered further into the waters of the pond. The bones noted at this scene did not display the level of thermal alteration noted on the patella and remains found in the ash lens near the garbage burn pit. Instead, the highest degree of burning noted was charring, the first of many color changes seen in thermally altered bones (Mayne-Correia, 1997; Symes et al., 2008). Upon closer inspection (without disturbing the remains), the skeletal elements at this location appeared to be relatively gracile, hinting that they may be those of a female individual.

Once all of the evidence was exposed through the denuding effort, a datum and baseline were set up over the remains. From there, the scene was documented in detail with a hand-drawn plan view map (see Figure 7.2). This provides an accurate record of the location, position, and orientation of the scattered elements *in situ*, with the relationship between elements and evidence highlighted. Although written, photographic, and georeferencing information should always be noted at a scene, creating a hand-drawn map enables closer analysis of the distribution of the elements (Dirkmaat & Cabo, 2016). Vegetation and other debris that can distract from or obscure evidence in photographs are not factors with the hand-drawn map. Especially at this complex scene with presumed multiple human bodies, presence of animal remains, multiple burning sites, and fire damage, careful documentation was essential to understanding the events that had transpired. Without information regarding the context at a scene, it is difficult to associate elements or to explain taphonomic alterations to the remains; certainly, determination of original orientation or deposition is impossible.

While the two recovery teams processed their respective sites, PSP continued a comprehensive search of the grounds and the structures on the property. They discovered

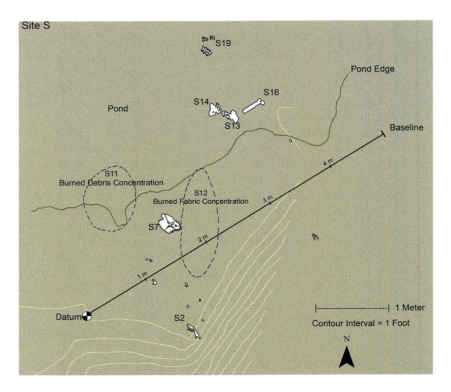

FIGURE 7.2   Hand-drawn plan view map of the pond scene.

a third potential scene near a large multi-car garage on the property. This third site consisted of an area of burnt ground, burnt wood, nails, and burnt human remains. Additionally, something glittered among the ashes – a pair of wedding rings. With evidence of human remains confirmed, a similar forensic archaeological approach was taken with this site as with the other two. A boundary was established along the outer periphery of the immediate scene, and team members denuded in a closing circle to expose all remains and evidence. Elements were documented *in situ*, then collected and bagged. This scene was designed as Site D for mapping purposes.

Simultaneous to the excavation of the three sites, PSP investigated the indoor aspects of the property. Inside a number of the out-buildings, including the garage, they found many empty gas cans. This discovery suggested that the remains had been burned over an extended period of time, using gasoline as an accelerant. Now with a site count of three and numerous fragments of human remains located, the question became: how do these locations all connect, and what was the sequence of events?

The human remains were transferred to the custody of the Mercyhurst Forensic Anthropology Laboratory for further documentation and analyses after the excavation of the three sites was complete and all skeletal elements and evidence were collected.

The laboratory analyses consisted initially of creating an inventory of the remains recovered from each of the three sites, most of which were significantly burned and fragmented. In addition, attempts were made to estimate the biological profile. These initial analyses indicated that the remains found along and in the pond were exclusively those of an adult female. The identifiable remains recovered in the ash lens near the house's trash burn pile were exclusively those of an adult male. The remains found in the burn area adjacent to the garage contained male and female remains.

Approximately four days after the initial scene processing, law enforcement contacted Dr. Dirkmaat to inform him that they had drained the contents of the pond to conduct an intense search of the pond floor for additional human remains. The M-FSRT returned to the scene for a third time. A tanker truck was called in to assist in the wet screening of the mud using 1/4" mesh sieves. Only a few additional human remains were recovered in this effort, although a particularly significant item of evidence was found: a partial dental appliance that included the upper right central and upper right lateral incisor models. The suspected female victim – a potential match for the gracile skull found at the pond the first day – was reported to have a full set of teeth. The male victim, however, was reported to have worn a dental appliance. If the appliance belonged to the late husband, it now appeared that remains of both the adult male and female were deposited at the pond location. With an interdisciplinary effort and the expertise of a forensic odontologist, a positive identification of the remains – those of the husband – was established via comparison of the dental appliance to antemortem dental records. With the positive ID of the husband, the M-FSRT could test a new hypothesis: that the male and female individuals were burned at the same place (the garage) and then dumped post-burning at two different sites (the pond and burn pile).

As more evidence was found at the couple's property, the story given by the surviving son began to alter. When questioned by law enforcement, the suspect's weak narrative about a fiery motor vehicle crash crumbled. When confronted with the fact that investigators had discovered human female remains on the grounds of the property, the suspect turned the blame to his father, claiming that perhaps he had killed his wife and fled town. Prior to the positive ID, there was no evidence to confirm that his story was not true, as all male remains were too burned to retrieve DNA. But this story, too, fractured with the positive identification (via the dental appliance) of the male adult remains as the suspect's father.

LABORATORY ANALYSIS

As law enforcement used the positive identification of the male individual to move forward with their investigation, Dr. Dirkmaat and his team focused on explaining the sequence of events at the scene as part of the effort to reconstruct the circumstances surrounding the deaths. This would require information from both the three-scene recovery and the laboratory. The team worked to reassociate the highly fragmented remains, differentiating when possible between the male and female individuals. Many fragments could not be sorted due to the highly fragmented and commingled state of the remains, but sex differences in other areas of the skeleton such as the humeral and femoral heads, in addition to the fragments of the skull and os coxae, could be attributed to a particular individual.

The final sorting of the remains in the laboratory indicated that two individuals were represented: an adult male and an adult female (see Figure 7.3). Further, lab analyses confirmed what had been observed on-site: two different burning patterns were represented. The male individual was burned to a much higher degree, as evidenced by the white, calcined, and highly fragmented appearance of the patella found on the road of the

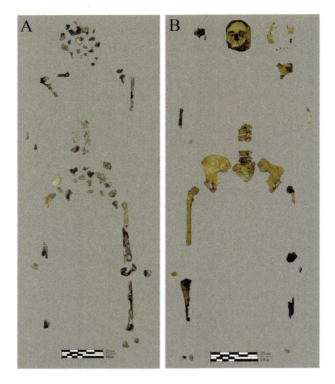

FIGURE 7.3   Skeletal elements attributed to both the male individual (A) and the female individual (B).

property (Mayne-Correia, 1997; Symes et al., 2008). The fragmentation and burn pattern suggested they were likely moved and purposely fragmented during the fire episode. In contrast, several elements of the female individual exhibited only minimal or no signs of burning, indicating that they likely were still associated with soft tissue at the time of burning (including the lower torso and pelvic region) and were exposed to a shorter duration of burning.

Careful examination of the remains also revealed perimortem wounds in the skulls of both victims. In both cases, a fragment of occipital exhibited internal beveling indicative of a gunshot entrance wound that traveled from back to front. The relatively intact base of the skull of the adult female exhibited evidence of an exit wound and permitted the reconstruction of the path of the bullet as back to front, somewhat superior to inferior, and slightly left to right (see Figure 7.4). The skull of the adult male was highly fragmented and incomplete, so an exit defect was not found (see Figure 7.5). The remains of the adult male were found in all three sites, whereas the remains of the adult female were found in only two of the sites.

The differential burning and pattern of dispersal suggested that the suspect burned the husband first at the site outside the garage. The process likely took a long time and required much fuel. Most of the burned (calcined) remains were then placed in a receptacle and dumped on the back dirt pile of the residence's garbage burn pit. This collection effort was incomplete, and some of the husband's burned skeletal elements were left in the burned area (see Figure 7.6).

FIGURE 7.4 Skull fragments of the female individual exhibiting differential burning.

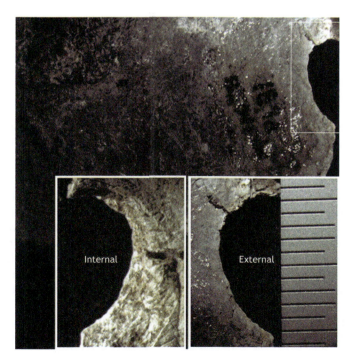

FIGURE 7.5 Posterior view of occipital fragment with close-up views of internal (left inset) and external (right inset) beveling.

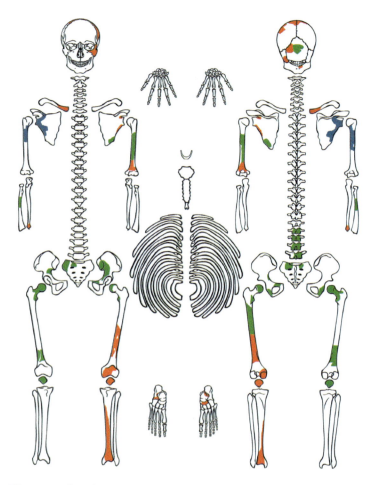

FIGURE 7.6 Homunculus diagram representing the skeletal elements of the male individual. Blue indicates remains from the pond site, orange indicates remains found at the garage site, and green indicates remains from the burn pile.

Next, the wife was burned, but not as extensively as the husband (perhaps the suspect ran out of fuel or patience). Her remains were then removed from the initial burn area near the garage and taken to the pond's edge (perhaps via a wheelbarrow) and dumped. It did not appear that much or any effort was made to move the remains from the pond's edge to deeper water. In addition, some of the remains of the husband's skeletal elements from the first burning were collected with the wife's remains (second burning) and also ended up deposited into the pond, verified by the discovery of the dental appliance (see Figure 7.7).

When confronted with this evidence, the suspect took the plea and a trial was not necessary. Without the forensic anthropology team assisting at both the scene with forensic archaeology techniques and in the laboratory to properly identify the individuals and sources of trauma, it is unlikely that the crime scene could have been reconstructed to such an extent. Forensic archaeology, and the use of hypothesis testing, is crucial at the crime scene for proper taphonomic interpretations.

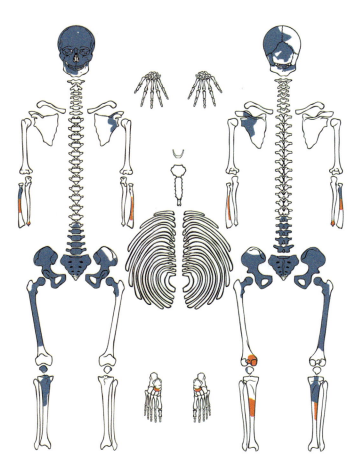

FIGURE 7.7 Homunculus diagram representing the skeletal elements of the female individual. Blue indicates remains from the pond site and orange indicates remains found at the garage site.

LESSONS LEARNED

- Using forensic archaeological techniques at the scene, such as hand-sorting the burned remains, constructing hand-drawn plan view maps, keeping elements associated with their designated sites, and using methods such as water-screening, maximum recovery of the evidence was possible, even when the bones were fire damaged, brittle, and highly fragmented. These recovery techniques, when paired with methodologies in the laboratory, enabled the forensic anthropology team to note the differential degrees of burning between the male and female remains, important in determining the sequence of events.
- Careful examination of each burned fragment also led to the discovery of the gunshot trauma, altered by the fire damage but still evident.
- Multiple levels of documentation and careful recovery processing, and analysis of each fragment enabled reconstruction of the circumstances surrounding the deaths of the two individuals, enabling law enforcement to confront the suspect

with enough evidence that a trial was not necessary. These taphonomic interpretations would not have been possible at such a complex scene without context and without continually testing hypotheses. This case is a particularly well-rounded argument for the importance of information gained both at the recovery and in the laboratory; only by considering all evidence and continually testing hypotheses can the most parsimonious explanation be uncovered.

DISCUSSION QUESTIONS

7.1 The authors argue that investigative information is important for hypothesis testing when conducting forensic archaeological searches and recoveries as well as forensic anthropological analyses. Other practitioners feel that due to potential biases, all analyses should be completed blindly. What are the pros and cons of both approaches?

7.2 Had a forensic anthropological team not been called to conduct the search and recovery in this case, and instead, law enforcement collected the remains and sent them to the forensic anthropology laboratory, what information may have been lost? How might that complicate analyses, interpretations, and the case outcome?

7.3 Do you see any patterns in the anatomical distribution of the remains as depicted in the homunculus images (Figures 6 and 7)? How might you explain those distributions?

REFERENCES

Dirkmaat, D. C., & Cabo, L. L. (2016). Forensic archaeology and forensic taphonomy: Basic considerations on how to properly process and interpret the outdoor forensic scene. *Academic Forensic Pathology*, 6, 1–16.

Dirkmaat, D. C., Cabo, L. L., Ousley, S. D., & Symes, S. A. (2008). New perspectives in forensic anthropology. *American Journal of Physical Anthropology*, 51, 33–52.

Dirkmaat, D. C., Olson, G. O., Klales, A. R., & Getz, S. (2012). The role of forensic anthropology in the recovery and interpretation of the fatal fire victim. In D. C. Dirkmaat (Ed.), *Companion to forensic anthropology* (pp. 113–125). New York: John Wiley & Sons.

Mayne-Correia, P. M. (1997). Fire modification of bone: A review of the literature. In: W. D. Haglund, & M. H. Sorg (Eds.), *Forensic taphonomy: The postmortem fate of human remains* (pp. 275–293). Boca Raton, FL: CRC Press.

Symes, S. A., Rainwater, C. W., Chapman, E. N., Gipson D. R., & Piper, A. L. (2008). Patterned thermal destruction of human in a forensic setting. In C. W. Schmidt, & S. A. Symes (Eds.), *The analysis of burned human remains* (pp. 15–54). Cambridge: Academic Press.

CHAPTER 8

Dismembered, Burned, and Dumped: But in What Order?

Lindsay H. Trammell

CONTENTS

RECOVERY

In late winter, burned and dismembered human remains were discovered by workers in a wooded conservation region near a river in Missouri. Law enforcement responded to secure the area and they flagged, photographed, and mapped the remains and items of potential evidentiary value. To aid in documentation, numbered placards were placed near items of evidentiary value and near general concentrations of remains. A site datum was established, and global positioning system (GPS) coordinates were taken of the remains as well as the nearby river's edge and parking lot. Officers and cadaver dogs completed extensive grid searches in multiple directions from the original point of discovery and found no additional human remains. The appropriate local authorities also searched upstream and downstream on the river via watercraft but were unsuccessful in locating further remains. Everything found was left *in situ*, and the medical examiner's office (MEO) was contacted for forensic anthropological assistance.

The dismembered human remains were incomplete and exhibited varying stages of thermal damage from fire exposure in addition to fresh, unburned tissue. They were located beneath trees with low-hanging, unburned branches and surrounded by fresh leaf litter (see Figure 8.1).

Due to the thermal damage and dismemberment trauma, a systematic inventory was difficult to complete at the scene; however, a general catalog of skeletal elements was documented during the recovery process. The first step of the recovery involved removal of the leaf litter and miscellaneous foliage debris. Notably missing were the axial skeleton and pelvis. The numbered placards were photographed in numeric order, tissue was

FIGURE 8.1   Burned and dismembered human remains *in situ*.

placed into a body bag, and evidence not representing biological material was collected by the law enforcement crime scene unit.

A right femur was identified at the scene, as the proximal end sustained no thermal damage, and (fresh) soft tissue was still present; evidence consistent with sharp force trauma was visible to the exposed femoral head. Osseous items exhibiting calcination or obvious fragility were wrapped in aluminum foil (following Dirkmaat et al., 2012) to preserve bony integrity during transport. Smaller osseous fragments were placed in brown paper bags. Everything was sealed, dated, and labeled with the recovery location.

After removal of the burned tissue and skeletal elements from the scene, the ash and surface soil were collected with a trowel and dustpan to ensure complete recovery of additional osseous fragments and potential evidentiary items. The ash and soil were secured in a large, clear plastic bag and placed in the body bag for sorting and examination in a more controlled environment.

Unburned nonhuman remains were also discovered among the human elements. Those easily identifiable as nonhuman were bagged separately and provided to the law enforcement agency. The secured human remains were transported to the MEO for further examination. Chain-of-custody was maintained and documented by the respective agencies, and an MEO case number was allocated upon arrival.

FORENSIC ANTHROPOLOGY EXAMINATION

A limited forensic pathological examination of the burned and dismembered remains was completed by the Chief Medical Examiner, and all material was photographed and x-rayed. Soft tissue was sampled from the right femur by the investigating law enforcement agency for DNA analysis. Forensic anthropological assistance was subsequently requested.

FIGURE 8.2 Layout of remains for forensic anthropological analysis including post-macerated right femur. Metric scale.

The remains sustained varying stages of thermal damage. The right thigh had fresh soft tissue with no obvious evidence of discoloration or decomposition and a lack of insect activity. The remains were consistent in anatomical representation and biological indicators with a single, incomplete adult individual (see Figure 8.2). Portions of the limbs were recovered but the axial skeleton (including the skull) was not found. Due to extensive thermal damage, a more detailed catalog of the highly fragmented portions could not be completed.

Biological profile analyses were completed to assist in identifying the unknown decedent. Sex was estimated as male based on postcranial osteometrics of the unburned portions of the right femur (Spradley & Jantz, 2011). The gross morphological and radiological observations of complete epiphyseal fusion on the non-fragmented skeletal elements indicated the individual was an adult (Scheuer & Black, 2000). Burned tissue was cut away from the intact right tibia to facilitate a length measurement for stature estimation. A stature 90% prediction interval of 67.1 to 74.0 inches was obtained using discriminant function analysis in Fordisc 3.1 (Jantz & Ousley, 2005) for "any" adult group. Due to the incomplete recovery and the state of the remains, ancestry was not estimated.

Based on forensic pathological examination findings and sharp force trauma observed during the exam (as well as on scene), a proximal portion of the right femur containing the suspected defects was cut from the burned lower portion of the midshaft using an electric bone saw. This was done so it could be macerated for detailed evaluation without further damaging the burned remains. Given the varying levels of thermal damage, with some skeletal elements being highly calcined, the decision was made to preserve the integrity of the remaining osseous portions and not attempt maceration beyond the right proximal femur.

NEW DEVELOPMENTS

Approximately two weeks after the location and recovery of the remains, a DNA match was obtained in the Combined DNA Index System (CODIS) database, and the individual was positively identified as a 50-year-old White male. The identification led law enforcement to a primary suspect who, in turn, provided police with a secondary site where they discovered the remainder of the decedent. The intact head, torso, and left brachium were recovered in a neighboring county. These remains displayed no thermal damage and were intact and encased in fresh soft tissue. The remains were conveyed to the same MEO (as the first accession) for further examination. Because this accession of remains could be identified visually as the same decedent who was identified previously by DNA testing, the same case number was utilized.

The second accession of remains was comprised of fresh soft tissue, with mild discoloration of the soft tissue and only slight decompositional odor. No insect activity was noted. A partially skeletonized incomplete rib cage, complete vertebral column, sacrum, and pelvis were also present among the remains. The sternal ends of the ribs displayed evidence consistent with carnivore damage.

The Chief Medical Examiner completed a partial forensic pathological examination. The decedent sustained four gunshot wounds: two to the head, one to the neck, and one to the chest. The projectiles were recovered during autopsy and released to the law enforcement agency. Given the complete nature of this second set of remains, the medical examiner did not feel an anthropological analysis of the gunshot wounds was necessary but requested forensic anthropological assistance to complete a skeletal trauma analysis on select elements with evidence of dismemberment. The following elements were cut away from the individual to facilitate this examination: portions of the left and right os coxae, distal left humerus, and right clavicle. The medial end of the right clavicle had sharp force damage due to the pathologist's postmortem examination. The resected bones were macerated using a mild detergent and warm water to remove adherent soft tissue and allowed to dry prior to further analysis.

TRAUMA ANALYSIS

Evidence of sharp force trauma consistent with dismemberment was noted on several skeletal elements. Figure 8.3 shows the generalized areas of dismemberment trauma and thermal damage.

Multiple superficial parallel cut marks were visible on the lateral epicondylar ridge of the left humerus. A series of superficial cut marks was noted slightly superior and lateral to the acetabulum of the left os coxa with additional cut marks on the lunate surface of the acetabulum (see Figure 8.4). The lunate surface of the acetabulum of the right os coxa exhibited one superficial cut mark. Multiple cut marks were present on the proximal left femur, with multiple cuts observed on the superior-anterior femoral head and another on the anterior femoral neck that extended through the greater trochanter. A series of superficial cut marks was visible on the superior aspect of the left femoral neck (see Figure 8.5). Multiple cut marks were visible on the right proximal femur, as well, including one through the anterior surface of the greater trochanter and several superficial cuts just superior to this defect. A series of cut marks was observed on the anterior aspect of the right femoral head and neck (see Figure 8.6).

Overall, evidence of dismemberment trauma was observed on multiple skeletal elements, primarily at joint articulations of the limbs. Further examination of the kerf walls and cut surfaces was hindered because the injuries were predominately through trabecular bone. No discernable striations were noted via gross or microscopic analysis. The lack of striation evidence and consistency in observed kerf-shapes suggests these dismemberment injuries were likely produced with a single tool. All observed kerfs were V-shaped, consistent with the use of a knife or similarly beveled instrument. To be conservative, a distinction between a serrated or non-serrated blade was not made given the lack of

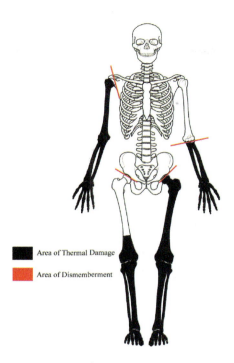

FIGURE 8.3 Illustration of generalized areas of dismemberment trauma and thermal damage.

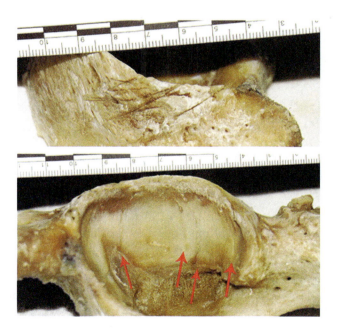

FIGURE 8.4 Series of superficial cut marks superior and lateral to the left acetabulum (above). Red arrows designate four cut marks to the lunate surface of the left acetabulum (below). Superior is left in image. Metric scale.

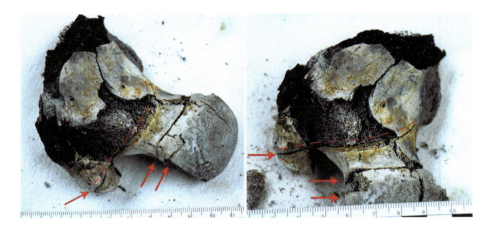

FIGURE 8.5   Two anterior views of at least three cut marks to the left proximal femur. The red arrows designate each obvious sharp force injury. The dotted red line shows the continuation of one cut through the anterior femoral neck and greater trochanter. Metric scale.

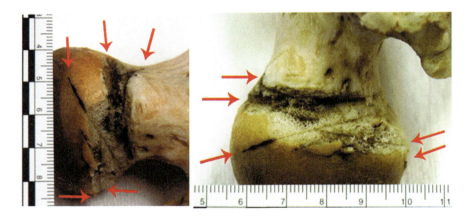

FIGURE 8.6   Anterior view (left) and superolateral view (right) of sharp force trauma to the right femoral head and neck. Red arrows show at least five cut marks. Metric scale.

visible striations. Similarly, a general lack of distinct features such as exit chipping or trailing scratches precluded observations regarding the direction of cuts.

Patterns of thermal damage were also recorded. The burned remains were sorted into the following categories: identifiable human; identifiable nonhuman; cortical bone fragments; cancellous bone fragments; and ash. All remains were photographed and radiographed. Thermal damage due to fire exposure was described and documented on the identifiable human osseous elements (see Figures 8.2 and 8.3).

The remains sustained varying stages of thermal damage, which ranged from no evidence of burning, to fragmentation and calcination of skeletal elements. As described by Symes and colleagues (2015), in a "typical" burn pattern areas of bone protected by more soft tissue are expected to be more protected from the fire (Fojas et al., 2015;

Symes et al., 2015). In an intact body, the pugilistic posture is induced as muscle fibers shrink and contract in reaction to fire. This leads to increased exposure of specific anatomical areas such as the anterior tibia, posterior elbow, and dorsal aspect of the wrist. However, the anterior portion of the elbow (cubital fossa) and the palm are more protected from damage due to joint flexion and soft tissue protection.

In this particular case, both typical and atypical burn patterns were noted. Burned soft tissue was present on the right arm and hand, left forearm, and the left and right legs and feet. Increased calcination was noted to proximal and distal portions of the burned radii and ulnae. Less tissue was present on these portions of the bones, consistent with a normal burn pattern – the epiphyseal portions burned prior to proceeding along the shafts. Also, the hands and feet exhibited the pugilistic response, indicating the wrists and ankles were intact at the time of burning.

Increased calcination was also noted on the joint surfaces of the right humeral head and left femoral head, with only charring and intact burned tissue on the shafts. This is not a normal burn pattern. Usually, these surfaces would be some of the last to burn as they are protected by the bony articulations and soft tissue of the shoulder and hip, respectively.

After detailing the sharp force trauma and thermal modification, the two sources of trauma were interpreted together. Though this case displays both expected and abnormal burn characteristics, the overall thermal alterations do not represent a normal pattern for burned human remains, which suggests that the dismemberment trauma occurred prior to the burning event. Scene context and evidence of sharp force trauma to the proximal femora, humerus, and os coxae indicate that the right arm and right and left lower limbs were dismembered at the shoulder and hip joints. This directly affected the typical burn pattern by exposing these generally protected regions to the fire. The complete absence of thermal damage on the proximal right femur in comparison to the left lower limb is also atypical. These atypical burn patterns indicate that the body was burned after it was dismembered.

INVESTIGATION RESULTS

Law enforcement investigation yielded a primary suspect who confessed to the shooting, dismemberment, and subsequent burning of the decedent. The individual was shot and dismembered in a residence and transported to two separate locations in steel drums. One barrel contained the first accession of burned and dismembered remains, and the second barrel contained the intact head, torso, and left brachium. Fire was set to one portion of the remains while still inside of the drum. This was done at the scene, and the remains were then dumped from the barrel when the fire had somewhat dissipated. The barrel was taken by the suspect. This explains the lack of unburned surrounding branches and fresh leaf litter on the scene as the fire was mostly contained. It also clarifies the atypical burn pattern observed on the remains. Bony regions exposed by the dismemberment process may have been more exposed to the fire, depending on their position within the barrel relative to the fire. Given that remains were in the barrel post-dismemberment, it is possible that the right proximal femur, although dismembered, may have ended up at the bottom of the barrel and remained relatively protected from the fire, while the left proximal femur was more exposed. The weapon used to dismember the individual was ascertained to be a partially serrated hunting knife. This is consistent with the observed sharp force trauma characteristics on the left humerus, left and right os coxae, and left and right femora.

LESSONS LEARNED

- This case study demonstrates the need to understand typical patterns of thermal damage in order to recognize and explain deviations from the expected pattern of burning. Scene documentation is also important, as the lack of burning of the surrounding leaf litter and vegetation suggested the body was not burned *in situ*. Understanding these patterns is vital for differentiating perimortem trauma from postmortem thermal damage and interpreting the evidence to deduce a sequence of events.

- It is also important for practitioners to consult the most current research and literature to apply up-to-date analytical and recovery techniques. Familiarity with recent scientific journal articles and publications facilitated the use of aluminum foil as a means of bony preservation during scene recovery and transport in order to protect elements and signs of potential trauma (Dirkmatt et al, 2012; Lewis & Christensen, 2016).

- This case also highlights the relationship between the medical examiner and forensic anthropologist. The medical examiner maintains jurisdiction and control over the case, while the forensic anthropologist provides consultation as requested by the medical examiner. The medical examiner decides when a forensic anthropologist is required at a scene and the extent of the forensic anthropological analysis they require for the case. In this particular instance, the forensic anthropologist was not requested at the scene of the second set of remains and was asked to analyze only those remains with suspected sharp force injuries, not those related to gunshot wound trauma. This was due to the presence of fresh soft tissue allowing the pathologist to accurately track each projectile. Also, only a limited trauma analysis was requested for the first accession of remains to identify tool class while preserving the integrity of the heavily calcined portions of bone. Other medical examiners or coroners may request a full osteological and trauma analysis of all remains; it depends on the medical examiner/coroner involved, their experience, and the analyses they deem necessary to provide the details they need to make (and defend) a determination of cause and manner of death.

DISCUSSION QUESTIONS

8.1 Besides using aluminum foil, what are additional suggestions for how the remains could have been adequately secured at the scene to ensure proper preservation during transport?

8.2 The weapon used for dismemberment was a partially serrated hunting knife. Given this, what other features may have been present on bone at the gross and microscopic levels? What else could be discerned about the weapon?

8.3 How would the pattern of thermal damage differ had the individual not been dismembered? Would burning in a barrel affect the pugilistic posture of remains that had not been dismembered?

8.4 Do you think the forensic anthropologist should have been requested to macerate and examine the full set of skeletal remains? What additional information might this provide? What are the pros and cons of more extensive processing and analysis?

REFERENCES

Dirkmaat, D. C., Olson, G. O., Klales, A. R., & Getz, S. (2012). The role of forensic anthropology in the recovery and interpretation of the fatal fire victim. In D. C. Dirkmaat (Ed.), *A companion to forensic anthropology* (pp. 113–135). Malden, MA: Blackwell Publishing.

Fojas, C. L., Cabo, L. L., Passalacqua, N. V., Rainwater, C. W., Puentes, K. S., & Symes, S. A. (2015). The utility of spatial analysis in the recognition of normal and abnormal patterns in burned human remains. In N. V. Passalacqua, & C. W. Rainwater (Eds.), *Skeletal trauma analysis* (pp. 204–221). Wiley Blackwell.

Jantz, R. L., & Ousley, S. D. (2005). FORDISC 3.1: Personal computer forensic discriminant functions.

Lewis, L. M. S., & Christensen, A. M. (2016). Effects of aluminum foil packaging on elemental analysis of bone. *Journal of Forensic Sciences*, 61(2), 439–441.

Scheuer, L., & Black, S. (2000). *Developmental Juvenile osteology*. Academic Press.

Spradley, M. K., & Jantz, R. L. (2011). Sex estimation in forensic anthropology: Skull versus postcranial elements. *Journal of Forensic Sciences*, 56(2), 289–296.

Symes, S. A., Rainwater, C. W., Chapman, E. N., Gipson, D. R., & Piper, A. L. (2015). Patterned thermal destruction in a forensic setting. In C. W. Schmidt, & S. A. Symes (Eds.), *The analysis of burned human remains* (pp. 17–59). Elsevier Ltd.

CHAPTER 9

Body in the Barrel: Complex Body Disposal and Recovery

Soren Blau and David Ranson

CONTENTS

BACKGROUND

A 23-year-old male was reported to police as a missing person, as he had not been seen for four months. Police intelligence led the investigation to a domestic property where an 80-cm high steel drum was located in the back of a vehicle. The vehicle and steel drum were transferred to the local police forensic science center where the contents were inspected (see Figure 9.1).

The forensic anthropologist identified fragments of burnt human bone in the upper levels of the debris in the drum. Initial inspection of the remainder of the drum's contents revealed burnt and unburnt decomposed human remains, clothing (including a balaclava, shoes, and plastic gloves), and pieces of glass, metal, and other burnt debris. The contents of the barrel were placed in a body bag and transferred to the local institute of forensic medicine for formal examination.

DOCUMENTATION AND SKELETAL ANALYSIS

A postmortem computed tomography (PMCT) examination of the body bag and its contents was performed as part of routine postmortem procedures. Although the body bag contained a considerable amount of non-bone debris, the PMCT enabled the identification and safe extraction of numerous potentially diagnostically important skeletal elements (see Figure 9.2).

The use of PMCT as a valuable initial "triaging" tool has been widely demonstrated in both domestic and disaster victim identification cases (Blau, Robertson, & Johnston, 2008; Brough, Morgan, & Rutty, 2015; Viner et al., 2015). In this case PMCT assisted

FIGURE 9.1   Steel drum and the contents laid out for inspection.

FIGURE 9.2   Virtual reconstruction of the soft tissue masses showing several identifiable skeletal elements in addition to quantities of debris.

the forensic anthropologist in locating the skeletal remains among large decomposed soft tissue masses. Identifiable bone elements were separated from the undiagnostic bone fragments, which ranged in size from 10–40 mm, and the decomposed muscle tissue was removed from the long bones to facilitate detailed examination of the bony remains. The analysis revealed the incomplete, decomposed, and partially burnt remains of one individual. The skull (cranium and mandible) and the majority of the axial skeleton (upper body) were absent (see Figure 9.3) (see below for an explanation about the missing skeletal elements).

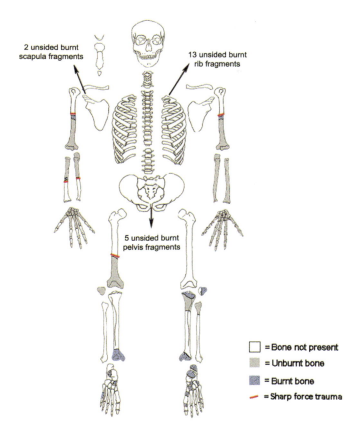

FIGURE 9.3 Schematic representation of the condition and preservation of the skeletal remains.

   The partial long bones (upper and lower limbs) were predominantly unburnt with incompletely burnt decomposed muscle tissue attached. The identifiable fragments of foot bones were all burnt. The left and right wrists and hands were not easily visually recognizable within the decomposed soft tissue (the bones were unburnt), but could be clearly located within the soft tissue masses on the CT scan (see Figure 9.4). Subsequent dissection revealed that the left wrist and hand were bound with tape and other material, and a bracelet was found around the left wrist.

   Commenting on the ancestry, sex, or estimated age of the individual using anthropological techniques was not possible due to the poor preservation of the remains and the absence of the skull and pelvis. For the same reason, dental identification was not possible. Given the hypothesis, based on intelligence from the criminal investigation process that the individual might be the missing person, a positive identification was ultimately achieved using direct DNA comparison.

   The left ulna, radius, and the bones of the left and right hands were complete and unburnt. However, most of the remaining surviving skeletal elements were incomplete and showed evidence of burning (see Figure 9.3). The proximal left and right humeri were absent, and the mid-shafts of both humeri were charred black from the fire. Examination of the sections of both humeri fragments revealed a series of

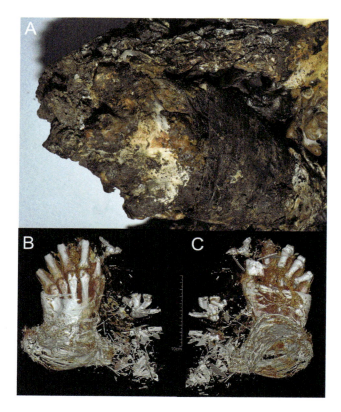

FIGURE 9.4   Dorsal view of the decomposed and partially burnt left hand and unburnt bones of the wrist with associated black tape (A); CT virtual reconstruction of the dorsal (B) and ventral (C) views of the left hand providing a clearer understanding of the relationship of the hand to the tape. (The VR images in B and C are reprinted from Blau et al. (2018, Fig 9.33, pg. 701) with permission provided by Elsevier.)

parallel grooves c. 2 mm in diameter with well-defined margins (see Figure 9.5). The distal right ulna and radius were also absent, and the mid-shaft of the right radius was slightly burnt. The striations and grooves observed on other long bone fragments were not as obvious on the ulna and radius. It is possible that following partial application of sharp force, blunt force was then applied to the arms. The proximal right femur was absent. The mid-shaft section of the right femur was unburnt and also displayed a series of non-uniform striations and exit chipping (see Figure 9.6). The left femur was not recovered.

The parallel striations were consistent with perimortem sharp force skeletal trauma typical of dismemberment (Konopka, Strona, Bolechala, & Kunz, 2007; Porta et al., 2016; Reichs, 1998; Ross & Cunha, 2019; Rutty & Hainsworth, 2014; S. A. Symes, Williams, J.A., Murray, E.A., Hoffman, J.M., Holland, T.D., Saul, J.M., Holland, F.P., Pope, E.J., 2002). The striations were morphologically similar to defects produced by an implement such as a saw (Hainsworth, 2017). The pattern of burning suggests the individual was dismembered prior to the attempt made to destroy the remains using fire. While some of the exposed bone cross-sections (i.e., sites of skeletal dismemberment) were exposed to the fire and charred, other regions showed no sign of burning. The

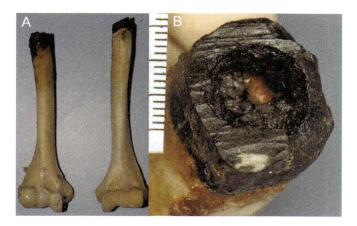

FIGURE 9.5 Anterior view of the distal left and right humeri showing the burnt proximal sections (A); detail of the burnt mid-shaft section of the right humerus showing a series of parallel grooves (B).

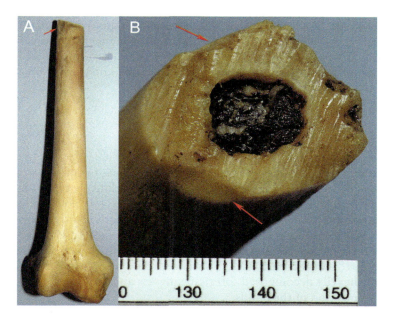

FIGURE 9.6 Anterior view of the right distal femur (A) and cross-section (B) showing evidence of non-uniform striations and exit chipping (red arrows).

absence of the burning on the femur may possibly be explained by the fact the bone was relatively more protected by large muscles, or that the position of the lower limbs within the barrel protected the femora from the effects of the fire.

Although the police recovered bullet fragments and a pistol from the vehicle located at the scene, the poor preservation of the remains and lack of soft tissue meant there was an insufficient evidence base for the forensic pathologist to determine an unequivocal medical cause of death. Therefore, the cause of death was unascertained.

COURT TESTIMONY

Both the forensic pathologist and the forensic anthropologist provided expert testimony in the Supreme Court. Both experts testified about the preservation and completeness of the remains and the actions used to dispose of the individual (specifically dismemberment and fire). However, the cross-examination by the defense team focused largely on a circular metal ring (bracelet) and the observation of plastic and some kind of tape that appeared to be around the deceased's left wrist. The reasons for this were not explicitly identified during questioning of the anthropologist or pathologist. Given that identification of the victim was not at issue during the trial (DNA returned positive results) and the prosecution had not advocated a particular medical cause of death, the defense team focused on the material evidence noted by the forensic anthropologist and pathologist. This is more than likely because it related to the circumstances surrounding the death that had been alleged by the prosecution.

It is common for experts, including forensic anthropologists and pathologists, to testify but never formally be informed about the outcome of the trial. This highlights the important role of forensic experts in providing independent evidence within their area of expertise. In this case, the outcome was obtained through media reports. The court heard that the accused individuals kidnapped and shot the victim, dismembered his body with a chainsaw and tried to dissolve it in acid. The accused then placed the victim's remains in a container and threw them into the ocean. The main offender was found guilty of drug trafficking, kidnapping, and murder and sentenced to 32 years in jail, while the other offender was found guilty of murder, drug trafficking, and gun possession and was jailed for 21 years.

LESSONS LEARNED

- While radiographic techniques have a long history of use in anthropology, routine application of PMCT in forensic anthropology is a relatively recent development (Franklin, Swift, & Flavel, 2016; Uldin, 2017). PMCT is now well-recognized as an important part of a forensic examination significantly contributing to the identification process (Brough et al., 2015; Davy-Jow & Decker, 2014) and interpretation of skeletal trauma (Christensen, Smith, Gleiber, Cunnignham, & Wescott, 2018). In this case, PMCT assisted in the recovery of skeletal elements among the heavily decomposed soft tissue masses and facilitated the documentation of the position of the hands.
- Despite relatively poor preservation and the absence of a cause of death, it was possible for the forensic anthropologist to provide evidence regarding the type of perimortem skeletal trauma that had been inflicted on the individual, including the burning and dismemberment, that is, actions that occurred around the time of death. This was important for the prosecution's case in relation to the alleged attempts to dispose of the body and hinder identification.
- It should be remembered that it is often not possible for the expert witness to assume the line of questioning that may occur during oral testimony. Indeed, trying to anticipate the direction of oral testimony can be dangerous for the forensic witness as it may lead to a restriction in the scope of preparation that they undertake. In any event, the value of a pre-trial conference does allow the

witness to more effectively prepare for "evidence in chief" with regard to the evidential matters that the prosecution wishes to elucidate through the witnesse's testimony (Henneberg, 2016).

DISCUSSION QUESTIONS

9.1 How did the preservation and condition of the remains in this case affect the analysis and conclusions of the forensic anthropologist and pathologist?

9.2 Forensic anthropologists vary in the depth of their description of sharp force trauma defects. How much detail would you provide in describing the sharp force trauma? Are there any potential consequences of providing too many details? What would you do if investigators brought you a specific saw and asked if it was the one used to dismember the body?

9.3 There are differing views about the use of terminology to describe the timing of skeletal trauma (Andrews & Fernández-Jalvo, 2012; Blau, 2017; Cattaneo & Cappella, 2017; Kemp, 2016). Some forensic anthropologists refer to fire modification and/or dismemberment as perimortem trauma because the modifications were made in fresh bone when it retained its elastic biomechanical properties (S. A. Symes, L'Abbe, Stull, LaCroix, & Pokines, 2013; Ubelaker, 2013). Others argue that in such cases, as the one presented here, it is clear that the dismemberment and subsequent fire modification occurred after the individual was deceased and thus should be referred to as postmortem (Blau, 2017). Which term would you use and why?

9.4 How much information can a forensic anthropologist provide about material artifacts associated with the skeletal remains before going beyond their expertise?

REFERENCES

Andrews, P., & Fernández-Jalvo, Y. (2012). How to approach perimortem injury and other modifications. In L. S. Bell (Ed.), *Forensic microscopy for skeletal tissues: Methods and protocols, methods in molecular biology* (Vol. 915, pp. 191–225). Berlin: Springer.

Blau, S. (2017). How traumatic: A review of the role of the forensic anthropologist in the examination and interpretation of skeletal trauma. *Australian Journal of Forensic Sciences*, 49(3), 261–280.

Blau, S., Ranson, D., & O'Donnell, C. (2018). *An atlas of skeletal trauma in medico-legal contexts* (p. 701). London: Academic Press, Figure 9.33.

Blau, S., Robertson, S., & Johnston, M. (2008). Disaster victim identification: New applications for postmortem computed tomography. *Journal of Forensic Sciences*, 53(4), 1–6.

Brough, A. L., Morgan, B., & Rutty, G. N. (2015). Postmortem computed tomography (PMCT) and disaster victim identification. *Radiol Med*, 120(9), 866–873.

Cattaneo, C., & Cappella, A. (2017). Distinguishing between peri- and post-mortem trauma on bone. *Taphonomy of Human Remains: Forensic Analysis of the Dead and the Deposition Environment*. In E. M. Schotsmans, N. Marquez-Grant and S. L. Forbes (Eds.) (pp. 352–368). Chichester, Wiley.

Christensen, A. M., Smith, M. A., Gleiber, D. S., Cunnignham, D. L., & Wescott, D. J. (2018). The use of X-ray computed tomography technologies in forensic anthropology. *Forensic Anthropology*, 1(2), 124–140.

Davy-Jow, S. L., & Decker, S. J. (2014). Virtual anthropology and virtopsy in human identification. In X. Mallet, T. Blythe, & R. Berry (Eds.), *Advances in Forensic Human Identification* (pp. 271–288). Boca Raton, FL: CRC Press.

Franklin, D., Swift, L., & Flavel, A. (2016). 'Virtual anthropology' and radiographic imaging in the forensic medical sciences. *Egyptian Journal of Forensic Sciences*, 6, 31–43.

Hainsworth, S. (2017). Identification marks – Saws. In S. Black, G. N. Rutty, S. Hainsworth, & G. Thomson (Eds.), *Criminal Dismemberment: Forensic and Investigative Analysis* (pp. 135–155). Boca Raton: CRC.

Henneberg, M. (2016). The expert witness and the court of law. In S. Blau & D. H. Ubelaker (Eds.), *Handbook of Forensic Anthropology and Archaeology* (2nd ed., pp. 635–641). London: Routledge.

Kemp, W. L. (2016). Postmortem change and its effect on evaluation of fractures. *Academic Forensic Pathology*, 6(1), 28–44.

Konopka, T., Strona, M., Bolechala, F., & Kunz, J. (2007). Corpse dismemberment in the material collected by the Department of Forensic Medicine, Cracow, Poland. *Legal Medicine (Tokyo)*, 9(1), 1–13.

Porta, D., Amadasi, A., Cappella, A., Mazzarelli, D., Magli, F., Gibelli, D., . . . Cattaneo, C. (2016). Dismemberment and disarticulation: A forensic anthropological approach. *Journal of Forensic and Legal Medicine*, 38, 50–57.

Reichs, K. J. (1998). Postmortem dismemberment: Recovery, analysis and interpretation. In K. J. Reichs (Ed.), *Forensic Osteology: Advances in the Identification of Human Remains* (2nd ed., pp. 353–388). Springfield: Charles C. Thomas.

Ross, A. H., & Cunha, E. (Eds.). (2019). *Dismemberments*. London: Academic Press.

Rutty, G. N., & Hainsworth, S. V. (2014). The dismembered body. In G. N. Rutty (Ed.), *Essentials of Autopsy Practice* (pp. 59–87). London: Springer-Verlag.

Symes, S. A., L'Abbe, E. N., Stull, K. E., LaCroix, M., & Pokines, J. T. (2013). Taphonomy and the timing of bone fractures in trauma analysis. In J. T. Pokines & S. A. Symes (Eds.), *Manual of Forensic Taphonomy* (pp. 341–366). Boca Raton: CRC Press.

Symes, S. A., Williams, J.A., Murray, E.A., Hoffman, J.M., Holland, T.D., Saul, J.M., Holland, F.P., Pope, E.J. (2002). Taphonomic context of sharp-force trauma in suspected cases of human mutilation and dismemberment. In W. D. Haglund, M. H. Sorg (Ed.), *Advances in Forensic Taphonomy: Method, Theory, and Archaeological Perspectives* (pp. 403–434). Boca Raton: CRC Press.

Ubelaker, D. H., Montaperto, K.M. (2013). Trauma interpretation in the context of biological anthropology. In C. Knüsel, Smtih, M. (Ed.), *The Routledge Handbook of the Bioarchaeology of Human Conflict* (pp. 25–38). Florence: Taylor and Francis.

Uldin, T. (2017). Virtual anthropology – a brief review of the literature and history of computed tomography. *Forensic Sciences Research*, 2(4), 165–173.

Viner, M. D., Alminyah, A., Apostol, M., Brough, A., Develter, W., O'Donnell, C., . . . Woźniak, K. (2015). Use of radiography and fluoroscopy in Disaster Victim Identification. *Journal of Forensic Radiology and Imaging*, 3(2), 141–145.

CHAPTER 10

Sealed for Your Protection: A Triple Homicide Involving the Use of a Corrosive Agent to Obscure Identity

Laura C. Fulginiti, Kristen M. Hartnett-McCann, and Detective Frank Di Modica

CONTENTS

INTRODUCTION

In the summer of 2001, a woman and her two children were scheduled to fly to Ohio to visit relatives, but they never arrived. The local law enforcement agency went to their home and questioned the husband, who was the stepfather to the children. The husband asked them to please search for his family. Three days later, the woman's vehicle was found. The vehicle had been wiped clean and yielded only five fingerprints, later matched to the missing woman. Investigators returned to the family home and discovered all new bedding, a new washer and dryer, freshly cleaned carpeting, and a strong odor of pine cleaner. During a search of the home, they found blood spatter and a large blood stain on the laundry room concrete slab. Rubber gloves, empty pine cleaner bottles, and two bloody knives were recovered from the trash. Even without the bodies of the missing family the husband was charged, tried, and convicted of capital murder in 2002. In 2004, the husband was sentenced to death for the three murders.

In October of 2005, work crews were preparing a large desert parcel for the foundation of a superstore. As part of the process, trenches were dug around the larger Palo Verde ("green stick") trees on the lot to prepare them for transplantation elsewhere. A rusted metal 55-gallon drum was uncovered near the bottom of one tree. When the backhoe struck the drum, an object that they thought was a large gnarled tree root was thrown out of it. They removed the drum, crated the tree, and moved it away for relocation. At the base of a second tree the backhoe struck and opened the top of a second rusted metal 55-gallon drum. When the men looked inside, they observed the nude body of an adult female. She was positioned head down in the drum with her buttocks and feet extending upwards (see Figure 10.1).

The police department with jurisdiction for the location where the body was discovered was contacted, and the recovery process began. An aerial survey was conducted via helicopter to obtain overall photographs of the vast scene. The forensic anthropologist received photos of the item from the first drum and confirmed that it was a human pelvis with proximal femora attached. The vertically oriented second drum was excavated *in situ* by the forensic anthropologist, and all human tissue was collected. Relatively early on in the investigation, a detective with the county sheriff's office with jurisdiction over the murder cases suggested that the decedents could be the woman and her children missing since 2001.

Since the remains were suspected to be two of the three members of the missing family, the search continued at the scene for a third set of remains while the first two were being examined at the Forensic Science Center in Phoenix, Arizona. The pathologist and forensic anthropologist worked together to assess the remains. The first pelvis was macerated completely to assess age-at-death and sex, and to determine if any tool marks associated with dismemberment were present. The adult female was mummified, and the contents of the drum had a peculiar odor that did not resemble decomposition or decay. The aroma was musty with a vague chemical overtone. The anthropologist initially thought the bodies had been in the drums for an extended period of time, decades even, based on the odd smell. Additionally, both bodies exhibited unusual patterns of decay. The muscle tissue was extremely dry and almost crumbly with a darker red/rust color that resembled that of historic or even prehistoric mummified humans. The ends of the bones of the pelvis were eroded rather than cut, and the adult female had multiple swaths of eroded and corroded tissue and bone.

Because the degraded appearance of the bone and soft tissue and the absence of a decomposition odor suggested an extended postmortem interval, the forensic anthropologist initially doubted that these two individuals could be part of the missing family. Only as the maceration process of the first pelvis continued and more of the bone structure was revealed did she begin to doubt her initial impression. The pelvis and femora

FIGURE 10.1 *In situ* view of the top of the 55-gallon drum that contained the adult female.

appeared more recent, as more of the bone was observed. The morphology appeared to be possibly female, although the remains were of a young person. Eventually, the age-at-death was estimated at 12–13 years old, which was consistent with the age of the female missing child. Concurrently, the forensic odontologist was performing a comparison of the antemortem and postmortem dental records and concluded that the adult female was consistent with the missing mother (more on this process later).

The county sheriff's office took over the investigation once identity was confirmed. The detectives were notified that the drums contained two of the three missing persons, and that one of them was only represented by the pelvis, at which point the search process was escalated. Cadaver dogs, metal detectors, and probes were brought to the scene in an effort to locate a suspected third drum or additional body parts. When news broke of the first two bodies, concerned citizens even offered their psychic abilities and dowsing rod services. The search area was increased and stretched approximately one-quarter of a mile at its greatest width. Unfortunately, the cadaver dogs were unable to detect any human remains in the area. Metal detectors were used to search for a third drum but proved ineffective; the high metallic content of the surrounding soil and rocks made it impossible to distinguish a true metal find from the general area itself (see Figure 10.2).

The search and recovery team consisted of members of two law enforcement agencies, the forensic anthropologist from the medical examiner's office, and a student intern. The large, multi-agency team sifted through tons of dirt that had been removed during the tree relocation efforts in search of evidence or other remains that may have been unearthed prior to the discovery of the bodies. The team designed and built a large sifting screen, measuring 12' × 12' and utilized a front-end loader to deposit scoops of dirt onto the screen (see Figure 10.3).

Approximately four to six persons at a time moved the dirt through the screen by hand to effectively handle the enormous quantities of dirt. After the recovery process, numerous piles of sifted dirt were scattered throughout the site, many as high as 10 feet. Despite these efforts, minimal human remains were recovered in the screen. It was noted

FIGURE 10.2   Use of a metal detector at the scene. The adult female was found adjacent to this crated tree.

FIGURE 10.3 Members of the local law enforcement, county sheriff's office, and forensic anthropology team work together to screen the previously disturbed soil as well as the soil excavated from around the trees. A large 12' × 12' screen was constructed specifically for this recovery.

that during the initial search process the investigators found several fragments of what they thought was bark from the native mesquite trees in the area. These pieces varied in size but were mostly linear and corrugated much like cardboard. The texture was coarse, and the color was grayish brown. Only a few exemplars were retained.

Given the close proximity (within 50 feet) of the female remains to one another, the search for the second child focused on the immediate vicinity of the remains. After a week of searching, and deploying every possible tool on hand, the third drum was discovered outside of the original search area. One of the homicide detectives, who coincidentally and somewhat ironically had an extreme fear of dead bodies, was standing under a small Palo Verde tree in a wash. A "wash" is a riverbed with no water in it. These channels only run after a heavy rain. The ground gave way underneath him and he realized he was standing inside a 55-gallon metal drum. Belatedly comprehending what he had discovered, he made an extreme attempt to scramble out of the drum as quickly as possible. Each of the 55-gallon drums had been buried vertically, with the tops being approximately four to six inches below ground level and covered with dirt. Over time, the lids rusted and became unattached, allowing first the backhoe and ultimately the detective to "remove" them.

Once the third drum had been discovered a more precise recovery took place. Discussion regarding the possible acceptable ways to recover the third body included (1) a full excavation of the drum and its contents, using a bobcat to remove dirt from one side of the drum to allow access to it, or (2) removal of the drum *en bloc* with excavation of the contents at the medical examiner's office. Given the badly corroded state of the metal drum, any attempt to move it intact would have caused it to disintegrate and resulted in potential loss of evidence. Ultimately, the decision was made to dig a pit and excavate one side of the drum from the pit, keeping the soil in place around the remaining circumference to reinforce and keep the drum intact. After one half of the drum was exposed, the forensic anthropology team used standard archaeological tools and methods to excavate the remains from the drum. Brushes, small picks, and trowels were used to access and expose the skeleton through a large eroded area at the bottom of the drum (see Figure 10.4).

One of the first items uncovered from the third drum was a femoral head epiphysis. The stage of development and fusion of this epiphysis suggested that the third missing person, a ten-year-old, had finally been located. The painstaking recovery process yielded approximately six fragments of long bone, a proximal femur, an unfused os coxae, and several fragments of mummified tissue. The bone was degraded and eroded in a similar manner to the females' remains. The tissue was dry, grainy, light in weight, and crumbly and was similar to the dozens of pieces of "tree bark" that had not been collected during the search process.

The recovery process also included a reevaluation of the location of the drum where the adult female was found. At the base of the drum a swath of rust-colored material extruded outward from the drum into the surrounding dirt. This material was dry, formed large clumps, and had a strong chemical odor. Samples were collected and tested. Numerous broken plastic fragments and a few safety seals for covering bottle mouths marked "SEALED for YOUR PROTECTION" were recovered from the soil surrounding this drum. These plastic fragments and safety seals were not immediately identified as being connected to the search, but their presence was noted. Nine additional "SEALED for YOUR PROTECTION" safety caps were found in the immediate area of the third drum. At this point, the connection was made that some liquid substance was likely poured into the drums with the remains prior to the burial of the drums. Research later conducted by the authors determined that the safety seals were consistent with seals on bottles of various corrosive agents designed to prevent splashing or spilling of the content (discussed further below).

Laboratory analysis of all three sets of remains was conducted in the Forensic Anthropology Laboratory at the FSC. The first set of remains was determined to be an adult White female who was approximately 5′ 3″ tall. The adult female was nearly complete, but exhibited taphonomic alterations on her legs, feet, abdomen, ribs, and head

FIGURE 10.4 The forensic anthropology team excavates the remains from the third drum using standard archaeological procedures and tools.

consistent with exposure to a corrosive agent while she was head down in the drum (see Figure 10.5). There was destruction of her natural teeth, although an anterior bridge containing porcelain fused to stainless steel crowns was unaffected. There were defects on the facial skeleton and an incised defect on the right side of the cranial vault with blunted edges and exposure of the diploe and internal table. While these defects had characteristics of blunt force impact to the face and a possible sharp force impact to the head, the corrosive agent altered the defects and reduced the validity of conclusive statements about the mechanisms that caused the injuries (see Figure 10.5). There were linear defects in the rib cage with similar alterations. The antemortem dental records of the missing adult female were copies of images that were upside down and backward on the film. The forensic odontologist was able to make a scientific positive identification by comparing the missing adult female records to the recovered adult female's dentition, with the caveat that the images were reversed.

The second set of remains consisted of a pelvis and partial femora and was estimated to be a female approximately 11–14 years of age. Due to the lack of a skull and intact long bones, ancestry and stature were not estimated. The third set of remains was incomplete and fragmentary, and consisted of six bone fragments from the pelvis and femora. Age was estimated to be 9–11 years of age at death based on long bone epiphyseal fusion, but sex was indeterminate. No trauma was visible on either of the children's remains,

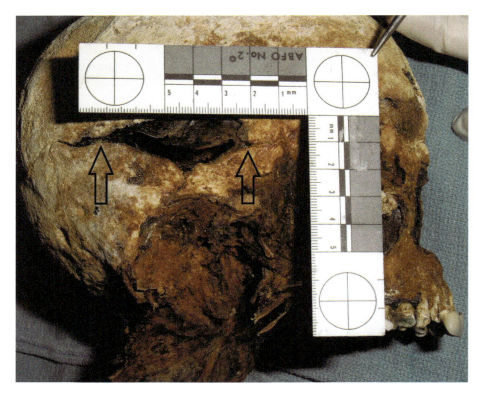

FIGURE 10.5 Anterior aspect and right lateral aspect of the adult female cranium. The bones and teeth are eroded. An incised defect is observed on the right side of the cranium, but the edges are eroded and blunted making it difficult to formulate conclusive statements about the mechanisms of trauma.

FIGURE 10.6 Partial left proximal femora from the male child. The bone is extremely light, brittle, and friable.

though taphonomic alterations from the substance poured into the drums severely compromised the bone fragments available for analysis. The female child was positively identified by DNA comparison. Because of the poor condition of the male child's remains (see Figure 10.6), identification was complicated. He was eventually identified with circumstantial evidence and through DNA comparisons, though the DNA statistics were fairly low due to the paucity of remains and degraded bone condition.

In 2008, after the bodies had been recovered, the state Supreme Court overturned the 2004 murder conviction of the husband. They argued that the fact that his semen was found in the teenage girl's bed should not have been allowed as an aggravating factor. He was re-tried in 2012, and the forensic anthropologist was called to testify. She explained the recovery process to the jury along with additional information regarding the corrosive substance that was used. She provided expert testimony on the defects on the adult female that could be ascribed to both blunt and sharp force trauma. She also explained that the corrosive agent had altered the typical elements used to establish postmortem interval to such a degree that the original estimate was much longer than the actual one (decades vs. four years). Ultimately, the husband was again convicted of all three murders in September of 2012.

The anthropology team was intrigued by the corrosive substance that nearly obliterated the skeletons of the 10-year-old male and 12-year-old female and caused extensive damage to the adult female. Safety seals were present at the locations of all three victims, and the suspicion was that the seals may have been removed from containers of liquid corrosive chemicals. Through research of various packaging practices, the seals were found to be consistent with those used on containers of various acids. The material found in the dirt outside of the adult female's drum appeared to have breached the bottom ring of the drum fairly early in the deposition period. Since the acid leached out of the base of the drum early on, the soft tissue and underlying bone were only exposed to the substance for a short period of time. The skeletons of the two children were likely less ossified due to their age, and a larger number of safety seals (nine seals) were recovered in the immediate area around the drum containing the male child than in the area around the other drums.

The anthropology team hypothesized that the husband used more agent on the youngest child (the boy), who would also be of smaller stature, than he used for the girl and the adult female. The recovery team also observed less leaching of rusty material around the barrel containing the boy, indicating that the drum integrity remained intact for a longer period of time and allowed the agent more time to affect the remains. The greater degree of leaching from the drum containing the adult female resulted in less contact with the acid and halted consumption of the remains. This preservation allowed for the determination of both identity and cause of death for the adult female.

Curiosity led to a detailed scientific study regarding the effects of corrosive agents on human bone, skin, teeth, hair, fingernails, and muscle tissue (Hartnett et al., 2011). This research evaluated if any substance could completely eliminate an entire human body. A human femur with attached muscle and skin was purchased from an anatomical supply company, teeth were donated from a local dentist, and a hair salon provided hair and fingernail clippings. Each of these human tissues was subjected to a variety of corrosive agents with water as a control. The liquids included the water control, carbonated beverage (soda), bleach, sulfuric acid, and muriatic acid (also known as hydrochloric acid). A powdered digestive enzyme and bacteria septic tank additive (brand name "Rid-X") was also tested.

Each specimen was weighed prior to the start of the test and then placed into four-ounce round glass jars with metal lids that were approximately three-quarters full. The specimens were evaluated every 15 minutes for alterations to consistency, reactions to the substance, weight, and degree of elimination. The most effective agent was hydrochloric acid, which at full strength consumed all tissue types in under 24 hours with only a small amount of soft, unrecognizable material floating on top of the acid. A nearly complete adult femur was also placed into 20 ounces of hydrochloric acid and left overnight. No recognizable pieces of bone were observable the next morning. Overall, muriatic acid seems the likeliest candidate for the destructive agent used to dissolve the remains of a mother and her two children in buried metal drums in the Arizona desert. Muriatic acid is inexpensive ($1.79/gal), readily available at local hardware, lawn, and garden stores (at least in pool friendly states), and exceptionally effective on all human tissue types; however, it corrodes metal with the same intensity. Some brands even come with safety seals on their bottles that read, "SEALED for YOUR PROTECTION."

LESSONS LEARNED

- Effective communication among all players is critical during a multi-jurisdictional investigation so that information and evidence are not overlooked or lost in the process. All materials of questionable origin recovered during a search and recovery operation (i.e. wood, rock, metal, etc.) should be examined by a forensic anthropologist before being discarded.
- Metal detectors are ineffective in areas with a high metallic soil and rock content.
- Acids and/or chemicals on bodies may eliminate or mask human decompositional odors, making cadaver dogs an ineffective search tool.
- Traditional search methods are usually effective in finding human remains, though luck can help!
- A large screen can be constructed rapidly and for a small cost and can be used in conjunction with heavy machinery to speed up the screening process significantly. Ingenuity is the key.
- Corrosive substances, especially muriatic acid, will damage and destroy human remains in a relatively short time span, making identification and analysis difficult. Recognizing the effects of acid etching on bone can provide important lines of evidence relating to body disposal, obfuscation of identify, and intent for the prosecution (see Hartnett et al., 2011).

DISCUSSION QUESTIONS

10.1 A scene recovery can be a complicated, evolving investigation with multiple players and events that can alter the course of the process. How does this case address those issues? List several factors a forensic anthropologist must consider when supervising a search and recovery operation.

10.2 Do you think a second search and recovery for the discarded "bark" should have been completed? What benefits and cost would come from such exhaustive search efforts?

10.3 In this case, the third drum was found by chance. It can be extremely difficult to locate buried remains in a potentially large search area. If the detective hadn't fallen into the drum by chance and you were the forensic anthropologist, how would you extend the search for the third body? Are there any other tools or resources you would consider using?

REFERENCE

Hartnett, K. M., Fulginiti, L. C., & DiModica, F. (2011). The effects of corrosive substances on human bone, teeth, hair, nails, and soft tissue. *Journal of Forensic Sciences*, 56(4), 954–959.

CHAPTER 11

Differential Diagnosis in Forensic Entomology: Mites versus Pathologies and Taphonomy

César Sanabria-Medina, Luz Elena Cifuentes, and Maria Alexandra Lopez-Cerquera

CONTENTS

INTRODUCTION

The medical examiner faces a complex challenge during the autopsy of skeletonized remains, as the soft tissues that could provide evidence of the death events are significantly decomposed or absent (Sanabria-Medina, 2017). This process is complicated further if remains are buried illegally or exposed to the environment, thereby introducing additional skeletal modifications and obscuring associated evidence. The medical examiner must rely on the expertise of an interdisciplinary team of specialists to answer questions about the identification of the remains and circumstances of death. This team of professionals usually involves an odontologist and a forensic anthropologist, but may also include other specialists, such as ballistic technicians, biologists, entomologists, geneticists, chemists, and physicists. In the following case, the advantages of an interdisciplinary approach between pathology, anthropology, and forensic entomology are discussed in the context of a case that presented unique cranial lesions of unknown etiology.

Historically, the main application of forensic entomology has been the estimation of the postmortem interval (Campobasso & Introna, 2001). It is possible to determine an approximate time of death through the life cycle and the rate of development of Diptera (flies) as the first colonizers of a body. Likewise, it is possible to arrive at the same information by relating groups of insects that sequentially arrive during decomposition. However, the contribution of forensic entomology to the reconstruction and understanding of the conditions of death is much broader (Cifuentes, 2017).

A decomposing body becomes a source of energy for various organisms. During decomposition fungi, bacteria, vertebrate and invertebrate organisms participate and play different roles. Thus, the decomposing body is a food resource and a temporary habitat for hundreds of organisms, including a diverse array of insects (Cifuentes, 2017). The presence of different groups of arthropods and their interaction with the body may cause alterations in tissues, which at the time of the autopsy may create difficulties in differentiating the timing of defects relative to the death and the etiology of defects. For example, a soft tissue wound caused by a knife can be modified from its original form by insect activity. The blood exposure from the defect creates an ideal location for female Diptera (flies) to lay their eggs, leaving the natural orifices, normally the first regions of the body to be colonized, as a secondary option. Once the active feeding period begins with the emergence of first stage larvae, a distortion of the wound occurs, increasing its size and making it difficult to recognize the wound or potentially obscuring all traces of a wound (Rodríguez & Bass, 1983; Campobasso & Introna, 2001; Byard, 2005). Other saprophagous and ominivorous organisms can alter remains and, depending on the species, may feed off remains anytime from initial body deposition to years after death. The analysis of resultant modifications, particularly the timing of the event (i.e., antemortem, perimortem, or postmortem) requires extensive experience with ample knowledge of the taxonomy, biology, ethology, ecology, and geographical distribution of the arthropods that intervene in the decomposition process (Haskell, et al., 1997). Using this knowledge, the entomologist can support the observations made by pathologists and forensic anthropologists, determining that certain particularities associated with death may or may not correspond to actions caused by the activity of different organisms on human tissues (Cifuentes, 2017).

ARTHROPODS THAT CAUSE ALTERATIONS IN HUMAN BONE TISSUES

As mentioned earlier, several groups of arthropods settle in a decomposing human body (Haglund & Sorg, 1997; Cerda, et al., 2008; Tibbett & Carter, 2008; Pokines & Symes, 2014). Both the arthropods and their remains provide information about the body and the context in which it was found. Likewise, arthropods can be agents of postdepositional alterations in taphonomic processes (Huchet, et al., 2013). Although Diptera (flies) are the most often discussed insect in relation to human remains, termites and mites have also been documented in taphonomic contexts.

Termites (Order Isoptera), for example, frequently live in underground habitats as well as in dry surface environments. They can penetrate remote depths (> 9 meters in some species) and establish tunnels between buried wood and the ground (Triplehorn & Johnson, 2005). Although osteophagy (i.e., consumption of bone) in these organisms has not been extensively studied, it has been reported in archaeological publications (Huchet et al., 2009; Lloveras et al., 2016; Wrobel & Biggs, 2018). These insects build tunnels through bones, potentially damaging these tissues significantly (Backwell, 2012).

Similarly, mites (Class Acari) are arthropod organisms that have colonized virtually all habitats on the planet and have been associated with decomposition and human remains (e.g., Braig & Perotti, 2009; Frost et al. 2009; Salona et al. 2010). More than a century ago, about 2.4 million mites of the genus *Tyrophagus* (Megnin, 1874) were estimated in the skull of a child in Paris. This finding was explained according to the phoretic behaviors that until then were known about these organisms (i.e., present due to their attachment to other insect hosts involved in the decompositional process, such as flies and beetles). More than a century later, a new analysis was conducted by Perotti (2009), who refuted this conclusion by studying the biology of these arthropods, denying their phoretic condition, and establishing that they are organisms with a strong presence on the ground (i.e., they are not necessarily associated with host insects). Although mites have been associated with human cadavers and decomposition, there are no known records of these organisms feeding directly on bone tissue.

CASE STUDY

The skeletonized body of a 19-year-old male who died due to multiple gunshot wounds to the thorax was recovered by law enforcement. For reasons unknown, no medicolegal autopsy was performed. The body was placed in a wooden coffin and buried. Twelve years later, the Office of the Prosecutor ordered an exhumation and a medicolegal autopsy in order to officially establish the manner, cause, and mechanism of death.

The remains were analyzed by an interdisciplinary team of six specialists: a forensic pathologist, two forensic anthropologists, a forensic odontologist, and two geneticists. An inventory and assessment of the biological profile indicated the presence of a single individual, consistent with the known profile of the decedent. DNA analyses confirmed the identity. Trauma analyses were carried out with a stereomicroscope and revealed perimortem injuries associated with gunshot trauma to the thorax. The medical examiner concluded that the cause of death was polytrauma due to gunshot wounds to the thorax and the manner of death was established as homicide. In addition, the skull had small rounded non-perforating injuries/bone lesions that varied in diameter. The edges of the lesions were thick, and their depth only reached the spongy tissue of the diploe. It was not immediately clear if these cranial defects were ante-, peri-, or postmortem in nature (see Figure 11.1).

The general morphology of bone lesions initially suggested they were taphonomic, but the specific agents and mechanism were unknown, leaving some uncertainty. Therefore, the pathologist and forensic anthropologists investigated additional hypotheses. The possibility of multiple myeloma was considered. Multiple myeloma is a type of cancer of the bone marrow with an abnormal proliferation of plasma cells. The myeloma cells also interfere with the normal production of osteoclasts and osteoblasts: the myeloma cells produce a substance that signals the osteoclasts to accelerate bone resorption, causing the bone undergoing remodeling to resorb without new bone formation to replace it (American Cancer Society, 2018). Macroscopically, the lytic processes of the myeloma

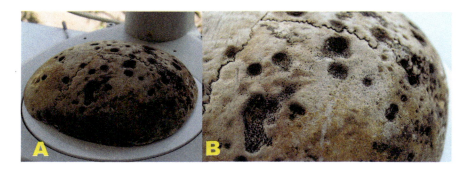

FIGURE 11.1   Overall and close-up view of circular defects observed on the skull.

tend to have sharp edges. In this case, however, they were rounded or polished, although it is not possible to establish if the polish was caused by taphonomic effects (Ortner, 2003). Multiple myeloma generally affects people over 40 years of age, but has been reported in younger individuals (Kasenda et al., 2011). Still, it would be exceptionally rare to see such disease progression in a 19-year-old male. Furthermore, according to his relatives, the man had no medical history of this disease. Samples were sent to the histopathology laboratory where the microscopic diagnosis definitively ruled out multiple myeloma.

Stereoscopic examination of the regions of bone loss revealed evidence of non-bony substances within the defects. Samples were taken and submitted to the forensic entomology lab, and results indicated that the unknown substances consisted of mites (see Figure 11.2). The samples sent to the lab included more than one thousand dead organisms, most of them highly deteriorated with only their shell remaining. The taxonomic identification determined that the mites belonged to the class Acari, the order Mesostigmata, and the suborder Uropodina (see Figure 11.3).

The Mesostigmata order involves a large and cosmopolitan group of parasitiform mites which live in an unusual variety of habitats. Most species are predators (Karg, 1993), while many others are parasitic or symbionts of mammals, reptiles, or arthropods

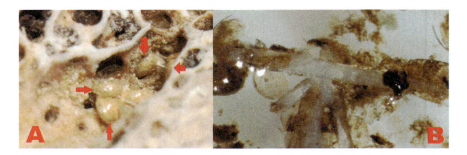

FIGURE 11.2 Stereomicroscopic view of mites found inside the circular regions of bone loss in the skull. (A) Overall view with shells noted with arrows. (B) Magnified image of mites.

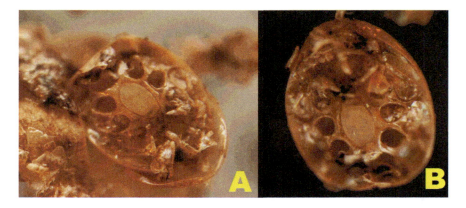

FIGURE 11.3 Stereomicroscopic view of an adult mite found within the circular defects and determined to be of the suborder Uropodina.

(Strandmann & Wharton, 1958). Relatively few species feed on pollen or nectar fungi (Walter & Proctor, 1998; Krantz et al., 2009). Mites of the order Mesostigmata have been found in association with soil, fungi, garbage, carrion, nests, and rotten wood, among other materials. The individuals of the suborder Uropodina are easily distinguished from the rest of members of the Mesostigmata order, due to the specific morphological characteristics of their legs and the chaetotaxy (the arrangement of bristles) of their palps. The biology of this cohort is relatively unknown, but they are known to use predatory feeding strategies. As supplementary diet, they feed on fungi and decomposing organic matter (Fassch, 1967). On the other hand, in immature stages they are associated with phoretic behaviors (Bajerlein & Bloszyk, 2004). That is, they use other organisms (often insects) to move to other habitats.

The entomological analyses could not determine the specific species of the mites, creating a challenge when trying to understand the exact etiology of the bone losses in the skull. A few hypotheses were generated to explain the presence of mites/defects in the cranium. First, it is not possible to confirm or rule out any osteophagic feeding activity of the mites on the bone tissue. Some of the circular regions of bone loss had dried and dark residues shaped like shells, which could correspond to fungi that previously colonized the bone. If so, the fungi may have attracted the mites (as a food source) to the skull. It is possible that other environmental factors could have caused bony degradation and nurtured fungal, bacterial, or other organismal accumulation on the skull, making it a suitable habitat and food resource by the mites. Phoretic behavior could have contributed to the colonization by the mites; however, since these organisms may inhabit the superficial strata of the soil, the mites could have extended their habitat to the inhumed body. Additional research by forensic entomologists, in conjunction with botanists and biologists, is needed to better understand such microscopic taphonomic agents, their interaction with one another, and their effects on bone.

LESSONS LEARNED

- An interdisciplinary approach is key to a successful analysis of any case. The forensic anthropologist should not extend themselves beyond their expertise and should be prepared to consult entomologists or other specialists as needed.
- Forensic practitioners should not assume their first hypothesis as true. Rather, they should generate alternative hypotheses and differential diagnosis. This methodology will give greater certainty to their conclusions. All contextual information should be considered when assessing the hypotheses, including scene information, time since death, and antemortem records of the decedent. For example, in the above case one of the potential hypotheses for the presence of the circular skull defects was multiple myeloma. However, microorganisms such as fungi and bacteria can also affect cortical and internal bone structures. This can create numerous tunnels, giving the bone a spongy appearance that may be misinterpreted as disease (e.g., multiple myeloma), when in reality it is a postmortem modification. Despite the superficial appearance of the lesions as similar to multiple myeloma, subsequent scientific analyses ruled out that possibility in this case. Given the decedent's life history, it was more likely that the defects were taphonomic in origin, which lead to the microscopic identification of the mites.

DISCUSSION QUESTIONS

11.1 Do you think the mites created the circular defects observed? Why or why not? What alternative hypotheses can you come up with for these bony lesions? Why do you think the defects were confined to the skull region?

11.2 If the observed defects are not related to the cause or manner of death of the individual, why put in the extra effort to identify the mechanism of the defect? Could there be any forensic implications?

11.3 What other specialists may have been able to provide information in this case study?

11.4 If the described defects are the result of mites, why hasn't this been well documented in the literature? How might this phenomenon be investigated in a controlled, systematic manner?

REFERENCES

American Cancer Society (2018) What is multiple myeloma? Retrieved October 1, 2018, from https://www.cancer.org/cancer/multiple-myeloma/about/what-is-multiple-myeloma.html

Backwell, L. R. (2012). Criteria for identifying bone modification by termites in the fossil record. *Palaeogeography, Palaeoclimatology, Palaeoecology, 337–338*, 72–87.

Bajerlein D., & Bloszyk J. (2004). Phoresy of *Uropoda orbicularis* (Acari: Mesostigmata) by beetles (Coleoptera) associated with cattle dung in *Poland. European Journal of Entomology, 101*, 185–188.

Braig, H. R., & Perotti, M. A. (2009). Carcases and mites. *Experimental and Applied Acarology, 49*(1–2), 45–84.

Byard R. W. (2005). Autopsy problems associated with post-mortem ant activity. *Forensic Science Medicine and Pathology, 37*, 37–40.

Campobaso, C. P., & Introna, F. (2001). The forensic emtomologist in the context of the forensic pathologist's role. *Forensic Science International, 120*(1–2), 132–139.

Cerda, M., Villalaín-Blanco, J., & Prósper E. (2008). Introducción a la tafonomía forense. Análisis del depósito funerario y génesis de fenómenos pseudopatológicos. In C. Sanabria (Ed.), 2008. *Antropología forense y la investigación médico legal de las muertes* (2nd ed.). Bogotá D.C., Colombia: ACAF.

Cifuentes, L. E. (2017). Manejo de la evidencia entomológica en circunstancias de necropsia: un enfoque para médicos rurales. In C. Sanabria-Medina (Ed.), (2017). *Manual de Medicina Legal y Ciencias Forenses Para Médicos Rurales* (pp. 329–352). Editorial SIGMA Editores & Fondo Editorial Universidad Antonio Nariño/Facultad de Medicina.

Fassch, H. (1967). Beitrag zur Biologie det einheimischen Uropodiden Uroobovella marginata und Uropoda orbiculris und experimentelle analyse ihres phoresieverhaltens. In W. Krantz & D. E. Walter (Eds.), 2009. *A manual of acarology. Tercera edición. Capítulo 11. Orden Mesostigmata. Estados Unidos de Norteamérica.* Texas Tech University Press.

Frost, C. L., Braig, H. R., Amendt, J., & Perotti, M. A. (2009). Indoor arthropods of forensic importance: insects associated with indoor decomposition and mites as indoor markers. In J. Amendt, M. L. Goff, C. P. Campobasso, & M. Grassberger (Eds.) *Current concepts in forensic entomology* (pp. 93–108). Springer, Dordrecht.

Haglund, W. D., & Sorg, M. H. (1997). *Forensic taphonomy: The postmortem fate of human remains*. Boca Raton, FL: CRC Press, Taylor & Francis Group.

Haskell, N. H., Hall, R. D., Cervenka, V. J., & Clark, M. A. (1997). On the body: Insects' life stage presence and their post-mortem artifacts. In W. D. Haglund & M. H. Sorg (Eds.) *Forensic taphonomy. The post-mortem fate of human remains*. Boca Raton, FL: CRC Press.

Huchet, J. B., Dverly, D., Gutierrez, B., & Chauchat C. (2009). Thaponomy evidence of a human skeleton gnawed by termites in a Moche-Civilisation grave at Huaca de la Luna, Peru. *International Journal of Osteoarchaeology. 10*, 11.

Huchet, J. B., Mort, F. L., Rabinovich, R., Blau, S., Coqueutgniot, H., & Arensburg, B. (2013). Identification of dermestidae pupal chambers on Southern levant human bones: Inferens for reconstruction of middle bronze age mortuary practices. *Journal of Archaeological Science, 40*, 3793–3803.

Karg, W. (1993). Raunbmilben: Acari (Acarina), Milben Parasitiformes (Anactinochaeta) Cohort Gamasina Leach. Tierwelt Deutsch. 59. In W. Krantz & D. E. Walter, (Eds.), 2009. *A manual of acarology*. (3rd ed.). Capítulo 11. Orden Mesostigmata. Estados Unidos de Norteamérica. Texas Tech University Press.

Kasenda, B., Ruckert, A., Farthmann, J., Schlling, G., Schnerch, D., Prompelier, H., Wasch, R., & Engelhardt, M. (2011). Management of multiple myeloma in pregnancy ¾ Strategies for a rate challenge. *Clinical Lymphoma Myeloma and Leukemia, 11*, 190–197.

Krantz, G. W. & Walter D. E. (2009) *A manual of acarology*, 3rd ed. Lubbock, Tex.: Texas Tech University Press.

Lindquist, E. E., Krantz, G. W., & Walter, D. E. (2009). Order mesostigmata. G. W. Krantz & D. E. Walter (Eds.) *A manual of acarology*, 3rd ed (pp. 124–232). Lubbock, Tex.: Texas Tech University Press.

Lloveras, L. l., Rissech, C. & Rosado, N. (2016). Tafonomía forense. In C. Sanabria-Medina (Ed.), *Patologíay antropología Forense de la muerte: la investigación científico-judicial de la muerte y la tortura, desde las fosas clandestinas, hasta la audiencia pública* (pp. 453–523). Bogotá D.C., Colombia: Forensic Publisher.

Mégnin P. (1874). Me´moire sur les hypopes. *Journal of Anatomy and Physiology, 1874*: 225–254.

Ortner, J. D. (2003). *Identification of pathological conditions in human skeletal remains* (2nd ed., pp. 481–501). San Diego, CA: Academic Press.

Perotti, M. A. (2009). Mégnin re-analysed: The case of the newborn baby girl, Paris, 1978. *Experimental and Applied Acarology, 49*, 37–44.

Pokines, J. T., & Symes, S. A. (2014). *Manual of forensic taphonomy*. Boca Raton, FL: CRC Press, Taylor & Francis Group.

Rodriguez, W. C., & Bass W. M. (1983). Insect activity and its relationship to decay rates of human cadavers in east Tennessee. *Journal Forensic Science, 28*, 423–432.

Sanabria-Medina, C. (2017). Necropsias en cadáveres esqueletizados: aporte a la estimación del intervalo *postmortem*. In C. Sanabria-Medina (Ed.) *Manual de Medicina Legal y Ciencias Forenses para Médicos Rurales* (pp. 329–352). Editorial SIGMA Editores & Fondo Editorial Universidad Antonio Nariño/Facultad de Medicina.

Saloña, M. I., Moraza, M. L., Carles-Tolrá, M., Iraola, V., Bahillo, P., Yélamos, T., ... & Alcaraz, R. (2010). Searching the soil: forensic importance of edaphic fauna after the removal of a corpse. *Journal of forensic sciences, 55*(6), 1652–1655.

Strandtmann, R. W., & Warton G. W. (1958). Manual of mesostigmatid mites parasitic on vertebrales. Ed. C. E. Yunker. College Park, MD: Contrib. 4, Institute of Acarology.

Tibbett, M., & Carter, D. O. (2008). *Soil analysis in forensic taphonomy. Chemical and biological effects of buried human remains*. Boca Raton, FL: CRC Press, Taylor & Francis Group.

Triplehorn, C. A., & Johnson, N. F. (2005). *Borror and delong's introduction to the study of insects*. Septima edición. Estados Unidos de América. Thompson. Books/Cole.

Walter, D. E., & Proctor H. C. (1998). Feeding behaviour and phylogeny: Observations on early derivative Acari. *Experimental and Applied Acarology, 22,* 39–50.

Wrobel, G. S., & Biggs, J. (2018). Osteophageous insect damage on human bone from Je'reftheel, a Maya mortuary cave site in west-central Belize. *International Journal of Osteoarchaeology, 28,* 745–756. Early View.

CHAPTER 12

Lacustrine Skeletal Taphonomy from Southeastern Tennessee

Murray K. Marks, Jonathan D. Bethard,
and Darinka Mileusnic-Polchan

CONTENTS

INTRODUCTION

Medicolegal investigations involving forensic anthropological consultation can take many forms and involve analyses of human remains ranging from recent events (e.g., skeletal trauma analysis where a decedent's identity is known) to circumstances where remains are not located or identified for months, years, or even decades after the death event (Hinkes, 2017; Kamenov, Kimmerle, Curtis, & Norris, 2014; Kimmerle, 2014). In these "cold case" contexts, investigators may have no idea as to a decedent's identity; however, in other instances, investigators may have been pursuing a particular case for years. For example, the National Missing and Unidentified Persons System (NamUs) (www.namus. gov) currently lists over 12,400 unidentified persons in its database, the earliest of which dates to 1915 (NamUS Case # UP8428). In this chapter, we present a resolved case of a woman who was missing for more than 50 years and whose recovery and identification required a multidisciplinary and multiagency collaboration.

CASE HISTORY

In February of 1956, a 26-year-old female motorist traveling an hour through Tennessee for a job interview disappeared. She was declared dead in 1975 after family and friends had no subsequent communication with her since her disappearance. Various scenarios concerning her disappearance were discussed by family, friends, law enforcement, and the local community, with no evidence as to the missing woman's whereabouts. The case remained unsolved for over 50 years, receiving only periodic attention when a missing person was searched for in area lakes. In 2006, via "cold case" resources, the county sheriff's office reopened the case, focusing on retracing the woman's last known activities.

After interviewing the original lead detective and other witnesses, interest turned to the route she might have taken that morning. Investigators traced the probable journey along a rural state highway, noticing the road's proximity to a deep 1,900-acre lake in Cherokee National Forest equidistant between her starting location and destination. During the summer months of 2007, a Tennessee Bureau of Investigation (TBI) diving team provided county detectives with information on submerged vehicles in the lake. A TBI diving team was later assembled in May 2008 to search the lake bottom adjacent to the sinuous road the victim may have traveled. At a depth of approximately 45 feet in near zero visibility, the divers discovered a vehicle matching the description of the missing woman's car. Upon searching the vehicle's interior (the roof had collapsed) divers located and recovered a partial skeleton, remnants of clothing, and other personal effects (see Figure 12.1). All items were transported by the county sheriff's office personnel to the Forensic Anthropology Laboratory in the Regional Forensic Center at the University of Tennessee Medical Center in Knoxville, Tennessee for analysis.

The skeleton was inventoried after a drying period of three days. Slightly less than 2 millimeters of shrinkage was recorded between the metric values on wet versus dry long bones. The prolonged water exposure of 50 years and a clouded, near zero visibility scene in lake sediment deposits produced deleterious effects on the analysis of the remains and recovery methods. Numerous cranial and postcranial elements were missing or incomplete. No facial bones or teeth were recovered, and the scapulae, ribs, and vertebrae were highly fragmented. The clavicles were nearly complete while humeri and the left femur displayed heavy cortical erosion at their proximal and distal ends. The right femur, right tibia, and os coxae were fragmentary (see Figure 12.2). No hand bones were discovered and only a few foot bones, which were contained in the foot region of a nylon stocking, were recovered. The radii and ulnae displayed fairly pristine condition.

FIGURE 12.1   Photograph of some of the personal effects recovered from the submerged vehicle.

FIGURE 12.2   Photograph of the right os coxae. Note the preserved auricular surface and greater sciatic notch. The discoloration and cortical exfoliation represent the taphonomic alterations caused by long-term submersion in the lake.

Taphonomic changes to the skeleton were remarkable and pronounced; the vocabulary for description here and in the discussion section follows Haglund and Sorg (1997). Most notably, those portions of the skeleton comprised primarily of exposed trabecular bone received heavy cortical erosion and dissolution. Of the six recovered vertebrae, no complete centra presented and all but one was represented exclusively by the neural arch. Pubes and ischia were absent while ilia presented a highly weathered and exfoliated appearance similar to that from long-term sun exposure (Behrensmeyer, 1978) (see Figure 12.2). Long bone diaphyses were discolored and weathered with macroscopic evidence of variously compromised cortical bone integrity (Behrensmeyer, 1978). With the exception of the forearm bones, all long bone epiphyses were highly eroded or absent altogether (Figure 12.3).

Though hindered by a difficult scene resulting in incomplete recovery and the long-term taphonomic effects of water exposure/movement, standard osteometric methods and a macroscopic evaluation was used to develop a biological profile. Intact portions of the os coxae (see Figure 12.2) and the overall gracile appearance of the intact radius and ulna corroborated the metric values for a "female" sex estimation. Age-at-death was estimated from the auricular surface and, though the retroauricular area was missing, an absence of billowing and presence of striae was consistent with a young adult age estimate. Ancestry was not assessed as the craniofacial skeleton, cranial base, and dentition were damaged or absent. Stature was calculated in Fordisc 2.0 using maximum radial and ulnar lengths of 20th century White females as a reference sample.

A section of the right femoral midshaft was excised for histological analysis and embedded following petrographic thin section procedures (Marks, Rose, & Davenport Jr, 1996). The purpose of histological analysis was to corroborate the auricular surface age estimate and to examine microscopic taphonomic changes. Thin section examination was achieved utilizing a Leica DMRX research light microscope at 28×, 88×, and 175× magnifications (see Figure 12.4). Histological analysis demonstrated preserved osteons and osteocyte densities, despite a high eroded macroscopic appearance caused by cortical exfoliation.

Though no dental or skeletal means of positive identification was possible, the findings indicated that the missing 1956 woman could not be excluded as a match. The femoral midshaft sample was sent to the Federal Bureau of Investigation's (FBI) Laboratory in Quantico, Virginia. The FBI DNA Analysis Unit compared the suspected decedent's nuclear DNA profiles with her now elderly male sibling. The results of the DNA analysis did not exclude the decedent as a possible match. Given the remains of a young adult female were recovered from the same type of automobile she owned and located in close proximity to a route she was known to travel, the osteological and genetic findings enabled the Regional Forensic Center's Medical Examiner and the county sheriff's office officials to state that a presumptive identification was acceptable. The Medical Examiner ruled that the death was due to accidental drowning and in February 2009, the decedent was buried after a graveside service was conducted with members of her family

FIGURE 12.3 Photograph of the right distal humerus. Note the discoloration and cortical exfoliation represent the taphonomic alterations caused by long-term submersion in the lake.

FIGURE 12.4 Undecalcified ground thin section (unpolished) at ×28, ×88, and 175 showing exfoliation and fragmentation from the femur midshaft (labeled A, B, and C, respectively). Note the discoloration of damaged region and unorganized osteonal structure in image A. Note the osteonal remnants in the discolored layer in image B. Note the normal appearing, unaltered osteons in image C. Osteocytes are visible in B and C.

in attendance. In May 2009, a member of the Tennessee House of Representatives com-memorated the life of the decedent, as well as the multiagency investigation which led to the identification with House Joint Resolution 398.

DISCUSSION

While forensic pathologists traditionally inspect and interpret more recent soft tissue deterioration as an estimation of time since death (see Marks, Love, & Dadour, 2009) anthropologists often measure prolonged decomposition in differing environments (see DiMaio & DiMaio, 2001; Payne-James, Jones, Karch, & Manlove, 2011). Haglund and Sorg (1997, 2001) brought initial attention to decomposition that has evolved into the cross-sectional and longitudinal study of taphonomy and its application to forensic case-work. This focus has become a near-cornerstone in the armature of the anthropologist with dedicated volumes and papers guiding that endeavor (see Dirkmaat, 2012; Pokines & Symes, 2013; Schotsmans, Márquez-Grant, & Forbes, 2017; Stuart & Ueland, 2017), alongside contributions from human and vertebrate paleontology (Fernandez-Jalvo & Andrews, 2016).

Moreover, few studies have described cases of long-term submersion in either fresh-water or marine environments, though cases studies have been published from disparate time periods ranging from the medieval period (Bell & Elkerton, 2008) to the 20th cen-tury (e.g., Dumser & Türkay, 2008). While several studies have investigated taphonomic effects of water on human skeletal tissues, few researchers have examined how long-term aquatic exposure influences the histological organization of skeletal and dental tissues (Bell & Elkerton, 2008; Cotton, Aufderheide, & Goldschmidt, 1987). Such exposure may obscure histological aging methods (Gocha, Robling, & Stout, 2019; Kerley, 1965; Kerley & Ubelaker, 1978) which are of interest to forensic anthropology; however, in the case presented here, long-term water exposure did not hamper DNA identification. We postulate that a primary reason for this positive result was the preserved density of osteocytes observed histologically. Recently, several studies have demonstrated that osteocyte density is critical to DNA survivability, regardless of skeletal element type (Antinick & Foran, 2019; Johnston & Stephenson, 2016; Mundorff & Davoren, 2014). Additionally, several peri- and postmortem environmental situations have prompted an experimental microscopic examination of the organizational structure of bone includ-ing burning (Bradtmiller & Buikstra, 1984; Holden, Phakey, & Clement, 1995; Neson, 1992; Shipman, Foster, & Schoeninger, 1984; Ubelaker, 2009), freezing (Tersigni, 2007) and water immersion (Bell & Elkerton 2008). Further, the dentition has been evaluated regarding to heat exposure (Fairgrieve, 1994; Schmidt, 2015) and water immersion (Bell et al. 1996).

Although this case study provides a taphonomic picture from an extraordinary context, it is not the first to call attention to the effects of aquatic environments and fluvial transport on human remains (Bassett & Manhein, 2002; Brooks & Brooks, 1997; Ebbesmeyer & Haglund, 2002; O'Brien, 1997). Research has investigated soft tissue changes in freshwater environments (O'Brien, 1997) as well as decomposition sequences produced by invertebrates (Haglund, 1993). Studies have also investigated the role marine environments play on initiation and spurring decomposition, adi-pocere formation, and skeletonization (Anderson & Hobischak, 2004; Dumser & Türkay, 2008; Hobischak & Anderson, 2002; Kahana et al., 1999; Lewis, Shiroma, Guenthner, & Dunn, 2004; Stojanowski, 2002). Here, researchers have drawn conclusions about the effects of marine environments on human tissues that have been exposed from a few days (Anderson & Hobischak, 2004) to over four centuries (Bell & Elkerton, 2008). This wide time expanse indicates that although case studies exist for remarkable isolated aquatic contexts, additional data are needed to clarify the effects of numerous variables from freshwater and marine environments. In a case of long-term aquatic exposure, Cotton and colleagues (1987) discuss taphonomic changes from a case where two individuals were submerged for a five-year period in Duluth (Minnesota) harbor. Though soft tissue changes are described from both macroscopic and histological perspectives, surface and trabecular skeletal changes are not discussed (Cotton et al., 1987). Last, a recent study examining the formation of adipocere in freshwater contexts has helped clarify this complicated process (Stojanowski, 2002) and Ubelaker and Zarenko (2011) have provided a comprehensive synthesis on the subject.

Although this case is an isolated context, both macroscopic and histological observations provide a remarkable opportunity to document taphonomic effects resulting from long-term exposure to a freshwater environment. Though taphonomic bone changes were pronounced, coupling macroscopic description with histological analysis proved effective. In typical experimental settings, long-term submersion intervals are typically not possible. Despite recent advances in estimating the postmortem submersion interval (PMSI) (Benbow, Pechal, Lang, Erb, & Wallace, 2015; Dickson, Poulter, Maas, Probert, & Kieser, 2011; Heaton, Lagden, Moffatt, & Simmons, 2010), these studies do not describe submersion intervals encountered in this case. Indeed, virtually no published literature describes a submersion interval similar to this case study and it is our hope that this contribution supplements the experimental literature.

LESSONS LEARNED

- Some long-term cold cases can be solved when forensic anthropologists collaborate in multidisciplinary investigations. Positive identification is often accomplished when multiple lines of evidence are examined in conjunction with the biological profile generated by forensic anthropologists.
- It is important to understand that taphonomic alterations to osseous remains are variable, particularly in aquatic environments, which are variable themselves (e.g., marine, lacustrine, riverine, freshwater, salt water, brackish, etc.).
- Despite instances when bone may appear macroscopically altered, DNA may still be preserved and used to generate identifications.

DISCUSSION QUESTIONS

12.1 How might forensic anthropologists utilize the NamUS database to work with other forensic professionals to reexamine longstanding cold cases?

12.2 Why is an understanding of taphonomic change critical for forensic anthropology?

12.3 Why might DNA have been preserved in the aquatic environment described in this case study?

ACKNOWLEDGMENTS

We hope this chapter helps a family memorialize a decedent missing more than fifty years. We thank Ms. Monica Datz, Detective, Bradley County, Tennessee, Sheriff's Office in Cleveland, Tennessee, for her investigative skills and permission to present this case. Special thank you to Dr. Christian Crowder of the Dallas County Medical Examiner's Office and Southwest Institute of Forensic Sciences for assessment of the femoral thin sections.

REFERENCES

Anderson, G., & Hobischak, N. (2004). Decomposition of carrion in the marine environment in British Columbia, Canada. *International Journal of Legal Medicine, 118*(4), 206–209.

Antinick, T. C., & Foran, D. R. (2019). Intra- and inter-element variability in mitochondrial and nuclear DNA from fresh and environmentally exposed skeletal remains. *Journal of Forensic Sciences, 64*(1), 88–97.

Bassett, H. E., & Manhein, M. H. (2002). Fluvial transport of human remains in the lower Mississippi River. *Journal of Forensic Sciences, 47*(4), 1–6.

Behrensmeyer, A. K. (1978). Taphonomic and ecologic information from bone weathering. *Paleobiology, 4*(02), 150–162.

Bell, L. S., & Elkerton, A. (2008). Unique marine taphonomy in human skeletal material recovered from the medieval warship Mary Rose. *International Journal of Osteoarchaeology, 18*(5), 523–535.

Bell, L. S., Skinner, M. F., & Jones, S. J. (1996). The speed of post mortem change to the human skeleton and its taphonomic significance. *Forensic Science International, 82*(2), 129–140.

Benbow, M. E., Pechal, J. L., Lang, J. M., Erb, R., & Wallace, J. R. (2015). The potential of high-throughput metagenomic sequencing of aquatic bacterial communities to estimate the postmortem submersion interval. *Journal of Forensic Sciences, 60*(6), 1500–1510.

Bradtmiller, B., & Buikstra, J. E. (1984). Effects of burning on human bone microstructure: A preliminary study. *Journal of Forensic Sciences, 29*(2), 535–540.

Brooks, S., & Brooks, R. (1997). The taphonomic effects of flood waters on bone. In W. D. Haglund & M. H. Sorg (Eds.), *Forensic taphonomy: The postmortem fate of human remains* (pp. 553–558). Boca Raton, FL: CRC Press.

Cotton, G. E., Aufderheide, A. C., & Goldschmidt, V. (1987). Preservation of human tissue immersed for five years in fresh water of known temperature. *Journal of Forensic Sciences, 32*(4), 1125–1130.

Dickson, G. C., Poulter, R. T., Maas, E. W., Probert, P. K., & Kieser, J. A. (2011). Marine bacterial succession as a potential indicator of postmortem submersion interval. *Forensic Science International, 209*(1–3), 1–10.

DiMaio, V. J., & DiMaio, D. (2001). *Forensic pathology*. Boca Raton, FL: CRC Press.

Dirkmaat, D. (2012). *A companion to forensic anthropology* (D. Dirkmaat, Ed.). Malden, MA: John Wiley & Sons.

Dumser, T. K., & Türkay, M. (2008). Postmortem changes of human bodies on the Bathyal Sea floor—two cases of aircraft accidents above the open sea. *Journal of Forensic Sciences, 53*(5), 1049–1052.

Ebbesmeyer, C. C., & Haglund, W. D. (2002). Floating remains on Pacific Northwest waters. In W. D. Haglund & M. H. Sorg (Eds.), *Advances in forensic taphonomy: Method, theory, and archaeological perspectives* (pp. 219–242). Boca Raton, FL: CRC Press.

Fairgrieve, S. I. (1994). SEM analysis of incinerated teeth as an aid to positive identification. *Journal of Forensic Sciences, 39*(2), 557–565.

Fernandez-Jalvo, Y., & Andrews, P. (2016). *Atlas of taphonomic identifications: 1001+ images of fossil and recent mammal bone modification.* Springer.

Gocha, T. P., Robling, A. G., & Stout, S. D. (2019). Histomorphometry of human cortical bone. In M. A. Katzenberg & A. L. Grauer (Eds.), *Biological anthropology of the human skeleton* (pp. 145–187). Hoboken, NJ: John Wiley & Sons.

Haglund, W. D. (1993). Disappearance of soft tissue and the disarticulation of human remains from aqueous environments. *Journal of Forensic Sciences, 38*(4), 806–815.

Haglund, W. D., & Sorg, M. H. (1997). *Forensic taphonomy: The postmortem fate of human remains.* Boca Raton, FL: CRC Press.

Haglund, W. D., & Sorg, M. H. (2001). *Advances in forensic taphonomy: Method, theory, and archaeological perspectives.* Boca Raton, FL: CRC Press.

Heaton, V., Lagden, A., Moffatt, C., & Simmons, T. (2010). Predicting the postmortem submersion interval for human remains recovered from UK waterways. *Journal of Forensic Sciences, 55*(2), 302–307.

Hinkes, M. (2017). Forensic anthropology in cold cases. In R. H. Walton (Ed.), *Cold case homicides* (2nd ed., pp. 381–400). Boca Raton, FL: CRC Press.

Hobischak, N. R., & Anderson, G. S. (2002). Time of submergence using aquatic invertebrate succession and decompositional changes. *Journal of Forensic Sciences, 47*(1), 142–151.

Holden, J. L., Phakey, P. P., & Clement, J. G. (1995). Scanning electron microscope observations of heat-treated human bone. *Forensic Science International, 74*(1), 29–45.

Johnston, E., & Stephenson, M. (2016). DNA profiling success rates from degraded skeletal remains in Guatemala. *Journal of Forensic Sciences, 61*(4), 898–902.

Kahana, T., Almog, J., Levy, J., Shmeltzer, E., Spier, Y., & Hiss, J. (1999). Marine taphonomy: Adipocere formation in a series of bodies recovered from a single shipwreck. *Journal of Forensic Sciences, 44*(5), 897–901.

Kamenov, G. D., Kimmerle, E. H., Curtis, J. H., & Norris, D. (2014). Georefrencing a cold case victim with lead, strontium, carbon, and oxygen isotopes. *Annals of Anthropological Practice, 38*(1), 137–154.

Kerley, E. R. (1965). The microscopic determination of age in human bone. *American Journal of Physical Anthropology, 23*(2), 149–163.

Kerley, E. R., & Ubelaker, D. H. (1978). Revisions in the microscopic method of estimating age at death in human cortical bone. *American Journal of Physical anthropology, 49*(4), 545–546.

Kimmerle, E. H. (2014). Forenisc anthropology in long-term investigations: 100 cold years. *Annals of Anthropological Practice, 38*(1), 7–21.

Lewis, J. A., Shiroma, C. Y., Guenthner, K., & Dunn, K. N. (2004). Recovery and identification of the victims of the Ehime Maru/USS Greeneville collision at sea. *Journal of Forensic Sciences, 49*(3), 539–542.

Marks, M. K., Love, J., & Dadour, I. R. (2009). Taphonomy and time: Estimating the postmortem interval. In D. W. Steadman (Ed.), *Hard evidence: Case studies in forensic anthropology* (pp. 165–178). New York: Taylor & Francis.

Marks, M. K., Rose, J. C., & Davenport, W. D., Jr (1996). Thin section procedure for enamel histology. *American Journal of Physical anthropology, 99*(3), 493–498.

Mundorff, A., & Davoren, J. M. (2014). Examination of DNA yield rates for different skeletal elements at increasing post mortem intervals. *Forensic Science International: Genetics, 8*(1), 55–63.

Neson, R. (1992). A microscopic comparison of fresh and burned bone. *Journal of Forensic Sciences, 37*(4), 1055–1060.

O'Brien, T. (1997). Movement of bodies in Lake Ontario. In W. D. Haglund & M. H. Sorg (Eds.), *Forensic taphonomy: The postmortem fate of human remains* (pp. 559–565). Boca Raton, FL: CRC Press.

Payne-James, J., Jones, R., Karch, S. B., & Manlove, J. (2011). *Simpson's forensic medicine*: CRC Press.

Pokines, J. T., & Symes, S. A. (2013). *Manual of forensic taphonomy*. Boca Raton, FL: CRC Press.

Schmidt, C. W. (2015). Burned human teeth. In C. W. Schmidt & S. A. Symes (Eds.), *The analysis of burned human remains* (2nd ed.) (pp. 61–81). San Diego, CA: Academic Press.

Schotsmans, E. M., Márquez-Grant, N., & Forbes, S. L. (2017). *Taphonomy of human remains: Forensic analysis of the dead and the depositional environment.* John Wiley & Sons.

Shipman, P., Foster, G., & Schoeninger, M. (1984). Burnt bones and teeth: An experimental study of color, morphology, crystal structure and shrinkage. *Journal of Archaeological Science, 11*(4), 307–325.

Stojanowski, C. M. (2002). Hydrodynamic sorting in a coastal marine skeletal assemblage. *International Journal of Osteoarchaeology, 12*(4), 259–278.

Stuart, B. H., & Ueland, M. (2017). Decomposition in aquatic environments. In E. M. J. Schotsmans, N. Marquez-Grant, & S. L. Forbes (Eds.), *Taphonomy of human remains: Forensic analysis of the dead and the depositional enviornment* (pp. 235–429). John-Wiley & Sons.

Tersigni, M. A. (2007). Frozen human bone: A microscopic investigation. *Journal of Forensic Sciences, 52*(1), 16–20.

Ubelaker, D. H. (2009). The forensic evaluation of burned skeletal remains: A synthesis. *Forensic Science International, 183*(1–3), 1–5.

Ubelaker, D. H., & Zarenko, K. M. (2011). Adipocere: What is known after over two centuries of research. *Forensic Science International, 208*(1–3), 167–172.

Trauma

Who Pulled the Trigger … First? Bone Biomechanics Recreate the Story Behind a "Police Shooting"

Steven A. Symes, Ericka N. L'Abbé, and Mark M. LeVaughn

CONTENTS

INTRODUCTION

The use of deadly force by the American police has recently been hailed as the most visible and controversial aspect of the country's criminal justice system (Klinger et al. 2015; Ross 2015). Police-involved shootings are frequently documented in the United States with deadly force often being questioned in the national and international media as well as in the courtroom (Ross 2015). Current awareness and sensitivity surrounding police-involved shootings in America emphasize the importance of inter-professional collaboration among pathologists, anthropologists, and law enforcement in evaluating gunshot wound (GSW) injuries based on the witness accounts leading up to and including the shooting event associated with the evidence obtained from soft and hard tissues observed during autopsy.

Forensic anthropologists contribute to the investigation of a shooting event by detailing fracture patterns in skeletal remains, as supplements to establishing orientation and direction of bullet trajectory, and sequencing the number of impacts (Symes et al., 2012; Madea & Staak, 1988; Dixon, 1984; Berryman & Symes, 1998; Langley, 2007; Smith et al.,1987). Despite the use of the same biomechanical principles to analyze cranial and postcranial fractures, the sequencing of fractures and minimum number of defects must be approached differently. Sequencing fractures and number of impacts in the postcrania is often more difficult due to fewer intersecting radiating fractures between GSW defects.

Additionally, GSW defects are often non-existent in postcranial bones and, depending on the position of the victim at the time of the shooting, a bullet may travel through various skeletal elements before exiting the body completely. Thus, definitive sequencing of GSWs may be difficult among postcranial elements without information on scene context and soft tissues (Langley, 2007).

Scene context and soft tissues are essential for interpreting skeletal trauma such that valuable information may be lost when: an anthropologist does not participate in the autopsy, the remains are completely skeletonized, or accurate context is unknown. In all, a loss of context limits a forensic anthropologist's interpretations of bone fractures and the overall contribution to the forensic pathologist's determination of cause and manner of death (Symes et al. 2012).

A forensic anthropologist needs to develop good relationships and establish lines of communication among various interdisciplinary medicolegal practitioners, particularly forensic pathologists, odontologists, and law enforcement. This allows them to learn the various procedures involved in an autopsy, the procedures to remove bone from autopsy for analysis, and the use of multiple sources of information (soft tissue and eye witness accounts, for example) to successfully and accurately interpret bone injuries (Symes et al. 2012).

This paper examines a police-involved shooting where a victim has multiple gunshot wounds to the head, chest, and limbs. An invasive postmortem examination at autopsy by the pathologist, anthropologist, and investigator, along with documentation and reconstruction of the processed cranial bones, was used to assist in clarifying the direction and sequence of gunshots for the eventual determination of cause (COD) and manner (MOD) of death.

THE SHOOTING EVENT

Police officers attempt to arrest a defiant and dangerous felon by rousting him at home during the night. The officers have three weapons trained on the front door of the suspect's home, including two 40-calibre police issue handguns and a .223 rifle. Overall visibility is less than desirable as the accused exits his home and walks toward his confronters, wearing only boxer shorts and a red kerchief around his neck. The police mediator spots an object that may be a handgun in the suspect's right hand. The officer gives immediate instructions for the suspect to "drop his weapon" and to "hit the ground." The suspect noticeably changes his posture and almost simultaneously a gunshot rings out. Instinctively, all three police officers open fire, hitting the suspect numerous times. As paramedics arrive at the scene, the deceased suspect is faceup and handcuffed. Near the body is a 9mm handgun.

EXTERNAL EXAM AND AUTOPSY

At autopsy, the State Anthropologist (SAS) and the investigating officer for the State accompany the Forensic Pathologist (MML). The officer begins to recount what he has learned from interviews with the mediator and each police shooter. The SWAT member with the .223 rifle said he could see the laser dot on the suspect's right chest when the shooting began, and he confidently said that his first shot was lethal.

The deceased presents at autopsy with multiple gunshot wounds in the head, chest (including the right to left high-velocity mid-chest perforating shot), and limbs. During the initial examination, it was found that two perforating GSWs entered the head and these were labeled A and B (see Figure 13.1). These occur on the right side of the skull and appear roughly consistent with the position of the poised police officers in relation to the suspect at the time of the shooting.

During the initial external exam at autopsy, the trajectories of the two wounds are assigned A→D and B→C (see Figure 13.1). Additional perforating and penetrating wounds of the chest and limbs were labeled with letters E to P, documenting numerous bullet impacts and exits. Bullet trajectories outside the body are at best hypothetical in anthropology (at least without sophisticated 3D mapping equipment). However, the trajectory of ballistic projectiles within the body can be tracked. Also, bullet trajectories occasionally indicate body posture at the time a victim is shot. If straight-line trajectories are accurately tracked in the body, they are recordable and may be of value in court.

As the external examination evolved into preparation for autopsy, wound B was found to be "complicated": it is actually two entrance wounds (see Figure 13.2), where the original frontal wound associated with GSW B exits via C above the left ear. A second penetrating wound on the frontal bone was also discovered and later labeled Q→R, exited forward and inferior, through the right frontal down to the right eye (see Figure 13.2 lower image). The entrance Q (obscured in Figure 13.1) is likely a by-product of a bullet entering a pre-existing skull fracture from a previous GSW. The skin was carried into the separate defect and created a gaping separation in the frontal bone. Once the tissue was extended, the two circular defects (entrance B and Q) were visually obvious

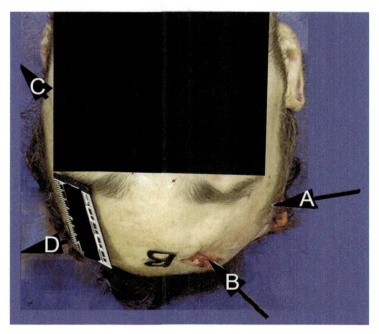

FIGURE 13.1 The anterior view at the autopsy table reveals two perforating head wounds (arrows). They appear consistent with lower speed bullets from handguns or middle-velocity rifles. Wound track B→C presents a trajectory from the upper right frontal down to left temporal bone, just above the ear. Wound A→D enters the right parietal bone above the ear with an upward and right to left trajectory, with exit D in the posterior left parietal.

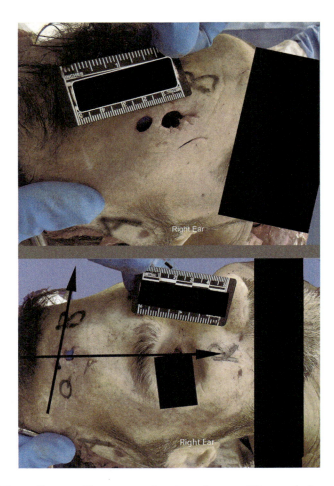

FIGURE 13.2   Upper Image: Closer examination of wound B revealed an additional GSW hidden by deformation of the skin covering fractured skull bone. In addition at least two skin lacerations are noted, probably due to bone fragmentation from high energy destruction. Lower Image: the new GSW is labeled Q→R. The wound progresses superior to inferior, and posterior to anterior. Arrows indicate the perpendicular progress of perforating B→C and the newly discovered Q→R, progressing through previously compromised bone. Therefore, three GSWs enter the right skull, each with a unique trajectory.

(see Figure 13.2). Keep in mind, multiple head shots to standing victims produce differing trajectories due to intentional and unintentional (gravity) movement of the victim.

While shaving scalp wound A for photographs, the pathologist discovered the wound was uncharacteristic to all other wounds in the victim. Wound A is a contact GSW wound with obvious soot on the skin, bone, and within the neural cranium, not to mention the suspected ejector rod imprint on the skin (see Figure 13.3).

How is it possible to have a contact gunshot wound when police officers allegedly opened fire from behind their cars at a respectable distance from the suspect? Additionally, the officers each opened fire (based on individual interrogation data) after hearing a close-range gunshot. The initial physical evidence is inconsistent with the police officers' testimonies. With this new scenario, what options need to be considered?

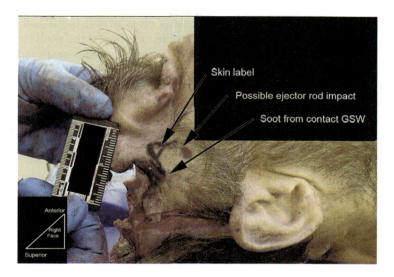

Skin label

Possible ejector rod impact

Soot from contact GSW

Anterior

Right
Face

Superior

FIGURE 13.3 Close examination of skin wound A after autopsy and skull cap removal. Shaving reveals a contact gunshot wound with a possible ejector rod imprint and obvious soot surrounding the wound, in the bone, and in the neurocranium.

Since the obvious COD is multiple GSWs, the missing link is in establishing MOD. The sequence of the cranial GSWs warranted closer examination of dry bone associated with the injuries to substantiate or refute the officer's final theory of "fatal police shooting."

The removal of the cranial vault from a fresh body is possible at autopsy and routinely performed by forensic anthropologists in MEOs in the United States. Yet the author (SAS) has found some individuals to express opinions that this act of tissue removal is unethical, a form of body desecration, and in violation of funeral traditions and religious requirements, particularly if the skeletal elements are retained and later used for research or instruction of professionals. Forensic experts are tasked with taking science to its highest level of diagnostic interpretation while preserving the dignity of the victim and being sensitive to mourning family and loved ones. Bone should be removed during autopsy if it provides evidence or support for cause and manner of death in an investigation. Care should be taken to obtain as much bone as possible from both sides of each fracture, so as to allow for determination of the exact location of failure. This often involves collecting bone that was displaced by energy, as well as bone beyond the fracture with recognizable anatomical landmarks, in order to maintain context (without which any interpretation of trauma is meaningless).

BASIC BIOMECHANICS OF BALLISTIC INJURIES TO SKULLS

Numerous authors provide fracture pattern descriptions for GSW injuries to the skull (e.g., Smith et al., 1987; Berryman & Symes, 1998; DiMaio 1999; Symes et al., 2012; Fackler, 1988) so only a brief review is provided here. Fracture morphology associated with cranial GSWs include plug-and-spall bone fragments, radiating fractures, and concentric heaving fractures, all of which can assist a forensic anthropologist in interpreting the direction of a shot as well as the number and sequence of impacts (Symes et al., 2012; Madea & Staak, 1988).

At impact, a bullet penetrates tissues or pushes plugs of soft tissue and bone into the body. Due to the high velocity associated with ballistic injuries, the skull resists the associated forces to a higher degree than it does blunt forces, causing bone to fail immediately with a plug and spall. The focused entrance defects may *approximate* bullet shape, but due to numerous extrinsic and intrinsic variables, such as velocity, projectile design, angle of impact, and bone strength, bullet impact defects only represent general features of the projectile, such as small, medium, or large caliber for handguns (Berryman et al., 1995; Ross, 1996). A high velocity projectile travels in a straight line, but once the bullet enters the body or impacts any solid object, it can destabilize gyroscopically and may tumble as it exits the body.

If a bullet's energy is not neutralized after punching through the skull, hoop fractures may form due to a build-up of internal pressure of the missile shocking its target. Hoop fractures spread from the point of impact, and are produced by continuous, internal shock waves from a fast-moving bullet. With sufficient energy inside the skull, the bone resists and then reactively splits from the entrance into hoop fractures (Symes et al., 2012).

If a penetrating bullet creates high internal pressure in the cranium and hoop fractures do not relieve pressure, pie-shaped pieces begin to lift off the cranium, initially near the plug. Eventually, and with enough energy, concentric heaving fractures may also occur in response to energy from the shock wave (Smith et al., 1987; Symes et al., 2012). The concentric fractures are a consequence of the 'heaving out' or 'lifting off' of bone from the cranium and often appear as arched or semi-arched fractures between preexisting radiating fractures at the bullet's entrance or exit (Langley, 2007; Hart, 2005; Symes et al., 2012). Concentric heaving fractures are the tertiary fractures in a GSW since the plug and spall, with enough energy, forms radiating fractures, and again with sufficient energy the radiating fractures may be intersected by concentric heaving fractures that connect two radiating fractures as the bone bows outward.

While the bulls-eye pattern formed with concentric heaving fractures is similar to severe blunt trauma to the skull, the heaving out of the fracture is used to distinguish between the two injuries (Langley, 2007; Hart, 2005). The energy of a high-speed bullet traveling in a mostly closed container will push the pie-shaped pieces outward. Given a concentric fracture and the internal pressure from a high-velocity projectile, the inner bone table of the skull vault fails in tension and eventually tears in compression on the outer bone table, lifting the bone. Concentric heave-out fractures are always beveled externally, irrespective of being associated with an entrance or an exit defect, since the origin of failure is pressure from inside the skull directed externally. The wedge-shaped piece of bone is displaced in the opposite direction of blunt force trauma, which, if covering enough surface pushes all bone inward as the low energy impact crushes tissues, usually contacting a larger area than a ballistic missile. The initial concentric failure in blunt trauma reveals tension on the external skull.

Exit wound fractures are often more irregular than entrance wound morphology on account of reduced speed, gyroscopic instability, deformation, fragmentation, tumbling of the projectile, and interruption of radiating fracture energy into preexisting fractures from the entrance defect (Symes et. al., 2012). Radiating fractures travel at the speed of sound in that material, i.e. thousands of feet per second in bone. Fractures from an entrance defect travel across the cranium before any bullet can exit. Under such circumstances, exit wound radiating fractures and concentric heaving fractures may terminate into preexisting entrance radiating and concentric heaving fractures (Berryman &

Symes, 1998; Smith et. al., 1987; Symes et. al., 2012). The axiom enables the potential of sequencing GSW trauma in the skull (Madea & Staak, 1988).

In a case of multiple gunshot wounds to the cranium, Puppe's Law of Sequence, or the phenomenon of intersecting fracture lines, can be applied to determine fracture order (Madea & Staak, 1988). Following Puppe's Law of Sequence, radiating fractures from a second impact site continue until they intersect a previous fracture, whereby most of the strain energy is dissipated (Smith et. al., 1987; Madea & Staak, 1988; Dixon, 1984; Berryman & Symes, 1998). In other words, a fracture from an exit impact may terminate into a previously existing fracture from an entrance. By carefully documenting all radiating fractures and identifying a few key attributes, a practitioner may determine the sequence of shots. However, the theory only works if enough kinetic energy is present for fractures to intersect. If no intersection is present, bevel direction is the main indicator of bullet progress and direction.

The cranial vault is round so many bullets do not impact the structure at a perpendicular angle. A keyhole defect is used to describe a bullet striking tangentially often leaving a wound with a partial oval entrance and a partial external spall and bevel. This characteristic morphology is due to the bullet, even at an angle, punching a partial plug and spall into the bone. However, the angled trajectory also puts the projectile on edge with the plane of the bone, as opposed to a perpendicular punch through, and thus, this angle of impact also forms an exit bevel (Dixon, 1982; Berryman & Symes, 1998).

With a tangential impact, a radiating fracture may form ahead of the bullet at the initial point of impact, and as the bullet encounters maximum resistance from the bone. At this angle to the bone's surface, a focused concentric heaving fracture may form between the two Y-shaped radiating fractures, creating an outward spalling of bone as the bullet goes "on edge" with the plane of bone and either completely penetrates, partially penetrates, or at the shallowest trajectory, skips off the bones' surface. The morphology of this defect resembles an old-fashioned keyhole, thus the "keyhole fracture" (Dixon, 1982; Berryman & Symes, 1998; Symes et. al., 2012). The beveled surfaces on entrance and exit wounds as well as keyhole defects can be used to establish the trajectory (direction) of the bullet. Likewise, the loss of kinetic energy between entrance and exit defects is useful in establishing wound sequence, with exit wounds commonly exhibiting shorter (per cm) radiating fractures and fewer concentric heaving fractures than entrance wounds (Symes et al. 2012).

APPLYING THESE PRINCIPALS TO THE POLICE SHOOTING CASE

From autopsy, three cranial GSWs were established with trajectories A→D, B→C, and Q→R. The forensic anthropologist was tasked with establishing the sequence of the GSWs from the bone. Advantages and disadvantages exist when analyzing specific tissues, but osseous tissues are essentially the indestructible "bare bone" indicators of trauma. In the reconstructed processed bone, wound entrances A, B, and Q are clearly evident and contribute to an accurate perspective of the death event.

GSW A represents a classic keyhole defect in the right parietal bone (see Figure 13.4). The bullet traveled from right to left and exited the left upper parietal. The trajectory of the bullet is evident from the bevel and spall on the entrance defect. Notice the inferior border of wound A is typical for an entrance GSW, with a curved impact site, while the superior border is externally beveled as the bullet goes on edge with the bone. Does that

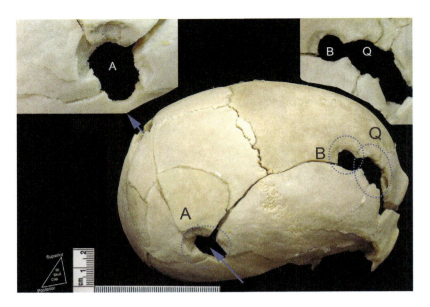

FIGURE 13.4   Reconstructed calotte with three entrance wounds circled. GSW A, a key-hole entrance, represents the greatest release of kinetic energy in the cranium. Wounds B and Q (right insert) each show only minor energy transfer compared to A (left insert) because entrance GSWs B and Q impacted a compromised skull and possibly struck pre-existing radiating fractures. The contact GSW A (see soot on upper border of wound A), destabilized much of the cranial vault (see also Figures 13.5 and 13.6).

mean some or all the projectile exited here? No, it just means the kinetic energy is releasing internally and externally.

Entrances B and Q are also visible in Figure 13.4. Notice entrance B has uniform chipping of outer table despite uniform internal bevel. This chipping is often associated with a perpendicular entrance to skull bone and is interrupted by what appears to be a preexisting fracture (see Figure 13.5, tracked by the two arrows on the anterior cranium). The fracture is likely a product of wound A, the wound designated as a contact wound. Wound Q appears to enter next to B, but the wound is poorly discernible due to the preexisting fractures.

The fracture pattern of wound A→D presents massive destruction to the bone with radiating fractures traveling across the cranium, followed by as many as three generations of concentric heaving failure (see Figures 13.4 and 13.5, arrows). A practitioner should compare the length of the radiating fractures for the entrance and exit defects. In GSW A, the length of the fracture from the defect can be used as a rough estimate of the loss of kinetic energy from the bullet (Smith et al., 1087). As a rule, entrance radiating fractures are longer than exit radiating fractures, as most kinetic energy is released upon entrance (Symes et al., 2012; Berryman & Symes, 1998). Similarly, the number of fractures are commonly greater with entrance than with exit defects. Any subsequent shots (second, third, etc.) also present with less kinetic energy release because the bone has already been compromised, diminishing the skull's resistance to failure.

In this case, the .223 rifle shooter reported that he only shot toward and hit mid-body mass, not the head. The variation in energy expression (cm of total bone fracture) for entrance A as opposed to the entrances B and Q (see Figures 13.4 and 13.5) can be

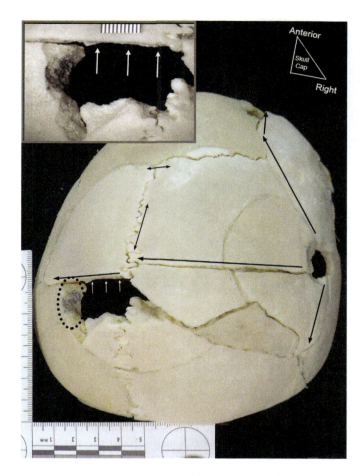

FIGURE 13.5   GSW A→D (superior view of reconstructed vault). Fractures from the entrance defect (right side) radiate across the cranium (elongated single pointed arrows). Further energy is released with as many as three generations of concentric heaving fractures. Shock energy is also being released through the sagittal and coronal sutures (double arrows). The associated exit GSW D (left side) presents with less energy than entrance A, as the bullet punches through bone except where a previous radiating fracture from A exists (white arrows). Extensive bullet wipe is also noted at exit GSW D (oval).

explained using Puppe's Law of Sequence (Madea & Staak, 1988). First, the contact GSW A→D destabilized much of the cranial vault such that wounds B→C and Q→R present a different fracture morphology on account of less energy transfer and bony resistance. Additionally, entrance GSWs B and Q appear to strike preexisting radiating fractures (see Figures 13.4 and 13.5). The loss of kinetic energy and the lessening of bone's resistance to higher velocity is also notable with entrance and exit wounds from the same bullet (trajectory) (Smith et. al., 1987).

The bullet trajectory A→D does reveal external spalling as the bullet exited (D), but some areas of bone around wound D are not beveled. Half of exit defect D exhibits classic external plug and spall with bullet wipe (probably lead in this case) on the bone where the projectile exited, while the other half of the wound simply resembles one side of a

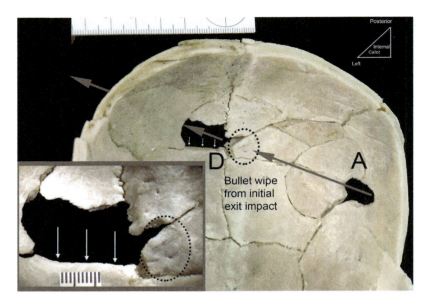

FIGURE 13.6 Endocranial view of the calotte reconstruction, presenting radiating and concentric fractures associated with GSW A→D. Note the radiating fracture traveling from A across the cranium (gray arrow), with GSW D exiting near the preexisting fracture of GSW A (white arrows). Preexisting fractures and destabilized bone provide resistance to massive failure from subsequent skull GSWs. Ovals show bullet's impact, with bullet wipe, on the endocranium before exiting the skull.

radiating fracture (see Figure 13.5). The observable morphology describes a bullet exit defect created *after* radiating fractures originating from the explosive entrance wound already compromised the skull near the exit, neutralizing all energy in this area. The bullet exit initial impact on the internal surface is indicated in Figure 13.6.

    This case illustrates the importance of assessing the internal *and* external surfaces of the cranium because the internal surface often provides more clues about the intersections of radiating fractures than does the external surface. Figure 13.6 shows the internal surface of the calotte and the trajectory of bullet trajectory A→D. As mentioned above, most of the cranium was destabilized from entrance GSW A, indicating that this wound resisted to an extremely high level, causing the cranium to shatter. The internal surface also illustrates the reduced energy associated with exit GSW D compared to entrance GSW A, as it impacted in the region of a preexisting radiating fracture from GSW A.

    In summary, an analyst can explain observed fracture patterns and establish the sequence of shots in the cranium using basic biomechanics of a ballistic injury and Puppe's Law of Sequence. In this case, the contact GSW A→D is the initial GSW to the head, followed by GSW entrance wounds B and Q. GSWs B and Q transferred minor energy to the skull compared to A→D, despite all wounds being associated with low- to middle-velocity weapons.

    The ultimate anthropological assistance in complex bone trauma cases is reconstructing the damaged bone for a "reconstruction of the crime scene." A reconstruction not only assists the pathologist in assessing the COD and MOD, but it may also assist in the overall investigation and in a court of law (see Figure 13.7).

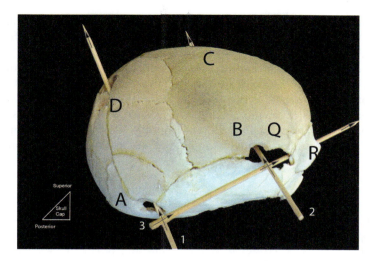

FIGURE 13.7   All trajectories and sequences of the skull: A→D, B→C, and Q→R. The first GSW A is consistent with a tight contact wound.

LESSONS LEARNED

- During the initial autopsy and upon discovering the contact GSW, the investigating officer recounted his interviews with the shooting officers. Each officer recalled only opening fire after thinking the suspect was shooting at them. The investigating officer summarized his description of events by stating, "Well, I guess we opened fire after the suspect committed suicide with that first shot." The evidence from hard tissues (bone) was used to assist in establishing a MOD and to corroborate police officer testimony.
- Forensic anthropologists can apply several principles to their analysis and interpretation of hard tissues to provide additional evidence with regard to bullet trajectory and sequence. The pattern of intersecting fractures can be used to determine the direction of fire, as well as the sequence of shots. Radiating fractures indicate the energy and direction of the various shots. Ballistic injury biomechanics are potentially recognizable in intrinsic bone failure, as it conveys the absorption of high-velocity kinetic energy (KE = $\frac{1}{2}mv^2$).
- Complicated fracture patterns may be unraveled using an understanding of bone dynamics and inter-professional teamwork to reconstruct the sequence of events. In the end, all law enforcement bullet impacts to the skull indicate they followed the initial contact head wound, which in this case was defined as a suicidal gunshot wound. While the forensic pathologist bears the responsibility of COD and MOD, additional expertise of anthropology and investigation may contribute useful information beyond that derived from the autopsy.

DISCUSSION QUESTIONS

13.1 The autopsy team made three significant discoveries during their analysis, namely the contact gunshot wound, the radiating fractures in bone, and police testimony which was pertinent to the forensic pathologist establishing the MOD. What are the benefits of inter-professional teamwork in the forensic sciences? Are there any cons?

13.2 In this case the forensic anthropologist works in a medical examiner's office and was called in to assist in a trauma analysis in a recent death. Other offices may not have a forensic anthropologist on staff and may only call in for their expertise in cases involving decomposed, skeletonized, or otherwise modified remains. When should a forensic anthropologist be utilized? Should they be used in every trauma case involving bones? Discuss the pros and cons.

REFERENCES

Berryman, H. E., Smith, O. C., & Symes, S. A. (1995). Diameter of cranial gunshot wounds as a function of bullet caliber. *Journal of Forensic Sciences*, 40(5), 751–754.

Berryman, H. E., Symes, S. A. (1998). Recognising gunshot and blunt cranial trauma through fracture interpretation. In K. J. Reichs, (Ed.), *Forensic osteology: Advances in the identification of human remains* (pp. 333–352). Springfield, IL: Charles C. Thomas.

DiMaio, V. J. M. (1999). An introduction to the classification of gunshot wounds. In *Gunshot wounds: Practical aspects of firearms, ballistics, and forensic techniques* (2nd ed., pp. 65–122). New York: CRC Press.

Dixon, D. S. (1982). Keyhole lesions in gunshot wounds of the skull and direction of fire. *Journal of Forensic Sciences*, 27(3), 555–566.

Dixon, D. S. (1984). Pattern of intersecting fractures and direction of fire. *Journal of Forensic Sciences*, 29(2), 651–654.

Fackler, M. (1988). Wound ballistics; as review of common misconceptions. *JAMA*, 259(18):2730–2736.

Hart, G. O. (2005). Fracture pattern interpretation in the skull: Differentiating blunt force from ballistics trauma using concentric fractures. *Journal of Forensic Sciences*, 50(6):1–6.

Klinger, D., Rosenfeld, R., Isom, D., Deckard, M. (2015). Race, crime, and the micro-ecology of deadly force. *Criminol Public Policy*, 15, 193–222. doi: 10.1111/1745-9133.12174

Langley, N. R. (2007). An anthropological analysis of gunshot wounds to the chest. *Journal of Forensic Sciences*, 52(3),532–537.

Madea, B., & Staak, M. (1988). Determination of the sequence of gunshot wounds of the skull. *Journal of the Forensic Science Society*, 28, 321–328.

Ross, A. H. (1996). Caliber estimation from cranial entrance defect measurements. *Journal of Forensic Sciences*, 41, 629–633.

Ross, C. T. (2015). A multi-level Bayesian analysis of racial bias in police shootings at the county-level in the United States, 2011–2014. *PLOS ONE 10*(11), e0141854. doi: 10.1371/journal.pone.0141854

Smith, O. C., Berryman, H. E., & Lahren, C. H. (1987). Cranial fracture patterns and estimate of direction from low velocity gunshot wounds. *Journal of Forensic Sciences*, 32(5), 1416–1421.

Symes, S. A., L'Abbé, E. N., Chapman, E. N., Wolff, I., & Dirkmaat, D. C. (2012). Interpreting traumatic injury from bone in medicolegal investigations. In D. C. Dirkmaat (Ed.), *A companion to forensic anthropology* (pp. 540–590). London, UK: Wiley-Blackwell.

CHAPTER 14

Forensic Anthropological Contributions to Manner of Death in a Case of Multiple Suicidal Gunshot Wounds

Diana L. Messer

CONTENTS

CASE BACKGROUND

On a winter day in the rural Midwest United States, squirrel hunters walking in the snowy woods near their house came across what appeared to be a human skull. Looking closer, they noticed a hole near the top of the head. Immediately they reported their finding to the local police department, telling them that the skull appeared to have a possible bullet hole. The police responded to the scene with the medical examiner's office to conduct a recovery of the skeletonized remains. A handgun and shell casing were found at the scene. The remains were recovered by law enforcement and transported to the medical examiner's office to find out who this person was and how they died.

A conventional autopsy of the recovered remains was precluded by lack of soft tissue. Without the pathology report to determine cause and manner of death, Dennis C. Dirkmaat, PhD, D-ABFA, Director of the Mercyhurst Forensic Anthropology Laboratory

(M-FAL) at Mercyhurst University was contacted by the medical examiner to conduct a comprehensive forensic anthropological analysis. They were particularly interested in the skeletal trauma assessment, hoping to find evidence to support the medical examiner's effort to ascertain identity and cause of death. It is important to note that forensic anthropologists do not determine the cause and manner of death. They do, however, contribute evidence for the coroner or medical examiner to make those official determinations. In cases such as the present study, where a conventional autopsy cannot be performed, information derived from skeletal trauma analysis may be the only evidence available to understand the cause and manner of death.

The person who legally determines cause and manner of death varies by state. Sixteen states and the District of Columbia have a centralized medical examiner's office, six have county or district-based medical examiner's offices, and 25 have a state medical examiner whose office is responsible for the determination of cause and manner of death (Center for Disease Control and Prevention, 2016). The rest of the states either have a coroner system or a mixture of the two (Center for Disease Control and Prevention, 2016) and overall, more jurisdictions have a coroner system (Hanzlick, 2006). Coroners are elected positions and in most states are not required to hold a medical degree (National Public Radio, 2011; Hanzlick, 2006). The coroner system historically began as a position to keep records of births and deaths, primarily as it pertained to taxes in early England (Jentzen, 2010; National Public Radio, 2011). These individuals were originally tasked to confirm identity, determine how the person died through inquiries, and assist with seizing of property. This gradually evolved as the coroners were requested to verify injuries and bear witness at the earliest version of trials (Jentzen, 2010; Gross, 1892).

The coroner or medical examiner makes the decision of when to involve a forensic anthropologist. Whether or not a forensic anthropologist is called to consult on a case may depend on a variety of factors. It could be hypothesized that coroners without a medical background may be more likely to call in experts to assist in the death investigation given their lack of anatomical expertise, while medical examiners may feel that their expertise is sufficient to conduct a skeletal trauma assessment. Conversely, medical examiners may appreciate fellow scientific experts more and be more apt to call in an anthropologist. Budgets may also play a role. The key factor, however, is that the coroner or medical examiner is aware of local forensic anthropologists, the work that they can provide to support death investigations, and that a good working relationship is established. Fortunately, Dr. Dirkmaat has been active in the region for over 30 years, and his forensic work is well known in the area.

DESCRIPTION OF REMAINS

The human remains found in the woods were brought to the M-FAL for analysis with associated clothing including a pair of large boots, hooded sweatshirt, winter hat, one glove, and a pair of glasses. The remains were mostly skeletonized, with only small portions of desiccated tissue remaining. The skeletal elements were in good condition, with no evidence of significant taphonomic modification due to weathering or animal activity. Prior to analysis, the skeletal elements were photographed, cleaned of dirt and debris with soft-bristled brushes, and lightly brushed with water.

Once dried, a skeletal inventory was conducted. Elements present included the cranium, right portion of the mandible, and most of the larger postcranial elements

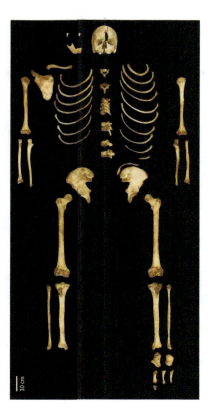

FIGURE 14.1 Skeletal layout in anatomical position. Photo courtesy of the Mercyhurst Forensic Anthropology Laboratory and Dennis C. Dirkmaat, PhD, D-ABFA.

(see Figure 14.1). Skeletal elements missing included the left half of the mandible, the hyoid, sacrum, several ribs, several vertebrae, all hand bones, the entire right foot, and the majority of the left foot. Portions of the cranium were also missing (see Figure 14.2). An offer by the M-FAL to search the area for the missing skeletal elements was declined.

FORENSIC TAPHONOMY ANALYSIS

Since detailed documentation of context at the scene was unavailable, the ability to comment on postmortem movement of the remains, absence of certain skeletal elements, or an estimate of postmortem interval was significantly limited. Had the Mercyhurst Forensic Recovery Team (M-FSRT) conducted an archaeological recovery of the remains, more knowledge may have been gained regarding the dispersal and distribution of skeletal elements, and more bones may have been recovered despite the light covering of snow at the time of discovery.

Postmortem modifications were evident on the remains. The skeletal elements had a medium to dark brown coloration with small portions of adherent tissue, particularly at or within joint surfaces. Small amounts of desiccated tissue were present on the femora, tibiae, fibulae, and humeri. Soft tissue has been noted to persist in the

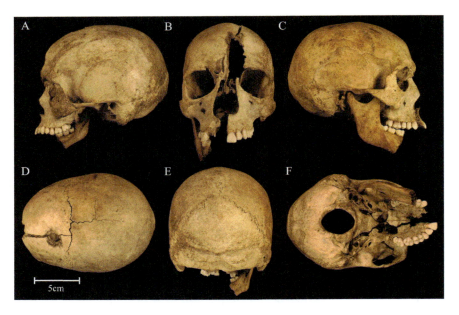

FIGURE 14.2 Six views of the skull. (A) Left lateral, (B) Anterior, (C) Right lateral, (D) Superior, (E) Posterior, and (F) Inferior. Photo courtesy of the Mercyhurst Forensic Anthropology Laboratory and Dennis C. Dirkmaat, PhD, D-ABFA.

Midwest region after one year; however, this depends on season of deposition, animal scavenging, insect activity, and a variety of other factors (Dirkmaat et al., 2014). Little to no evidence of carnivore damage or animal scavenging was evident on these remains, perhaps explained by the date of deposition and subsequent recovery. The degree of skeletonization and lack of soft tissue was consistent with a postmortem interval of approximately one year in an outdoor wooded location before discovery in December.

BIOLOGICAL PROFILE ANALYSIS

Prior to a positive identification, a blind analysis of the biological profile was performed.

Ancestry and Sex Estimation

Analysis of cranial non-metric indicators of ancestry was limited by the extensive trauma to the facial region and mandible. Ancestry estimation was performed through metric analysis. Thirty-nine postcranial measurements were entered into Fordisc 3.1.301 (Jantz & Ousley, 2005) and compared to White and Black individuals in the Forensic Data Bank. Discriminant function analysis of 10 lambda-stepwise selected variables assigned the remains to the White male group, with posterior and F-ratio typicality probabilities of 0.919 and 0.922, respectively. This model correctly classified 89% of the postcrania in the reference database, after correcting for overfitting through leave-one-out cross-validation (Jantz & Ousley, 2005).

Age-at-Death Estimation

The spheno-occipital synchondrosis displayed a visible fusion scar, corresponding to an age interval greater than 13.3 years (Shirley & Jantz, 2011). The right and left medial clavicle epiphyses exhibited a billowed surface with no epiphyseal flake present, which corresponds to an age less than or equal to 17.4 years old (see Figure 14.3A) (Langley-Shirley & Jantz, 2010). Several long bones exhibited evidence of epiphyseal non-fusion, including the right and left humeri, radii, ulnae, femora, tibiae, and fibulae. Most of these, such as the unfused proximal humerus, indicate an age between 16 and 20 years old (see Figure 14.3B) (Scheuer & Black, 2004). Notably, the epiphyseal growth cap of the right and left iliac crest were not fully fused, though union was beginning in the anterior half (see Figure 14.3C). This union typically begins prior to 17 years of age (McKern & Stewart, 1957) and continues fusion through 17 to 19 years in males (Scheuer & Black, 2004). In addition, the epiphyses of the ischial ramus and tuberosity were still in an active state of fusion (see Figure 14.3D).

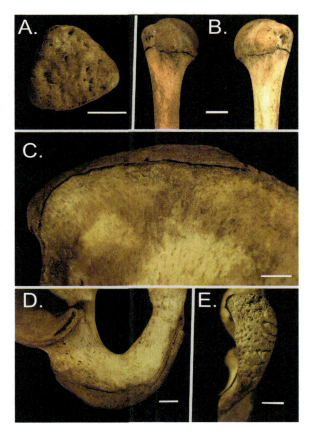

FIGURE 14.3   Anatomical areas evaluated to estimate age-at-death. (A) Unfused medial end of right clavicle, (B) Right and left proximal humeri with epiphyseal line, (C) Unfused right iliac crest, (D) Unfused right ischiopubic ramus and tuberosity, and (E) Right pubic symphyseal face. Photo courtesy of the Mercyhurst Forensic Anthropology Laboratory and Dennis C. Dirkmaat, PhD, D-ABFA.

The pubic symphyseal faces of the right and left innominates exhibited prominent billowing, no delimitation, and no distinct pubic tubercle (see Figure 14.3E). These features are consistent with Phase 1 of the Brooks and Suchey (1990) age estimation method, corresponding to an age range of 15 to 23 years, with a mean of 18.5 years (+/–2.1).

Transition analysis (Boldsen et al., 2002), an adult age-at-death estimation method that statistically combines information from independently scored features of the cranial sutures, pubic symphyses, and iliac auricular surfaces, produced an age estimate of 15–20.1 years, with a maximum likelihood of 15 years, using the reference sample for White males and a 95% confidence interval. Combining these age indicators and methods of age-at-death estimation provides a narrow age range of 15–20 years.

Stature Estimation

The most accurate stature estimation from these remains was derived from the maximum lengths of the calcaneus, femur, and tibia. This estimate was derived using Fordisc 3.1.301 (Jantz & Ousley, 2005), which utilizes a reference sample of 20th century White males from the Forensic Data Bank, with a 95% prediction interval. This equation produced a stature range of 65.7 to 73.5 inches (95% interval), with a point estimate of 69.6 ± 3.9 inches.

The skeletal analysis indicated that the skeletal remains belonged to a White male between the ages of 15 and 20 years with a stature of between 65.7 and 73.5 inches. This was consistent with a local missing person, a white 16-year-old male with a stature of 5 feet 9 inches (69 inches). A forensic odonotologist made a positive dental identification.

SKELETAL TRAUMA ANALYSIS

In this case study, the term "perimortem" refers to the interval at or around the time of death when bone retains its organic biomechanical properties and exhibits no evidence of healing. This analysis documents only skeletal trauma; however, it is possible that additional soft tissue injuries were present that did not impact bone.

Two perimortem circular defects were noted on the hard palate. One defect was located posterior to the central incisors (see Figure 14.4, Defect A, green inset) and the other defect was located posterolaterally on the left palatine bone (see Figure 14.4, Defect B, green inset). Two additional perimortem circular defects were noted on the cranial vault. One defect was located in the middle of the frontal bone (see Figure 14.4, Defect C, yellow inset) and the other defect was located to the left posterolaterally, anterior to the coronal suture (see Figure 14.4, Defect D, blue inset).

The morphology of these circular defects was consistent with two separate high-velocity projectile impacts (i.e., gunshot trauma). It should be noted that while the term "ballistic trauma" has been traditionally used to describe gunshot trauma, it is no longer considered accurate. "Ballistic" is a term that generally refers to the study of the motion of projectiles such as bullets and in forensic science sometimes refers to the practice of matching ammunition to the firearm it was fired from (Christensen et al., 2014).

The gunshot wound located more anteriorly will be described as the *anterior gunshot* wound. The other gunshot wound located more posteriorly will be described as the *posterior gunshot*. Defects A and C were associated with the anterior gunshot event (see Figure 14.4). Defect A was comprised of a circular defect with sharp margins

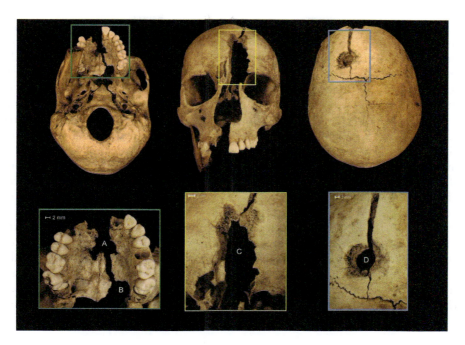

FIGURE 14.4 Three views of the skull showing two perimortem defects. Upper left: inferior view with close-up in green inset. Upper middle: frontal view, with close-up in yellow inset. Upper right: superior view with close-up in blue inset. Photo courtesy of the Mercyhurst Forensic Anthropology Laboratory and Dennis C. Dirkmaat, PhD, D-ABFA.

and radiating fractures, which is consistent with an entrance wound (DiMaio, 1999; Berryman & Symes, 1998; Kimmerle & Baraybar, 2008 and references therein). There was no internal beveling noted, likely due to the low cortical thickness of the hard palate. Defect C exhibited external beveling with one radiating fracture running directly posterior until it joined with the coronal suture (see Figure 14.4). The force from the projectile impact likely diffused into the suture lines from this radiating fracture, creating diastatic fracturing indicated by the expanded space along the suture lines and a hairline fracture extending laterally from the sagittal suture on the right side. The external beveling and radiating fracture indicate that Defect C is an exit wound (DiMaio, 1999; Berryman & Symes, 1998; Kimmerle & Baraybar, 2008 and references therein). The trajectory of this projectile is interpreted to have entered bone at the anterior portion of the hard palate (at Defect A), traveled superiorly through the facial region and through the frontal sinus before exiting through the anterior central portion of the frontal bone (at Defect C) (see Figure 14.5 and 14.6, green arrow). The trajectory of this gunshot wound was interpreted to be inferior-to-superior, slightly anterior-to-posterior, and slightly right-to-left.

Defect B and Defect D are associated with the posterior gunshot event. Defect B exhibits sharp margins and is located on one of the radiating fracture lines from Defect A. The sharp, roughly circular margins are consistent with a gunshot entrance wound (DiMaio, 1999; Berryman & Symes, 1998; Kimmerle & Baraybar, 2008 and references therein). Defect D is located on the left frontal bone immediately anterior to the coronal suture and a few centimeters to the left of bregma. This defect exhibits external beveling, concentric fractures around the defect margin, and is located on a radiating fracture from Defect C. These characteristics are consistent with an exit wound (DiMaio,

1999; Berryman & Symes, 1998; Kimmerle & Baraybar, 2008 and references therein). The anterior portion of this radiating fracture has a widened appearance, likely due to distributed force from the second gunshot wound. The trajectory of this projectile was interpreted to have begun at the posterior aspect of the hard palate (at Defect B), traveled superiorly through the sphenoidal sinus and cranial vault, and exited at the top of the cranium (at Defect D) (see Figure 14.4). The path of the posterior gunshot event is roughly parallel to the path of the anterior gunshot event (see Figure 14.5 and 14.6, red arrow). The trajectory of this gunshot is interpreted as inferior-to-superior, slightly anterior-to-posterior, and slightly right-to-left.

Analysis of these defects indicates that the anterior gunshot event occurred prior to that of the posterior gunshot event. This is evidenced by the distribution of force seen from the posterior gunshot event and the intersection of the radiating fracture from the initial exit wound with the second exit wound, using Puppe's law (Madea & Staak, 1988; Viehl et al., 2009). Both the entrance (Defect B) and exit wound (Defect D) are located on radiating fractures that resulted from the anterior gunshot event. Defects A and C do not have any radiating fractures of their own. Instead, there is an expansion of the radiating fractures that connect the two entrance and exit wounds, in comparison to other radiating fractures that exhibit a smaller width. This is likely due to the force of the second, posterior gunshot event diffusing along these already preexisting radiating fractures. In

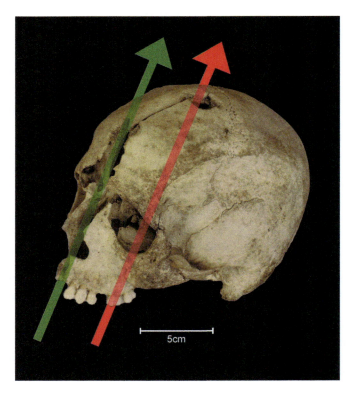

FIGURE 14.5 Left oblique view of cranium showing approximate gunshot trajectories. Anterior defect (green arrow) and posterior defect (red arrow). Note that these trajectories are approximate illustrations. Photo courtesy of the Mercyhurst Forensic Anthropology Laboratory and Dennis C. Dirkmaat, PhD, D-ABFA.

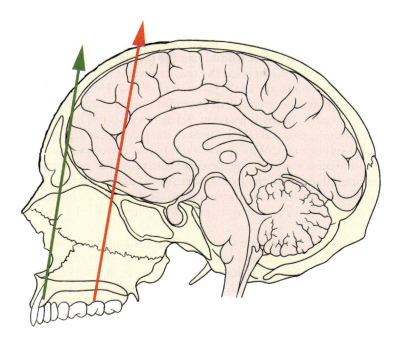

FIGURE 14.6 Midsagittal illustration of bullet trajectories through cranium with associated neural tissue. Anterior defect (green arrow) and posterior defect (red arrow). Note that these trajectories are approximate illustrations. Adapted image: "Skull and Brain Sagittal" © Patrick J. Lynch, medical illustrator, C. Carl Jaffe, MD, cardiologist. From Wikimedia Commons. https://creativecommons.org/licenses/by/2.5/

addition, the force from the posterior gunshot event's exit wound (Defect D) likely also diffused into the suture lines, contributing to the diastatic fractures noted. Both gunshot events occurred in an inferior-to-superior direction, from either a submandibular or intraoral gunshot. Without the entirety of the mandible (not recovered by law enforcement), a distinction between a submandibular and intraoral entrance could not be made.

Forensic anthropologists are sometimes asked to comment on bullet caliber in cases with gunshot trauma. However, in this case, the intersection of the second gunshot wound through the radiating fracture of the first gunshot wound contributed to dispersal of forces through preexisting fractures that might have rendered measurements inaccurate at best and, at worst, misleading and confusing for the medicolegal investigation. In addition, the entrance wounds were associated with several areas of missing bone that prevented a complete measurement from being taken.

DISCUSSION

In this case study, after the forensic anthropological analysis and report were completed, the medical examiner and law enforcement agency needed to consider whether the gunshot wounds described were the result of a homicide or a suicide. In the case at hand, this was not immediately clear. Though more than one gunshot wound in cases of suicide are not inconceivable, multiple gunshot wounds generally call for a close examination to rule out homicide.

In a review of suicidal gunshot wounds conducted in 1981, 74% were located in the head, with 13% of those located in the mouth, chin, or submental region (Eisele et al., 1981). Though a single gunshot wound has the potential to be lethal, multiple gunshot suicides where the individual pulled the trigger several times have been documented (Karger et al., 1997; Introna et al., 1989; Jacob et al., 1989; Hejna et al., 2012). The projectile trajectory and potential areas of the nervous and/or cardiovascular system damaged in such cases are particularly important. Karger et al. (1997) noted cases of multiple suicidal gunshot wounds, particularly wounds to the face or eye region, where the initial wound(s) did not result in a lack of incapacitation. In this review, Karger et al. (1997) also noted that gunshot wounds in such suicide cases generally did not impact cerebral tissue or resulted in focused damage to the frontal lobe, avoiding the primary motor cortex and other areas of the brain related to motor function. Of the cases examined in their review, 70% passed above the anterior cranial fossa and damaged only the frontal lobe (Karger et al., 1997).

Specifically, evidence of submental and transoral handgun or low energy gunshot wounds that produce serious but non-fatal brain injuries when the path of the bullet is limited to the frontal area has been documented (Kreit et al. 2005). This type of injury is consistent with the path of first gunshot event in this case study, which passed through the nasal region of the face and traveled through the frontal sinus, likely impacting only a small portion of the anterior frontal lobe if any cerebral tissue at all. Incapacitation of the individual in cases where only a small portion of the frontal lobe is wounded is highly unlikely because the motor cortex is not impacted, thus would not impair the ability for physical activity (Karger et al., 1995; Karger et al., 1997). If the bullet trajectory does not impact the central nervous system and passes only through the facial region, such as in the present case study, the ability of the individual to act following the gunshot event is possible.

In this case, the forensic anthropologist's anatomical knowledge, detailed description of bullet trajectories, and ability to sequence the gunshot wounds provided the information necessary for the medical examiner to determine the manner of death as a suicide.

LESSONS LEARNED

- Forensic archaeological recovery would most likely ensure recovery of most major skeletal elements. In this case, failure to utilize forensic archaeological methods resulted in an incomplete recovery of remains that limited the skeletal trauma analysis. In particular, the absence of the other half of the mandible resulted in an inability to determine the exact entrance site for both gunshot wounds.
- Skeletal trauma interpretations require advanced anatomical knowledge. When analyzing skeletal trauma and preparing a report, the forensic anthropologist should consider how their report will be used and provide information that is relevant to the coroner and/or medical examiner's determination of cause and manner of death. Incorporating information about surrounding anatomical structures may provide information relevant to the medical examiner's interpretation. It is important to reconstruct remains and document the number and sequence of defects, as this may contribute to determination of cause and manner of death.

DISCUSSION QUESTIONS

14.1 Do you think the medical examiner's manner of death ruling of suicide would have changed if the skeletal trauma analysis had been unable to provide an opinion regarding the order of the GSWs?

14.2 Can you think of another circumstance involving skeletal remains where *manner* of death might rely solely on evidence from the forensic anthropologist?

14.3 What other anatomical correlations with skeletal trauma might contribute to a determination of *cause* of death?

14.4 How important is understanding the death investigation system to the forensic anthropologist? How might the state, district, and/or county specific systems affect the forensic anthropologist?

ACKNOWLEDGMENTS

The author would like to acknowledge Dennis C. Dirkmaat, PhD, D-ABFA as well as the graduate students of Mercyhurst University (classes of 2018 and 2019) for their assistance with the forensic anthropological analysis. Thanks also to Luis Cabo Perez for his thoughtful comments and edits. Additionally, thanks to the editors, Dr. Heather Garvin and Dr. Natalie Langley, for the invitation to contribute this case study.

REFERENCES

Berryman, H. E., & Symes, S. A. (1998). Recognizing gunshot and blunt cranial trauma through fracture pattern interpretation. In K. J. Reichs (Ed.), *Forensic osteology: Advances in the identification of human remains* (2nd ed. pp. 333–352). Springfield, IL: Charles C. Thomas.

Boldsen, J. L., Milner, G. R., Konigsberg, L. W., & Wood, J. W. (2002). Transition analysis: A new method for estimating age from skeletons. In R. D. Hoppa & J. W. Vaupel (Eds.), *Paleodemography: Age distributions form skeletal samples* (pp. 73–706). Cambridge, UK: Cambridge University Press.

Brooks, S., & Suchey, J. M. (1990). Skeletal age determination based on the *Os pubis*: Comparison of the Acsadi-Nemeskeri and Suchey-Brooks methods. *Human Evolution*, 5, 227–238.

Center for Disease Control and Prevention. (2016). Death investigations systems. https://www.cdc.gov/phlp/publications/coroner/death.html. Accessed on August 21, 2018.

Christensen, A. M., Passalacqua, N. V., & Bartelink, E. J. (2014). Chapter 13. Analysis of Skeletal Trauma. In *Forensic anthropology: Current methods and practice* (pp. 341–375). Kidlington, UK: Elsevier, Inc.

DiMaio, V. (1999). An introduction to the classification of gunshot wounds. In *Gunshot wounds: Practical aspects of firearms, ballistics, and forensic techniques* (2nd ed., pp. 65–122). Boca Raton, FL: CRC Press.

Dirkmaat D. C., Cabo, L. L., & Fredette, S. M. (2014). Refining postmortem interval estimates in the northeast. *Proceedings of the 66th Annual Meeting of the American Academy of Forensic Sciences*, Seattle, WA.

Eisele, J. W., Reay, D. T., & Cook, A. (1981). Sites of suicidal gunshot wounds. *Journal of Forensic Sciences, 26*(3), 480–485.

Gross, C. (1892). The early history and influence of the office of coroner. *Political Science Quarterly, 7*(4), 656–672.

Hanzlick, R. (2006). Medical examiners, coroners, and public health. *Archives of Pathology and Laboratory Medicine, 130,* 1274–1282.

Hejna, P., Miroslav, S., & Zaptopkova L. (2012). The ability to act – Multiple suicidal gunshot wounds. *Journal of Forensic and Legal Medicine, 19,* 1–6.

Introna, F., & Smialek, J. E. (1989). Suicide from multiple gunshot wounds. *American Journal of Forensic Medicine and Pathology, 10*(4), 275–284.

Jacob, B. Barz, J., Haarhoff, K., Sprick C., Worz, D., & Bonte, W. (1989). Multiple suicidal gunshots to head. *American Journal of Forensic Medical Pathology, 10,* 289–294.

Jantz, R. L., & Ousley, S. D. (2005). *FORDISC 3.0: Personal computer forensic discriminant functions.* Knoxville, TN: University of Tennessee.

Jentzen, J. M. (2010). *Death investigation in America: Coroners, medical examiners, and the pursuit of medical certainty.* Cambridge, MA: Harvard University Press.

Karger, B. (1995). Penetrating gunshots to the head and lack of immediate incapacitation. II. Review of case reports. *International Journal of Legal Medicine, 108*:117–126.

Karger, B., & Brinkman, B. (1997). Multiple gunshot suicides: Potential for physical activity and medico-legal aspects. *International Journal of Legal Medicine, 110,* 188–192.

Kimmerle, E. H., & Baraybar, J. P. (2008). Chapter 7: Gunfire injuries. In Kimmerle & Baraybar (Eds.), *Skeletal trauma: Identification of injuries resulting from human rights abuse and armed conflict* (pp. 321–400). Boca Raton, FL: CRC Press, Taylor and Francis Group.

Kriet, J. D., Stanley, R. B., & Grady, M. S. (2005). Self-inflicted submental and transoral gunshot wounds that produce nonfatal brain injuries: Management and prognosis. *Journal of Neurosurgery, 102,* 1029–1032.

Langley-Shirley, N., & Jantz, R. L. (2010). A Bayesian approach to age estimation in modern Americans from the clavicle. *Journal of Forensic Science, 55*(3), 571–583.

Madea, B., & Staak, M. (1988). Determination of the sequence of gunshot wounds of the skull. *Journal of the Forensic Science Society, 28,* 321–328.

McKern, T. W., & Stewart T. D. (1957). *Skeletal age changes in young American males.* Quartermaster Research and Developmental Command Technical Report EP-45, Natick, MA.

National Public Radio. (2011). Coroner's don't need degrees to determine death. https://www.npr.org/2011/02/02/133403760/coroners-dont-need-degrees-to-determine-death. Accessed on August 23, 2018.

Scheuer, L., & Black, S. (2004). *The juvenile skeleton.* Cambridge, UK: Academic Press.

Shirley, N. R., & Jantz, R. L. (2011). Spheno-occipital synchondrosis fusion in modern Americans. *Journal of Forensic Science, 56*(30), 580–585.

Viel, G., Gehl, A., & Sperhake, J. P. (2009). Intersecting fractures of the skull and gunshot wounds. Case study and literature review. *Forensic Science, Medicine, and Pathology, 5,* 22–27.

CHAPTER 15

A Unique Case of Skeletal Trauma Involving Scissors

Alexandra R. Klales

CONTENTS

CASE BACKGROUND AND FORENSIC ARCHAEOLOGICAL RECOVERY

A housing authority employee discovered suspected decomposing human remains within a trash receptacle outside of a residential housing unit during the hot summer months. Law enforcement was contacted, and a forensic anthropology recovery team traveled to the scene to assist state law enforcement and the local coroner's office with the recovery the following day.

Upon arrival at the scene, the forensic anthropology recovery team discovered three metal trashcans and seven plastic bags stacked outside a housing unit (see Figure 15.1). An initial inspection of the scene revealed a skeletonized human mandible, previously discovered by the housing employee. The mandible was found sitting atop a blanket and garbage in one of the metal trashcans (see Figure 15.1, middle bin). The team quickly deemed the remains to be forensically significant human bone and a forensic archaeological recovery commenced.

The trashcans were determined to be a secondary deposition site given that the remains were out of anatomical order (e.g., mandible separate from the cranium) and were contained within two separate containers, without any evidence of dismemberment. This indicated that significant time had lapsed between the death event and substantial decomposition took place prior to deposition in the trash receptacles. With this information, law enforcement decided to search the suspect's apartment unit where it was believed the remains had been stored and decomposed. No human bones or soft tissue were discovered in the apartment; however, a large quantity of blood and decompositional fluid was located within the kitchen closet/pantry. An abundance of living and dead flies were present throughout the unit despite having been fumigated recently by

FIGURE 15.1   Overview of the scene.

housing authority employees. The primary scene was photographed, and entomological evidence was collected.

Due to the exposed nature of the secondary scene and lack of privacy from the public and media, the decision was made to relocate the trash receptacles to a more private and secure location where their contents could be documented in detail to preserve any contextual information. Prior to disturbing the scene, however, photographs were taken, and a sketch of the placement of trashcans and bags was made. The trash receptacles were then individually contained, and each unit was labeled with a unique identifier, which was also labeled on the scene sketch so no contextual information would be lost. Two of the metal trashcans (the third was empty, Figure 15.1 bin on right) and the seven plastic bags were transported to a secure off-site location to document, sort, and recover the contents of each can/bag. A nearly complete, partially skeletonized set of remains was found wrapped within a blanket in the trashcan containing the mandible. The lower extremities and lower thorax were primarily fleshed (although in a moderate state of decomposition), while the skull, upper thorax, and upper extremities were mostly skeletonized. A secondary, smaller concentration of eleven bones was found in a separate trash bag. Following the recovery, the remains were transported to the local coroner's office for an autopsy and subsequently transported to the forensic anthropology laboratory for a complete osteological analysis.

LABORATORY ANALYSIS OF THE REMAINS

When remains are partially covered in soft tissue, it is imperative to conduct a thorough external and radiographic examination prior to any processing to identify and document any soft tissue defects before that information is lost as well as any potential skeletal trauma. Based on what is observed, the anthropologist may alter the maceration process (e.g., decide to use no sharps if there's potential sharp force trauma, or alter procedures to preserve fragile elements/evidence). It also helps to be able to show jury members that

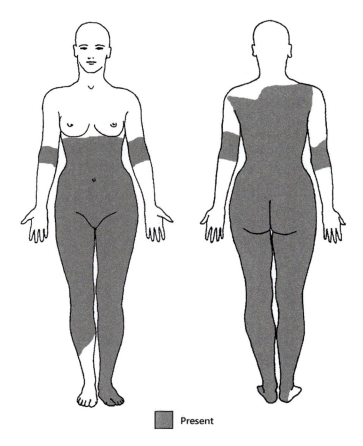

Present

FIGURE 15.2 Homunculus illustrating the distribution of the remaining soft tissues.

trauma was present prior to maceration to definitively rule out the possibility that it was caused during processing. In this case, signs of suspected traumatic defects were noted on several skeletal elements during initial documentation in the laboratory and prior to the removal of the soft tissue. Entomological activity and soft tissue defects were also noted throughout the remains (Figure 15.2), especially on the lower extremities and thorax. Given the amount of soft tissue, scalpels were used to remove large portions of the soft tissue but were not used in the regions surrounding the possible trauma noted during initial inspection of the remains. All elements were then macerated through immersion in warm water to remove the remaining soft tissue.

Both qualitative and quantitative forensic anthropological methods were used to blindly estimate the biological profile parameters (i.e., age, ancestry, sex, stature) of the decedent following maceration and inventory. The generated biological profile was consistent with the known demographic information of the decedent, who had been positively identified through dental records by a forensic odontologist. A suspect was in custody, but a potential murder weapon had yet to be located. Investigators were interested in what type or class of weapon was used in the commission of the crime. The focus of the requested forensic anthropological report, therefore, was specifically on the trauma analysis.

TRAUMA ANALYSIS

Antemortem (i.e., healed) trauma was noted on the left scapula, which showed signs of bony remodeling (see Figure 15.3, orange circle and 15.4, blue box). Perimortem trauma (i.e., at or around the time of death) was more pervasive on the remains and was analyzed both macroscopically and microscopically. The morphology of the traumatic defects indicated they likely occurred while the bone still retained its organic components, i.e., while the bone was still fresh and elastic. Defects were noted on the cranium, left scapula, left and right humerus, left radius and ulna, left and right hand elements, several thoracic elements (i.e., ribs and vertebrae), left and right innominates, sacrum, left and right femur, left and right tibia, left and right fibula, and right talus (see Figure 15.3). Each defect was photo-documented macro- and microscopically, charted, and described with anatomical locations and measurements.

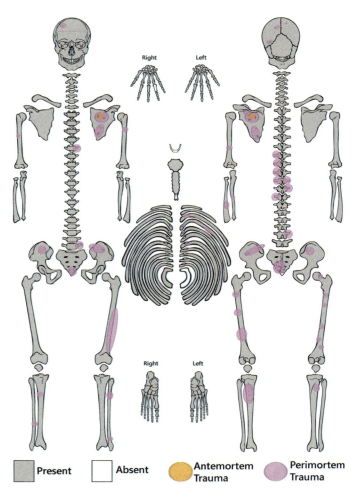

FIGURE 15.3 Homunculus indicating bones present and areas of ante- and perimortem skeletal trauma. Left: anterior view. Right: posterior view.

The bones predominantly exhibited characteristics of blunt force trauma with puncturing, penetration, scoring, and fracturing of the cortical bone, yet there was also evidence of sharp force trauma (incised defects) on several skeletal elements. Blunt force trauma is defined from a biomechanical standpoint as a slow-loaded impact to bone and can result in abrasions, contusions, lacerations, and/or fractures (Symes et al., 2012). Punctures refer to defects with a clear entrance and an exit through the bone, while penetrating defects refer to those with an entrance and no exit through the bone. Sharp force trauma, defined from an osteological perspective, is a narrowly focused, dynamic, slow-loaded, compressive force with a sharp object that produces damage to hard tissue in the form of an incision (Symes et al., 2012).

A pattern in the shape of the blunt force penetrating defects was noted on the scapula, innominates, and sacrum, although the defects were somewhat unusual in contour (see Figure 15.4). The defects exhibited damage at the site of impact including flaking, fragmentation, and multiple fractures caused by the penetrating pressure of the weapon into the bone. These defects also exhibited varying degrees of crushing along the edges. The impacts were made with sufficient force to cause blunt force trauma (slow-loading trauma) and were directed from posterior-to-anterior on the scapula, often at oblique angles, and lateral-to-medial on the innominates. The defects are consistent with a blunted object with a small focal area causing penetrating and puncturing wounds. The puncturing defects were mostly rectangular or square in shape with a thickness of 3mm at one end at 3.5mm at the other end (in the case of the rectangular defects). The length of the defects was also consistent at 11mm for the rectangular defects and 3mm for the square defects. Multiple defects were also present on the calotte of the skull, the thoracic region, and the extremities (see Figure 15.3 and 15.5). Penetrating defects were present throughout these skeletal regions, as were pairs of parallel linear incised defects measuring approximately 3mm apart (see Figure 15.5), consistent with the puncturing defects described above. Most of the rib defects appeared to be caused by blunt trauma with crushing and flaking of the cortical bone, but there were also areas of sharp force trauma

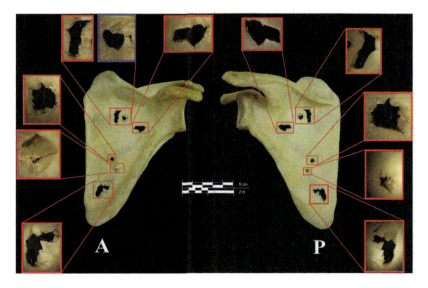

FIGURE 15.4   Example of the penetrating defects on the left scapula in anterior view (left) and posterior view (right). Antemortem outlined in blue and perimortem in red.

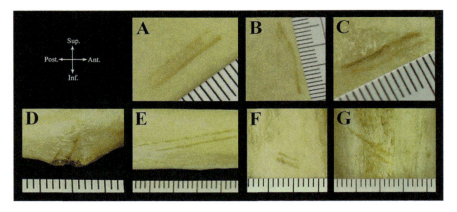

FIGURE 15.5 Photographs showing the consistent parallel linear defect trauma throughout the skeleton. (A) Left parietal defect near the landmark asterion;(B) Right parietal defect near the landmark bregma; (C) Right parietal defect near the temporal lines; 9D) Left sixth rib defect on the superior aspect of the head; (E) Left ninth rib defect on the internal surface near mid-shaft;(F) Left humerus defect; (G) Left femur defect.

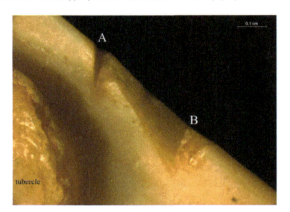

FIGURE 15.6 Close-up view of the two linear incised defects on the inferior margin near the tubercle of the left seventh rib. (A) acute angle; (B) obtuse angle.

(i.e., incised) on several ribs. Upon magnification of the parallel linear defects on the seventh and ninth ribs, it became apparent that one incision was wide (obtuse) and one was more narrow (acute angle) (see Figure 15.6). On the long bones, defects were more prevalent on (1) the posterior surface of the bones, (2) the left side, and (3) the lower limbs. Both hands and the talus of the right foot also had minor defects. The defects on the extremities mirrored the patterns on the thorax and calotte.

The combination of blunt force and sharp force defects, coupled with the consistent measurements and patterning of the shape of the defects seemingly would make the identification of tool class easier, however, this case proved challenging due to the unconventional murder weapon. The dearth of defects, variety of expression, and uniqueness of the morphology of the defects presented specific challenges in identifying specific tool classes. I turned to Petraco's (2010) *Color Atlas of Forensic Toolmark Identification* and began flipping through the chapter on "Common Hand Tools Seen in Casework."

The blade impression of a standard slotted screwdriver matched the length and width dimensions of the rectangular defects on the scapula, innominates, and sacrum. However, a screwdriver lacks a beveled edge and could not have created the parallel incised defects seen elsewhere on the remains or explain the angled edge on the scapula (see Figure 15.4, most lateral defect). Furthermore, could the flattened, wedge-shaped tip of the screwdriver explain the square defects and the parallel scoring defects throughout the remains? To test this, I experimentally inflicted defects on clay, soft wood, and chalk to see if the parallel striations and squared defects could be replicated – to no avail. At this point, I considered the possibility of perhaps two different weapons, one that created the blunt force defects, like a slotted screwdriver, and perhaps a second that created the sharp force defects, but what sharp force implement would consistently create parallel scoring and incised bone?

Thoroughly perplexed by this case and getting increasing pressure from the investigators to identify a potential weapon class, I wandered down to a colleague's office to discuss the case. We went through each defect, bone by bone. I explained some of the literature review I had conducted and the various experimental defects created. In a moment of ingenuity, my colleague looked around her office, her wheels spinning, and settled on a pair of scissors on her desk. She opened the scissors and dragged the blade and honed edges along the wood, chalk, and clay. In each instance, the blade produced a series of approximately 3mm parallel lines. Next, we stabbed the clay with the scissors open and closed, thereby replicating both the squared 3mm defects and the rectangular defects with the angled edge (11mm by 3mm). A quick follow-up literature search on screwdriver and scissor trauma mainly produced forensic pathology and soft tissue case studies; however, the dimensions described in the scissor reports were consistent with those in this particular case. Could we say for certain that the defects were created by a pair of scissors and not a similarly shaped implement? No. But we could deduce that the defects could have been created by one implement rather than two. The summary of the trauma section of the final case report is presented below to demonstrate how the tool class description was worded and presented to investigators.

> Perimortem trauma was identified in multiple locations on the skeleton. The bones exhibit predominantly blunt force trauma with puncturing, penetration, scoring, and fracturing of the cortical bone, yet there is also evidence of sharp force trauma (incised defects) on several skeletal elements. The parallel linear defects found throughout the body are consistent in dimensions and measure approximately 3mm apart. Given the associated nature of the blunt and sharp force trauma, a single blunt force tool with a sharp edge cannot be ruled out. The rectangular and linear defects are consistent with a tool that exhibits a thin, rectangular end or tip. For example, the defects are consistent with a pair of scissors; however, it is not possible to exclude other similarly shaped tools. Nearly all the defects are going in the posterior-to-anterior direction (i.e., from back to front) and more defects are present on the left side than on the right side of the skeleton.

CASE RESOLUTION

It was later learned that the perpetrator had been instructed to vacate his unit so local housing authority staff could fumigate his housing unit in response to complaints of odor and increased insect activity from neighbors. Fearing detection, the perpetrator removed the remains from the kitchen pantry, where they had been decomposing for a number of weeks, to the trashcans outside his unit. At the time of the forensic archaeological recovery, the investigators had the perpetrator in custody and had reasonable cause for a presumptive identification of the remains discovered outside the perpetrator's rental unit. The perpetrator later plead guilty to third-degree murder charges after admitting to stabbing the victim multiple times with a pair of scissors and was sentenced to 30 years in state prison. In many cases, the murder weapon may not be found, or the perpetrator does not admit to the commission of the crime. In this unique case, we were able to confirm our hypothesis that the weapon that caused the soft tissue and skeletal defects discovered throughout the remains was, in fact, a pair of scissors.

LESSONS LEARNED

- In cases of unique or unusual trauma, turning to literature and case reports outside of forensic anthropology can be helpful.
- The analyst must keep in mind that a single tool may create a variety of shaped defects, and care must be taken to not get too focused on the first potential tool that may match some of the characteristics. It is important to think like a defense attorney and be your own worst critic. Ask yourself, "is there any other weapon, object, or implement that could create the same defect?" Often the answer will be yes, so while you may have a good idea about the potential weapon or you've been provided information by the investigators, it's important not to overstate your conclusions while still trying to relay helpful information to investigators.
- It is also important to think outside of the box. Perkins (2003) study of violent crimes in the US indicated that about 6% are commissioned with "sharp objects," of which in 15% of these cases (88,000 per year) the weapon consisted of a sharp object other than a knife, including scissors and ice picks.
- Lastly, it never hurts to reach out to colleagues when you are struggling with a particular case or even a research project. Two heads are always better than one and much of what we learn about trauma is based on exposure to casework, so consulting your peers with different experiences and backgrounds from your own is often beneficial. Some may not have the luxury of working down the hall from a fellow forensic anthropologist, and in those cases, it may be beneficial to have a colleague perform an external review of the case materials and conclusions or at the least an external review of the case report, as once the report is submitted, the anthropologist must be prepared to defend its contents in court (a time at which even small typos can become embarrassing).

DISCUSSION QUESTIONS

15.1 When you encounter a case of unique trauma, what steps can you take to potentially identify a tool class? How should you word a forensic case report when you are unable to narrow the case down to a specific weapon, tool, or tool class?

15.2 Why is it important to note in the report the direction of the impact (in this case, that most were from back to front)?

15.3 If you were called to court for this case and the defense argued that all of the sharp force trauma defects were made by you when you were processing the remains with a scalpel, how would you counter this argument?

REFERENCES

Perkins, C. (2003). Weapon use and violent crime. *Bureau of Justice Statistics Special Report, 20.* September 1948.

Petraco, N. (2010). *Color atlas of forensic toolmark identification.* Boca Raton, FL: CRC Press.

Symes, S. A., L'Abbe, E., Chapman, E. N., Wolff, I., & Dirkmaat, D. C. (2012). Interpreting traumatic injury to bone in medicolegal investigations. In D. C. Dirkmaat (Ed.), *A companion to forensic anthropology.* New York: Wiley.

CHAPTER 16

Sharp Force Trauma with Subsequent Fire Alteration: A Complicated Case Study

Erin N. Chapman

CONTENTS

BACKGROUND

The local police and fire department were called to the scene of a house fire. After suppressing the fire, a deceased individual assumed to be the homeowner was found lying supine in the remnants of a chair at the bottom of the stairs (see Figure 16.1A). A window above the decedent was open at the time of discovery. The remains were charred with significant damage to the anterior portions of the body, including exposure of abdominal organs. After the body was removed from the scene, firefighters examined and removed the debris that was surrounding the body in an effort to locate any missing elements or physical evidence. The debris was removed with shovels and tossed out of the open window. Because the homeowner, well known to the authorities, had a history of heavy smoking and drinking, the police and fire department assumed the fire and subsequent death was accidental.

FIGURE 16.1   Photographs of the decedent at the scene and during autopsy. (A) Photograph of the decedent at the scene lying supine on the remnants of a chair at the base of a set of stairs. (B) Overall photographs of the decedent at autopsy showing his left and right sides as well as his back. (C) Photographs of the decedent's face before debris was removed (top) and after debris was removed (bottom).

AUTOPSY AND INITIAL EXAMINATION

The morning following the recovery, an autopsy was performed by a local medical examiner. Initial external examination of the body revealed injuries to the head and face indicative of a suspicious death rather than an accident (see Figure 16.1B and C). The medical examiner also noted the smell of kerosene on the body and clothing of the decedent. The autopsy confirmed substantial sharp and blunt force injury to the head and face. Thermal damage in the form of extensive charring and variable soft tissue loss was noted, primarily on the anterior aspect of the decedent's body. Examination of the pharynx revealed no soot in the trachea, and a rapid test to determine carbon monoxide saturation in the blood came back negative. These results verified the individual was not breathing when the fire occurred. The medical examiner informed local law enforcement that the death appeared to be a homicide necessitating further investigation. The office's forensic odontologist later confirmed the decedent's identity as the middle-aged male who owned the home.

POLICE AND FIRE DEPARTMENT INVESTIGATIONS

With the knowledge that the death was ruled a homicide, the police began an investigation. From interviews and surveillance video from neighboring homes, the police were able to reconstruct the events that led to the death of the homeowner and the fire. Around 9:00 pm, the homeowner was seen getting out of a taxi, staggering into his house. At approximately 10:00 pm, a neighbor called 911 to report a house fire. Police were able to obtain video footage of the fire starting that captured an individual running out of the front door with his clothing on fire and entering a neighboring home. After identifying the fleeing individual, a search warrant was obtained to search his home. In their search, police found a roofing hammer with traces of what appeared to be blood in a dresser drawer. The tool was sent to the police crime laboratory for DNA analysis to confirm and identify the blood residue. The DNA analysis confirmed the presence of the decedent's blood as well as fingerprints from the suspect. Police interviews with the suspect yielded a confession soon after the inflicting tool was located. The assailant admitted that he was drinking with the victim in his home. At some point, the victim began to make offensive anti-Semitic remarks to the suspect. The suspect admitted to picking up the roofing hammer and hitting the victim in the head. According to the suspect's story, the victim began making gurgling noises and he continued to hit the victim until he stopped making noise.

In addition to the police investigation, the fire department's investigation into the cause of the fire revealed that it was started near the decedent and evidence of accelerants were found. The fire, originally assumed to be accidental, was now ruled an act of arson. Based on the results of the autopsy, and the police and fire investigations, it became clear that the fire was started in an attempt to conceal a homicide.

ANTHROPOLOGICAL ANALYSIS

Condition and Inventory of Remains

The forensic anthropologist consulted with the medical examiner regarding the retention of remains for trauma analysis. In the autopsy room, the hands and forearms were examined for defensive wounds. No injuries were noted in the gross examination, although extensive thermal damage was present in the area. The anthropologist decided to remove the head for tool mark examination. In addition, due to the heat alteration of the hands, it was deemed necessary to remove and process them to identify any potential defense wounds not identified in the soft tissues.

The head (cranium and mandible), three cervical vertebrae (C1, C2, and partial C3), hyoid, and both hands (including carpal bones) were retained for anthropological examination. After the remains were processed free of soft tissue, it became apparent that the majority of the mid-face, especially on the left side, was highly fragmented and would require extensive reconstruction. Figure 16.2 depicts the condition of the skull after soft tissue removal. The anterior view of the skull in Figure 16.2B (in blue) highlights the reconstructed portions of the facial region. Numerous small fragments from this region were recovered but due to their size, were not able to be reconstructed. After reconstructing the skull, the anthropologist noted a large area of the forehead that was not recovered from the scene.

Taphonomic Effects of Fire

The implications of heat modification to human remains and forensic analysis are numerous and varied (Stewart, 1979; Symes, Rainwater, Chapman, Gipson, & Piper, 2015). One of the major goals in this case was separating fire alteration (taphonomy) from perimortem injuries. In cases where fire is used to cover up a homicide, anthropologists must answer questions in court as to how damage created by a fire is distinguished from trauma signatures related to the death event. The ability to distinguish between antemortem,

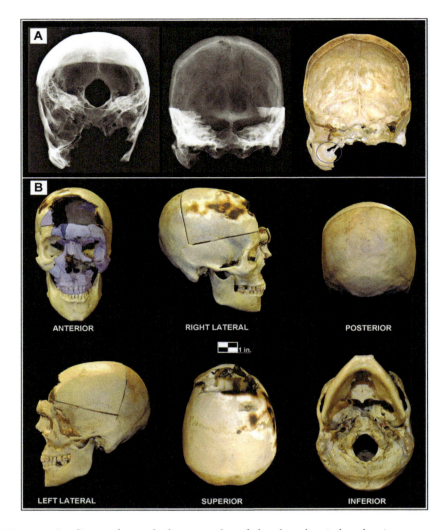

FIGURE 16.2   Radiographs and photographs of the decedent's head prior to processing and after reconstruction. (A) Superior and AP radiographs and anterior photograph of the cranium after processing (without the skullcap). The blue circle highlights the location of the right eye orbit. (B) Six views of the skull after reconstruction was completed. The blue (anterior view) highlights the portions of the face that were reconstructed.

perimortem, and postmortem damage based on pattern recognition is imperative to the analysis of remains subjected to heat alteration (de Gruchy & Rogers, 2002; Gonçalves, Thompson, & Cunha, 2011; Herrmann & Bennett, 1999). The destructive forces of fire on a body and physical evidence create a recognizable pattern that assists in the interpretation of the scene and the body itself (Symes et al., 2015). Three process signatures should be evaluated in these cases (Symes et al., 2015, p. 34): (1) body position and tissue shielding in bone; (2) color change in thermally altered bone; (3) burned bone fracture biomechanics.

In this case the areas of the body with the most significant heat alteration were consistent with the most common areas to burn first (forehead, dorsum of hands, knees, shins, etc.) (Mayne Correia & Beattie, 2002; Symes, Dirkmaat, Ousley, Chapman, & Cabo, 2011). Charring was most pronounced on the mid-face, portions of the scalp, right upper extremity, left distal forearm and hand, right side of the torso, and several patchy areas on the lower extremities. Several areas of the body were relatively spared and showed evidence of soot deposition. Pugilistic posture was noted with flexion at the elbows, wrists, fingers, hips, knees, and ankles (see Figure 16.1). Heat-related fractures were noted in the wrists and fingers. The exception to the normal burn pattern in this case was the region of the mid-face. Areas covered with thin layers of tissue (i.e., little to no underlying muscle) such as the nose, forehead, and chin typically exhibit the most extensive burning or heat alteration (Symes et al., 2011; Symes et al., 2015). However, the inconsistencies in the burn pattern of the mid-face in this case are due to the perimortem trauma.

Skeletal Trauma Analysis

The retained skeletal remains were examined both grossly and microscopically for the presence of trauma. All damage to the hands was consistent with postmortem heat modifications. Numerous sharp and blunt force injuries were present in the soft and osseous tissues of the head (see Figures 16.3 and 16.4). The majority of the injuries noted are on the anterior portion of the skull. Additionally, sharp force injuries were noted on the endocranial surface of the left side of the calvaria. All perimortem injuries were consistent with chopping-type wounds, classified as sharp-blunt or "cortocontundente" (Fundacion de Antropologia Forense Guatemala, 2009; Pinheiro, L'Abbé, Symes, Chapman, & Stull, 2014), where injuries were created by a tool with a sharp edge with associated blunt impact injuries due to the weight and force transferred from the tool to the victim.

Figure 16.3A highlights the sharp force injuries (blue) with associated blunt force injuries (red). Numbers designated in Figure 16.3B are used to denote the injuries referenced herein. These numbers do not correspond to the sequence in which the blows or injuries occurred.

There were numerous superficial sharp force impact wounds on the mid-face and mandible measuring between 1/10″ to 7/8″ in length and approximately 1/8″ wide (see Figure 16.2– 16.4). The incised bone and narrow kerf of the superficial impact wounds suggests a sharp implement with edge bevel. In addition, a minimum of seven deep wounds was observed in the skull (see Figure 16.3). Two of these blows (impacts #16 & 17) penetrated deep into the mid-face and struck bone in the region superior to the spheno-occipital synchondrosis at the base of the cranium. The depth of these wounds was estimated to be four to five inches. The remaining five deep impacts were located on the endocranial surface of the right side of the frontal and parietal bones. The five impacts range from approximately two and a half inches to four inches deep and range in length from one-third of an inch to half an inch. Figure 16.4 highlights soft tissue (16.4A) and osseous evidence of sharp force trauma (16.4B) with associated blunt fracturing (16.4C).

Figure 16.3B illustrates the minimum number of impacts observed in the skull. Twenty-six probable impacts or blows were noted from the tool mark evidence on the skull. Caution should be exercised with this estimate because it is difficult to speculate what may or may not have occurred after the blows to the mid-face compromised the bony structures. Additional blows may have created additional tool marks on the bones of the face, which were not in proper position after they were crushed inward. A conservative estimate of a minimum of 20 blows were inflicted to the head of the victim.

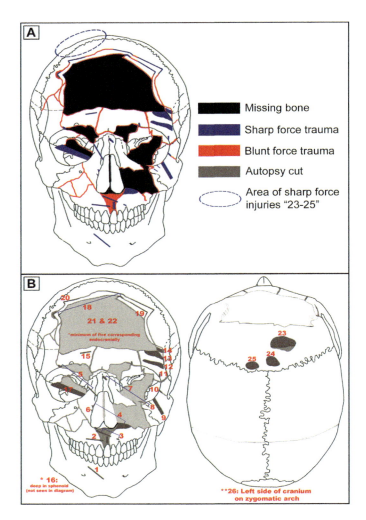

FIGURE 16.3 Diagrams of the sharp and blunt force injuries observed in the skull. (A) Diagram of the anterior skull highlighting sharp force (blue) injuries, associated blunt force (red) injuries, missing portions of the skull (black) and postmortem autopsy modifications (gray). (B) Anterior and superior hand-drawn diagrams of the location and minimum number of impacts observed (numbers do not indicate a sequential order). The blue brackets represent one impact.

Thermal Alteration versus Perimortem Trauma

During pretrial preparation, prosecutors asked for an explanation of how anthropologists differentiate skeletal trauma from damage created by fire. This is an important point as defense attorneys often argue that all damage to a deceased individual could have been caused by thermal alteration. In addition, it is important that anthropologists and medical examiners be prepared to show evidence and explain their conclusions on whether or not a person died because of a fire (determined by a medical examiner) or whether or not perimortem trauma occurred prior to the fire (determined by an anthropologist or

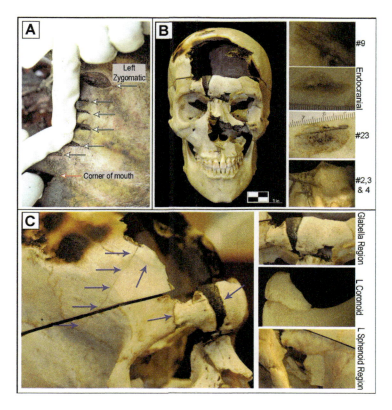

FIGURE 16.4 Illustrates sharp and blunt force injuries in soft and osseous tissues. (A) Six incised soft tissue injuries (white arrows) noted during autopsy on the left side of the mid-face. These soft tissue injuries are associated with injuries to the underlying bone. (B) An overall photograph of the anterior skull highlighting sharp and blunt force injuries. The close-up images provide examples of incised bone associated with a tool with edge bevel. (C) Examples of several blunt force injuries observed. The blue arrows indicated perimortem blunt force fractures observed on the right lateral and anterior skull. Close-up photographs of other areas of blunt force trauma.

medical examiner). Defects have to be excluded as taphonomic in nature before they can be considered to have occurred perimortem (Symes et al., 2015).

Figure 16.5 highlights several areas in the skull that indicate that perimortem trauma was inflicted prior to the fire. The first process signature (body position and tissue shielding) can be applied to the sparing of the soft tissues of the back. The tissue on the decedent's back has been shielded from the heat and is consistent with his found position at the scene. However, an abnormal pattern appeared in the forehead area of glabella. In Figure 16.5A, a fracture through the region of glabella was observed (small blue arrows in inset). The right side of the fracture exhibited fire alteration while the left side appeared to be unaltered. The pattern observed in the fragments highlighted that a fracture (inset – blue arrows) occurred prior to the heat alteration. The fracture borders lined up perfectly but only the right side was heat altered. A closer look at Figure 16.1C revealed the fragment of the forehead sticking out away from the soft tissues and exposed to the heat.

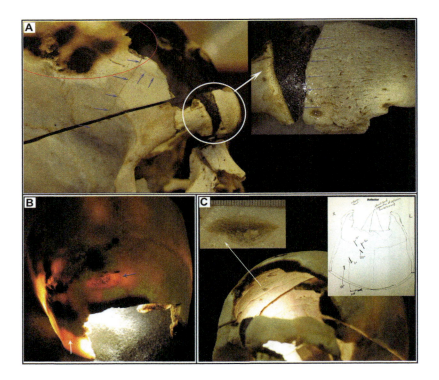

FIGURE 16.5   Illustrates the presence of perimortem trauma prior to heat alteration. (A) Right side of forehead region showing perimortem fractures (blue arrows) outside of the area of heat alteration (red oval). Perimortem fracture of the glabella region (white circle and inset) showing area of altered and unaltered borders (small blue arrows). (B) Backlit skullcap highlighting areas of altered (opaque) and unaltered (light shining through) bone. Note, the sharp force injury (impact #23 – blue arrow) as well as a perimortem linear fracture (white arrow) in areas not altered by fire. (C) View from the anterior left side of the skull illustrating the sharp force impact sites on the endocranial surface of the skull. Insets show unaltered sharp force impact site and a diagram of the five impact sites observed.

The second process signature (color change in thermally altered bone) was observed in the sharp force injury on the right side of the frontal (see Figure 16.5B [blue arrow]) (normal); in the heat line along the right side of the frontal region (see Figure 16.5A [area of red oval]) (normal); and the heat line observed along the right side of the glabella fragment (abnormal) at the junction of burn and unburned bone (see Figure 16.5A). Figure 16.5B depicts the skullcap with oblique light shining behind it. The light was used to reveal the areas of heat modification. The areas in which light shined through indicate unaltered bone, including the sharp force injury indicated by the blue arrow (impact #23). Conversely, the opaque areas were areas of heat-altered bone in which dehydration had occurred.

The third process signature (burned bone fracture biomechanics) was evaluated by the observation of the hands and fractures of the skull. The fractures observed on the dorsal metacarpals were consistent with canoeing in which the cortical bone is compromised. These fractures were associated with fire modification. Conversely, the linear fractures observed on the right frontal (see Figure 16.5A [blue arrows] and 16.5B [white arrow]), fracture of the left coronoid process of the mandible (see Figure 16.4C), and fracture of the left sphenoid region (see Figure 16.4C) are consistent with wet bone perimortem fractures and are located in areas outside of the heat alteration. In addition, sharp force impacts were observed on the unaltered, endocranial surface of the right front and parietal bones (see Figure 16.5C).

Suspect Tool Examination

As mentioned previously, a search of the suspect's home yielded a roofer's hammer with a wooden handle and brown leather straps in a dresser drawer. After anthropological analysis and documentation of the skeletal remains was completed, the anthropologist and the medical examiner inspected the suspect tool at the police crime laboratory. Photographs, measurements, and silicone molds were made of the tool (see Figure 16.6). The cutting/sharp edge of the tool appeared well used with several defects. The tool edge

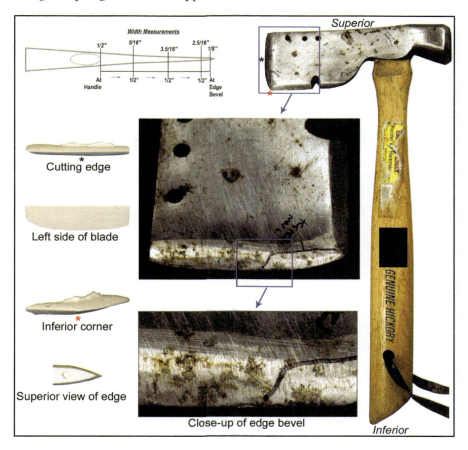

FIGURE 16.6 Illustrates some of the information recorded from the examination of the suspect tool. Measurements of the width of the blade were taken along the superior edge of the tool. The superior and inferior corners (red star) of the cutting edge differ in shape. The inferior edge of the hatchet portion of the tool is beveled in a similar manner as the lateral edge (black star). Note the striations visible in the close-up images of the edge bevel. In addition, changes in striation direction suggest the tool was likely re-sharpened at some point.

also exhibited signs of possible re-sharpening of the edge bevel, which slightly changed the striation pattern and angle ("Close-up of edge bevel"). On the opposite side of the hatchet is a hammer. Although an exact weight was not obtained, the tool itself is classified as a medium weight hand tool. This is a common feature of tools that the operator swings because the weight of the tool assists in driving a nail (for instance) into another material.

A comparison between the observed bony injuries and the dimensions of the tool was completed (see Figure 16.7). In general, when doing a tool mark examination in bone, the analyst looks for class characteristics of the tool based on evidence in the bone. These class characteristics consist of general information regarding size of the tool, shape of the tool, shape of the cutting edge, and other characteristics of dimensionality and condition.

When comparing the dimensions of the tool and the tool marks in the bone the anthropologist noted several areas of consistency in the height of the blade, the width of the blade at edge bevel with measurements taken from the injuries in bone (see Figure 16.7; Table 16.1).

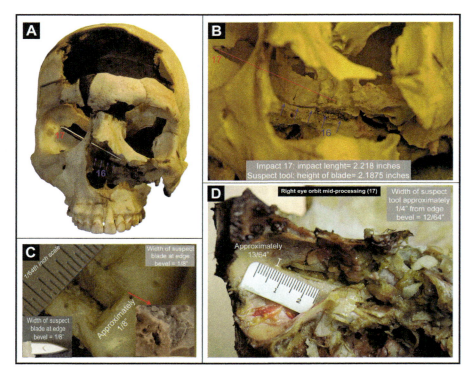

FIGURE 16.7 Comparison of injuries in the skull with dimensions of the suspect tool. (A) Overall anterior photograph of the skull with deep impacts #16 (blue) and #17 (red) highlighted. The position and orientation of the injuries can be visualized. (B) Close-up photograph of impacts #16 (blue) and #17 (red). Note the impression of the tool left in the inferior end of impact #17. (C) Microscopic image of the inferior edge of the tool mark created by impact #17. Note the squared shape. Insets of the superior edge of the suspect tool (bottom left) and a negative mold of the impact site (bottom right). (D) Photograph of the superior end of impact #17 in mid-processing. Photographs were completed mid-processing due to the fragile nature of the bone fragments in this area.

TABLE 16.1   Congruencies between Suspect Tool Measurements and Bone Injury Measurements

Tool Area	Tool Measurement	Impact #	Bone Measurement
Blade Height	2 3.5/16″	17	2 7/32″
Blade Height	2 3.5/16″	18	2 ¼″
Width of Blade 2″ from handle (Thickness)	3.5/16″	17 (superior)	13/64″
Width at Edge Bevel (Thickness)	1/8″	5	1/8″
Width at Edge Bevel (Thickness)	1/8″	17 (inferior)	1/8″

Several impacts exhibited striations in the bone. There were some irregularities or lack of consistency and a lack of sharpness in the striation pattern in the bone. This is suggestive of the inflicting tool being well used and dull. In addition, there was a change in direction of the striation pattern in impact #9 (see Figures 16.3B and 16.4B), this may be suggestive of a change in the angle of the striae on the blade. The directionality of the blows to the face of the victim was somewhat variable. For the majority of injuries the tool marks were consistent with the blade traveling in a downward arc, impacting the left side of the head and traveling upward and to the right, going through the frontal bone, incising the brain and hitting the endocranial surface of the right frontal and parietal bones. The inferior edge of impact #17 confirms that the handle was positioned on the right side of the victim's head at impact (see Figure 16.7B and C). In general, there are a number of changes in direction. Due to the massive amounts of facial trauma, both blunt and sharp, it is difficult to reconstruct the position of the handle for some of the injuries. Once the bones of the mid-face were fractured, accurate sequencing and directionality become more difficult to assess.

CONCLUSIONS AND SUMMARY

The injuries present on the skull exhibit characteristics of sharp-blunt force or "corto-contundente" trauma with subsequent heat alteration. Significant fragmentation, with evidence of incised bone, is suggestive of a medium weight hatchet-type tool with a sharpened blade. Nearly all of the injuries exhibit characteristics of chopping wounds (essentially perpendicular to the bone) except a few in which the tool grazes the surface of the bone at an acute angle. The injuries are concentrated on the mid-face and left side of the face. The majority of the tool marks suggest the handle was on the right side of the victim's head, with the blade traveling from inferior to superior, and anterior to posterior.

Several impacts to the skull suggest different directionality. However, based on the osseous material available, there is no evidence that any other part of the tool was used to inflict the injuries. Conservatively speaking, a minimum of 20 or more (26 probable) impacts to the head were noted (see Figure 16.3). Some injuries in the mid-face appear to be superficial with the edge of the blade penetrating the bone only slightly; other impacts penetrate into the deep portions of the skull base. The impacts to the deep portions of the base of the skull and the impacts on the endocranial surface penetrated the victim's head between two and five inches and required significant force. Measurements and observations of the tool examined at the police crime laboratory and the tool marks in the bone suggest similar class characteristics of size, shape, and wear.

The suspect initially pled "not guilty" by reason of "extreme emotional distress." However, the day before trial began the defendant changed his plea to "guilty" when faced with the overwhelming evidence against him (DNA evidence, anthropological evidence of excessive trauma given the number of blows to the victim, and autopsy evidence showing the victim was deceased before the fire). He was sentenced to 20 years to life for second-degree murder and 5–15 years for third-degree arson.

LESSONS LEARNED

- Shoveling and racking through debris often increases the risk of displacing, concealing, or destroying fragmented remains and other evidence. In some cases, these methods may also inflict further modifications and damage to the bone (Dirkmaat, Cabo, Ousley, & Symes, 2008). As anthropologists, we must continue to push for fatal fire victim recovery to be executed by an anthropologist using forensic archaeological techniques. A large portion of the frontal was not recovered from this scene. These bone fragments likely were not consumed by the fire but rather were missed during the recovery effort. Any remains or contextual evidence not documented or recovered is lost.
- Anthropologists can contribute a wide range of important information to a medicolegal investigation. Detailed information regarding the type of tool inflicting injuries to bone, the mechanism of those injuries, and the minimum number of blows can assist in the investigation and in some cases can contribute to forcing confessions from perpetrators. Information regarding the context and mechanism of skeletal injuries can lead investigations and assist investigators in uncovering the truth.

DISCUSSION QUESTIONS

16.1 What additional valuable information might have been gained if a trained anthropologist recovered the remains from the house fire rather than the police and fire department?

16.2 Should anthropologists frame their analyses around the potential questions a defense attorney could ask? How do anthropologists reduce bias in their analyses, while simultaneously attempting to anticipate questions they may be asked in a court of law?

16.3 How do anthropologists address the lack of validated skeletal trauma studies when approaching cases such as this?

ACKNOWLEDGMENTS

A special thanks to Chief Medical Examiner, Dr. Tara Mahar and all of the staff at the Erie County Medical Examiner's Office. Thank you to Drs. Steven Symes (formerly of) and Dennis Dirkmaat of Mercyhurst University for the use of microscopic equipment in the documentation of this case. A sincere thank you to Drs. Kathryn Grow Allen and Alexandra Klales for proofreading and feedback on the chapter. Finally, I am grateful to the editors of this volume for including me in their project and providing me with valuable feedback.

REFERENCES

de Gruchy, S., & Rogers, T. L. (2002). Identifying chop marks on cremated bone: A preliminary study. *Journal of Forensic Sciences*, *47*(5), 933–936.

Dirkmaat, D. C., Cabo, L. L., Ousley, S. D., & Symes, S. A. (2008). New perspectives in forensic anthropology. *Yearbook of Physical Anthropology*, *51*, 33–52.

Fundacion de Antropolgia Forense Guatemala. (2009). Standard Operating Procedures. D. Cortocontundente (TCC). Fundacion de Antropologia Forense Guatemala-FAFG, Guatemala [in Spanish].

Gonçalves, D., Thompson, T. J. U., & Cunha, E. (2011). Implications of heat-induced changes in bone on the interpretation of funerary behavior and practice. *Journal of Archaeological Science*, *38*(6), 1308–1313.

Herrmann, N. P., & Bennett, J. L. (1999). The differentiation of traumatic and heat-related fractures in burned bone. *Journal of Forensic Sciences*, *44*(3), 461–469.

Mayne Correia, P. M., & Beattie, O. (2002). A critical look at methods for recovering, evaluating, and interpreting cremated human remains. In W. D. Haglund & M. H. Sorg (Eds.), *Advances in forensic taphonomy: Method, theory, and archaeological perspectives* (pp. 435–450). Boca Raton, FL: CRC Press.

Pinheiro, J. E. S., L'Abbé, E. N., Symes, S. A., Chapman, E. N., & Stull, K. E. (2014). New approach to the traditional English classification of trauma and bone implications. Presented at the 66th annual meeting of the American Academy of Forensic Sciences, Seattle, WA.

Stewart, T. D. (1979). *Essentials of forensic anthropology, especially as developed in the United States*. Springfield, IL: Charles C. Thomas Publishing.

Symes, S. A., Dirkmaat, D. C., Ousley, S. D., Chapman, E. N., & Cabo, L. L. (2011). Recovery and interpretation of burned human remains. Funded by the National Institute of Justice Award No. 2008-DN-BX-K131. Final Technical Report prepared for the National Institute of Justice. Department of Justice.

Symes, S. A., Rainwater, C. W., Chapman, E. N., Gipson, D. R., & Piper, A. L. (2015). Patterned thermal destruction of human remains. In C. W. Schmidt & S.A. Symes (Eds.), *The analysis of burned human remains* (pp. 17–59). New York: Elsevier Press.

CHAPTER 17

Forensic Anthropology's Role in Clarifying Cause of Death in the Appeal of a No Body Homicide Conviction

Ashley E. Kendell, Eric J. Bartelink, and Turhon A. Murad

CONTENTS

INTRODUCTION

In 2007, a man was convicted of first-degree murder of an elderly male from California. The suspect, a previously convicted felon, was hired by the decedent to trim trees and perform other work on his property. When the decedent went missing in 2006, the handyman became the primary suspect in the case. An investigation revealed that he had stolen the decedent's checkbook and forged his signature on several checks. This finding led law enforcement to surmise that the motive for the murder came when the decedent confronted the handyman regarding the stolen checkbook. Law enforcement believed that, fearing a return to prison, the handyman murdered the decedent, leaving behind a bloody crime scene, a spent cartridge casing, and a handprint, but no body. Without a body, the prosecution set out to convince the jury that the handyman had committed the murder, using an arterial spurt pattern as their primary evidence of homicide. This case represents only the second "no body homicide" conviction in the county since the 1980s.

CASE HISTORY

The decedent was reported missing in 2006. An uncharacteristic lack of contact with family and friends prompted the local sheriff's office to investigate further. Law enforcement began their search at the decedent's residence. An investigation of the interior of the home revealed a number of factors suggesting foul play, including evidence of arterial spurt on a bedroom wall and a spent .22 caliber shell casing found on a bedroom dresser. In addition, a handprint was located within the residence that was matched to the decedent's handyman. The handyman quickly became the focus of the investigation as the prime suspect. The suspect was later found in possession of the decedent's travel trailer and a number of personal items, including a passport, checkbook, and a box of .22 caliber ammunition. Additionally, four of the decedent's firearms were recovered from friends of the suspect (.22 caliber revolver, 9mm handgun, 357 Magnum, .22 caliber pistol), who stated that he had asked them to hide the firearms on his behalf. Finally, a beer box stained with the decedent's blood was located within the travel trailer. With a case quickly building against the suspect, law enforcement was faced with one notable missing link: the victim's body. Law enforcement conducted an extensive search of the property with the assistance of cadaver dogs but failed to turn up any human remains. The suspect was subsequently arrested and held in custody pending trial. Without a body or murder weapon, the district attorney's office began to build their case based largely on circumstantial evidence.

The suspect was faced with a jury trial for the murder in 2007. Without a body, the prosecution built their case on the victim's disappearance, in tandem with the excessive blood documented at the crime scene. The evidence of arterial spurt in a bedroom led the prosecution to speculate that the victim's throat had likely been cut. The prosecution also revealed the motive to the jurors. The suspect was found to be in possession of the victim's checkbook and was known to have forged his signature on a number of checks made out to himself. After being confronted over the stolen checks, the suspect was said to have killed the victim to avoid returning to prison. The suspect was ultimately convicted of a "no body" homicide. In addition to the homicide charges, he also faced two counts of being a felon in possession of a firearm and one count of grand theft. He was sentenced to 25 years to life.

Following sentencing, the suspect's attorneys appealed the case. While incarcerated, however, the suspect revealed the location of the victim's body to his cellmate. This information was turned over to law enforcement, who then returned to the victim's residence for an additional search of the premises. Upon locating a potential clandestine grave on the property in 2008, the sheriff's office contacted the CSU, Chico Human Identification Laboratory (CSU, Chico HIL) to assist in a recovery effort. The forensic anthropology team arrived at the residence later that day to assist with the recovery. The gravesite, located alongside the residence, was determined to contain human remains and was excavated using standard forensic archaeological methods.

The excavation revealed a shallow pit containing articulated human remains of a single individual. The shallow grave pit was oriented in an approximate north-to-south direction (head-to-toe) along the south side of the residence. The decedent's remains were primarily skeletonized, although small amounts of soft tissue adhered to the thoracic area, vertebral column, pelvic area, and upper thigh region. The skeleton lay supine and extended within the grave. The left arm was positioned over the sternum, while the right arm appears to have been flexed at the elbow so that the hand was beneath the right side of the head. A zip-up blue sweat jacket was found along the southeastern grave wall in

FIGURE 17.1   The decedent's remains in an extended and supine position within a shallow grave (note the presence of an electrical cord and blue sweat jacket).

contact with the left leg. Also present was a bed sheet with a severed electrical cord (plug end intact) wrapped around it. The "corded" bag appeared to contain other clothing items (later identified at the coroner's office). The free end of the electric cord extended along the length of the grave, with the "plug end" near the decedent's right arm. The deteriorated remnants of underwear were recovered, including an elastic waistband that was still attached to the pelvic girdle (see Figure 17.1).

ANTHROPOLOGICAL ANALYSIS

Following the recovery of the human remains from the residence, the sheriff's office requested that the remains be transported back to the CSU, Chico HIL for a forensic anthropological analysis. The anthropological analysis included the establishment of a biological profile and a trauma analysis.

RESULTS OF BIOLOGICAL PROFILE ASSESSMENT

The biological profile suggested the remains belonged to a White male, over the age of 65 years, and approximately 68.5 ± 2.7 inches tall (5 feet, 6 inches–5 feet, 11 inches). Ancestry and stature were assessed through metric analyses conducted in Fordisc 3.0 (Jantz & Ousley, 2005). The ancestry assessment was consistent with a white male with a posterior probability of 0.960 and typicality probability of 0.620, when compared to American Indian, Black, Hispanic, and White samples. Age was estimated primarily on the morphological assessment of the pubic symphysis (Brooks & Suchey 1990; Phase VI with 95% CI (confidence interval) of 34–86 years and Mean of 61.2 years). The pubic symphyses, combined with other aging indicators including the right 4th sternal rib morphology (Işcan et al., 1984), auricular surface morphology (Lovejoy et al., 1985), cranial suture closure (Meindl & Lovejoy, 1985), significant degenerative changes observed on the thoracic and lumbar spine, and advanced dental attrition led the third author to determine the decedent's age to be greater than 65 years. All components of the biological profile aligned with the demographic information obtained for the missing decedent, who was reported to be a 72-year-old White male, 5 feet, 8 inches in height. The remains were later positively identified using antemortem dental records.

RESULTS OF TRAUMA ANALYSIS

Antemortem Trauma

The trauma analysis revealed a number of antemortem conditions, including evidence of surgical intervention on the proximal and distal ends of the right tibia, and the distal end of the right femur (see Figure 17.2). The decedent had undergone a total knee replacement in life, demonstrated by the presence of a right knee prosthesis. A surgical screw was visible in the medial malleolus of the right tibia (see Figure 17.2C). The left ankle also showed evidence of a healed fracture of the distal tibia, the talus, and calcaneus (see Figure 17.2A, B). Of particular note is the presence of a complete fracture through the talus, which failed to heal back together, resulting in two separately healed segments.

In addition to the high level of consistency between the biological profile and the victim's demographic information, the antemortem conditions recorded also corresponded to injuries documented throughout the victim's lifetime (a mountain-climbing accident in his twenties), thereby strengthening the presumptive identification of the victim.

Perimortem Trauma

Multiple perimortem defects were observed on the skull, all consistent with small-caliber gunshot trauma (Berryman & Haun 1996; Berryman & Symes 1998; Berryman et al., 2013). Gunshot wounds were noted on the left lateral and posterior aspects of the cranium, as well as the anterior aspect of the decedent's mandible. Four entrance wounds

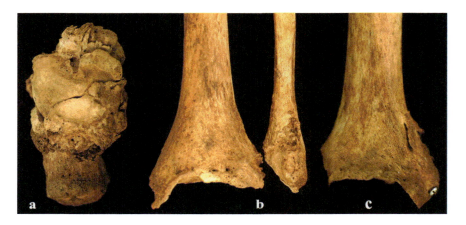

FIGURE 17.2   Bilateral healed fractures to both ankles (a); left talus articulated with calcaneus; (b) left tibia and fibula; (c) right tibia with surgical screw in the medial malleolus).

and two exit wounds were recorded. One projectile entered the superior aspect of the left mandibular ramus and exited the cranium through the right anterior maxilla, just inferior to the right eye orbit (see Figure 17.3). A second projectile entered the left lateral aspect of the skull at the base of the temporal arch, near the supramastoid crest and posterior to the external auditory meatus (see Figure 17.4). This projectile did not exit and was recovered from within the cranium. A third projectile entered the cranium posterior to the foramen magnum and exited through the right eye orbit (see Figure 17.5). Finally, a fourth entrance defect was recorded to the right of the midsagittal plane on the mandible, where it remained embedded in the bone (see Figure 17.6). The presence of four gunshot

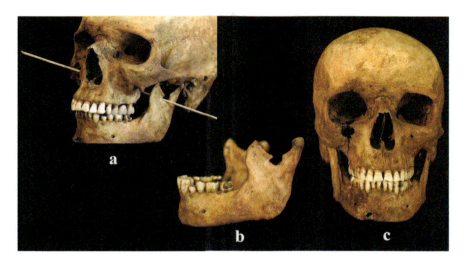

FIGURE 17.3   Trajectory of the first projectile wound (a) with bullet entrance on the left mandibular ramus (b) and corresponding exit wound on the right maxillae inferior to the right eye orbit (c).

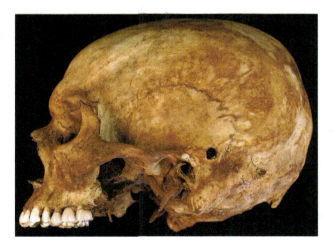

FIGURE 17.4   A second entrance wound in the left temporal bone, near the supramastoid crest (projectile recovered from within the cranium).

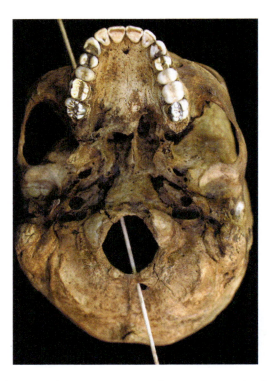

FIGURE 17.5   A third projectile entrance wound posterior to the foramen magnum that exited through the right eye orbit.

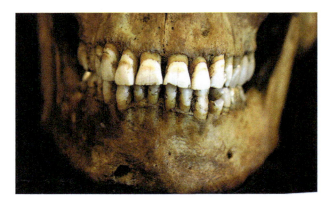

FIGURE 17.6   A fourth projectile entrance just right of the sagittal plane through the mandible (projectile embedded within the bone at the time of analysis).

wounds on the decedent's skull and lack of evidence of sharp force trauma to the neck area, clarified interpretations of the cause of death. The bullet trajectories suggest that the decedent was shot once from anterior to posterior in the lower jaw, twice at a lateral angle from posterior to anterior on the right side of the cranium, and once from posterior to anterior near the base of the skull.

Perimortem Trauma Revisited: The Actual Cause of Death

The most interesting component of the anthropological analysis, and the most integral component for the prosecution, was the perimortem trauma recorded on the remains, consisting of at least four gunshot wounds on the decedent's skull, three of which were potentially fatal. The .22 caliber antique pistol found to be in the possession of one of the handyman's friends was later determined to be the weapon used in the homicide. Also worthy of note, the decedent's prescription eyeglasses were found in the house and located near the arterial spurt. The right eye glass was shattered, likely the result of the bullet that passed through the right eye orbit as it exited the skull, an unexplainable finding until the cause of death was amended by the new trauma evidence.

While the prosecution was able to convict the suspect based on the presence of arterial spurt and significant blood loss, and circumstantial and physical evidence at the crime scene and in the victim's travel trailer, interpretation of the cause of death was largely presumptive and speculated to be from sharp force trauma to the neck. Following the recovery and anthropological analysis, the cause of death was clarified. These findings ultimately led the defense to drop the appeal, and the suspect's conviction was upheld.

LESSONS LEARNED

- Forensic anthropologists are trained to recognize potential clandestine burials based on disturbances in soil and differences in vegetation and likely could have helped law enforcement discover the remains earlier if called into the investigation in advance.
- In any anthropological analysis, it is important to avoid confirmation bias. Although it had previously been suspected that the decedent may have died as a result of sharp force trauma, due to the presence of arterial spurt in the decedent's bedroom, the forensic anthropologists still completed a comprehensive osteological analysis. This analysis revealed no evidence of sharp force trauma, but instead four gunshot wounds to the head.
- Forensic anthropological analyses can provide integral information to assist the medical examiner in their interpretation of cause and manner of death.

DISCUSSION QUESTIONS

17.1 Do you think the presence of the arterial spurt misguided the investigation in this case? If the arterial spurt blood was not present, would more emphasis have been placed on the guns the suspect had asked friends to hide? How can investigators and forensic anthropologists avoid confirmation bias?

17.2 The suspect was initially convicted under what appears to be incomplete information about the circumstances of death and without a body. Do you feel such a conviction is fair? Why or why not?

17.3 Why is it important to document antemortem trauma even if the decedent is identified through other means (e.g., dental comparisons)?

REFERENCES

Berryman, H. E., & Haun, S. J. (1996). Applying forensic techniques to interpret cranial fracture patterns in an archaeological specimen. *International Journal of Osteoarchaeology*, 6, 2–9.

Berryman, H. E., Shirley, N. R., & Lanfear, A. K. (2013). Basic gunshot trauma interpretation in forensic anthropology. In *Forensic anthropology: An introduction* (pp. 291–306).

Berryman, H. E., & Symes, S. A. (1998). Recognizing gunshot and blunt cranial trauma through fracture interpretation. In K. J. Reichs (Ed.), *Forensic osteology: Advances in the identification of human remains* (2nd ed.). (pp. 333–352). Springfield, IL: Charles C. Thomas.

Brooks, S., & Suchey, J. M. (1990). Skeletal age determination based on the os pubis: a comparison of the Acsádi-Nemeskéri and Suchey-Brooks methods. *Human Evolution*, 5(3), 227–38.

Işcan, M. Y., Loth, S. R., & Wright, R. K. (1984). Metamorphosis at the sternal rib end: A new method to estimate age at death in white males. *American Journal of Physical Anthropology*, 65, 147–56.

Jantz, R. L., & Ousley, S. D. (2005). *FORDISC 3.0: Personal computer forensic discriminant functions*. Knoxville, TN: University of Tennessee.

Lovejoy, C., Meindl, R., Pryzbeck, T.,& Mensforth, R. (1985). Chronological metamorphosis of the auricular surface of the ilium: a new method for the determination of adult skeletal age at death. *American Journal of Physical Anthropology*, 68(1), 15–28.

Meindl, R. S., & Lovejoy, C. O. (1985). Ectocranial suture closure: A revised method for the determination of skeletal age at death based on the lateral-anterior sutures. *American Journal of Physical Anthropology*, 68(1), 57–66.

Human Rights and Mass Disasters

A Multiyear Forensic Archaeological Recovery of Human Remains from a C-124 Aircraft Crash on Colony Glacier, Alaska, USA

Gregory E. Berg, Owen L. O'Leary, and Kelley S. Esh

CONTENTS

The views expressed here are those of the authors and do not reflect an official opinion or position of the Department of Defense or United States Government.

INTRODUCTION

A routine transportation flight turned into a disaster on the morning of November 22, 1952. A C-124 personnel transport aircraft left McCord Air Force Base in Washington bound for Elmendorf Air Force Base, Alaska; the last contact with the aircraft was approximately 20 minutes before its scheduled arrival. Due to poor weather and possible navigational errors, the aircraft crashed into the side of Mount Gannett, resulting in the loss of 52 Americans. Search and rescue procedures were delayed for several days. On November 28, 1952, a two-person team in a small Cessna aircraft found the crash site on the western side of Mount Gannett. In their opinion, the crash occurred higher on the mountain, triggered an avalanche, and this resulted in a debris field covering two acres of snow pack on "Surprise Glacier." The debris field was on a 40° slope, and snow depths were estimated at 18 to several hundred feet thick. Bad weather forced a larger recovery team to take a roundabout route to the crash location, arriving 11 days later. Due to another eight feet of new snow, no human remains and little debris were found. Because of the severe risks associated with a more detailed winter recovery, the site was abandoned for the year. Later the following spring and summer, recovery attempts met similar results – the crash site was essentially gone.

Sixty years later, on June 9, 2012, an Alaska National Guard helicopter crew spotted wreckage, including a large aircraft wheel, on Colony Glacier (note: Surprise glacier was renamed to Colony Glacier) approximately 70 km east of Anchorage, Alaska. During the following days, an on-ice survey located aircraft wreckage, personal identification media, and human remains. The Alaska Command contacted the Joint POW/MIA Accounting Command (now the Defense POW/MIA Accounting Agency or DPAA) for its aircraft recovery expertise. The DPAA fielded an archaeological recovery team to Colony Glacier in June of 2012 for a site assessment and a human remains recovery of the missing personnel.

Every year since 2012, a summer field season has been conducted on Colony Glacier (DPAA recoveries 2012–2015, Air Force Mortuary Affairs Office [AFMAO] recoveries 2016–2018) to recover personal effects and human remains. Over that time, the ever-changing surface of the glacier has produced multiple recovery loci, located from the toe of the glacier to the first major icefall (see Figure 18.1). This chapter discusses the first five years of archaeological recovery, detailing the complex challenges that were overcome. Since very few large-scale archaeological recovery efforts on glaciers or ice fields are detailed in the literature, we describe our preparations for working in this

FIGURE 18.1   An aerial view of Colony Glacier; the red circle indicates general recovery locations. Inner Lake George is in the foreground, and the firn line is up-glacier to the left. Above and left of the circle is the icefall (rugged area of ice that resembles ice-mountains).

environment, present field methods and improvements employed each year, synthesize an overall approach for future glacier work, and detail the lessons learned from our experiences. All data compiled in this chapter are taken from internally generated yearly field reports by the authors, as well as Dr. Derek Congram (for 2012).

PREPARATION

Understanding the environment is critical to planning and undertaking an archaeological recovery on a glacier, in part to comprehend the risks involved and to properly prepare a team for the challenges ahead. There are two main glacial surface environments and they are demarked by the firn line; the lower region is solid ice and the upper region is ice permanently covered by snow. Both zones have unique challenges. For example, the snow-covered crevasses above the firn line present deceptively safe passages where climbers may fall and become trapped, and the wide-open network of crevasses below the firn line may necessitate hundreds of meters of travel to avoid them, tiring the climber.

Glaciers produce dynamic, localized weather patterns that necessitate safety plans for ground and air (helicopter) operations. Glacial weather effects range from dropping cloud ceilings with increased precipitation to massive and/or erratic wind shifts in direction and intensity. Glacial wind shifts are fairly unpredictable, as they can change on an hourly basis and produce sustained winds of 25 knots or higher. In addition, the surface environment is a constantly changing, moving, slick, and watery landscape. Crevasses can open and grow quickly; one day a crevasse may be easy to step over, the next it may require a ten-minute traverse around. Water streams can grow from a trickle to a gushing, roaring river in hours, carving out new paths across the glacier's surface. Moulins can change from small, half-meter in diameter holes to massive ten-meter in diameter shafts with no bottom. Finally, the glacier's toe is typically floating, which means it can calve unexpectedly, removing large portions of seemingly "safe" ice. All of these conditions need consideration when planning for helicopter landing zones, evacuation routes, equipment storage locations, work priorities, and team safety. As summer exacerbates the rapid changes, June was deemed the most acceptable timeframe for on-ice operations.

The DPAA recovery team undertook a weeklong training course about glacial traversing, ice climbing, and mountaineering to mitigate some of the working hazards. The training was designed to equip personnel with the necessary skills, equipment, and procedures in case of an emergency, but not to make them become "experts." Each year, prior to fielding the team, another round of training was offered to train new team members and provide a refresher for returnees. The AFMAO recovery seasons included a more stringent four-week military mountaineering school.

Planners from DPAA, Alaska Command, and Alaska Air National Guard outlined the glacier recovery in detail, from creating emergency extraction procedures to implementing a local weather team encamped near the glacier with an on-site weather station. Communications included satellite, radio, and cellular devices. All safety procedures were outlined (e.g., emergency call = immediate activation of a stand-by Air Force Pararescue evacuation helicopter; glacier surface winds above 30 knots sustained = pull the team off the glacier; cloud ceilings below 500 feet = no flight operations; early morning rain and low ceilings = no work that day). These robust preparations and rules produced a work environment in which redundant safety procedures were ready for execution, maintained team confidence and peace of mind in the dangerous environment, and ensured the day-to-day archaeological recovery was not hampered.

A forensic anthropologist and forensic archaeologist jointly led the on-ice recovery effort. Additional team members were mountaineering qualified military personnel (six to eight individuals), a medic (Special Forces), a pathfinder (military individual specializing in helicopter operations), a forensic photographer, and frequently, a Pararescue jumper. Each morning, a detailed flight plan, a weather briefing, and a safety briefing were conducted by the helicopter squadron commanders. The forensic anthropologist and archaeologist listed the day's goals and work assignments. This discussion of the daily work plan reinforced safety, potential weather impacts, and work priorities.

2012 FIELD SEASON

To start, several helicopter surveys of the entire glacier identified three discrete aircraft wreckage concentrations associated with the glacier's toe (loci A, B, and C in Figure 18.2). South and up-glacier of the main concentration was a large icefall (a substantial drop in elevation creating the look of a frozen waterfall). The firn line was approximately one to two km further south from the icefall, up glacier, and it obscured any visual identifications of plane debris. The overflights estimated the site size to be ~ 600–900 meters in length, over multiple ice formations. Two helicopter landing zones were identified sufficiently away from the site so the helicopter rotor wash would not impact site integrity. An emergency shelter (a 3×3×3 m shipping conex) also was airlifted and placed near the site.

The first portion of the on-ice recovery strategy was a pedestrian survey of each locus using parallel transects across the glacier surface to establish boundaries and locate areas for equipment storage and staging. For loci B and C, these transects were easily accomplished, as the ice was not heavily crevassed and a moraine (small- to medium-sized rocks and boulders) was present on the ice surface. Loci B and C were relatively sparsely populated with aircraft wreckage and human remains; therefore the recovery strategy was simply GPS point position recordings.

The majority of the crash debris and human remains was found at locus A, which was also the most glacially complex area of the site (see Figure 18.3). Locus A was bisected in regular intervals with east-west running crevasses (perpendicular to glacier flow) that were anywhere from small cracks to 2.5 meter-wide chasms, 10–30 meters deep. The crevasses created natural "units" of ice shelving between them. Prior to recovery, several experimental site datum emplacements were tried – rebar into the ice surface, deep ice screws, and wooden poles. All melted out of the ice in a matter of hours due to sun-generated heat. The need to devise a technique that would allow for comparable measurements (e.g., provenience and context) to be captured in a moving environment was paramount. We decided to use large flat rocks as a site datum, supplemented with nine subdatums (numbered secondary datums with known GPS coordinates); the additional subdatums increased the chance of finding at least two known points in following year(s). All were painted with consecutive numbers and were placed in approximate 20-meter intervals; grid north was 50° east of magnetic north (mN).

Each ice platform between two crevasses was treated as a collection unit, searched on hands and knees in a line-abreast fashion, and mapped (see Figure 18.3B). All incident related wreckage was removed from each unit and transported to collection points away from the main site area to be removed at the recovery season end. When possible, wreckage and remains were "excavated" out of crevasses by removing all of the moraine materials from the base of the crevasses (see Figure 18.3C). Due to the fast melt of the ice

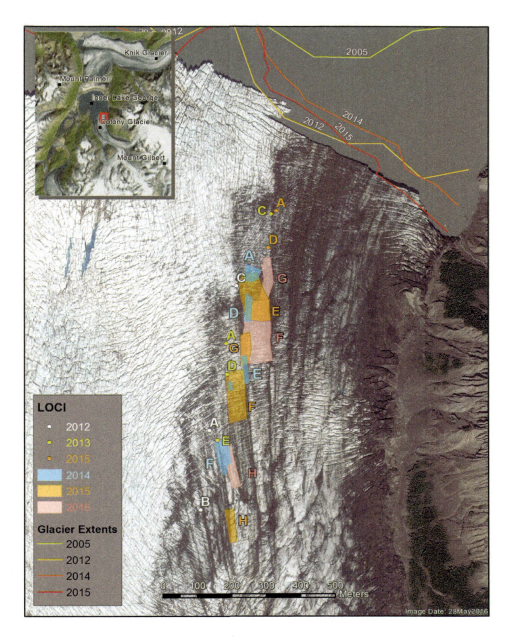

FIGURE 18.2 Aerial photograph of Colony Glacier showing the changing loci locations of the loci (A–H) over the five yearly recovery efforts as well as the glacier toe over a 10-year span.

(in some cases, about 10–20 cm of ice per day), each unit was searched three times over the two-week recovery period.

Site mapping was accomplished using measuring tapes laid parallel to and along the edge of a crevasse for each unit and used as a baseline. For the first unit, the location of probative evidence was recorded in two dimensions: along the baseline (i.e., the distance

FIGURE 18.3   Site conditions. (A) Example of the typical environment with remains and wreckage melted out together on the edge of a deep crevasse. (B) Excavators' "hands and knees" approach to the intensive collection of the glacier surface looking for human remains and material evidence. Yellow tape is a mapping sub-baseline, and the painted rocks are subdatum 26. (C) Example of crevasse excavation. The team member is secured to the glacier via an ice screw for safety. Note: yellow material in the crevasse base is a life-preserver.

from the subdatum) and the perpendicular distance from the baseline. Although this produced very precise spatial locations, the abundance of remains, the fast-changing climatic conditions, and the difficult nature of the recovery area warranted streamlining the recovery process of subsequent units. Additional units were collected in one-meter blocks along the baseline (i.e., all remains between 11m and 12m were placed in the same bag). This method increased collection speed, while sufficiently maintaining possible spatial associations.

Each day, all collected evidence was placed in resealable plastic bags, labeled with provenience information, and placed in plastic buckets packed with ice for preservation. Items were transported back to the Joint Base Elmendorf – Richardson hospital morgue, where the evidence was stored in a secure refrigerator maintained near freezing for preservation. For the flight from Alaska to Hawaii, the evidence bags were wrapped in newspaper, placed in coolers with dry ice, sealed with duct and evidence tape, and placed in a locked transfer case. At the conclusion of the journey, all remains were still cold with no deleterious effects.

In total, during the 2012 field season, approximately 5,250m^2 of the glacier surface was surveyed and approximately 1,400m^2 was intensively collected. The recovered
human remains were often small, fragmented pieces, and only small amounts of soft
tissue were present. Frayed, splintered long bone ends, and cracked, split ends of ribs
and thin bones were identified as a hallmark of glacial taphonomy (Pilloud et al., 2016).
The tight spatial collection allowed for precise associations of remains to each other
(Congram and Berg, 2013, Pilloud et al., 2016), and the first field season resulted in the
identification of 17 individuals.

2013 FIELD SEASON

In March 2013, another fly-over showed new wreckage and remains on the glacier. The
2013 archaeological recovery followed the same format as the 2012 effort in terms of
preparation and site set-up. By late June 2013, the archaeological recovery team was
recovering four distinct loci on the glacier. The site datum and several subdatums from
locus A were relocated; the loci were designated A, C, D, and E (loci A and C were the
same as the previous year, and locus E likely correlates to locus B from the previous
year). A new locus was found between locus A and E, labeled locus D (see Figure 18.2).
The official site datum traveled 270 meters downslope and dropped some 25 meters in
elevation.

Since the use of painted subdatums was successful in relocating site loci despite the
drastic glacier movements, the same method was used. For each locus, additional subdatums were established, typically spaced approximately 20 meters apart, and placed on a
site map with GPS coordinates. A N-S baseline was then established for each locus, from
the two furthermost subdatums along a 330° azimuth (mN). This established the first
permanent baseline for the entire site and was used consistently between the loci. Units
were also redefined to represent the area between subdatums, rather than the area on an
ice shelf between crevasses.

For each collection site, 100m tapes were used to create a secondary baseline, orthogonal from the site baseline. Items were then piece plotted on the overall site map utilizing
both the secondary baseline and a line orthogonal to it, per locus, and this proved to be
a much more effective mapping strategy over 2012. The collection strategy remained
the same. Any evidentiary items recovered outside of these mapping areas were recorded
via GPS coordinates. All aircraft wreckage was removed from the glacier except for several oxygen tanks and landing tires, as these created additional reference points easily
observed during aerial reconnaissance.

In locus A, multiple linear arrangements of rocks, aircraft wreckage, and human
remains were found between crevasses (see Figure 18.4). These alignments could represent the collection of materials trapped into crevasses from the year before. It was
hypothesized that the freeze/thaw yearly opening and closing action of the crevasses
"pushed" these materials to the surface again, much like a child forces ice up from the
bottom of an ice-pop. Alternatively, these accumulations could represent the base of a
crevasse after the glacier surface has melted down to the crevasse base. In either case,
it was hypothesized that the greatest concentrations of human remains and wreckage would be found in conjunction with these linear rock arrangements for the lower
glacier area.

Collection activities covered an area greater than 9,100m^2 (six times more than in
2012). Additional pedestrian surveys were conducted to the base of the icefall, yielding

FIGURE 18.4   Aerial photograph of a linear wreckage arrangement surrounded by rocks
and sediment (red circle) at locus A, prior to 2013 recovery. The tire at left (yellow arrow)
was at the bottom of a crevasse at the end of the 2012 recovery. The same uplifted crevasse also revealed a high concentration of possible human remains.

no incident materials. A larger volume of human remains was recovered from this year,
including larger limb and torso portions. Additionally, soft tissue was more often present in association with the bony elements. Based on the increase of both human remains
and personal effects, it was believed that a more substantial piece of the crash site had
been recovered. Given the appearance of locus D in between previously observed loci, we
suspected the site could continue to yield remains and material evidence as the ice moved
closer to the terminal end of the glacier. Furthermore, the appearance of locus D indicated that incident related material was distributed vertically through the glacier. Four
individuals were identified from this fieldwork.

2014 FIELD SEASON

In early March 2014, aerial recognizance indicated wreckage deposits on the glacier's
surface; training and planning commenced. Helicopter imagery and visual observations
showed four independent loci on the ice based on the subdatums and a new locus, up-
glacier, near the icefall. On-ice archaeological recovery started on June 6, 2014. Locus C
was no longer present and was assumed to have calved into Lake George. Locus A was
positioned where Locus C was the previous year, and Locus D and E were approximately
200 meters down-glacier of their previously recorded locations (see Figure 18.2). A new
locus, F, was located up slope of locus E. Both Locus D and E expanded in size and
material density from the previous year. The site datum in locus A was located ~200m
downslope from the previous year.

The recovery strategy remained the same, but several changes to site mapping occurred. First, rather than referring to the blocks between subdatums as "units," this designation was dropped, and every substantial item of evidentiary value was piece plotted using the baseline and orthogonal method described above (2013). Second, subdatums were established every 20m in the new loci along a baseline. Third, the southernmost subdatum per locus was given the arbitrary origin reference point of N100/E100, and all point plots were then calculated from it (e.g., a humerus found in the middle of locus D has the point plot of locus D, N145/E121, surface). Fourth, highly aggregated areas of evidence were called "concentrations" and human remains were bulk gathered and given a center point coordinate of the concentration. Independent loci maps annotated all relevant provenience data and site boundaries (see Figure 18.5).

The intensive site collection spanned a total of 9,472m² of surface area. Within loci A and D, the subdatums were in the same relative position to one another, indicating minimal rearrangement of the loci as they advanced down slope. A large portion of locus E was further to the east and outside of the area defined as Locus E in 2013, indicating a new section of this locus melted out in the intervening year. Substantial evidentiary materials and human remains were recovered from *all* loci, even those that had been intensively collected for the two years prior (even larger portions were recovered this year). This pattern suggested an extensive vertical distribution of material throughout the ice. Finally, the highest concentrations of human remains were associated with linear rock formations representing the glacial till accumulations from the closed or melted out crevasses, confirming the prior year's hypothesis.

2015 FIELD SEASON

Substantial changes to the site were evident from the air in March 2015, and extensive new deposits of material were present on the glacier's surface. The DPAA team collected nearly 23,000m² during the three-week field season. Two new loci, G and H, were identified (see Figure 18.2). Locus G fell between the existing loci E and F. Locus H was against the foot of the icefall. The recovery techniques remained unchanged. The site datum in locus A was once again located, providing four years of comparative positional data. The glacier moved downslope approximately 200–300m per year with a 50m total change in elevation.

Surprisingly, locus A was still safe to approach near the glacier toe, and a final collection effort produced a small amount of human remains, suggesting that deposits were either fully exposed by the time a locus reached the glacier toe, or materials embedded further in the ice would not be recovered prior to calving into Lake George. Given a 50m elevation drop between the northern and southern aspects of the site, we believe the former is true.

The successive location of the 2012–2014 subdatums was critical in reestablishing the glacier grid system and retaining provenience between field seasons. Multiple subdatums established in locus F during 2014 were not located in 2015; subdatums not found in 2014 from earlier years were found again in 2015 (Locus A, D, and E). As the glacier moved downslope, the subdatum rocks likely fell into open crevasses, but were then pushed back or the crevasses melted enough to reexpose them. A dramatic example of the opening and closing of the crevasses as the glacier travels downslope is demonstrated in the comparison of Locus F between 2014 and 2015 (see Figure 18.6). Thus, the utility of establishing many subdatums was proven to be effective, and it was the most reliable way to relocate the various loci from year to year.

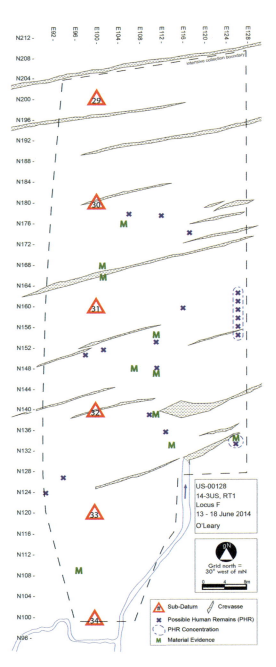

FIGURE 18.5   Example of a site map, detailing locus F, from 2014.

The unexpected appearance of Locus G between Locus E and F in 2015 indicated it was situated deeper in the glacier ice than either Locus E or F. This suggests site formation and modification processes created a "layered" or "pocketed" site. A dramatic example of the site layering was found in an outcrop of ice that had incident materials on its surface, a "sterile" four-meter zone of no materials, then more incident materials

FIGURE 18.6   Locus F, final photograph from 2014 (top), view to the west, and 2015 (bottom), view to the east. Comparison between the two years shows the dramatic changes in the ice surface as the loci move downslope. The upper zone had few crevasses, while the middle zone was heavily crevassed, suggesting topographic changes in the substrate beneath it.

below that. The other new locus, H, appeared farther south and closer to the icefall than the previous new "Locus F" in 2014. Its appearance so far upslope might have been a direct consequence of an atypically hot summer the year before, enabling more glacier melt.

Eight distinct concentrations of remains were found in 2015, three in Locus F, four in Locus G, and one in Locus H. These areas were distinguished based upon the high density of the remains, to include a nearly complete individual, and many personal effects (e.g. wallets, clothes, rings, shoes). Altogether, 2015 represented a substantial increase in incident materials and remains over the previous three years.

2016 FIELD SEASON

Due to changing jurisdictional authority, the 2016 field season was led by the AFMAO instead of DPAA. All prior to fieldwork preparations were undertaken, though the team was created and staffed by the AFMAO. Only one forensic anthropologist was present at the site in a solely advisory capacity. It is unknown if provenience information was retained from the 2016–2018 field seasons.

Unlike previous years, 2016 saw a general stagnation in the overall growth of the site, but the amount of recovered remains increased. Several more near complete individuals and large portions of individuals were recovered, and six remains concentrations (four in Locus F and two in Locus H) were processed. Locus A was either too close to the glacier toe for safe operations or had calved into the lake. Locus D and E moved the usual approximate 200m downslope. Locus G and F essentially grew together (no intervening space clearly demarking an end to either one). Locus H slightly expanded and slid downslope approximately 150m. No additional loci up-glacier of Locus H were found. All archaeological techniques were applied and the site mapped per the 2014/2015 conventions. Approximately 23,275m² were intensively collected during that field season, remaining fairly consistent with the 2015 field season.

DISCUSSION

As of this writing, 39 of the 52 individuals have been identified from the incident. The glacier has revealed the crash site slowly, with loci of concentrated incident material exposed over the course of multiple field seasons (Table 18.1 summaries site size/collections per year). We believe the initial discovery year (2012) represented about two to three years of site exposure prior to being "found." The initial appearance was not the total horizontal and vertical extent of the site; those limits were finally realized in 2015/16. The 2015/16 field seasons might represent the apex of the site size and extent, but we are uncertain without additional data.

The site underwent three main transitions in appearance below the icefall. The first zone was the relatively flat surface north of the icefall with some crevassing and a few streams. Evidentiary materials were relatively sparse and relatively "light" in weight. The glacier next dropped steeply in response to a change in underlying bedrock, producing an area of increased crevasse growth and no surface water (see Figure 18.6). Light to moderate amounts of materials were recovered in this zone. The underlying bedrock then leveled, and the ice transitioned to a flatter surface with few crevasses, linear rock alignments, and surface water streams and pools. The vast majority of evidentiary materials were recovered in this final zone.

The horizontal distribution of artifacts and remains remained consistent within the glacier – no area moved faster or in a significantly different direction than the areas surrounding it. Elevation differences between loci were likely due to the underlying bedrock formation and not differential melt rates between areas on the glacier. Melt rates in the summer were, however, quite incredible. During the average two to three week field season, the glacier lost approximately two vertical meters of surface ice while the teams were on site.

The data presented here suggest that the crash debris was buried in the ice in layers and discrete pockets, and not in a single, thin, linear mass all at the same level. This is evidenced by significant distances between loci and elevational differences throughout the site. We hypothesize that the original crash dynamics, the ensuing avalanche, movement of these materials onto and across the original glacier surface, the nature of the wreckage

TABLE 18.1   Intensive Surface Collection Coverage for Each Locus per Year. Total Field Days Is in Parentheses After Each Year. Empty Cells Had No Collection for the Year

Locus	2012 (6)[1]	2013 (16)[1]	2014 (18)[1]	2015 (22)[1]	2016 (22)[1]
A	1,361m²	1,841m²	2,368m²		
C		2,230m²			
D		1,378m²	1,824m²		
E		3,683m²	1,936m²	8,800m²	
F			3,344m²	8,900m²	14,720m²
G				2,400m²	5,000m²
H				2,500m²	3,552m²
Total	1,361m²	9,132m²	9,472m²	22,600m²	23,272m²

itself (i.e., heavy vs. light material), and glacier actions/movements created these pockets and layers. Some heavier materials, such as the engines, have yet to be found and may be stuck in the original bergschrund or may have penetrated deeply into the original glacier and will not be found prior to calving into Lake George. Our observations reveal complex relationships that we could not have begun to understand without a thorough, yearly archaeological observation and mapping program of Colony Glacier.

LESSONS LEARNED

- Preparation is key. The project's success lies greatly with the due diligence completed before setting foot in the field – especially given the limited field season available. Having the right skill sets, establishing safety rules, practicing evacuation drills, and avoiding unnecessary risks enabled successful yearly recoveries without any catastrophic injuries.
- Because of the continuously evolving nature of this archaeological site, our methods and techniques also had to evolve. Glaciers are moving conveyor belts of ice and stone, and each season yielded additional and larger recovery areas. Datum points fixed in space were untenable. Allowing subdatums to travel with and through the glacier while staying in the same relative fixed location *to the site itself* provided an innovative solution to maintaining provenience and tracking changes in the glacier movement.
- At the end of each field season, we were constantly asked, is that all? Are you done yet? This simple yet complex question could really only be understood after the fourth year of fieldwork. At that point, we were confident that new loci would appear at the base of the icefall, and then take a minimum of four years to travel the 1.3 km to the glacier's toe, all the while dropping in elevation and ice thickness, enabling the full vertical scope of a locus to manifest. If no new loci appeared at the base of the icefall in a given year, it could be that the project would only have four more years. This prediction was only possible after completely understanding the travel, direction, elevation changes, and behavior of the evidence in the ice due to the archaeological techniques we employed at this site.
- The difficult nature of this recovery operation resulted in significant media attention. Despite media restrictions and team efforts to keep the site location and activities confidential, issues arose during fieldwork, to include an individual who kayaked across the lake and walked across the ice "just to see what we were doing." It was not uncommon to see aerial "low-level" fly-bys of the site as well, and it was reported that a helicopter company landed paid tourists on the site after the team had left for the field season. Some crash-related materials also showed up on auction sites for sale. All efforts should be made to prevent such activities.

DISCUSSION QUESTIONS

18.1 What other types of potential recovery sites present dynamic environments? What aspects of the archaeological methods utilized in this glacier recovery could be applied in these other environments?

18.2 How and why would the recovery outcome have differed if the effort had been limited to one or two seasons versus five seasons? At present, 13 individuals remain unidentified from the crash. Is there a point that the risks outweigh the benefits?

18.3 Lessons were learned and excavation methods evolved over the five years of recovery. If given a "do-over," how would you approach the preparation, mapping, and collection of evidence from the start of the first season?

ACKNOWLEDGMENTS

While the authors would like to personally acknowledge a host of individuals across multiple organizations that made this recovery possible, this section would be nearly as long as this book chapter. Therefore, in brief, we would like to thank Alaska Command, Alaska Air National Guard, Alaska Air National Guard 212th Rescue Squadron, Elmendorf Airforce Base Hospital Morgue, the Armed Forces DNA Identification Laboratory, the Armed Forces Medical Examiner System, Pacific Command, and the Defense POW/MIA Accounting Agency. Without the support of these agencies, commands, and organizations, the remains of these US Service personnel would not have been recovered.

REFERENCES

Congram, D., & Berg, G. E., (2013). *Archaeological recovery at an aircraft crash site on a glacier [podium presentation]*. Society for American Archaeology. Honolulu, Hawaii.

Pilloud, M. A., Megyesi, M. S., Truffer, M., & Congram, D. (2016). The taphonomy of human remains in a glacial environment. *Forensic Science International*, 261, 161.e1–161.e8.

CHAPTER 19

Quadrilateral Defects in the Tuskulenai and Leon Trotsky Cases: Skeletal Trauma Associated with Soviet Violence in Two Different Contexts

Cate E. Bird and Rimantas Jankauskas

CONTENTS

INTRODUCTION

As one of the principal organizers of the Bolshevik revolution, Leon Trotsky represented a celebrity personality of the global Communist movement. His assassination by a suspected Soviet agent in 1940 in Mexico City was not entirely unexpected provided his sour relations with Joseph Stalin and the multiple failed assassination attempts he survived. However, his death by means of an ice axe was surprising given its general lack of a utilitarian function in a subtropical region. His mode of death may be construed as anomalous, save for similarities to another case of violence inflicted by Soviet agents in Lithuania following the Second World War.

The study of human skeletal biology, specifically of perimortem trauma on osseous remains, permits researchers to understand the experiences of victims and reflect on the perpetrators of violence. This includes state principals who order violence from above, as well as agents or workers who inflict violence locally on behalf of the state. This study employs a forensic anthropological approach to examine skeletal trauma in the Trotsky case in Mexico and prisoners executed and interred at the Tuskulenai Estate (aka Tuskulenai case) from 1944 to 1947. In particular, it evaluates characteristics of quadrilateral wounds (n = 24) on the crania of 21 executed prisoners from two burial pits in the Tuskulenai case and compares them to skeletal trauma observed in the Trotsky case. The similarity of quadrilateral defects observed in these two vastly different contexts suggests they were caused by a similar class of object. Given that the Tuskulenai executions and Trotsky assassination were performed by Soviet agents, we discuss reasons why this object was selected for violence.

THE TUSKULENAI CASE

Situated on the Baltic Sea coast, the Republic of Lithuania represents one of the geographical, cultural, and political intersections between Western Europe and Russia. As a result of its unique geo-political location, it has served as a battleground for various European groups since the tenth century. Lithuania was formally recognized as an independent state following the First World War (1918), but its sovereignty lasted less than 20 years. The Non-Aggression Treaty (aka Molotov–Ribbentrop Pact) between the Union of Soviet Socialist Republics (USSR or Soviet Union) and the National Socialist German Workers' (aka Nazi) Party in 1939 partitioned Eastern Europe. By June 1940, all three Baltic States were annexed by the USSR and assimilated as Soviet states, including the formation of the Lithuanian Soviet Socialist Republic (LSSR) (Damusis, 1998). The first Soviet annexation of Lithuania lasted until June 1941, when Nazi Germany pushed the Soviets east and occupied Lithuanian territory until July 1944. This period in Lithuanian history had devastating results for the local populations, especially the Jewish communities. However, by the middle of 1944, the Soviets once again annexed Lithuania and governed until 1990.

During the period after the Second World War, Soviet authorities increased mass arrests and deportations in Lithuania (Applebaum, 2003). Ethnic minorities, particularly those located in borderlands (e.g., Poles, Germans, Lithuanians, Latvians) were targeted for repression in greater numbers due to their perceived alliances with foreign enemies and their strong sense of nationalism (Gregory, 2009). However, subjugation during this second occupation in Lithuania was met with strong opposition, and a long partisan war ensued from 1944 to 1953 (Kiaupa, 2005). Although partisans waged anti-Soviet campaigns from the forests of Lithuania, Soviet authorities quelled explicit opposition by the end of the first decade of their second occupation. Both partisans and noncombatants (civilians) were targeted regarding counter-revolutionary activities (Pocius, 2006). Between 1944 and 1949, approximately 350,000 Lithuanians were deported (Lane, 2001).

Prisoners were transported to the People's Commissariat for State Security (NKGB)-Ministry of State Security (MGB) basement prison in Vilnius and either executed or deported to Gulag prisons throughout the Soviet Union (Rudienė & Juozevičiūtė, 2006). According to historical records, approximately 767 people were sentenced to death between 28 September 1944 and 16 April 1947, generally under Article 58 of the Russian SFSR Penal Code (counter-revolutionary activity) (Vaitiekus, 2011). Executions during this period were performed in the NKGB-MGB basement execution chamber, and based on eyewitness accounts, the protocol for managing condemned prisoners appears fairly standardized. Condemned prisoners were transported from prison cells to a room adjacent to the execution chamber. Here, they confirmed their biographical information to security agents, and their death warrants were signed by NKGB-MGB Commanders. Then they were led into the execution chamber where the door would close. According to

Vaitiekus (2011), prisoners were often distracted or hit on the head with a heavy object to lose orientation, before being shot in the head.

After executions, the bodies of prisoners were transported on trucks to the Tuskulenai Estate and were interred in clandestine graves. Secrecy was paramount during the trials, executions, and interments of prisoners, and these activities were concealed until the end of the Soviet occupation in Lithuania. However, with the discovery of state security documents in 1994, the Lithuanian President established a working group of archaeologists, anthropologists, and forensic experts to investigate these activities (Jankauskas, 2009). Due to the immense secrecy under Soviet administration, the national revival in Lithuania during the 1990s focused on the establishment of truth and historical justice concerning the occupation period (Vaitiekus, 2011). This included the search, recovery, and identification of prisoner remains at the Tuskulenai Estate.

Jankauskas et al. (2005) provide the most comprehensive record of archaeological and skeletal data regarding the Tuskulenai case. Archaeological excavations undertaken at the Tuskulenai Estate in 1994, 1995, and 2003 revealed a series of 45 pits containing a total of 724 individuals. Remains were discovered in two general locations: inside a former garage and north of the garage. Jankauskas et al. (2005) note the presence of 720 males and 4 females, aged 19 to 66 years at death. Researchers argue that groups of prisoners were likely executed during one night and subsequently buried in one pit together (Jankauskas et al., 2005). Thus, each pit represents one discrete night of executions. By examining the skeletal, archaeological, and historical data in concert, Jankauskas and colleagues have chronologically sequenced at least 25 pits based on specific dates of execution and burial. Anthropologists continue the process of identification of executed prisoners, identifying 55 executed individuals to date and returning 7 of these individuals to surviving family members or the Catholic church for reburial. The skeletal remains now reside in an accessible columbarium at the Tuskulenai Memorial Complex in Vilnius.

Jankauskas et al. (2005) report that 97% of individuals in the Tuskulenai case exhibit trauma inflicted around the time of death (i.e., perimortem). The majority of perimortem wounds represent gunshot trauma, with upward of 95% of individuals demonstrating single and multiple gunshot wounds of the head (see Figure 19.1). However, the remains of prisoners also exhibit other types of perimortem trauma, including sharp and blunt force. One unique pattern of perimortem trauma observed in the remains of Tuskulenai prisoners are quadrilateral defects (see Figures 19.2 and 19.3). These square or rectangular-shaped wounds have four straight sides and four corners. In the Tuskulenai case, quadrilateral defects have been documented in the remains of over 100 prisoners (Jankauskas et al., 2005).

In this study, quadrilateral defects were examined in the skeletal remains of prisoners from two burial pits in the Tuskulenai case located within the former garage: pits 23 and 26. Quadrilateral defects in crania were analyzed using a number of variables, including size, shape, and the presence of secondary fractures and edge damage. Each quadrilateral defect was digitized and superimposed on a cranial homunculus to illustrate the approximate placement, size, and shape of all defect entry wounds for each pit (see Figure 19.3).

Burial pit 23 contained 11 individuals who were executed on August 26, 1946. Analyses of these individuals demonstrated ten males and one female who were young or middle-aged adults. All individuals exhibited major trauma on the cranium, specifically gunshot entry wounds predominantly located on posterior crania. One individual exhibited blunt force trauma on the face. Nine individuals (82%) from pit 23 exhibited a total of 13 quadrilateral defects.

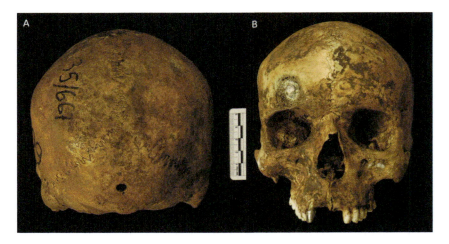

FIGURE 19.1 Example of gunshot trauma in the Tuskulenai case. (A) Posterior view illustrating a gunshot entrance wound to the occipital bone. (B) Anterior view of same cranium illustrating a gunshot exit wound with a bullet lodged in the right frontal bone.

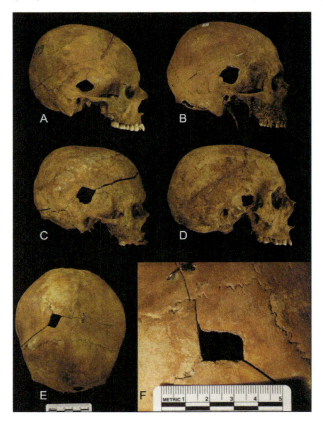

FIGURE 19.2 (A–D) Examples of quadrilateral wounds in right temporal region of four different crania. (E) Quadrilateral wound to the right parietal of a cranium, with a close-up illustrating linear edges with delamination and radiating fractures in (F).

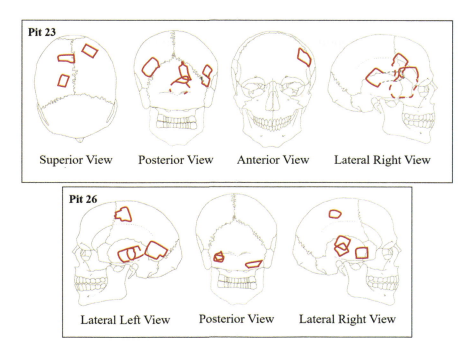

FIGURE 19.3   Composite images of quadrilateral defects from crania in pits 23 and 26.

Burial pit 26 contained 13 individuals executed on November 18, 1946 (only 10 individuals were available for analysis at the time of study). Examination of remains demonstrated the presence of ten males; most individuals were young or middle-aged adults. Only eight individuals demonstrated major trauma, while trauma was absent in two individuals. Only one individual in pit 26 exhibited a gunshot wound (of the left side of the head), and one individual exhibited blunt force trauma on the face. Seven individuals (70%) from pit 26 had a total of 11 quadrilateral entry defects.

A total of 15 individuals from these two burial pits exhibited 24 quadrilateral defects. Composite images of quadrilateral defects in pit 23 demonstrated that defects cluster on the superior, posterior, and right sides of crania. Only one quadrilateral defect entrance wound was observed on the anterior aspect, while no entries were observed on the left side or inferior of any crania in pit 23. In pit 26, quadrilateral defect entries were observed on the right side, left side, or posterior of crania, while no defects were observed on the superior, inferior, or anterior aspect of any crania.

The maximum dimensions of all quadrilateral sides were between ten and 30 millimeters long. Square defects were the most commonly observed shape (46%), followed by irregular shapes (29%), rectangular shapes (21%), and incomplete (4%) (see Figure 19.4). Differences in defect shape, particularly irregular and rectangular shapes, likely represent varying angles or depths of penetration of the forceful object (although the use of a second tool cannot be ruled out). Additionally, one case of an incomplete quadrilateral defect on the occipital bone was observed in close proximity to a complete square defect to the right (see Figure 19.5). A closer view of the defect revealed that it was likely caused by a pointed object, which slid across the bone surface but did not penetrate the endocranium. Radiating fractures, as well as a wedge of bone lifted outward was present, which likely corresponds to damage associated with removal of the object. This incomplete defect was particularly helpful in elucidating the shape (pointed end) of the object.

Radiating fractures were also observed in association with quadrilateral defect entry wounds. These ranged from zero to three fractures, with two radiating fractures being the most common. Additionally, a number of quadrilateral defect entry wounds demonstrated lipped or raised edges and "wedges" of bone pulled outward between radiating fractures (see Figure 19.6). Internal and external beveling was also observed; however, internal beveling tended to be much more extensive than external beveling. This edge damage was likely associated with the object penetrating the cranium and then being extracted.

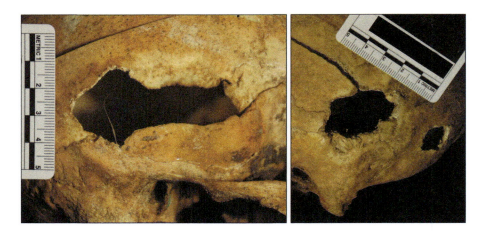

FIGURE 19.4   Example of irregular quadrilateral defects.

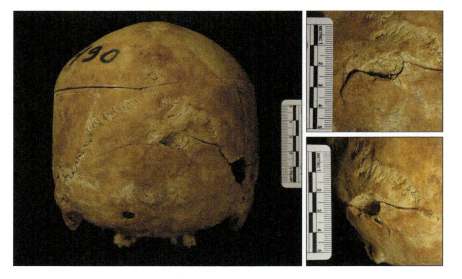

FIGURE 19.5   Incomplete quadrilateral entry wound to occipital bone. Note also a gunshot entrance wound (inferior portion of the occipital bone) and complete quadrilateral wound (right asterion region).

FIGURE 19.6   External view of quadrilateral entry wound with edge damage suggestive of extraction of the tool. The arrow highlights the externally lipped bone and the star indicates an externally displaced wedge of bone.

Quadrilateral exit wounds were less common. Two cases were observed in burial pit 23: one individual demonstrated a possible incomplete exit wound on the inferior of the cranium, just right of the foramen magnum. This exit defect was directly opposite from a quadrilateral entrance wound, which was located on the superior aspect of the cranium. The exit defect manifested as bone pushed outward (like a reverse depression fracture), but the instrument did not completely penetrate the outer table. The second individual demonstrated a square-shaped possible exit wound on the right squamosal suture, which laid directly opposite from a much larger entry wound on the left side of the cranium. Consistent with an exit wound, external beveling was noted, but no internal beveling was observed. Furthermore, one radiating fracture was noted as extending superiorly.

Based on the skeletal defect characteristics, we contend that these quadrilateral defects are likely caused by long objects with sufficient mass to penetrate bone. They are square in cross-section, taper from base to tip, have a pointed tip, and are long enough to span the width of a cranium (due to the presence of exit wounds on some crania). They likely represent a combination of sharp and blunt force trauma, similar to trauma caused by machetes or axes (Kimmerle & Baraybar, 2008). Previous researchers (Jankauskas et al., 2005) have speculated that these quadrilateral defects may have been caused by a pick, ice axe, or bayonet. While this unusually patterned defect is not common in other skeletal cases of violence, Leon Trotsky's assassination may help with the interpretation of the square or rectangular-shaped wounds in the Tuskulenai case.

LEON TROTSKY'S ASSASSINATION

Leon Trotsky represented a principal organizer of the Bolshevik revolution, an effective military leader, and one of the architects of the USSR (Thompson, 1998). During the early years of the Soviet state, Joseph Stalin and Trotsky fundamentally disagreed regarding the direction of the party's policies. As Vladimir Lenin's power and health waned during the early 1920s, Stalin's power increased. Following Lenin's death in January 1924, Stalin isolated Trotsky by removing him as the Commissar for War in 1925, from the Politburo in 1926, and expelling him from the party in 1926. By 1928, Stalin had exiled Trotsky from the Soviet Union (Thompson, 1998).

Trotsky eventually settled in Mexico in 1937. However, Stalin was notoriously suspicious of many of his rivals and allegedly ordered their assassinations from afar, including Trotsky. As early as 1931, Stalin secretly noted to other Soviet leaders that, "Trotsky, this criminal gang-boss and Menshevik charlatan, has to be bumped on the head …he has to know his place" (Volkogonov, 1996, p. 439).

Trotsky survived numerous assassination attempts throughout his life. However, on August 20, 1940, he was attacked in his home in Mexico City by Ramon Mercader, who had befriended Trotsky the previous year. As Trotsky and Mercader met in Trotsky's study, Trotsky directed his attention away from Mercader who hit Trotsky on the top of the head with an ice axe, an implement which he had brought with him (Volkogonov, 1996). Upon hearing a commotion, Trotsky's bodyguards rushed into the room and detained Mercader for questioning. While Trotsky lived more than a day and was lucid for some of that time, he eventually died of blood loss on August 21, 1940.

Following Trotsky's death, an autopsy was performed, and a cast of his skull was made which demonstrated his injury (see Figure 19.7). This cast is currently on display at the International Spy Museum in New York City. The wound is located on the right parietal bone, indicating a superior-to-inferior direction of force consistent with Mercader's testimony. The quadrilateral defect, which is approximately 10 × 23 mm in size, has four straight sides and four corners. Linear projections extend from the

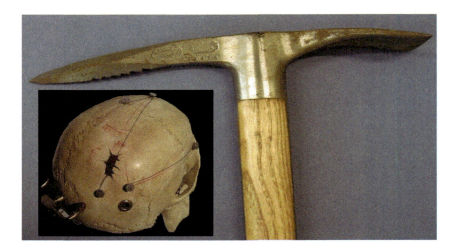

FIGURE 19.7   Cast of Trotsky's cranium illustrating a quadrilateral entry wound, on display at the International Spy Museum in New York City. Image of an ice axe similar to the one used by Roman Mercader during the assassination of Leon Trotsky. Ice axe image obtained from Auckland Museum [CC BY 4.0 (https://creativecommons.org/licenses/by/4.0)], via Wikimedia Commons.

defect, but these likely represent artifacts of the casting process. While the quadrilateral defects in the Tuskulenai case are slightly larger than the one seen in the Trotsky case, defects in both cases exhibit similar characteristics. The similar patterning of defects suggests that a comparable class of object, such as a pointed axe, may have been employed in both cases.

DISCUSSION

While skeletal trauma often falls into well-defined biomechanical typologies, differences in forces applied by specific objects can span these categories. In bone, sharp force trauma generally refers to fractures produced by the slow-loading, localized application of objects with edges, points, or beveled surfaces, which produces punctures, gouges, clefts, or incised alterations. Conversely, blunt force trauma generally refers to fractures caused by the slow-loading application of objects with a blunt surface, which produces various fracture types, plastic deformation, and delamination (Passalacqua & Fenton, 2012; Wedel & Galloway, 2013). Based on the skeletal characteristics observed in our examination, we argue that quadrilateral defects in our samples represent a mixed mechanism category of trauma, with aspects of both sharp and blunt force.

While not well documented, quadrilateral defects have been observed in skeletal remains from other contexts (Murphy et al., 2010; Novak, 2007), including the Battle of Towton from AD 1461. Associated with the War of the Roses and fought in Yorkshire, England between soldiers from the Houses of York and Lancaster, this battle was reportedly one of the deadliest fought on English soil (Novak, 2007). Skeletal remains associated with the Battle of Towton were first discovered in 1996 during the construction of a garage and included a minimum of 38 individuals, 29 of which had crania present. Novak (2007) documented the battle-related trauma associated with this case, noting four individuals with seven quadrilateral defects similar to those found in the Tuskulenai case (Novak, 2007). These patterned wounds are attributed to the beak of a medieval war hammer or poleaxe, which is pointed and square in cross-section.

While fusillade, or firearms, was a state-mandated means of execution throughout much of the Stalinist period in the Soviet Union, it is evident from the Tuskulenai case that other mechanisms of force were employed by executioners implementing violence on behalf of the state (Bird, 2013). Beatings and stabbings can be non-generic modes of execution, but the presence of patterned quadrilateral wounds in over 100 prisoners executed by two different execution squads in the Tuskulenai case suggest a certain convention of acceptable use with this weapon. While ice picks were common in cold climatic regions such as Lithuania, their availability as a utilitarian item in Mexico City (a temperate, subtropical environment) is perplexing.

Historically, ice axes (aka "ice picks") were first manufactured by Grivel in 1840 which modified the traditional alpenstock (a long wooden pole with a spike at the end) with a sharpened blade opposed to a flattened blade for the purposes of mountaineering in cold climatic regions (Ingram, 2001). Ice axes permitted climbers to scale steep terrain by penetrating ice and creating holds for hands and feet. During the early part of the 20th century, the size and shape of the ice axe evolved into shorter, lighter axes with small wooden handles that could be swung with one hand in confined spaces. The ice axe used during the Trotsky assault had a wooden handle (approximately one foot long) and two metal projections, including a long, pointed end and a short, flat end (see Figure 19.7).

The pointed end tapered from base to tip, was square in cross-section, and produced a similar wound to those found in the Tuskulenai case. Mercader testified that he brought the ice axe with him, hidden within his raincoat, to assault Trotsky but he never stated *why* he chose that particular object for the assassination.

The climate in Mexico varies significantly with its topography. Although much of Mexico contains arid deserts and tropical environments, mountaineering is practiced, particularly on the high-altitude volcanoes in the inner part of the country, some of which are within 60 miles of Mexico City. These volcanoes are crowned with snow and ice and require appropriate mountaineering equipment (e.g. ice axes) for successful ascents. While it is possible that Ramon Mercader chose an ice axe to assassinate Trotsky because he was a mountaineer, owned the object, and was skilled in its use, no evidence of his participation in this alpine activity has surfaced. Rather, Mercader's association with the Soviet security apparatus may shed light on his choice of weapon for the assassination.

Ramon Mercader was born in Barcelona, Spain in 1913 and raised in France. Following in the footsteps of his parents, Mercader embraced Marxist–Leninist ideology and participated in leftist organizations in Europe during the 1930s. In 1938, he moved to Mexico City under the assumed name, "Jacques (or Jacson) Mornard." Befriending Trotsky as a sympathizer, Mercader gained access to his heavily guarded Coyoácan home where he struck Trotsky on the head with the ice axe in 1940. Mercader was convicted in a Mexican court and served 20 years in prison. Upon his release, Mercader traveled between the Soviet Union and Cuba and eventually died in Havana in 1978 (Volkogonov, 1996).

During his trial, Mercader maintained the identity of "Jacques Mornard," and contended that his motivation for killing Trotsky stemmed from a dispute over his desire to marry Sylvia Ageloff, a friend of Trotsky's. However, historical evidence now supports that Mercader was a Soviet agent who acted on behalf of the Kremlin. Mercader was reportedly recruited during the Spanish Civil War by NKVD agent, Col. Leonid Eitingon, and trained as a Soviet agent in Moscow (Volkogonov, 1996). Following his release from prison, Mercader was awarded the Hero of the Soviet Union by Alexander Shelepin, the head of state security. Ramon Mercader's training as a Soviet agent, coupled with the Tuskulenai data, is potentially informative for further understanding the Soviet security apparatus.

In order to retain authoritarian control, the USSR developed a centralized security apparatus that enforced ideological conformity, limited citizen autonomy, ensured social and political control, and punished transgressions (Shelley, 1996). The centralized nature of state control was mirrored in the security apparatus: orders pertaining to state security originated in Moscow and were then disseminated down the bureaucratic hierarchy to be implemented at a local level. While the state security apparatus experienced a number of tortuous splits and merges between the 1920s and 1940s, it emerged stronger following the Second World War. Not only did the apparatus become more centralized during the war, but it had *well-trained* agents at its disposal (Levytsky, 1972). One mode of building an efficient bureaucracy is formally incorporating agents into the institution: training is one method employed by higher authorities to monitor and modify the character of agents. The expert training of agents ensures that individuals are versed in the official conduct of the bureaucracy which is regulated by formal rules and documents. While not much is known about the training of violence workers in Soviet state security, the expert execution of prisoners in other cases of state violence such as Vinnytsia, Ukraine (1937–1938) and Katyn, Russia (1940) indicate that Soviet executioners were skilled in maximizing the fatal potential of small-caliber weapons and may have received training to do so (Bird, 2013; Cienciala et al., 2008; Kamenetsky, 1989; Sanford, 2005).

While it is possible that objects used for violence in the Trotsky and Tuskulenai cases represent unassociated objects of convenience, this explanation seems less likely given the specialized function of ice axes in Mexico. Given the allegedly specialized training of Soviet agents in other instances of violence, it may be that violence workers in both cases had training in the application of these objects. Finally, imitation is the highest form of compliment, and it is possible that local violence workers in the LSSR emulated the mechanism of violence in the Tuskulenai executions based on a historical and well-known assassination: that of Leon Trotsky.

LESSONS LEARNED

- *Individual* cases of skeletal trauma can benefit from a larger, *aggregate* approach, which reveals biomechanical characteristics useful for understanding unique mechanisms of trauma. Based on findings in the Tuskulenai and Trotsky cases, quadrilateral defects represent patterned wounds likely caused by long objects that taper from base to tip, are square in cross-section, and have a pointed tip.
- While the random use of comparable objects for execution between the Tuskulenai and Trotsky cases are possible, the similarly patterned skeletal trauma in both cases suggests intent by either the state or violence workers.
- This study elucidates that signs of violence on the body reveal not only what decedents experienced around the time of death, but also provide insight into how and why violence workers choose a particular method of execution. While the scientific understanding of skeletal trauma and perpetrators of violence is important, we also contend that research into these cases should be couched within the historical context and the local communities who experienced this trauma, and coupled with the unwavering humanitarian intent to identify victims and return remains to families or their communities.

DISCUSSION QUESTIONS

19.1 All quadrilateral wounds were observed in crania. What are possible reasons for this observation? Do you think the shape of crania and the manner in which force dissipates may play a role in the ability of anthropologists to recognize any patterned wounds in postcranial elements?

19.2 When working a case, forensic anthropologists generally do not relay information about the specific objects used in blunt force injuries. What makes this case different? How did the authors remain conservative in their assessments while still providing important information about the potential object that created the skeletal defects?

19.3 What similarities and differences were found between the remains in the two burial pits examined? What information does such a comparison provide?

ACKNOWLEDMENTS

We would like to thank the following Lithuanian institutions for their support: Tuskulenai Memorial Complex, the State Forensic Medicine Service, and The Genocide and Resistance Research Center of Lithuania.

REFERENCES

Applebaum, A. (2003). *Gulag: A history of the Soviet concentration camps.* New York: Doubleday.

Bird, C. E. (2013). *State-sponsored violence in the Soviet Union: Skeletal trauma and burial organization in a post-world war II Lithuanian sample* (Doctor of Philosophy Dissertation). Michigan State University.

Cienciala, A. M., Materski, W., & Lebedeva, N. (2008). *Katyn: A crime without punishment.* Yale University Press.

Damusis, A. (1998). *Lithuania against Soviet and Nazi aggression.* American Foundation for Lithuanian Research, Inc.

Gregory, P. R. (2009). *Terror by quota: State security from Lenin to Stalin (An Archival Study).* Yale University Press.

Ingram, S. (2001). All tooled up. Retrieved from https://www.thebmc.co.uk/all-tooled-up

Jankauskas, R. (2009). Forensic anthropology and mortuary archaeology in Lithuania. *Anthropologischer Anzeiger, 67*(4), 391–405.

Jankauskas, R., Barkus, A., Urbanavičius, V., & Garmus, A. (2005). Forensic archaeology in Lithuania: The Tuskulėnai mass grave. *Acta Medica Lituanica, 12*(1), 70–74.

Kamenetsky, I. (1989). *The tragedy of Vinnytsia: Materials on Stalin's policy of extermination in Ukraine during the Great Purge, 1936–1938.* New York: Ukrainian Historical Association.

Kiaupa, Z. (2005). *The history of Lithuania.* Vilnius, Lithuania: Baltos Lankos.

Kimmerle, E. H., & Baraybar, J. P. (2008). *Skeletal trauma: Identification of injuries resulting from human rights abuse and armed conflict.* CRC press.

Lane, T. (2001). *Lithuania: Stepping westward* (1st ed.). Routledge.

Levytsky, B. (1972). *The uses of terror: The Soviet secret police 1917–1970.* Coward, McCann & Geoghegan.

Murphy, M. S., Gaither, C., Goycochea, E., Verano, J. W., & Cock, G. (2010). Violence and weapon-related trauma at Puruchuco-Huaquerones, Peru. *American Journal of Physical Anthropology, 142*(4), 636–649.

Novak, S. (2007). Battle-related trauma. In V. Fiorato, A. Boylston, & C. Knüsel (Eds.), *Blood red roses: The archaeology of a mass grave from the Battle of Towton AD 1461* (pp. 90–102). Oxford, UK: Oxbow books.

Passalacqua, N. V., & Fenton, T. W. (2012). Developments in skeletal trauma: Blunt-force trauma. In D.C. Dirkmaat (Ed.), *A companion to forensic anthropology* (pp. 400–411). Wiley Blackwell.

Pocius, M. (2006). *1944–1953 metų partizaninio karo Lietuvoje istoriografija.* Istorija. *Lietuvos aukštųjų mokyklų mokslo darbai, 64*(4), 52–64.

Rudienė, V., & Juozevičiūtė, V. (2006). *The museum of genocide victims: A guide to the exhibitions.* Museum of Genocide Victims.

Sanford, G. (2005). *Katyn and the Soviet massacre of 1940: Truth, justice, and memory*. London, UK: Taylor and Francis Group.

Shelley, L. (1996). *Policing Soviet society: The evolution of state control*. Routledge.

Thompson, J. M. (1998). *Russia and the Soviet Union: A historical introduction from the Kievan state to the present* (4th ed.). Westview Press.

Vaitiekus, S. (2011). *Tuskulenai: Victims of execution and their henchmen (1944–1947)*. Genocido ir Rezistencijos Centras.

Volkogonov, D. (1996). *Trotsky: Eternal revolutionary*. Simon and Schuster.

Wedel, V. L., & Galloway, A. (2013). *Broken bones: Anthropological analysis of blunt force trauma*. Charles C Thomas Publisher.

CHAPTER 20

Sexual Offense in Skeletonized Cadavers: Analysis, Interpretation, Documentation, and Case Report

*César Sanabria-Medina, Jorge Andrés Franco
Zuluaga, and María Alexandra Lopez-Cerquera*

CONTENTS

INTRODUCTION

Article 5 of The Universal Declaration of Human Rights establishes that no human being can be subjected to torture or other cruel, inhumane, or degrading treatment. This is a non-derogable right, which cannot be suspended or altered under any circumstance. Despite the efforts of international agencies, the practice of torture and other cruel and inhumane treatments persists in more than half of countries worldwide (United Nations, 2016). Although these phenomena are complex in the legal context, they are even more so in the forensic context. Victims may be recovered months or years after death, in varying states of decomposition and skeletonization, and many times displaying natural or man-made modifications, making postmortem examinations difficult. It becomes a challenge for forensic practitioners and criminal investigators to establish how, when, and why these deaths occurred. These questions are key, as victims' families and other stakeholders demand answers from the investigative institutions to ensure appropriate application of the law and the delivery of justice. In addition, answering relatives' inquiries in the forensic context is part of the comprehensive reparation process to victims in the context of international human rights law (Sanabria-Medina, Quiñones, Osorio, & Barraza, 2016).

SEXUAL VIOLENCE IN SKELETONIZED CADAVERS:
THE ROLE OF FORENSIC EXPERTS

In cases of skeletonized remains, evidence of sexual violence may be overlooked under the mistaken assumption that the probability of finding signs indicative of sexual violence is low or nonexistent. This is a mistake associated with the "culturally-shaped" idea that a sexual crime almost exclusively refers to the act of penetration (oral, vaginal, or anal) and the presence of biological fluids (sperm, saliva) inside or on the victim, which at first glance would be impossible to demonstrate in skeletonized remains. However, other evidence of sexual violence may be elucidated with appropriate strategies, especially with the support of forensic experts who do not often participate in these types of cases. The case presented here required consultation with the forensic physics laboratory. Analysis of trace evidence, corroboration of witness accounts, evaluation of scene contextual information, and comparison to patterns documented in previous cases of sexual violence all contributed to the ultimate conclusions.

As indicated by Morales (2011a,b), in many cases of torture and sexual violence, victims are silenced with death. In such cases, forensic practitioners and investigators are left to fill the gaps. "The absence of evidence is not evidence of absence,"[1] is perhaps the gold standard of the forensic anthropological analysis in such cases. Cases of skeletonized or decomposing cadavers face a paradox: the greater and more serious the crime, typically the less physical evidence available for scientific analysis. The search for evidence of sexual violence in skeletonized cadavers is a complex exercise of "cautious diagnosis." Still, to indicate in a forensic report that the findings associated with acts such as sexual violence are indeterminate based solely on the fact that the remains are decomposed or fully skeletonized is irresponsible if an in-depth analysis of all potential evidence has not been conducted. In such circumstances, both the scene and cadaver must be approached by forensic experts familiar with these crime modalities and the circumstances that may have surrounded them and, above all, an understanding of the implications of their work to the investigation. Knowing international protocols is also useful in cases of presumed sexual violence, especially when forensic practitioners are aware of the regulations on attacks against personal dignity, in particular, the humiliating and degrading treatments referred to by the Additional Protocols I and II (Articles 3 and 75 and Article 4, respectively) of the Geneva Conventions.

Both the European Court of Human Rights and the United Nations Office of the Special Rapporteur indicate that rape and other forms of sexual violence can be identified as torture and abuse (United Nations, 2016, paragraph 51). Per the Special Rapporteur, "Rape is tantamount to torture when it is inflicted by a public official, at his/her instigation, or with his/her consent or acquiescence" (paragraph 51). Furthermore, in many cases of armed conflict, both state agents and non-state actors perpetrate acts of sexual violence, which constitute "violations of international humanitarian law (IHL) and correspond to torture according to the jurisprudence of international criminal law" (paragraph 52). In addition, a diversity of criminal activities in situations of detention in armed conflicts are considered torture, cruel, inhumane, and degrading treatment, regardless of the perpetrator (civilian or state agent). An example is forced nudity which, when combined with homicide, may be overlooked during the autopsy. The evidence may indicate a violation of personal dignity, which may have been preceded by physical forms of sexual violence. Forms of sexual offense (and their legal implications) may vary among countries and depend on the act (e.g., carnal vs. other sexual acts), the age of the victim, and the ability of the victim to resist.

CASE STUDY

The forensic anthropological literature on sexual violence is notoriously scarce. For this reason and with the aim of making a small contribution to the field, this chapter presents the forensic analysis of a case involving multiple skeletonized female individuals with signs of sexual violence examined in the framework of the investigations carried out by the Prosecutor General's Office in Colombia. These deaths were attributed to paramilitary groups.

Stemming from the peace agreement signed between paramilitary groups and the Colombian government in 2005 (Law 975, 2005), hundreds of judicial investigations were conducted to recover the remains of persons assassinated by these illegal armed organizations during the armed conflict. One of these investigations resulted in the recovery of the skeletonized remains of four young sisters subjected to forced disappearance by a paramilitary group.

Investigative information led to the search of a rural sector in Colombia. Multiple boreholes were dug, searching for changes in soil consistency (indicative of a grave) or contact with remains. Eventually, this process revealed two grave features: one containing a single skeletonized individual and a second containing three skeletonized individuals (see Figure 20.1). The remains were commingled, naked, and presented signs of dismemberment. Inside the collective grave, intimate garments (underwear bottoms and bras) were found adjacent to the cadavers (see Figure 20.1, arrows). The remains were carefully excavated by the Prosecutor's Office and sent to the National Institute of Legal Medicine and Forensic Science for a postmortem evaluation. An interdisciplinary team of eight specialists worked collaboratively on the case: a forensic pathologist, three forensic anthropologists, a forensic odontologist, two geneticists, and one forensic physicist.

The first step was to separate the skeletal elements by individual. Macroscopic techniques of individualization were used according to the methodologies proposed by L'Abbé (2005) and Adams and Byrd (2006) along with information from forensic archaeological recovery. Structures were paired by morphological similarities, including conformity at the level of joint areas, rearticulation of fracture edges and breaks, degree of taphonomic

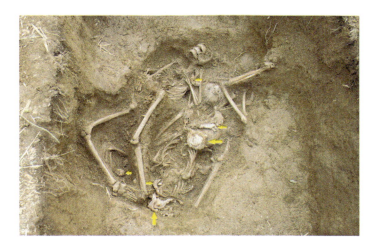

FIGURE 20.1   Mass grave containing three skeletonized individuals. Arrows point at intimate clothing. (From Photographic Archive Technical Investigation Unit, General Prosecutor's Office of the Nation.)

preservation between adjacent structures, and similarities in sex and age characteristics. The reassociation process confirmed the presence of at least three individuals in the mass grave (see Figure 20.2) and confirmed that the single grave contained remains from a single individual (see Figure 20.3).

The biological profiles (sex, biological age, and stature) were estimated from traditional skeletal markers utilizing Scheuer and Black (2000) and Buikstra and Ubelaker (1994), as well as the dental criteria provided by the forensic odontologist. Subsequent genetic comparisons with family members from the missing sisters confirmed the four identities. Two of the sisters were twins and their individualization was possible because one of them had an antemortem clavicle fracture.

Trauma analyses were performed with the support of a stereomicroscope, facilitating close-up examination of the angle, shape, color, and morphology of fracture edges to determine their ante-, peri-, or postmortem origin (Galloway 1999; Kimmerle & Baraybar, 2008; and Wieberg et al., 2008). Perimortem sharp force trauma consistent with dismemberment was observed in multiple joint areas across all four individuals. One individual showed signs of decapitation (see Figure 20.4). Overall, the trauma patterns indicated that the injury mechanism was sharp force trauma with a heavy object, such as a machete.

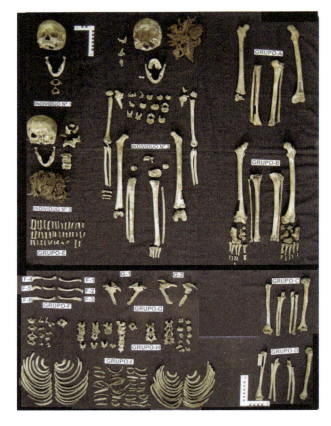

FIGURE 20.2   Reassociation process of remains contained within the mass grave. (From Mónica Chapetón, National Institute of Legal Medicine and Forensic Sciences, Regional Bogotá, Forensic Anthropology Laboratory.)

FIGURE 20.3 Body from individual grave. (From Mónica Chapetón, National Institute of Legal Medicine and Forensic Sciences, Regional Bogotá, Forensic Anthropology Laboratory.)

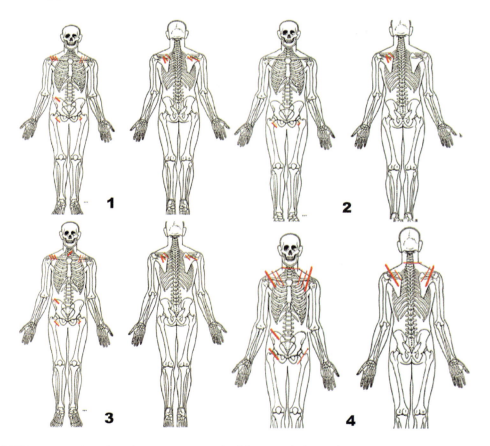

FIGURE 20.4 Graphic documentation (red lines) of sharp force trauma injuries associated with dismemberment in all four individuals. Note the potential decapitation of individual 4. (From Jorge Andrés Franco, National Institute of Legal Medicine and Forensic Sciences, National Group of Forensic Pathology.)

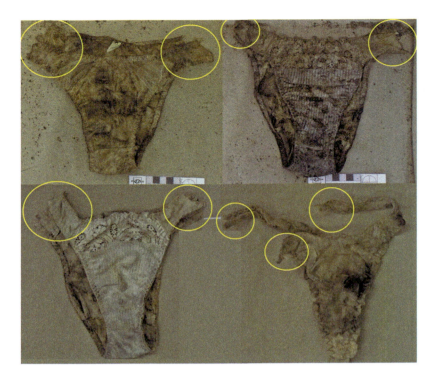

FIGURE 20.5 Intimate garments (underwear bottoms) with lateral tears at the points of least tension (circles). (From Mónica Chapetón, National Institute of Legal Medicine and Forensic Sciences, Regional Bogotá, Forensic Anthropology Laboratory.)

The clothing was analyzed by the forensic physicist. Three underwear bottoms were found with irregular lateral tears at the points of least tension, that is, not at the sewing areas (see Figure 20.5). A bra was also found with irregular tears in the anterior central region (see Figure 20.6). Note that these clothing items were not found in correct anatomical locations within the graves (see Figure 20.1), appearing instead to have been tossed in with the bodies. The tears were thus not likely to be the result of natural degradation or bloaty expansion during decomposition.

The case analysis included the following: (1) the study of national laws on sexual violence; (2) a consultation of the international protocols mentioned above; (3) the analysis of the scene report; (4) the study of the versions of the events provided in the case file; (5) the analysis of skeletal trauma; (6) the taphonomic analysis of bone and clothing; and (7) the analysis of undergarments with the support of the forensic physics lab.

CASE CONCLUSIONS AND DISCUSSION

The case involved four skeletonized bodies recovered from an individual and a mass grave, all female, one of them a minor (<18 years). All remains showed signs of dismemberment, and one evidenced signs of decapitation. The bodies were found naked with the exception of intimate clothing items, including three underwear bottoms and a bra recovered from within the mass grave out of anatomical context. Considering the national and international regulations mentioned earlier, as well as the recommendations by national

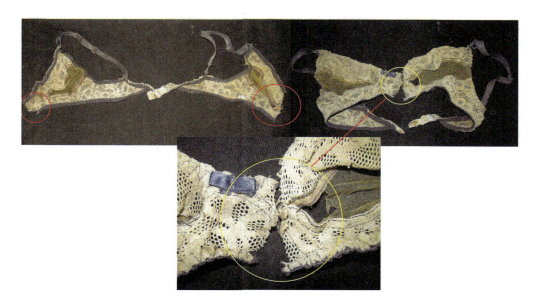

FIGURE 20.6   Irregular tear in anterior central area of the bra. (From Mónica Chapetón, National Institute of Legal Medicine and Forensic Sciences, Regional Bogotá, Forensic Anthropology Laboratory.)

and international protocols and similar cases, the following variables compatible with sexual offense were obtained:

- Two clandestine graves containing only cadavers of young women.
- The women were naked.
- The intimate garments were disassociated from the cadavers.
- The underwear bottoms presented a pattern of bilateral tears, which according to the report offered by the forensic physics experts, indicates they were pulled while the fibers were in tension (i.e., the victims were wearing the undergarments when they were torn). This hypothesis led to the conclusion that forced nudity prior to death was perpetrated by a third party. A similar situation occurred with the bra, whose breaking point was located at the point of least tension in the anterior region. Taphonomic agents were ruled out as a causative factor since other areas of the fabrics were in a good state of preservation.

The cause of death was determined as perimortem sharp force trauma from a heavy weapon associated with a dismemberment pattern. "Dismemberment" as cause of death was proposed in the scientific literature by Morcillo and Campos (2012). It was not possible to rule out additional trauma to soft tissues on account of the skeletonized state of the remains. Examination of the garments suggested that the victims were stripped and their intimate garments torn apart while being worn by the victims. From the experts' point of view, these findings are related to forced nudity. Likewise, it was not possible to confirm or rule out that penetration or another type of sexual abuse occurred. The manner of death was established as violent and homicidal from the medicolegal opinion.

To reach the final conclusion of the present case, elements were considered that may be overlooked in the analysis of remains found in clandestine graves. For example, the nakedness of the bodies and the fact that they were all female, which alone suggests several types of probable sexual violence (e.g., violation of privacy and public nudity by coercion of third parties which is compatible with humiliation and degradation of a person), or other cruel, inhumane, or degrading treatment or punishment indicated in the four Geneva Conventions that make up IHL.

Likewise, tangible physical evidence such as the pattern of tears in undergarments suggested the sisters were dressed when the underwear was torn apart. The evidence led the experts to conclude that this was a case of sexual violence, not necessarily the typical case of sexual violent penetration with which sexual offense is usually identified (however, no soft tissues and/or additional evidence were available to test such a hypothesis). Yet, it was clear that forensic practitioners were dealing with a case of forced nudity, which does not mean this is legally considered a less serious crime. Additionally, although it is outside the competence of forensic experts to express themselves officially in this regard, this case is characterized by the perpetration of conduct classified as crimes against humanity (i.e., forced disappearance, torture, sexual crime, homicide, concealment of the body and, of course, feminicide) insofar as they were committed in the context of an armed conflict. That is, the crimes were systematic and deliberate.

A number of national and international documents were useful in concluding this case. Those sources included the Declaration of Human Rights (1948); the Colombian Code of Criminal Procedure in the chapters that address Crimes against Persons and Protected Property by the IHL, including Title IV, Crimes Against Sexual Freedom, Integrity and Education and Abusive Sexual Acts; the Inter-American Convention on the Prevention, Punishment and Eradication of Violence against Women, known as the Convention of Belém do Pará (site of its adoption in 1994); the sentence of 2009 on the case *González y Otros ("cotton field") vs. Mexico* and publications addressing the problem of torture and sexual crime in complex cadavers (Droege, 2007; Morales, 2011a,b 2011a and Sanabria-Medina et al., 2016, among others).

For all the above, the study conducted by forensic anthropologists as support to the medicolegal autopsy cannot be limited to the estimation of a biological profile and the documentation of trauma. This may produce an incomplete and inefficient expert report ultimately absent of the burden of proof provided by potential evidence, which would be detrimental to the surviving victims. Furthermore, as stated by Morales (2011a. pp. 24)

> an expert opinion of this kind, which is inadequate, insufficient, and removed from the context, will not be considered as an expression of conceptual poverty [by the forensic practitioner], but as a scientifically supported truth, endorsed by the signature of an expert and may well contribute to the filing of a case due to the of evidence. (Translated from Spanish)

LESSONS LEARNED

- It is imperative that medical examiners, forensic anthropologists, and other forensic experts analyze the case "in context." In addition to examining the scene, the remains, and physical evidence, as in the present case study, the experts must read, interpret, and if necessary apply the following aids to the analysis of the case(s): (1) national and international laws, (2) international treaties, (3) previous judicial decisions – national and international – that have been taken in cases with a similar context, and (4) cases reported in the literature.

- The myth that sexual violence can only be documented in cases of fresh cadavers should be abandoned. The absence of clothing in the victim, clothing removed, a naked or partially dressed cadaver, are all signs of a possible sexual offense. Likewise, the condition of the garments must be analyzed, including possible tears and patterns of intentional damage.
- The location of graves containing only female cadavers could be an indicator of sexual violence. Investigators should take into account the position of the bodies and their belongings; the state of the clothing or the lack of it; the presence of fractures in pelvic and/or leg bones in order to immobilize the victim; the presence of ties on female's arms using the victim's clothing or a different material, and the lack perimortem of dental pieces.
- A key factor in clandestine grave analysis is having organized and clear hypotheses about the events so experts can confirm or rule out the possibility of sexual violence prior to death. The rigorous application of protocols can make it possible to clarify the existence of more than one crime (e.g., sexual violence and homicide), and the commission of misconduct prior to death such as torture and/or sexual violence.

DISCUSSION QUESTIONS

20.1 This case illustrates the collaborative efforts of an interdisciplinary forensic team to draw conclusions regarding potential sexual and humanitarian violence in skeletonized remains. Whose role is it to make a final determination about whether the evidence supports sexual or inhumane violence?

20.2 Would the conclusions of the case have been different if the clothing artifacts had not been recovered and documented *in situ*? If so, how would they have differed?

20.3 Why do the authors argue that it is important for forensic practitioners to know and understand the laws surrounding criminal violence and humanitarian rights? How does knowledge of such regulations impact forensic analyses?

NOTE

1. Although this phrase is usually attributed to Carl Sagan, it seems that it actually belongs to Tsiolkovski (1857–1935).

REFERENCES

Adams, A., & Byrd, J. (2006). Resolution of small-scale commingling: A case report from the Vietnam War. *Forensic Science International*, 156, 63–69.

Buikstra, J., & Ubelaker, D. (1994). Standards for data collection from human skeletal remains. In *Proceedings of a seminar at the field museum of natural history*, 68(1), Arkansas Archaeological Survey.

Droege, C. (2007). El verdadero Leitmotiv: la prohibición de la tortura y otras formas de malos tratos en el derecho internacional humanitario. *International Review of the Red Cross*, 89(867), 515–541.

Galloway, A. (1999). *Broken bones: Anthropological analysis of blunt force trauma.* Springfield, IL: Charles C Thomas.

Kimmerle, E. H., & Baraybar, J. P. (2008). Sharp force trauma. In *Skeletal trauma* (pp. 263–319). Boca Raton, FL: CRC Press, Taylor & Francis Group.

L'Abbé, E. A (2005). Case of commingled remains from rural South Africa. *Forensic Science International, 151,* 201–205.

Organización de las Naciones Unidas (ONU). (1948). Declaración de los Derechos Humanos. Retrieved from http://www.un.org/es/documents/udhr/index_print.shtml. Law Number 975. Diario Oficial de la República de Colombia, Bogotá, Colombia, July 25th, 2005.

Morales, R. M. (2011a). La necropsia medicolegal en víctimas de desaparición forzada: Documentación de la tortura y la violencia sexual. Editado por Instituto Nacional de Medicina Legal y Ciencias Forenses y Programa de las Naciones Unidas para el Desarrollo (PNUD) / Programa Fortalecimiento a la Justicia en Colombia. Bogotá D.C., Colombia.

Morales, R. M. (2011b). Violencia sexual y tortura en desaparición forzada. Editado por Instituto Nacional de Medicina Legal y Ciencias Forenses y Programa de las Naciones Unidas para el Desarrollo (PNUD) / Programa Fortalecimiento a la Justicia en Colombia.

Morcillo-Méndez, M., & Campos, I. Y. (2012). Dismemberment: Cause of death in the Colombian armed conflict. Retrieved from https://www.ncbi.nlm.nih.gov/pubmed/22948397

Sanabria-Medina, C., Quiñones, E., Osorio, R. H., & Barraza, S. M. S. (2016). Tortura y delito sexual en cadáveres esqueletizados: Análisis, interpretación y documentación. In C. Sanabria-Medina (Ed.), *Patología y antropología forense de la muerte: la investigación científico-judicial de la muerte y la tortura, desde las fosas clandestinas, hasta la audiencia pública* (pp. 719–746). Bogotá D.C., Colombia: Forensic Publisher.

Scheuer, L., & Black, S. (2000). *Developmental juvenil osteology.* Academic Press.

Wieberg, D. A., & Wescott D. J. (2008). Estimating the timing of long bone fractures: Correlation between the postmortem interval, bone moisture content, and blunt force trauma fracture characteristics. *Journal of Forensic Sciences, 53*(5), 1028–1034.

CHAPTER 21

Making the Best of Limited Resources and Challenges Faced in Human Rights Investigations

Eugénia Cunha, Maria Teresa Ferreira, Cristina Cordeiro, and Duarte Nuno Vieira

CONTENTS

INTRODUCTION

This chapter presents a forensic anthropology case study in the context of an investigation of crimes against humanity and humanitarian forensic actions in Africa. It involves two African missions in which the authors were called by an international agency to identify victims, document potential human rights violations, and determine cause and manner of death of a group of 25 persons who went missing. The dates and country are being kept anonymous for security purposes. Purportedly, the individuals were executed after being tortured, and their bodies allegedly were inhumed in a mass grave, although it was also possible that some bodies were inhumed in a local cemetery. The crime occurred about seven months before the first mission and one year before the second one. All victims were males (the majority young individuals) from the country in question and almost all were military. The forensic team included two forensic pathologists, two forensic anthropologists, and one geneticist. In the first mission, the work had to be done during one week in a small room at the local hospital, hereafter designated as the workroom. The second mission, which occurred five months after the first one, was accomplished in four days.

THE FIRST MISSION

When we arrived, the mass grave had already been excavated by the local authorities who were convinced they had done a good job and that our task was appreciably easier after their work. We arrived to the workroom to find 21 small paper bags, sealed with red tape, arranged one-by-one on the floor. The chain of custody seemed to have been respected during the excavation and transport. Several larger black plastic bags contained clothing and various personal belongings, one contained ribs, another contained vertebrae still encased in soft tissues, and others contained commingled long bones with varying degrees of adherent soft tissues. It became immediately apparent that the excavation was not completed to forensic archaeological standards, and there were numerous misconceptions to clarify. Further examination revealed that 19 of the paper bags each contained a single skull which had been isolated from the remaining body, one paper bag had two skulls, and another paper bag had a shirt and a plastic bag with a military identification card. The excavators had expended considerable time and effort removing the clothes of the 21 individuals and putting them all together in plastic bags. They had similarly removed the 12 pairs of ribs of each of the individuals, which must have been difficult given that the remains were not fully skeletonized, retaining tendons and some muscles.

Given the now commingled nature of the remains, the similar profile of the individuals (males of same ancestry and similar age) and lack of association with material artifacts, it was clear that the individualization of each skeleton (i.e., association of all remains) would not be possible from gross examination. A strategy was needed to save and infer the maximum data possible, and the focus shifted to determining the identity of the deceased and documenting any remaining evidence regarding the circumstances of death. Isolated bones and groups of elements that were still somewhat articulated by soft tissues were separated by types and side, when appropriate. The ribs were counted and separated into two large piles of lefts and rights; attempts were not made to assign ribs to individuals. Efforts were made to separate out individual vertebral columns from the 500 vertebrae. Once right and left os coxae were separated, several pairs were matched on the basis of size, morphology, and taphonomic changes. After that, the sacra were allocated using congruency with the sacroiliac joints. Once the pelvic girdle was complete, the next step was to allocate the femora. We found the hip articulation to be one of the more reliable elements when assembling the bones of an individual. Given the cavitation of the acetabulum and close-fit with the femoral head, exclusions could be made when the femoral heads did not fit. Nevertheless, in some cases cases it was challenging to decide the best match. At the end of the process, five bones were associated with a skeleton (two os coxae, one sacrum, and two femora) for all individuals.

It is important to bear in mind that the majority of the femora and other elements were fragmented due to the violent nature of the deaths (discussed later). The fragmentation precluded the assemblage of some elements to individuals. Some of the skulls were particularly damaged due to trauma, but it was thought important to return these to the families. Associating the skull with the infra-cranial skeleton is especially difficult, as the only articulation is via the vertebral column, which had been disassociated with the remains. The forensic anthropologists completed a biological profile for each skull. Two molars were removed from each skull and a section of femur was removed from each set of assembled infra-cranial skeletons and sent to another country for genetic/DNA analysis. Time and cost precluded DNA analysis of every skeletal element but by analyzing the skull and group of five associated postcranial elements, we maximized the return of elements to families based on available resources.

The MNI – minimum number of individuals – was estimated based on the inventory of the skulls, humeri, radii, ulnae, femora, tibiae, fibulae, vertebrae, ribs, ossa coxae, and sacra. Bones of the extremities were not completely recovered since almost all phalanges were missing. In every case, the MNI achieved was 21 individuals. That is, all the skeletons, albeit completely disarticulated, were mostly complete, suggesting the burial was a primary deposition site and the bodies had been buried when they were relatively fresh. The completeness of the bodies and general anatomical positioning was confirmed by the team who performed the excavation.

Identification Process

The identification process began by assessing all possible biological profile information from the skulls, ossa coxae, humeri, and femora. We applied metric and non-metric approaches for sex estimation (following Asala, 2001; Beauthier, 2011; Işcan & Steyn, 2013). Age indicators included the 4th rib, pubic symphysis, auricular surface, sternal end of the clavicle, vertebral bodies, degenerative changes, and root transparency of monoradicular teeth (again following Asala, 2001; Beauthier, 2011; Işcan & Steyn, 2013). Ancestry was assessed using non-metric features of the skull, namely of the face (Hefner, 2009; Işcan & Steyn, 2013), and, whenever possible, craniometric data were used to estimate ancestry. Stature was estimated on the basis of femoral lengths (Lundy, 1983). Invariably, all 21 individuals from the mass grave were young adult males of African ancestry, providing no assistance in discerning the individual identity of the missing individuals. Efforts then focused on individualizing traits. In the face, the presence of anterior diastemas, antemortem fractures to the anterior dentition, dental prostheses, and dental treatments were among the most diagnostic features. For the postcrania, preexisting pathological changes in bones included mild degenerative osteophytic changes on the vertebral bodies, especially the lumbar region.

The families searching for their loved ones were involved. Several relatives of each victim went to a nearby room to provide the necessary antemortem data for the comparative phase of identification. We received only one antemortem x-ray, but families provided records of medical exams, pictures (especially those of the face and smiling), notation of any known individualizing features, and an estimation of stature. DNA samples were collected from close relatives of the victims. The relatives were interviewed by the forensic pathologists while DNA samples were gathered by the geneticist.

Cause and Manner of Death

Given that two of the authors are medical doctors, cause and manner of death were evaluated (this would be outside the scope of anthropology). Skeletal elements were separated from soft tissues and dried, followed by a macroscopic examination. Unfortunately, imaging means were not available. Nine skulls displayed perimortem traumatic injuries. In seven, clearly demarcated entry and exit wounds indicated gunshot wound trauma (GSW) (see Figure 21.1), while two others displayed fracture patterns more consistent with blunt force trauma (BFT). As the skulls were not in association with the infra-cranial skeleton, cause of death could only be determined in the cases of severe BFT or GSW to the head, which was approximately half of the individuals. Postcranially, perimortem injuries varied by type and region. A GSW to the thorax was distinguishable through entrance wounds to the

FIGURE 21.1 Right superior-oblique view of a skull displaying an exit defect consistent with a GSW. The entrance was through the anterior aspect of the frontal, in the left orbit, near nasion, while the exit was through the right side of the frontal (indicated by external beveling). The exit caused extensive damage to the right frontal and upper face. The radiating fracture lines at the exit indicate the magnitude of energy transmitted during the injury.

ribs and left marks on clothing (facilitating association with clothing for this individual). Some vertebrae exhibited severe fractures that may have been associated with death. Some scapulae exhibited fractures with plastic deformation consistent with perimortem trauma. A GSW was also observed on an os coxae where the entrance and exit defects indicated a back to front bullet trajectory. Three humeri displayed comminuted fractures consistent with GSW trauma. The femora were among the most damaged elements; while some were reduced to fragments (comminuted fractures compatible with GSW trauma), others had fractures more consistent with BFT or a mixture of blunt and sharp force trauma consistent with an impact with a heavy instrument with a cutting edge, such as an axe. One sacrum had a severe longitudinal alar fracture consistent with a fall from height, and the fracture continued into the lumbar vertebrae allowing the reallocation of those vertebrae to the respective sacrum. Despite the severity of the postcranial traumatic injuries, only those affecting the vertebrae could be directly related to the cause of death. Overall, the cause of death was considered as consequent to multiple and extensive injuries; the manner of death was considered "Unnatural / Homicidal injury."

Documentation of Human Rights Violation

All skulls had blindfolds in association (see Figure 21.2) suggesting a violation of human rights. Moreover, iron chains were recovered, which produced a green discoloration on some bones indicating their positioning in contact with those elements in the grave. Some forearms were also tethered with ropes or fabric (see Figure 21.3). Undoubtedly, these individuals were blindfolded and tied up, and thus could not defend themselves while they were severely beaten and/or shot to death. In one case an antemortem injury is suggestive of torture. A body of a scapula displayed macroscopic signs of a healing injury,

FIGURE 21.2   Skull with a blindfold still *in situ*.

FIGURE 21.3   Right radius and ulna tethered with fabric.

including periosteal reaction. Since some of the victims were in prison before they were killed, it is probable that the injury to the scapula occurred while the victim was arrested. Histological and imaging analyses would have been useful to evaluate the post-traumatic interval (PTI); however, these resources were not available. Since previous maltreatment is forbidden per article 5 of the Universal Declaration of Human Rights this information is important in establishing the circumstances surrounding the case in question.

The Skeletons Exhumed from the Local Cemetery

As previously mentioned, there was suspicion that the remaining missing individuals were inhumed in the local cemetery to hide and preclude their identification. As a result, four bodies were exhumed from the cemetery by a local team. Each individual was completely skeletonized and in an individual grave. Anthropological analyses revealed that the remains belonged to two females and two males whose biological profile, namely sex and age-at-death, indicated they were not the young males who went missing.

THE SECOND MISSION

The excavation team was alongside us during all work on the remains during the first mission. We took that opportunity to explain what should and should not be done in such excavations, providing practical examples of what information was lost because of the errors in the exhumation process. The local excavation team was dedicated and wanted to be helpful. Gradually they understood that separating the skulls from the remaining bodies and removing the clothes and putting them all together in a bag hindered our forensic analyses. Although this learning process was too late for the first mission, they were able to put to use what they learned in a second mission. This second mission occurred five months after the first one, when a new mass grave site was located. They properly exhumed six victims and the evidence remained *in situ*. Indeed, the bodies analyzed from this second mission were not only completely dressed, but they also kept the ropes *in situ*. Given that all the bodies were dressed, clothing could be used as circumstantial evidence. One individual was tied with a rope from the feet to the head (see Figures 21.4 and 21.5) indicating he was completely unable to defend himself. Damage to the clothing matched the wounds found on the bones. During this second set of analyses,

FIGURE 21.4   Example of one of the individuals of the second mission with clothes in association and a rope tied around the remains (particularly the neck region).

FIGURE 21.5   Skull of the individual of Figure 21.4. There is extensive fragmentation of the right temporal and parietal, as well as fractures of the maxillary, frontal and right zygomatic arch. The fracture pattern was consistent with BFT.

we followed the same methods to identify the individuals and determine the cause and manner of death. Once again, the six individuals represented young adult African males who were beaten and/or shot to death.

CONCLUSIONS

The history of the incident was more or less known, and authorities acknowledged that the victims were severely beaten with several instruments, including firearms, with no way to defend themselves. Deaths occurred during or soon after the physical assaults. DNA analyses resulted in the positive identification of all victims and facilitated the association of the skulls with the five associated postcranial elements. Unfortunately, we never heard back about the results of the trial. Although there was no obligation to tell us about what happened, we feel it is important to keep forensic experts aware of the legal proceedings to support and attest to the scientific validity of the forensic work and avoid any risk of rejection of evidence. Nonetheless, the forensic team's work is particularly rewarding in situations where our knowledge contributed to the return of the victims to their families, providing justice to them and to society. Without forensic anthropology, the victims' identities and the stories of their deaths would not have been unveiled.

LESSONS LEARNED

- The local excavation team members weren't the only ones to learn something from these missions. This experience illustrated that a great majority of forensic tasks are too specific to be performed by individuals without appropriate train-ing, especially given what is at risk when information is lost. Forensic recovery is an invasive process where it is impossible to undo events. During the forensic process, a single misstep may limit and compromise all subsequent phases and potentially jeopardize evidence and results in a court of law. Forensic training to agencies in countries where human rights violations have occurred is needed urgently so that when a situation arises, it is handled properly. If the forensic evidence is collected and handled appropriately throughout the recovery process, perpetrators are more likely to be convicted, and the deaths of victims won't be forgotten and left unpunished.

- Not all situations are ideal, but we must work with what we are presented. In many human rights scenarios, resources are limited. Radiographs may not be an option, DNA analyses may be limited, time and costs are restricted, and forensically educated personnel may not be available. We must use our training and expertise to employ the methods and best practices at our disposal. In this case that meant trying to associate a good portion of each skeleton using macroscopic techniques, which could be reconciled with limited DNA analyses. In an ideal situation, the forensic team would have been called in to perform the excavation themselves, complete with extensive and written documentation of remains *in situ*, provenience would have been maintained through the recovery, and then DNA analyses combined with the association of the remains would have permitted all skeletal elements to be returned to the appropriate families and a more extensive trauma analyses to be conducted per individual.

- It is important not to get frustrated when outside agencies or individuals take erroneous steps. Their intentions are usually good, and they do not realize the implications of their actions. It is our responsibility to educate these individuals and agencies to prevent mishaps and strengthen future collaborative efforts.

DISCUSSION QUESTIONS

21.1 It is not uncommon for forensic anthropologists to never hear back about case resolutions. Do you think some notification of trial proceedings should be required? How would knowledge about case outcomes impact forensic anthropological procedures?

21.2 How could the excavation missteps of the first mission affect related court trials?

21.3 In what ways do human rights cases differ from other (e.g., domestic) forensic anthropological cases?

REFERENCES

Asala, S. A. (2001). Sex determination from the head of the femur of South African whites and blacks. *Forensic Sciences International, 117*, 15–22.

Beauthier, J. -P. (2011). *Traité de Médicine Légale* (2nd ed.). Bruxelles, Belgium: De Boeck.

Hefner, J. T. (2009). Cranial nonmetric variation and estimating ancestry. *Journal of Forensic Sciences, 5*, 985–995.

Işcan, M. Y., & Steyn, M. (2013). *The human skeleton in forensic medicine* (3rd ed.). Springfield, IL: C. C. Thomas.

Lundy, J. K. (1983). Regression equations for estimating living stature from long limb bones in the South African Negro. *South African Journal of Science, 79*, 337–338.

CHAPTER 22

Search for Spanish Civil War Victims in the Cemetery of Sant Ferran, Formentera (Spain): Oral Witness Testimonies, Secondary Deposition Site, and Perimortem Trauma

Almudena García-Rubio, Juanjo Marí Casanova,
Glenda Graziani, Francisca Cardona, Pau Sureda,
Sergi Moreno, and Nicholas Márquez-Grant

CONTENTS

INTRODUCTION

The Spanish Civil War (1936–1939) resulted in a high number of civilians killed as a result of the conflict between the Fascist or Nationalist Regime of Francisco Franco and the Republican faction. Thousands of civilians were killed as a result of violence behind the frontline. The figures of those killed during the Civil War and the first ten years of the dictatorship that followed have been estimated at 49,272 from the Republican Repression and 130,179 from the Francois repression (Preston, 2012). Excluding those killed by bombings in a number of towns and cities, as well as those who died during exile or through battle, there were those summarily executed during *paseos* ("walks") where they were walked out to fields or forests, during *sacas* (mass killings of prisoners) and those killed near cemeteries (Juliá, 1999).

After the Spanish Civil War, a number of exhumations were carried out by those of the victorious side (Francoist regime). These exhumations are a result of the Government ordinances of 1939 and 1940 that stated that those who wanted to exhume the remains of their relative killed by the Marxists (Republicans) to later inter them in a cemetery can do so within six months of the publication of those ordinances and at no cost (BOE 130:3207). Between the death of Francisco Franco in 1975 and the *coup d'état* of 1981, relatives of Republican deceased undertook a number of exhumations; however without following any scientific (archaeological) protocols (see Ferrándiz, 2011; Etxeberría et al., 2016 for more information). Since the year 2000 with the exhumation of 13 Republican individuals from the mass grave at the village of Priaranza del Bierzo (León), an increasing number of mass grave exhumations in Spain were brought to the international stage in the context of searching for victims of human rights abuses. This mass grave at Priaranza del Bierzo was the first to be exhumed following archaeological methods and techniques and where the deceased were identified through DNA analysis (Prada et al., 2003; Silva, 2006; Ríos & Etxeberría, 2016). Between the years 2000 and 2018 there have been more than 600 exhumations in Spain and over 8000 individuals recovered (Aranzadi Society of Sciences). The primary aim of these investigations, after a number of failed attempts to seek justice and therefore legal action, is humanitarian in nature (Martín, Pallín, & Escudero, 2008; Moreno, 2016).

The Spanish government has created legislation to regulate these exhumations.[1] Aside from the existing national laws, there are also regional legislation in most of the autonomous communities. For example, in the Basque Country there is an agreement since 2003 between the Human Rights Authority of the Basque Government and the Aranzadi Society of Sciences, which aims to search for those missing in this region during the Spanish Civil War and the dictatorship that followed, and as of November 2017 Valencia approved a law that covers the tasks of search, exhumation, and identification of victims (BOE-A-2017-15371). Within this national trend, the government of the Balearic Islands passed in June 2016 the law regarding the search and recovery of those missing as a result of the Civil War and the Francoist regime with the unanimous support of all parliamentary groups. Since this law was approved in the Balearic Islands there have been a number of searches, exhumations, and identifications of victims of the Francoist regime in various cemeteries on the island of Mallorca totaling 70 individuals exhumed and 14 positive identifications; and in the islands of Ibiza (cemetery of Figueretas) and Formentera (cemetery of Sant Ferran). The latter is the aim of the current paper.

The case study we present here is from the island of Formentera and it provides a good example of how archaeology and anthropology have assisted in the search and identification of human remains from the Spanish Civil War and in the reconstruction of the sequence of events surrounding and following deposition. In addition, this case study aims to illustrate how physical/forensic anthropological knowledge of skeletal trauma can assist in locating individuals in a human rights context, especially when their remains are not located in their primary deposition site.

THE CEMETERY OF SANT FERRAN IN FORMENTERA

Formentera is the smallest of the Spanish Archipelago of the Balearic Islands located in the Western Mediterranean, with a land surface of 83.2 km^2 (see Figure 22.1A). While today the population is of over 12,000 inhabitants, in 1930 the population was 3,328 inhabitants (www.ine.es).

On July 18th, 1936 there was a military rebellion and all the Balearic Islands except Minorca were taken by the Nationalists. Previously, Formentera had been politically dominated by the Nationalist Confederation of Labor (CNT in its Spanish acronym) and the Salt Syndicate. After several weeks of Republican redomination, the Nationalist forces occupied Ibiza and Formentera. All over Spain, the Nationalists under Franco's leadership occupied the political power. This Fascist political party (*Falange Española*) was the one to execute the first phase of repression characterized by extrajudicial executions.

In Formentera, the memory of Francoist repression was kept alive by numerous survivors (Parrón, 2015). A total of five (mass) graves have been located through archival research and witness testimony, and this information includes the exact number of victims in each grave and their names, totaling 12 people who faced an extrajudicial execution. In addition to these, there is a communal grave in the new cemetery at Sant Francesc, also in Formentera, which includes the remains of at least 58 individuals who died in captivity at the island's prison of la Savina. The case of Formentera according to a number of authors was a very cruel and bloody repression, which had a great impact on an island with a small population and where families were interrelated (Parrón, 2015).

It is the cemetery of Sant Ferran (see Figure 22.1B) which is the focus of this chapter. The cemetery, which belongs to the church, was founded in 1903 and is still in use today, although since the opening of the new cemetery in 1940, its actual use is minimal. It is today a protected historical site (*Bien de Interés Cultural*). Here, according to witness accounts, on the March 1st, 1937, five men were buried in a mass grave after being detained by the Fascists and executed outside the cemetery walls. These five executed male individuals included: a 42-year-old sailor; a 37-year-old sailor; a 24-year-old sailor;

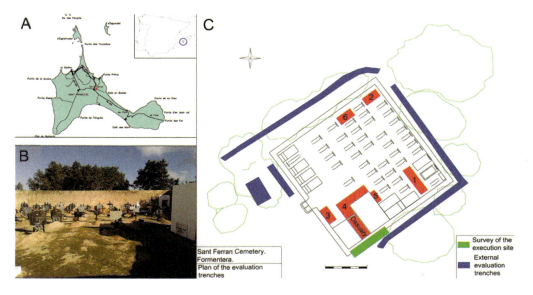

FIGURE 22.1   Overview of cemetary. (A) Map illustrating island of Formentero and cemetery location. (B) Overview of the Sant Ferran Cementery. (C) Map of the cemetery illustrating marked graves, survey areas, external evaluation trenches (blue), and internal evaluation trenches (red).

an 18-year-old (occupation unknown) and another male individual (age and occupation unknown). The archaeological and anthropological investigation aimed at locating, recovering, and identifying these individuals.

ARCHAEOLOGICAL AND ANTHROPOLOGICAL INVESTIGATION

The cemetery of Sant Ferran covers a small area of 250m² with graves and, adjacent to the southern wall, a small shed for tools, an ossuary and a columbarium which was built in the 1950s (see Figure 22.1B and C). The excavation team was comprised of seven archaeologists (three of whom are also physical/forensic anthropologists) and carried out their work between the November 29th and theDecember 3rd, 2017, at the request of the association *Forum per la Memória d'Eivissa i Formentera*, with the consent and support of relatives of the deceased and the financial support of the local council and government (*Consell Insular de Formentera* and *Govern de les Illes Balears*).

In the absence of any official cemetery register or any plans showing the location of the clandestine graves, it was the oral witness accounts that pointed to two possible locations for the mass grave. These locations were areas of the cemetery with an absence of tombstones. The archaeological strategy targeted these two areas first (see Figure 22.1C). Trial or evaluation trenches were opened and dug by hand (see Figure 22.2A).

The first trench to be dug (Trench 1) initially measured 2 × 1m but was later expanded to 3.5m in length and a 1m depth. During this excavation, three ordinary coffined single graves were found (see Figure 22.2B). These were excluded from being the missing individuals after assessing the decomposition of the coffin, the presence of any artifacts, the position of the remains, and the biological profile according to the anthropological

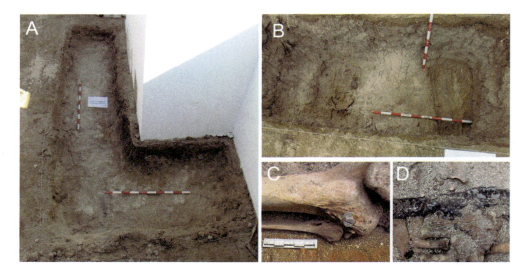

FIGURE 22.2   Initial excavations. (A) Example of evaluation trenches dug to look for burial features. (B) Overview of remains discovered within degraded coffins. (C) Example of remains containing surgical device. (D) Example of remains of a female holding rosary beads. These remains were able to be excluded as the missing individuals.

assessment at the site. The anthropological analysis of the human skeletal remains took into account the internationally recognized methods recommended by the Spanish Association of Forensic Anthropology and Odontology (AEAOF) and published in 2013 (Serrulla, 2013). One of the individuals was an adult female, with a rosary in her hands (see Figure 22.2D); another was an adult male; and the third coffin had a relatively intact coffin so it was decided not to intefere further. Moreover, the trench section showed a clear cut for these graves within the natural sediment, showing that there was no disturbance or a previous cut for a larger grave of five individuals.

Following the above, a second trench was opened (Trench 2). The location for this one was at the northeast corner of the cemetery and the dimensions of the excavated area were 2×1m and 1m in depth. A first coffin grave was identified which corresponded to a middle-aged adult male with a denture and evidence of surgery on the left tibia as a result of a fracture (see Figure 22.2C). It would appear that this type of surgery started in the main Spanish hospitals in the 1960s, so that this individual would not have died in the 1930s. Beneath this burial was another adult male individual, also in a coffin lying on the natural and undisturbed sediment. This latter individual was missing its skull and right lower limbs elements, likely due to truncation from the previous burial.

The above trenches, which were initiated from witness testimonies, were thus negative in the search for those killed during the Spanish Civil War. Therefore, four additional trenches were excavated. One was located near the entrance to the cemetery (Trench 3), two others surrounded the columbarium built in the 1950s (Trenches 4 and 5) and the other targeted another area of the cemetery which had no tombstones (Trench 6); but all of these yielded negative results.

As a result of the above, the search expanded to the outside of the cemetery taking into account the knowledge that the executions had taken place outside the southeastern wall of the cemetery. With this in mind, an evaluation trench was opened alongside the perimeter of the cemetery using a mechanical digger to remove the top layer (see Figure 22.3A). The trenches were 70–80cm wide and 40–60cm in depth and covered an area of 54.30m. They all proved negative with no undisturbed natural soil. In addition, a metal detector survey was undertaken in the area where the executions had presumably taken place. The latter yielded various 7mm caliber projectiles attributed to a Spanish Mauser rifle (Martínez Velasco 2008), a weapon used during the Spanish Civil War. One complete bullet, two fragments, and part of the full metal jacket of a fourth one were found (see Figure 22.3B and C).

Having explored the surrounding and taking into account the investigation within the cemetery walls, it was noted that Trenches 2 and 6 evidenced the reuse of graves. This led to the conclusion that perhaps the space where the five individuals had been buried may have been used for later interments and thus their remains deposited in the ossuary. In order to test this hypothesis, it was decided to assess the thousands of bone elements within the ossuary for any evidence of perimortem trauma which may relate to the events of 1937.

THE OSSUARY

The ossuary is a small rectangular structure measuring 1.30×3m (see Figure 22.4A). There were two different layers or contexts which were clearly distinct. The most superficial layer (strata 04 in Figure 22.4B) comprised mainly fragments of cemetery material such as flower wreaths, plastic, and tombstone debris. Below that was a 55cm deep

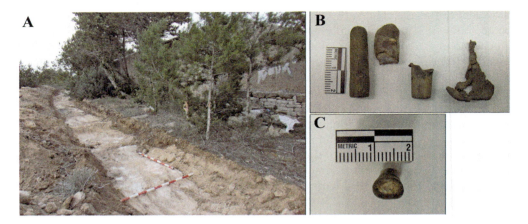

FIGURE 22.3　(A) Example of evaluation trench outside the cemetery. (B) Spanish Mauser rifle projectiles found near the cemetery wall. (C) 7mm caliber bullet.

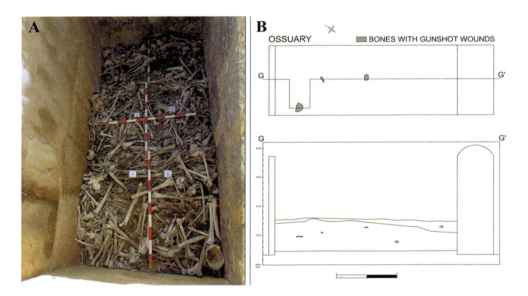

FIGURE 22.4　The ossuary. (A) Aerial view of the ossuary, with strata 05 exposed and the division into quadrants for a more systematic documentation. (B) Profile maps from the ossuary, illustrating location of bones with gunshot wounds.

layer which included human remains and sediment (strata 05 in Figure 22.4B). In order to approach this systematically and have some spatial reference, the space was divided into quadrants. Each bone was recovered by hand and assessed for perimortem trauma, particularly related to gunshot or ballistic trauma, and this process continued until all bones were recovered from the ossuary. All the bones were disarticulated, commingled, and derived from individuals once interred in their primary graves but whose remains were later exhumed and deposited here to create space for new burials. Many of the bones were incomplete, fragmented, and some had significant weathering and other taphonomic alterations. Nevertheless, each bone was assessed by anthropologists on site and an attempt was made to identify those with perimortem trauma. Three incomplete bones were found to evidence perimortem trauma. These were located in the northern half of the ossuary and are described below.

The first was an incomplete cranium with occipital, right parietal, and right temporal. This fragment was found in the western corner in quadrant 1 and at a depth of 137cm. The assesment of the mastoid process and the posterior zygomatic arch were consistent with a male morphology. On the temporal squama there was a perforating circular 7mm defect with internal beveling, consistent with an entry wound as a result of high velocity projectile or ballistic trauma (see Figure 22.5A and B).

A second cranial fragment with part of the right parietal and frontal also displayed signs of ballistric trauma. This was found at the intersection of the quadrants at a depth of 122cm. At the edge of the frontal fragment is the presence of a defect penetrating the outer and inner cranial vault which is semicircular. This defect has a diameter of 7mm and internal beveling. Once again, these traits are characteristic of a entry wound resulting from high velocity projectile or ballistic trauma (see Figure 22.5C and D). Externally there is a clear radiating wet bone or perimortem fracture likely associated to this entry wound.

The third element displaying ballistric trauma was a proximal third of an adult left humerus. This incomplete bone was found in the eastern corner of quadrant 1. The vertical and transverse diameters of the head (40.6mm and 42.1mm, respectively)

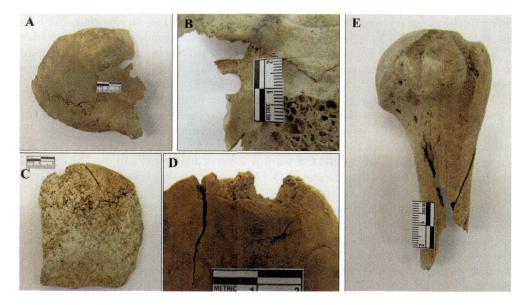

FIGURE 22.5 Ossuary bones displaying perimortem trauma. (A) An incomplete cranium with a circular 7mm defect on the temporal squama consistent with an entry wound as a result of high velocity projectile or ballistic trauma. (B) Internal beveling in the same bone. (C) Cranial fragment with a semicircular defect in the frontal consistent with an entry wound as a result of high velocity projectile or ballistic trauma. (D) Internal beveling on the frontal bone. (E) An adult left proximal humeral shaft with a small bevel from which there are radiating fractures, consistent with ballistic trauma.

are consistent with male dimensions following Spanish discriminant functions for sex estimation (Alemán et al., 1997). In the diaphysis there was a small bevel from which there are radiating fractures. These radiating fractures appear as a double buttefly shape steming from an initial point of contact as a result of gunshot or ballistic trauma (see Figure 22.5E).

The characteristic of these lesions, regarding both their typology and location are similar to those extensively documented for executed victims during the Spanish Civil War (Ríos et al., 2014; Congram et al., 2014). In summary, the investigation of the ossuary and the anthropological assessment of the human remains on site, identified three bones with perimortem trauma compatible with high velocity projectile or ballistic trauma. These bones belong to at least two adult individuals, at least one of whom appeared to be male. The diameter of those defects, especially the cranial ones, would also be compatible with the projectiles found by the outer wall of the cemetery. With this in mind, it could be hypothesized that those bones belonged to at least two of the individuals being searched who were executed and buried during the Spanish Civil War.

IDENTIFYING OF THE DECEASED AND RECONSTRUCTION OF THE EVENTS

To test the hypothesis that the bones with perimortem trauma were the remains of at least two individuals from the group of five men executed in 1937, two lines of enquiry were followed. The Civil Registers were checked to explore whether or not any of the people buried in the cemetery had died as a result of gunshot trauma. In addition, the bones were submitted for DNA analysis alongside DNA reference samples from living relatives.

The Civil Register in Formentera includes every death (excluding those executed during the Spanish Civil War) with the name of individual, date, burial place, and cause of death.[2] Archival research of this Register from 1903 onwards when the cemetery of San Ferran had its first burial did not find any cause of death associated with gunshot wounds or ballistic trauma. The only case that was to be considered further was a male who died due to a 'brain injury' (*traumatismo cerebral* according to the Register) but further research found that this was caused by a fall from a cliff. With regard to the DNA samples, selected relatives from the five victims submited reference samples; however, the bone fragments did not yield sufficient mtDNA (there was no nuclear DNA) and therefore the genetic testing was unsuccessful.

How then to test the hypothesis that the three bone fragments correspond to at least two of the five individuals? Firstly, the information obtained through oral testimony clearly indicated that the five men were detained, executed outside the cemetery, and buried in the cemetery of Sant Ferran in Formentera. Secondly, the cemetery ground was investigated as well as the outside perimeter with archaeological methods and the absence of a mass grave and the presence of individuals with perimortem trauma compatible with high velocity projectiles were found in the ossuary. Moreover, these fragments with trauma belonged to at least two individuals, at least one of them male, with defects compatible with bullets from a Mauser rifle such as those found with the metal detector outside the cemetery wall. We believe that if this information is taken into account along with the fact that no person was buried in the cemetery who had died from a gunshot wound, this provides strong evidence that those bones belong to the Spanish Civil War victims that the team was searching for; however, there appears to be no means to obtain a postive identification.

LESSONS LEARNED

Since 2000, the number of Spanish Civil War exhumations by professional archaeologists, anthropologists, forensic pathologists, and other specialists have increased in order to provide a dignified burial and closure for relatives. This case of Sant Ferran is one of such interventions to search, attempt to locate, recover, and identify those who were executed and buried in clandestine graves during the Spanish Civil War and the dictatorship that followed. The case presented here to locate and identify five men who died in 1937, provides scientists with a number of lessons that may help in future investigations.

- It cannot be stressed enough the importance of having a profesional team of historians to undertake the archival research, a number of archaeologists to systematically search the cemetery and eliminate areas with confidence, the use of different methods such as metal detector to identify projectiles, and in particular a strong team of anthropologists to eliminate areas where individuals have been found buried there. In particular for this case, it was the anthropological assessment of human remains with perimortem trauma that yielded results.
- Although faced with numerous limitations, including the volume of commingled bones and bone fragments in the ossuary, the alteration of bones due to various taphonomic factors, and the lack of DNA results, by reconstructing the remains and eliminating others; some answers were provided. Areas believed by witnesses to contain the remains were excluded and a potential scenario for the disposition of the remains was constructed. Although definitive answers and positive identifications could not be obtained, the relatives of the deceased have more information and potential evidence that the individuals were there and subsequently were placed in the ossuary, and these answers may provide at least a little closure to the families.
- There are limits to forensic science, and much of it depends on the preservation of context and associated evidence. Not all cases can be solved with 100% confidence.

DISCUSSION QUESTIONS

22.1 Do you believe that the authors have enough information to determine whether the remains belong to the missing individuals from the Civil War? How confident are you and how would you present the results in a case report?

22.2 Are there any additional methods that the team could carry out to further prove that the remains belong to those individuals?

22.3 The remains with perimortem gunshot trauma represent a minimum of two (and a maximum of three) individuals. Where then are the remains of the individuals who were executed that same day? Are there any other methods you would employ to search for additional remains?

ACKNOWLEDGMENTS

We would like to express our gratitude to Dr Escribano (Madrid) and Dr Galtés (Barcelona), for providing information on the type of surgery and its chronology associated with the surgical device discovered during excavations.

NOTES

1. BOE 2007: Ley 52/2007 of December 26th, regarding a number of measures to recognise those who died violently during the Civil War and the dicatorship that followed; BOE 2011: protocol for the exhumation of victims from the Spanish Civil War and the dictatorship; BOGC2017: reform of the previous law 55/2007 of 26th December 2007.
2. From 1994 onwards there is no specific information on the cause of death (BOE 13489 ORDEN of June 6th, 1994) but the Institute of Legal Medicine who has Formentera under its jurisdiction have indicated that there have been no deaths as a result of gunshot wounds or ballistic trauma (Juan Ramón Sancho-Jaráiz, personal comment).

REFERENCES

Alemán, I., Botella, M., & Ruiz, L. (1997). Determinación del sexo en el esqueleto postcranial. Estudio de una población mediterránea actual. *Archivo Español de Morfología, 2,* 69–79.

Aranzadi Society of Sciences. (n.d.). Retrieved October 1, 2018, from http://www.aranzadi.eus/?lang=en

Congram, D., Passalacqua, N., & Ríos, L. (2014). Intersite analysis of victims of extra- and judicial execution in Civil War Spain: Location and direction of perimortem gunshot trauma. *Annals of Anthropological Practice, 38*(1), 81–88.

Etxeberría, F., García-Rubio, A., Herrasti, L., Jiménez, J., & Márquez-Grant, N. (2016). Mass graves from the Spanish Civil War: Exhumations, current status and protocols. *Archaeological Review from Cambridge, 31*(1), 83–103.

Ferrándiz, F. (2011). Guerras sin fin: guía para descifrar el Valle de los Caídos en la España contemporánea. *Política y Sociedad, 48*(3), 481–500.

Juliá, S. (1999). *Víctimas de la Guerra Civil.* Madrid, Spain: Temas de Hoy.

Martín Pallín, J. A., & Escudero, R. (2008). *Derecho y Memoria Histórica.* Madrid, Spain: Trotta.

Martínez Velasco, A. (2008). Breve introducción a la cartuchería para arqueólogos. *Sautuola, XIV,* 383–398.

Moreno Gómez, F. (2016). *Los Desaparecidos de Franco. Un Estudio Factual y Teórico en el Contexto de los Crímenes Internacionales y las Comisiones de la Verdad.* Madrid, Spain: Alpuerto.

Parrón i Guasch, A. (2015). La represión franquista en Ibiza y Formentera. *Memòria antifranquista del Baix Llobregat.* 34–38.

Prada, E., Etxeberría, F., Herrasti, L., Vidal, J., Macias, S. & Pastor, F. (2003). Antropología del pasado reciente: una fosa común de la Guerra Civil Española en Priaranza del Bierzo (León). In M. P. Aluja, A. Malgosa, & R. M. Nogues (Eds.), *Antropología y Biodiversidad Volumen I* (431–446). Barcelona, Spain: Ediciones Bellaterra.

Preston, P. (2012). *El Holocausto Español.* Debate. Barcelona.

Ríos, L., & Etxeberría, F. (2016). The Spanish Civil War forensic labyrinth. In O. Ferrán & L. Hilbink (Eds.), *Legacies of violence in contemporary Spain: Exhuming the past, understanding the present* (pp. 174–198). New York: Routledge Studies in Modern European History.

Ríos, L., García-Rubio, A., Martínez, B., Herrasti, L., & Etxeberria, F. (2014). Patterns of perimortem trauma in skeletons recovered from mass graves from the Spanish Civil War (1936–1939). In C. Knüsel & M. J. Smith (Eds.), *The Routledge handbook of the bioarchaeology of human conflict* (pp. 621–640). London, UK: Routledge.

Serrulla, F. (Ed.). (2013). *Recomendaciones en Antropología Forense.* Asociación Española de Antropología y Odontología Forense.

Silva, E. (2006). *Las fosas de Franco.* Madrid, Spain: Temas de Hoy.

CHAPTER 23

Excavation and Analysis of Human Remains from Mass Graves in the Western Sahara

*Francisco Etxeberria, Lourdes Herrasti,
and Carlos Martin-Beristain*

CONTENTS

BACKGROUND

In 2013, the Asociación de Familiares de Presos y Desaparecidos Saharauis (Unknown, 1970), under the guidance of the University Institute Hegoa, contacted Basque University for assistance in the exhumation and analysis of two mass graves containing human remains. The graves were previously discovered by relatives of the victims in the Sahara Occidental (Western Sahara). Their deaths were related to conflicts which ensued during the decolonization of the Sahara by Spain and the invasion of the Western Sahara by Morocco. A multidisciplinary team was formed by specialists in forensic archaeology, forensic anthropology, forensic pathology, and forensic genetics from Basque University and Aranzadi-Society of Sciences. The formation of such a diverse and experienced team of specialists was of great importance given the multiple objectives of the investigation: victim identification, determination of cause of death, and documentation of the medical-legal significance of the circumstances around their disappearance, death, and interment with consideration of the judicial process of crimes against humanity in Spain.

EXCAVATION AND ANALYSIS

In accordance with international protocols used in matters of forced disappearances,[1] the following procedures were carried out: protection of the site, observation and analysis *in situ*, collection of material evidence, and transfer of material evidence and DNA samples to the appropriate laboratories using chain of custody for supportive analyses; all of which was continuously documented via photographs and video. The excavation was performed carefully, preserving as much evidence as possible for anthropological and pathological analyses and DNA sampling. At the same time, and essential in any forensic investigation of forced disappearances, oral testimonies from family members were recorded on video. DNA from living relatives was also collected for comparison.

We began with the inspection of the two mass graves, designated Grave (Fosa) 1 and Grave (Fosa) 2, in Fadret Leguiaa, Sahara Occidental. The sand was extracted with a shovel until skeletal remains were reached, whereupon wood instruments and brushes were used to carefully excavate the remains and avoid damages to the bones. The remains were exposed, but kept *in situ*. Both mass graves and the surrounding areas were checked with a metal detector, resulting in the discovery of ten cartridge cases from firearms issued in the years 1963, 1964, and 1975.

After the burials were exposed, the cases were documented in a field report with the following data observed *in situ*: the position of the individual, the sex, the estimated age, any individualizing skeletal characteristics, any pathological indicators (e.g., signs of trauma), and associated material evidence. Each grave was documented with photography, video, and hand drawings (Mays et al., 2018; Campillo & Subira, 2004). Methods outlined in Buikstra and Ubelaker (1994), Ubelaker (2007), and Cardoso (2008a, 2008b) were used to estimate age and sex. The reference tables proposed by Trotter and Glesser in Ubelaker (2007) and Mendonça (2000) were used to estimate stature. Pathology was assessed according to features and characteristics outlined in Etxeberria and Carnicero (1998), Isidro and Malgosa (2003), Kimmerle and Baraybar (2011), Berryman and Symes (1998), DiMaio (1985), and Galloway (1999).

GRAVE (FOSA) 1

The mass grave was located on a gentle slope in a sandy ravine of low uniform depth, consisting of very fine sand (see Figure 23.1A). The maximum length of the grave feature measured 8.90m and the width varied between 80cm and 120cm. Human remains, however, were dispersed, with the furthest elements discovered more than 30m from the grave in the area of greatest decline of the ground surface. This dispersion is likely a consequence of their shallow inhumation using scarce coverage of sand, with a natural displacement of the remains over time, as wind, gravity, and occasional rain carried the bones toward the lower end and western side of the burials. In the center and bottom of Grave 1 only a few elements were found in their original anatomical position: the left fibula and calcaneus, and the left upper extremity. Most of the bones displayed some taphonomic changes, including sun bleaching and cortical bone degradation due to environmental factors (e.g., fluctuations in temperature and humidity). All skeletal analyses were conducted at the site because of legal and logistical issues in transporting the remains out of the country for analyses and then back to this remote area to family members.

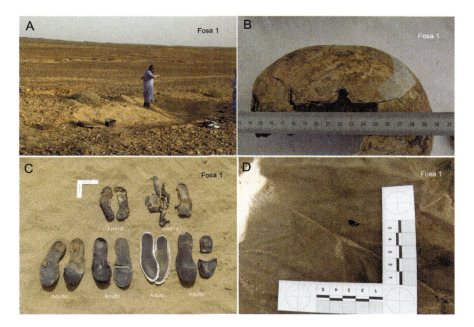

FIGURE 23.1   Grave (Fosa) 1. (A) Photograph of the ravine where Grave 1 was discovered and an investigator using a metal detector to look for evidence. (B) Cranium recovered from Grave 1 with a circular lesion consistent with a gunshot wound entry to the left side of the cranium. (C) The soles of six pairs of shoes found in the grave. On top, the soles of smaller size are those of juveniles, with the larger soles below representing adult individuals. (D) Article of clothing with a circular perforation produced by a firearm projectile.

The inventory of the remains confirmed the presence of six individuals: four adult males and two subadults, probably male, between 15 and 20 years of age based on bone and teeth development. Four of the crania exhibited traumatic lesions with related fractures. Two of them showed clear evidence of gunshot wounds, with characteristic circular defects, radiating and concentric fracture lines, and beveling (Kimmerle & Baraybar, 2011) (see Figure 23.1B). Clothing was also recovered from the scene, including the soles of six pairs of shoes, two of which were smaller and presumably associated with the subadult remains (see Figure 23.1C). Some clothes also exhibited defects consistent with an impact from a firearm projectile (see Figure 23.1D). In addition, 7.62 caliber cartridge cases were recovered from the immediate proximity of the grave, with manufacturing dates marked prior to 1976 (see Figure 23.2).

Teeth were extracted and sent off for DNA analyses; otherwise the remains did not leave the scene. Ideally, the remains would have been reinterred in the grave *in situ* following all analyses, but given the dispersed and exposed nature of the remains upon discovery, all of the dispersed remains were interred in a single location near the grave. The bones were buried at a depth of 1m at a well-documented location, and a grave marker was placed to demarcate the location and identify the victims. This ensured that the grave could be revisited again if additional data were needed and also provided the families a location to visit and memorialize the victims. The personal artifacts recovered from the grave were handed over to the Saharauis authorities, along with the chain of custody.

FIGURE 23.2   Cartridge cases of a 7.62 caliber rifle with markings from the years 1963 and 1964 recovered from the scene of Grave 1.

GRAVE (FOSA) 2

Grave 2 was located 30m from Grave 1. The site was a small elongated depression with a sandy floor. A small median elevation with overlying loose stones demarcated the burial (see Figure 23.3A). Two sun-bleached femora were observed exposed on the surface of the grave. The mass grave measured 2.20m long by 0.85m wide. As before, the sand was carefully excavated to expose the skeletal remains, revealing two complete skeletons in anatomical position wearing traditional clothes of the Bedouins (a nomadic population of the Sahara) and accompanied by numerous personal effects (see Figure 23.3B and C). All evidence was documented with photographs and in written form.

Anthropological analyses indicated that Individual 1 was an adult male of mature age (>40 years) at the time of death. Along with the associated personal artifacts and clothing, a National Identity Document was found with the decedent's name. The document was enveloped in white paper and plastic. The left pelvis displayed missing bone and a fracture pattern consistent with a gunshot injury to the left ilium (see Figure 23.3D). The beveling on the posterior side of the defect was consistent with an exit wound, indicating a gunshot wound to the abdomen, with an anterior-to-posterior descending trajectory and exit on the posterior side of the left ilium. The pathologist suspected the lesion caused rapid and massive bleeding, resulting in death by hypovolemic shock.

The remains of Individual 2 corresponded to a young adult male. Again, a National Identity Document with the presumed decedent's name was found in a plastic wallet among the personal artifacts and clothes. This individual displayed a gunshot injury to the head with the entrance defect on the left side of the frontal-temporal area and the externally beveled exit defect in the mastoid region of the right temporal bone (see Figure 23.4). The trajectory was from anterior to posterior, from left to right, and from superior to inferior. Further, the mandible displayed a complete fracture on the left side of the mental eminence consistent with direct trauma to that area.

When all documentation and analyses were complete, the perimeter of the burial was delimited with stones and covered with sand to secure the preservation of the site.

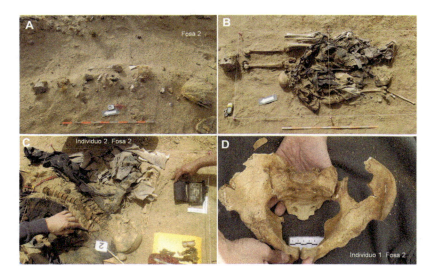

FIGURE 23.3   Grave (Fosa) 2. (A) Grave 2 before the excavation. A few sun-bleached skeletal elements can be observed among the stones on the surface. (B) The grave with the remains of two individuals *in situ*. Note that the stone to the left of the picture was kept in place as a marker, and the sun-bleached femur observed prior to excavation is evident at the top of the photograph. (C) Close-up of Individual 2 illustrating anatomical position of remains and the documentation and collection of associated personal artifacts. (D) Defect and radiating fractures consistent with a gunshot wound to the left ilium.

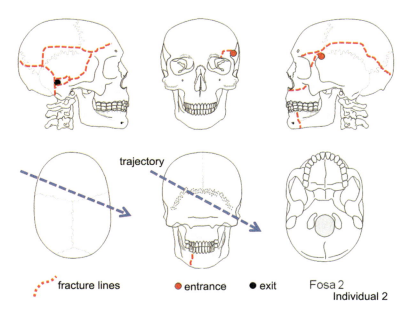

FIGURE 23.4   Schematic representation of the defects caused by the gunshot wound to the cranium of Individual 2 (Grave 2).

DNA ANALYSES

The 37-year-old grave sites were located in a desolate area of the Sahara Desert that presented logistical challenges and limited time and resources for recovery. Nonetheless, all the decedents were identified following certified international standards. First, the personal documentation discovered with the human remains in both mass graves, combined with the historical facts of 1976 and corroborating witness testimonies, enabled a quick narrowing down of the identities of the victims. Second, the relatives present recognized several personal objects of the victims during the excavation. In circumstances with a closed group of decedents, circumstantial evidence holds more weight in victim identification. For the purpose of a genetic comparative study with the ultimate goal of identification, oral swab samples (unambiguous samples) of the relatives were taken upon request. Additionally, a permission form was completed providing information about the familial relationships between the donors and victims. The genetic samples from the relatives (unambiguous) and those from the bones and teeth sampled from the skeletons (ambiguous) were analyzed in the laboratory of Forensic Genetics of the Basque University, which confirmed the presumed identification of all victims.

CONCLUSIONS

The following conclusions were drawn from the excavations and analyses:

1. Two mass graves (Grave 1 and Grave 2) were discovered and investigated in Fadret Leguiaa, in the region of Samra (Sahara Occidental), with a total of eight male victims, six adults and two juveniles.
2. Both the circumstantial evidence (testimonies and physical evidence found on site) and the genetical analysis confirmed the identities of all victims.
3. Cartridge cases of 7.62 caliber firearms were discovered in the immediate proximity of both mass graves.
4. Characteristic lesions, including radiating and concentric fracture lines and beveling, were consistent with gunshot injuries observed on the preserved human remains and associated clothing.
5. The medical-legal etiology of death was due to the injuries incurred during violent homicide, with a direct medical cause of death of hypovolemic shock and/or a traumatic destruction of vital organs.
6. In accordance with the circumstantial evidence and the analyzed evidence, the previously collected witness testimonies confirm the violent death of these people in February 1976, associated with the decolonization of the Sahara by Spain and the invasion of the Western DelSahara by Morocco.

In spite of the remote desert location with no infrastructure, successful completion of this project was possible thanks to an experienced multidisciplinary team capable of adapting to the situation at hand.

LESSONS LEARNED

- The recovery of human skeletal remains from potential human rights violations commonly takes place in remote settings, where interment into mass graves is possible without raising suspicion. This presents challenges to the excavators given the logistics of transporting personnel and equipment to the site, health and safety concerns during the excavation process, and housing accommodations for the team. Teams must be flexible and make the most of what is available to them.

- Although in domestic forensic cases it is common to document the scene, excavate the remains and then transport them to a laboratory for further analysis, this is not always a possibility or necessarily the best practice in certain human rights situations. There are logistics and legalities to consider when transporting remains across country boundaries, and anthropologists need to be cognizant of the desires of the victims' families and their cultural and religious beliefs. The pros and cons must be weighed in each scenario.

DISCUSSION QUESTIONS

23.1 The violent deaths of the individuals described in this case study occurred in 1976, 37 years prior to their discovery and excavation. What effects did time have on the discovery, analysis, and interpretation of the remains? How can time affect the prosecution of such crimes?

23.2 What do the bullet trajectories in this case tell you about the circumstances surrounding the death of these individuals?

23.3 How does working with human remains in a human rights situation such as this one differ from working with remains in a domestic forensic case? How do procedures vary? Are there different objectives? Do certain types of evidence hold more weight?

NOTE

1. For example: (1) Principles on the Effective Prevention and Investigation of Extra-legal, Arbitrary and Summary Executions Recommended by Economic and Social Council resolution 1989/65 of 24 May 1989 (Union Nations). (2) Annex 1989/65. Effective prevention and investigation of extra-legal, arbitrary and summary executions (Resolution. 1989/65 of 24 May 1989). (3) Manual on the Effective Prevention and Investigation of Extra-Legal, Arbitrary and Summary Executions of Union Nations (New York, 1991). (4) Declaration on the Protection of All Persons from Enforced Disappearance of Union Nations Resolution 47/133 of 18 of December of 1992.

REFERENCES

Unknown. (1970, January 01). AFAPREDESA. Asociación de Familiares de Presos y Desaparecidos Saharauis. Retrieved October 1, 2018, from http://afapredesa. blogspot.com.es/

Aranzadi- Society of Sciences. http://www.aranzadi.eus/?lang=en

Berryman, H. E., & Symes, S. A. (1998). Recognizing gunshot and blunt cranial trauma through fracture interpretation. In K. J. Reichs (Ed.) *Forensic osteology. Advances in identification of human remains* (pp. 333–352). Springfield, IL: CC Thomas Publisher.

Buikstra, J., & Ubelaker, D. (1994). *Standards for data collection from human skeletal remains*. Arkansas Archeological Survey.

Campillo, D., & Subira, M. E. (2004). *Antropología física para arqueólogos*. Edit. Ariel.

Cardoso, H. F. V. (2008a). Age estimation of adolescent and young adult male and female skeletons II, epiphyseal union at the upper limb and scapular girdle in a modern Portuguese skeletal sample. *American Journal of Physical Anthropology, 137*, 97–105.

Cardoso, H. F. V. (2008b). Epiphyseal union at the innominate and lower limb in a modern Portuguese skeletal sample, and age estimation in adolescent and young adult male and female skeletons. *American Journal of Physical Anthropology, 135*, 161–170.

DiMaio, V. J. M. (1985). *Gunshot wounds: Practical aspects of firearms, ballistics, and forensic taphonomy*. New York: Elsevier.

Etxeberria, F., & Carnicero, M. A. (1998). Estudio macroscópico de las fracturas del perimortem en Antropología Forense. Study macroscopic of the fractures made in the perimortem of Forensic Anthropology. *Revista Española de Medicina Legal, 84–85*, 36–44.

Galloway, A. (1999). *Broken bones. Anthropological analysis of blunt force trauma*. Springfield, IL: CC Thomas Publishers.

Isidro, A., & Malagosa, A. (2003). *Paleopatología. La enfermedad no escrita*. Edit. Masson.

Kimmerle, E. H. & Baraybar, J. P. (2011). *Traumatismos óseos. Lesiones ocasionadas por violaciones a los Derechos Humanos y conflictos armados*. Lima, Peru: EPAF.

Mays, S., Brickley, M., Dodwell, N., & Sidell, J. (2018). *Historic England 2018. The role of the human osteologist in an archaeological fieldwork project*. Swindon, Historic England.

Mendonça, C. (2000). Estimation of height from the length of long bones in a Portuguese adult population. *American Journal of Physical Anthropology, 112*, 39–48.

Ubelaker, D.H. (2007). Enterramientos humanos. Excavación, análisis, interpretación. *Munibe*. Suplemento, (24), 22–163.

SECTION **V**

Other Considerations

CHAPTER **24**

The Use of Human Skeletal Remains in Palo Rituals in Orange County, Florida

John J. Schultz, Ashley Green, Ronald A. Murdock II, Marie H. Hansen, Joshua D. Stephany, and Jan C. Garavaglia

CONTENTS

INTRODUCTION

The ritual use of human and nonhuman skeletal remains may be misinterpreted when skeletal remains are discovered out of context. Even when found in context, establishing medicolegal significance during forensic investigations can be problematic. Correctly interpreting and determining the forensic significance of ritualistic human remains requires an understanding of the religions utilizing skeletal remains, their beliefs, and their practices. This chapter focuses on the use of human bones in *Palo* rituals through multiple forensic anthropology case studies from Orange County, Florida.

Palo is an Afro-Caribbean syncretic religion with four distinct branches (*Palo Mayombe, Palo Monte, Palo Kimbisa, Palo Briyumba*), each originating from African tribes in different regions (Wetli and Martinez, 1983; Canizares, 1999; Dodson, 2008; Murrell, 2010). Palo is often confused with Santería, which is also an Afro-Caribbean syncretic religion, but does not typically involve human remains in its rituals (Wetli & Martinez, 1981; Wetli & Martinez, 1983). Both religions have African origins and some elements of Christianity adopted during times of slavery. They share some similarities in beliefs and practices and have borrowed concepts of symbolism from one another over time – making it easy for those unfamiliar to confuse the two religions (Kail, 2008; Murrell, 2010). Palo specifically involves the use of human remains, so this will be the primary focus of the rest of the chapter.

Ochoa (2004) asserts that foremost, Palo is the "practice of apprehending and working with the dead" (p.251), as Palo practitioners enter into a familial relationship with the dead. It originates from the African Bantu belief that spirits of the deceased (*nfumbe*) can be called upon to interfere with the lives of the living, or to perform acts for the practitioner (Kail, 2008). Palo incorporates the basic belief in a supreme being that is commonly referred to as *Nsambi* (Wetli & Martinez, 1983; Gonzales-Whippler, 1989; Canizares, 2002; Kail, 2008; Murrell, 2010). Palo priests, or paleros, rely on the power of the *nganga*, an assemblage of sacred objects typically contained in an iron cauldron, to aid in their workings (Martinez & Wetli, 1982; Gonzalez-Whippler, 1989; De La Torre, 2014; Kail, 2008; Murrell, 2010; Ochoa, 2010).

The palero harnesses the nfumbe and places them inside of the nganga, along with objects endowed with the spiritual power of Nsambi to draw power. These objects are referred to as *nkisis* and their associated spirits/deities as *mpungo* (Perlmutter, 2005; Dodson, 2008; Kail, 2008). Each mpungo is associated with certain material objects, and colors (see Figure 24.1 for colors that may represent a mpungo), and is summoned from the nganga by the palero to participate during a ritual (Murrell, 2010). Examples of objects found within the nganga may include human skeletal remains, cemetery dirt, dirt from crossroads, wooden branches, spices, animal carcasses, shells, herbs, grasses, stones, items made of iron such as railroad spikes and bladed weapons, mercury, bamboo canes, stones, and blood (Wetli & Martinez, 1981; Wetli & Martinez, 1983; Canizares, 2002; Dodson, 2008; Kail, 2008; Murrell, 2010). Each of the objects placed in the nganga is added for specific reasons and to direct the intentions of the nganga. For example, machetes symbolize the memory of the history of violence endured and give the nganga a weapon to protect itself and its owner (De Mattos Frisvold, 2010). Seashells are added to bring about money and wealth, while the feathers of specific birds provide spiritual wings (De Mattos Frisvold, 2010).

The nganga must reside in a sacred space, typically outside of the house in a private area such as a shed, garage, or small building (Kail, 2008). The sacred space and remains within are often painted with sygills, or cosmograms, known as the *firma*, in colors of white, black, and red (see Figure 24.1) (Canizares, 2002; Kail, 2008; Gill et al., 2009). The firma represent the language of Palo and act to communicate specific important religious aspects such as spells, deities, or prayers, the most common elements of which are the cross, circle, arrow, "X," and skull (Canizares, 2002; Gill et al., 2009; Kail, 2008). The sacred space for the nganga is often constructed with elements of the forest such as greenery and preserved animals remains (Dodson, 2008).

The nganga is of interest to the forensic anthropologist, as this is the vessel in which the palero traps and controls the nfumbe through the use of human skeletal remains. Human skeletal remains are commonly procured from a cemetery, but may also be obtained from a botanica (folk medicine retail store) or the internet (Martinez & Wetli, 1982; Gonzalez-Whippler, 1989; Kail, 2008; Winburn et al., 2017). When describing the nganga construction, Canizares (2002) suggests that the skull, fingers, toes, tibia, and ribs are the specific human bones that should be procured. It is common to recover other human long bones instead of a tibia associated with the nganga, such as the femur. The skull is normally placed at the center of the nganga and the

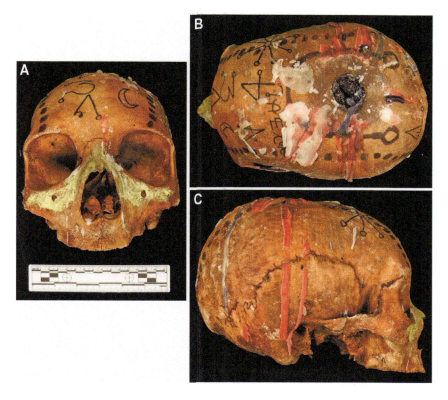

FIGURE 24.1   A cemetery cranium consistent with a grave robbery that has been tapho-nomically modified for Palo ritual use. Note the firmas on the superior surface of the cranial vault (A–B) and the red, white, black, and yellow coloring (A–C). Also note the depressed, circular burnt area on the superior surface for mounting a candle with adhered red and white colored wax (B).

tibia (or other long bone) is located either within or near the nganga (Kail, 2008). The uses of the nganga are numerous, and the palero invokes the nfumbe with an animal horn or with his tibia scepter (*kisingue*) (Wetli & Martinez, 1983; Gonzalez-Whippler, 1989). The palero sprays the nganga with rum as a type of salute and uses gunpowder to coerce the spirit in the nganga to do his bidding, whether of benign or ill intentions (Gonzalez-Whippler, 1989). Animal remains may be found within and surrounding the nganga, as animal sacrifice is of primary importance to the Palo religion. It is through sacrifice and presentation of blood and animal carcasses that a palero "feeds" the nganga (Gonzalez-Whippler, 1989; Wetli & Martinez, 1983; Dodson, 2008; Kail, 2008).

CASE EXAMPLES

Over a 12-year period, six cases of Palo were documented in Orange County, Florida, where at least one cauldron was found in association with either a human cranium or skull. Each case is presented briefly (see Table 24.1), focusing on the scene and origin

TABLE 24.1 Summary of the Six Cases Presented

Case	Active or Discarded Nganga	Location	Skull/Cranium #	Postcranial Bones	Bone Origin	Documentation of Origin
1	Active	Backyard Shed	Four skulls	Os coxa and 20 ribs	Teaching	Price tags on 2 skulls that are affixed with string to zygomatic arches, string affixed to a zygomatic arch of third skull; receipt for three skulls; taphonomy consistent with being cleaned and lightened for teaching
2	Active	Warehouse	Two crania	Humerus, 3 tibiae, 2 articulated hands, and three ribs	Teaching	Metal hardware articulating hands; price tag affixed to 1 cranium on zygomatic arch with string and only string present on other cranium; receipt for human bones documented by law enforcement; taphonomy consistent with being cleaned for teaching
3	Discarded	River	One skull	None	Most likely cemetery	Uniform dark brown soil staining, bone erosion, and fragmentary areas
4	Discarded	Wooded area behind parking lot	One partial cranium	Distal femur fragment	Cemetery	Uniform dark brown soil staining, extremely light bones, and eroded and fragmented bone quality
5	Discarded	Buried in backyard	One fragmentary skull	Partial tibia	Cemetery	Uniform dark brown soil staining, localized iron staining, extremely light bones, and eroded and fragmented bone quality
6	Discarded	Backyard shed; possible original sacred space	One cranium	Femur shaft fragment; right and left ilium fragments	Teaching	String affixed to left zygomatic arch without price tag; mother of palero confirmed material purchased over internet; taphonomy consistent with being cleaned and bleached for teaching

of the human skeletal remains to determine forensic significance. Criteria used to determine the origin of skeletal material was based on biological information (e.g., ancestry) and the unique suite of taphonomic modifications (e.g., Dupras & Schultz, 2014; Paolello & Klales, 2014; Pokines & Baker, 2014; Schultz, 2012; Schultz et al., 2003). For example, anatomical/teaching material was categorized based on the presence of a homogenous lightened and bleached coloration (white hue) from cleaning, drill holes or hardware attached in predictable locations to articulate bones, and glued single-rooted teeth. Modifications due to frequent handling during teaching, including patina, a dirty appearance, writing, broken teeth, and postmortem fractures, also suggested teaching specimens. Cemetery material may be derived from either above ground or below ground mortuary contexts, and may be from recent or historic archaeological contexts historic archaeological contexts. Cemetery remains have taphonomic modifications associated with coffins (e.g., coffin wear; preservation of coffin fabric adhered to the bones; adhered desiccated tissue; adhered scalp and facial hair; embalming artifacts) or extended burial (e.g., uniform dark brown soil staining; eroded and fragmented bones; root etching; localized copper and iron staining).

Case 1

Law enforcement responded to a residence in a suburban neighborhood in reference to gunshots being fired. They searched the residence for victims and found none. While searching the backyard, they observed fired cartridge casings on the ground near an air-conditioned shed. Upon further inspection of the shed, blood spatter was observed on the bottom of the door and on the ground outside the shed. Music was playing and the shed contained candles, four large pots against the back wall (the second from the left contained a skull), a large wooden statue, and a large horn hanging on the wall (see Figure 24.2). The foreground included three skulls in metal tubs, a number of plates, figurines, and urns with animal parts (see Figure 24.2). Animal blood spatter was on the walls and floor. A metal shed with caged roosters, four chained up large dogs, caged birds, and a small goat were also discovered. Two fire pits were present; one contained a burnt rooster, the other contained burnt nonhuman bones. A smaller shrine containing no human remains was discovered near the door of the residence. A receipt the purchase of three human Chinese skulls and additional bones from a natural history store in California that sells human and animal skeletal remains was also recovered.

Four human skulls were removed from the shed. Two skulls collected from the large metal tubs had a merchant price tag attached to the left zygomatic arch; a third had a piece of string tied to the left zygomatic arch (likely a remnant from the price tag). Numerous ritual taphonomic modifications were noted, including fresh and dried blood. The skull recovered from the nganga clay flower pot (see Figure 24.2A) had a gold metallic crown affixed to the superior aspect of the cranial vault and was covered in a black residue, possibly wax, with blood and adhered feather remnants (see Figure 24.2D). Additionally, an adult human right os coxa with a drill hole through the acetabulum and 20 human left ribs were recovered. The skeletal modifications were consistent with being cleaned and lightened, and the Asian ancestry affinities of the skulls were consistent with teaching material from China.

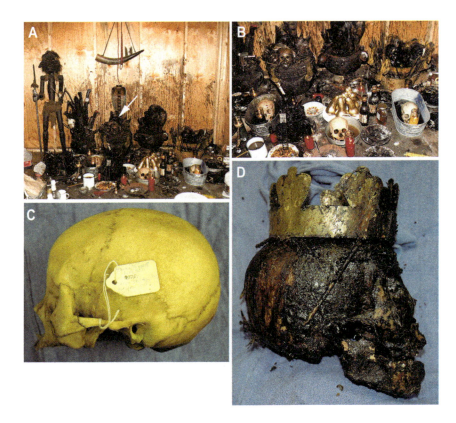

FIGURE 24.2 (A) Overview of sacred space inside of shed exhibiting a teaching skull (white arrow) within the nganga clay flower pot in the background. (B) Close-up showing nganga and three skulls in metal tubs in foreground. Note the anatomically prepared teaching skulls in the tin tubs. (C) The price tag attached to the left zygomatic arch of one of the skulls indicates $375 and a Chinese origin. (D) Note the ritual taphonomic modifications to one skull that includes dried blood, adhered feather remnants, dried black material (possibly wax), and an affixed gold colored crown.

Case 2

The fire department responded to a fire within a business warehouse and found a Palo sacred space. The sacred space was partitioned from the main warehouse with walls covered with firmas (see Figure 24.3A), and it was believed the fire was started by candles in the space. At least three cauldrons, a large metal tub, three large ceramic plates, and a large wooden statue were recovered. A small shrine was located in a corner of the warehouse containing no human bones. Law enforcement was called when a human cranium was discovered.

Two crania, a partial left humerus, three tibiae (one of which was screwed to a large horn), three ribs, and two articulated anatomical hands held together with metal hardware were removed from the scene. One cranium (see Figure 24.3B) exhibited adhered feather fragments, dried blood, and two metal spears (approximately 175cm long) placed through the nasal aperture and extending to the posterior aspect of the foramen magnum. A small merchant price tag was attached to the left zygomatic arch via string, but the

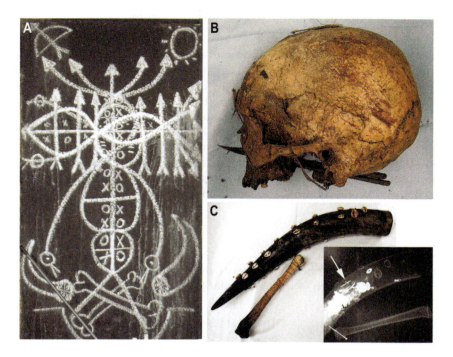

FIGURE 24.3   (A) Panel from the outside of the sacred space for Case 2 with chalk firmas (approximately seven feet high). (B) Note the taphonomic ritual modifications to one teaching cranium: that includes dried blood, adhered feather fragments, two metal spears positioned through the anterior and posterior nasal apertures, and a merchant price tag attached to the left zygomatic arch (black arrow). (C) Animal horn with attached cowrie shells and a human tibia (teaching specimen) attached to the horn with a screw. (C) The inset x-ray highlights the material stuffed into the horn that includes very small swords (white arrow).

writing was not legible. The second cranium exhibited dried blood, an unidentified black residue adhered to the outer surface of the bone, and a silver colored chain inserted into the nasal cavity. The right zygomatic arch was fractured during the postmortem period, and a short piece of string was attached to the arch. Although a price tag was not attached to the end of the string, the type of string and length was consistent with the price tag on the other cranium. The animal horn was decorated with cowrie shells, and the attached human tibia had been placed in a large ceramic plate on the floor of the sacred space. The human tibia was wrapped in copper-colored wire, and the open end of the horn had been sealed with a hard substance. An x-ray indicated the horn was filled with rocks and metallic material, including a number of small swords (see Figure 24.3C).

The skeletal material preparation modifications were consistent with being cleaned, with a number of bones being initially, bleached. The Asian ancestry affinities of the skulls were also consistent with teaching material from China. The owner informed law enforcement that he purchased the human bones over the internet and receipts were provided. Further, the book entitled *The Book on Palo: Deities, Initiatory Rituals, and Ceremonies* by Baba Raul Canizares (2002) was noted at the warehouse. It was clear that the book was used as a guide for the palero who constructed the sacred space.

Case 3

A group of friends were fishing on a river and spotted a large metal cauldron partially sticking out of the water along the bank. They pulled four metal stakes, a crucifix, and a large key from the cauldron before pulling the cauldron from the bank and emptying it out on the ground (see Figure 24.4A). They alerted law enforcement upon finding a skull among the contents. Further analysis of the cauldron's contents indicated various items associated with ngangas such as sticks, knives, beads, railroad spikes, handcuffs, stakes, cloth dolls, and animal bones. It was unknown how long the nganga was in the river. While the skull was the only human skeletal material associated with the nganga, taphonomic modifications (e.g., uniform dark brown soil staining, bone erosion, and fragmentary areas due to prolonged burial) were most consistent with a cemetery origin.

Case 4

Two men were drinking beer behind a grocery store when they noticed a rusted metal tub lying on its side near the curb of the parking lot (see Figure 24.4B–D). The tub contained

FIGURE 24.4   (A) The scene for Case 3 on the bank of the river after the contents of the nganga were dumped out. Note the cranium (white arrow) in the foreground that most likely originated from a cemetery context. (B) The scene for Case 4 with the nganga ritual items scattered in the parking lot near the curb and edge of wooded area; note the cauldron in the foreground and partial cemetery cranium in the background (white arrow). (C) Close-up of the cauldron containing the railroad spikes, including one inserted in the medullary cavity of a distal femur fragment (white arrow) from a cemetery origin. (D) Close-up showing the cemetery partial cranium that was deposited in the trash area.

knives, railroad spikes, a small wooden cross, and bones. They called law enforcement upon noting a partial human cranium and a knife in the nearby wooded area. Additional items associated with the metal cauldron were also located in the vicinity and included horseshoes, a silver chain, at least one animal horn, and cut sticks. The partial cranium and a distal right femur fragment with a railroad spike inserted in the medullary cavity were the only human bones present. The taphonomic modifications (e.g., uniform dark brown soil staining, extremely light bones, and eroded and fragmented bone quality due to prolonged burial) of the human skeletal remains from the nganga were consistent with a cemetery origin.

Case 5

A homeowner was trimming trees and bushes in the back yard of their residence and discovered some knives and handcuffs coming up from the ground along the back fence. The homeowner dug and removed a cauldron and other buried items (see Figure 24.5). Upon arrival at the scene. law enforcement noted a large hole in the ground along the fence and a rusted cauldron with an attached chain near the hole. A white bucket was noted with the rusted handcuffs, rusted stakes, a small doll, and other items from the cauldron. Other materials from the cauldron, including cloth and bones, were also lying on the ground and appeared to represent both human and nonhuman bones (see Figure 24.5). The only human bones present were a fragmentary skull and tibia. The taphonomic characteristics (e.g., uniform dark brown soil staining, localized iron staining, extremely light bones, and eroded and fragmented bone quality due to prolonged burial) were consistent with a cemetery origin.

FIGURE 24.5   Scene for Case 5 after the homeowner dug up the nganga. Note the fragmentary skull and tibia in the background against the fence (white arrows) that were derived from a cemetery origin.

Case 6

Kids entering a backyard through a missing section of fence found a cranium inside an open shed (see Figure 24.6). They removed the cranium and placed it on the east side of the house before contacting law enforcement. Law enforcement noted other bones inside the shed, which appeared to be some type of shrine or place of worship. Three cauldrons, two large flower pots, and a large, wooden statue were also present. Inside the cauldrons and flower pots were various items typically associated with ngangas such as railroad spikes, shells, statues, dirt, animal bones, etc. The property owner was contacted and stated that the material belonged to her son and was used for religious purposes until his death approximately two years prior. She unsuccessfully tried contacting the religious group with which he was involved to remove the material. She believed her son purchased the human bones over the internet.

The cranium was consistent with a juvenile of Asian ancestry. Further, the cranium had been modified with dark staining, the left mastoid process was damaged, and animal feather remnants and dried blood droplets were present on the cranial vault. These taphonomic changes were due to ritual use of the cranium and storage in an open shed. However, the cranial vault and face still exhibited an overall light coloration that was

FIGURE 24.6 The open shed (A) and close-up (B) showing nganga material associated with a sacred space that was haphazardly piled in the shed for Case 6. (C) The location where kids deposited the teaching cranium behind the house (white arrow) after removing it from the shed. (D) Teaching cranium with string attached to the left zygomatic arch (white arrow) that had previously contained a price tag.

consistent with being cleaned and most likely bleached. A short piece of string tied around the left zygomatic process (see Figure 24.5) was consistent with string used to attach a price tag to human skulls sold for teaching purposes, as observed in Cases 1 and 2. Also, a fragment of the proximal femoral diaphysis was recovered along with matching right and left ilia that were cut cleanly and horizontally across the lower aspect of the ilium between the anterior superior and inferior iliac spines. These partially cut bones appear to have been cleaned and exhibit relatively fine cut marks consistent with a powered saw with a small-toothed blade. Although it is common to find skeletal elements representing teaching material in Palo scenes, we previously had not observed cut human bone pieces from teaching materials. However, it is possible to purchase cut bones originating from a teaching context for cadaver dog training over the internet. The scene and remains were consistent with the practices of Palo, and the human skeletal remains were consistent with a teaching material origin.

DISCUSSION

Knowledge of Palo ritual practices can help the forensic anthropologist interpret the ritual taphonomic modifications and context of skeletal remains, and ultimately determine the medicolegal significance of remains. The six cases presented here reveal similar Palo trends, as well as some variations (see Table 24.1).

Although all cases included at least one skull or cranium, the two cases involving active sacred spaces included at least two crania (Case 1 and 2), with four skulls recovered from Case 2. Four of the six cases contained a skull or cranium along with postcranial material. This is consistent with the literature that summarizes the most commonly found human bones in the sacred space or nganga as the skull, hands, feet, tibia, and femur (Gonzalez-Whippler, 1989; Canizares, 2002; Perlmutter, 2004; Perlmutter, 2005; Kail, 2008). The tibiae and femora found in these cases were likely used as a bone scepter by the palero to invoke the nfumbe (Wetli & Martinez, 1983; Gonzalez-Whippler, 1989). While a tibia was not recovered from Case 1, a large animal horn was hanging on the wall behind the nganga, suggesting that the horn may have been used instead of a tibia as a method of invoking the nfumbe. Interestingly, while Case 2 was the only case where all human bone types (skull, fingers, toes, tibia, and ribs) specified by Canizares (2002) were recovered, this was also where a copy of the Palo book by Canizares (2002) was noted.

Of the six cases presented, two active sacred spaces were located outside the paleros homes (a backyard shed and a business warehouse), which is consistent with locations of the sacred spaces of previous cases in the literature (Kail 2008; Winburn et al., 2016). Case 6 involved sacred space items that appeared to be piled haphazardly in a shed, which was likely the original location of the sacred space for the nganga. In the other three cases, the nganga objects were discarded in different contexts. While the sacred space within which the nganga resides will provides specific contextual information to enable identification of a Palo ritual, ngangas are often discarded at secondary locations. It has been suggested that a palero's nganga is dismantled when he dies, and the elements are returned to nature by being buried in the woods or wrapped in burlap/plastic and deposited in a cemetery (Wetli & Martinez, 1981; Wetli & Martinez, 1983; Martinez & Wetli, 1982). Only Case 5, which contained skeletal material that most likely originated from a cemetery, was found buried. Case 4 appeared to have been dumped in the vicinity of a wooded area, and

Case 3 involved the disposal of the nganga in a river. The secondary locations of the discarded ngangas for the cases presented are consistent with common depositional contexts of other reported cases such as near water, buried or deposited in yards, and associated with wooded areas (Gill et al., 2009; Pokines, 2015; Winburn et al., 2016).

Similarities among the cases suggest that local paleros may communicate or consult similar sources. The presence of the Canizares (2002) book at Case 2 (a guide to the practices of Palo Monte) indicate this may have been the specific Palo branch represented by a majority of these cases. One commonality noted among 50% of the cases (Case 1, 2, and 6) included preservation of receipts and merchant price tags (only the string remained on the zygomatic arch of Case 6) to document the legal purchase of anatomically prepared teaching specimens. While Winburn et al. (2016) note that the majority of specimens examined in their Florida sample were products of grave robbing or a historic/archaeological origin, only half of the cases in this study were likely cemetery specimens that were potential grave robberies. While Palo paleros participate in grave robbing to obtain human remains for their nganga, this study, and various others, have documented that paleros also obtain teaching specimens legally (Gonzalez-Whippler, 1989; Martinez & Wetli, 1982; Kail, 2008; Winburn et al., 2016; Winburn et al., 2017). It is, however, becoming more difficult to purchase human skeletal remains through internet sources, and Winburn et al. (2017) noted that the online auction site eBay no longer allows the sale of human skeletal remains. The limited availability and high cost of legally obtaining remains for Palo could possibly lead to increased grave robbing scenarios.

LESSONS LEARNED

- When human skeletal remains are discovered outside of the context of the sacred space and/or nganga, it is important to recognize the ritual taphonomic indicators they exhibit as being related to the Palo ritual, as well as indicators of a teaching or cemetery origin of the remains.
- While it is common for cemetery or teaching specimens used for Palo rituals to be retained at medical examiner offices, bones from grave robberies can be reinterred if they can be associated with the original gravesite.

DISCUSSION QUESTIONS

24.1 What are examples of taphonomic modifications indicative of Palo ritual use? How about cemetery or teaching origins?

24.2 What analyses might you conduct to support a suspicion that a skull was a teaching specimen originating from China?

24.3 Although remains used in Palo rituals may not be associated with a homicide or otherwise suspicious death, they are at times obtained illegally. In such a case, are the remains of forensic significance?

24.4 If the remains were obtained legally, are there any laws being broken by the practitioners? Does the fact that this is a religious practice change any perspectives about the rituals?

REFERENCES

Canizares, R. (1999). *Cuban Santeria: Walking with the night.* Rochester, VT: Destiny Books.

Canizares, B. R. (2002). *The book on Palo: Deities, initiatory rituals, and ceremonies.* Old Bethpage, NY: Original Publications.

De La Torre, M. A. (2014). *Santería: The beliefs and rituals of a growing religion in America.* Grand Rapids, MI: Wm. B. Eerdmans.

De Mattos Frisvold, N. (2010). *Palo Mayombe: The garden of blood and bones.* London, UK: Scarlet Imprint.

Dodson, J. E. (2008). *Sacred spaces and religious traditions in Oriente Cuba.* Albuquerque, NM: University of New Mexico Press.

Dupras, T. L., & Schultz, J. J. (2014). Taphonomic bone staining and color changes in forensic contexts. In J. T. Pokines & S. A. Symes (Eds.), *Manual of forensic taphonomy* (pp. 317–324). Boca Raton, FL: CRC Press.

Gill, J. R., Rainwater, C. W., & Adams, B. J. (2009). Santeria and Palo Mayombe: Skulls, mercury, and artifacts. *Journal of Forensic Sciences, 54*(6), 1458–1462.

Gonzalez-Whippler, M. (1989). *Santeria: The religion, faith, rites, magic.* Saint Paul, MN: Llewellyn Publications.

Kail, T. M. (2008). *Magico-religious groups and ritualistic activities: A guide for first responders.* Boca Raton, FL: CRC Press.

Martinez, R. & Wetli, C. V. (1982). Santeria: A magico-religious system of Afro-Cuban origin. *American Journal of Social Psychiatry, 11*(3), 32–38.

Murrell, N. S. (2010). *Afro-Caribbean religions: An introduction to their historical, cultural, and sacred traditions.* Philadelphia, PA: Temple University Press.

Ochoa, T. R. (2004). Aspects of the dead. In M. A. Font (Ed.), *Cuba today: Continuity and change since the 'Periodo Expecial'* (pp. 387–420). New York: Bildner Center for Western Hemisphere Studies.

Ochoa, T. R. (2010). Prendas-Ngangas-Enquisos: Turbulence and the influence of the dead in Cuban-Kongo material culture. *Cultural Anthropology, 25*(3), 387–420.

Paolello, J.M. & Klales A.R. (2014). Contemporary cultural alterations to bone. In J. T. Pokines & S. A. Symes (Eds.), *Manual of forensic taphonomy* (pp. 118–199). Boca Raton, FL: CRC Press.

Perlmutter, D. (2004). *Investigating religious terrorism and ritualistic crimes.* Boca Raton, FL: CRC Press.

Perlmutter, D. (2005). Ritualistic crime. In J. Payne-James, R. W. Byard, T. S. Corey, & C. Henderson (Eds.), *Encyclopedia of forensic and legal medicine* (Vol. 3, pp. 547–564). Oxford, UK: Elsevier.

Pokines, J. T. (2015). A Santeria/Palo Mayombe ritual cauldron containing a human skull and multiple artifacts recovered in western Massachusetts, U.S.A. *Forensic Science International, 248,* e1–e7.

Pokines, J. T., & Baker J. E. (2014). Effects of burial environment on osseous remains. In J. T. Pokines & S. A. Symes (Eds.), *Manual of forensic taphonomy* (pp. 73–114). Boca Raton, FL: CRC Press.

Schultz, J. (2012). Forensic significance of skeletal remains. In D. Dirkmaat (Ed.), *A companion to forensic anthropology* (pp. 66–84). Boca Raton, FL: CRC Press.

Schultz, J. J., Williamson, M. A., Nawrocki, S. P., Falsetti, A. B., & Warren M. W. A. (2003). Taphonomic profile to aid in the recognition of human remains from historic and/or cemetery Contexts. *Florida Anthropologist, 56,* 141–147.

Wetli, C. V., & Martinez, R. (1981). Forensic sciences aspects of Santeria, a religious cult of African origin. *Journal of Forensic Sciences, 26*(3), 506–514.

Wetli, C. V., & Martinez, R. (1983). Brujeria: Manifestations of Palo Mayombe in South Florida. *Journal of the Florida Medical Association, 70*(8), 629–634.

Winburn, A. P., Martinez, R., & Schoff, S. K. (2017). Afro-Cuban ritual use of human remains: Medicolegal considerations. *Journal of Forensic Identification, 67*(1), 1–30.

Winburn, A. P., Schoff, S. K., & Warren, M. W. (2016). Assemblages of the dead: Interpreting the biocultural and taphonomic signature of Afro-Cuban Palo practice in Florida. *Journal of African Diaspora and Archaeological Heritage, 5*(1), 1–37.

CHAPTER 25

"To Understand the Parts, It Is Necessary to Understand the Whole": The Importance of Contextualizing Patterns in Forensic Anthropology Casework

Laura C. Fulginiti, Andrew Seidel, and Katelyn Bolhofner

CONTENTS

INTRODUCTION

Pattern recognition is a fundamental component of the practice of forensic anthropology. The patterns of osteological growth, development, remodeling, and senescence form the basis for the most widely used methods for the estimation of age-at-death. Patterns are evident in the different conformations of osteological features that enable practitioners to estimate sex from skeletal remains. The estimation of ancestry is based largely on patterns embedded within the architecture of the skull, and the interpretation of skeletal trauma often relies on the predictable patterning produced by both fracture dynamics and the processes of healing. Patterns pervade case work, and their recognition is essential for the accurate interpretation of osteological remains.

Whether due to the vagaries of casework or to differences in the sensitivity and training of forensic pathologists, anthropologists often must formulate a biological profile or interpret osteological trauma based on partial skeletons, isolated bones, or even bone fragments. This situation, combined with a tendency to favor those patterns that underlie preferred analytical techniques, often results in circumstances where larger-scale patterns that have bearing on a case are obscured. When analyzing excised portions of bones or bone fragments recovered during a search, the focus should be on the fact that each of these elements was once integrated within a living person and part of a dynamic biological system. In the casework experiences shared here, making a conscious effort to place skeletal elements back into their original context – the living body – whether through reconstruction or other means, often yielded new lines of inquiry, as well as a glimpse of the broader patterns surrounding the death of the decedent. Unappreciated biases that can be brought to bear on an investigation by external parties often are minimized by the hypothesis testing of the forensic anthropological analysis. By using data generated from an integrated view of the decedent, the anthropologist can provide an enhanced perspective. This understanding of these larger-scale patterns impacts how a case is investigated and strengthens the working relationship between law enforcement and the medicolegal community.

Each of the following four brief case descriptions provides a slightly different example of how considering the body as an integrated and living whole, even when only partial remains are available for analysis, changes the understanding of a case and/or provides new and relevant information concerning the circumstances surrounding an individual's death. After the case descriptions, the authors explain how this hypothesis testing approach altered the outcome or produced beneficial effects outside the context of the case. One of the objectives of this chapter is to demonstrate the forensic anthropologist's role during this type of hypothesis testing; the analyzes can support a working theory put forward by law enforcement or a forensic pathologist, or provide an alternative interpretation of the evidence based on observations in the skeleton.

Case 1

A man took his dog to a local lake to test a newly refurbished motor on his power boat. Later in the day the boat was found, still running, with its nose up against a sheer cliff face. The dog was recovered unharmed from the bow. After the boat motor was shut down, a search commenced and the man was located facedown in the water, deceased. Numerous sharp and blunt force injuries were present on the head, torso, and legs of the decedent. The soft tissue wounds resembled long, linear lacerations and abrasions, and the underlying bones exhibited multiple fractures and misaligned segments.

At autopsy, multiple scenarios were considered for the events surrounding the man's death. There were suggestions that he may have fallen out of the boat (multiple open and unopened beer bottles in the boat suggested he might have been drinking); that he may have run into the cliff either purposefully or accidentally and was ejected during the impact; or that he suffered a medical event and fell overboard. Depending on the specific events, the manner of death could be ruled accident, suicide, natural, or undetermined. The strike marks on the man's torso and legs were thought to be caused by the propeller of the boat, and the forensic anthropology team was asked to assess the damage and determine how the patterning could be interpreted in the context of the death. While standard osteological analysis would not have been able to differentiate whether the injury itself occurred peri- or postmortem, a contextualized reconstruction of the patterning of injury could help to elucidate the timing and nature of the events that caused the injuries.

The team attended the autopsy and examined the injuries to the bones of the head, torso, lower limbs, and feet. At the conclusion of the exam, the bones of the lower limbs and right foot were removed and macerated for further study. The right distal femur, tibia, fibula, and foot were reconstructed, with the foot glued together for a more detailed assessment (see Figure 25.1). There were oblique fractures and incised wounds on the medial aspect of the distal right femur, the entire right tibial shaft, and the superior surfaces of the right foot. Parts of the superior aspects of the tarsals were shaved off, and the right metatarsals were fractured at their mid-shafts.

Lake patrol detectives provided a propeller exemplar so experiments could be conducted to determine how the strikes in the bone may have been created. Long sheets of butcher paper and carbon paper were laid out in the lab and the propeller was turned along them, marking the butcher paper as the propeller blades were drawn against the carbon paper (see Figure 25.2).

At the conclusion of these experiments, the forensic anthropology team was able to match one specific pattern on the butcher paper with the injuries observed in the bones. Based on the design and movement of the propeller and the skeg (a sternward extension of the keel that makes up part of the propeller assembly), the angles of the cuts on the lower extremity could have been produced in only one way.

The ultimate conclusion based on this analysis was that the decedent was facedown prior to being drawn feet first and from the right side of the boat through the propeller and skeg apparatus. The man had significant cardiovascular disease, a slightly elevated ethanol level, and had fallen into the water. There was no determination regarding whether the man fell from the boat prior to or as a result of the boat striking the cliff face. However, the forensic anthropology reconstruction that he was likely floating facedown when pulled

FIGURE 25.1 Overall view of the reconstructed and rearticulated right lower limb of a man struck by the propeller of a small motorized water craft.

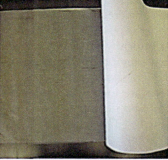

FIGURE 25.2 The experimental process used to recreate the pattern of incised injury observed on the lower leg and foot of this decedent a) rolling the propeller across a two sheets of white butcher paper sandwiching a carbon paper (b) the carbon paper in between two white sheets of butcher paper (c) the marks made by the propeller turning (stationary, full throttle).

into and struck by the propeller blades provided data that he was probably unconscious or deceased prior to being impacted by the propeller. As a result, the medical examiner listed the manner of death as accident, with cardiovascular disease contributing.

Case 2

During the overnight hours of Halloween, several suitcases were discovered in and next to a dumpster. When the finder opened one of them he saw parts of a dismembered body inside. Law enforcement subsequently emptied the dumpster in an attempt to locate all of the body segments. At the conclusion of the search, portions of the limbs were still missing. Following autopsy, the forensic anthropology team examined the decedent and determined that a minimum number of one adult male was represented, provided a biological profile, evaluated the osseous traumata, and examined the cut ends of the bone segments for tool marks. The individual was subsequently identified but as there were multiple limb segments unaccounted for during the original search and recovery, multiple exemplars representing each cut surface were macerated and retained for further study. Upon evaluation following maceration of the remains, it was determined that the squared cross-section of the tool marks and the uneven striations evident on the cut surfaces of the bones were consistent with dismemberment using a manual saw. Ultimately, these specimens were retained for evidentiary purposes.

Almost a year later, construction workers clearing brush at the end of a street discovered mummified body parts. The recovered segments included the distal two-thirds of a left humerus articulated with the left radius and ulna and a complete left hand; the left distal femur, patella, and proximal tibia and fibula; the distal left tibia and fibula articulated with the left foot minus the distal phalanges; the right metatarsals and phalanges and one right cuboid, which rearticulated with the right forefoot. The shafts of the long bones exhibited cut marks consistent with dismemberment.

Given that the newly found remains represented the skeletal elements missing from the previous dumpster case, there was immediate suspicion that they may be from the same individual. There was concern that lifting usable fingerprints for comparison and confirmation of this tentative identification may prove time-consuming and challenging due to mummification. Thus, the anthropologist was tasked with comparing the elements and cut surfaces with those recovered previously and archived at the medical examiner's office to more quickly explore the possible identification. The proximal left humerus from the dumpster, could be anatomically fitted onto the cut surface of the mummified distal left humerus, indicating that they originated from the same decedent (see Figure 25.3). The immediate anthropological assessment of these disparate elements recovered ten months apart and from separate locations obviated the need for dedicating resources to a new death investigation and demonstrated the value of archiving specimens when an incomplete recovery is made. The tentative identification ultimately was confirmed through fingerprint comparison.

Case 3

Hikers photographing the desert discovered a burned, decomposed, and skeletonized body in a wash (i.e., a dry river). The decedent was supine with his wrists bound in front of his waist, his hips externally rotated, and his legs bent at the knees (see Figure 25.4).

FIGURE 25.3   Rearticulated segments of the left humerus.

FIGURE 25.4   Partially burned and decomposed adult male at the scene of recovery.

The distal legs and ankles appeared to have sustained significant thermal damage. The forensic anthropology team was asked to attend the autopsy and provide a biological profile and trauma assessment. At the conclusion of the exam, the hyoid, partially ossified thyroid cartilage, pubic symphysis, sternal ends of the fourth ribs, and the right and left distal legs and feet were removed and macerated for further study.

After maceration, the distribution of thermal damage apparent on the bones of the feet and legs appeared to be non-random. The distal ends of both tibiae and fibulae exhibited significant charring. The lateral aspect of the left calcaneus was fragmented, with portions of bone missing and blackening evident at the margins of the missing areas. The superior two-thirds of the left talus exhibited significant charring, with the superior aspect of the trochlea beginning to whiten as a result of calcination. The majority of the left tarsals were significantly blackened, with thermal alteration limited to the superior surface of the cuboid. Charring was also evident on the dorsal aspect of the bases of the metatarsals and exhibited clear lines of demarcation between adjacent portions of burned and unaffected bone. The right distal tibia, fibula, tarsals, metatarsals, and phalanges were more fragmented than the left, but exhibited a nearly identical pattern of thermal alteration.

The bones of the feet were rearticulated and then associated with the distal tibiae and fibulae (see Figure 25.5). Upon rearticulation, the pattern of thermal alteration was confined to a clearly demarcated band including the superior aspects of the tarsals and bases of the metatarsals, as well as the distal portions of the tibiae and fibulae, while the inferomedial aspects of both calcanei were unaffected. Missing portions of bone were presumed to be cremated or not recovered/recognized at the scene. Although charring was also evident on the skin of the decedent's thighs as well as his clothes, the pattern of thermal alteration to the bone described above suggests that his ankles had been bound together and that these bindings were subject to more prolonged and/or more intense combustion than the remainder of his body, perhaps representing the initial point of ignition for the fire.

Reconstruction and rearticulation also demonstrated that the observed relationships between thermal alteration and lines of fracture could be divided into three categories. In the first category, thermal alteration crossed fracture lines, involving the broken surfaces of bone as well as the external aspects of both segments of bone surrounding the fracture. In the second category, thermal alteration was constrained by the line of fracture, affecting the distal fragment up to the fracture line but leaving the proximal fragment unaffected. The third category included fractures that were entirely unaffected by thermal alteration. The latter two categories suggest that some of the fracturing of the distal tibiae and fibulae that was initially attributed to thermal injury was, in fact, precedent and perimortem.

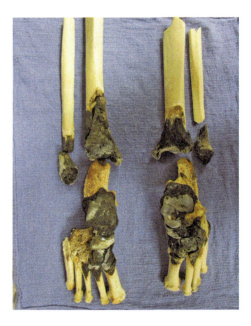

FIGURE 25.5 Thermal pattern on legs and feet, post-maceration and after reconstruction and rearticulation.

Case 4

At the autopsy of an elderly male who allegedly had expired as a result of injuries sustained in a fall in the care facility to which he had been admitted for dementia, the medical examiner noted complex skull fractures that appeared inconsistent with the received narrative. As a result, she asked the forensic anthropology team to assess the calvaria and provide a detailed analysis. The calvaria was macerated to facilitate further study. The injuries discovered following maceration were even more extensive than those noted by the medical examiner, exhibiting characteristics of both perimortem and antemortem fractures (see Figure 25.6).

Acute fractures involving both the external and internal tables of bone, were arrayed across the right and left parietal bones, forming a network of comminuted fracturing across the vault. The patterns in which these fracture lines intersected were consistent with a minimum of four distinct points of impact. In addition to these acute injuries, the calvaria exhibited two ovoid areas in which a combination of lytic and proliferative bony reactions could be observed. These two areas also coincided with localized discoloration of the bone that was consistent with hemorrhage and, as a result, were interpreted as a lytic response to the pooling of blood with subsequent bony remodeling. The combination of acute fracturing and macroscopically visible remodeling suggested that the decedent had experienced at least two traumatic episodes. The overall pattern of these injuries and their location on the superior aspect of the cranium rendered them inconsistent with the narrative that had been reported to law enforcement: a single fall resulting in an impact to the back of the head.

As a result of the anthropological examination, the medical examiner ruled the case a homicide. Local law enforcement was initially skeptical and requested a consultation with the pathologist and the anthropology team to review their findings. After reviewing detailed photographs and receiving an explanation of the injuries on the macerated calvaria, they agreed that a more intensive investigation into the death was warranted.

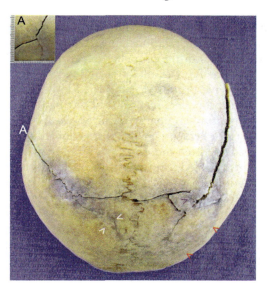

FIGURE 25.6 Overall view of cranial vault after maceration with impacts: A (inset) – converging fractures left side of vault; white arrows – triangular shaped fracture; purple arrow – convergence of fractures with lytic response; red arrows – extension of fractures (difficult to see in this view), creating a square-shaped pattern.

LESSONS LEARNED

- Although forensic anthropologists frequently are limited to working with isolated portions of a skeleton, the practitioner should never forget that these elements were once part of a person. Keeping this fact in focus helps us remember the humanity inherent in our profession, and enables us to discern patterns that were not immediately evident and that sometimes alter our conclusions. The process is consistent with the scientific method; develop a hypothesis, test the hypothesis, confirm or alter the hypothesis. In Case 1, for example, the rearticulation of the skeletal elements removed at autopsy revealed a pattern of injury that was replicated through experimentation with a boat propeller, bringing clarity to the events surrounding the decedent's death and contributing to the determination of manner of death. Similarly, in Case 3, reconstruction and rearticulation revealed new details pertaining to the perimortem period for this decedent, providing evidence that the nature of the crime was perhaps darker than initially perceived and suggesting that law enforcement should consider the possibility of multiple scenes. Case 4, although not entailing physical reconstruction of the remains, provides an example of how carefully considering the totality of evidence can falsify the narrative that accompanies a death. In this case, the pattern of injuries was inconsistent with a single fall and provided evidence for multiple traumatic episodes as well as neglect, prompting the designation of the manner of death as homicide. By moving beyond the fragments of bone available for analysis and considering their integration into a living and dynamic body, the outcomes and/ or process of investigation of these cases were changed. The forensic anthropologist is charged not only with describing skeletal modifications, but also interpreting them, which cannot be done appropriately if each defect is considered only in isolation.
- A less direct, but no less important, result of considering cases in this fashion is the effects of a strong working relationship with law enforcement. The participation of Lake Patrol officers in the propeller experiment in Case 1 inspired their interest and cooperation. As a result of their involvement with the anthropological analysis, they have proposed using road-killed deer to conduct further experiments with boat propellers in a natural setting to document the resultant patterns of injury for future comparative purposes. In Case 2, the archiving of dismembered remains for evidentiary purposes enabled their rearticulation with skeletal elements recovered ten months later. While the reapproximation of the dismembered remains enabled for a characterization of the implement used, of more immediate importance is that the anthropological association of these two sets of remains to the same individual prevented unnecessary expenditure of time and resources by the medical examiner office and law enforcement, who had been preparing to initiate a new search and investigation. Presented with the raw description of the injuries in Case 4, law enforcement remained unconvinced of the need for additional investigation. During their consultation, the team demonstrated that it was unlikely for a living person to fall in such a way as to produce that pattern and location of fracturing on multiple occasions, resulting in the initiation of a homicide investigation. In the experiences of this team, and as illustrated by several of the cases above, the result of such collaborations has been an increased understanding of and respect for the role of forensic anthropologists within medicolegal investigations.

- Finally, these cases highlight the forensic anthropologist's role in hypothesis testing. In each of the described cases there were initial speculations regarding the remains or circumstances of the death. Could the trauma be attributed to boat propellers? Did the remains match a previous case? Were the legs bound? Was the trauma representative of a single fall? Forensic anthropologists must use various resources to test hypotheses and remove bias from our analyzes, which may require thinking outside the box. In some cases, the results may corroborate initial suspicions, but in other cases they may refute the case histories provided. It is the forensic anthropologist's role to document all potential evidence and provide an unbiased and scientifically sound interpretation of their meaning, which in turn assists the medical examiner/coroner in deciding the manner and cause of death.

DISCUSSION QUESTIONS

25.1 In Case 1 (boat propeller case), the medical examiner ultimately ruled the manner of death an accident believing that the man was floating face down and unconscious at the time of impact with the propeller. It is the medical examiner's job to use the available evidence to ultimately rule on the *most probable* scenario. Can you think of any other scenario(s) that would match the results? If the medical examiner ruled the death a suicide, homicide, or undetermined, what kind of legal repercussions might that have?

25.2 In a dismemberment case, remains can be spread in different geographic regions and may ultimately end up at different medical examiner's office (and at different times). If the remains in Case 2 didn't end up at the same office, do you think they would have been linked together? What if they hadn't archived the remains from the previous case? Consider various options for archiving remains (CT scans, photographs, casts of specimens, actual specimens) – what are the advantages and disadvantages of these various options in a case like this? What steps could forensic anthropologists take to increase the likelihood that remains in such cases are more easily associated to one another?

25.3 Give two additional examples (beyond those in the chapter) where anthropological assessments could benefit from considering the overall pattern of injury (or disease, or biological profile constructions).

CHAPTER 26

Perpetrators, Pack Rats, and Postmortem Disturbances: A Case Study Involving Multiple Contexts, Jurisdictions, and Identities

Angela Berg, Kent J. Buehler, and Carlos Zambrano

CONTENTS

INTRODUCTION

Early one dreary rainy holiday weekend, forensic anthropologists from the Office of the Chief Medical Examiner (OCME) met a sheriff's deputy in a convenience store parking lot in rural Oklahoma to examine the contents of a shoebox. The deputy informed them that the item in the box was discovered by a family who carried this item home believing it to be a bowl. The shoebox contained a partial human cranium; an anthropological safari to the scene was now in order. The deputy led the anthropologists along a maze of gravel and red-dirt covered country roads to a narrow 200-yard drive where the cranium had been discovered on a cleared oil well pad.

The site displayed heavily disturbed soils in the clearing and was surrounded by dense trees and vegetation (see Figure 26.1). During the initial evaluation of the partial cranium we concluded that the decedent was likely a male given the robust nuchal markings, and the circumstances surrounding his death were suspicious given the fracture pattern. The appearance of the remains indicated a relatively short postmortem interval of less than one year as

FIGURE 26.1   Google Earth Pro satellite imagery dated 14 days prior to the recovery and excavation.

suggested by the presence of desiccated soft tissues and the faint presence of decomposition odors. The absence of embedded soils suggested we were dealing with a surface scatter. A brief investigation of the site revealed no further skeletal elements. The weather and the holiday weekend limited the availability of additional personnel; therefore, the law enforcement agency and anthropologists agreed to conduct a more thorough search at a later date.

Four days later, the sheriff coordinated a search party that included multiple state and local agencies. The sheriff also formally requested that a state law enforcement agency assume full responsibilities of the crime scene and investigation. After a briefing, an assessment of the site was conducted by a representative of each agency, as well as the OCME anthropologists and the Oklahoma Archaeological Survey forensic archaeologist. The well site and several animal trails were examined to determine the manpower needed to search such a large area. While returning to the county road, the forensic archaeologist noticed black plastic bags on the edge of the dirt road. A brief discussion regarding the potential significance of the bags ensued, until the forensic archaeologist and an anthropologist peered into the woods and noticed something of further interest. In the woods, just west of the trail and plastic bags was a depression and disturbed area in the red clay with fabric protruding from within. Upon closer examination, a partially exposed human tibia and femur were recognized, and the fabric was identified as a fitted bedsheet. It was now evident that a clandestine burial had been disturbed by animal activity.

FORENSIC ARCHAEOLOGY/RECOVERY

Examination of the feature indicated it was a shallow grave containing human skeletal elements that had been disturbed by animal scavengers, probably coyotes. The original fragmentary cranium was found on the surface more than 50 yards north of the grave.

That fact, and considerable previous experience with scavenged and scattered human remains, suggested more bones could be spread over a large area. We were thus presented with two major considerations in processing the scene, one being the grave and the other being a potentially large area of land surrounding the grave. Law enforcement personnel not assisting with the grave excavation were tasked with conducting the surface search. Line searches were conducted over an area approximately 200 yards wide from the road on the south to approximately 100 yards north of the oil well pad, more-or-less centered around the grave site. No human remains were found during the line searches, which were hampered by extremely dense brush.

The forensic archaeologist and forensic anthropologists processed the grave, with law enforcement personnel assisting as needed. The grave consisted of an elongated hole in the ground roughly five feet in length. Width varied from ~2.3 to 3 feet (see Figure 26.2) with a maximum depth of approximately one foot. The walls of the grave sloped inward with increasing depth, particularly at the northwest end of the grave.

Two human lower limb bones, a right femur and a right tibia, protruded from the grave fill. Both bones showed evidence of carnivore modification. The proximal half of the femur and distal end of the tibia remained buried. The presence of leg bones at the northwest end suggested that the head had been to the southeast. The sheet's position suggested it had been dragged partially out of the grave, likely by animal scavengers digging into the grave.

A large section of tree trunk and several portions of tree branches lay on the surface immediately adjacent to the grave; their condition indicated they had been dead for several years. Their position suggested they may have been used to cover the grave, but were subsequently moved by scavengers attempting to gain access to the buried body. A considerable "halo" of soil was present on the grave margins and was likely due to both animal disturbance and failure of the perpetrator(s) to return all backfill into the grave.

Processing the grave and the area immediately around it was done in four general stages using forensic archaeology procedures. During Stage 1, a simple grid system was

FIGURE 26.2 Overview of grave facing east prior to processing. Note the skeletal remains (north end) and sheet protruding from grave (west edge).

established incorporating the grave area. The investigating state agency used a total station to map the grave and adjacent features. Because the grave had been substantially disturbed by scavenging animals, Stage 2 began by carefully walking the area looking for scattered skeletal elements and other potential evidence. All items were flagged with pin flags to be recorded with the total station and given a unique evidence number. The area was also scanned with a metal detector; then it was processed a second time by personnel on their hands and knees using small hand rakes and trowels to scrape through the leaf litter on the forest floor down to the ground surface. The leaf litter and other debris were collected in buckets and then screened by law enforcement personnel through ¼ inch mesh archaeological sifting screens to recover any small items that might have been overlooked. All items found *in situ* were left in place, flagged, photographed, numbered, recorded, and mapped.

During Stage 3 the attention turned to processing the grave. The "halo" of backdirt around the grave was scraped down to the original ground surface using hand trowels. This dirt was collected and processed through the sifting screens. All items found were left in place, flagged, photographed, numbered, recorded, and mapped; any items found during screening were documented by the specific area of recovery. The depth of the "halo" ranged in thickness, up to two inches in places. Stage 4 consisted of the excavation of the interior of the grave using small hand tools such as trowels, spoons, and scoops. Due to the high clay content of the soil, a fire department water truck was brought in and all dirt was water screened. Items encountered within the grave were left *in situ* when possible and documented as described above prior to removal.

Excavation continued until the bottom of the grave was reached, a depth varying from 1.5 to 1.9 feet below the surface (see Figure 26.3). The bottom was detectable by the soil being considerably more compact than the disturbed grave fill, resulting in an accumulation of decomposition fluids, as well as the presence of large, uncut tree roots. At this point, all exposed items were photographed for a final time, mapped, assigned a number, extracted, and bagged. The following items were recovered from within the grave: left zygomatic, maxillary fragments and teeth, right femur, right tibia, right fibula, fabric fragments, a bedsheet containing an unidentified bone fragment and a finger or

FIGURE 26.3   Overview of grave facing west after processing.

FIGURE 26.4   Pack rat nests from various scenes encountered by the authors.

toenail, a belt fastened around the sheet, hair, clear plastic consistent with adhesive tape, and a copper-jacketed projectile.

The pedestrian, "hands and knees" search, and "halo" excavations produced skeletal and other potential evidentiary items. The items recovered from the surface around and near the grave included: multiple cranial fragments and teeth, right scapula, right humerus, left femur, left tibia, left fibula, two metatarsals, fabric fragments, fragments of a t-shirt and boxer shorts, and three black plastic sacks. The general pattern of distribution of items lead northeast away from the grave.

Expansion of the search in that direction resulted in the discovery of a large, active packrat nest approximately 26 yards northeast of the grave with human remains embedded in the nest. The nest was approximately four by five feet horizontally and two feet high, consisting of a dense pile of small branches, twigs, and leaves (see Figure 26.4, upper right image). The entire nest was dismantled by hand and all debris screened. Seven bone fragments were recovered, including the left scapula, left ulna, half of the mandible, one unidentified bone fragment, and three nonhuman bones.

INTERPRETATION OF THE ARCHAEOLOGICAL DATA

The condition of the remains (skeletonized with minor amounts of attached desiccated tissues, slight odor of decomposition, relatively minor evidence of weathering, and significant animal scavenging) suggested a postmortem interval (PMI) of less than one year. This interpretation was strengthened by botanical evidence at the scene. The grave was in a heavily wooded area containing large numbers of deciduous trees. The limited amount of leaf litter contained inside the grave and on top of the backdirt halo surrounding the grave represented leaf fall from a single autumn season, suggesting the grave was likely dug prior to the previous autumn but not before the autumn of two years prior. The act of digging and refilling a grave damages and kills existing vegetation and delays replacement growth (Caccianiga, Bottacin, & Cattaneo, 2012). The grave lacked evidence of foliage from previous growing seasons, and growth of new vegetation in the grave and backdirt halo was limited, supporting the hypothesis that the grave and subsequent disturbance by scavengers occurred recently enough that replacement growth had not yet had time to become established, likely within one year or less.

The grave was typical of clandestine graves – shallow, hand-dug, irregular in size and shape, and barely large enough to contain the body. A length of approximately five feet at the surface, with sloping sides to about four feet long at the bottom, suggests the body was probably buried in a partially flexed position as the grave was not long enough to contain a fully extended adult male (in our experience, graves of bodies buried in a fully flexed or fetal position tend to be circular). The relative positions of the right femur, tibia, and fibula, which were still in the grave, support this hypothesis. Their positions suggest the right lower extremity was flexed at the knee; however, subsequent disturbance by scavengers makes it difficult to determine exact positioning. The central and southern portions of the grave contained only very small skeletal items, all related to the cranium. Since no elements relating to the pelvis, torso, or arms were found in the grave, it is plausible that these parts were originally in the now empty portions of the grave.

The presence of the bedsheet suggests it played a role in the transportation of the decedent to the burial scene. The association of the buckled belt around a portion of the sheet and not in association with pants reinforces this interpretation. An intact, rifled, minimally deformed copper-jacketed projectile was recovered from inside the grave, suggesting the decedent had suffered a gunshot wound. Because of the disturbance of the grave by animal scavengers, it is not possible to correlate the projectile to a specific part of the body. However, its location near the grave's southern end suggests it may have been associated with the torso.

A prominent feature of the scene was the substantial disturbance of the grave and distribution of skeletal elements (see Figure 26.5). Following the burial of the body, scavengers removed most of the body, leaving behind only the right lower extremity (minus foot), a few cranial fragments, and some teeth. The complete lack of ribs, vertebrae, and pelvic elements suggest these elements may have been removed as one or more articulated units, as were the left lower limb and both upper limbs. Support of this interpretation was provided by the close association of the right femur, tibia, and fibula within the grave as well as the right scapula and humerus that were found scattered just outside the burial site. Scavengers routinely detach and carry off large, articulated skeletal segments, such as the spinal column and limbs (Haglund, 1997a; Haglund et al., 1989). Virtually all of the recovered human remains showed moderate to severe scavenger modification consistent with carnivores (Pokines, 2014).

Although a large amount of skeletal material was absent from the scene, examination of the 14 elements identified in the field showed that most major portions of the body were at least minimally represented: the skull (cranium and mandible), the right and left upper and lower extremities, and at least one foot. The small and easily consumed or destroyed bones of the hands, ribs, vertebrae, and the pelvic girdle were missing. The most likely explanation was that these elements were removed from the area by the scavengers.

Some of the partially consumed bones pulled from the grave were also secondarily foraged by rodents. Three human bone fragments and four nonhuman bone fragments were recovered from a large packrat nest over 70 feet from the grave. Since this initial

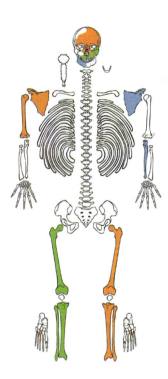

FIGURE 26.5   Homunculus depicting recovery location of each skeletal element. Green indicates grave recovery; orange indicates surface scatter; blue indicates packrat nest.

encounter with a packrat nest during a field recovery, we have continued to examine and process such nests in subsequent scenes with the recovery of a large amount of skeletal and evidentiary material. Based on our experience the potential of scene disturbance, modification, and distance of transport by rodent activity, likely has been underestimated by the literature and merits further evaluation.

ANTHROPOLOGICAL EVALUATION

After recovery, all skeletal material and evidentiary items were transported to the OCME facility for further evaluation; all nonskeletal evidence was then released to law enforcement. The 14 recovered skeletal elements were consistent with one individual based on their size, symmetry, and development. Morphological traits and discriminant function analyses suggested a male with an undetermined ancestry. Only the cranial sutures and periarticular surfaces were available for age estimation; the combination of these age-related indicators suggested a middle-aged adult. Forensic stature was determined using maximum femur length in the linear regression equations for any US population within Fordisc 3.0 (Jantz & Ousley, 2005).

Antemortem conditions included a well-healed, depressed defect above the orbit just to the right of glabella, multiple dental restorations, impacted third molars, and marginal lipping and osteophytes suggestive of osteoarthritis.

A fragmentary skull is always suspicious and requires thorough radiographic and visual examination before and after reconstruction. Via radiographs, radiopaque

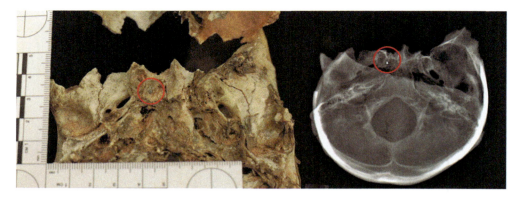

FIGURE 26.6   Inferior view of cranium. Radiopaque fragment noted to the right of vomer, via visual and radiograph assessment.

fragments were noted to the right of the vomer on the ectocranial surface of the sphenoid (see Figure 26.6), in the maxillary bone superior to the right molars and premolars, and along the defect of the right anterior mandible. The skull fragments and dentition were reconstructed using a temporary adhesive cement with the assistance of a forensic dentist. Much of the right side of the cranium and the left mandible were missing; however, radiating fractures, external beveling of the superior left parietal, fracturing of the mandible, and internally embedded radiopaque fragments were suggestive of a gunshot wound. Since the cranium displays multiple radiopaque fragments and a fracture pattern consistent with a gunshot wound, then this would imply that the associated projectile would also display fragmentation. Since the recovered projectile is intact and minimally deformed, then this projectile was not associated with the gunshot wound to the head. The intact and minimally deformed projectile suggests that the impact velocity was reduced, either by great distance or absorption of the energy by the soft tissues (Haag, 2015).

IDENTIFICATION

Several weeks after the scene recovery, the OCME anthropologists and forensic archaeologist were working on another skeletal recovery in a central Oklahoma county. Since the suspected decedent at this site was missing from another jurisdiction, the agency from that jurisdiction was invited by local law enforcement to assess and assist at the scene. During scene processing, the secondary agency's detectives noted that they had recently received a tip that a missing person was buried in a rural area of central Oklahoma. The closest city to the site they referenced was also the closest city to the recovered burial discussed above. This was the first lead to the identity of the remains.

The missing person's information was also known to the anthropologists. Eight months prior, the sister of the missing person contacted the OCME attempting to locate her brother; a search of the OCME database yielded no persons by name or unidentified persons by identifiers who matched her brother. Her brother's case was already listed in NamUs.gov but only included limited biometric information, such as hair and eye color, height, and weight. A phone interview with the sister provided more information about his physical traits, as well as information about medical treatments, dental facilities, and fingerprints. She also noted that his birth name (Identity A) was different than the name listed in the NamUs profile (Identity B).

At our request, a state agency searched for fingerprints and a CODIS sample; fingerprints in IAFIS were located but a CODIS DNA sample was not. The Interstate Identification Index (III) report revealed that the missing person had 16 aliases and seven dates of birth that were used throughout multiple states. The fingerprints from the birth name (Identity A) matched the name entered into NamUs (Identity B). The Oklahoma CODIS administrator contacted the other states to locate a potential CODIS sample; that sample was eventually found in another state under yet another alias (Identity C) where he had been previously incarcerated for ten years.

Federal Statute (45 CFR § 164.512(g)(1)) allows medical examiners and coroners to access medical records to aid in the death investigation and decedent identification, so with this new information, we searched for medical records and radiological imaging at all local hospitals. Most facilities did not balk at searching their records for all combinations of names and birth dates. We were able to locate medical records and antemortem CT scans of his skull under his local name (Identity B). Since an antemortem defect was noted on the frontal bone of the decedent, we requested photographs of the missing person from law enforcement (Identity B). The subject displayed a scar in the same region of the defect, which supported the tentative identification.

The state penitentiary that collected the CODIS sample was also contacted for medical or dental records and they were able to locate dental records and one panoramic dental radiograph taken 21 years prior during his incarceration under Identity C. The forensic dentist was tasked with a dental comparison of the decedent to the 21-year-old dental records for Identity C, as well as dental comparison to the CT imaging of the skull for Identity B. Dental comparison established positive identification for Identity C; the dental CT imaging could not exclude the identity of Identity B as being the same individual.

The skull features from the CT image series for Identity B was compared to the cranial features of the decedent. There were multiple points of concordance, including the depressed frontal bone lesion and frontal sinus pattern, which provided positive identification for Identity B (see Figure 26.7).

We established positive identification of the decedent as being both Identity B and Identity C; these names were linked in IAFIS via fingerprints as all belonging to the same individual, including to his birth name (Identity A). Based on this identification, the law enforcement agency who was handling the missing person case assumed complete jurisdiction of this case from the state law enforcement agency who had been requested by the sheriff; from this point on, they processed the evidence and pursued prosecution.

Due to the confusion over the decedent's name as well as new information about the suspect and the event leading up to the death, the law enforcement agency handling the case and the district attorney's office handling the prosecution decided to pursue genetic analyses for evidentiary comparison. They chose a DNA lab who could process bone and provide them with a CODIS profile of the decedent. Approximately 11 months after recovery of the decedent, the unidentified remains were matched positively to the family reference DNA samples (Identity A). Two months later, we were also notified that the decedent had a CODIS match to Identity C.

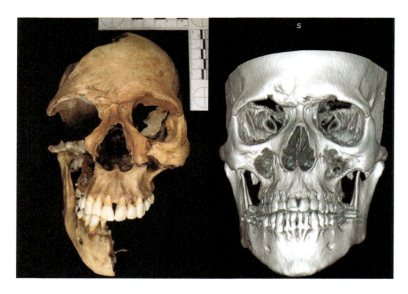

FIGURE 26.7   Anterior view of skull, postmortem reconstruction vs. antemortem CT 3D volume rendering. Note the consistencies of bone structure, dentition, and right frontal bone lesion.

CONCLUSION

The ability to work with multiple law enforcement agencies, complete an extended but thorough two-day recovery, and communicate with the family to address all identification issues enabled us to provide comprehensive forensic archaeological and anthropological analyses. Lack of identification of a decedent can limit law enforcement's ability to investigate, so it is imperative to follow all leads that may identify or exclude the individual, even if there are multiple identities. Once identification was established, law enforcement was able to locate a suspect who confessed to shooting the victim in the abdomen, transporting him to a wooded area wrapped in fabrics, digging a shallow burial pit, then shooting the decedent in the head when the suspect discovered he was still alive before covering his body with soil. Despite all of this corroborating information from the scene and skeletal evidence, the suspect initially desired a trial, but eventually took a plea deal.

LESSONS LEARNED

- Our initial evaluation of a surface scatter was incorrect; it was a scatter, but from the disturbance of a clandestine burial not a surface deposition. This highlights the importance of being prepared and able to adapt during the search and recovery process.
- Rodents, notably packrats, are more actively involved in the taphonomic changes than we suspected. Previous studies of rodents as taphonomic agents have, while briefly mentioning their proclivities for removing small bones from a scene, focused on the modification of skeletal elements principally through gnawing activities (Galloway, 1997; Haglund, 1997b; Klippel & Synstelien, 2007).

However, in the years following our initial packrat nest encounter, we have excavated several dozen more nests from about a half dozen scenes. These nests have yielded multiple skeletal elements and evidence a substantial distance from the primary concentration/deposition of remains. These nests may be a treasure trove for locating the missing items at a scene and require thorough evaluation. At one scene, nearly 30 items were recovered from a single nest, while another productive nest was located approximately 76 yards from the surface scattered elements. Nor were the remains confined to small skeletal elements such as hand and foot bones; we have recovered fibulae, scapulae, mandibles, vertebrae, and other large elements from nests. Additionally, rodents are constantly moving elements and remains may be found weeks later in new nests as well as previously searched sites.

- The identity of the decedent is crucial to any investigation; confusion about the identity can cripple law enforcement's ability to secure search warrants and follow leads, and a family is left without answers. In this case, we used every methodology of identification at our disposal – fingerprints, dental radiographs, CT imaging, and genetic analyses – to link the multiple identities of one person in order to establish a solid identity.
- This case's resolution received a jumpstart due to the good relationships built over years between the archaeologist, anthropologists, and law enforcement in the state. The agency that took over the case was initially skeptical when they inquired about the future search for a buried body. We informed them we already recovered one from the area; however, their willingness to communicate was a big first step in identifying this individual. Relationship-building is further strengthened by the fact that the deputy who held that box took a forensic anthropology/osteology course with an author, less than a month prior to the discovery of the cranial fragment – because of that class he knew that the cranium was probably a male and that "heads don't magically fall apart." He was concerned enough to get permission to request the assistance of anthropologists and additional agencies to investigate. From that point, the sheriff decided to use all available resources, even those outside of his jurisdiction. Because of the involvement of multiple forensic experts and the communication between all law enforcement jurisdictions, the investigation led to an identification, a suspect, and a successful prosecution.

DISCUSSION QUESTIONS

26.1 Knowing that rodents and carnivores can take skeletal elements a long distance from the primary scene, how far should you extend a search if a single skeletal element is discovered? What contextual or geographic information might you use to guide the search in certain directions? What if you still are unable to find additional elements (as was the case here for the pelvis and vertebrae)?

26.2 After the scene has been thoroughly searched and released, how would you explain to the family or law enforcement when additional skeletal or evidentiary items are found at the same site at a later date?

26.3 If the decedent has multiple names or dates of birth due to aliases or misdocumentation, which name do you provide for the decedent and why? What are the possible implications of using an incorrect (or alias) name in medicolegal documentation (e.g., vital statistics, insurance benefits, prosecution)?

26.4 Why do you think the authors have chosen to document this case study in English units of measurement instead of metric?

REFERENCES

Caccianiga, M., Bottacin, S., & Cattaneo, C. (2012). Vegetation dynamics as a tool for detecting clandestine graves. *Journal of Forensic Sciences, 57*(4), 983–988.

Code of Federal Regulations, Security and Privacy, (45 CFR. § 164.512(g) (1)). (2000).

Galloway, A. (1997). The process of decomposition: The model from the Arizona-Sonoran Desert. In W. D. Haglund & M. H. Sorg (Eds.), *Forensic taphonomy: The postmortem fate of human remains* (pp. 139–150). Boca Raton, FL: CRC Press LLC.

Haag, L. C. (2015). Base deformation of full metal-jacketed rifle bullets as a measure of impact velocity and range of fire. *American Journal of Forensic Medical Pathology, 36*(1), 16–22.

Haglund, W. D. (1997a). Dogs and coyotes: Postmortem involvement with human remains. In W. D. Haglund & M. H. Sorg (Eds.), *Forensic taphonomy: The postmortem fate of human remains* (pp. 367–381). Boca Raton, FL: CRC Press LLC.

Haglund, W. D. (1997b). Rodents and human remains. In W. D. Haglund & M. H. Sorg (Eds.), *Forensic taphonomy: The postmortem fate of human remains* (pp. 405–414). Boca Raton, FL: CRC Press LLC.

Haglund, W. D., Reay, D.T., & Swindler, D. R. (1989). Canid scavenging/disarticulation sequence of human remains in the Pacific Northwest. *Journal of Forensic Sciences, 34*(3), 587–606.

Jantz, R. L., & Ousley, S. D. (2005). *FORDISC 3.0: Computerized forensic discriminant functions.* Version 3.1. Knoxville, TN: The University of Tennessee.

Klippel, W. E., & Synstelien, J. A. (2007). Rodents as taphonomic agents: Bone gnawing by brown rats and gray squirrels. *Journal of Forensic Sciences, 52*(4), 765–773.

Pokines, J. T. (2014). Faunal dispersal, reconcentration, and gnawing damage to bone in terrestrial environments. In J. T. Pokines & S. A. Symes (Eds.), *Manual of forensic taphonomy* (201–248). Boca Raton, FL: CRC Press LLC.

CHAPTER 27

The Case of the … Cases: The Flow of the Ordinary into a Medical Examiner's Office

James T. Pokines

CONTENTS

INTRODUCTION

The chance to learn about and solve exciting cases is likely what draws many students to the field of forensic anthropology: one can never be certain what puzzle will show up next. Many cases offer unusual provenience histories of remains (Chapter 10). Others offer difficult perimortem trauma for analysis (Chapter 13). Complex causes of death also are often presented in cases studies (Chapter 17). Some cases offer a series of subtle clues regarding the identity of the individual involved that must be pieced together to develop a biological profile (Chapter 1). Each of these examples can present an important caveat to guide future investigations, providing one possible explanation for a given set of observations and allowing a viable hypothesis to be formed. Unusual cases educate students and professionals about the ever-expanding boundaries of possible case resolutions.

However, many cases do not: the bulk of cases received for forensic anthropological examination, while certainly of great interest to the practitioner, make for unremarkable case studies. These include cases where the primary question is whether the remains are of forensic interest (i.e., recent remains requiring identification). An overreliance upon case studies as an indication as to what one will encounter when practicing forensic anthropology may lead to unrealistic expectations about the nature of this career, as most case studies are chosen precisely because they are atypical.

The following data come from the author's experience as Forensic Anthropologist for the Commonwealth of Massachusetts, working at the centralized office of the Chief Medical Examiner in Boston. While Massachusetts has multiple satellite Medical Examiner offices, all cases requiring forensic anthropological analysis are forwarded to the Boston office, including all skeletonized or partially skeletonized cases. The cases presented in this chapter were originated from November 2011 to October 2018, excluding (extensive) earlier cold cases in storage that required additional forensic anthropological analysis during this period. These were excluded due to the biased nature of the sample, as many cold cases consist of former cemetery remains that usually cannot be identified to an individual (Pokines et al., 2016), or cases that have been in storage for multiple decades and often lack complete provenience information. These older cases receive a full forensic anthropological analysis, including DNA sampling to achieve positive identification whenever possible, but they tend to accumulate to a greater degree than do more identifiable cases. Cases where only radiographic identifications or trauma analysis was performed were also excluded from the present analysis. The latter type of cases normally derive from individuals whose identity is known, but the cause and manner of death are in doubt, and include many cases of dismemberment. Radiographic identifications and trauma analyses may represent a substantial portion of a forensic anthropological case load, but they often represent different types of analytical procedures. The present sample does include some field cases that did not result in any human remains being returned to the OCME for analysis, such as cases where Native American burials were identified as such in the field, and temporal jurisdiction was turned over to the Massachusetts Historical Commission (MHC; i.e., the state archaeology office). Artifact-only cases were omitted from the sample (e.g., empty urns being found washed ashore or buried). Cases where there was only consultation by the forensic anthropologist with the medical examiners which also resulted in no formal report were omitted, and the amount of these is substantial.

The parameters of the sample of human cases are presented in Table 27.1. These are grouped by taphonomic background (i.e., history of postmortem changes). The author employs a taphonomic framework in forensic anthropological analyses, whereby the ultimate case disposition is in large part determined by its depositional and postdepositional history (Pokines, 2018). These taphonomic changes largely derive from the main environment to which the remains were exposed long-term after death (cemetery burial, other burial, terrestrial surface, marine/river, former anatomical teaching specimen, ritual, or trophy; these categories are discussed below). The relative age (postmortem interval) of the remains also was tabulated: archaeological, which includes Native American burials (17.5% of the sample); historical, which are close to or over 100 years old (primarily coffin burials in cemeteries; 17.5%), and modern (those ranging from a few days/weeks to a under 100 years old; 65.5%), with the latter category often including very recent remains. Skeletal representation was scored as isolated elements ($\leq$5 elements; 47.45% of the sample), partial skeleton (25.0%), or complete (or nearly complete) skeleton (27.5%). Skeletal cases also included those where significant soft tissue was present at the time of initial examination, but enough skeletal exposure had occurred that forensic anthropological analysis was requested/required.

Previous research in this jurisdiction has determined that most bone cases enter into the system passively, i.e., no active search by law enforcement was underway at the time of discovery, usually from accidental encounter by pedestrians, dog-walkers, or (in the

TABLE 27.1   A Sample of Forensic Anthropological Case Types (n = 120) at the OCME Boston, Massachusetts, November 2011 to October 2018 (Seven Years), Broken Down by Relative Age (Postmortem Interval) of the Case and Overall Skeletal Representation

TaphonomicBackground	Age					Representation		
	Archaeological/Native Amer	Historical	Modern	Total	%	Isolated	Partial	Complete
Anatomical	0	0	8	8	6.7	5	2	1
Burial – Cemetery	0	19	10	29	24.2	18	7	4
Burial – Other	20	0	5	25	20.8	7	8	10
Marine/River	1	1	18	20	16.7	17	3	0
Indoor	0	0	5	5	4.2	0	0	5
Terrestrial	0	0	31	31	25.8	8	10	13
Trophy	0	0	1	1	0.8	1	0	0
Ritual	0	1	0	1	0.8	1	0	0
Total	21	21	78	120		57	30	33
%	17.5	17.5	65.0			47.5	25.0	27.5

case of buried remains) construction workers (Pokines et al., 2017). After the initial find of human remains, law enforcement personnel then search the area for additional remains and associated evidence. A constant stream of bones in need of identification is the norm.

NONHUMAN AND NON-OSSEOUS CASES

In the practice of forensic anthropology, it is common to have to rule out remains as nonhuman or non-osseous (i.e., not bone). These cases have not been tabulated directly here, as previous analysis for this jurisdiction has determined around 90% of the forensic anthropological case load following the parameters discussed above are determined to be nonhuman or non-osseous (Pokines, 2015a; Woods and Pokines, 2013). The human cases examined here, therefore, are only around 10% of the total number of cases, although they easily exceed 90% of analytical time expended: nonhuman/non-osseous cases can be disposed of with a brief analysis and a single-page form. The proportion of nonhuman/non-osseous cases here is generally higher than that reported by other forensic anthropologists among case work (Bass & Driscoll, 1983; Falsetti, 1999; Marks, 1995). In Massachusetts, the nonhuman species include most commonly cattle (*Bos taurus*), white-tailed deer (*Odocoileus virginianus*), and pig (*Sus scrofa*), with lesser amounts of seal species (Phocidae), birds, sheep/goat (*Ovis aries/ Capra hircus*), domestic dogs (*Canis familiaris*), and many other taxa in small amounts (Pokines, 2015a). While many of these species overlap in size with adult humans, many do not and are highly dissimilar morphologically, such as birds; these possibly are mistaken for human infant remains. Efficiency of case work is greatly increased, including through decreased transportation and scene-securing time of law-enforcement personnel, if the majority of these remains can be ruled out through electronic photographs that include multiple views and a scale (Pokines, 2018). Non-osseous remains were only a small amount of the overall cases but included multiple instances of plastic skeletal remains being turned in as possibly human.

BURIED REMAINS

Former cemetery remains (i.e., those starting off in a coffin in what was usually a demarcated cemetery, even if the coffin has completely decomposed since burial) accounted for a significant portion (24.2%) of the case load (Pokines et al., 2016) (see Figure 27.1). These rarely can be identified to a specific grave and hence to a specific individual, although often the cemetery of origin is known. Colonial cemeteries in Massachusetts often have a long history of use (as far back as the 1620s), and many early graves received only perishable markers or no markers at all. Incomplete record-keeping also led to subsequent graves disturbing previous, no longer marked remains. A history of small, family cemeteries on private property also contribute to the forensic problem. It is possible in some cases to return the remains to the cemetery for reburial, and cases over 100 years of age become the temporal jurisdiction of the MHC. Three cases were more recent in origin, as determined through artifact associations, the condition of the remains, or the age and use history of the cemetery. These include one set of partial remains from an amputation that apparently had been buried, shallowly and unauthorized, in a cemetery plot. The majority (18 of 29 cases) of cemetery remains were isolated remains, which in most instances were disturbed by digging. All of the complete (n = 4) and all but one of the partial (n = 6)

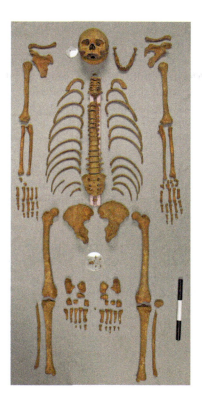

FIGURE 27.1 An example of former cemetery remains, in this case a nearly complete skeleton excavated by the author. Note the overall brown staining from contact with the soil.

cemetery cases were recovered through excavation by the author and/or the MHC. In each case, the remains had already been partially disturbed by construction or new burial excavation and required additional forensic anthropological assessment.

Other burials did not start out in coffins but in direct contact with the soil, and these tend to group as older (archaeological) or more recent (clandestine burials). All of the older buried remains were identified through a combination of ancestry estimation, dental wear, taphonomic state, known proximity to previously identified Native American sites, and contextual evidence as Native American, although European early Colonial remains that were not originally coffin burials certainly could be encountered in this jurisdiction. In Massachusetts, Native American remains get turned over to the MHC for repatriation, and cases with the remains still partially buried have the option of staying in place for reburial.

Recently buried remains away from cemeteries were only a small portion of the sample (n = 5). Each of these was a clandestine burial under suspicious circumstances, requiring significant investigation by law enforcement. These ranged from very recent, largely undecomposed remains to fully skeletonized remains that had been in the ground for several years. All of these were complete/nearly complete remains except one case, where the majority of remains had been placed in a secondary clandestine burial, and an isolated element had been left behind at the primary clandestine burial location in the basement of a house.

TERRESTRIAL (GROUND SURFACE) AND INDOOR REMAINS

A significant number of forensic cases came from outdoor forensic contexts (Pokines, 2016). While remains from other contexts sometimes end up deposited temporarily on the ground surface, such as accidentally disturbed cemetery remains, the taphonomic categories discussed here refer to the environments in which they spent the majority of their history and which altered them most significantly. The terrestrial remains therefore are all of recent modern origin, which is due to long-term exposure of bones on the ground surface (decades to centuries) normally leading to the destruction of bone through taphonomic processes, including flaking apart from subaerial weathering (Junod & Pokines, 2014). Terrestrial cases totaled 25.8% of the overall sample, and the majority of these (n = 23) were complete or partial skeletons. In this environment, it is common for at least partial disarticulation to occur, primarily due to large vertebrate scavengers (Pokines, 2016).

Relatively fewer (4.2%) indoor cases were examined in the present sample, and all were recent. This is likely due to the on-average earlier find of indoor remains: most come from the deceased's own residence, and long-term lack of contact with that individual or decomposition smells often lead to the discovery of indoor remains earlier in the postmortem interval before skeletonization has occurred than outdoor remains. Consequently, indoor cases usually are more easily identified and less likely to need forensic anthropological analysis. The cases examined here include private residences and one dry-docked boat.

MARINE AND RIVER REMAINS

Given the large amount of coastline, it is no surprise that many Massachusetts cases derive from the ocean, either netted while fishing or other ocean harvest activity, or washed ashore (Pokines & Higgs, 2015). These accounted for 16.7% of the sample and consisted only of isolated remains or partial skeletons (see Figure 27.2). Dispersal of remains in the ocean is common, as limbs shed skeletal elements during the decomposition process, with the most distal elements usually lost first (Haglund, 1993; Sorg et al., 1997). This dispersal means that it is common for an isolated skeletal element to wash ashore. In some cases, the partial remains were recovered over multiple instances: one case was compiled from four separate finds of bones washing ashore in the same location over a span of months, and one was compiled from five separate finds over a broader area in a span of years. Nuclear DNA testing can be used to identify isolated remains, which was necessary in the two cases mentioned above to reassociate some of the remains with each other. Recent remains from marine sources often yield positive DNA results despite the exterior changes that the bones have undergone. All marine cases were of recent

FIGURE 27.2   A tibia bleached and rounded from extended time in a marine environment.

FIGURE 27.3   Proximal end of a radius battered from river transport, with traces of dark brown staining likely from former burial still visible.

origin except one, where an isolated skeletal element was recovered near the site of a known, submerged Native American occupation site. Its taphonomic condition indicated long-term marine immersion, including mineral staining, marine organism growth, and battering and rounding of exposed margins.

A single human case was derived from a freshwater context, although these were more common among the many nonhuman cases examined over the same time over which the present sample was accumulated. This consisted of a single radius that likely derived from a historical burial that was eroded from a known cemetery site upstream, based upon the deep brown soil staining of portions of its surface, with the remaining surfaces removed through fluvial battering (see Figure 27.3).

ANATOMICAL TEACHING SPECIMENS

Former anatomical teaching specimens are a common source of forensic anthropological casework, either because they are mistaken for remains from another source or have been seized by law enforcement as a precaution. Sometimes, they are also knowingly turned over to law enforcement by individuals who no longer wish to keep them and do not know what else to do with them, in the hopes of achieving proper disposal. While some remains of this type are easy to diagnose due to the presence of anatomical mounting hardware (e.g., springs to secure the mandible to the cranium, screws and hooks to attach the calotte to the inferior cranium, and pins to keep the calotte from shifting), anatomical labeling, and sectioning, these indicators are not present on all specimens (see Figure 27.4). The postmortem interval of these remains normally spans a few decades, given that most show signs of significant use wear and/or derive from countries that have banned the export of teaching skeletons in recent decades, including India (Pokines, 2015c; Pokines et al., 2017). These can be returned to the source in some cases, if they were legally obtained.

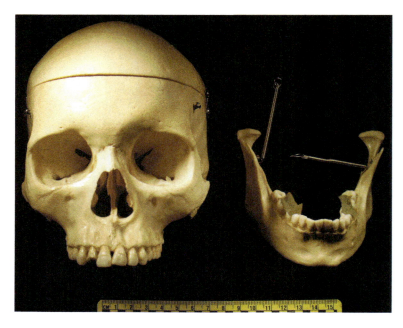

FIGURE 27.4   An example of former anatomical teaching specimens, in this case a skull with clear indications of metal hardware and calotte sectioning.

The majority of former anatomical teaching specimen cases were isolated remains, commonly skulls. The two partial cases were collections of postcranial elements used for osteology training that were turned over to law enforcement voluntarily, as the owners did not know how else to dispose of them and did not want to dispose of them illegally. Only one complete skeleton was examined during the sample period; these are less common, as they are typically still mounted/wired together, and their origin presents no mystery. One anatomical case was unusual in that it was a cranial vault fragment from a teaching specimen lost in a fire over 100 years previously, unearthed during archaeological testing of a college campus. Despite being calcined, this fragment retained its even machine cutting from the sectioning of the vault. Its taphonomic state therefore was consistent with its suspected origin, although sectioning from previous autopsy or practice for surgery could not be ruled out completely.

TROPHY SKULLS

In some military conflicts body parts of slain enemies are collected, and these are broadly termed trophy skulls. The collection of trophy skulls was popular during the Vietnam War and the Second World War in the Pacific Theater, and many were returned to the US (Yucha et al., 2017). After the respective wars, the veterans often hid the remains away (Harrison, 2006). Veterans sometimes tried to repatriate these skulls years later, or they were sometimes rediscovered after the death of the veteran by surviving family members. In either case, they are often reported to law enforcement or sometimes to the relevant foreign embassy/consulate. Forensic anthropological assessment is needed to confirm the origin of these remains, and contextual information including newspaper clippings, diaries, letters, photographs, and military records from the time help establish provenience.

The present sample included only one trophy skull (described in Yucha et al., 2017), although other trophy skulls were still in storage and underwent additional analysis. This skull had an unusual history, in that it had spent decades buried in a back yard, and the veteran who had obtained it during the Second World War requested prior to his death that his son excavate it and turn it over to law enforcement so that it could be repatriated to Japan (repatriation has occurred in this case, via the Japanese consulate in Boston). The sample of cemetery remains also includes one isolated element (a temporal) that had been reported as part of a trophy skull from Okinawa during the Second World War. The development of this bone, however, indicated that the age at death was around one year, and the taphonomic condition was most consistent with previous burial. While the bone may in fact derive from Okinawa, it could not be classified as a trophy skull.

RITUAL REMAINS

Human remains used in modern ritual contexts have the potential to highlight interesting case studies, as they are often associated with unusual artifacts that may be unfamiliar to members of law enforcement in a given jurisdiction. Multiple religious traditions with differing uses of human remains also may intersect with law enforcement, and these need to be distinguished from each other. Santería and Palo Mayombe (see Chapter 24) are West African-derived religions/sects with components of Catholicism, and both involve the ritual use of nonhuman skeletal remains, making them an increasing object of forensic interest. Palo Mayombe also involves the use of human skeletal remains placed within ritual cauldrons (*ngangas*) along with multiple other artifacts, including stones, keys and locks, wooden sticks from different species and other plant remains, beads, animal remains, knives, coins, beads, mercury, and sometimes railroad spikes (Gill et al., 2009; Paolello & Klales, 2014). Santería practices rarely include small amounts of human remains, but Palo Mayombe typically includes whole crania/skulls and a human tibia to be used as a scepter, hence the interest of these cases to law enforcement. Even larger sets of remains, including whole bodies from coffins, may be taken for use in rituals (pers. obs.). The present sample includes a single case of ritual remains, the unusual nature of which made it (quite hypocritically) suitable for publication as a case study (Pokines, 2015b). A *nganga* was recovered from a periodically drained watercourse in western Massachusetts, and it contained typical Palo Mayombe artifacts. It also contained the skull of an adult male, hence the need for forensic anthropological involvement.

The origins of ritual remains themselves, however, may be more mundane. Palo Mayombe calls, ideally, for the acquisition of a human skull and tibia from a cemetery and from an individual who was a criminal or mentally unstable in life so that the spirit of this person will be easier to control through use of the *nganga*, to which it is symbolically bound. In practice, ritual economy (Metcalf, 1981) may overrule these considerations, and the human remains may be obtained from expedient sources, including the purchase of former anatomical teaching specimens, which are unfortunately available for internet purchase. The remains in this case were likely from a cemetery, although it was not possible to link them with any known cemetery disturbance in the area.

LESSONS LEARNED

- The term "forensic interest" can be a misnomer. The author feels that if the case has been sent to or obtained by the forensic anthropologist for analysis, that alone makes it of "forensic interest," as it is only through forensic anthropological analysis that its origin and final disposition are determined. While many cases are recent and may include situations of homicide or suicide and searches for missing persons, these must be separated from other cases where there is no real (or prosecutable) question of foul play, including archaeological, cemetery, trophy, and ritual remains and the constant influx of nonhuman remains. A better term therefore might be "of police interest." The more mundane types form a large proportion of the cases tabulated here and require considerable time and just as thorough documentation in most cases as complex homicides.

- These case proportions are not necessarily typical across the US, in that the subject jurisdiction may have particular regional differences from other states. The time depth for cemeteries from European settlement combined with the population density and degree of new earthmoving construction, for example, may greatly increase the amount of cemetery remains entering into the medical examiner system in Massachusetts (Pokines et al., 2016). The presence of substantial coastline also affects the case load in a way that landlocked jurisdictions will lack; here, marine cases are frequent and often difficult to identify, especially in the case of isolated skeletal elements. Substantial forest cover also may contribute to the number of bodies deposited outdoors, as these areas offer cover for clandestine activities. Massachusetts also differs from other jurisdictions in that it has a centralized medical examiner system covering the entire state, as opposed to some large metropolitan areas where the system is county-based. All forensic anthropological cases in Massachusetts, across a variety of cities, towns, suburbs, and more rural areas, are required to come under analysis in a centralized location. It is therefore unknown if the forensic anthropological case load in this jurisdiction is typical of the US as a whole. Other institutions, such as the Central Identification Laboratory on Oahu, Hawaii, for example, specialize their forensic efforts to the recovery and identification of the remains of US servicemembers from previous wars, including aircraft crashes, battlefield burials, ship wrecks, and exhumations of cemetery unknowns. Their forensic case profile therefore differs from the one outlined here, so multiple sources should be consulted regarding the types of cases typical to a given laboratory.

- It is also important to keep in mind that while forensic anthropology can be an exciting career, it also requires mastery of many practical skills; while the former often gets most of the attention, the latter is the reality for which one must train.

DISCUSSION QUESTIONS

27.1 Do you believe that all forensic anthropology students should be required to take courses on nonhuman osteology (e.g., zooarchaeology, vertebrate comparative anatomy, or comparative osteology)? Why or why not?

27.2 How does taphonomic analysis separate cases into their dispositional categories? How is this approach helpful in case resolution?

27.3 How does the depiction of forensic anthropologists' roles in television shows, in which every case is an exciting riddle that is always solved by the end of the episode, parallel depictions of police and medical professions in popular media? What effect might these portrayals have on juries?

REFERENCES

Bass, W. M., & Driscoll, P. A. (1983). Summary of skeletal identification in Tennessee: 1971–1981. *Journal of Forensic Sciences, 28*(1), 159–168.

Falsetti, A. B. (1999). A thousand tales of dead men: The forensic anthropology cases of William R. Maples, Ph. D. *Journal of Forensic Sciences, 44*(4), 682–686.

Gill, J. R., Rainwater, C. W., & Adams, B. J. (2009). Santeria and Palo Mayombe: Skulls, mercury, and artifacts. *Journal of Forensic Sciences, 54*(6), 1458–1462.

Haglund, W. D. (1993). Disappearance of soft tissue and the disarticulation of human remains from aqueous environments. *Journal of Forensic Sciences, 38*(4), 806–815.

Harrison, S. (2006), Skull trophies of the Pacific War: Transgressive objects of remembrance. *Journal of the Royal Anthropological Institute* (New Series), *12*, 817–836.

Junod, C. A., & Pokines, J. T. (2014). Subaerial weathering. In J. T. Pokines & S. A. Symes (Eds.), *Manual of forensic taphonomy* (pp. 287–314). Boca Raton, FL: CRC Press.

Marks, M. K. (1995) William M. Bass and the development of forensic anthropology in Tennessee. *Journal of Forensic Sciences, 40*(5), 741–750.

Metcalf, P. A. (1981). Meaning and materialism: The ritual economy of death. *Man* (New Series), *16*(4), 564–578.

Paolello, J., & Klales, A. (2014). Contemporary cultural alterations to bone. In J. T. Pokines & S. A. Symes (Eds.), *Manual of forensic taphonomy* (pp. 181–199). Boca Raton, FL: CRC Press.

Pokines, J. T. (2015a). Identification of nonhuman remains received in a medical examiner setting. *Journal of Forensic Identification, 65*(3), 223–246.

Pokines, J. T. (2015b). A Santería/Palo Mayombe ritual cauldron containing a human skull and multiple artifacts recovered in western Massachusetts, U.S.A. *Forensic Science International, 248*, e1–e7.

Pokines, J. T. (2015c). Taphonomic characteristics of former anatomical teaching specimens received at a Medical Examiner's office, MA. *Journal of Forensic Identification, 65*(2), 173–195.

Pokines, J. T. (2016). Taphonomic alterations to terrestrial surface-deposited human osseous remains in a New England, U.S.A. environment. *Journal of Forensic Identification, 66*(1), 59–78.

Pokines, J. T. (2018). Differential diagnosis of the taphonomic histories of common types of forensic osseous remains. *Journal of Forensic Identification, 68*(1), 87–145.

Pokines, J. T., Appel, N., Pollock, C., Eck, C. J., Maki, A. G., Joseph, A. S., Cadwell, L., & Young, C. D. (2017). Anatomical taphonomy at the source: Alterations to a sample of 84 teaching skulls at a medical school. *Journal of Forensic Identification, 67*(4), 600–632.

Pokines, J. T., Eck, C. J., & Sharpe, M. E. (2017). Sources of skeletal remains at a chief medical examiner's office: Who finds the bones? *Journal of Forensic Identification, 67*(2), 278–299.

Pokines, J. T. & Higgs, N. (2015). Macroscopic taphonomic alterations to human bone in marine environments. *Journal of Forensic Identification*, 65(6), 953–984.

Pokines, J. T., Zinni, D. P., & Crowley, K. (2016). Taphonomic patterning of cemetery remains received at the Office of the Chief Medical Examiner, Boston, Massachusetts. *Journal of Forensic Sciences*, 61(S1), S71–S81.

Sorg, M. H., Dearborn, J. H., Monahan, E. I., Ryan, H. F., Sweeney, K. G., & David, E. (1997). Forensic taphonomy in marine contexts. In W. D. Haglund & M. H. Sorg (Eds.), *Forensic taphonomy: The postmortem fate of human remains* (pp. 567–604). Boca Raton, FL: CRC Press.

Woods, K. N., & Pokines, J. T. (2013, February 20). Analysis of nonhuman skeletal material received in a medical examiner setting. Poster presented at the 65th annual meeting of the American Academy of Forensic Sciences, Washington, DC.

Yucha, J. P., Pokines, J. T., & Bartelink, E. J. (2017). A comparative taphonomic analysis of 24 trophy skulls from modern forensic cases. *Journal of Forensic Sciences*, 62(5), 1266–1278.

Index

Cities of The Future

Towards Integrated Sustainable Water and Landscape Management

Cities of The Future
Towards Integrated Sustainable Water and Landscape Management

Editors

Vladimir Novotny
Northeastern University, Boston, MA

Paul R. Brown
CDM, San Diego, CA

Proceedings of an International Workshop held July 12–14, 2006 in Wingspread Conference Center (Racine, WI)

Sponsored by
National Science Foundation, Washington, DC, USA
The Johnson Foundation, Racine, WI, USA
CDM, Cambridge, MA, USA
Northeastern University, Boston, MA, USA
International Water Association, London, UK

Publishing

Published by IWA Publishing, Alliance House, 12 Caxton Street, London SW1H 0QS, UK

Telephone: +44 (0) 20 7654 5500; Fax: +44 (0) 20 7654 5555; Email: publications@iwap.co.uk
Website: **http://www.iwapublishing.com**

First published 2007
© 2007 IWA Publishing

Printed by Lightning Source
Cover design by www.designforpublishing.co.uk
Typeset by Aptara, New Delhi, India

Cover photo provided courtesy of Aqua-Tex Scientific Consulting Ltd. with permission from the
Windmill Developments and Vancity, and architect Busby Perkins and Will.

British Library Cataloguing in Publication Data
A CIP catalogue record for this book is available from the British Library

Library of Congress Cataloging-in-Publication Data
A catalog record for this book is available from the Library of Congress

ISBN: 1843391368
ISBN13: 9781843391364

Contents

Acknowledgement

This material is based upon work supported by the National Science Foundation under Grant No. 0544161. Any opinions, findings, and conclusions or recommendations expressed in this material are those of the author(s) and do not necessarily reflect the views of the National Science Foundation.

Wingspread Workshop Organizing Committee

Vladimir Novotny, CDM Chair Professror, Northeastern University, Boston, MA, USA (workshop organizer)

Paul R. Brown, President, Public Services Group, CDM, Carlsbad, CA, USA

Kate Bowditch, Charles River Watershed Association, Weston, MA, USA

Lee Breckenridge, School of Law, Northeastern University, Boston, MA, USA

Jiri Marsalek, Canada Centre for Inland Waters, Burlington, ON, Canada

Peter Shanahan, Massachusettes Institute of Technology, Cambridge, MA, USA

Ex-officio:

Patrick Brezonik, Division of Bioengineering and Environmental Systems, National Science Foundation, Washington, DC, USA

Carole Johnson, The Johnson Foundation, Racine, WI, USA

Andrew Speers, International Water Association, London, UK

Introduction to the book

It is my great pleasure to introduce this monograph of edited contributions presented July 12–14, 2006 during the Wingspread International Workshop "Cities of the Future – Bringing Blue Water to Green Cities". The mission and the reputation of Wingspread conferences and workshops are in convening smaller meetings of thoughtful and rigorous inquiry in an atmosphere of candor and purpose to advance the knowledge and reach consensus and solutions. Candor comes from trust which develops when participants get to know each and interact. Sustainable development and environment is one of three major focus areas of Wingspread conferences.

In the terminology of sustainable development of cities, blue water means clean water suitable and safe for human consumption, swimming and for protecting aquatic life. Green cities are those communities that are based on the principles of sustainable landscape ecology and do not damage or overuse their water resources. In spite of billions spent we have not yet reached these goals. The chapters presented in this book investigate the fundamentals of sustainable urban development, vulnerability of cities to the stresses by extreme hydrologic events such as hurricanes and tsunami, population increase, changes of urban landscape, and pollution. Other chapters deal with solutions to these problems, how to make the cities and their landscapes resilient to these stresses, what are the barriers and how to overcome them.

The invited workshop participants/delegates and writers of the chapters are the leading scientists and environmental specialists devoted to research and protection of urban ecologic systems. The compositions of delegates and chapters are interdisciplinary, cutting across the board of science and engineering. Delegates arrived from eight countries and included members of the National Academy of Engineering, several endowed chair professors, experts, and distinguished leaders of the environmental community. Their specializations ranged from urban hydrology,

hydrology, ecology, management, economics, landscape architecture, and law. The sponsorship of the National Science Foundation, which I now represent, signifies and underlines the importance NSF is putting on solving scientifically the problems the cities are facing such as aging and ecologically unsustainable infrastructure, insufficient supply of good quality water, threats of pollution, and impairment of integrity of urban waters. Urban water bodies represent the lifeline of the cities and their sustainability is a prerequisite to protection of public health and ecology. The aftermath of the hurricane Katrina in a historic City of New Orleans and destruction of urban areas in the coastal Louisiana and Mississippi revealed a high vulnerability of coastal cities to extreme climatic events that is expected to increase with global warming. In the same time, urban water bodies will have to receive increased effluent flows and loads of pollution form urban landscape that are then later reused by downstream cities for water supply and irrigation. Urban waters are also stressed by overuse resulting in diminished or even disappearing flows.

The workshop presentations and deliberations contained in this book outline the paths towards the hydrological and ecological sustainability of cities. The workshop also forged alliances and cooperation between universities, agencies, consultancies, and nongovernmental environmental organizations which is hoped to increase with time as one of the most important part of engineering and ecological science.

Special thanks go to the Johnson Foundation that provided sponsorship for the workshop and the facility designed by the most famous US architect, Frank Lloyd Wright. Wright was a master whose architectural vision was focusing on letting buildings be a part of nature and surroundings as exemplified not only by Wingspread but also by his other architectural marvels, such as Falling Waters in Pennsylvania, Taliesin in Wisconsin or Imperial Hotel in Japan. Sponsorship of CDM, a leading environmental engineering consultancy, is also acknowledged and appreciated. The International Water Association, the largest international professional organization dealing with water, sponsored the workshop and its publishing arm, IWA Publishing, has produced this book.

Prior to my current position at NSF I was a Dean of the College of Engineering at the Northeastern University in Boston. The Wingspread workshop was organized and this monograph has been produced by the Center for Urban Environmental Studies of the Northeastern University. The university is striving towards being one of the top urban research universities and the Center for the Urban Environmental Studies is an important part of this effort. NEU has been ranked in the US as a number one institution of higher learning in cooperative engineering education and internships.

Allen L. Soyster
Director of Division of Engineering Education and Research Centers
National Science Foundation
Washington, DC, USA

Preface
Cities of the future: The fifth paradigm of urbanization

V. Novotny

Northeastern University, Boston, MA 02115, USA
E-mail: Novotny@coe.neu.edu

P. Brown

CDM, Carlsbad, CA
E-mail: BrownPR@cdm.com

Summary: This preface outlines the history of paradigms of urban water resources and drainage and defines the fifth paradigm of sustainability for the Cities of the Future. It also summarizes the results of the Wingspread Workshop: Cities of the Future – Bringing Blue Water to Green Cities held in Wisconsin in July 2006 and the recommendations of the workshop.

HISTORICAL PERSPECTIVES – PARADIGMS

The physical connections (both structural and natural) between cities and their water resources have changed through the centuries. At the same time, our conceptual models of these systems and our understanding of how they should function and relate to one another have changed as well. There are at least four recognizable historical models or paradigms that reflect the evolution and development of urban water resources. The emergence and broad acceptance of a new model, or "fifth paradigm," is documented in the monograph that follows. First, a brief review of historical perspectives.

In the *first paradigm*, ancient cities relied on wells for water supply; surface water bodies for their functional needs like transportation, irrigation, and washing;

and streets for waste disposal and drainage (both rain water and snow melt). The well-preserved Roman city of Pompeii had stone paved narrow streets equipped with stepping stones for pedestrian crossings that, during rainfalls, served as drainage channels. Almost all waste, including fecal matter from people and animals, was disposed onto streets where street sweepers, night soil collectors, and stormwater were the primary means of getting rid of detritus. In medieval Paris, solid waste accumulations in the streets often reached a meter in depth, emitting nauseating odors. Only main streets were paved with semi-permeable cobblestone surfaces. Side streets were unpaved. Some waste was recycled. For example, Romans used urine as bleach in their laundries. This first paradigm is characterized by the utilization of shallow groundwater, the exploitation of easily accessed surface water bodies for many purposes, and the shared use of streets for the conveyance of people, waste products, and rain water. These conditions were prevalent into the middle ages in cities in Europe, China, Japan, and other countries.

As water demands increased and easily accessed local groundwater and surface supplies were insufficient to support life and commerce, a *second paradigm* emerged in growing ancient (Rome is said to have had a population approaching one million) and medieval cities – the engineered capture, conveyance and storage of water.

There are many examples of water brought to ancient cities from long distances. For example, the aqueducts of ancient Rome brought water to fountains and villas from mountains as far away as fifty kilometers. In many cities and even rural castles, rain water was collected and stored with drinking water in underground cisterns. The Basilica Cistern in the ancient east Roman capital of Constantinople (present Istanbul) built in 6th century by Byzantine Emperor Justinian could hold $80\,000\,m^3$ (21 million gallons) of rainwater and spring water collected 19 kilometers from the city.

To handle the increased urban runoff, the invention of sewers allowed polluted street flows to be conveyed underground. The Roman sewer Cloaka Maxima has been functioning for more than two thousand years. Sewers were installed much later, however, in other European cities, in most cases during the industrial revolution in the eighteen and nineteen centuries, and in Japan after the World War II. Gradually, domestic wastewater was introduced into these storm sewers following the introduction of flushing toilets in Europe and the United States (while toilets that flushed were common in the baths and villas of ancient Rome; it would be centuries before ordinary citizens would avail themselves of such luxuries). With the introduction of wastewater the so-called "combined" sewer was created. And while combined sewers solved the immediate problem of getting wastewater and stormwater out from under foot, it overloaded receiving waters with raw sewage and created new problems downstream.

The third paradigm for urban water and wastewater added a massive investment in the control and treatment of point sources of pollution resulting from combined and separated sewer systems and provided increased treatment of potable water supplies. These improvements emerged at the beginning of the twentieth century and were driven by epidemics of waterborne diseases caused by the contamination of water supplies with raw sewage discharges and by leaking sewers contaminating wells. Beyond the deadly health risks, the smell of anoxic receiving waters overloaded with biodegradable organic and fecal sludge was overpowering. At the beginning of the twentieth century, the British Parliament could not meet during the summer because of the smell of the Thames River.

At the same time, impervious surfaces in cities were also increasing, resulting in higher volumes of stormwater runoff and more frequent flooding. To address the problem of increasing runoff and urban flooding, as well as poor water quality, many rivers were lined and occasionally buried underground. The aim of these *fast conveyance* urban drainage systems was to remove large volumes of polluted water as quickly as possible, protecting both public safety and property and discharging these flows without treatment into the nearest receiving water body. The third paradigm introduced first primary and then secondary wastewater treatment but did not address the overall, uncontrolled water-sewage-water cycle (see chapters by Lanyon and Novotny in this monograph). In the United States, the period culminated in the passage of the Water Pollution Control Act Amendments of 1972 (Clean Water Act) making end of pipe treatment mandatory.

In our *fourth paradigm*, attempts to control pollution originating from diffuse, non-point sources were added to the growing complex of structural water management infrastructure. This paradigm could also be called the *"end-of-pipe control"* because the predominant point of control of both point and diffuse pollution is where the polluted discharge enters the fast conveyance system (sewer or lined channel) or the receiving water body.

Pollution from urban runoff and other diffuse sources was recognized as a problem only about thirty to forty years ago. The proponents of the Clean Water Act noted this type of pollution and included in the Act provisions for unregulated voluntary controls of nonpoint pollution. At the same time, Safe Drinking Water Act was introduced to protect the quality of drinking water and its sources that were also polluted, primarily by diffuse (nonpoint) pollution. At the end of the twentieth century the European Parliament enacted the Water Framework Directive. The period between the enactments of the Clean Water Act in the United States and the Water Framework Directive in Europe until the beginning of this century has comprised this *fourth paradigm* of urban water management and protection in which both point and increasingly diffuse sources of pollution were considered (see also a chapter by Yamada and Muhandiki in this monograph) and addressed in many separate and discreet initiatives.

While vast sums of money have been spent in developed countries on point and diffuse pollution controls, many have realized that no matter how much money is spent to reduce controllable regulated sources of pollution, the integrity of overused urban water bodies has been severely impaired and will remain so if the fast conveyance, end of pipe treatment paradigm alone continues to be the prevailing model.

The fast-conveyance drainage infrastructure conceived of in Roman times to eliminate unwanted, highly-polluted runoff and sewage has produced great gains in protecting public health and safety. And yet, in spite of billions spent on costly "hard" solutions like sewers and treatment plants (see a chapter by Brown in this monograph) water supplies and water quality remain a major concern in most urbanized areas. A large portion of the pollution is caused by the typical characteristics of the urban landscape: a preference for impervious over porous surfaces; fast "hard" conveyance infrastructure rather than "softer" approaches like ponds and vegetation; and rigid stream channelization instead of natural stream courses, buffers and floodplains. Because the hard conveyance and treatment infrastructure under the fourth paradigm (sewers, combined and sanitary sewer overflow facilities, treatment plants) was designed to provide only five to ten year protection these systems are usually unable to safely deal with the extreme events and sometimes failed with serious consequences.

A new paradigm is emerging from the successes and failures of efforts to control pollution that offers the promise of adequate amounts of clean water for all beneficial uses. Urban waterways are the historic core of our cities' economies and have the potential to be rich sources of biological diversity, contributing to the quality, economy and health of urban life.

TOWARD HYDROLOGICAL AND ECOLOGICAL SUSTAINABILITY OF FUTURE CITIES – THE FIFTH PARADIGM

Sustainable development has been defined as "development that meets the needs of the present without compromising the ability of future generations to meet their own needs" (Brundtland et al., 1987; see also a chapter by Speers in this monograph). Mays (2007) presented several definitions of water resources sustainability that comply with Brundtland more general definition, for example

> *Water resources sustainability is the ability to use water in sufficient quantities and quality from the local to the global scale to meet the needs of humans and ecosystems for the present and the future to sustain life, and to protect humans from the damages brought about by natural and human-caused disasters that affect sustaining life.*

Mays (2007) then presented several historic examples of civilizations that vanished, most likely because of extreme natural events such as drought and flooding. This occurred in regions of Middle East and also in pre-Columbian Americas, including the Hohokam and Anasazi cultures in Southwestern United States. However, catastrophes on a similar scale have also occurred in recent history. The flood damage in New Orleans and destructive tidal surge along the Mississippi coastline resulting from Hurricane Katrina (see chapters by VanHerdeen and Englande et al.) may have caused similar irreparable damage and population disappearance.

The need for ecological sustainability of urban watersheds and water resources and their resilience to extreme events (floods, storm surges, tsunamis) and/or excessive pollution leads us to the *fifth paradigm* of urban water management, a model of sustainable and resilient urban waters and watersheds. This paradigm adopts a holistic, systems approach to the urban watershed, rather than a functionally discrete focus on individual components (drinking water, sewage, stormwater) characteristic of earlier models.

The main drivers of change to sustainable urban aquatic and terrestrial ecological systems are:

1. **Population growth and migration**
   In the next 50 years the world population is expected to increase by 50 percent, magnifying already overpopulated urban centers, especially in developing countries. In about 40 years, the population of the United States will increase by 100 million or 33 percent. Nearly 75 percent of the United States population today is served by municipal water supply and wastewater systems, which is expected to increase to 90 percent in the future. This growth will no doubt increase pressure on already overstressed water resources (see a chapter by Marsalek et al).

2. **Global climatic changes**
   There is general consensus among the scientific community that the earth's atmosphere will be, on average, warmer by several degrees by the end of this century. The impact on water resources may be significant, not only on quantity but also on quality and water temperature. For one, more heat will be released by urban cooling systems. Overheated pavements will carry warmer runoff into receiving waters that may have less flow during warm periods and already higher temperatures due to more solar radiation. Global warming is expected to increase the frequency and strength of hurricanes and coastal flooding and may result in rising sea levels. It may also stimulate shift of phytoplankton composition of surface water resources from diatomic and green alga population to prolific and resistant cyanobacteria that produce harmful toxins and cause bad taste and odor in water.

3. **Land stewardship and ethics**
   The work of Aldo Leopold, a professor at the University of Wisconsin and land steward fifty years ago, defined the notion of land ethics and nature sustainability that includes both waters and their watersheds. People generally put high value on protecting nature and desire sustainability. This demand for attaining and protecting integrity of aquatic ecosystems has been incorporated into the environmental protection laws of many countries. It galvanizes citizens and will drive the urban engineering and ecologic efforts to green and sustainable development and retrofitting.

4. **Multiple use and overuse of urban waters**
   Urban waters and their corridors provide benefits for many, often conflicting uses. Withdrawal of water for drinking and other uses diminishes natural flows and the used water is often transferred via sewers to a discharge point many kilometers downstream. Hydraulic modification of streams to accommodate increased flood flows damages or destroys habitat and reduces recharge of groundwater.

5. **Hardening of the urban landscape**

   High imperviousness and sewers built in the last hundred and fifty years in United States cities have dramatically changed the hydrology of urban areas. Imperiousness increased peak high flows in urban streams by a factor of 4 to 10 and diminished the base flow. Increased variability and higher frequency of bank overtopping has resulted in unstable eroding channels and a loss of habitat in many locations.

6. **Long distance transfer of water and sewage**

   As discussed above, historically, local water supply and sewerage systems were replaced by long distance transfers of water and sewage in regional systems (see chapter by Heaney in this monograph). The long transfers of raw water and wastewater deprived water bodies in the source area of flow and increased the effluent content of receiving waters supplied by large regional treatment plants (see a chapter by Novotny in this monograph) are the consequences.

7. **Improved efficiency of wastewater and storm runoff treatment**

   The level of treatment of urban wastewater achieved in many communities has reached such efficiency that effluent quality is similar to and occasionally better than that of upstream receiving waters. Some utilities have justifiably changed their names from "wastewater" or "sewage" treatment to "water reclamation" agencies. The treatment efficiencies for removing biodegradable organics, nutrients, and toxics are rapidly improving. Small package and highly efficient and automated treatment plants for subdivisions and small communities have been developed and are economically affordable in many applications.

   New and old technologies have emerged for best management practices dealing with the quantity and quality of urban runoff. Urban rainwater has been collected and used for millennia throughout the world. Technologies are available and within the reach that will enable rediscovery of the vast potential of these natural sources and the redesign of urban drainage systems to maximize the benefits of reuse (see a chapter by Heaney).

8. **Limits have been reached and developing remedies is urgent**

   In most major urban areas, ecological systems, both terrestrial and aquatic, have reached limits or thresholds that will require rethinking water management concepts. Groundwater levels in Tucson (arid west), and downtown Boston, and Mexico City, and cities throughout the world have been drawn down to levels of serious concern – threatening the availability of supplies for all uses and undermining buildings and structures as a result of land subsidence (see chapter by Shanahan). The City of Boston is using large volumes of drinking water for replenishment of groundwater to prevent subsidence resulting from high imperviousness, dewatering of underground tunnels, and old fast-conveyance drainage.

   Urban streams have lost their base flow, yet the channels cannot accommodate the increased floods. Many urban streams have been covered and put out of sight or converted to concrete lined channels. Temperature of some receiving urban water bodies used for cooling (e.g., the Lower Des Plaines River in Illinois or Charles River in Boston) and/or summer flows from overheated black urban impervious surfaces can reach high levels lethal to fish. Riparian corridors and floodplains have been built over.

When looking towards the future of cities, the evolving paradigm is a model of integration of both new and older urban developments with the landscape, drainage, transportation and habitat infrastructure that will make cities resilient to extreme hydrological events and pollution, while providing an adequate amount of clean water for sustaining healthy human, terrestrial and aquatic lives; and an optimal balance among recreation, navigation and other economic uses.

Sustainable cities of the future will combine concepts of "smart green" development (see chapters by Ahern and Hill); interconnected ecotones (parks, river riparian zones); and the control of diffuse and point source pollution from the landscape. They will be based on reuse of highly treated effluents and urban stormwater for various purposes including landscape and agricultural irrigation; groundwater recharge to enhance groundwater resources and minimize subsidence of historic infrastructure; environmental flow enhancement of effluent-dominated and flow-deprived streams; and ultimately for water supply (see chapter by Furumai).

This fifth paradigm will evolve from the concept of the total hydrologic water and mass balance where all the components of water supply, stormwater, and wastewater will be managed in a closed loop (Figure 1.1). It will incorporate landscape changes including less imperviousness, more green space used as buffers and for groundwater recharge (see chapters by Ahern and Hill), and it will restore the landscape's hydrological and ecological functions (see chapter by Maimone et al or Yamada and Muhandiki). It will rely on greatly enhanced removal of organic chemicals, nutrients and endocrine disruptors from effluents and will promote the application of best management practices that provide treatment, water conservation, and storage of excess precipitation for reuse (chapters by Furumai and Yamada and Muhandiki). Closing the water loop may require decentralization of some components of the urban water cycle in contrast to the current highly centralized regional systems employing long distance water and wastewater transfers (see chapter by Heaney).

One of the goals of the fifth paradigm is to develop an urban landscape that mimics but not necessarily reproduces the processes and structures present in a predevelopment natural system. Eco-mimicry includes hydrological mimicry, where urban watershed hydrology imitates the predevelopment hydrology, relying on reduction of imperviousness, increased infiltration, surface storage and use of plants that retain water (e.g., coniferous trees). It will also include interconnected green ecotones around urban water resources that provide habitat to flora and fauna, while providing storage and infiltration of excess flows and buffering pollutant loads from the surrounding urban surfaces (chapters by Ahern, Hill and Furumai).

The new or retrofitted urban landscape will be based on storage-oriented drainage with less reliance on underground conduits and more surface storage, infiltration and flow retardation. It will be based on ecological principles to sustain urban terrestrial and aquatic biota. In coastal areas susceptible to extreme weather,

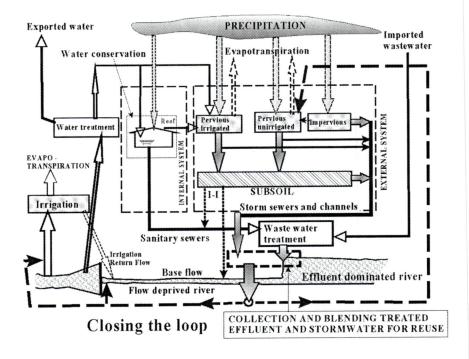

Figure 1.1. Recycle and reuse of wastewater and urban runoff (dashed line) closes the loop in the total hydrologic cycle in urban areas.

the landscape of the city and surrounding areas (including barrier islands) will provide resiliency and protection of humans from the catastrophic consequences of these occasional events (see chapter by VanHerdeen).

Currently, elected officials, community and business leaders, and environmental interests in many major cities are promoting "green" and "smart" development – ideas and programs that incorporate ecological principles into urban development. Their "quality of life" goals are broad; frequently, ecologically-balanced aquatic systems are not recognized as offering the enormous potential for improvement that is possible. There is a need to unify our thinking on these subjects and propose "soft" and "hard" approaches that work together to reverse the adverse effects of urbanization; repair ecologically damaged or even destroyed urban water resources; restore natural hydrology of streams; and recharge groundwater with collected rainwater, increased water conservation and reuse of treated effluents.

In most cases, green developments are small piecemeal patches created by individual developers rather that an integrated effort of the entire community to incorporate interconnected functioning ecotones into the urban area (see chapter by Hill). The pioneering and visionary concepts developed by the famous nineteenth-century landscape architect, Frederick Law Olmstead, of building interconnected

Figure 1.2.   Wingspread conference center.

urban landscape ecotones (parks), often surrounding urban streams, was abandoned by most United States cities in the latter half of the twentieth century.

While ecologically and hydrologically sustainable urban areas provide a significant means of protecting public health (the original and principal goal of civil engineering), they offer the potential of addressing other issues and problems affecting urban public health. Issues such as human exposure to toxic pollutants in the atmosphere, legacy pollution in paint, in soils, and food contaminated by irrigation water are also being addressed in the drive towards ecologically and hydrologically sustainable urban systems. Eliminating public health catastrophes caused by extreme hydrologic and pollution events will be the most obvious area of a concentrated focus of the fifth paradigm. These problems can be addressed in parallel, coordinated research efforts by public health scientists and professionals.

Despite the potential benefits offered by a holistic, systems-based perspective, there are significant economic and institutional barriers to the sustainable, ecologically and hydrologically functioning green development and retrofitting of cities (see chapter by Clune and Braden). Correcting and eliminating these barriers will be a challenging task and will require education, legislative actions by the central and local governments and citizens stakeholder involvement (see the chapter by Adler). Innovative financing mechanisms must be proposed, investigated and implemented (see chapter by Bowditch).

The grand challenge now facing the environmental and urban planning and design professions is to create a set of tools that can be used by government agencies and industry to develop and implement plans for restoration of impaired urban watersheds and maintain sustainable management of multiple-use urban water bodies and landscape. These tools are essential as communities "retrofit" their water infrastructure in older cities and design new systems for expanding or revitalizing urban centers.

In order to develop the tools, models, and guidance manuals, researchers must collect and analyze data provided by new "urban observatories." Such observatories are now in place in Baltimore (Maryland) and Phoenix (Arizona). More observatories for interdisciplinary investigations of urban ecological and hydrological systems funded by the National Science Foundation and other agencies/partners will be installed by universities and other organizations to investigate water flows and mass balances of pollutants in the urban environment (see chapters by Welty and Baker and Brezonik). International cooperation, exchanges of scientists and students with countries of Europe, Asia, and with Australia has to be established on a large scale. The results of the research and experience with implementing the fifth paradigm will need to be extrapolated to megacities in developing countries, knowing that the solutions may not be exactly the same as those to be implemented in the developed countries.

WINGSPREAD WORKSHOP

A two-day international interdisciplinary workshop of experts was convened in the Wingspread Conference Center (Racine, Wisconsin), operated and managed by The Johnson Foundation. The workshop developed visionary concepts on how to ensure that cities and their water resources become ecologically sustainable and are able to provide clean water for all beneficial uses. With urban waters as a focal point, the experts at the workshop explored the links between urban water quality and hydrology, and the broader concepts of green cities and smart growth, and addressed the current vulnerability to extreme hydrologic conditions. The workshop also addressed legal and social barriers to urban ecological sustainability and proposed practical ways to overcome those barriers.

The workshop was organized by the Center for Urban Environmental Studies of the Northeastern University in Boston and sponsored by the National Science Foundation (Washington, DC), CDM (Cambridge, Massachusetts), The Johnson Foundation (Racine, Wisconsin), and the International Water Association (London, UK).

The workshop papers presented in this book are aimed at establishing an agenda for achieving ecologically and hydrologically balanced water use, drainage and wastewater disposal systems, as well as the remediation of damaged water bodies and watershed landscape – all contributing to the quality of life and economic vitality of cities.

Each of the twenty-five invited international expert speakers prepared a chapter on an assigned interdisciplinary discussion topic, summarizing the state of the art and developing a vision of research and actions that would lead to sustainable urban landscape and receiving waters. The workshop participants also included invited experts who participated in the discussion, commented on the presentations, reviewed the chapters, and participated in the formulation of the final workshop synthesis. The total number of invited participating delegates arriving from seven countries (USA, Botswana, Canada, China, France, Japan, and Sweden) was forty.

The chapters in these book/proceedings are organized around the following themes:

- Keynote introductory presentations
- Extreme events
- Urban observatories and total mass balance of pollution in cities
- Hydrologic linkages, sustainable drainage
- Urban pollution stresses and reclamation of urban waters
- Technological solutions
- Sustainable urban landscape concepts – green infrastructure
- Implementing future urban hydrological and ecological systems

RECOMMENDATIONS

In addition to listening to and discussing the presentations, delegates were engaged in extensive discussions. This is the core of Wingspread programs. Some key recommendations proposed by the delegates in the discussion session were:

1. Elevate the importance of water as a central and essential organizing element in a healthy and sustainable urban ecosystem.
2. Develop and implement new approaches to the management of urban water systems, including:

   - Moving towards an integrated system approach based on the total hydrologic cycle that addresses all of the uses of water in the urban environment;
   - Building multiple benefits into all projects and programs that contribute to the economic, social, and environmental health of cities;
   - Promoting new, innovative design concepts that incorporate structural solutions and natural system restoration, replication, and enhancement; and
   - Adopting adaptive management focused on prototype development, implementation, monitoring, and performance evaluation.

3. Increase the resilience and redundancy of urban water systems to resist and rebound from extreme events.
4. Create multifunctional urban landscapes that would be hydrologically and ecologically functional.

The overriding recommendation of the workshop participants is to support the formation of a national and international coalition of researchers, stakeholders, agencies, and government officials dedicated to establishing the fifth paradigm and enlarging the movement towards sustainable urban cities and their life-giving water resources.

REFERENCES

Brundtland, G. (ed.). (1987). *Our Common Future: The World Commission on Environment and Development*, Oxford University Press, Oxford.
Mays, L.W. (2007). *Water Resources Sustainability*, McGraw – Hill, New York, and WEF Press, Alexandria, VA.

PART ONE

Urban Water Sustainability

1

1

The importance of water infrastructure and the environment in tomorrow's cities

Paul R. Brown, AICP

President, Public Services Group, CDM, San Diego, CA
E-mail: BrownPR@cdm.com

Summary: Public sector municipal government and utility leaders responsible for providing reliable water, wastewater, and stormwater management are confronted by several important trends affecting the future of cities. These trends include the need to increase the social and economic benefits created by urban infrastructure, improving collaboration among overlapping agencies and jurisdictions, making the transition from "fast conveyance" to "closed-loop" systems, introducing public stakeholders into decision-making and program implementation, and preparing for extreme events. All of these issues suggest changing priorities within urban communities, elevating environmental and energy issues to a higher level than mobility and economic development. This key note chapter concludes with a discussion of the implications of that change to those involved in planning, designing, and implementing traditional civil, environmental, and hydraulic infrastructure projects.

INTRODUCTION

CDM is an interdisciplinary consulting, engineering, construction, and operations firm whose corporate mission is "to improve the environment and infrastructure." The company works for public and private sector clients from over 100 offices worldwide. Our public sector clients include municipal government and utility leaders directly responsible for providing urban infrastructure, protecting the urban environment, and delivering reliable services in the areas of water, wastewater, and stormwater management.

This monograph provides a brief summary of several issues and trends confronting our public sector clients. Many of these issues are addressed in the work and contributions of the experts who gathered for the Wingspread workshop. They reflect the challenges faced by those on the front line of urban infrastructure management. In the end, they have important implications regarding the role that water resources and the environment play in providing for healthy and sustainable cities of the future.

Increasing the social and economic benefits provided by environmental infrastructure

More than ever, CDM engineers and scientists are working with clients to address the social and economic needs of their communities, in addition to providing reliable services and keeping them in compliance with regulatory requirements. Reliability and compliance are the minimum expectations of every public agency and utility. Many communities are asking for more, however. They are asking that every urban infrastructure project be viewed as a multi-purpose, multi-benefit opportunity.

Advocates for sustainability challenge corporations to consider the social and environmental returns they produce, in addition to the traditional economic bottom line performance that drives decision-making. Our challenge is to create greater economic and social returns from infrastructure projects designed primarily to protect and improve the environment and public health. How can we take the billions of dollars that will be spent on controlling combined sewer overflows, for example, and use those dollars to do more than simply hollow-out caverns underground. Delivering bottom line environmental, economic, and social returns (the so-called "triple bottom line") should be expected from every institution and agency in our society. There is a real need for creativity and innovation in delivering those benefits in traditional civil, sanitary, and hydraulic infrastructure projects.

Improving collaboration among agencies and jurisdictions

Every day we are addressing the convergence of urban utility functions. The most obvious example is the overlap and, in some cases, consolidation of water, wastewater, and stormwater utilities. But convergence goes beyond that. The desire for sustainability in every aspect of urban development heightens the environmental and energy aspects of all urban infrastructures – particularly in buildings and transportation systems. There is a need for better inter-jurisdictional collaboration that goes beyond building a fence around the "silos." This means more integrated planning efforts, improved system modeling capabilities, and a sustained commitment to joint project planning, implementation, monitoring, and accountability for results.

In a similar vein, we are seeing the erosion of governmental borders and legal property lines in favor of the softer natural transitions that define the boundaries of topography and ecosystems. Looking for watershed-based approaches in densely urbanized, multi-jurisdictional settings introduces conflicts and incongruities in the way individual stakeholders literally "see the world." We need to recognize and address the inherent conflicts that exist between "bright-line" legal borders and the borderless continuum of natural systems.

Making the transition from fast-conveyance to closed-loop systems

Almost everywhere, we are attempting to transition from "fast-conveyance" systems to more closed-loop, self-sufficient systems, and that is not easy. The increased demands for water reclamation and reuse creates concerns about water quality degradation and public health, while at the same time offering increased sustainability and greater independence from over-committed sources of supply. This public policy debate is far from over, and we need much better tools for presenting and discussing health and safety risks in open public meetings.

Introducing public stakeholders into decision-making and implementation

As we transition from depending solely on large-scale structural solutions to those that restore or mimic natural systems, we must rely on the most dynamic force in the urban landscape – its human inhabitants – to help. Those institutions and utilities that have met constituent and rate-payer needs as "invisible infrastructure" will need to reacquaint themselves with the partners needed to function in an "integrated" world.

Whether we state it explicitly or not, we are asking for more understanding, engagement, and stewardship from ourselves and our customers. We are looking for supportive leadership and volunteers in every community. We don't need everyone everywhere, but we do need someone everywhere. The new world of conservation, reduced development impacts, and long-term sustainability is going to force us into partnerships with the people we serve. We are going to do it with our customers as partners and not simply for them.

We need structured, documented, and transparent decision-making combined with improved communication, simulation, and visualization tools for public stakeholder dialogue and education.

Preparing for extreme events

In spite of the politics, we are seeing the wide acceptance of the notion that something is changing with the weather. In most cases the discussion has evolved from "is it real?" to "what are we going to do about it?" That reality has led to a much greater emphasis on: (1) minimizing emissions of green-houses gases, and (2) planning for and adapting to extreme events. What are the implications on our engineered systems? As we respond to the global implications of altered weather and changes in sea-level, the public looks to us to rise to the occasion – not necessarily with bigger and bolder structural solutions (although these are no doubt part of the answer) but with fundamental re-thinking of the relationships between human settlements and natural ecosystems – a challenge beyond any we have undertaken to date.

LEADING IN A TIME OF RAPIDLY CHANGING PRIORITIES

It is now becoming apparent that a significant shift in the priorities placed on urban infrastructure is occurring, with the environment and energy moving ahead of mobility and economic growth in terms of their relative importance to the public. These priorities have been well documented in the United States by Gallup and other public opinion surveys (Center for American Progress, 2005).

> "Americans continue to favor the environment when asked to choose between environmental protection and economic growth. After dipping slightly below 50% last year [2004], a majority (53%) once again says that protection of the environment should be given priority, when environmental protection conflicts with economic growth."

Are we keeping up with the challenges reflected in those changed priorities? It is not clear that we are and much more can be done. Let me over-simplify to make a point.

Historically, urban water infrastructure has been something of an enlightened afterthought. After the shelter, after the roads, after the commerce, and after the disease and squalor, the plumbing followed. Hydraulic and sanitary systems were developed from antiquity through the modern age largely in response to the demands and problems created by the evolution of cities themselves. They are a reaction to increasing populations and urbanization. In this respect, they have been a lagging technology in the urban environment – and maybe in some respects they still are.

Cities grew at locations offering opportunities for transportation, housing, agriculture, and industry – that, of course, frequently led to the establishment of cities along waterways. But those waterways were almost always viewed as a means of transportation, a raw water supply, a source of energy, and a convenient location for waste disposal. Our growing cities have beneficially exploited rivers in almost every conceivable way, always looking at the river as a means to an *end* – rarely an end in itself.

Our core business of environmental engineering lagged behind those roads and buildings, providing the "plumbing" – often after-the-fact – in response to public health crises, flooding, and deadly levels of pollution and environmental degradation.

It's not that the contributions resulting from better plumbing haven't been appreciated for their public health and quality of life benefits. Lewis Mumford (1961) stated it well:

> "Perhaps the greatest contribution made by the industrial town was the reaction it produced against its own greatest misdemeanors; and, to begin with, the art of sanitation or public hygiene.... Nineteenth-century achievements in molding large glazed drains and casting iron pipes, made possible the tapping of distant supplies of relatively pure water and the disposal, at least as far as a neighboring stream, of sewage; while the repeated outbreaks of malaria, cholera, typhoid, and distemper served as a stimulus to these innovations, since a succession of public health officers had no difficulty in establishing the relation between dirt and congestion, of befouled water and tainted food, to these conditions."

In the development of cities, we have responded heroically to the failures of economic success. For the future, our clients are looking for something more proactive and preemptive in avoiding that kind of failure. It is often referred to as "sustainability."

Today, urban planners are working urgently to redefine themselves in this rapidly urbanizing world. In a joint position paper entitled, "Reinventing Planning: A New Governance Paradigm for Managing Human Settlements," (American Planning Association, 2006) leaders from the American Planning Association, the Canadian Institute of Planners, the Commonwealth Association of Planners, the Royal Town Planning Institute, the Council of European Spatial Planners, and UN-Habitat joined together to address "the challenges of rapid urbanization, the urbanization of poverty and the hazards posed by climate change and natural disasters."

What do they identify as the most important contributions that this reinvention can produce? First "Reduce vulnerability to natural disasters" and second "Create environmentally-friendly cities."

Have we been equally ambitious in reinventing our role in shaping the future of rapid urbanization worldwide? Will we remain leaders in lagging technologies – following the parade with brooms and shovels, cleaning up environmental damage and compensating for the impacts of economic development? There is clearly an opportunity for us to reinvent our role in the future of sustainable urban development. To help environmental decision-makers incorporate economic and social ends in their pursuit of environmental and public health protection. We cannot be accused of ignoring the environment. We may be guilty, however, of being isolated from the economic and social issues related to urbanization and land use.

If it is fair to say that virtually all the problems associated with water quantity and quality in urban watersheds are significantly impacted by land use, doesn't it follow that we could have a huge influence on the future by directly engaging as a stakeholder in the planning and decision-making surrounding those land use decisions?

This would not put the environmental engineering community in charge. On the contrary, it would merely establish parity with the other drivers affecting land use. What would change if the aquatic ecosystem in the urban watershed served as the starting point for planning tomorrow's cities? Those scientists, planners, engineers who have followed development with sophisticated plumbing would have to take into consideration many new issues that are currently handled by others.

Of course, the process isn't linear and no one really leads in the complicated dance of urbanization. And yet, if for a moment, the urban watershed came first and every other profession, institution, agency, and law was designed to protect its long-term integrity (while allowing for increasing population and economic growth) would we see more green roofs, porous pavement, solar energy, recycled water, rapid transit, and innovations in technology and behavior too numerous to quantify?

If there was ever a time to step forward and contribute to our understanding of what "sustainability" in urban infrastructure means, now is it. Again, this doesn't mean "taking over" from the developers, architects and planners who largely drive the form or our urban landscape. It means joining with them as leaders (not followers) in the creation of something brand new.

If there was ever a time when we should be challenging ourselves to be as bold in vision, confident in one another's intelligence and values, and tireless in our search for innovation, this is the moment. The Wingspread workshop provides many important examples of where these changes are already well underway. There remains much to learn and more to do in shaping a sustainable future for our rapidly urbanizing world.

REFERENCES

Center for American Progress. (2006). "Public Opinion Watch," (quoting from 2005 Gallup Poll). http://www.americanprogress.org/issues/2005/04/b596339.html (Accessed July 10, 2006).

Mumford, Lewis. (1961). *The City in History*. San Diego: Harcourt, Inc., pp. 474–475.

American Planning Association. (2006). "Reinventing Planning: A New Governance Paradigm for Managing Human Settlements." http://www.planning.org/knowledge/reinventingplanning.htm (Accessed July 10, 2006).

2

Developments towards urban water sustainability in the Chicago metropolitan area

Richard Lanyon

General Superintendent, Metropolitan Water Reclamation District of Greater Chicago
E-mail: Richard.Lanyon@mwrdgc.dst.il.us

Summary: This introductory key note article describes history of Chicago's drainage system, its impact on and protection of the city's water supply that lead to the reversal of the Chicago River and building of the Chicago Sanitary and Ship Canal (CSSC) and the Tunnel And Reservoir Plan (TARP). New steps in the twenty first century towards sustainability, including green city developments, are now being implemented throughout the Metropolitan Water Reclamation District of Greater Chicago. The agenda towards hydrological and ecological sustainability is outlined.

HISTORICAL PERSPECTIVES

19th Century Chicago

The City of Chicago was incorporated in 1837 and at that time had a population of about 4,100 people. The population grew rapidly because it was strategically located where the Great Lakes and the inland waters could be easily traversed. The City undertook many projects before and following the Great Chicago Fire of 1871 to make a sustainable city. One of the first steps occurred in 1855 with the adoption of a program to install sewers. It was necessary to raise the grade of the streets of this topographically flat and poorly drained landscape so that sewers could be installed to drain stormwater and wastewater from low areas. This was done by raising buildings and streets to the new grade level to facilitate drainage and make streets and sidewalks safer and easier to maintain. Improved drainage moved wastewater to the river and out into Lake Michigan, thus making the river a nuisance and polluting the lake.

The next major step was to keep wastewater out of the lake, the source of the drinking water supply. A sub-continental divide ten miles west of the Lake Michigan shoreline in Chicago and little topographic relief, made the excavation of a canal diverting sewage carrying flows of the Chicago River away from Lake Michigan a feasible remedy. This canal could allow for the diversion of water and sewage out of the Lake Michigan watershed into the Des Plaines River watershed, a tributary of the Illinois and Mississippi Rivers. To make this affordable, a new taxing body was needed with a sufficiently large area of authority. In 1889, the State of Illinois adopted legislation that allowed for the creation of such a body and the Sanitary District of Chicago was created by referendum in the same year with the purpose of keeping pollution out of Lake Michigan.

Before and after the reversal of the river

The immediate task facing the new Sanitary District of Chicago was to cut off the main source of pollution to the lake. To do this, the District engineers reversed the flow of the Chicago River. The Chicago Sanitary and Ship Canal was constructed in the 1890s to connect the South Branch of the Chicago River with the Des Plaines River and so carry wastewater away from the city and dilute it with lake water as it flowed downstream. In 1910, the North Shore Channel was completed to divert more lake water to aid dilution in the North Branch and in 1922; the Calumet-Sag Channel reversed the flow of the Calumet River. With other improvements, including intercepting sewers along the lakefront, this 125-kilometer (78-mile) long system of man-made waterways immediately improved the quality of life for everyone who depended on Lake Michigan for drinking water. There are three points where lake water enters the waterway system and one outlet control at the discharge to the Des Plaines River. Figure 2.1 shows the Chicago waterways before and after the reversal of the rivers and diversion of sewage flows from Lake Michigan.

MWRD waterways and facilities

The reversal of the rivers protected Lake Michigan but the waterway system was virtually an open sewer until intercepting sewers and sewage treatment plants were built in the 1920s and 1930s. Today, the Metropolitan Water Reclamation District has seven water reclamation plants in Cook County, each located along a stream or river. These plants treat the wastewater from 5.25 million people, the industrial

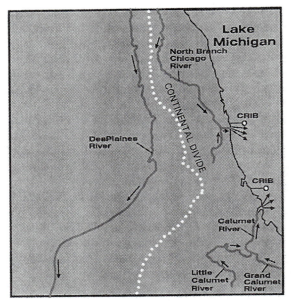

Before Construction of the Canals

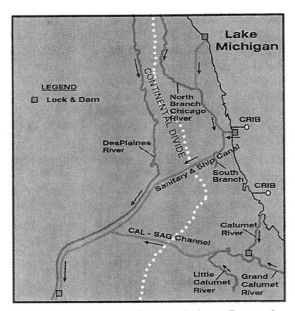

After Construction of the Canals

Figure 2.1.   Chicago waterways before and after the reversal of the Chicago River flow in early 1900s.

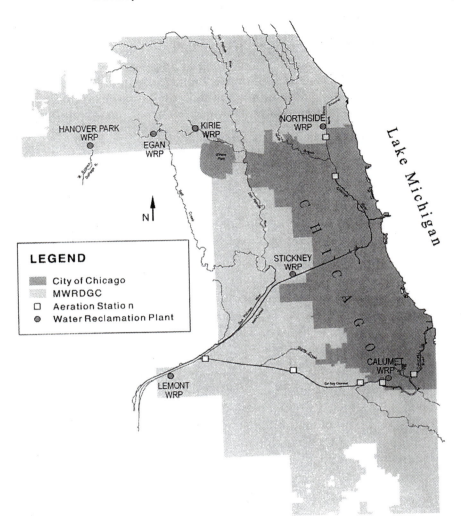

Figure 2.2.   Metropolitan Water Reclamation District of Greater Chicago – Waterways and facilities.

equivalent of another 4.5 million people and a combined sewer overflow equivalent of 0.6 million people, for a total average daily flow of 5.2 million m³ (1.4 billion gallons). The effluent from these water reclamation plants accounts for most of the flow in the waterway system. To improve the dissolved oxygen level in the slow-moving waterways, the MWRD has seven supplemental aeration stations along the waterway system (Figure 2.2). More of these are likely to be constructed if current water quality standards are upgraded.

TARP system

To address the problem of combined sewer overflows, the MWRD adopted the
Tunnel and Reservoir Plan in 1972 (Figure 2.3). The tunnels range in size from
4.5- to 10-meters (15- to 33-feet) in diameter and are located in bedrock between 61
and 91 meters (200 and 300 feet) underground. The project has been recognized by
the American Society of Civil Engineers and others as an outstanding engineering

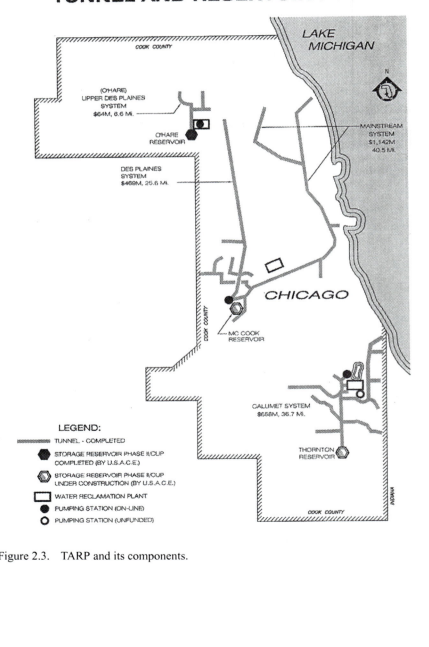

Figure 2.3.   TARP and its components.

achievement. All 175 kilometers (109 miles) of tunnels are now complete and operational, capable of storing 9 million m³ (2.4 billion gallons) of combined sewer overflows. Since the first part of the TARP system was finished in 1985, more than 3 billion m³ (800 billion gallons) of combined sewer overflow have been captured, and conveyed to the nearest water reclamation plant for full secondary treatment. The TARP system includes three storage reservoirs to provide an eventual total of 70 million m³ (18.5 billion gallons) of combined sewer overflow storage. The smallest reservoir, the O'Hare Reservoir, was finished in 1998. An interim reservoir in the south suburbs already provides 11.6 millions m³ (3.1 billion gallons) of stormwater storage while work continues to complete the two large reservoirs. The cost of the TARP system will be more than three billion U.S. dollars.

AGENDA TOWARDS ECOLOGICAL SUSTAINABILITY

The Chicago Wilderness region

The Chicago Region Biodiversity Council is comprised of 200 public and private organizations in a regional network dedicated to the protection, restoration and stewardship of biological diversity in the Chicago Wilderness region. This region, located around the southern end of Lake Michigan, is noted for its richness in diverse species of fauna and flora. The topography, formed by the receding continental glaciers, was a tapestry of dunes, lake plains, prairie and woodlands. Urban commercial and industrial development over the past 150 years has taken a toll on the biodiversity and natural features of the landscape. Participants in Chicago Wilderness are dedicated to preserving and restoring natural features to benefit present and future generations of people and businesses in the region. This work is carried out through individual and collective efforts. Governmental agencies, such as, the MWRD and City of Chicago, are examples of leadership and pioneering in carrying out the goals of Chicago Wilderness. Local governments are committing resources and leading by example in initiating preservation and restoration activities.

Guide to stormwater best management practices

This guide is an example of local government leadership. In Mayor Richard M. Daley's introduction to this guide, he says the use of Best Management Practices for stormwater can be a cost-effective means to protect our water resources. The city is demonstrating and promoting innovative alternatives to managing stormwater. The MWRD was recently granted the authority for stormwater management in Cook County by the State of Illinois and will be implementing Best Management Practices and other innovations throughout Cook County through regulatory programs, watershed planning and capital improvement projects. In the past decade, the MWRD has begun a program to convert traditional turf landscaping at its facilities with native prairie landscaping to improve stormwater management, wildlife habitat and biodiversity; and to develop wetlands for these same benefits and nutrient reduction.

Mayor Daley's message highlights eight Best Management Practices. These are 1) green roofs; 2) downspout disconnection, rain barrels and cisterns; 3) permeable paving; 4) natural landscaping; 5) filter strips; 6) biofiltration rain gardens; 7) drainage swales; and 8) naturalized detention basins. Use of these practices for stormwater runoff from developed land will correct many of the problems caused when stormwater is not effectively or efficiently managed.

City of Chicago's water agenda

This agenda was developed to guide water related decisions for many years. It provides a strategy for keeping our water safe, clean and plentiful, and for improving the infrastructure that keeps the area's homes and businesses supplied with water. The Great Lakes represent 20 percent of the earth's and 95 percent of the nation's fresh surface water supply, so we must not take this resource for granted. The Water Agenda outlines a strategy for caring for our water resources as a whole, complex and connected system. The City of Chicago has been aggressive in water conservation programs and has reduced per capita water consumption by more than 20 percent (Figure 2.4).

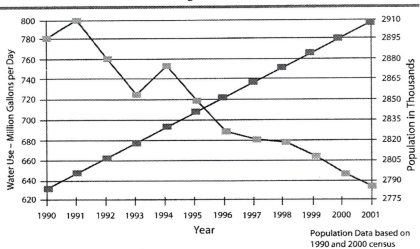

Figure 2.4. Population increase and water conservation impacts (to convert from million gallons per day to m^3 /day multiply by 0.00378).

To reduce consumption, the city conducts audits and recommends upgrades for large industrial water users. Currently, industrial or commercial water users, new residential users and residential buildings with more than three units all have water meters. The Department of Water Management will develop a plan to meter all residential users so that everyone pays for the water they actually use. The Department of Water Management, has invested in infrastructure improvements that have reduced water consumption by 453 600 m^3/day (120 million gallons per day) even as the city's population has increased. This has occurred through outreach to water consumers and water main leak detection, repair and replacement.

The city also supports the international Great Lakes Governors and Premiers in their efforts to establish a governance system for future withdrawals of Great Lakes water. Leaders of cities located around the Great Lakes joined forces in recent years because of their concern for stagnation in the work of both federal governments in preserving and protecting this great resource. Local initiatives often set the pace and tone of programs that state, provincial and federal governments can follow.

Chicago river agenda

The City of Chicago, Office of Mayor Daley, has prepared an agenda for improving water quality in the Chicago River. The report sets forth a set of bold initiatives for the river and incorporates many of the MWRD's priorities for improving the quality of the river. The MWRD staff worked with Mayor Daley's staff assisting in the development this agenda and preparing this report. The four central goals of the agenda are improving water quality, protecting nature and wildlife in the city, balancing river use and enhancing neighborhood and community life.

The location of the City of Chicago on the Calumet and Chicago Rivers at their mouths on Lake Michigan is a key to understanding the growth and development of the city. The river has improved dramatically over the past three decades, but more work needs to be done to further improve water quality and the quality of the river corridor. Intense storms result in combined sewer overflows; pollutants and debris are carried by stormwater runoff into the river from alleys, streets and parking lots; and the legacy of historic pollution still remains in the sediments on the river bottom in some areas.

The agenda calls for the completion of TARP, consideration of disinfection of treated effluents, additional technologies for water quality improvement, river bank protection and naturalization, improving aquatic and terrestrial habitat along waterways, developing linear parks and trails along the waterways and increasing access to waterways for recreational use.

Fish species in the Chicago river system

Fish in the rivers are the best indicators of improved water quality. Fish cannot survive if the water is polluted and deprived of oxygen. As a result of improvement in wastewater treatment, removal of polluted discharges to waterways, operation of the TARP tunnels and discontinuation of effluent disinfection with chlorine, there

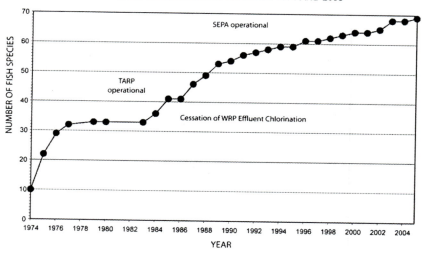

CUMULATIVE TOTAL NUMBER OF FISH SPECIES COLLECTED FROM THE CHICAGO AND CALUMET RIVER SYSTEMS BETWEEN 1974 AND 2005

Figure 2.5.   Effect of water quality improvement by MWRDGC on fish species abundance.

has been a dramatic increase in the number of fish species and the abundance of fish found in the waterways. Since 1974, when there were only ten species of fish found in the waterways, the number of species has climbed to 68 in 2005 (Figure 2.5). Increases are specifically attributed to the cessation of effluent chlorination in 1984 and initial portions of the TARP tunnels becoming operational in 1985. Additional fish diversity and improvement was achieved after installation of Side Elevated Pool Aeration (SEPA) systems.

Environmental action agenda, building the sustainable city

Implementation of sustainable practices for stormwater runoff in a city the size and age of Chicago requires dynamic leadership. Nowhere is this more evident than in the departments and facilities of city government. Installing a green roof on City Hall, opening a green technology center, greening of city boulevards and thoroughfares has provided the incentive for others to follow. Each department of city government has developed an environmental agenda, outlining specific measures to take for a greener and sustainable Chicago. Once outlined, each department must follow through on their measures and is held accountable by the Mayor's Office. In so doing, city government is leading by example and, as a result, businesses and citizens are beginning to follow and implement similar practices.

Information sources

More information on all these topics can be found at the following web addresses:

www.chicagowilderness.org
www.cityofchicago.org
www.mwrd.org
www.ChicagoAreaWaterways.org

3

Water and cities – overcoming inertia and achieving a sustainable future

Andrew Speers

International Water Association, 12 Caxton Street, London SW1HOQS, UK

Summary: There is clear evidence that many urban systems are unsustainable. Water systems, while protecting public health and in many cases providing safe, high-quality water services may be energy intensive and place stress on ecosystems. In numerous cities, the situation is likely to be exacerbated by climate change. Rapid responses to emerging crises would be valuable but often our decision-making is slow and incremental. This paper examines, briefly, these decision-making processes and explores recent developments which, when applied to water services, could move society more rapidly to a sustainable future.

INTRODUCTION

Urban water systems provide healthful and life-giving services to millions of city dwellers and are arguably the most vital of urban services. However, it is evident that water, sewerage and drainage services are often unsustainable: more water is consumed than will sometimes be available; ecosystems suffer as a result of constrained or polluted flows; some communities suffer economically or in other ways for the benefit of others. Where such problems do exist, they are likely to be exacerbated if climate change comes to pass and they may emerge in regions that are rapidly urbanising. Thus, any strategy to improve the sustainability of cities must include a focus on water management.

Many urban systems improve (or perhaps sometimes only change) slowly. There may be very good reasons that precipitous change should be avoided. Yet faced with constrained economic and natural resources, increasing populations and possible climate change effects the maintenance and expansion of water and wastewater systems may require more than incremental improvement.

How this might be achieved, and the key stimulants to more rapid change, form the key discussion points of this paper. Overall, the intention of the paper is to highlight the potential for improvement in urban water systems and the areas in which the greatest potential exists. The flow of argument is that:

- Cities are dynamic and 'messy' things, difficult to manage and subject to a myriad of social, economic and other forces. While we would wish to plan cities, in many ways our attempts at planning represent and art, not a science
- It is nevertheless necessary to have in mind a goal. Making cities more sustainable is a goal and a well considered one, albeit the means of its achievement is uncertain and the pace of change can be frustratingly slow
- Some key developments that may stimulate more rapid change and the relevance of these to the better management of urban water systems is therefore explored.

Managing in a 'Fuzzy' environment

It is always best to start with a goal in mind but cities are agglomerations of millions of individual decisions taken over many decades, if not centuries, where an overall strategy cannot be gleaned. While our broader institutions affect they way cities develop, decisions on where and how to live, where to work or recreate, when to invest, where to worship, are essentially private ones. Despite this anarchy, on many levels cities have been a success, providing satisfactorily high standards of living for many citizens.

Equally, though, cities have failed their citizens. Slums have developed, the environment has been decimated, crime has risen and examples of inefficiencies abound. We have attempted to correct these failures by creating departments of planning and many other institutions to limit what may be built where and how. Transport plans, traffic management systems, land use controls, zonings, greenbelts and height restrictions are all instruments that we use to impose order on chaos. Some of these have produced spectacular results, the introduction of sewerage and potable water systems perhaps being the most beneficial. But such is not universally so: road systems we thought led to the future, became clogged almost as soon as they were opened (witness the Périphérique in Paris), public transport systems have been abandoned and then rebuilt (light-rail systems in California), dream settlements of post-war urban planners have worked against the sense of community for which they strove; poor zoning decisions have everywhere led to the abandonment of formerly vibrant inner city areas.

Thus, we have a paradox. Centralised control fails citizens because the will of the state does not satisfy their individual desires in what is a very dynamic system constantly buffeted by economic and social change. Anarchy fails because individual decisions do not necessarily maximise social wellbeing. There must always be planning; it will not always work. While this is true of any human-created system; cities present a particularly challenging environment in which to attempt a reconciliation.

On accepting the 2005 Balzan Prize for the Social and Cultural History of Cities Sir Peter Hall, Professor of Planning at the Bartlett School of Architecture and Planning, University College London remarked that, "Sigmund Freud famously said that he had spent his life asking "What do women want?" but had never found out. I've spent my academic life asking "How do cities work?" and I don't think that I've yet cracked my mystery either".[1] While we would not give up on the challenge of understanding better how our cities operate, I suggest that we've reached level of sophistication sufficiently high to accept that there is no answer; that cities are muddled, fuzzy, complex and dynamic systems and that we must accept these conditions.

If that is so – and surely it must be – we are presented with a particular challenge: How, in a fuzzy and uncertain environment, do we move a city from a position of unsustainability to one of long-term viability within the sort of time frames many practitioners feel are essential?

What is sustainability?

In addressing this challenge, we need to consider what sustainability means. Despite the difficulty of positing specific objectives in 'fuzzy' environments, it is nevertheless worthwhile to understand what the key conceptions of sustainability might be.

Over the past decades awareness of the importance of balancing human needs with those of the ecosystems that support us has risen. It is clear that people ascribe a high value to protecting the environment both from a position of self interest (they enjoy and benefit from the services and conditions provided by the environment) and for the environment's own sake (other species have a right to exist). By many objective assessments, however, numerous ecosystems and environments are declining in quality and function under human influence. This is acutely felt in cities, where ecosystems have been substantially and comprehensively altered.

Faced with this reality, international efforts have been directed to mitigating the impacts of human activities without inhibiting economic development, at least not unduly, and while meeting our social needs and equity goals.

In 1983 the World Commission on Environment and Development was established. The report of the Commission, *Our Common Future* (Brundtland, G. ed., 1987) – which is also known as *'The Brundtland Report'*, after its Chairperson, Gro Harlem Brundtland – defined the concept of 'Sustainable Development'

---

[1] http://www.balzan.com/documents/preisverleihung05/rede-hall-en.pdf

recognising the need to reconcile competing environmental and human priorities. The Commission's definition was "[development] which meets the needs of the present without compromising the ability of future generations to meet their own needs". Associated text also makes clear that societal needs are culturally determined – goods 'needed' by citizens of less developed nations will be significantly fewer in number, volume and refinement than those 'needed' by citizens of wealthy nations – and that the ecosystems on which we depend should be protected.

Numerous definitions of sustainability have been proposed since, many of which are quite detailed and subtle. The *Brundtland Report* definition nevertheless continues to be a touchstone for those active in implementing and debating the concept. Concentrating on its core elements, the concept of sustainable development requires social, economic and environmental factors to be considered in decision-making so that human society continues to develop, while spreading the benefit of this development more widely within and across generations and making sure that ecosystems continue to function.

Muddling through

We have, therefore, a goal of sorts, albeit one that is generalised. Each of us, however, will have different views as to the policies needed to move us to achieve that goal and the speed with which change is needed. A person seeing nature as ephemeral and significantly damaged by relatively minor pressures would, for example, advocate for rapid and radical change, whereas someone viewing nature as resilient would likely believe that time is not so much of the essence. Dispute is not, however, the only reason policy development and implementation might be slow, nor is incrementalism of itself negative. The following discussion analyses, briefly, the reasons a more incremental approach to policy development is generally taken.

The "rational-comprehensive" approach to public administration is often presented as the ideal method for developing public policy. According to this approach policy makers begin addressing a particular policy issue by ranking values and objectives. Next, they identify and comprehensively analyse all alternative solutions, making sure to account for all potential factors. In the third and final step, administrators choose the alternative that is evaluated as the most effective in delivering the highest value in terms of satisfying the objectives identified in the first step.

This approach seems to make perfect sense. But bureaucrats and administrators don't work this way in the real world, according to Charles Lindblom (Lindblom 1959).

First, defining values and objectives is very difficult. There are always trade-offs in public policy. It is difficult to say with certainty, for example, that it is better to spend less on education in order to balance the budget, or that building more roads is a better way to reduce traffic congestion than raising gasoline taxes.

Second, separating means from ends (policy recommendations from the objectives of those policies) is impossible. Instead, the policy solution is always bound up with the objectives. The problem of reducing traffic congestion could involve building either highways or mass transportation. But for many interested parties each of these potential "solutions" to the problem of congestion is likely to be a policy goal in its own right.

Third, it is impossible to aggregate the values and objectives of the various constituencies of the executive bureaucracy – citizens, private organizations, legislators, and appointed officials, among others – to determine exactly which preferences are most important. The virtue of a policy is indicated by its ability to achieve broad support, not by some assessment that it is most efficient according to some abstract criteria.

Finally, it is inefficient to identify and analyse every policy option. For all but the most narrow policy choices it takes too much time and too many resources. Administrators are very busy and the volumes of detail on even relatively simple issues would be overly burdensome to analyse.

Instead of comprehensive analysis of every policy option, a much more constrained process of "successive limited comparison" is really how policies are developed, insists Lindblom. According to this "branch" method, administrators usually look only at policies that differ in relatively small degree from the policies currently in effect, thereby reducing the number of alternatives to be investigated while simultaneously narrowing the scope of investigation. In other words, they look at two nearby branches, not the whole tree, roots and all.

Successive limited comparison – or muddling through – is thought to be the primary cause of the tendency toward incrementalism in policy development. Only rarely are dramatically different new policies developed. Instead, administrators in both the public and private sectors tend to build on existing policies, tweaking them here and there in a continuous, evolutionary process.

This sometimes causes frustration on the part of citizens and other interested parties, who feel that the government is sluggish and unresponsive. But, Lindblom thinks that such incremental "muddling through" is a good thing. It is efficient (it analyses practical options much more quickly than the root method) and in the end it is responsive to the goals of a sufficiently broad set of constituents.

Since Lindblom wrote new techniques of decision-making and policy comparison have been developed. Multi-criteria analysis, for example, a means of weighting and comparing options that can actively involve stakeholders in the process of identifying and weighting the desirability of certain outcomes, can be used to great effect. Alternatives which aid comparison and decision making, such as 'Ecological Footprinting' – a process of comparing options according to the land area needed to provide resources or consume/recycle waste products – and 'Factor X' analysis – though which the impact of a percentage reduction of '$x\%$' in, say, pollution or the consumption of natural resources can be compared, have also emerged in response to the perceived need to become more sustainable. These are also potentially valuable aids. However, the complexity of applying such techniques to all but the most significant policy decisions must necessary limit their application. 'Muddling through' is still a much used and practical approach to policy development.

Accelerating change?

Day to day then, our policy options are usually reduced to the more immediately achievable. However, as noted, sustainability might require more direct and radical action. Those urban areas lacking basic city services, such as water and sewer, and in which sustainability equates to mere survival require, for example, more urgent and fundamental solutions that 'muddling through' would generally reveal. Heavily eutrophic lakes, diminishing aquifers, gross reductions in the volume of water available per capita, reduction in species diversity are problems that may not wait for an incremental approach.

How might we leapfrog then to a condition whereby more radical solutions might be considered? While Lindblom's arguments are compelling, there are numerous examples from the real world where huge steps have been taken in a short time. In particular, recent developments in three critical areas have driven society in new and dramatic directions within very short time frames. While successive limited comparison may be the general methodology of policy development, the 'policy space' may be reduced if those developments producing the most significant developments in society generally were applied to water specifically. I would avoid a suggestion that the developments discussed below are the *only* critical developments driving society forward, I would suggest that they are fundamental and have potential to be applied within the water industry (or in many cases are already being applied) in ways that will lead to more rapid steps toward more sustainable outcomes.

These developments are:

Technological Developments

Technological development in our time has driven society in ways that are clearly not incremental. The internet barely existed 10 years ago, nor the fibre optic networks to support it 20 years ago. Landline services are becoming obsolete, or the marginal cost of them is approaching zero. Data storage is becoming almost limitless and computer-based modelling has taken huge leaps forward. The impact of these developments is profound. Huge efficiencies in communications, data analysis and computing have been achieved, and the rate of change is continuing unabated. City infrastructure is responding to these changes. Decentralisation is more feasible, new industries have developed, notably IT, and knowledge workers become valuable commodities. In particular, technological developments have given us powers of analysis way beyond that which was imaginable 30 years ago.

Education, Knowledge and Learning

Awareness-raising is a category of change that would include both an increase in the standard of education and the provision of new information where previously there was none available. In the developed world literacy is almost 100% and even in the developing world it is now 85% for those 15 years or younger, and 74% for older groups (up from 74% and 57% respectively in 1980) (UNESCO 2002). Piggybacking on the power of computers and remote sensing technology there is also an increasing array of information available on environmental conditions and on the interrelationship between those conditions. For example, the relationship between ocean temperature and climate fluctuation (e.g. the 'El Niño' effect) is now better understood and the community better informed of its likely impact. It could be suggested that the rise in environmental awareness in the early 1990s was a response to increased information and mounting evidence of human-induced environmental change.

Growth of Market-based Economies
We also see the almost complete victory of what is variously known as 'economic rationalism', 'supply-side economics', 'globalisation' or the concept that "markets usually provide more satisfying answers to questions of choice, consumer preference and so on – and in doing so, provide a more rapidly advancing level of total well-being for all concerned – than decisions by diktat, whether those be by politicians, bureaucrats or controllers generally" (Stone 1992). Regardless of one's opinion of the benefits of such approaches, the victory of the market has had a fundamental impact on society in a short period.

Water systems and rapid change

If these examples are legitimate, they would seem to suggest that there are societal forces that can propel us forward in a manner that is beyond incrementalism. Harnessing these forces for the benefit of water systems may bring about significant and rapid change. To explore this suggestion future, it is worthwhile identifying parallels between the categories of development outlined and the modern provision of urban water services. Such discussion will also move us from the realm of the philosophical to the practical.

Harnessing technology

Water systems are, fundamentally, transport systems. The goal of early installations, admirably met, was to keep the source of water separated as far as possible from the disposal point of wastewater. Thus, we take water from dams, lakes, rivers and aquifers and discharge waste products, primarily human faecal matter and urine – at a point at which people are unlikely to re-ingest contaminants. The rate of cholera, typhus and other water borne communicable diseases plummeted as a result – indeed the success of some systems was measured by these criteria. This means of transportation, however, is remarkably inefficient. We use tonnes of water to move what becomes very dilute waste (less than 1% of sewage is faecal matter) and we have diminished the quality of the transported effluent through the introduction of modern industrial chemicals (although it is acknowledged that we have also improved treatment processes accordingly). The advent of technology has, however, made it completely possible to produce very high quality water from poor quality sources. In particular membrane micro- and ultra-filtration technology is becoming widespread and improving in efficacy and efficiency rapidly.

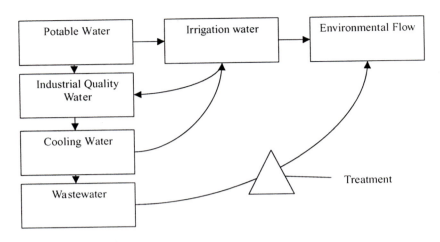

Figure 3.1.   Cascading uses of water.

Such a development makes it possible to locate sources of water close to sources of wastewater and, in particular conditions make one the input to the other. Our water 'loop' therefore, moves from being kilometres in extent and possibly thousands of years in duration to being smaller, more immediate and more controllable. Additionally, the pressure on natural resources, rivers, lakes, aquifers, is reduced because there is significantly more reuse within the system. Of course, not all systems need be closed loop. Potable, small scale treatment technology also makes it possible to 'cascade' uses. Figure 3.1 below illustrates one view of the way in which water might be used and reused at different stages in its journey from first quality potable water to wastewater, the central message being that first quality water is not needed for all purposes. While transport might still be required to move second quality water from source to user, in new development areas land use zonings can be used to encourage collocation of industries such that the waste from one could become the input from another. In particularly well designed systems, the output from one facility can be the input to another without there being a need for treatment. For example, cooling water from an electricity generator, could be irrigation water for a farm or feedlot. Figure 3.1 illustrates this scenario.

Treatment options are obviously a technological breakthrough but in some senses they still represent an incremental step, building on previous concepts of treatment, albeit with new materials. Combined with information technology, however, new options are emerging. In particular, it is possible to schedule flows within water systems in a more sophisticated way. Thus, pressures, flows, the replenishment of reservoirs, discharges to the environment and so on can be scheduled to take advantage of peak conditions (e.g. scheduling flows during off peak times, or when environmental conditions are such that a receiving water body is able to cope best with discharges of treated effluent). It has not been utilised yet to its fullest extent. Particular opportunities exist with regard to the design of systems such that new treatment technologies can be incorporated in systems to maximise transport

A. Speers

Configuration without Flow Balancing

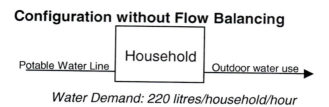

Water Demand: 220 litres/household/hour

Configuration with Recycling and Flow Balancing

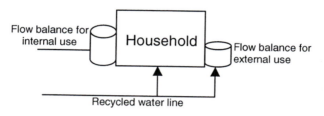

Water Demand: Potable: 15 litres/household/hours
                       Non-potable 45 litres/household/hour

Figure 3.2.   Impacts of flow balancing.

efficiency and the scheduling of flows can become very innovative, maximising the recovery of nutrients.

Figure 3.2, below, shows two water systems. The first, a traditional configuration, the second including the use of recycled water and flow balancing – or bringing of water onto a site during off peak periods or over an extended time - to reduce significantly the size of infrastructure required to deliver water. As is illustrated, the volume of water required per hour is significantly reduced. This would greatly reduce the size and extent of infrastructure required to serve a particular development.

Localised water treatment could be incorporated in such designs ensuring the quality of water supplied. A second example, not illustrated, would see urine, which contains 90% of the phosphorus found in domestic strength effluent, separated from so-called blackwater (faecal matter and other household waste) and treated separately. The same infrastructure could be used for transport of both urine and blackwater through the use of IT to schedule flows. This could significantly improve the recovery of phosphorus, which could then be reused as a fertiliser, rather than being discharged to sensitive receiving environments or expensively separated at 'the end of pipe'.

Obviously, any infrastructure will improve as technology improves. However, something fundamental can potentially occur with regard to water systems as new understanding and better control and treatment allow water systems to be more contained (smaller loops) and more intelligently managed. The benefits of such developments will be revealed through cost reduction and a reduction in dependence on natural resources.

Utilising Awareness

Turning to the issues of education and awareness-raising, along with most environmental issues, water management now attracts broad community attention. Issues such as water shortages, drought and climate change effects, pollution of rivers lakes and oceans, over-abstraction of aquifer water and the like are the subject of wide debate and anxiety. As such, the desire to move beyond step by step improvement is strong. While Lindblom would suggest that such desire is often thwarted by bureaucratic incrementalism, our improved knowledge, and our ability to rapidly build new knowledge widens the range of potentially successful choices available. For example, remote sensing technology and sophisticated computer modelling have led to quantum leaps in our understanding of the dynamics of river systems and the importance of environmental flows, the impact of abstractions and the relative importance of point and non-point source discharges. We are armed, therefore, not just with a knowledge that phosphorus discharges to a river can be limiting, but the points at which those discharges have most effect. It would therefore be possible – as has been the case on the Hawkesbury-Nepean river system on which the city of Sydney, Australia depends for its water supply – to demand specific treatment processes be installed in vulnerable areas but allow discharges to continue in less sensitive reaches (NSW Environment Protection Authority 1995).

Increases in our knowledge obviously come with an increase in complexity and this can work against a narrowing of our policy options by muddying the waters. Better knowledge will not, therefore, lead to better decision-making and, of course, statistics are subject to manipulation. However, improved understanding has significant potential to lead to an awareness of problems not previously identified and a better understanding of cause and effect. An outcome of particular benefit here is the development of more effective environmental indicators and improved understanding of the relationship between aspects of environmental performance. Indicators to monitor the environment and the impacts of policy and physical change have been widely used for at least the past 30 years. However, frequently these have measured the implementation of administrative steps, rather than environmental change. Furthermore, indicator use and measurements themselves were often poorly coordinated nor common across locations.

Now days, remote sensing, improved river, aquifer and ocean measuring and modelling and a better understanding of the impact of events and materials in particular environments has led to the provision of very good quality data that is widely available. Most OECD countries produce State of the Environment Reports, and the OECD itself has played a key role in harmonising indicators and fostering widespread dissemination of readily accessible data (OECD 2003). The outcome is more targeted policies and interventions and better knowledge widely spread among stakeholders.

Economic Reform

Thinking finally about changes in the underlying economic philosophy of the liberal democracies (and indeed others such as China) the way in which we price and cost water and water services has undergone a revolution in the past 15–20 years. Water services, traditionally, have been the domain of governments. This reflects historic circumstances in which the high capital cost of water services made them natural monopolies. The growth of capital, greater efficiencies, and innovative approaches to creating competition have led to the introduction of private capital. Equally fundamentally, the pricing and costing of water services has changed to recognise better the true cost of water and to pass it on to consumers. Thus, in many jurisdictions the cost to consumers of water services reflects the volume of water used in some way. Pay-for-use pricing properly implemented reflects the cost to the water service provider of water service capital and operating costs, plus a return on investment. Pay-for-use pricing has helped to rationalise consumption as water is no longer seen as a 'free good'.

Pricing water to reflect capital and operating costs plus return on investment is only part of the picture, however. Increasingly recognised is the importance of externalities. An externality exists when a third party, including the environment bears a cost (or receives a benefit) for which they are not compensated (or do not pay). For example, an externality would exist if an angler were fishing downstream of a sewage treatment plant, and had their catch reduced because of the plant's discharges and was not compensated for that loss. A recent study by CSIRO Australia suggests that the cost of repair to a water main on a moderately busy road would be more than double were externalities (business loss and delays in travel time) taken into account (CSIRO 2001). However, one could argue that calculation of opportunity costs is more important in setting the true cost of water services than the incorporation of externalities. However, I suggest that for the immediate future incorporate of externalities is of more immediately practical use as most water systems are already installed and that significant new opportunity costs arise only when important new infrastructure is required.

Presently, externalities are poorly dealt with, not least because they are difficult to identify and quantify. However, as long as we allow externalities to remain unconsidered, the goods and services that are offered are under-priced. This leads to over-consumption of those goods, and where this leads to consumption of environmental goods, to a rate of degradation that may be unsustainable or even catastrophic. If, for example, we accept climate change as a reasonable hypothesis, the costs – which will be in the billions – of that change should be factored into the cost of fossil fuels. Fossil fuels would costs more and there would be a reduction in use of energy overall, or a switch between fossil fuels and renewables. Viewed another way, we might say that the costs of the clean up of contaminated sites – funded by devices such as the Superfund in the United States and other mechanisms in other nations – represents an externality now being paid for by this generation, that was the legacy of a previous generations.

Despite the challenges, getting the price of water right would produce significant benefits by rationalising the consumption of water and the production of wastewater. If, for example, the cost of trade waste discharges were increased to compensate for the cost of treatment or the impact on the environment, there would be a financial incentive for companies to redesign their industrial processes, pre-treat wastes prior to discharge or invest in recycling technology (or the relationships with other companies as described earlier in relation to 'cascading uses'). Of course, suggesting that we can reduce our policy space by universally agreeing to cost water services more accurately is somewhat naïve. The Dublin Statement on Water and Sustainable Development (UNCED 1992) included this statement:

> *"Past failure to recognize the economic value of water has led to wasteful and environmentally damaging uses of the resource. Managing water as an economic good is an important way of achieving efficient and equitable use, and of encouraging conservation and protection of water resources."*

This statement has beeen criticised by some Islamic nations as, those countries view water as a free gift from God. While meaning no disrespect to those who hold those views, the economic arguments in favour of accurately pricing water services are strong and are reflected now in a number of seminal documents in countries around the world (see, for example, the Bonn Charter for Safe Drinking Water (IWA 2004)) the Dublin Statement on Water and Sustainable Development (developed through the participation of 500 hundred delegates, including government-designated experts from a hundred countries and representatives of eighty international, intergovernmental and non-governmental organizations) and the World Health Organisation's Guidelines for Drinking Water Quality (3rd Ed.) (WHO 2004). Furthermore, the rising costs of the installation of new infrastructure and of maintenance of existing services have raised issues to do with future financing. Whether one agrees with the commoditisation of water or not, it is evident in developed and developing nations that governments will play an increasingly small role in the provision of capital to fund infrastructure renewal and expansion. Pricing of water services to reflect these rising costs ensures a regular stream of monies with the concomitant benefit of rationalising water consumption.

New technologies and awareness have more rapidly and less controversially narrowed the policy space, than will the commoditisation of water. Nevertheless, we should at least be able to agree that correct water service pricing is an issue that will need to be carefully considered (and which in some jurisdictions where private capital is already deeply involved in infrastructure provision will be fundamental).

CONCLUSION

Cities are fuzzy, complex entities, the form of which frequently reflects merely the cumulative effect of millions of individual welfare maximising decisions. We seek to bring order to this randomness through planning mechanisms, because we recognise that individual decisions will not maximise community welfare; inequalities, bottlenecks and the like will exist unless some centralised decision-making takes place. Perversely, the impact of individual decisions, changing preferences and styles will thwart centralised planning (as, of course, will poor policy development which neglects to take into account community preferences).

We have sought, at times, to significantly alter the face of our cities though large scale planning or policy development. Some of the schemes implemented have been successful, others have failed. Lindblom suggests that most public administration is no better than 'muddling though' – progressing policy development through incremental decision-making where the policies chosen are limited to the familiar and the immediately possible – but also suggests that this is an acceptable form of decision-making given that the cumulative impact of small-step decisions generally works in the right direction without getting too much catastrophically wrong. Yet the populace often feels frustrated and, more fundamentally for the issues we are now considering, we will not be able to progress as rapidly as we need to in the face of increasing evidence environmental decline. We need to be sustainable, and we need to become so sooner rather than later.

If 'muddling through' is not good enough for us when faced with the urgent need to move beyond our presently unsustainable circumstances, we should seek means of narrowing the policy space. These means can be identified through examples drawn from recent history. Three critical developments – advances in technology (particularly IT), increased awareness and understanding of environmental conditions and the adoption of an economic philosophy based on the dominance of markets – each go well beyond incrementalism. These developments can be applied in the water sector and examples have been provided in the text.

My conclusion then is that it is possible to make rapid strides toward sustainability. However, a word of caution is required because mere application of these developments will not necessarily lead to a well implemented outcome. Two precursors are necessary to minimise the risk of failure. The first is to recognise that not all water systems will suit all circumstances. We have adopted blanket approaches to water systems in the past; one size fits all. However, the more flexibility we create in design of systems the more likely we will produce a 'good fit' taking advantage of local conditions. Secondly, the community must be involved in decision-making, even though that will prove difficult at times.

No system of planning will result in completely successful outcomes, but nor will reliance on individual welfare-maximising decisions. However, there are steps that we can take to make outcomes more successful and more far sighted. In this period in human and environmental history, far-sightedness will not go astray.

REFERENCES

Brundtland, G. (ed.) (1987) *Our Common Future: The World Commission on Environment and Development*, Oxford University Press, Oxford.

CSIRO Australia (2001) Setting and Evaluating Customer Service Standards.

International Water Association (2004) The Bonn Charter for Safe Drinking Water Thompson, Michael (2002) Understanding Environmental Values; A Cultural Theory Approach Carnegie Council on Ethics and International Affairs.

Lindblom, Charles (1959). "The Science of Muddling Through." Public Administration Review 19, (2) 79–88.

New South Wales Environment Protection Authority (1995) State of the Environment Report (see Catchment Case Studies: Hawkesbury-Nepean).

Organisation for Economic Cooperation and Development (OECD) OECD Environmental Indicators, Development, Measurement and Use (2003).

Stone, J. (1992) "The future of clear thinking", Quadrant, 36 (1/2) 56–63

United Nations Commission on Environment and Development (1992) Dublin Statement on Water and Environmental Development (see http://www.wmo.ch/web/homs/documents/english/icwedece.html)

United Nations Educational, Scientific and Cultural Organisation (UNESCO) Institute for Statistics (2002) Special Estimates and Projections of Adult Illiteracy for Population Aged 15 Years Old and Above, By Country and By Gender. Paris: UNESCO. Available online at: http://www.uis.unesco.org/en/stats/stats0.htm

World Health Organisation (2004) Guidelines for Drinking Water Quality (3rd Ed).

PART TWO

Impact of Extreme Events

4

Hurricane realities, models, levees and wetlands

I.L. van Heerden, G.P. Kemp and H. Mashriqui

Louisiana State University, Baton Rouge, LA 70803, USA
E-mail: ivor@hurricane.lsu.edu

Summary: Hurricane Katrina made landfall as a fast-moving Category 3 storm. Thereafter 85 percent of greater New Orleans was flooded, 1,500 lives were lost and approximately 500,000 left homeless. The hurricane protection system failed catastrophically with over 50 breachings or breaks reflecting that levee designs did not account for poor soil foundations, and/or because of unsuitable levee material. Kilometers of levees lacked armoring to protect from waves. Surge elevations were exacerbated by the loss of coastal wetlands which exceeds 4,000 km^2 since 1930. With global warming accelerating, 'smarter' planning is needed for many coastal cities and communities. Surge defenses making the full use of natural as well as man-made components need to be augmented with sustainable development and some retreat from low-lying coastal regions. For coastal Louisiana, an east-west major levee system needs to be built across the mid-coast, complimented and protected by aggressive coastal wetland and barrier island restoration.

INTRODUCTION

The coastal wetlands and estuaries of Louisiana are one of the world's great ecosystems. For millennia, the Mississippi River has supplied the coast with an immense resource of freshwater, nutrients, and sediment to build a vast expanse of marsh and swamp land. These lands have been altered by natural erosional processes. The dynamic interplay of land and water, where new lands are continuously built and old lands changed and lost, has produced an environment rich in natural habitats and extraordinary biological productivity. Millions of people rely directly or indirectly on the marshes for their livelihood and for protection against hurricanes and storms. The delta also is of enormous economic importance in ways indirectly related to wetlands, especially because it produces some 15–20% of the nation's oil and almost 30% of its natural gas and because the Mississippi River ranks as the country's most important inland navigational waterway.

In the last several decades humans have impacted this ecosystem in many ways, especially by controlling rivers so natural floods are no longer a part of wetland maintenance and creation, and by dredging channels that expose freshwater marshes to salt water at an unnatural rate. As a consequence the natural surge buffeting offered by the wetlands has been severely reduced in the last 50 years. Louisiana is thus very vulnerable to hurricanes and major flooding as 70 percent of the population lives on 36 percent of land at or below sea level (van Heerden, 1994). That includes New Orleans (Figure 4.1).

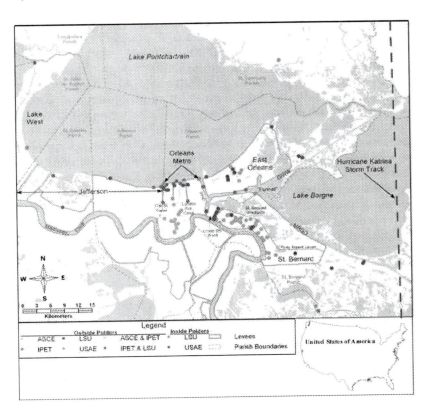

Figure 4.1. The Greater New Orleans area showing principal features that define flood defenses on the "east" bank, the locations of high water marks surveyed outside and inside of the polders, and the five subareas for assessing surge (Lake West, Jefferson, Orleans Metro, East Orleans and St. Bernard).

Recognizing that coastal Louisiana is super susceptible to storm surges, researchers from the Center for Study of Public Health Impacts of Hurricanes at the Louisiana State University Hurricane Center (LSU) provide emergency managers with storm surge forecasts generated by the ADvanced CIRCulation hydrodynamic model (ADCIRC) (Westerink et al., 1994, Luettich and Westerink, 1995). As Katrina approached the northern Gulf Coast, model output was made available in near-real-time (www.hurricane.lsu.edu/floodprediction) based on the National Oceanic and Atmospheric Administration National Hurricane Center (NHC) standard tropical cyclone advisories (Knabb et al., 2005).

LSU ADCIRC graphics were carried by numerous media outlets including in the New Orleans Times-Picayune newspaper on the day before the storm struck as well as on several local television stations. They have subsequently been featured in many reports and documentaries. Despite the tragic outcome, it is clear that ADCIRC saved lives by encouraging many to leave before too late. ADCIRC is also a great planning tool.

Hurricane Katrina made landfall in Louisiana as a fast-moving Category 3 hurricane at 6:10 am on Monday 29th of August, 2005. Of the populated areas that constitute Greater New Orleans (GNO), 80 percent of Orleans Parish, 99 percent of St. Bernard Parish, and approximately 40 percent of Jefferson Parish were flooded, in some cases for weeks. This flooding cost the lives of more than 1,500 residents. Over 100,000 families were rendered homeless, the great majority of whom had heeded evacuation orders, making this disaster the worst since the 1927 Mississippi River flood in terms of homes destroyed. The 1927 flood prompted a complete redesign of the Nation's strategy for managing Lower Mississippi River flooding. That response led to construction of the Mississippi River & Tributaries Project (MR&T) that has successfully prevented a repeat of the 1927 disaster. In contrast, many observers have noted similarities between the patterns of surge-induced flooding that occurred during Katrina and the previous storm of record, Hurricane Betsy in 1965. The GNO hurricane protection system (HPS) was, in fact, intended to prevent a repeat of Betsy disaster, but this system failed during Katrina with a twenty-fold increase in loss of life. It is hoped that knowledge gleaned form the Hurricane Katrina catastrophe will lead to development of a new HPS that will prevent future hurricane flooding as effectively as the MR&T has stopped uncontrolled flooding of the Lower Mississippi.

STATE OF THE ART-CURRENT KNOWLEDGE
The coastal wetlands dilemma

Over the past 7000 years, the Mississippi River has created seven deltas (Kolb and Van Lopik, 1958) by a process of delta switching in which the river abandoned its main channel approximately every 1000 years in favor of a shorter route to the Gulf. The abandoned delta eroded and the locus of sediment deposition shifted to the new active watercourse. By this process the distributary ridges and extensive interdistributary wetlands that comprise the vast deltaic plain of South Louisiana were created. Until human intervention reversed the trend about 100 years ago, the delta switching process, coupled with the shorter-period spring flood sedimentation, accounted for a net gain of land of the magnitude of 3 km2/yr (Templet and Meyer-Arendt, 1988).

When the French began settling the region in 1699, the Mississippi mouth was not confined to the narrow bird-foot delta. Rather the river flowed sweet and wild into the Gulf through a mouth which extended for more than 260 km along the coast (Condrey, 1993). This vast mouth was characterized by a series of bays, bayous, tree lined cheniers, dense levee breaks, and impenetrable marshes. Actually, the river had two main forks. The right fork is now called the Mississippi River and the left fork Bayou Lafourche. The area between these forks was a vast island when the Mississippi was in flood. At a larger system scale, the land was built up as a consequence of delta switching; however, smaller scale and more frequent river flooding events were responsible for some wetland creation and certainly responsible for wetland aggradation and maintenance.

To early explorers, control of the Mississippi in a political jurisdiction sense was the prime issue. Flood control was not an issue until the earliest settlements were established. Europeans had developed the technology of constructing levees along the Po, Danube, Rhine, Rhone and Volga rivers, prior to the establishment of New Orleans (Davis, 1991). So it was only natural that Louisiana's new immigrants would start immediately building levees. Thus as early as 1717 the Mississippi River Levee System was born. While flooding was of concern to Louisianans' along the river, national interest in the first half of the 19th Century was focused on enhancing river navigation. Levee construction thus increased in pace such that by 1851 a continuous levee existed from Baton Rouge to 32 km (20 miles) downstream of New Orleans. These levees were generally fairly low and not very wide and were frequently breached during floods.

In 1849, as a consequence of a series of severe floods, the flooding problem became a National issue with the enactment of the Swamp Land Act (Davis, 1991). As a consequence of this initiative, and the establishment of levee districts and boards by the state, by 1858 more than 3218 km of 2.4 m. to 3.0 m. high (with a base of 15.2 to 21.3 m.) levees lined the Mississippi River. However, these levees were too often breached. The flood of 1927 was considered the "greatest peace time disaster in our history" (Simpich, 1927, p. 245). It was the product of abnormally high rainfall in the Mississippi's drainage basin. Approximately 800,000 individuals were driven from their homes and in excess of 59,570 km^2 were inundated (Simpich, 1927). The severity of the 1927 flood resulted in passage of the 1928 Flood Control Act and the U.S. Army Corps of Engineers began to construct the present Mississippi River guide levees. Construction of the levees has effectively eliminated overland flooding, crevassing and also sealed all distributaries from the Mississippi River south of Baton Rouge. Natural wetland accretion was terminated by the artificial levees.

The 20th century exploration and mining of oil and gas sealed Louisiana's wetlands fate. Thousands of kilometers of access canals, pipeline canals and navigation channels were dredged enhancing saltwater intrusion into sensitive fresh water wetlands sealing their doom; as well as the canals, channels and associated levee spoil piles being extremely disruptive to the natural wetland hydrology.

As the twentieth century progressed, the Louisiana coast lost its wetlands at an increasing rate to reach about 103 km^2 (40 square miles) per year in the 1970's. Recently, the rate has slowed, but losses of tens of km^2 per year still occur (approximately, 0.5 km^2/day). Total wetland loss since the turn of the century has been over 4,000 km^2, an area 1 1/2 times that of Rhode Island (Templet and Meyer-Arendt, 1988). Loss of wetland means more than loss of the prevailing resource base. Life and property are increasingly threatened as the populated, low-elevation natural levee lands become more exposed to the Gulf of Mexico. The loss of storm-buffering protection by wetlands and barrier islands not only jeopardizes the safety of isolated bayou communities but also the New Orleans metropolitan area. The potential impact of a hurricane directly striking the city is much more serious today than in decades past because of the increased storm surge levels expected with adjacent open-water conditions (Templet and Meyer-Arendt, 1988).

Storm surge modeling

LSU began issuing experimental ADCIRC results during the 2002 hurricane season. By 2005 emergency managers were seeking operational surge forecasting within each 6-hour NHC advisory window. Some simplifications are required for the rapid turnaround necessary in this degree of operational support, particularly the exclusion of astronomical tides and setup due to waves. Astronomical tides in the northern Gulf are less than 0.5 m, but wave setup can be more significant under some storm scenarios (Weaver and Slinn, 2004). The most important ADCIRC output provided to emergency managers shows the maximum elevation of water reached at every point in the relevant portion of the model domain (Figure 4.2). In some cases, animations are also produced to study surge development in more detail. LSU issued five surge forecasts over the two days prior to landfall at 1200 (UTC) based on NHC Advisories 16, 17, 18, 22 and 25, (Table 4.1).

High water mark (HWM) elevations outside of the GNO hurricane surge protection system were surveyed to assess the accuracy of near-real-time surge forecasts. Similar data inside the polders was also acquired to permit calculation of the volume of water that entered the GNO flood defense perimeter. A good hurricane track prediction is essential to accurate surge forecasting. The NHC track prediction for Katrina was exceptionally consistent and accurate for more than 2 days prior to landfall (Knabb et al, 2005). The center of the eye was forecast to pass between 13 and 59 km east of downtown New Orleans Superdome for ADCIRC simulations beginning with NHC Advisory 16 issued 51 hours prior to landfall (Table 4.1). The forecast track shifted gradually to the east over this period within this narrow corridor. Other characteristics of the storm forecast at the GNO latitude changed, however, from 51 to 24 hours before landfall, indicating an expected increase in intensity with a lower central pressure and greater sustained wind speeds (Table 4.1).

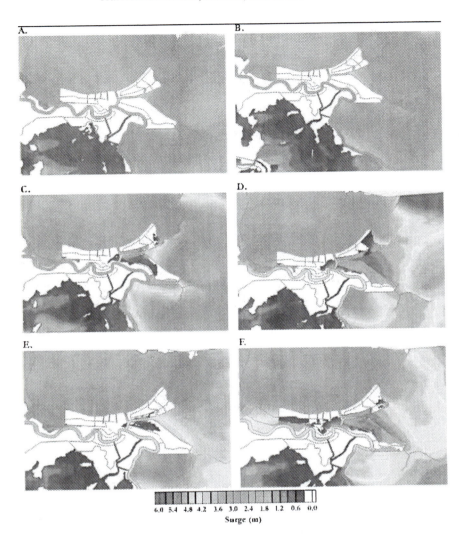

Figure 4.2.   The LSU ADCIRC forecast sequence for Hurricane Katrina based on NHC advisories with hours to landfall: (A) Adv. 16–51h, (B) Adv. 17–45 h, (C) Adv. 18–39 h, (D) Adv. 22–24 h, (E) Adv. 25–9 h, (F) Adv. 31–(−27 h).

This trend in the forecast was reversed, however, in an advisory issued 9 hours before landfall (Adv. 25) that predicted substantial weakening and an increase in forward speed. The post landfall forecast (Adv. 31) showed less weakening than expected earlier but an even greater forward speed (Table 4.1). The final NHC track has the eye center passing 10 km to the east of the St. Bernard levee defenses, so that only the western, weaker eye wall winds affected any populated portion of

Table 4.1. National Hurricane center forecasting and LSU surge forecasting for Hurricane Katrina.

	National Hurricane center advisory information			Predicted storm characteristics at closest CNO approach 30° N				Surge analysis complete	
Advisory number	Advisory date: Time (UTC)	Time to landfall (h)	Eye center location (km)	Central pressure (mb)	Max sustained wind (m/s)	Speed (m/s)	Date: Time (UTC)	Elapsed time (h)
16	8/27/03 0900	31	13.0 East	968	47	3.3	8/27/03 1950	10.3
17	8/27/05 1500	45	27.8 East	957	54	5.8	8/27/05 2000	5.0
18	8/27/03 2100	39	18.3 East	941	51	3.8	8/27/05 2700	6.0
??	2/28/05 1200	74	33.3 East	920	59	5.8	8/28/05 2000	8.0
25	8/29/05 0300	9	48.2 East	963	50	6.7	8/29/05 0950	6.5
31	8/30/05 1500	37	159.3 East	937	156	8.6	Post Storms	

GNO (Figure 4.1). On the other hand, much of the storm remained over water-prior to the second landfall on the Louisiana-Mississippi border, and experienced relatively little weakening. Katrina dealt GNO a glancing blow when it crossed Louisiana as a relatively fast-moving Category 3 storm (Knabb et al., 2005).

Beginning with the advisory issued 39 hours before landfall, ADCIRC predicted a surge that would enter GNO through the artificial funnel formed by the convergence of the Gulf Intra Coastal Waterway (GIWW) and Mississippi River Gulf Outlet (MRGO) levees. This surge then sequentially overtopped levees along the connected channels of the MRGO, GIWW and Inner Harbor Navigation Canal – Industrial Canal (IHNC). Less significant overtopping was forecast along the southern St. Bernard levee and none was predicted on the Lake Pontchartrain shoreline. ADCIRC predicted a potential for overtopping along the western flood-wall protecting the Jefferson polder and along the southern St. Bernard levee. Only one water level gage at the southern terminus of the IHNC recorded useful data through the peak of the surge (IPET, 2006), but HWMs acquired outside the flood protection levees and floodwalls permit assessment of the accuracy of the LSU ADCIRC forecasts. Peak surge elevations in the five sub-areas (Figure 4.1) show an increase from west to east, ranging from 2.2 m in Lake West to 4.9 m outside the St. Bernard polder (Table 4.2). The surge forecast associated with the most intense storm prediction (Adv. 22) produced the best overall match to observed values, resulting in a predicted GNO mean peak elevation of 3.65 m, 0.18 m lower than the observed (3.83 m). The RMS error overall was 0.60 m, resulting from over-prediction to the west (Lake West and Jefferson) and under-prediction for the East Orleans and St. Bernard polders on the east (Table 4.1). The post-storm simulation (Adv. 31) produced a mean elevation (3.75 m) only 0.08 m lower than the observed mean, but with a larger RMS error (0.72 m) than the Advisory 22 forecast.

Most observed HWMs contain an added, but unknown, surge increment associated with waves. This is more likely for those surveyed on the eastern side of GNO where wave action was most significant (IPET, 2006). If this is true, then the difference between a mean surge elevation of 4.5 m predicted by ADCIRC and 4.9 m observed in St. Bernard, for example, may be largely attributable to waves. Waves can damage unarmored levees like those that collapsed along the MRGO, but do not advect much water over intact structures (Seed et al, 2005). Surge forecasts as accurate as those produced during the Katrina approach show not only where levees are challenged but can also provide an estimate of the volume of water that would have been introduced if the levees and floodwalls had not failed (Table 4.3).

Overtopping predicted by ADCIRC would have delivered far less water than what actually entered the city, ranging from 11 percent of that measured in St. Bernard to 37 percent in East Orleans (Table 4.3). Just under 60 percent of the nearly 23,000 ha that were submerged on 29 September would not have experienced standing water. Moreover, mean flood depths caused by overtopping would have been less than 0.7 m in all three polders. The surge predicted by ADCIRC exceeded existing levee crests by no more than 1 m anywhere for a period of less than two

Table 4.2. Peak surges of the hurricane Katrina.

Zone West to East	No. Obs.	HWM (SE) m	Forecasts: Hours to Louisiana Landfall [NHC Advisory Number]					
			51 h [16]	45 h [17]	39 h [18]	24 h [22]	9 h [25]	−27 h [31]
Lake West	5	2.20 (0.21)	2.16	1.66	2.30	2.24	1.94	2.71
		RMSE	0.34	0.62	0.37	0.30	0.38	0.66
Jefferson	4	2.45 (1.67)	2.49	2.10	2.94	3.14	2.60	3.73
		RMSE	0.37	0.34	0.28	0.23	0.26	0.34
Orlean: Metro	20	3.63 (0.31)	1.99	2.28	2.73	3.40	2.77	3.49
		RMSE	1.69	1.39	0.99	0.35	0.92	0.33
East Orleans	9	4.8 (0.55)	2.76	2.90	3.43	4.06	3.35	4.18
		RMSE	1.57	1.41	0.99	0.45	1.02	0.63
St. Bernard	14	4.16 (0.51)	3.11	3.05	3.96	4.44	3.58	4.52
		RMSE	2.00	1.98	1.52	0.80	1.48	0.72
All	52	3.13 (0.14)	2.46	2.51	3.14	3.65	2.90	3.75
		RMSE	1.68	1.50	1.16	0.60	1.08	0.72

Table 4.3. Observed flooding and forecast overtopping for New Orleans during Hurricane Katrina.

Zone (no. obs.)	Mean high water (SE) (m)	Mean elevation of flooded land (m)	Mean depth of flooding (m)	Flooded area (ha)	Area flooded (%)	Flood volume (m³)	Flood volume (%)
		Observed flooding from overtopping and breaches					
Orleans Metro (22)	0.70 (0.02)	−0.59	1.36	8,156	100	117,604,500	100
East Orleans (12)	−0.52 (0.03)	−1.77	1.23	5,951	100	66,071,800	100
St. Bernard (58)	3.21 (0.09)	0.75	2.46	8,036	100	177,241,000	100
Total (92)				22,143	100	360,017,300	100
		Forecast flooding from overtopping					
Orleans Metro	−1.02	−1.61	0.59	2,455	30	14,888,742	13
East Orleans	−1.54	−2.22	0.68	3,608	61	24,361,620	37
St. Bernard	0.36	−0.21	0.57	3,031	38	18,861,091	11
Total Overtopping				9,094	41	58,111,453	16
		Estimated flooding from breaches					
Orleans Metro			0.77	5,701	70	102,715,758	87
East Orleans			0.57	2,343	39	41,710,180	63
St. Bernard			1.89	5,006	62	158,379,909	89
Total Breaches				13,038	59	302,805,847	84

hours. An overtopping event of this type would not have caused the extensive damage to interior drainage and pumping infrastructure that, in fact, took place. Instead, breaches scoured deeply into the levee base to establish a direct connection with the sea outside that rendered pumping futile (Seed et al., 2005). On average, breaches are estimated to have transmitted nearly 84 percent of the observed GNO flood volume. Because breaching played such a dominant role, processes that led to the initiation and expansion of breaches greatly influenced the volume of water introduced into each polder.

The ADCIRC model has proved to be an accurate prediction tool for emergency managers and lends itself for use in other hurricane prone areas. Additionally, it has a real value as a planning tool.

The levee failures

"Team Louisiana" was commissioned by the Louisiana Department of Transportation and Development to undertake an investigation of the numerous levee and floodwall collapses observed in the GNO after Katrina (van Heerden et al., 2007). Failure can occur when a system is stressed beyond what it was designed to resist or failures can take place because a system is under-designed. This situation can arise when risks and failure mechanisms that should have been known at the time are either ignored or underestimated.

It is important in any forensics investigation to avoid an appearance of omniscience in 20:20 hindsight. Designs are developed at a particular time when less is known. It is not necessarily appropriate to expect those who designed structures in the past to know everything that we do today. An understanding of what was known at the time, what tools were available then and what tools were used is critical (van Heerden et al., 2007). On the other hand, all engineers – then or now –must deal with uncertainty and follow accepted practice to account for unknowns that could increase the risk of failure. This is generally done for structures like levees and floodwalls by inflating the known stress to be resisted by a "factor of safety" that is sufficient to account for unknowns, or for the unlikely cumulative effects of multiple processes operating simultaneously.

Five surge events

The GNO metropolitan area sequentially faced five surge events, distinct in time, space and intensity.

(1) The first surge was coincident with landfall (0600) when the storm created a dam at the mouth of the Mississippi River and reversed flow in a bore that moved upstream, causing overtopping of the mainline Mississippi River levees on both banks as far north as Belle Chasse.

The remaining surge events involved water that flowed west from the Mississippi and Alabama coasts and was trapped against the bulge formed by Louisiana and the southeast trending Mississippi River levee system.

(2) The second surge collected against the Mississippi River levees in Plaquemine Parish south of New Orleans. This surge sent water over all of the parallel levees that follow the River, affecting communities on both banks. A relatively small amount of water overtopped the hurricane protection levee that runs east from Caernarvon, wetting St. Bernard for the first time.

(3) The third surge event was associated with Lake Borgne 'funnel' grew prior to landfall and reached a maximum elevation, exclusive of waves, of approximately 5.2 m at 0700. The Lake Borgne surge was conveyed to the Inner Harbor Navigation Canal where it reached about 4.6 m by the western leg of the Mississippi River-Gulf Outlet. This is a part of the Gulf Intracoastal Waterway that was vastly enlarged for ocean shipping. The Lake Borgne surge caused more flooding than all other events combined, destroying kilometers of levees and floodwalls and inundated the Lower 9th Ward, East Orleans, St. Bernard and much of Gentilly.

Two surge events affected Lake Pontchartrain, one on the south shore followed by one on the north.

(4) The first Pontchartrain surge occurred following an initial buildup as water streamed into the lake from the Mississippi coast over a couple of days. Apparently, the 1.8 m (6 ft) lake level elevations at this time were enough to initiate movement on some of the floodwalls. As the storm passed over Lake Borgne at around 0900, the winds over Lake Pontchartrain shifted from easterly to northerly and westerly and pushed a surge of between 3 to 3.6 meters against GNO levees and floodwalls. The Orleans Canal spillway began flowing, and three I-Wall reaches along the London and 17th St. Canals gave way.

(5) The second rise in Lake Pontchartrain occurred along the north shore at about 1200 and took place after the storm moved onshore as winds diminished and the huge surge that hit Mississippi relaxed and spread west. This surge diminished in to the west, but put more than 3.6 m into Slidell and flooded significant portions of St. Tammany parish.

Besides the aforementioned channels, others of interest on the west side of the city are the London Avenue Canal, the Orleans Canal, the 17th Street Canal and the Parish Line Canal (also called Duncan Canal).

General findings

- The hurricane protection system (HPS) relied upon by the residents of New Orleans failed catastrophically wherever surge waters reached elevations of 3 m or greater on canal levees or floodwalls.
- An estimated 84 percent of the floodwater volume that entered the city came through breaches, rather than over the tops of levees that were too low.
- Catastrophic levee and floodwall failures were confined to artificial waterways connected to Lake Borgne or Lake Pontchartrain, or both.

- Levees along the Lake Pontchartrain lakefront and those that were protected by extensive wetland systems east and west of the greater New Orleans area did not fail even if they were overtopped
- Levees and floodwalls constructed in close proximity to deep channels tended to fail catastrophically.

Specific findings

- 17th Street Canal, London Ave Canal and the Industrial Canal failures reflected a lack of foundation support for the I-wall design, generally exacerbated by subsurface water movement.
- I-Wall failures on the London Ave. and 17th St. Canal at surge levels of between 9 and 10 ft should have been predicted by geotechnical engineers working with the data and tools available to them when the walls were designed.
- No overtopping has been identified in any of the Orleans Metro drainage canal systems except in one section of the Orleans Canal, where a lower elevation 'spillway' was present because a section of the floodwall on the east side near the terminal pump station was missing.
- I-Wall sections that form the west side of the Jefferson Parish HPS along the Parish Line Canal did not fail but experienced significant movement as a result of the storm.
- Some floodwalls along the IHNC failed before they were overtopped, while others survived the limited overtopping.
- Levees built of sand along the MRGO were destroyed primarily by wave action.
- MRGO levees were destroyed well before they could have been overtopped.

The floodwall failures associated with the drainage canals have attracted the attention of the soils and foundation engineering community, and have been subjected to detailed analyses. Indeed, the USACE conceded last month that these I-wall failures were predictable when shortcomings in the original design were coupled with a poor characterization of the properties of the underlying soil strata. Nothing more is required to set these failures in motion than 2.4 to 3 m of static head in the canal. In contrast, the processes that precipitated breaching in the levees and floodwalls of the Industrial, GIWW and MRGO canals are more complex in an oceanographic sense. Analysis has included the additional forces and loadings associated with high velocity flows and waves (van Heerden et al., 2007). The soils that make up the levees, rather than only those that comprise the foundation, assume greater importance. Waves and currents operate primarily on levee faces rather than foundations, and have received less attention to date (van Heerden et al., 2007) but are the dominant forces leading to collapse of the MRGO levees.

ASSESSMENT

A long-term Hurricane Protection System (HPS) is needed for coastal Louisiana that focuses on levee systems coupled with wetland creation, restoration, and maintenance and has as its central theme, delivering sediment and fresh water from the Mississippi (van Heerden and Bryan, 2006). Any big-picture surge protection system for this state must include major barrier levees and flood gates in conjunction with creating wetlands and barrier islands. The best procedure is to build the hard structures where they can be protected by an existing "platform" of wetlands, no matter how fragile, and where there's a reasonable supply of good soil building material for levees. The wetland platforms seaward of the barrier levee and floodgates would be the target sites for future wetlands creation projects, with barrier islands seaward of the expanding wetlands base. Barriers—wetlands—barrier islands: This progression assures the survival of the estuarine bays necessary to keep the commercial and recreation fishing industries alive, as well as to supply the breadth of natural habitats that make up this unique ecosystem.

Here are the main elements of the major barrier system (Figure 4.3):

- At the main entrance to Lake Pontchartrain, build a flood control structure with multiple gates. Such a structure will need at least one shipping lock. These floodgates in the entrance to the lake would protect all communities around the Lake Pontchartrain and communities as well as all the way up to Baton Rouge.
- The pump stations must be moved to the head of the 17th Street, London Avenue, and Orleans Avenue canals, blocking any storm surge from the lake from entering the canals.
- During Katrina we saw how water poured from the Industrial Canal into Lake Pontchartrain. A gated structure with either a large lock or a surge barrier plus a lock needs to be built where the Ted Hickey Bridge crosses the Industrial Canal, a hundred yards or so from the lake.
- Eastward from the Lake Pontchartrain flood gate all levees need to be substantially raised. East of New Orleans a butterfly-gate type of flood control structure is needed in the area of the Funnel. Substantial levees are needed from this butterfly gate linking with the levees along the Mississippi River that protect from river floods. This could be a rebuilt version of the 40 Arpent Canal levee. All levees should have armoring that saves them from both wave attack and erosion should they be overtopped.
- From the Mississippi River westward we need to build one giant barrier levee that basically follows a bulging curve seaward, taking in Houma and some of its surrounding communities (some smaller curves may be necessary to incorporate some communities). With careful location, this levee, about 135 km (84 miles long), can balance the protection of the communities "inside" and wetlands fronting the system on the "outside." This inland levee would include numerous gated structures at key navigation channels and where river/sediment/freshwater diversions will be built. The barrier levee would then tie into the East Guide Levee of the Atchafalaya River at a point close to Morgan City. This levee can then be extended from the West Guide Levee of the Atchafalaya River all the way to Texas, with navigation gates at the required locations. This giant levee needs to be on the seaward flank of a large navigation channel, parts of which would include the existing Intracoastal Waterway. This channel would allow

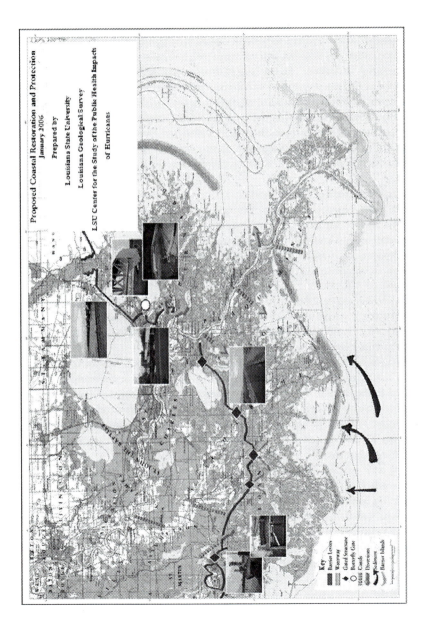

Figure 4.3.   Conceptual Plan for Hurricane Protection System (Based on van Heerden and Bryan 2006).

access for heavy equipment and barges with soil, rock, and other levee-building materials. To some extent this new levee would replace the network of levees now in place in this area south and west of the Mississippi River. The existing network creates some funnel effect problems of its own.

- This whole system—barrier levee with navigation/sediment distribution gates—must be accompanied by legislation that stops development in the wetlands anywhere in the newly protected areas. Many communities are going to be outside this new levee system. Some retreat has to take place. It should also be required that all new construction within the new barrier be elevated a given number of feet as determined by the design specifics of the whole system. This would provide an extra margin of protection in the event of some levee overtopping, or to reduce flood damage should there be a rainfall flood that exceeds the capacity of the pumps.

In order to protect the levee system, and as a hedge against a rising sea level and climate change, we must concentrate on building and restoring wetlands seaward of the barrier, using the sediment resources of the Atchafalaya and Mississippi rivers. Even with the major barrier system, these wetlands are still the outer defense and must be maintained for this purpose, as well as for their commercial, recreational, and environmental uses. Natural systems and some developed areas within the core should also be maintained, in order to accommodate expected overtopping and to avoid the degradation of water quality that the Dutch have experienced. The major elements necessary for large-scale wetland creation include:

1. Divert the Mississippi River into Breton and Chandeleur Sounds Approximately 20 km^2 of new wetland will be created in this stable basin every year. Increased sediment and freshwater input will greatly aid the St. Bernard marsh complex.
2. Reconnect Bayou Lafourche to the Mississippi River and divert about 5% of the Mississippi River down the bayou. From here it can be diverted over a wide area of wetlands.
3. Increase the discharge from the Mississippi down the Atchafalaya River from 30% to 40% and disperse this extra flow in the coastal marshes as widely as possible.
4. Long-term restoration of Louisiana's wetlands also depends on the restoration of Louisiana's barrier islands. Thus, management of the Mississippi River for wetland restoration should be simultaneous with barrier island restoration.

The above four point plan will not be cheap; however, it will ensure the continued existence of coastal Louisiana, its inhabitants, and unique culture. Additionally, it presents many new opportunities to expand the regions' job base and hence improve its economy. However, by way of ending, we quote van Heerden from "CNN Reports: Katrina – State of Emergency":

"So now is the time to put politics, egos, turf wars, and profit agendas aside; now is the time to effectively reconstruct New Orleans; engineer

proper levees and restore the coast. Now, because nature will not give us a second chance. We owe it to the 1,000 who died; we especially owe it to their survivors. Maybe more important we owe it to future generations"

ACKNOWLEDGEMENTS

This research was supported by the Louisiana Department of Transportation and Development under the leadership of Secretary Johnny Bradberry, as well as by the Center for the Study of Public Health Impacts of Hurricanes, a Louisiana Board of Regents Health Excellence Fund Center, and by the McKnight Foundation. Dane Dartez, Young S. Yang, Kate Streva and Ahmet Binselam of the LSU Hurricane Center provided data collection and analysis assistance. Independent consultant Billy Prochaska, Lou Capozzoli and Art Theis contributed significantly to the Team Louisiana effort.

REFERENCES

Condrey, R.E. (1993). The Early Explorers' Views of the Barataria-Terrebonne system: The Fork in the River and its Island.

Davis, D.W. (1993) Crevasses on the Lower Course of the Mississippi River. Coastal Zone '93, vol. 1, p. 360–378, July 19–23, 1993, New Orleans, Louisiana.

Interagency Performance Evaluation Task Force (2006) Performance Evaluation Plan and Interim Status, Report 1 of a Series: Performance Evaluation of the New Orleans and Southeast Louisiana Hurricane Protection System. (U.S. Army Engineers MMTF 00038-06).

Knabb, R.D., J.R. Rhome, D.P. Brown (2005) Tropical Cyclone Report: Hurricane Katrina. (NOAA National Hurricane Center).

Kolb, C. R., and J. R. Van Lopik (1958) Geology of the Mississippi river deltaic plain, southeastern Louisiana, v. 1. United States Army Engineer Waterways Experiment Station, Vicksburg, Mississippi.

Luettich, R.A. Jr., J.J. Westerink (1995) in Quantitative Skill Assessment for Coastal Ocean Models, D. Lynch, A. Davies, Eds. (American Geophysical Union Press, Washington, D.C), vol. 48, pp. 349–371.

Seed, R., et al. (2005) "Preliminary Report on the Performance of the New Orleans Levee Systems in Hurricane Katrina on August 29, 2005" (Report No. UCB/CITRIS – 05/01, November 17).

Simpich, F. (1927) The great Mississippi flood of 1927: The National Geographic Magazine, v. 52, no. 3.

Templet, P.H., and K.J. Meyer-Arendt (1988) Louisiana Wetland Loss: A Regional Water Management Approach to the Problem. Environmental Management, vol. 2, No. 2, p. 181–192.

Van Heerden, I. Ll. (1994) A long-term, comprehensive management plan for coastal Louisiana to ensure sustainable biological productivity, economic growth, and the continued existence of its unique culture and heritage. NSMEP-CCEER Report, Louisiana State University, 49pp. (http://www.publichealth.hurricane.lsu.edu/ Adobe%20files%20for%20we bpage/CCEER%201994.pdf).

Van Heerden, I.L. and M. Bryan (2006) "The Storm" Publ. Penguin, New York, New York, 308pp.

Van Heerden, I. L., G. P. Kemp, H. Mashriqui, R. Sharma, B. Prochaska, L. Capozzoli, A. Theis, A. Binselam, K. Streva, and E. Boyd (2007) The Failure of the New Orleans Levee System during Hurricane Katrina. TEAM LOUISIANA Final report for Louisiana Department of Transport and Development.

Weaver, R.J., D.N. Slinn (2004) Coastal Eng. 2004, Proc. 29th Int. Conf. 2, 1532.

Westerink, J. J., R.A. Luettich, Jr., J. Muccino (1994) Tellus 46a, 178.

5

Water quality and public health – case studies of Hurricane Katrina and the December 2004 tsunami in Thailand

A. J. Englande[1], R. Sinclair[1], and P. Lo[2]

[1]*Tulane University, New Orleans, LA 70112, USA*
E-mail: aenglan@tulane.edu
[2]*Materials Management Group, New Orleans, LA 70114, USA*
E-mail: paullo@mmgnola.com

Summary: The South East Asian Tsunami in Thailand and Hurricane Katrina in the United States were natural disasters of different origin but of similar destruction and response. Both disasters exhibited synonymous health outcomes and similar structural damage from large surges of water, waves, and flooding. A systematic discussion and comparison of the disasters in Thailand and the Gulf Coast assess the: environmental health planning before the disaster, the environmental health impacts, and future planning in expectation of a similar event in coastal areas. By considering both calamities to be similar types of disaster in different coastal locations, valuable comparisons can be made for improvements in response and preparedness. Recommendations are made to: (1) improve disaster response time in terms of needs assessments for public health and environmental data collection; (2) development of an access oriented data sharing policy; (3) and prioritization of natural geomorphic structures such as barrier islands, mangroves, and wetlands to help reduce the scale of future natural disasters.

INTRODUCTION

On December 26, 2004 at 7:58 in the morning, an undersea earthquake west of Sumatra Island in Indonesia initiated a strong tsunami first hitting nearby beaches with the most intense force, then other countries including India, Malaysia, Maldives, Myanmar, Somalia, Sri Lanka, and Thailand. The waves in Thailand were first reported on the Phuket Province coastline at 9:38am; when, three – meter high tidal waves hit the shore followed fifteen minutes later by a three to ten meter giant wall of water which extended for several kilometers pushing floodwaters up to one kilometer inland. A third series of five-meter-high waves followed the water wall (Sinthuvanich and Boonprakub, 2005b). This is one of the most deadly natural disasters in modern history with a total estimated 229,866 persons lost, including 186,983 verified as dead and 42,883 missing. In Thailand, 8,212 people are verified dead or missing with over 2,448 of those as non-Thai from thirty-six different countries (UN, 2005). Losses in Thailand are estimated at 1.6 billion U.S. dollars with a minimum repairing cost of 482 million US dollars (Figure 5.1).

Nine months after the tsunami in Asia, Hurricane Katrina made landfall on the U.S. Gulf Coast on August 29, 2006 at 6:10 am (see also the preceding chapter). The disaster was characterized by heavy rains, a devastating storm surge up to ten meters high in Mississippi, and strong winds reported at a maximum of 215 km/h near Buras, Louisiana. The storm surge's wall of water caused wide-spread tidal damage to coastal areas to Gulf Coast states with many areas in Mississippi and Louisiana completely destroyed. It also induced breeches in the New Orleans levee system, flooding 80% of the city, limiting government access and isolating many families and individuals, many in poor health status, without any electricity,

Figure 5.1.   Tsunami destruction in Thailand (photo author)

Figure 5.2.   Destruction by Hurricane Katrina (author's neighborhood in Chalmette, LA)

food, or drinking water (Daley, 2006). The remaining population of New Orleans had to live multiple days in a harsh urban landscape where little was known about security, safety of the ubiquitous floodwater, or the future in general. The immediate destruction, flooding, and related issues account for a total of 1,695 deaths in all affected areas with 1,464 in Louisiana and 231 in Mississippi (Franklin, 2006; LDHH, 2006). The total damage from Katrina is estimated at 75 billion US dollars (Figure 5.2).

The S.E. Asian Tsunami in Thailand and Hurricane Katrina in the U.S. were natural disasters of different origin but of similar destruction and response. Both disasters exhibited synonymous health outcomes, similar population targets including low socioeconomic status groups, and similar structural damage from large surges of water and waves. These events demonstrate great learning opportunities for better preparedness for future disasters and a need of improved: coherent public health response and prioritization, coordinated government response, a coherent rapid data collection plan for health and environment indicators, and more emphasis on natural environmental protection such as mangroves or wetlands. By considering both calamities to be similar types of disaster in different coastal locations, valuable comparisons can be made for improvements in response and preparedness. It is suggested that other coastal areas in the world should heed the advice and lessons learned and be proactive by developing plans in anticipation of the next disaster rather than responding after the fact.

Scope and objectives

This paper will discuss both the Thailand Tsunami and the Louisiana/Mississippi Hurricane Katrina in reference to environmental health planning before the disaster, the environmental health impacts, and future planning in expectation of a similar event in comparable locations. Though emphasizing water and environmental health, the paper will also address: response by the community, government, and volunteer aid groups; environmental health issues of concern; commonalities between the two disasters; and recommended actions to improve response to future disasters in similar coastal areas.

CURRENT KNOWLEDGE

Pre-storm Thailand

The areas affected by the storm in Thailand included many tourist beach resorts on the water front. A few national forests, a number of mangrove areas, and some undeveloped areas were hit by the tsunami. Behind the tourist resorts, there was a variety of housing ranging from middle class villas, to temporary squatter huts built by migrant Myanmar workers or nomadic Thai *Moken* fishermen. Many of the poor communities around the tourist resorts or shrimp farms had pre-existing disease burdens of acute respiratory illness or chronic diarrhea, both characteristic of global poverty. Previous to the storm, the Royal Thai government had some resettlement disputes with a few of the affected local communities. The exchange rate, poverty descriptors, and cost of living in Thailand are much different than anywhere on the Katrina affected U.S. Gulf Coast.

Disaster plans existed among government agencies and hospitals. They were regularly rehearsed for general types of disasters, most likely mass causalities associated with the many isolated terrorist threats in the southern provinces. Being quick in emergency response and post-impact recovery, these plans were applauded by many international observers. The plans lacked longer term preparedness and mitigation due to natural disasters but efforts by the royal family before the disaster have emphasized the protection of coral and mangroves for ecological, and unknowingly, natural defensive justifications (Weerapong, 2004). Anecdotal evidence suggests that the close knit nomadic Moken fisherman had fewer fatalities and destruction. This is allegedly due to their consistent awareness of their surroundings, native understanding of the marine environment and their "collective memory" of tsunami warning signs (Holland, 2005).

Pre-storm New Orleans

In August, 2005 metropolitan New Orleans and surrounding areas were served by five large scale waste treatment facilities and sewerage systems. As the city is built on elevations sometimes lower then the surrounding lake or river, the separate storm water is pumped into Lake Pontchartrain during and following a rain event. The sewerage system and drinking water distribution system were in need of repair.

The New Orleans Sewerage and Water Board (NOSWB), responsible for more than 400,000 customers was implementing a one billion dollar capital improvement program to address long-term maintenance and repair needs, including compliance with a 1998 court-ordered sewer system consent decree. Because the infrastructure was built more than seventy-years ago, most of the leaking before and after the storm is underground in unverified locations for both the water supply and sewerage systems.

Much of the city's development occurred on land previously wetland or marsh. This is illustrated with detailed maps of the city during the 1850's which, in a recent conference presentation, was overlaid with maps showing the Katrina flooded neighborhoods; an obvious correlation exists where houses in the 1800s were built on land that did not flood in the 2006 Hurricane Katrina (Meffert, 2006). Communities 150 years ago in New Orleans realized the vulnerable flood areas and built their flood resistant homes on higher elevations. The 1.6 billion dollars of damages to New Orleans by Hurricane Betsy in 1965 prompted many city planners and politicians to emphasize protection of the city from hurricanes. One proponent of this new emphasis on planning for disaster was Governor John McKeithen who pledged that his administration would "see that nothing like this occurs in our state again" (Colton, 2006). This era of planning emphasized infrastructure improvements; more canalization and pumping stations which may have contributed to a false sense of security in some areas. Consequently, some home owners built structures on highly vulnerable flat concrete slabs. Fortunately, recent pre-Katrina hurricanes and better environmental awareness have prompted national and local groups to raise alarm to these unnecessary risks. The U.S. Army Corps of Engineers had even initiated a wetlands restoration study to "restore the natural geomorphic structures and processes" which would protect the land from major storm events (USACE, 2004).

Pre-storm Mississippi

In August 2005, the Gulf Coast area of Mississippi was served by wastewater treatment facilities approaching overload conditions from significant population increases due to rapidly developing subdivisions and casino development within the affected coastal counties. Rural mobile home parks, using on-site wastewater treatment, were also increasing. Similarly, the many small water treatment facilities in these counties were approaching capacity prior to the storm. The seven counties most severely affected by Hurricane Katrina (i.e., Jones, Pearl River, Stone, George, Jackson, Harrison and Hancock) received drinking water from 210 public water supplies serving a populations of about 495,000. Only seven of these systems served a population base greater than 10,000.

The State of Mississippi had an early warning system and hurricane evacuation routes set up, but other logistics such as shelter locations, volunteer management, and response to potential hazardous waste contamination from industrial facilities was limited. This is evidenced by the many vulnerable houses, structures, and industrial facilities located in coastal Mississippi.

Figure 5.3.   Post-Katrina recovery and assessment.

Impact Thailand

The Thai Ministry of Public Health responded rapidly to the tsunami disaster through organizing clinicians, medical supplies, treatment of injuries, burying, and identification of the dead. A rapid assessment tool was used to collect health data on syndromic illness, morbidity, mortality, and environmental health concerns/needs (TMOH et al., 2005). The health needs assessment found that an increase in acute diarrhea was observed, but still much less than outbreaks in other disaster settings. A large number of enteric pathogens were cultured from wounds which suggested surface contamination with organisms cultured as *A. hydrophila* and *Vibrio vulnificus,* similar to some polymicrobial infections detected during hurricane Katrina (Engelthaler et al., 2005). The Thai relief efforts were complemented by foreign governments such as the United States Center for Disease Control (CDC) collaboration, the government of Germany, and many others, but the Thais maintained that most programming of the disaster response was organized by the government itself (Prabahorn, 2005) through its central command in Bangkok with command centers in each of the six impacted provinces. Tulane University faculty, students, and Thai alumni coordinated with the Pollution Control Department (PCD) and a provincial health department to conduct a needs assessment on environment related issues.

Environmental health data was collected by the Pollution Control Department (PCD) but not made public in the months after the disaster. Because flooding associated with the Tsunami drained quickly away from affected areas, the environmental data focused on wells and surface water bodies. In addition to environmental parameter testing, government officials and the public were quick to declare most surface water unsuitable for drinking or treatment if rumors suggested that a human or animal corpse was in the water. Even if not true, these perceptions took time to disprove; fortunately, drinking and bathing water were readily available from well-stocked government supplies and many natural springs from nearby mountains. Additionally a system of dual water supply was enacted using uncontaminated reservoir or mountain spring water, depending on location. A membrane technology was used for drinking and food preparation and stored and labeled at a central location in each temporary camp; while, untreated source water was also stored and labeled to be used for other purposes.

Storm impact New Orleans

The first needs assessment was conducted by the CDC on October 14–17, approximately seven weeks after the hurricane, and therefore no longer considered "rapid." Residents in Orleans Parish were still without basic utilities and services such as running water (20.2%), electricity (24.5%), and garbage and debris removal. The statistics from Jefferson and Orleans Parishes were not typical of post-hurricane recovery; where, even after seven weeks, most households had basic environmental health needs which had not improved since two weeks after the hurricane (Norris et al. 2006). The needs assessment determined residents and relief workers in the New Orleans area had diarrhea (3.5%), acute respiratory infections (12.1%), and skin or wound infections (15.4%). It was also determined that almost half (49.8%) of the adult residents in the New Orleans area exhibited levels of emotional distress indicating a need for mental health services, while over half of the households (55.7%) had a member with chronic health conditions (Daley, 2006).

Data collection during the Katrina disaster was organized by various groups including the U.S. Environmental Protection Agency (EPA), the Natural Resource Defense Council (NRDC), Louisiana State University, and local researchers such as Louisiana Environmental Action Network, and Louisiana Bucket Brigade (NRDC, 2005). Although much data was not made immediately available due to copyright concerns of researchers and data processing time, many data summaries are now available on the internet. The NRDC website attempts to present data from all sources with little interpretation. Studies conducted at area universities have discussed these data with varying interpretations.

Although the news media claimed the flood waters were a "toxic soup", researchers reported the waters as comparable to typical storm water. Early reports claimed an absence of a major toxic effect in the flood water due to a massive dilution, the absence of refineries or chemical plants in the levied areas of New Orleans, gasoline stations in short supply as the hurricane approached, the strong alkalinity in the flood waters buffering the potential acidic effect of the many submerged car batteries, and the less than normal amount of hydrocarbons in the flood water (Pardue et al., 2005).

Figure 5.4.   The New Orleans East Bank wastewater treatment facility one week after
Hurricane Katrina

The infamous aid response to New Orleans Katrina disaster is known through
the news media globally. Although wind damage and flooding due to rain occurred
in the New Orleans area, the flooding due to levee breaks introduced a different
sort of destruction than that seen in Mississippi or the Thai tsunami. Over 80%
of the city's homes and buildings were flooded and access was a major issue for
stranded residents and emergency workers.

All of the major wastewater and drinking water treatment plants around New
Orleans were impacted by flooding, wind damage, electrical problems, or other
associated issues. In the eastbank wastewater treatment plant, the surrounding
waters rose to levels above the facility's earthen levees (Figure 5.4). A Federal
Emergency Management Agency (FEMA) task force was assigned to mitigate
these issues and within six-nine weeks, many of the areas facilities were back online
(Bankston, 2006). Since the hurricane, the NOSWB had to double the pumping
capacity of drinking water and increase the chlorine dose to super-chlorination
levels which increased consumer concern over taste and odors and disinfection
byproducts. Low water pressure handicapped fire fighting activities. Another major
environmental concern after the storm was huge piles of debris, and its removal as
nearby landfills approached their limits.

Storm impact Mississippi

All of the twenty-six wastewater facilities in Mississippi counties severely affected by Hurricane Katrina were damaged with many infrastructure and electrical problems. Some of these treatment facilities suffered from process failures preventing them from producing compliant effluent discharges. Infrastructure problems including sewage lift stations, grinder pumps, sewer plugging, electrical damage and breaks in the sewer system resulted in sewage contamination (MSDEQ, 2005). Thousands of on-site treatment facilities, particularly above ground systems, were also damaged. Similarly, most water treatment facilities and distributions systems sustained significant infrastructure damage to electrical components, storage tanks and broken and inaccessible water mains and valves. This prompted a blanket boil water alert for all six coastal counties.

On September 14th a rapid community Needs Assessment was conducted using U.S. census blocks and randomly selecting households in the worst hurricane affected Hancock County. In regards to water and sanitation, multiple households lacked critical services such as electricity (41%), functioning indoor toilets (37%), and running water (21%), while most households used bottled water (68%) or relief agency water (26%) (McNeil et al., 2005). This was conducted over one month before the New Orleans rapid Needs Assessment.

As many residents of these affected counties could not evacuate, emergency health services and housing shelters were developed by the National Guard and many private charities. Besides large amounts of debris in roadways, access in most cases was not as serious of an issue as it was in New Orleans. Federal, state, and volunteer emergency services were quick to arrive, coordinate and provide relief with minimal turf issues as evidenced in other areas. Thirty thousand FEMA mobile homes and RVs were placed in the affected area initially with ground disposal of waste followed by localized wastewater treatment at each park. Private drinking water wells were utilized, but mostly bottled drinking water was transported to the site in the early weeks following the hurricane. By October 10, 85% of drinking water facilities and 95% of wastewater facilities in Louisiana and Mississippi were operational, but the boil water recommendation was still recommended until all treatment plants could be tested (Copeland, 2005). Some membrane treatment systems were used for desalinization of drinking water; while, in rural areas the State Department of Health initiated a program for individual well testing and instruction to owners for disinfection and sampling of their private wells. Various other government agencies and state groups assisted in the testing and community drinking water awareness efforts in Mississippi. These included mobile labs and technicians from groups such as the U.S. Public Health Service, the CDC, the EPA, the U.S. Army Corps of Engineers, agencies from other states including North Carolina, Kentucky, Florida, California, and others through an interstate compact.

Environmental data were collected by the EPA in the storm surge impacted portions of Hancock, Harrison, and Jackson Counties. Soil and sediment samples were collected from industrial facilities including DuPont, SONFORD, CHEMFAX,

Chevron, Omega Protein, Polychemie, Ershigs Fiberglass and several others. The samples were assessed using the EPA's RMP (Risk Management Plan), Tier II, TRI (Toxic Release Inventory) for various contaminants of concern such as dioxins/furans, Acetophenone, Acetone, PCBs, Agent Orange, and other site related compounds. Based on the sampling results, the EPA does not believe that most sites were impacted by Hurricane Katrina. The data and summaries are available on the EPA website(USEPA, 2005).

Current and future solutions – Thailand

The most urgent problem facing the Thai Ministry of Natural Resources after the tsunami was spread of contamination from "seawater, wastewater, mud, sand and decomposing bodies" (Sinthuvanich and Boonprakub, 2005a). Although these may not have had serious pathogenic consequences as originally assumed, the Thais did prioritize these issues and minimized them in only a few weeks during the time following the disaster.

To mitigate and prepare for another tsunami, the Thai government released official reports from the Ministry of Natural Resources and Environment to prioritize the marine and coastal ecology including mandates to further protect, enhance, acknowledge, and beautify the naturally occurring coral, sea grass beds, and mangrove forests. The ministry also recommends more eco-tourism promotion, more natural resources for drinking water such as artesian wells, and coordinated work with local communities to fulfill these tasks (Sinthuvanich and Boonprakub, 2005b). Lastly, the early warning system is stressed where the Thai can have up to an hour's warning for complete evacuation. The published documents make no mention of how these preparedness steps can apply to natural disasters besides tsunamis. The objectives in protecting mangroves, coral, and other natural resources are strongly promoted through a tsunami preparation logic. The inherent value in these natural resources was only lightly addressed and the importance of community environmental education was mentioned only as a side note.

Current and future solutions – New Orleans

The priorities outlined in the mayor's "Bring New Orleans Back" (BNOB, 2006) plan (health and social services committee) include: (1) building the resiliency, preparedness, and sustainability of New Orleans in light of future hurricane seasons; (2) re-designing the system of health care based on other successful national models; (3) emphasizing the importance of environmental health within the health department and centralizing safe drinking water, food safety, and air quality data collection efforts to be easily available to the public; (4) increasing the cities emphasis on promoting healthy lifestyles and chronic disease prevention; (5) revamping medical records and databases to a useable, holistic, and rapidly accessible format; (6) establishing a regional healthcare collaborative; (7) and developing an orderly reliable census information system which is updated more frequently. These proposed solutions attempt to continue working towards complete recovery and reduce vulnerability through improving resiliency in health status, databases, and disaster response. The recommendations acknowledge that various environmental health related data has been obtained and interpreted (Mielke et al., 2006; Pardue et al., 2005; Presley et al. 2006) and that many contaminant levels in the city are similar to background levels before the storm, while some parameters such as lead and arsenic are in excess of EPA recommended levels. It also acknowledges that there are many conflicting interpretations to similar data and findings available from groups such as the NRDC and the EPA (NRDC, 2005; USEPA, 2005), and there is a need to have government representatives as specialists in risk communication to decipher available data. Most importantly, the report emphasizes expedient release of scientific data to all stakeholders. This is in response to the often encountered problem in disasters where agencies wait to release vital data due to publication concerns, fears of misinterpretation, or no funding for making data public.

The report recommends that there is still a need to: identify data gaps; assess the health risk of mold; assess the soil, sediment, and surface water samples for bacterial contamination and chemical compounds; and stress that data be released to all stakeholders with culturally sensitive interpretations to be utilized by policy makers who consider science and other social/economic issues. Although much verbiage included in the BNOB proposal is similar to plans discussed in Mississippi and Thailand, there appears to be a lack of discussion or synthesis of the municipal water and wastewater situation or related biological/chemical contamination concerns.

Since Hurricane Katrina, the New Orleans Sewerage and Water Board is running a deficit of $ 80,000 per day due to drinking water seeping out of the distribution system. The city has to pump more than twice as much drinking water into the system to keep the water pressure high to reach all customers as far out as the Ninth Ward and surrounding neighborhoods. The large amount of underground flooding also necessitates super chlorination which has the potential for public complaint due to odor and taste, in addition to potential disinfection byproducts. The NOSWB has applied to FEMA's public assistance program for reimbursement of the damages, but the details of this have yet to be negotiated (Krupa, 2006).

To address the major Katrina problems associated with storm drainage, drinking water and wastewater treatment facilities in the New Orleans area, the Federal Emergency Management Association (FEMA) has formed a task force with various local and national government groups to provide comprehensive engineering restoration/repair. The group was able to re-start two wastewater and one drinking water facility within six to nine weeks following Katrina which is regarded as one of the most challenging and successful engineering feats in New Orleans history. They also have initiated long-term repair operations on the city's drainage system, sewerage, and drinking water distribution systems (Bankston, 2006). Although highly useful in engineering solutions, the efforts by this group only addressed the basic hygienic need of sewage collection and drinking water treatment and did not address other environmental contamination issues.

Current and future solutions – Mississippi

While many people still live in temporary housing, the Mississippi Department of Health (MDH) coordinates with the MS Department of Environmental Quality and the EPA for continuous evaluations of rural and FEMA trailer park wastewater treatment and drinking water supplies. The state in coordination with the EPA has continued monitoring potential exposure to hazardous materials and chemicals in industrial areas. Recently, the state has received approval for a 500 million dollar grant from the U.S. Housing and Urban Development (HUD) to improve drinking water supply and sewerage infrastructure in the six affected MS counties (Spengler, 2006). The Gulf Coast Regional Infrastructure Program will develop areas where: (1) many of the displaced population have moved which do not have adequate water, wastewater, and storm water infrastructure; and (2) where existing infrastructure has been damaged by the storm.

Although there are limited published plans on future steps towards preparedness of hurricanes in the Gulf region of Mississippi, many homeowners have moved to developing rural areas farther away from the coastline. The public health infrastructure is also adapting by shifting to new developed areas and moving vital health data to a central location in Jackson (MDH, 2006). New priorities towards mental health efforts and risk communication strategies are also in development.

ASSESSMENT

Major similarities

Overall, the cost of the tsunami in Thailand was estimated at 1.6 billion US dollars, while Hurricane Katrina's cost was 75 billion US dollars across the Gulf Coast. The trend is reversed for mortality; Thailand had 8,212 fatalities while 1,836 occurred in the New Orleans/Gulf Coast areas. The difference is due to the lack of an early warning system in Thailand which has since been rectified.

Health

Table 5.1 shows the biggest health similarity is that most deaths were due to drowning across all income levels. However, the loss of life during Katrina approaches the scale of what may be expected in developing countries; in both cases many fatalities were among the elderly and poor. These communities are often marginalized by lack of transport, education, and poor housing locations in such areas as squatter huts behind beach front tourist resorts or areas which received previous floods like the infamous Ninth Ward (Mutter, 2005).

New studies in Thailand report that the opportunistic *Aeromonas spp.* were isolated from 22.6% of infected wounds following the tsunami (Hiransuthikul et al., 2005). Although largely unstudied in Katrina, these bacteria represent an emerging global health threat (Presley et al., 2006) and were found in many of

Table 5.1. Environmental and public health comparison after the disaster in the New Orleans area, Coastal Mississippi, and the Thailand tsunami

	H. Katrina New Orleans Area	H. Katrina Coastal Mississippi	Tsunami Thailand
Sanitation	Sewerage system damaged Use of portable latrines	Sewerage system damaged Use of portable latrines	Relocation to relief camps and pit latrines dug
Water supply	Water pressure in some areas but treatment not certain	Water pressure in some areas but treatment not certain	No well water
	Boiling recommended	Boiling recommended	No fresh water
	Bottled water supplied	Bottled water supplied	No water pressure Bottled water supplied
Socio-Economic groups affected	All groups	All groups	All groups
Days before CDC needs assessment	46 days	18 days	12 days
Availability of environmental test data	EPA & NRDC websites, publications	EPA website	Thai PCD did not publicly release data
Most common cause of mortality	Drowning	Drowning	Drowning
Man-made vulnerabilities	Canalization, drainage & pumping, navigation canals, reclaimed wetlands	Reclaimed wetlands Coastal development	Shrimp and prawn farming. Development of tourist resorts. Un monitored fishing
Source of structural damage	Standing flood water and surge in some areas	Storm surge	Wave surge
Natural solutions	Wetlands/barrier islands	Wetlands/barrier islands	Mangroves/ coral/islands

the flood waters. Some human cases were also reported (Engelthaler et al., 2005). It is assumed that poor hygiene and drinking water contributed to the elevated diarrhea problem in Thailand; diarrhea was measured in the New Orleans needs assessment (3.5%), but it is not clear if this is above endemic levels. Twenty four cases of *Vibrio vulnificus* and *V. parahaemolyticus* were reported with six deaths. For all three cases, no significant disease outbreaks grew beyond expected levels.

Structural damage

Wind and rainfall-induced flooding typically account for the majority of damage during a hurricane, but the storm surge associated with Katrina produced effects which resembled the 2004 Thailand tsunami. Structural damage in both areas was due to a huge wall of water, sometimes thirty feet high in areas of Thailand and the Gulf Coast. In both Thailand and the Gulf Coast, engineered structures suffered similar types of damages through "hydrostatic and hydrodynamic loads – both lateral and vertical – debris impact and damming loads, and scour of supporting soil. These were not anticipated in either event and extensive documentation of what happened during Katrina may help prepare for future disastrous events (Robertson et al., 2006). This storm surge had a different effect in New Orleans where only the Ninth Ward and St. Bernard Parish's Arabi/Chalmette neighborhoods reported surges of water. Most flooding in New Orleans was caused by levee failures and subsequent rising waters which remained in place for two to three weeks. Wind damage was also significant in all areas affected by the hurricane (Pace, 2005).

Environmental damage

Many areas in Thailand had thick mangrove systems extending for at least several hundred meters on land. Communities located behind these forests were often protected from the tsunami. The story was different in Sri Lanka where many Mangroves and other vegetation had been cut down, leaving only a thin weakly-protective layer of vegetation near the beach front; whereas, the fatalities from the tsunami were much more (Kathiresan and Rajendran, 2005; Wabnitz et al., 2005). Mangroves are shown not only to protect life in catastrophic flooding events, but also to lower groundwater salinity as it flows from ocean to land areas (Ridd and Sam, 1996).

Similar to Mangroves, the coral reefs in Thailand were under assault from poor water quality and destructive fishing practices. Effort to protect them was justified to protect marine biodiversity and maintain large fish populations to sustain livelihoods (Allen and Stone 2005); little consideration was given towards their protective effects. After the tsunami, the wave buffering effect by coral was witnessed in many countries; the Maldives and Thailand were relieved to discover that their policies of coral protection had also served as coastal protection from the tsunami (IEMA, 2005a; IEMA, 2005b).

In Thailand, the conservation of mangroves translates to less production of one of its largest exports. Thailand is the world's largest shrimp exporter and the U.S. is its biggest buyer. The flat land and brackish water in mangroves make the best environment for shrimp farming; often leaving many chemicals and antibiotics which may impact the surrounding marine environment (IEMA, 2005a). Fortunately, the royal family recognized these issues around mangroves and healthy coral; strategies to mitigate mangrove destruction and also provide alternative livelihoods for shrimp farmers were in place many years before the tsunami (Weerapong, 2004). Like most disasters, the tsunami prioritized the need for natural defenses; the government is now allocating more land for mangroves and allowing only a few of the impacted and displaced shrimp farmers to return.

Similar to how mangrove and coral acted as protective defenses against the tsunami, wetlands and barrier islands are the U.S. Gulf Coast's natural defenses. Unfortunately, these have been significantly degraded over the years. Since the 1930s coastal Louisiana has lost over 900,000 acres of coastal wetlands with a current a rate of 16,000 acres per year; where, 70% of this loss in coastal wetlands derives directly or indirectly from human activity (Galloway, 2003).

As a response to this, the U.S. Army Corps of Engineers (USACE) initiated an ecosystem restoration study to determine the approach to restore the natural geomorphic structures and processes unique to the Gulf Coast, and needed for physical protection against hurricanes (USACE, 2004). The report describes that at least five to ten years are needed to fully develop a large-scale and long-term restoration plan. The actual restoration of the destroyed areas could take a public commitment many decades (Vasisihth, 2006).

Both the Thai and U.S. governments officially recognized the protective benefits of the natural environment before the Katrina and tsunami disasters struck. Both governments had also designated budgets and plans to restore these features. The unfortunate disasters may have focused efforts on relief operations temporarily, but a positive outcome is that awareness of the issue has now increased many-fold among the lay population in Thailand and the U.S. Gulf Coast with a hopeful renewed emphasis in government and donor budgets.

Water and wastewater

With most water and waste water treatment systems suffering damages during the flooding and storm surge associated with Katrina, recovery of these facilities took more than six weeks before functionality was restored. In Thailand, the few local drinking water treatment facilities, which sustained damages, were restarted within a similar time frame. Some were not repaired due to relocation of communities. Before the tsunami, up to 20% of Thailand's population relied on untreated surface water for drinking and bathing (UN, 2006). This situation is likely to have dropped after the tsunami as the government provided improved water sources in all relief camps and new settlements.

Other environmental health concerns

Mold is the single most extensive environmental health issue in the Gulf Coast area. Most buildings in New Orleans rely on air conditioners for ventilation and prevention of mold growth during the hot season. Traditional Thai houses and communities are naturally adapted to mold and most use a building style with much ventilation, light, and few dark spaces. Mold was a concern more in the New Orleans area since most water did not drain away until weeks after the storm.

Mud, sediment, and debris were major environmental health concerns in the Katrina disaster. During the Thai tsunami, the government relocated families to military run shelters which did not suffer from this type of problem. The Thai infrastructure destruction situation, in general, was much smaller in magnitude than the Katrina disaster and much of the debris was gone only a few months after the tsunami.

Data sharing

In the Asian tsunami and the Katrina disasters, rapid data and other relevant information were collected by various groups including local and national government health/environment authorities, relief agencies, and academic groups. These emergency fact-finding expeditions were often given great priority for appropriate response to evidenced needs, but too often the data is shared only internally. In the immediate weeks and months following both the tsunami and Katrina, release or interpretation of collected environmental data was problematic. A frequently updated publicly accessible, searchable, and comprehensive database is needed to disseminate information to avoid duplication of efforts and avoid bad decisions in emergency contexts which are often resource limited (Mills et al., 2005). Currently there are data accessible to the public through organizations such as Relief Web (www.reliefweb.int) which had up to date information on the tsunami and other current disasters, the NRDC and EPA websites for domestic disasters, the Brookings Institute (www.brookings.com), the Cochrane review (www.cochrane.org/evidenceaid/project.htm) and large international organizations which sometimes post data from other groups on their website. A new unique GIS and satellite image effort is synthesizing ground based photos and videos with satellite images of destruction after disasters. This technology to visualize the impacts of earthquakes with satellites (VIEWS) was used in both Katrina and the tsunami. With proper funding and interest, this technology will grow to be a valuable resource in disaster situations (Ghosh et al., 2005). Unfortunately, the websites are not sufficiently comprehensive in specific health, environment, or spatial data and often not acknowledged by most relief workers in the first few days following any given disaster. The Mills article (2005) recommends a comprehensive database be available for all groups and all types of data (Table 5.2).

RECOMMENDATIONS AND NEEDS

Although the two disasters were different in type, scale, and destruction, the Thai government's response appears more efficient than what occurred in New Orleans. However, the scale of the affected area in Hurricane Katrina (the size of the UK) was much larger than the land affected in Thailand and the two governments, cultures, living standards, and political systems are much different. The biggest difference being that New Orleans/Gulf Coast had an early warning system of the disaster to come while Thailand did not. Why therefore was the response of the New Orleans/Mississippi areas slower? Quick decisive action by the Thai military and Public Health and other public utilities had little political boundaries to administer aid. There was little room for political bickering and turf battles in the

Table 5.2. Aims and challenges of a comprehensive database. Edited by authors (Mills et al. 2005)

AIMS
- Help the people who are making decisions by giving them fast access to the good quality information
- Facilitate systematic reviews to summaries and synthesize information for readers of all abilities
- Avoid unwarranted duplication of efforts
- Encourage collaboration across agencies and within governments
- Provide ready access to the public directly in a unique website and through the media
- Improve before and after evaluations of conflicts, disasters, and interventions
- Identify gaps in knowledge
- Facilitate the development of standard measures and methods to evaluate relief and development
- Offer low cost environmental monitoring suggestions for relief teams with limited resources.

CHALLENGES
- Creating a culture of responsible participation and collaboration
- Minimizing threats to agencies on issues of contention and threats to staff from host nations
- Encouraging academics to release findings before journal publication
- Reporting data of relevance and accuracy.

authoritarian disaster zone after the tsunami. The Mississippi response was also uniquely different from the New Orleans area response. The state is known for offering more coordination and participation with federal government aid groups, but the situation was also much different as standing water was not a major issue. Decentralized water and sewerage facilities in Thailand and Mississippi offered flexibility and more resilience than centralized systems. It is suggested that other coastal areas in the world heed the advice and lessons learned and be proactive by developing plans, structures, and societal changes in anticipation of the next disaster rather than responding after the fact.

ACKNOWLEDGMENTS

We acknowledge the support and logistics planning of Luksamee Promakasikorn and the Pollution Control Department of Thailand, the Faculty of Maihdol University School of Public Health, the Provincial Health Department and Hospitals in Phuket and Phang Nga provinces, Thailand and the Tulane University School of Public Health and Tropical Medicine.

REFERENCES

Allen, G.R., Stone G.S. (2005). Rapid Assessment Survey of Tsunami-affected Reefs of Thailand. In: Final Technical Report: Global Marine Programs Office: New England Aquarium

Bankston, W. (2006). FEMA public assistance: Katrina – an inside look at the recovery efforts. Power Point Presentation. Harahan: FEMA: Area Field Office.

BNOB. (2006). Bring New Orleans Back Plan: Health and Social Services Committee. New Orleans.

Colton, C.E. (2006). From Betsy to Katrina: Shifting Policies, Lingering Vulnerabilities. In: The Third Annual Magrann Research Conference: The Future of Disasters in a Globalizing World 2006, Newark, New Jersey:Rutgers University.

Copeland, C. (2005). Hurricane-Damaged Drinking Water and Wastewater Facilities: Impacts, Needs, and Response. Congressional Research Service: The Library of Congress. Available: www.ncseonline.org/nle/crsreports/05oct/RS22285.pdf [accessed June 2006].

Daley, W. (2006). Public health response to Hurricanes Katrina and Rita – United States, 2005. Morbidity Mortality Weekly Report 55(9):229–231.

Engelthaler, D., Lewis, K., Anderson, S., Snow, S., Gladden, L., Hammond, R., et al. (2005). Vibrio illnesses after Hurricane Katrina – multiple states, August–September 2005. Morbidity Mortality Weekly Report 54(37):928–931.

Franklin, M. (2006). Columbia geophysicist wants 'full' Katrina death toll. Associated Press (New York) October 28, 2006.

Galloway, G. (2003). America's WETLAND: Campaign to Save Coastal Louisiana: . New Orleans, Louisiana:Available: http://www.lacoast.gov/news/press/2004-04-13/report27Jan042.pdf [accessed June 2006].

Ghosh, S., Huyck, C.K., Adams, B.J., Eguchi, R.T., Yamazaki, F., Matsuoka, M. (2005). Preliminary Field Report: Post-Tsunami Urban Damage Survey in Thailand, Using the VIEWS Reconnaissance System Buffalo, New York:MCEER. Available: http://mceer.buffalo.edu/research/Reconnaissance/tsunami12-26-04/05-SP01.pdf [accessed June 2006].

Hiransuthikul, N., Tantisiriwat, W., Lertutsahakul, K., Vibhagool, A., Boonma, P. (2005). Skin and Soft-tissue Infections among Tsunami Survivors in Southern Thailand. Clinical Infectious Diseases 41:e93–e96.

Holland, J.S. (2005). Moken Sea Gypsies: Tsunami Update: Saved by Knowledge of the Sea. Available: http://www7.nationalgeographic.com/ngm/0504/feature4/online_extra.html [accessed June 2006].

IEMA. (2005a). Coral Reefs and Mangroves May Have Helped Saved Lives in Tsunami Area. Available: http://www.iema.net/news/envnews?aid=4972 [accessed June 2006].

IEMA. (2005b). WWF Issues Tsunami Paper. Available: http://www.iema.net/news/envnews?aid=4974 [accessed June 2006].

Kathiresan, K., Rajendran, N. (2005). Coastal mangrove forests mitigated tsunami. Estuarine, Coastal and Shelf Science 65(3):601–606.

Krupa, M. (2006). Hard-hit S&WB swamped in red ink. The Times-Picayune (New Orleans) June 10:A–8.

LDHH. (2006). Hurricane Katrina: reports of missing and deceased Available: http://www.dhh.louisiana.gov/offices/page.asp?ID=192&Detail=5248 [accessed October 2006].

McNeil, M., Goddard, J., Henderson, A., Phelan, M., Davis, S., Wolkin, A., et al. (2005). Rapid Community Needs Assessment After Hurricane Katrina — Hancock County, Mississippi, September 14–15, 2005. Morbidity Mortality Weekly Report 55(9):234–236.

MDH. (2006). Mississippi's Part C State Performance Plan 2005–2010: First Steps. Mississippi Department of Health: Office of Health Services: Bureau of Child and Adolescent Health: Early Intervention Division. Available: http://www.msdh.state.ms.us/msdhsite/_static/resources/1565.pdf [accessed June 2006].

Meffert, D. (2006). The Katrina Environmental Research and Restoration Network. In: Rebuilding for Health, Sustainability, and Disaster Preparedness in the Gulf Coast Region 2006, Tulane University, New Orleans, Louisiana.

Mielke, H.W., Powell, E.T., Gonzales, C.R., Mielke, P.W., Jr., Ottesen, R.T., Langedal, M. (2006). New Orleans soil lead (Pb) cleanup using Mississippi River alluvium: need, feasibility, and cost. Environonmental Science and Technology 40(8):2784–2789.

Mills, E.J., Robinson, J., Attaran, A., Clarke, M., Singh, S., Upshur, R.E., et al. (2005). Sharing evidence on humanitarian relief. British Medical Journal 331(7531):1485–1486.

MSDEQ. (2005). Healthcare Priorities: Environmental Health.

Mutter, J.C. (2005). When Disaster Strikes, What Makes the Poor Vulnerable? New York, New York:Columbia University. Available: www.earth.columbia.edu [accessed June 2006].

Norris, H., Speier, A., Henderson, A., Davis, S., Purcell, D., Stratford, B., et al. (2006). Assessment of Health-Related Needs After Hurricanes Katrina and Rita — Orleans and Jefferson Parishes, New Orleans Area, Louisiana, October 17–22, 2005. Morbidity Mortality Weekly Report 55(2):38–41.

NRDC. (2005). New Orleans Environmental Quality Test Results. Available: http://www.nrdc.org/health/effects/katrinadata/contents.asp [accessed June 2006].

Pace, D. (2005). National Trust to survey Gulf Coast historic buildings. Associated Press September 15.

Pardue, J.H., Moe, W.M., McInnis, D., Thibodeaux, L.J., Valsaraj, K.T., Maciasz, E., et al. (2005). Chemical and microbiological parameters in New Orleans floodwater following Hurricane Katrina. Environmental Science and Technology 39(22):8591–8599.

Prabahorn, L. (2005). Personal communication on public health response of government to tsunami. (Englande AJ, Sinclair RG, eds).

Presley, S.M., Rainwater, T.R., Austin, G.P., Platt, S.G., Zak, J.C., Cobb, G.P., et al. (2006). Assessment of Pathogens and Toxicants in New Orleans, LA Following Hurricane Katrina. Environmental Science and Technology 40:468–474.

Ridd, P.V., Sam, R. (1996). Profiling Groundwater Salt Concentrations in Mangrove Swamps and Tropical Salt Flats. Estuarine, Coastal and Shelf Science 43(5):627–635.

Robertson, I.N., Riggs, H.R., Yim, S., Young, Y.L. (2006). Lessons from Katrina. Civil Engineering 76(4):56–63.

Sinthuvanich, D., Boonprakub, S. (2005a). Unified Assisstance to Affected Areas. Bangkok, Thailand:Ministry of Natural Resources and Environment: The Office of Natural Resources and Environmental Policy and Planning. Available: http://www.onep.go.th/download/1y_after_tsunami/index.html [accessed June 2006].

Sinthuvanich, D., Boonprakub, S. (2005b). Rehabilitation Actions Taken. Bangkok Thailand:Office of Natural Resources and Environmental Policy and Planning Ministry of Natural Resources and Environment Available: http://www.onep.go.th/download/1y_after_tsunami/index.html [accessed June 2006].

Spengler, S. (2006). Gulf Coast Regional Infrastructure Program Report. Mississippi Development Authority. Available: www.deq.state.ms.us.

TMOH, W.H.O, USCDC. (2005). Rapid health response, assessment, and surveillance after a tsunami – Thailand, 2004–2005. Joint response from Thailand Ministry of Health (TMOH); World Health Organization (WHO); U.S. Centers for Disease Control (CDC). Morbidity Mortality Weekly Report 54(3):61–64.

UN. (2005). United Nations Office of the Special Envoy for Tsunami Recovery Website. New York:Available: http://www.tsunamispecialenvoy.org/default.aspx [accessed June 2006].

USACE. (2004). Louisiana Coastal Area (LCA), Louisiana Ecosystem Restoration Study. New Orleans:US Army Corps of Engineers. Available: http://www.mvn.usace.army.mil/prj/lca/ [accessed June 2006].

USEPA. (2005). Environmental Assessment Summary for Areas of Jefferson, Orleans, St. Bernard, and Plaquemines Parishes Flooded as a Result of Hurricane Katrina.

Environmental Protection Agency Response to 2005 Hurricanes. Available: http://www.epa.gov/katrina/testresults/katrina_env_assessment_summary.htm [accessed June 2006].

Vasisihth, A. (2006). Getting Humans Back Into Nature: A Scale-Hierarchic Ecosystem Approach to Adaptive Ecological Planning. Evoking An Ecosystem Approach to Disaster Planning: Nested Systems, Multiple Functional Boundaries and the Ecological Re-contextualization of New Orleans. Excerpted from Draft Dissertation. California State University, Northridge.

Wabnitz, C., Alder, J., Change, S. (2005). A tale of two landscapes. British Columbia:UBC. Available: http://www.fisheries.ubc.ca/publications/fishbytes/ [accessed June 2006].

Weerapong, D. (2004). New site for nature lovers in Bangkok. Bangkok, Thailand:World Wildlife Foundation Newsletter. Available: http://www.panda.org/about_wwf/what_we_do/marine/news/successes/index.cfm?uNewsID=14713 [accessed June 2006].

PART THREE

Monitoring, Urban Observatories and Total Mass Balance of Pollution in Cities

6

Design of an environmental field observatory for quantifying the urban water budget

Claire Welty[1], Andrew J. Miller[1], Kenneth T. Belt[2], James A. Smith[3], Lawrence E. Band[4], Peter M. Groffman[5], Todd M. Scanlon[6], Juying Warner[1], Robert J. Ryan[7], Robert J. Shedlock[8], and Michael P. McGuire[1]

[1] *University of Maryland Baltimore County, Baltimore, MD 21250, USA*
[2] *USDA Forest Service, Baltimore, MD 21227 USA*
[3] *Princeton University, Princeton, NJ 08544 USA*
[4] *University of North Carolina, Chapel Hill, NC 27599 USA*
[5] *Institute of Ecosystem Studies, Millbrook, NY 12545 USA*
[6] *University of Virginia, Charlottesville, VA 22903 USA*
[7] *Widener University, Chester, PA 19013 USA*
[8] *US Geological Survey, Baltimore, MD 21237 USA*
E-mail: weltyc@umbc.edu

Summary: Quantifying the water budget of urban areas presents special challenges, owing to the influence of subsurface infrastructure that can cause short-circuiting of natural flowpaths. In this paper we review some considerations for data collection and analysis in support of determining urban water budget components, with a particular emphasis on groundwater, using Baltimore as an example study area. We review selected data collection techniques and applications, including use of airborne thermal infrared imagery to determine locations of groundwater inputs to streams; use of seepage transects and tracer tests to quantify surface-subsurface exchange; estimating groundwater recharge rates; analysis of base flow behavior from stream gages; mining of public agency records of potable water and wastewater flows to estimate leakage rates and flowpaths in relation to streamflow and groundwater fluxes; and considerations for building a geodatabase that includes information relevant to urban hydrologic data.

INTRODUCTION

In this paper we outline design elements of an urban hydrologic observatory, using Baltimore as an example application area, with an emphasis on quantifying the influence of the components of the built environment that affect flowpaths, fluxes, and stores of groundwater. While certain aspects of interactions between the built environment and the hydrologic cycle have received a great deal of attention in an applied sense (stormwater management, sanitary engineering), there have been few efforts that have taken a comprehensive approach to evaluate how urban infrastructure affects groundwater on a watershed scale. Motivation for this effort stems from the recent emphasis by National Science Foundation to establish a national network of environmental observatories to address fundamental scientific water resource questions on a watershed scale, where data are collected in a coherent, coordinated fashion and the use of sensor networks and cyber-infrastructure are key system elements.

Precipitation, evaporation, transpiration, infiltration, runoff, streamflow, and groundwater flow are the main components of the hydrologic cycle; this is typically illustrated by a sketch of the natural landscape showing cycles and subcycles of water flow. The importance of the built environment is generally acknowledged by indicating how the percent impervious area affects the shape of the storm hydrograph, but the cumulative impact of infrastructure (buildings, roads, parking lots, culverts, storm drains, detention ponds, leaking water supply and wastewater pipe networks) on the hydrologic cycle and aquatic ecosystem function is poorly understood. Among the components of the urban hydrologic cycle, groundwater is of key importance, yet it has received relatively little attention from hydrologists in the U.S. in situations where groundwater is not used for water supply directly. The majority of the world's population lives in urban areas and is dependent on the complete hydrologic cycle for water supply. Therefore, understanding the impacts of urbanization on the hydrologic cycle and on groundwater in particular should be better integrated into the study of this science. From the engineering perspective, urban water has been addressed historically in terms of designing facilities for water supply, wastewater treatment, and stormwater management. However, this work has focused on mitigation of specific engineering problems, not on understanding and planning for the integrated functioning of the natural and built environment.

Figure 6.1 depicts the elements of the urban water cycle. Arrows connect components of both the natural system (lower, upper and right portions of the figure) and of the urban infrastructure (left and center), including:

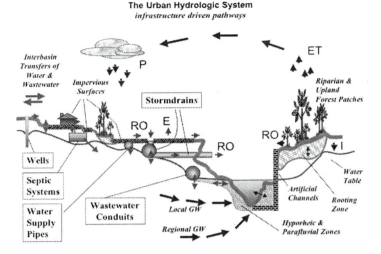

Figure 6.1.   The urban hydrologic cycle.

- water leaking from pressurized distribution pipes into the subsurface;
- infiltration and inflow into and exfiltration/overflows from both sanitary sewer lines and stormwater pipes;
- septic system discharge to groundwater;
- water routed through constructed stormwater ponds and basins;
- water supply import from or export to neighboring basins;
- wastewater import from or export to neighboring basins;
- point-source discharges to rivers from industrial operations;
- effects of impervious surfaces and hardened landscapes (turf, compacted soil, concrete stream channels) on runoff;
- influence of residential and commercial irrigation practices (lawn watering) on groundwater levels and base flow to streams in summer;
- influence of incision of urban stream channels on groundwater levels;
- interaction between urban vegetation and evapotranspiration processes;
- groundwater withdrawals for water supply; and
- preferential flow paths created by subsurface infrastructure.

Many of the pathways associated with the built environment traverse the subsurface. Urban aquifers are therefore not only natural reservoirs but also media through which piped water flows, with multiple impervious barriers, hollow conduits, and sharp transitions in hydraulic conductivity associated with excavation and fill. All pathways and exchanges that affect the mass balance on the urban watershed cannot be fully understood without baseline information on groundwater.

A water budget can be calculated for any time increment Δt for a chosen control volume. If an annual water budget is computed beginning and ending in winter when the soil moisture is at field capacity, the change in soil moisture is zero and does not need to be included. A control volume of interest can be defined laterally by chosen watershed boundaries and vertically from the top of buildings to bedrock. An example water budget for such a control volume and annual period is given as

$$P + I + W - S - ET = \Delta S_G \qquad (1)$$

where P = precipitation [L]; I = water imported to (+)/exported from (−) the basin via water distribution systems [$L^3 L^{-2}$]; W = wastewater imported to (+)/exported from (−) the basin [$L^3 L^{-2}$]; S = streamflow at the basin outlet [$L^3 L^{-2}$]; S_G = change in groundwater storage [L]; and ET = evapotranspiration [L], where L^2 is the watershed area and the time period evaluated must be the same for all terms The annual change in groundwater storage is determined by multiplying the change in average well water levels over the year by specific yield. Closing the water budget would imply that all terms are measured independently. While closing the urban water budget provides estimates of total volumes or stores of water, this calculation does not address fluxes or flowpaths, which are both critical to understanding the exchanges of groundwater and associated contaminants with the atmosphere, streams, and piped flow.

BACKGROUND
Urban water budget studies

Attempts at comprehensive long-term field studies of the urban water budget are few. Grimmond et al. (1986) and Grimmond and Oke (1986) focused on quantifying the relationship between lawn-watering practices and evapotranspiration in a 21-hectare area in Vancouver. The control volume selected did not include the saturated subsurface, nor did it attempt to address infrastructure leakage, and the study area was explicitly selected because it was devoid of streams. A major effort jointly sponsored by the European Union and Australia, "Assessing and Improving the Sustainability of Urban Water Resources and Systems" (AISUWRS, http://www.urbanwater.de), has a goal to develop fully coupled flow and transport models of the unsaturated and saturated subsurface and piped flow to track sources, sinks, flowpaths, and fluxes of contaminants. As part of this effort, Mitchell et al. (2001) have set forth a water balance approach to account for inflows and outflows of all piped water (stormwater, wastewater, potable water) at the watershed scale to document the potential for harvesting stormwater and wastewater as a potential resource; in further development a contaminant tracking capability has been added to this model (Mitchell et al., 2005). Horn (2000) has described a methodology using publicly available data to develop a water budget including basin use and interbasin transfer of water and wastewater. Sharp et al. (2003) have suggested that the effects of urban infrastructure on the shallow subsurface are similar to the effects of karstification and provide detailed simulations of the movement of water around pipes modeled as preferred flow channels. Paulachok and Wood (1984) took advantage of the opportunity to open hundreds of old wells in Philadelphia to create a water table map, showing the highly variable nature of the water table under the city. These studies provide a foundation from which to build a comprehensive assessment of the urban water budget using state-of-the-art methodologies and frameworks.

Baltimore as a model study area

Baltimore's urban watersheds are some of the most studied and densely instrumented in the world, stemming back to early work at Johns Hopkins University gaging storm sewers and quantifying water use (Geyer and Lentz, 1966) and to studies under the EPA Nationwide Urban Runoff Program (US EPA, 1983). More recent work includes stormwater and water resources investigations by local governments (e.g., Baltimore City Department of Public Works, 1999, 2001; Baltimore County, 2000), research by the NSF-funded Long-Term Ecological Research (LTER) Baltimore Ecosystem Study (BES, http://www.beslter.org), and mandated monitoring under EPA consent decrees with Baltimore City and County to address sanitary sewer overflow problems. The many data collection efforts currently being conducted, which are similar to those found in many urban areas, can be utilized as starting point for determining components of the urban water budget.

A primary BES study area is the Gwynns Falls watershed, a 171-sq km basin (Figure 6.2) that lies within the Patapsco River drainage to the Chesapeake Bay. This watershed was selected as an urban ecological study site because it is characterized by a gradient of urbanization from downtown Baltimore to suburban and remnant agricultural areas at the headwaters. Baisman Run and Pond Branch, in the adjacent Gunpowder River drainage (the main water supply of the Baltimore Metropolitan region) serve as forested reference watersheds. As part of the BES ecological studies, nine stream gages have been installed in the study area (Figure 6.2) by USGS (Doheny, 1999), in a nested watershed design that spans different types of land use and land cover, with two new stations added to the official BES/USGS network in 2005. In the Dead Run subwatershed there are five additional sites with water-level recorders upstream of the USGS gage that were monitored during the summers of 2003 through 2005 as part of an NSF-funded study of urban flood dynamics conducted by Miller and Smith (Nelson et al., 2006; Smith et al., 2005).

A network of rain gages and meteorological stations is indicated in Figure 6.2. In addition, a number of spatial-data products from airborne surveys are available that capture landscape physical features at a very fine scale. LIDAR (Light Image Detection and Ranging) data of the landscape were obtained in 2002 that quantify the urban/suburban topography on a 1-m horizontal and 10-cm vertical resolution. Baltimore County has recently acquired a new set of LIDAR data flown in 2005. EMERGEâ airborne color-infrared imagery provides a high-resolution aerial photography record. Also available are 1-ft resolution ortho-imagery for the years 1996, 2000, 2002, 2004, and 2005 from Baltimore County and from Baltimore City 3-in resolution ortho-imagery from 2000 and 1-ft resolution ortho-imagery from

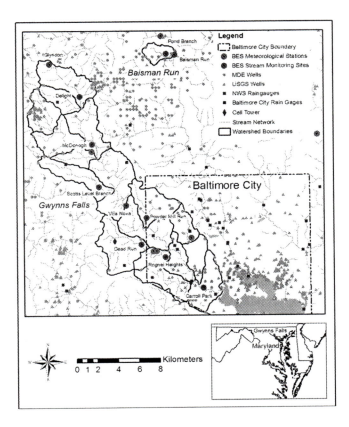

Figure 6.2.   Stream monitoring sites (also USGS streamflow gages), and rain gage/meteorological stations, and wells in the Gwynns Falls watershed.

2004. Four-m. multi-spectral/1-m panchromatic IKONOS imagery for Baltimore captured in 2000 is also available. GIS coverages of potable water, sewer, storm sewer and septic system locations for Baltimore County and Baltimore City are also available, as well as continuous flow records for potable water and sewerage flows at a great many locations.

Extensive monitoring programs in both Baltimore City and County, related to the identification of problems and rehabilitation of sanitary sewers, are being implemented as a result of mandates under consent decrees from EPA. The City effort is comprehensive in nature, with deployment of rainfall, groundwater and wastewater flow sensors. These include up to 150 monitoring stations operating simultaneously in pipes with diameters ranging in size from 8 inches to 12 feet, over a period of 18 months. This network includes about 30 groundwater wells that will aid our efforts greatly. The work will include pump station flow measurements and wireless telemetry for real time collection of data. The objectives of this city-wide flow monitoring program are to (1) obtain good rainfall and sewer flow records, (2) create baseline flow data for future sewer rehabilitation work, (3) provide information to develop a field inspection program for dry/wet weather flow characterization, and (4) support the development of a calibrated hydraulic model of dry and wet weather sewer flows.

Baltimore County's efforts are more geographically focused on isolated "trouble spots" than the City's approach (the City network has many more older sewers). The County is installing 45 new rain gages and new flow gages throughout the Gwynns Falls with an open time period for monitoring that will end when enough data have been obtained to evaluate infiltration and inflow problems and to suggest areas where sewers need to be rehabilitated, repaired or replaced. The County plans on monitoring sewage flows in the Dead Run watershed in 2007 and long-term plans include extensive cooperation with Baltimore City, as many streams/sewers traverse jurisdictions.

Although the original purpose of installation of the USGS gages in the BES was to provide data for hydro-ecological studies, this rich data resource has spawned additional hydrologic analyses. Most notably, the detailed computations of Miller and Smith in Moores Run and Dead Run have led to very fine-scale characterization of precipitation fields and the surface drainage systems in efforts to model rainfall-runoff processes. A recent GIS representation of the drainage system for Dead Run, completed by Meierdiercks et al. (2004) for input to a stormwater hydraulic model, is shown in Figure 6.3. This figure shows surface drainage and the major storm drains. A characteristic that is not widely recognized regarding these kinds of urban drainage systems that is well illustrated by this figure is that the drainage density is actually *increased* compared to a natural system, owing to the fine scale

Figure 6.3.   Dead Run drainage network. Dark grey lines represent open channel drainage; lighter grey lines represent the larger components of the buried storm drain system (Meierdierks et al., 2004).

at which the drainage structure is imposed onto the landscape (Turner-Gillespie et al., 2003). The dissection of the urban landscape by dense networks of pipes may exert great influences on urban water budgets that go beyond simple consideration of surface runoff volumes.

Considerations of regional hydrogeology

The Baltimore metropolitan region straddles the fall zone, where the Piedmont Plateau meets the Atlantic Coastal Plain. The fall zone is a region of locally steepened stream gradients marking the transition from the rolling upland of the Piedmont to the tidewater areas of the coastal plain and is aligned parallel to the Atlantic coast, across which a number of major cities in the US (Trenton NJ, Philadelphia, Baltimore, Washington DC, Richmond VA, Raleigh NC, and Augusta GA) have been settled to take advantage of abundant waterfalls for hydropower and milling operations at locations near the limits of navigation. Insights from hydrogeological investigations of the Baltimore region are therefore transferable to other major east-coast urban environments.

In the Piedmont and fall zone, groundwater flows through unconsolidated saprolite (weathered bedrock) and alluvium, and fractured bedrock that may play an important role in the urban water budget. The saprolite has a high porosity (20-30%); the porosity of the fractured rock is only 0.01–2%. The unconsolidated and fractured formations can be considered separate but interconnected flow systems (Greene et al., 2004). Due to its high porosity and storage characteristics, the unconsolidated system is not very responsive to recharge. It feeds the underlying low-porosity fractured bedrock, which owing to its low porosity is extremely responsive to recharge, resulting in significant changes to water levels in fracture-bedrock wells. The thickness of the saprolite is highly variable and can extend to depths on the order of 15–20 m; response of well levels and baseflow in streams to cycles of wetting and drying suggests that the saprolite may play an important role in buffering the response of the groundwater reservoir to these cycles, but that role is not well understood. Increases in impervious surfaces would be expected to reduce recharge in urban areas, which would be expected to reduce base flow to streams, but the behavior of the saprolite in this regard has not been quantified. If the saprolite is not sensitive to recharge, the surficial groundwater reservoirs may also be relatively insensitive to hydrologic extremes. Understanding the behavior of the saprolite is therefore essential for determining the water budget in Baltimore and in similar hydrogeologic environments.

QUANTIFICATION OF URBAN WATER BUDGET COMPONENTS

Base flow comparative studies

A high density of streamflow gages allows comparison of base flow characteristics and runoff ratios covering a broad range of urban development patterns.

Studies in Dead Run watershed (Smith et al., 2005) indicate that there are significant seasonal patterns in the storm-event water balance that can be related in part to antecedent moisture and (presumably) groundwater storage. There are also significant differences between runoff ratios in adjacent subwatersheds that may be related to impervious cover and to differing patterns of development. Rose and Peters (2001) found that a highly urbanized Atlanta watershed had lower baseflow discharge, shorter storm recession periods, higher 2-day recession constants and lower seasonal baseflow recession constants than in other less-urbanized watersheds. Streamflow records can be examined from the large network of gages identified above, focusing on the frequency distribution of dry-weather flows and on recession curves following storm events. Comparison of unit discharge at base flow for different seasons and antecedent moisture conditions can allow preliminary inferences to be made about differences in groundwater fluxes among watersheds, and comparative analysis of seasonal baseflow trends can yield some insight into seasonal trends in groundwater storage. Using available information compiled in GIS databases allows comparison of amount and spatial arrangement of impervious cover among watersheds, "natural" drainage density and the additional elements of drainage density associated with storm-drain networks and roads, percent of drainage area captured by stormwater retention facilities, and spatial extent of sanitary sewers in the riparian zone; information on differences in soil type is also available and can be included in the comparison. Statistical tools can be applied to determine whether there are identifiable correlations between patterns of urban infrastructure and trends in baseflow response. Using the nested watershed design (Figure 6.4) also allows examination of whether baseflow characteristics and recession curves follow trends with increasing drainage area that are consistent with simple scaling assumptions, or whether there are changes in baseflow response with increasing watershed scale that require alternative explanations. The potentially confounding effect of baseflow augmentation from leaking infrastructure can be addressed using information from leak detection studies while surveys of baseflows in stormdrains provide insight with respect to how networks of these pipes facilitate the removal of shallow groundwater to streams.

Regional groundwater characterization and modeling

The geologic framework in urban environments is often not well documented due to lack of exposed bedrock and outcrops for geologic mapping. However, there is a plethora of underutilized data in many locations, including Baltimore, that is collected as a matter of routine by highway departments in the form of plan and profile drawings and borings removed for geotechnical analysis any time a road, bridge, waterline, or sewer is constructed. This information can be used to determine the depth of overburden and also as a record of the depth to water table at the time the core was taken and to provide geologic context for urban groundwater studies.

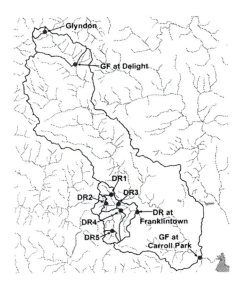

Figure 6.4.   Stream gauge locations and subwatershed boundaries of nested study sites in older suburban (Dead Run – DR) and newer suburban (Glydon and Gwynns Falls at Delight) development.

Information on the urban groundwater flow system is also often lacking, owing to the low spatial density of wells in this urban area that relies on surface water distribution for potable water supply. Nonetheless, locations of existing wells can be obtained from USGS and state data bases (Figure 6.2), as well as from other sources (e.g., county-level data). Selected existing wells can be opened to obtain a snapshot of the deep and shallow potentiometric surface elevations under wet (late winter/early spring) and dry (late summer) conditions. These measurements can be used to contour heads for shallow (e.g., 15 m) and deep (e.g., 100 m) well elevations. Comparison of the two potentiometric surfaces can provide an indication of regions of downward vs. upward flow exchange from the deep bedrock to the overlying saprolite and alluvium. An approximate indication of shallow lateral flow directions can be determined from the potentiometric head map derived from the shallow wells. The regional depiction of hydraulic head is expected to be too coarse to be very meaningful at the small subwatershed scale; additional methods can be employed to refine the groundwater flow field as needed.

The extent to which deep groundwater is contributing to stream base flow can be determined through model calibration to measured heads and stream-groundwater flux rates (see section below) plus any available information on hydraulic conductivity from aquifer pumping tests. A three-dimensional numerical groundwater model can be used in a screening mode where the flux rates are adjusted to match stream discharges measured with seepage tests. This makes it possible to determine the range of groundwater depths and the range of average aquifer hydraulic conductivities that will yield fluxes of the appropriate order of magnitude.

Groundwater recharge

In addition to describing the regional groundwater system, quantification of local recharge rates is desirable; this information can also be incorporated into a groundwater model. Opportunities for infiltration of precipitation in urban areas are reduced by the presence of impervious surfaces, hardened soils, and even turfgrass in some cases. Even where infiltration does occur, this may not be directly correlated to the amount of water reaching the saturated zone as recharge. Infiltrating groundwater is subject to high evaporation rates in areas lacking canopy vegetation, and to shortcuts and shortcircuiting through and around buried pipes that can serve as infiltration galleries and subsurface preferential flowpaths.

One method of estimating recharge on a scale of several square meters that does not rely on quantifying processes in the unsaturated zone but rather that focuses on mass reaching the water table, is the water-table fluctuation method (Healy and Cook, 2002). This simple method evaluates short-term changes in water levels in shallow wells, assuming that a short-term intense storm event causes a change in water table elevation. The water level change is multiplied by specific yield to calculate recharge. Determination of a field value of specific yield to utilize in the calculation at the appropriate scale is not straightforward, but Healy and Cook recommend a simple procedure. Soil moisture characteristic curves based on soil texture (percent of sand, silt, and clay) can be determined from databases of soil properties (e.g., Leij et al., 1996). Specific yield is then taken as the difference between specific retention and total porosity. SSURGO 1:24,000 soils coverages (http://www.ncgc.nrcs.usda.gov/products/datasets/ssurgo/) are available in digital form for the Baltimore area; these can be utilized to determine spatially variable soil texture to the extent possible. (It should be noted that disturbed urban soils are often not well represented by these maps, i.e., soils are designated as "urban complex". This issue is one example of the fundamental challenges in working with urban systems as opposed to natural systems; approximation procedures must be derived as needed.) Automatic water level recorders can be placed in selected shallow wells identified in the groundwater synoptic survey to record temporal changes in water levels. Recharge rates can be determined in these limited locations based on the above-described method. Locations where additional shallow wells should be drilled to fill information gaps can be identified.

Airborne thermal infrared imagery

Infrared imagery techniques, such as aerial infrared thermography and color infrared photography, are increasingly being used by regulatory and natural resources agencies to detect sewage and septic system discharges to streams. These technologies are included in a recent guidance manual produced by USEPA/CWP (Brown et al., 2004) supporting efforts under the federally mandated NPDES stormwater permit program, which requires municipalities to conduct illicit discharge detection and elimination (IDDE) programs in their storm drainage networks and streams. Thermal infrared imagery has also been used to determine concentrated locations of groundwater inputs into surface water systems based on temperature differences. The technique has proven to be useful in slow-moving shallow waters such as lakes (Anderson et al., 1995), estuaries (Portnoy et al, 1998), and wetlands (Olsen 2003, pp. 3–4); the technology has also been shown to work in fast moving waters (Torgerson et al., 2001; Loheide and Gorelick, 2006).

Airborne and ground level surveys of streamwater temperatures can be undertaken to identify concentrated points of groundwater discharge to streams and location of hillside seeps/springs, but at the same time sewage discharges may be incidentally located. Small-scale investigations can be done using a hand held IR instrument and both continuously recording and instantaneous temperature sensors. A TIR imaging device can be used to detect groundwater inputs to streams using methods described by Torgeson et al. (2001). The TIR data can be used as a guide to determine where significant fluxes of groundwater are entering the stream and where the optimal location would be to conduct seepage transects and tracer tests as described below. Also, this can be used as a guide for determining locations to deploy temperature probes, which can be used to record the dynamic signal of groundwater temperature inputs, whereas the TIR only provides a snapshot in time.

Seepage transects and tracer tests

For selected stream reaches, groundwater contributions to streamflow can be evaluated by utilizing the velocity gaging method (Rantz et al., 1982) in combination with the dilution-gaging method (Kilpatrick and Cobb, 1985) as described by Harvey and Wagner (2000). The combination of these methods allows determination of the net inflow (inflow minus outflow) as well as the gross inflow and outflow components for a stream reach. A tracer such as a solution of sodium bromide can be released at the upper end of a reach for the dilution-gaging method. Velocity measurements can be made using one of a variety of velocity meters. Groundwater fluxes to the stream calculated by this method can be compared to the crude approximation using Darcy's law applied to estimated hydraulic gradients from the shallow regional potentiometric surface map combined with published hydraulic conductivity data for the area.

A survey of dry weather flow rates in storm drain systems can be undertaken to further delineate the general extent of groundwater-streamflow interactions. This can include measurement of discharge at storm drain outfalls as well as at selected points in the upstream network (via access through manholes). Discharge measurements can be undertaken through measurement of velocities and areas (or by measures of volumes directly.) A limited suite of water quality constituents (e.g., fluoride, specific conductivity, temperature, and turbidity) can be determined at these points to enable flux calculations and to facilitate the determination of sources (e.g., Pitt 1993; Brown et al., 2004). These data can be combined with the seepage studies to address how much groundwater discharge to streams occurs in riparian vs. engineered settings.

Precipitation data analysis

Given the rapid and efficient delivery of runoff from urban watersheds, it is imperative to have accurate estimates of precipitation inputs. Daily rainfall fields at a spatial scale of 1 km^2 can be estimated from rain gage and radar reflectivity observations. For the Baltimore site, analyses are based on National Weather Service rain gages supplemented by rain gage networks maintained by Baltimore City, Baltimore County and BES (Figure 6.2), and with radar reflectivity observations from the Sterling, Virginia WSR-88D radar. "Volume scan" reflectivity data from the Sterling WSR-88D radar at a time resolution of 5-6 minutes and spatial resolution of 1 km in range by 1 degree in azimuth are archived on a routine basis at Princeton University. Rainfall estimates can be computed at 5-minute time intervals using algorithms that combine gage and radar observations (Baeck and Smith, 1998) and then aggregated to longer durations (15 minute, hourly and daily). The gage-radar rainfall estimation algorithm has been used for a range of applications, including storm-event water balance analyses in the Gwynns Falls watershed (Smith et al., 2005). Radar rainfall estimates are computed using a Z-R relationship of the form, $R = a\ Z^b$, where R is rainfall rate (mm h^{-1}), Z is radar reflectivity factor (mm^6 m^{-3}) and the Z-R parameters, a and b, are empirical coefficients. The "convective" Z-R relationship (for which a = 0.0174 and b = 0.71) is used for operational rainfall estimation by the National Weather Service (Fulton et al., 1998). Rain gage observations are incorporated into the analyses through a local multiplicative bias correction (Smith et al., 1996).

Evapotranspiration

Evapotranspiration can be explicitly calculated using any of several equations. The Penman-Monteith equation (see e.g., Maidment, 1996; Drexler et al. 2004) is popular because most standard meteorological stations collect data needed as input to this model – i.e. net all-wave radiation, wind speed, vapor pressure, air temperature, and surficial heat flux. Required leaf-area index data and reference vegetation data for Baltimore are available from local USDA Forest Service personnel. However, the inherent spatial heterogeneity of the urban environment places constraints on the accuracy of Penman-Monteith estimates developed for larger areas because of factors that influence canopy conductance (e.g. vapor pressure deficit, root zone soil moisture). Other factors might also affect local ET rates, such as the drainage efficiency of the urban infrastructure in controlling the amount of standing water on the surface. Presumably the aerodynamic conductance is high, so surface fluxes are expected to be highly coupled with the atmospheric properties. In order to address these sources of uncertainty, direct calculations of ET can be made using an eddy correlation station (Beringer and Tapper, 1996).

Pipe flows

GIS coverages of sanitary sewers, storm drains, septic systems, and potable water supply networks are available for many urban areas, including Baltimore; however the data are typically not compiled in one database and often lack key attributes needed for hydraulic modeling. Municipal water and wastewater records for a variety of scales are a rich but normally underutilized source of data for hydrologists. For example, water distribution system records are long-term and typically feature water meters at the household level, pumping and storage data for many points in the landscape, and gaging stations for key networks. Notably, these kinds of records are often indispensable for urban work, since "engineered" water can dominate the urban water budget. For example in the Baltimore watersheds, potable water is imported, no well water is used, and wastewater is exported from the watersheds but there are extensive exchanges with ground and surface waters. Records of the import/export volumes, extensive water quality data for dry weather and storm flows in streams and numerous infiltration and inflow studies are available from the Dept. of Public Works.

A data model for hydrologic information systems in urban areas

Modeling hydrologic processes in urban areas requires input data from numerous sources that are collected and archived at various temporal and spatial scales and in a variety of units and formats. There is a national movement to bring disparate data sources together for the purpose of facilitating analysis and visualization and an increased focus on data management techniques that allow researchers to focus their time on analysis rather than searching for data and using ad-hoc methods for converting data sets to units and formats required for a chosen model. Examples include CUAHSI (Consortium for the Advancement of Hydrologic Science, Inc., http://www.cuahsi.org); SEEK (Science Environment for Ecological Knowledge, http://seek.ecoinformatics.org); and KNB (The Knowledge Network for Biocomplexity, http://knb.ecoinformatics.org)). The emphasis in water-related fields has been on bringing together standard large-scale spatial data sets such as National Hydrography Data (NHD), National Elevation Data (NED), National Land Cover Data (NLCD), and temporal data such as those provided by USGS streamflow gages, NOAA weather stations, and Ameriflux towers. While great strides have been made for applications to natural landscapes, special requirements for urban areas have been virtually unaddressed. For example, in one recent application, Tenenbaum et al. (2006) used high resolution LIDAR elevation data with sampled soil moisture to determine that suburban catchments required order of magnitude finer resolution to adequately resolve topographic influence on water redistribution. In addition, the NHD does not account for piped flow, and the NLCD and NED datasets are available at too coarse a scale to capture the fine-grain spatial features of urban landscapes. Detailed data sets of the type required for water resources management in urban areas are available (LIDAR, high resolution aerial photography, GIS coverages of pipe networks, 5- or 15-minute streamflow data) for some locations, but these have not been systematically incorporated into organized national databases.

Related to this is the explosion in development and deployment of wireless sensors and sensor networks to deliver real-time data for environmental monitoring (e.g., Arzberger et al., 2005). One goal is ultimately to interface real-time data with mathematical models to provide dynamic predictions of water and contaminant movement through the terrestrial environment. Also, new developments in data warehouse and online analytical processing (OLAP) systems allow users to explore data across a large number of dimensions. Spatial OLAP, the integration of geographic information systems (GIS) and OLAP, can improve knowledge discovery from spatial distributions and relationships by allowing users to explore multidimensional data through spatial visualization. The application of SOLAP can prove to be a very powerful tool in exploring relationships between spatially and temporally dynamic environmental systems.

RECOMMENDATIONS AND FUTURE RESEARCH

Given the significance of the interaction between urban infrastructure and the water cycle, in this paper we aimed to review some relevant considerations in data collection and analysis, with an emphasis on the groundwater component. Future work will include implementation in Baltimore of many of the methods discussed. One important application is quantifying delivery of nutrients and contaminants from groundwater beneath cities to surface waters, and in the case of Baltimore, to the Chesapeake Bay. To aid in evaluating urban water resources problems, the national effort for developing a unified geodatabase can be built on by developing a prototype system for collecting, transmitting, storing, mining, manipulating, communicating, and visualizing hydrologic data sets pertaining to urban areas. This will involve fusing data streams from on-ground sensor networks including meteorological stations and USGS gages; fine-grid spatial data sets (LIDAR, EMERGE imagery, SSURGO soils, geology); city and county storm sewer and sanitary sewer systems; and census, health, and tax-parcel data.

ACKNOWLEDGMENTS

This material is based upon work supported by the National Science Foundation under Grant No. 0414206 and by the US Environmental Protection Agency under Grant CR83105801. Although the research described in this paper has been funded in part by the US Environmental Protection Agency, it has not been subjected to the Agency's required peer and policy review and therefore does not necessarily reflect the views of the Agency and no official endorsement should be inferred.

REFERENCES

Anderson, J.M., R.W. Duck, and J, McManus. 1995. Thermal radiometry: A rapid means of determining surface water temperature variations in lakes and reservoirs. J Hydrology: 173:131–144.

Arzberger, P., J. Bonner, D. Fries, A. Sanderson. 2005. Sensors for Environmental Observatories. Report of the NSF-Sponsored Workshop, December 2004.

Baltimore City DPW. 1999. City of Baltimore NPDES Storm Water Permit Program 1999 Annual Report. Baltimore, Baltimore, Baltimore City Department of Public Works Water Quality Management Office.

Baltimore City DPW. 2001. Reservoir water quality assessment for Loch Raven, Prettyboy and Liberty reservoirs. Baltimore, Baltimore City Department of Public Works Water Quality Management Office

Baltimore County DEPRM. 2000. Gunpowder study monitoring report, Baltimore County Dept. of Environmental Protection and Resource Management.

Baeck, M. L. and J. A. Smith. 1998. Estimation of heavy rainfall by the WSR-88D, Weather Forecasting, 13, 416–436.

Beringer, J and N. Tapper. 1996. The use of the Campbell Scientific Bowen ratio system for evapotranspiration measurement, IN Proceedings of the 2nd Australian Agricultural and Meteorology Conference, Brisbane, Australia.

Brown, E., D. Caraco and Pitt, R. 2004. Illicit Discharge Detection and Elimination - A Guidance Manual for Program Development and Technical Assessments. Center for Watershed Protection, Ellicott City, Maryland, 195 pp.

Doheny, E. J. 1999. Index of Hydrologic Characeristics and Data Resources for the Gwynns Falls Watershed, Baltimore County and Baltimore City, Maryland. USGS Open File Report 99–213, 17 pp.

Drexler, J. Z., R. L. Snyder, D. Spano, and K. T. P. U. 2004. A review of models and micrometeorological methods used to estimate wetland evapotranspiration. Hydrological Processes 18:2071–2101.

Fulton, R., J. Breidenbach, D.-J. Seo, D. Miller, T. O'Bannon, 1998. The WSR-88D rainfall algorithm. Weather and Forecasting, 13, 377–395.

Geyer, J. C. and J. J. Lentz 1966. An Evaluation of the problems of sanitary sewer system design. Journal of the Water Pollution Control Federation, 38(7).

Greene, E.A., A.M. Shapiro, and A. E. LaMotte. 2004. Hydrogeologic controls on groundwater discharge to the Washington METRO subway tunnel near the medical center station and crossover, Montgomery County, MD. USGS Water Resources Investigations Report 03–4294

Grimmond, C.S.B., T.R. Oke, DG Steyn 1986. Urban Water Balance: 1. A Model for daily totals. Water Resources Research, 22(10), 1397–1403

Grimmond, C.S.B. and T.R. Oke. 1986. Urban Water Balance: 2. Results from a suburb of Vancouver, British Columbia. Water Resources Research, 22(10), 1404–1412.

Harvey, J. W., and B. J. Wagner. 2000. Quantifying hydrologic interactions between streams and their subsurface hyporheic zones, in Streams and Ground Waters, edited by J. B. Jones and P. J. Mulholland, pp. 3–44, Academic, San Diego, Calif.

Healy, R.W. and P.G. Cook. 2002. Using groundwater levels to estimate recharge, Hydrogeology Journal, 10: 91–109.

Horn, M.A. 2000. Method for estimating water use and interbasin transfers of freshwater and wastewater in an urbanized basin: U.S. Geological Survey Water-Resources Investigations Report 99–4287, 34 p.

Kilpatrick, F.A., and E.D. Cobb. 1985. Measurement of Discharge using Tracers. Tech. Water-Resources Investigations, Book. 3, Chapter. A16. United States Government Printing Office.

Leij, F.J., W.J. Alves, M.T. van Genuchten, and J.R. Williams, 1996. Unsaturated zone hydraulic database. UNSODA 1.0 Users Manual, report EAP/600/R-96/095, USEPA, Adam OK 103 pp.

Loheide, S.P., and S.M. Gorelick. 2006. Quantifying stream-aquifer interactions through the analysis of remotely sensed thermographic profiles and in-situ temperature histories. Environmental Science and Technology, 10.1021/es0522074, published on the web 4/13/2006.

Maidment, D.R., ed. 1996. Handbook of Hydrology, McGraw Hill,.

Meierdiercks, K.J., Smith, J.A., Miller, A.J. and Baeck, M.L. 2004. The urban drainage network and its control on extreme floods. Trans. AGU 85(47), Abstract H11F-0370.

Mitchell, VG, RG Mein, and TA McMahon 2001. Modelling the urban water cycle, Environmental Modeling and Software. 16:615–629.

Mitchell, VG, C. Diaper, 2005. Simulating the urban water and contaminant cycle. Environmental Modeling and Software, in press.

Nelson, P. A., J. A. Smith and A. J. Miller. 2006. Evolution of channel morphology and hydrologic response in an urbanizing drainage basin, Earth Surface Processes and Landforms, 31, 1063–1079.

Olsen, LD. 2003. Thermal Infrared Imaging, pp. 3–4, in Selected applications of hydrologic science and research in Maryland, Delaware, and Washington, DC 2001–2003. USGS Fact Sheet FS-126-03, 8 pp.

Paulachok, G.N. and C. R. Wood. 1984. Water table map of Philadelphia, Pennsylvania, 1976–1980. US Geological Survey, Hydrological Investigations Atlas, HA-676.

Pitt, R., M. Lalor, R. Field, D. Adrian, and D. Barbe. 1993. Investigation of inappropriate pollutant entries into storm drainage systems: a user's guide. Washington, D.C., USEPA Office of Research and Development. Risk Reduction Engineering Laboratory, USEPA. Cincinnati, OH. EPA/600-R-92-238.

Portnoy, JW, BL Nowicki, CT Roman, and DW Urish. 1998. The discharge of nitrate-contaminated groundwater from developed shoreline to marsh-fringe estuary. Water Resources Research, 34(11), 3095–3104.

Rantz, S.E., et al. 1982. Measurement and computation of streamflow: volume 2. Computation and discharge: U.S. Geological Survey Water Supply Paper 2175.

Rose, S. and N.E. Peters. 2001. Effects of urbanization on streamflow in the Atlanta area (Georgia, USA): A comparative hydrological approach. Hydrological Processes, 18, 1441–1457.

Sharp, J.M., Jr., J.N. Krothe, J.D. Mather, B. Garcia-Fresca, and C.A. Stewart. 2003. Effects of Urbanization on Groundwater Systems. In Heiken, G., R. Fakundiny, J, Sutter, eds., Earth Science in the City: A Reader. American Geophysical Union, Washington, DC.

Smith, J. A., M. L. Baeck, M. Steiner, and A. J. Miller, 1996. Catastrophic rainfall from an upslope thunderstorm in the Central Appalachians: the Rapidan Storm of June 27, 1995, Water Resources Research, 32(10): 3099–3113.

Smith, J. A., M. L. Baeck, K. L. Meierdiercks, P. A. Nelson, A. J. Miller, and E. J. Holland, 2005. Field studies of the storm event hydrologic response in an urbanizing watershed. Water Resources Research, 41, W10413, doi:10.1029/2004WR003712.

Tennenbaum, D.E., L.E. Band, S. T. Kenworthy, and C.L.Tague. 2006. Analysis of soil moisture patterns in forested and suburban catchments in Baltimore, Maryland, using high-resolution photogrammetric and LIDAR digital elevation datasets. Hydrological Processes. 20, 219–240.

Torgersen, C.E., R.N. Faux, B.A. McIntosh, N.J. Poage, and D. J. Norton. 2001. Airborne thermal remote sensing for water temperature assessment in rivers and streams. Remote Sensing of the Environment: 76, 386–398.

Turner-Gillespie, D. F., J. A. Smith and P. D. Bates. 2003. Attenuating reaches and the regional flood response of an urban drainage basin, Advances in Water Resources: 26, 673–684.

U.S. EPA. 1983. Results of the Nationwide Urban Runoff Program, Final report. NTIS PB84-185552.

7

Ecosystem approaches to reduce pollution in cities

L.A. Baker[1] and P.L. Brezonik[2]

[1] Water Resources Center, University of Minnesota, 173 McNeal Hall, 1985 Buford Ave., St. Paul, MN 55108
[2] Department of Civil Engineering, University of Minnesota, 500 Pillsbury Drive S.E. Minneapolis, MN 55455-0116
E-mail: Baker127@umn.edu

Summary: Further reductions in pollution will require new management approaches based on ecosystem concepts. In this paper we examine three case studies that analyze flows of nutrients in urban ecosystems to identify new approaches for pollution management. An analysis of the food chain for the Twin Cities (Minneapolis-St. Paul, Minnesota, USA) is used to examine generation of nutrient pollution in the pre-consumption, consumption, and post-consumption phases. Several scenarios of reduced protein consumption are analyzed for their impacts on agricultural fertilizer requirements. Reductions of up to 40% of fertilizer N are possible with small to moderate dietary adjustments. We then examine phosphorus inputs to urban residential landscapes to show how a newly enacted lawn P fertilizer restriction has reduced P inputs by 90%. The third case study examines fluxes of carbon, nitrogen and phosphorus fluxes through suburban households. This study shows that "high-consumption" households may have 3–4 times higher carbon and nitrogen fluxes than "low-consumption" households. Developing pollution reduction policies based on ecosystem concepts will require new research that integrates social and biophysical sciences.

INTRODUCTION

With the exception of bans on certain toxic chemicals (e.g., lead and many chlorinated hydrocarbons) modern pollution reduction practice has mostly dealt with treating pollutants at the "end of the pipe". This has worked to a large extent to reduce pollutants discharge from pipes, like city sewers and industrial discharges. However, thirty-four years since passage of the Clean Water Act and the expenditure of more than \$500 billion to construct municipal wastewater treatment plants alone, we have achieved only limited success. The amount of oxygen-consuming material and bacterial contamination in rivers below cities has been reduced (EPA, 2000), but large portions of the surface and groundwater resource of the country remain polluted, particularly with nutrients (Mueller and Helsel, 2000; Nolan and Stoner, 2000).

By the early 1990s, it became apparent that nonpoint sources of pollution were responsible for much of the water pollution throughout the United States (Smith et al., 1987; Baker, 1992). At that time, cropland and pastures accounted for 43% of the N and 35% of the P entering U.S. surface waters (Carpenter et al., 1998). Nitrate contamination of groundwater underlying agricultural areas is now widespread (Neilsen and Lee, 1987; Mueller and Helsel, 2000). Urban stormwater is generally contaminated with nutrients, several metals, sediments and coliform bacteria (USEPA, 1983) and now is regulated under the National Pollution Discharge Elimination System (NPDES). Salinity has emerged as a major contaminant for reclaimed water in the southwestern U.S. (Thompson et al., 2006).

When the focus of water pollution shifted to control of nonpoint source pollutants, the "end-of-pipe" paradigm was brought along. Engineers designed stormwater detention basins, wetlands, buffer strips and other structural devices to remove pollution. Whereas structural treatment systems operate in a predictable manner when used to treat well-defined, steady flows such as municipal wastewater, operation of such systems is far less predictable when used to treat nonpoint source runoff, with variable, poorly-defined composition and flashy flows. Further, most pollutants are not truly removed (like BOD in a wastewater treatment plant); they simply are stored, eventually requiring removal. Structural controls are expensive to build; they require land that may not be available; they often are not maintained, or when maintained, maintenance costs often exceed expectations. With the notable exception of erosion reduction, there are few examples of long-term reductions in nonpoint source pollution in either agricultural or urban settings as the result of deliberate water quality management.

One reason for our failure to control nonpoint source pollution is that we have continued to rely upon the end-of-pipe paradigm. In this paper we propose that ecosystem-based approaches have great utility for reducing nonpoint source pollution and illustrate the potential with several case studies. The focus here is on ecosystem mass balance analysis, variously termed "ecosystem models", "box models" (or multicompartment box models), and "ecological flow networks" (Suh, 2005). These have been widely used for analysis of lakes, forests and other natural ecosystems. Over the past half dozen years we and others have used the concept of ecosystem network analysis to analyze urban pollution (Boyd et al., 1981; Decker et al., 2000; Baker et al., 2001a; Faerge et al., 2001), leading to a better understanding of nutrient flows through "human ecosystems" – the cities where most of us live and the farms that sustain us.

In this paper we examine nutrient flows through urban systems at three scales to examine potential pollution reduction efforts. First we examine nutrient flows through the human food chain of the Twin Cities (Minneapolis-St. Paul, Minnesota). We then examine P inputs to suburban landscapes to illustrate the potential impacts of a newly enacted lawn fertilizer P restriction. Finally, we examine nutrient movement through household ecosystems to show how consumption patterns have a major impact on fluxes of C and N.

CASE STUDIES OF NUTRIENT FLOWS

Case study 1: Flow of N through an urban food chain

Our first case study is nitrogen flow through the Twin Cities food chain. Adopting the terminology of industrial ecology, we refer to the agricultural system as the "pre-use" system, eating as the "consumption" system and disposal via wastewater and landfill waste as the "post-use" system. The total population of humans and dogs is represented. We estimated human food consumption using data from the Continuing Survey of Foods (Borrud et al., 1996), mapping nutrition intake by age and sex into the Twin Cities population using U.S. Census data. Food intake for dogs was based on an equation to estimate metabolic energy from weight (IAMS Corporation) and the average composition of dog foods (Baker et al., 2001a; Baker et al., 2007). We assumed 0.6 dogs per household based on an estimate of the U.S. dog population in the U.S. population (Patronek, 1995) and the average number of individuals per household in the 1990s.

In the United States, two-thirds of human protein consumption is derived from meat and one-third from crops. We assumed most dog food is derived from vegetable protein. In addition to food actually consumed, there is considerable food waste during industrial processing and usage. Here we use a value of 30% for overall food waste (Kantor et al., 1997), recognizing that this value has a high degree of uncertainty.

We then estimated N inputs and outputs needed to supply this amount of protein to the Twin Cities. Despite the complexities of the global food system that sends food to the Twin Cities, we can make reasonable estimates of major N transfers based on typical on "transfer efficiencies" from fertilizer-to-crop and from crop (feed)-to-livestock. Issues associated with estimating fertilizer-to-crop efficiencies are discussed by Cassman et al. (2002). Date compiled by Stuewe (2006) for a high-rate agricultural system in southwestern Minneosta, yield a value of 57% for all forms of N (fertilizer + soil mineralization + fixation + manure + atmospheric deposition, with about 80% coming from the first two sources). Feed-to-livestock transfer efficiencies vary from 15% to 50% (Ensminger, 1993; Van Horn et al., 1994; Lin et al., 1996; Baker et al., 2001b; Stuewe, 2006). Data from Stuewe (2006) yield a value of 34%. This means that 34% of the N in livestock food becomes meat, milk or eggs; the remaining 64% becomes manure. We assumed that 30% of manure N is volatilized before the manure reaches the field (Stuewe, 2006). Loss of N via leaching and runoff associated with farming is estimated at 15% of input

(Stuewe, 2006). Other transfers in the agricultural system include denitrification, volatilization of fertilizer and manure and change in soil storage (Baker et al., 2001b; Stuewe, 2006).

In the consumption phase, we assumed that excretion is approximately equal to consumption of N for both humans and dogs. Human excretion in the Twin Cities enters the wastewater system and is treated in eight municipal wastewater treatment plants. The sum of computed N excretion for the Twin Cities population is nearly identical to the sum of N inputs to the wastewater treatment plants (Baker et al., in prep.). The overall treatment efficiency for N (sum of inputs – sum of outputs/sum of inputs) was 79%. Based on Lauver and Baker (2000) we assumed that 10% of the N in a plant designed for denitrification become incorporated into sludge and the remainder of the N loss was denitrification.

Figure 7.1 shows that 44 Gg N/yr of N from all sources (fertilizer + soil mineralization + fixation + atmospheric deposition) is needed by the hypothetical

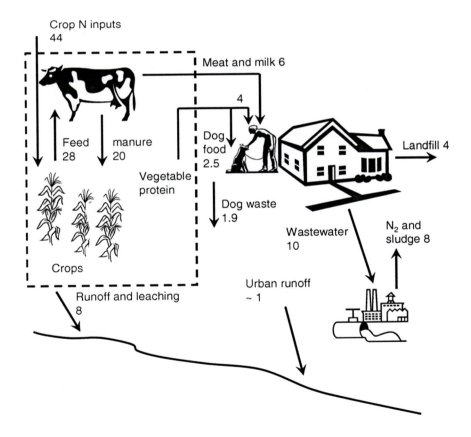

Figure 7.1.   Flows of N from agriculture (pre-consumption), through urban environment (consumption phase) and to ultimate disposal (post-consumption phase). Units are Gg/yr.

agricultural system to support food production for the Twin Cities. Only 16 Gg N/yr is exported from the agricultural system to the urban system in the form of animal and vegetable protein. About 8 Gg/yr is lost via leaching and runoff, and the rest is lot via volatilization and denitrification.

At the consumption phase, dog food is 16% of the total food N supplied to the Twin Cities and human food is 84% of the total. These totals include food waste. Food N is the second largest source of N to the Twin Cities, with the largest being abiotic fixation of N_2 by combustion (Baker et al., 2005). For typical suburban households in the Twin Cities, human and dog food comprise $\sim$40% of total N input.

In the post-consumption phase, about one-fourth of the food N enters landfills and three-fourths enters the sewage system. Eighty percent of wastewater N is "removed", becoming sludge (9%) or N_2 (91%).

Nitrogen flows through the urban food chain illustrate how modifications of urban consumption might alter the amount of fertilizer needed to supply this food. As a whole population, residents in the Twin Cities consume about 30% more protein than is required, based on the "recommended daily allowance" (RDA). This calculation is based on age- and sex-stratified RDA values extrapolated to the Twin Cities population, in comparison to actual nutrient intakes obtained from the Continuing Survey of Foods, also age- and sex stratified. Current consumption requires 44 Gg of N inputs to agriculture (Figure 7.1). Reduced consumption scenarios (Figure 7.2) show that reduced consumption of food by the urban population would have a significant impact of agricultural N inputs. Reducing human protein intake by 33%, maintaining a 2:1 ratio of meat:vegetable protein, would reduce the required N input by 30%. If the meat:vegetable protein ratio also were shifted to 1:1, the required agricultural N input would be 36% lower than the current requirement. Reducing the dog food input by 50% (by reducing the population of dogs, or the size of dogs or some combination of the two) alone would reduce the N requirement by 5%. All of these steps (30% reduction in human protein; 50:50 meat:vegetable protein; 50% reduction in dog protein) would reduce the required farm N input by 41%.

In the post-consumption phase, about 5 Gg N/yr becomes sewage sludge or landfill waste. In the Twin Cities, sewage sludge is burned and the ash is landfilled. In the future, sludge may become regarded as a valuable resource, as fertilizer prices increase due to energy costs (for N fertilizer) or exhaustion of mineral sources (for P fertilizer).

The use of a food-chain analysis to identify opportunities for pollution reduction is a dramatic departure from the current pollution reduction paradigm, which relies almost exclusively on end-of-pipe treatment, either on the farm (through agricultural BMPS, which have rarely been applied successfully over large areas) or through sewage treatment. Reductions in N fertilizer requirement that could be accomplished with decreased consumption – to levels well within current health guidelines – are greater than could be accomplished with conventional agricultural BMPs. Such a dietary shift might also be healthful. We note that the percentage reduction in N inputs to supporting agricultural systems would vary

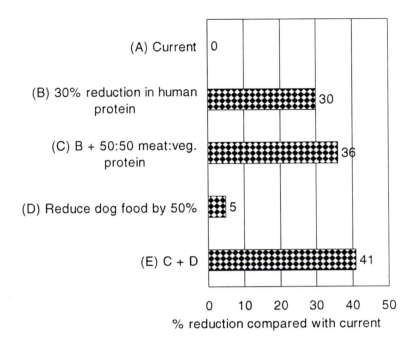

Figure 7.2.   Reductions of total N requirements to produce food for an urban population.

somewhat depending on the type of agricultural system providing this support, but in general, altering urban diets would have a significant effect on agricultural N requirements.

Case study II: Phosphorus in urban stormwater

In the United States, urban stormwater in cities with populations >10,000 (and some smaller ones) is now regulated under the National Pollution Discharge Elimination System (NPDES). This has compelled great interest in reducing pollution in urban stormwater. The main response has been a flurry of construction of end-of-pipe BMPs, such as wet and dry detention ponds, wetlands, infiltration basins, rain gardens and a plethora of proprietary devices. Education, public participation, and pollution prevention are mandated components of stormwater management programs, but there is an overwhelming reliance on constructed BMPs.

Here again we can use ecosystem mass balances to develop novel approaches for pollution reduction. For urban stormwater, we can establish several boundaries to establish frameworks for analysis. The boundary for the Level 1 analysis is the watershed itself. Excluding movement of P into the human food system, which is generally separated from the landscape system, common inputs include fertilizer P, dog food, polyphosphate used in municipal water systems for corrosion control, road sand and atmospheric deposition (Figure 7.3).

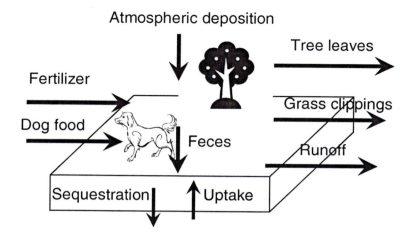

Figure 7.3.   Schematic of nutrient flows through an residential lawn.

The boundary for Level 2 analysis is the pervious landscape. The outer boundary is the watershed and the inner boundary is the edge of the street. Most P inputs to the watershed enter the pervious landscape, either directly (like fertilizer P) or indirectly (like P in municipal water, which enters landscapes through irrigation). Main P exports include runoff (soluble and particulate forms), tree leaves and grass clippings.

The boundary for Level 3 is impervious surfaces in the watershed. Most inputs come from pervious landscapes, with the exception of atmospheric deposition and road sand, which are probably minor sources. Outputs are stormwater and street sweepings. The upper boundary for Level 4 is the stormwater grate; the lower boundary is the discharge to a stream. Level 4 is the level used for designing structural stormwater BMPs.

We have analyzed the effect of Minnesota's lawn fertilizer P restriction on total P inputs to a hypothetical 5 km^2 residential watershed. This is a Level 1 analysis. The impervious surface was 20% of total area, and we assumed that 80% of pervious surfaces were fertilized at 1 lb P$_2$O$_5$ /1000 ft^2, following the University of Minnesota guidelines for medium fertility lawns. We assumed 0.6 dogs per household, with an average dog weighing 20 kg. Polyphosphates are commonly added to drinking water in Minnesota, most often at levels $\sim$1 mg PO$_4$ L^{-1} (Rezania, 2005). We assumed 0.5 m irrigation per year. An atmospheric deposition rate of 0.25 kg ha^{-1} yr^{-1} was used (Barr, 2004).

The Level 1 calculation shows that lawn fertilizer constituted 90% of total P input to the watershed prior to the fertilizer regulation, and the total P input was 5,702 kg yr^{-1}. Assuming 100% compliance with the regulation, total P input would have been reduced by 90%, to 569 kg yr^{-1}. The overwhelming dominance of fertilizer P inputs suggests that minor departures from stated assumptions would have little effect on the conclusion. In the post-regulation watershed, the main input

Table 7.1. Effect of Minnesota's lawn fertilizer P restriction

Source	Input rate	Before fertilizer P restriction			After fertilizer P restriction (all other inputs unchanged)	
		Total, kg/yr	%		Total, kg/yr	%
Fertilizer P	23 kg ha^{-1} yr^{-1} (1 lb P$_2$O$_5$/1000 ft^2)	5,133	90		0	0
Dogs	0.9 kg dog^{-1} yr^{-1}	21	0		21	4
Irrigation	1 mg PO$_4$ L^{-1} phosphate; 0.5 m/yr	422	7		422	74
Deposition	0.25 kg ha^{-1} yr^{-1}	125	2		125	22
Total	–	5,702	100		569	100

of P to lawns was phosphate from irrigation water, which accounted for 74% of total input. This analysis shows that lawn fertilizer was overwhelmingly the major source of P to our hypothetical residential watershed.

The impact of the P fertilizer restriction will not be an immediate 90% decline in stormwater P. Immediate losses of lawn fertilizer P to runoff range from 1% to 20%, depending largely on timing relative to irrigation or precipitation events (Baker, 2006, in prep.). Following cessation of fertilizer P inputs, grass and trees will "mine" stored soil P. Most lawn soils in Minnesota are enriched with adsorbed P (as measured by "Bray" or "Olsen" P) (Barten and Jahnke, 1997). Soluble P is released when grass is mowed and decomposes; this soluble P moves into streets during precipitation or irrigation events. Grass clippings and tree leaves also are deposited directly to streets, where they decompose and release P. Over the period of many years, available soil P will become depleted, reducing the P content in plants, which in turn will reduce the quantities of P released during decomposition. At some point, soil P could become depleted to the point that plant growth is inhibited. The rate of depletion of soil P depends largely on the magnitude of other P inputs (e.g., dog feces) and the rate at which P is exported via grass clippings and tree leaves. We predict that high maintenance lawns, in which grass clippings and tree leaves are bagged and removed from the property (thereby exporting P), will exhibit faster decline of runoff P concentrations than low maintenance lawns, where tree leaves and grass clippings are recycled, thereby returning P to the soil pool.

Case study III: Nutrient balances for households

The third scale of study for urban systems that may be useful is the individual household. This scale of analysis is important for several reasons: (1) households account for a major fraction of the C, N and P fluxes for cities; (2) understanding these fluxes could lead to novel methods for reducing pollution; (3) focusing on households enables us to elucidate the ultimate, rather than proximate causes of human biogeochemical perturbations; (4) the family household is a socially meaningful and practical unit of measurement; and (5) a model of household fluxes could become a valuable pedagogical tool to enable citizens to understand the impacts of their activities on their surrounding environment (Baker et al., 2007).

We recently developed a "household flux calculator" (HFC) to compute fluxes of C, N and P through households and have used the HFC to develop scenarios for "low consumption", "typical consumption" and "high consumption" households. All three scenarios were intended to represent four-person (two adults and two teenagers) living in a single-unit house located at an average commuting distance from work (8.2 miles in the Twin Cities). The "typical" household was developed using data from various government databases (OEA, 2000; EIA, 2001; USDA, 2001; Census, 2004), plus information from a pilot survey of 35 homes in the St. Paul suburb of Falcon Heights. High and low consumption scenarios were developed to be reasonable circumstances, not wild extremes. For example, the low and high consumption household energy scenarios were based on the 10th and 90th percentiles for household energy use in the Midwest climate region (EIA, 2001). Our high consumption household had two SUVs and one mid-size car and drove a total of 39,100 miles, whereas our low consumption household had two high mpg vehicles and drove a total of only 15,100 miles (see Baker et al., 2007 for other details).

The order of major C fluxes were transportation > household energy > air travel ≫ food. For N, the order of the top four fluxes was human food > transportation (NOx emissions) > fertilizer > dog food. The modeling analysis revealed that household consumption patterns had a very large effect on C and N fluxes (Figure 7.4). Modeled C flux was 3.5 times higher for the high consumption

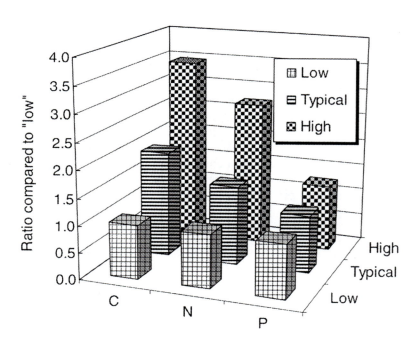

Figure 7.4.   Ratio of C, N and P fluxes through low, typical and high consumption scenario households, compared with the low consumption scenario.

household than the low consumption household. For N and P the ratios were 2.7 and 1.2. The ratio of P fluxes was small because we assumed that no fertilizer P was used in any of the households.

For our high consumption household, we estimated that reasonable modifications of behavior could result in a 20% reduction within one year (mainly through reduced driving and air travel), a 40% reduction in 3–5 years (replacing SUVs with more efficient cars, etc.), and a 70% reduction in 10 years (moving to a smaller home, etc.). We are currently analyzing C, N and P fluxes through each of the 35 homes we surveyed in Falcon Heights. Preliminary analysis suggests that the range of C and N fluxes is ~3–5-fold. These scenario calculations and preliminary analysis of data from real homes both suggest that elemental fluxes are very sensitive to household behaviors.

POLICY IMPLICATIONS AND RESEARCH NEEDS

These three case studies of ecosystem nutrient flows point to new directions in pollution management, with an emphasis on reducing the production of pollution rather than treating pollution at the end of the pipe. There are several reasons for moving in this direction. First, the end-of-pipe treatment approach that works well for point source pollution does not work as well for non-point source pollution. Source reduction also offers significant economic benefits through avoided waste (e.g., less cost for fertilizer; gasoline) and by lowering the cost of remaining pollution. Finally, source reduction is fairer, moving responsibility from the community-at-large to the potential polluter.

Developing pollution management policies based on ecosystem approaches raises many new research questions. Some of these include:

(1) *Disproportionality.* There is good evidence that consumption and pollutant generation/export is highly skewed, with a small fraction of the population (individuals, household or farms) producing a disproportionate fraction of the consumption or pollution. Examples include household water consumption (Mayer et al., 1999), tailpipe exhaust (Calvert et al., 1993), household energy use (unpublished analysis of data from the Energy Information Agency) and farmer fertilizer use (Birr, 2005). Nowak et al. (2006) argued that edaphic characteristics of landscapes accentuate this skewness for pollution production for agricultural systems and that the intersection of farmers' behaviors and landscape characteristics accentuates disproportionality. We hypothesize that this "second order" disproportionality exists for lawn runoff also. If so, targeting source reduction involves identifying both the behaviors (e.g., excessive fertilization) and landscape features (e.g., high slopes) that lead to pollution hotspots. Policies based on careful targeting could be highly effective and possibly more widely accepted.

(2) *Linkage between agricultural and urban systems.* Tracking nutrients through the complex food system will require new approaches that meld techniques from industrial ecology, ecosystem ecology, and environmental engineering

with recent and ongoing developments in sensor technology and cyberinformatics. This combination of techniques may make it possible to identify clear linkages between consumers in a given city and the geographic source of their food. For example, it should be possible to identify how much nutrient pollution in the Minnesota River, which drains a large agricultural watershed and empties into the Mississippi River, is caused by consumption within the Twin Cities. We hypothesize that this linkage would establish a sense of "ownership" of the specific agricultural systems which provide food to a city, encouraging reductions in consumption.

(3) *Feedback loops.* The potential for creating feedback loops for environmental management is growing rapidly. New technologies for imbedded and remote sensors and "smart" or auto-deployable sensor networks are expanding rapidly. Paralleling these developments (and essential for taking advantage of these developments) is an exponential growth in our ability to acquire, store and process in real time (or near real time) the massive quantities of data that can be generated by sensors. There is an unprecedented opportunity for harnessing this new technology to create information feedback loops and develop adaptive management strategies, including better ways to encourage source reductions. These concepts are the basis for an environmental observatory initiative in the planning and development stage in the Engineering and Geosciences directorates at the National Science Foundation: WATERS Network and its predecessor CLEANER (NRC, 2006).

(4) *Adaptive management.* As the concept of human ecosystems becomes established, we postulate that adaptive management concepts could play a major role in solving water quality problems. We know of one case study of adaptive management that resulted in documented improvement of water quality (Baker et al., 2006), but adaptive management concepts could prove valuable for solving regional water quality problems where many "actors" are involved. There is growing evidence that changes in behavior of urban dwellers could improve the quality of urban runoff, but there is a serious knowledge gap regarding factors that control the environmental behaviors. In our research on household nutrient flows we are using the Theory of Planned Behavior (TPB; Ajzen, 1991) as a framework for examining controls on behavior, because it breaks behavioral controls down into components – attitudes, social norms and "control beliefs". TBP allows researchers to identify the critical limitation(s)(the weak link) that control changes in environmental behavior. Once these limitations are identified, additional research is needed to identify appropriate types of feedback, types of communication that are effective for various groups and approaches to develop informal collaborative efforts.

REFERENCES

Ajzen, I. (1991). The theory of planned behavior. *Organizational Behavior and Human Decision Processes* 50:179–211.

Baker, L.A. (1992). Introduction to nonpoint source pollution in the United States and prospects for wetland use. Ecological Engineering 1:1–26.

Baker, L.A. (2006, in prep). Source reduction. In: Minnesota Stormwater Assessment Manual, edited by J. Gulliver and J. Anderson, University of Minnesota, St. Paul.

Baker, L.A., Brezonik, P.L., Mulla, D. and Stuewe, L. (2005). Nitrogen cycling in urban and agricultural watersheds in Minnesota (USA). Proc. Third International Conf. Nitrogen Biogeochemistry (N2004), Nanjing, China, October4 12–16, 2004.

Baker, L.A., Hartzheim, P., Hobbie, S., King, J. and Nelson, K. (2007). Influence of consumption choices on C, N and P fluxes through households. Urban Ecosystems 10:97–117.

Baker, L.A., Hope, D., Xu, Y. and Edmonds, J. (2001a). Multicompartment ecoystem mass balances as a tool for understanding and managing the biogeochemical cycles of human ecosystems. Proc. Second International Nitrogen Conference (N2001), Baltimore, Oct. 14–18, 2001.

Baker, L.A., Westerhoff, P. and Sommerfeld, M. (2006). An adaptive management strategy using multiple barriers to control tastes and odors. Journal of the American Water Works Association 98: 113–126.

Baker, L.A., Xu, Y., Hope, D., and Edmonds, J. (2001b). Nitrogen mass balance for the Phoenix-CAP ecosystem. *Ecosystems* 4:582–602.

Barr (2004). Phosphorus Sources to Minnesota Watersheds. Prepared by Barr Engineering for the Minnesota Pollution Control Agency, St. Paul, MN.

Barten, J., and Jahnke, E. (1997). Stormwater Lawn Runoff Water Quality in the Twin Cities Metropolitan Area, 1996 and 1997. Suburban Hennepin Regional Park District, Maple Plain, MN.

Birr, A.S. (2005). Paired Watershed Studies for Nutrient Reductions in the Minnesota River Basin. University of Minnesota, St. Paul.

Borrud, L., Wilkinson, E. and Mickle, S. (1996). What we eat in America: USDA surveys food consumption changes. Foods Review:14–19.

Boyd, S., Millar, S., Newcombe, K. and O'Neill, B. (1981). The Ecology of a City and Its People: The Case of Hong Kong. Australian National University Press, Canberra, Australia, Australian National University Press, Canberra, Australia.

Calvert, J.G., Heywood, J.B., Sawyer, R.F. and Seinfeld, J.H. (1993). Achieving acceptable air quality: some reflections on controlling vehicle emissions. Science 261:37–45.

Carpenter, S.R., Caraco, N.F., Correll, D.L., Howarth, R.W., Sharpley, A.N. and Smith, V.H. (1998). Nonpoint source pollution of surface waters with phosphorus and nitrogen. Ecological Applications 8:559–568.

Cassman, K.G., Dobermann, A. and Walters, D.T. (2002). Agroecosystems, nitrogen-use efficiency, and nitrogen management. Ambio 31:132–140.

Census. (2004). American FactFinder. U. S. Census Bureau, Washington, DC.

Decker, E., Elliott, H.S., Smith, R.A., Blake, D.R. and Rowland, F.S. (2000). Energy and material flow through the urban ecosystem. Annu. Rev. Energy Environ 25:685–740.

EIA (2001). Annual Energy Review. Energy Information Administration, U.S. Department of Energy, Washington.

Ensminger, M.E. (1993). Dairy Cattle Science, 3rd ed. Interstate Publishers, Danville, IL.

EPA (2000). Progress in Water Quality: An Evaluation of the National Investment in Municipal Wastewater Treatment (Executive Summary). U.S. Environmental Protection Agency, Washington, DC.

Faerge, J., Magid, J. and Penning de Vries, W.T. (2001). Urban nutrient balance for Bangkok. Ecological Modeling 139:63–74.

Kantor, L.S., Lipton, K., Manchester, A. and Oliveira, V. (1997). Estimating and addressing America's food losses. Food Review 20 (1): 1–11.

Lauver, L., and Baker, L.A. (2000). Nitrogen mass balance for wastewater in the Phoenix-Central Arizona Project ecosystem: implications for water management. Water Research 34: 2754–2760.

Lin, J.G., Hutjens, M.F., Shaver, R., Otterby, D.E., Howard, W.T., and Kilmer, L.H. (1996). Feeding the Dairy Herd. University of Minnesota Agricultural Extension Service, St. Paul. http://www.extension.umn.edu/distribution/livestocksystems/DI0469.html,

Mayer, P W, DeOreo, W B, Optiz, E M, Kiefer, J C, Davis, W Y, Dziegielewski, B and Nelson, J O (1999). Residential End Use of Water. Prepared for the American Water Works Association, Denver.

Mueller, D.K. and Helsel, D. (2000). Nutrients in the Nation's Waters–Too Much of a Good Thing? Circ. 1136, USGS, Washington, DC.

Neilsen, E.G. and Lee, L.K. (1987). The magnitude and costs of groundwater contamination: a national perspective. Report AGES870318, National Resources Economics Division, U.S. Department of Agriculture, Washington (DC).

Nowak, P., Bowen, S. and Cabot, P. (2006). Disproportionality as a framework for linking social and biophysical systems. Society and Natural Resources 19:153–173.

NRC. (2006). *CLEANER and NSF's Environmental Observatories.* Water science and Technology Board, National Research Council. The National Academies Press, Washington, D.C., 63 p.

OEA (2000). Minnesota Municipal Solid Waste Composition Study. Minnesota Office of Environmental Assistance, St. Paul, MN.

Patronek, G.J. (1995). Determining dog and cat numbers and population dynamics. Anthrozoos VIII:199–205.

Rezania, L. (2005). Optimizing phosphate treatment to minimize lead/copper seasonal variations. Waterline 12 (3): 2–3, Minnesota Department of Health.

Smith, R.A., Alexander, R.B. and Wolman, M.G. (1987). Water quality trends in the nation's rivers. Science 235:1608–1615.

Stuewe, L.A. (2006). Agricultural nitrogen and phosphorus balances in south-central Minnesota. M.S. Thesis, Department of Soil, Water and Climate, University of Minnesota, St. Paul.

Suh, S. (2005). Theory of materials and energy flow analysis in ecology and economics. Ecological Modeling 189:251–269.

Thompson, K., Christofferson, W., Robinette, D., Curl, J., Baker, L.A., Brereton, J. and Reich, K. (2006). Characterizing and Managing Salinity Loadings in Reclaimed Water Systems. Prepared for the American Water Works Research Foundation, Denver, CO.

USDA (2001). Briefing Room: Food Consumption: Individual Food Consumption (web page). Economic Research Service, U.S. Department of Agriculture.

USEPA (1983). Results of the Nationwide Urban Runoff Program. WH-554, Water Planning Division, U.S. Environmental Protection Agency, Washington, DC.

Van Horn, H.H., Wilkie, A.C., Powers and Nordstedt, R.A. (1994). Components of dairy manure systems. Journal of Dairy Science 77:2008–2032.

8

Field data requirements for monitoring and modelling of urban drainage systems

J.-L. Bertrand-Krajewski

URGC, INSA de Lyon, 34 avenue des Arts, 69621 Villeurbanne cedex, France
E-mail: jean-luc.bertrand-krajewski@insa-lyon.fr

Summary: Urban drainage systems are a significant part of global urban water systems. In order to reach and guarantee sustainable water management in cities, representative information is needed for both monitoring and modelling purposes. To account for the high variability of phenomena and processes occurring during storm events, especially regarding pollutant loads, large data sets are required. The paper illustrates these requirements by means of two case studies. The first one illustrates how the evaluation of the efficiency of a stormwater settling tank depends on the number of measured storm events; the second one indicates how the calibration and verification of even simple wet weather pollutant load models strongly depend on available field data. As collecting large data sets by means of traditional sampling and laboratory analyses is difficult and expensive, on-line continuous measurements by means of *in situ* sensors is suggested as an alternative.

INTRODUCTION

After their first appearance in 1980 in a IUCN report (International Union for the Conservation of Nature and Natural Resources) (Allen and Scott, 1980), the concepts of sustainability and sustainable development were clearly promoted and popularized in the Brundtland Report published in 1987 (Brundtland, 1987). Since that time, the concept of sustainability, despite the lack of consensus about its detailed and practical definition, has been widely disseminated and re-appropriated by individuals and institutions in many fields, and especially in the fields of water and of urban water systems. Reaching and ensuring ecologically sustainable urban water management is now a key objective for most municipalities, environment agencies, governments, NGOs, etc. Progress towards sustainability should be able to be evaluated and measurable, by means of monitoring allowing deriving e.g. performance indicators.

In order to reach and guarantee sustainable water management in cities, urban drainage systems (UDS), which are an important component of urban water systems (UWS), shall be better designed and monitored to evaluate their true performance. Compared to the traditional obligation of means, an obligation of results is nowadays more frequently requested, for technical, economical, environmental and social reasons. Improving both the design of new urban drainage facilities and the retrofit of existing infrastructures should be based on i) a better evaluation of existing facilities by means of monitoring and modelling and ii) a better knowledge of phenomena and processes occurring in UDS during storm events, and especially of their variability. Conception, design, modelling, management, evaluation of performance and related decision making shall be based on better *in situ* information and knowledge.

To illustrate the above needs, two case studies are presented in this paper. The first one examplifies how the evaluation of the efficiency of a stormwater settling tank depends on the number of measured storm events; the second one indicates how the calibration and verification of even a simple wet-weather pollutant load model strongly depend on available field data. In both case studies, information was obtained by means of samples and laboratory analyses. As this approach has obvious inherent limitations, additional short examples will be given to illustrate how on-line continuous sensors may offer an alternative. Lastly, some recommendations and research needs are suggested.

EVALUATING THE EFFICIENCY OF A SETTLING TANK

Within UDS, storage and settling tanks are very common facilities used to decrease pollutant loads discharged into receiving waters, in both separate and combined sewer systems. Stormwater effluents are temporarily stored in tanks, where settling occurs. As many pollutants in stormwater effluents are in the particulate phase, the removal of particles leads to the removal of pollutants, like e.g. COD (chemical oxygen demand), heavy metals, hydrocarbons, etc. The monitoring of such tanks is a critical issue in order to i) evaluate their true efficiency at both short and long time scales, ii) improve their conception, and iii) check if the objectives they had to reach are actually reached.

Monitoring is based on measurements, which are affected by uncertainties. In settling tanks, due to technical and practical difficulties, uncertainties may be very high and should be accounted for in performance evaluation. Additionally, storm events are random and show a great variability in terms of volumes and pollutant loads. As a consequence, a correct performance evaluation should be based on sufficiently large data sets. The two above aspects are illustrated in the following paragraphs.

Short time scale: event settling efficiency and its uncertainty

The Charbonnier storage tank, in the municipality of Vénissieux, France, is located at the outlet of a separate stormwater sewer system. Catchment and tank characteristics are summarised in Table 8.1. There are two inlet pipes: one C1200 circular pipe (diameter 1200 mm) at the outlet of the 178 ha sub-catchment, one T200 type egg-shape pipe (height 2 m) at the outlet of the 202 ha sub-catchment. The tank outlet is composed of a floating device which restricts the maximum outflow to 150 L·s^{-1} before it enters into a short length C400 circular pipe (diameter 400 mm). At the outlet of the C400 pipe, the runoff is infiltrated into the ground by means of a large infiltration tank (Bardin, 1999). In 1995, a measurement campaign was carried out to get knowledge about the functioning of the tank and to evaluate its pollutant removal efficiency. Both inlet pipes C1200 and T200 have been equipped with flowmeters (joint measurements of water depth h by means of piezo-resistive sensors and of mean flow velocity U by means of Doppler sensors). The outlet flow rate in the C400 pipe has been evaluated from the water depth above the floating device by using the relationship between flow rate and water depth given by the manufacturer, the water depth being measured by means of an aerial ultrasonic sensor. All data have been recorded with a constant time step $\Delta t = 6$ min. One hundred rainfall events have been observed. Automatic samplers have been installed within the inlet pipes and close to the surface weir of the floating device. Samples have been taken on regular time steps, and then combined proportionally to the measured flow rates and volumes in order to get event mean flow-weighted representative samples for each pipe. Many pollutants have been measured, but only TSS (Total Suspended Solids) will be considered in this paper. During the 1995 measurement campaign, seven rainfall events were completely monitored, with mean TSS concentrations.

The question to be answered is: what is the TSS removal efficiency ρ of the tank and its uncertainty? In order to answer this question with a reasonable level of confidence, the law of propagation of uncertainties (NF ENV 13005, 1999) and chronostatistics, associated with estimates of standard uncertainties, have been applied. Indexes e and s correspond respectively to tank inlet and outlet volumes V and TSS concentrations C. Detailed methods of calculation of all variables and of their uncertainties corresponding to 95% confidence intervals have been given elsewhere (Bertrand-Krajewski and Bardin, 2001, 2002).

Table 8.1. Charbonnier catchment and settling tank main characteristics.

Catchment area (ha)	380 (178 + 202)
Urbanisation (% of the area)	77% industries, 20% agriculture, 3% housing
Tank volume (m^3)	32 000
Tank mean depth (m)	2.10

Table 8.2. Measured values and associated uncertainties for the set of seven rainfall events.

date dd/mm/yy	V_e (m³)	$\Delta V_e/V_e$ (%)	C_e (mg.L⁻¹)	$\Delta C_e/C_e$ (%)	C_s (mg.L⁻¹)	$\Delta C_s/C_s$ (%)	ρ (-)	$\Delta\rho/\rho$ (%)
17/02/95	21400	6.7	39	29.0	21	37.7	0.462	56
21/02/95	1940	7.9	67	28.1	57	37.7	0.149	276
28/03/95	2850	8.1	64	28.3	47	37.7	0.260	138
15/05/95	5960	7.1	51	28.3	23	37.7	0.549	40
07/09/95	17600	9.8	135	28.3	43	37.7	0.681	23
03/10/95	1770	7.0	128	28.1	35	37.7	0.727	18
12/11/95	11500	7.6	37	28.5	15	37.7	0.595	33

The results for the set of seven rainfall events with measured TSS concentrations are given in Table 8.2. Relative uncertainties in total inlet volume V_e range from 6.7 to 9.8%. Relative uncertainties in concentrations C_e and C_s are quite constant. Depending on the rainfall event, the TSS removal efficiency ρ ranges from 0.149 to 0.727, with relative uncertainties $\Delta_{\rho/\rho}$ ranging from 18 to 276%. The highest values of $\Delta_{\rho/\rho}$ correspond to the lowest TSS removal rate values and $\Delta_{\rho/\rho}$ decreases when ρ increases.

The seven events are presented in chronological order in Table 8.2. Considering only the first three events, the settling efficiency of the tank appears as rather low, from 0.149 to 0.462, with very large uncertainties. On the contrary, considering the last three events, the efficiency appears as rather good, from 0.595 to 0.727, with a reasonable level of uncertainty. Depending on the sub-data set used for its evaluation, the settling efficiency may be very different. If the results are used e.g. to evaluate the strategy of pollutant interception, the perception of the municipality would be very different as well. In the first case, the existing strategy would not be validated (the settling tank would be considered as inefficient, or badly designed). In the second case, a more optimistic perception would be given, confirming the existing strategy and the design of the settling tank.

If one considers globally the seven rainfall events, the global removal efficiency ρ_7 is given by the following expression:

$$\rho_7 = 1 - \frac{\sum_{k=1}^{7} V_{sk}C_{sk}}{\sum_{k=1}^{7} V_{ek}C_{ek}} = 1 - \frac{1823}{4482} = 0.593 \tag{1}$$

Assuming that all rainfall events are independent, the application of the Law of Propagation of Uncertainties (LPU) to Eq. (1) gives $u(\rho_7) = 0.053$ and consequently $\frac{\Delta\rho_7}{\rho_7} = 17.8\%$.

The global TSS removal efficiency ρ_7, based on seven measured rainfall events, is equal to 0.59 and is known with a relative uncertainty equal to 18%, i.e. the true value of ρ_7 has a 95% probability to be between 0.49 and 0.70. Increasing the number of measured rainfall events would modify both above values.

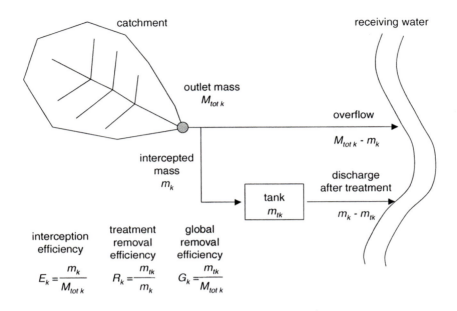

Figure 8.1.   Scheme of functioning of a storage-settling tank downstream of a catchment.

Long time scale: annual pollutant interception efficiency and its uncertainty

Consider now a storage and settling tank located at the outlet of a sub-catchment, as shown on Figure 8.1. During any rainfall event with index k, one can calculate:

- the interception efficiency E_k, which is the ratio between the mass m_k entering into the treatment facility and the total mass $M_{\text{tot} k}$ at the catchment outlet;
- the treatment removal efficiency R_k, which is the ratio between the mass m_{tk} trapped in the facility and the intercepted mass m_k (equivalent to the efficiency ρ calculated in the above paragraph);
- the global removal efficiency $G_k = E_k R_k$, which is the ratio between the trapped mass m_{tk} and the total mass $M_{\text{tot} k}$.

As the treatment removal efficiency R may vary significantly (it depends on the technology used, on the operation and maintenance of the facility, on the resuspension of antecedent deposits, etc.), this paragraph deals only with the interception efficiency E, considering obviously that only intercepted pollutants may have a chance to be removed.

The TSS annual interception efficiency E (%) for the N rainfall events observed during one year is given by:

$$E = 100\frac{\sum_{k=1}^{N} m_k}{\sum_{k=1}^{N} M_{\text{tot} k}} = 100\frac{m}{M} \qquad (2)$$

where $M_{\text{int } k}$ mass of TSS intercepted in the tank (kg) for the event nr k; $M_{\text{tot } k}$ total event TSS load (kg) for the event nr k; k index ranging from 1 to N for rainfall events; m sum of all m_k (kg) during one year; M sum of all $M_{\text{tot } k}$ (kg) during one year.

Two independent sources of uncertainty affect the value of E:

- the sampling uncertainty, denoted with the index s, due to the fact that not all N rainfall events, which constitute the population, are measured every year: only a sample of $n < N$ rainfall events is measured;
- the analytical uncertainty, denoted with the index a, due to the fact that the n sampled events are measured with uncertainties due to sensors, sub-sampling of water and pollutants, analytical procedures and techniques, etc. Evaluating the analytical uncertainty for each rainfall event is carried out as explained in Bertrand-Krajewski and Bardin (2002).

Applying the LPU to Eq. (2) gives:

$$\left(\frac{\Delta E}{E}\right) = \sqrt{\left(\frac{\Delta E}{E}\right)_s^2 + \left(\frac{\Delta E}{E}\right)_a^2} \tag{3}$$

If all hydrographs and concentrations for the N rainfall events are measured, there is no uncertainty linked to sampling and $(\Delta E/E)_s = 0$. Only the analytical uncertainty $(\Delta E/E)_a$ should be evaluated for all rainfall events.

The question to be answered is then: how many rainfall events should be sampled among N rainfall events per year in order to estimate the annual interception efficiency E with a given total uncertainty? The answer to the above question requires first a statistical analysis of the variables and then the application of the LPU. Details of hypotheses and calculations have been given in Bertrand-Krajewski et al. (2002).

Assuming (like in Lyon, France) that there are approximately $N = 80$ events per year generating runoff, Figure 8.2 shows $\Delta E/E$ as a function of the number n of measured rainfall events among N rainfall events per year.

If $n = 7$ events are measured among $N = 80$ events per year, the annual interception efficiency E is known with a relative uncertainty of 100%. To reach a relative uncertainty of 20%, one should measure $n = 50$ events among $N = 80$.

A further comparison of sampling and analytical uncertainties shows that the main source of uncertainty is the sampling representativity. In other words, it is more important to devote an effort to decrease the sampling uncertainty by increasing the number n of events measured per year, rather than decreasing the analytical uncertainty for each measured event. Similar results and orders of magnitude have been obtained e.g. by Rossi (1998) with other hypotheses applied to the distribution of pollutant concentrations and loads. This conclusion is not surprising for environmental fields like urban hydrology.

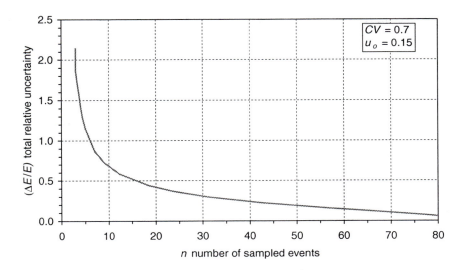

Figure 8.2.   Total relative uncertainty in the annual interception efficiency E versus the number n of rainfall events measured among $N = 80$ rainfall events per year with $u_o = 0.15$ (relative uncertainty in $M_{tot\,k}$ and m_k) and $CV = 0.70$ (coefficient of variation of TSS loads) (from Bertrand-Krajewski et al., 2002).

FIELD DATA AVAILABILITY AND MODEL CALIBRATION

This section aims to illustrate how stormwater quality model (SWQM) calibration and verification strongly depend on available field data (Bertrand-Krajewski, 2006). Among the various models proposed to simulate stormwater quality, regression models are frequently used to calculate EMC values (Event Mean Concentration) (see e.g. Servat, 1984; Driver and Tasker, 1990; Hoos and Sisolak, 1993; Saget, 1994; Hoos, 1996; Irish et al., 1998). In this section, one of the regression models implemented in the French software Canoé (Insa/Sogreah, 1999) is used as an example. It is given by Eq. (4):

$$C = a H^b I_{\max 5}^c ADW P^d \qquad (4)$$

with $ADWP$, antecedent dry weather period (hours); C, event mean concentration (mg.L^{-1}); H, rainfall depth (mm); $I_{\max 5}$, rainfall maximum intensity in 5 five minutes (mm.h^{-1}); a, b, c, d, parameters to be calibrated.

Seventeen storm events have been monitored in a 12-ha urban catchment equipped with a combined sewer system. All data (event mean TSS concentration, rainfall depth, intensity and antecedent dry weather period) are presented in chronological order in Table 8.3. These $N = 17$ events show the typical variability of all variables.

J.-L. Bertrand-Krajewski

Table 8.3. Values of C, H, $I_{max\ 5}$ and $ADWP$ measured for 17 storm events.

No. event	C (mg TSS.L^{-1})	H (mm)	I_{max5} (mm.h^{-1})	$ADWP$ (h)
1	834	3.2	2.4	141.2
2	602	3.4	3.2	374.2
3	707	4.6	8.8	7.3
4	625	4.0	5.2	11.0
5	288	6.4	3.2	16.4
6	410	6.0	3.2	48.3
7	402	4.8	3.2	44.0
8	914	15.0	6.4	48.0
9	632	3.4	4.8	26.6
10	399	3.0	3.2	0.6
11	743	3.6	4.8	43.5
12	474	2.8	2.4	56.0
13	331	4.6	4.0	0.7
14	625	5.2	10.4	14.8
15	548	10.6	2.4	21.6
16	508	5.6	4.8	86.1
17	1760	14.0	31.2	30.1

Table 8.4. Events used for calibration and verification.

Calibration no.	No. of the events used for calibration	No. of the events used for verification
1	1 to 6	7 to 17
2	12 to 17	1 to 11
3	1 to 17	None

Three different calibrations are presented in Table 8.4. Calibration 1 corresponds to a short measurement campaign, during which only the $n = 6$ first events have been measured. Calibration 2 is similar, but with the $n = 6$ last events. Calibration 3 is based on all $n = 17$ events, in order to use all available information. This 17 events data set is rather large compared to the usual operational practice revealed by a French enquiry: in 50% of the cases, up to 5 events are available for calibration, and up to 10 events in more than 80% of the cases (Gromaire et al., 2002).

The calibration has been made simply by means of the ordinary least squares method after logarithmic transformation of Eq. (4):

$$Ln(C) = Ln(a) + bLn(H) + cLn(I_{max\ 5}) + dLn(ADWP) \tag{5}$$

and by minimizing the sum S:

$$S = \sum_i (C_i - C_{mi})^2 \tag{6}$$

with C_i and C_{mi} respectively the observed and calculated EMC values for the event n° i, and i the index corresponding to the event numbers given in Table 8.4. This

Table 8.5. Calibration results.

Calibration	Parameter	Optimum	95% min. value	95% max. value
1	a	3220	5.62	1843510
	b	−1.308	−3.359	0.743
	c	0.209	−1.360	1.777
	d	−0.030	−0.591	0.530
2	a	128	22.12	743
	b	0.326	−0.703	1.355
	c	0.365	−0.241	0.972
	d	0.110	−0.183	0.402
3	a	193	107.64	347
	b	0.006	−0.332	0.345
	c	0.482	0.216	0.748
	d	0.110	0.018	0.202

method gives directly the optimum values of the four parameters a, b, c and d, and also their 95% confidence intervals.

For the three calibrations, the results obtained are:

– the values of the four parameters and their 95% confidence intervals (Table 8.5)
– the graphs showing C_m versus C, with circles for the events used in calibration and diamonds for the events used for verification (Figure 8.3)
– the total mass calculated for the 17 events (Figure 8.4).

According to Figure 8.3, both calibrations (i.e. circle dots) 1 and 2 appear satisfactory and not very different one another. The regression explains respectively 87% and 91% of the total variance for calibrations 1 and 2. This difference can be seen on the graphs, where the circle dots are a little bit closer to the diagonal for calibration 2 than for calibration 1. However, for verification shown by diamond dots, the results are significantly different: the dispersion is higher for calibration 1 than for calibration 2, especially for the 4 events marked by an arrow. This indicates that, despite a rather equivalent level of calibration, verification levels are very different. As a consequence, the total masses predicted by calibrations 1 and 2 are very different, as shown in Figure 8.4: respectively 5032 and 8656 kg of TSS, i.e. a relative increase of 72% for calibration 2. Compared to the observed total mass equal to 9099 kg of TSS, it appears clearly that calibration 2 is better. But, frequently, practitioners have only a limited data set and such comparisons are rarely made. This example shows that model calibration is very sensitive to the observations used: for the same amount of observations (6 events), results may differ dramatically. The influence of the event n°17, with a high concentration equal to 1760 mg.L^{-1}, appears as very important.

The results given in Table 8.5 show that the parameters a and b are the most sensitive ones to the data set used for calibration. They vary significantly between calibrations 1 and 2. For example, a changes from 3220 to 128, with very large confidence intervals. Similarly, b changes from −1.308 to 0.326, also with very

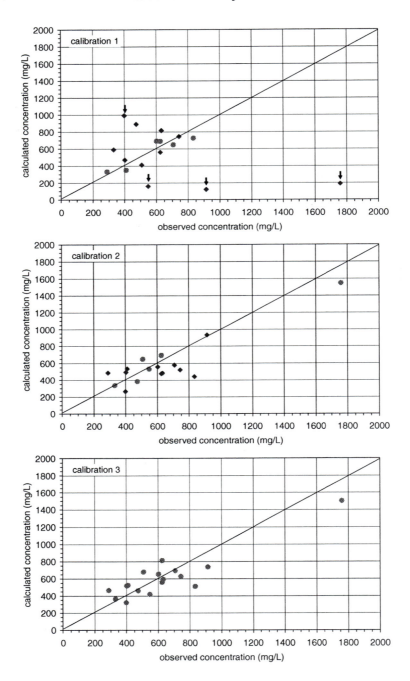

Figure 8.3. C_m versus C for all 17 events (arrows in the top graph correspond to inaccurately simulated events: see main text).

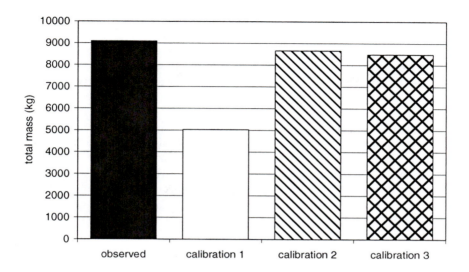

Figure 8.4.   Total TSS mass observed and calculated for the 3 calibrations.

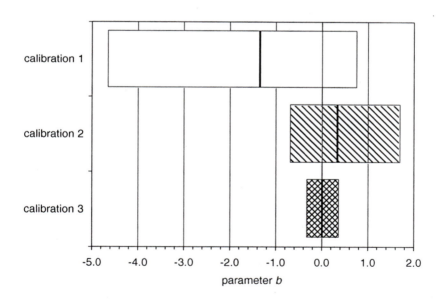

Figure 8.5.   Values of *b* and of its confidence intervals for the 3 calibrations.

large confidence intervals. The fact that *b* can be either negative or positive dramatically affects its role in the model. A further analysis indicates that confidence intervals for *b* decrease from calibration 1 to calibration 3, as shown in Table 8.5 and in Figure 8.5. In calibration 3, with 17 observations instead of 6 in calibrations

1 and 2, the optimum value of b becomes 0.006, i.e. very close to zero, and is located in a rather narrow confidence interval $[-0.332, 0.345]$. One may infer that b could be likely taken as equal to zero, which means that the rainfall depth H has no influence for this catchment and this data set, and that, consequently, it could be removed from Eq. (4). One observes here very clearly how increasing the number of observations has a strong effect on the width of confidence intervals.

Calibration 2 and calibration 3 presented in Table 8.5 and in Figure 8.3 appear as rather similar. The calculation of the total mass is also equivalent: respectively 8656 and 8485 kg of TSS, i.e. a relative difference of -2%, as shown in Figure 8.4. Both are close to the observed value of 9099 kg, with differences respectively equal to -5 and -7%, which is smaller than the uncertainty in the observed (i.e. measured) total mass estimated at approximately 25–30%. In this particular case, it seems that increasing the number of observations from 6 to 17 does not significantly change the results. But this conclusion is erroneous. There is a significant difference: in calibration 3, confidence intervals are narrower for all parameters (see Table 8.5), which allows the user to have a higher confidence in the model results.

This case study with a simple model shows clearly how calibration is sensitive to the data sets. For a given number of observations, very different parameter values are obtained. If the calibration may be similar (which is not always the case), the verification may be different. Increasing the number of observations may improve calibration and verification, and may also decrease the width of the parameters confidence intervals, but this is not systematically observed.

Similar results have been obtained with other regression models and also with detailed physically based models, with serious consequences when models are applied e.g. to design stormwater facilities (see Bertrand-Krajewski, 2006; Mourad, 2005).

USE OF CONTINUOUS MEASUREMENTS

All field data about pollutant loads shown in the above sections have been obtained with traditional sampling and laboratory analyses. Collecting large data sets in this way is technically difficult and very expensive. This is why, since more than two decades, attempts have been made to collect on-line and continuous information by means of sensors. However, on-line measurements in sewer systems are difficult (hard environment, difficult access for maintenance and checking, etc.). Approximately ten years ago, sufficiently robust technologies and sensors appeared on the market, allowing their use *in situ* in UDS. Among these sensors, turbidimeters and UV-visible spectrometers appear to be the most frequently used (Gruber et al., 2005; Ruban et al., 2005). In two experimental catchments of the OTHU (Field Observatory on Urban Hydrology) in Lyon, France, such sensors are used. The following two examples show recent results obtained with turbidimeters (Béranger, 2005). The sensors are initially calibrated with formazine and the measured turbidity expressed in NTU (Nephelometric Turbidity Unit) is then correlated with TSS or COD concentrations measured in samples by means of traditional laboratory analyses, including a specific correlation method accounting for uncertainties in both variables (Bertrand-Krajewski, 2004).

In the first example, two turbidimeters are installed respectively at the inlet and at the outlet of the settling tank of Chassieu, France. The data acquisition time step is 2 minutes. The settling tank is located at the outlet of a 185-ha industrial catchment, with a residual dry weather discharge of cooling and process water. Data obtained in February 2005 are shown in Figure 8.6. During the month, the total inlet TSS load was 2730 kg. During dry weather periods, approximately 480 kg of TSS entered the tank and 120 kg of TSS left the tank, which corresponds to a settling efficiency of 75%. During rainfall events, 2250 kg of TSS entered the tank and 340 kg of TSS left the tank, which corresponds to a settling efficiency of 85%. The results shown in Figure 8.6 could not have been obtained with traditional sampling and analyses. The sensors are installed in the Chassieu settling tank since September 2004, and continuous information is now available to evaluate the performance of the facility.

In the second example, one turbidimeter is installed at the outlet of the 245 ha residential catchment of Ecully, equipped with a combined sewer system. The data are stored with a 2 minutes time step. During the complete year 2004, 71 rainfall events occurred and 20 events among them generated overflows discharged in the small creek at the outlet of the catchment. All events have been monitored continuously. For the 20 events generating overflows, the mean TSS concentration $(mg.L^{-1})$ estimated from turbidity during the overflow periods and the

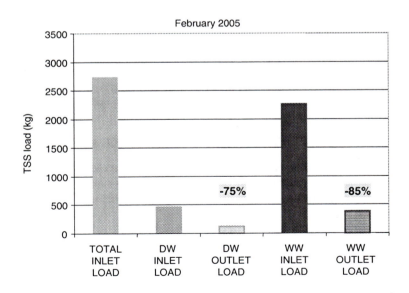

Figure 8.6.   TSS loads in dry and wet weather periods in the Chassieu settling tank, estimated by means of turbidimeters in Feb. 2005.

J.-L. Bertrand-Krajewski

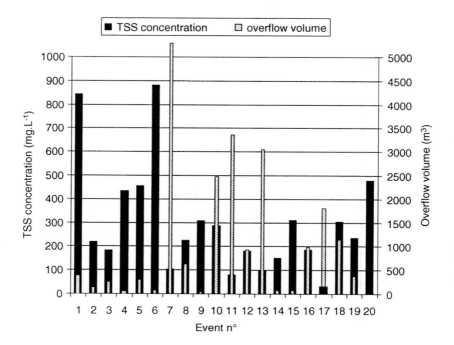

Figure 8.7.   Mean TSS concentrations estimated from turbidity for the 20 overflows observed at the Ecully catchment outlet in 2004.

corresponding volumes (m^3) are shown in Figure 8.7. Mean TSS concentrations in overflows show a very high variability, ranging from less than 50 to more than 850 mg.L^{-1}. Discharged volumes are also very variable, from a few to more than 5500 m^3. As in Chassieu, such dense information revealing the great variability of the phenomena is not obtainable with traditional sampling and analyses.

RECOMMENDATIONS AND FUTURE RESEARCH

Sustainability of urban water systems is a great objective. As a significant component of urban water systems, urban drainage systems need specific attention, especially regarding stormwater that transports large pollutant loads into the aquatic environment.

The progress towards sustainability should be evaluated and measurable. Compared to the past, this ambition means that urban drainage systems should be better known and modelled. This better knowledge is obtainable by means of monitoring and measurements which should be strongly promoted. The obligation of measurable and measured results should be the rule.

Due to the great variability of the phenomena occurring during rainfall events, large data sets are necessary to get representative and well based information, either for monitoring or modelling purposes. Limited data sets may lead to wrong results, interpretation, model calibration and verification, and consequently to wrong knowledge and decisions.

The above great variability should also be included in regulations and legal permits or authorizations, by replacing traditional single value design by design (and consequently performance evaluation) accounting for probabilities and statistics.

For each objective, a specific analysis should be carried out to define the appropriate time and space scales of measurement (geo- and chronostatistics).

Uncertainties in measurements and in knowledge should be taken into account, especially when comparison of results, technical solutions or investment strategies is concerned.

Traditional sampling and laboratory analyses are not the best solution. Some sensors already exist which, if calibrated and used on-line and continuously under appropriately controlled conditions, may deliver very dense and rich information giving a better knowledge and understanding of the phenomena. However, technological progress is still necessary before all pollutants of interest will be measurable on-line by sensors, especially e.g. for hazardous substances. This is why an adequate combination of sampling and on-line measurements, associated with a constantly improved modelling effort, is probably a valuable approach for the next years.

REFERENCES

Allen, R. and Scott, P. (1980). How to save the world: strategy for world conservation. London (UK): Kogan Page, 150 p. ISBN 0-85038-315-3.

Bardin, J.-P. (1999). Contribution à une meilleure connaissance du fonctionnement qualitatif des bassins de retenue soumis à un débit traversier permanent et à la prise en compte des incertitudes. PhD thesis: INSA de Lyon, France, January 1999, 341 p. + appendices.

Béranger, Y. (2005). Mesure en continu des flux polluants (MES, DCO) par turbidimétrie, dans les réseaux d'assainissement. Villeurbanne (France): URGC, INSA de Lyon, Research MSc report, September 2005, 127 p. Unpublished.

Bertrand-Krajewski, J.-L. (2004). TSS concentration in sewers estimated from turbidity measurements by means of linear regression accounting for uncertainties in both variables. Water Science and Technology 50(11): 81–88.

Bertrand-Krajewski, J.-L. (2006). Influence of field data sets on calibration and verification of stormwater pollutant models. Proceedings of the 7th International Conference on Urban Drainage Modelling, Melbourne, Australia, 2–7 April, 2: 3–20. ISBN 0-646-45903-1.

Bertrand-Krajewski, J.-L. and Bardin, J.-P. (2001). Estimation des incertitudes de mesure sur les débits et les charges polluantes en réseau d'assainissement: application au cas d'un bassin de retenue–décantation en réseau séparatif pluvial. La Houille Blanche 6/7: 99–108.

Bertrand-Krajewski, J.-L. and Bardin, J.-P. (2002). Uncertainties and representativity of measurements in stormwater storage tanks. Proceedings of the 9th International Conference on Urban Drainage, Portland, Oregon, USA, 8–13 September, 14 p.

Bertrand-Krajewski, J.-L., Barraud, S., and Bardin, J.-P. (2002). Uncertainties, performance indicators and decision aid applied to stormwater facilities. Urban Water 4(2):163–179.

Brundtland, G.H. (1987). Our common future. New York (USA): United Nations, Report of the World Commission on Environment and Development, August 1987, 374 p.

Driver, N.E. and Tasker, G.D. (1990). Techniques for estimation of storm-runoff loads, volumes, and selected constituent concentrations in urban watersheds in the United States. Washington, DC (USA): US Geological Survey, U.S.G.S Water-Supply Paper 2363.

Gromaire, M.-C., Cabane, P., Bertrand-Krajewski, J.-L., and Chebbo, G. (2002). Operational use of urban drainage pollutant load models – Results from a French survey. Proceedings of the SOM 2002 "Sewer Operation and Maintenance" International Conference, Bradford, UK, 26-28 November 2002, 8 p.

Gruber, G., Bertrand-Krajewski, J.-L., de Bénédittis, J., Hochedlinger, M., and Lettl, W. (2005). Practical aspects, experiences and strategies by using UV/VIS-sensors for long-term sewer monitoring. Proceedings of the 10th International Conference on Urban Drainage, Copenhagen, Denmark, 22–26 August 2005, 8 p.

Hoos, A.B. (1996). Improving regional-model estimates of urban-runoff quality using local data. Water Resources Bulletin 32(4): 855–863.

Hoos, A.B. and Sisolak, J.K. (1993). Procedures for adjusting regional regression models of urban-runoff quality using local data. Reston (USA): U.S. Geological Survey, Open-File Report 93–39, 39 p.

Insa/Sogreah (1999). Canoé User Manual. Villeurbanne (France): Alison – INSA de Lyon.

Irish, L.B., Barrett, E.B., Malina, J.F., and Charbeneau, R.J. (1998). Use of regression models for analyzing highway stormwater loads. Journal of Environmental Engineering 124(10): 987–993.

Mourad, M. (2005). Modélisation de la qualité des rejets urbains de temps de pluie: sensibilité aux données expérimentales et adéquation aux besoins opérationnels. PhD thesis: INSA de Lyon, Villeurbanne, France, 295 p.

NF ENV 13005 (1999). Guide pour l'expression de l'incertitude de mesure. Paris (France): AFNOR, August 1999, 113 p.

Rossi, L. (1998). Qualité des eaux de ruissellement urbaines. PhD thesis n° 1789: Ecole Polytechnique Fédérale de Lausanne, Switzerland, February 1998, 416 p.

Ruban, G., Bertrand-Krajewski, J.-L., Chebbo, G., Gromaire, M.-C., and Joannis, C. (2005). Précision et reproductibilité du mesurage de la turbidité des eaux résiduaires urbaines. Actes de la conférence "Autosurveillance, diagnostic permanent et modélisation des flux polluants en réseaux d'assainissement urbains", SHF-GRAIE-ASTEE, Marne-la-Vallée, France, 28–29 juin 2005, 191–200. ISBN 2-906831-62-X.

Saget, A. (1994). Base de données sur la qualité des rejets urbains de temps de pluie: distribution de la pollution rejetée, dimensions des ouvrages d'interception. PhD thesis: Ecole Nationale des Ponts et Chaussées, Paris, France, 227 p + annexes.

Servat, E. (1984). Contribution à l'étude des matières en suspension du ruissellement pluvial urbain à l'échelle d'un petit bassin versant. PhD thesis: Université des Sciences et Techniques du Languedoc, Montpellier, France, 182 p. + annexes.

PART FOUR

Hydrologic and Pollution Stresses, Response of Receiving Waters

9

Ground water and cities

P. Shanahan[1] and B. L. Jacobs[2]

[1] *Department of Civil and Environmental Engineering, Massachusetts Institute of Technology, Cambridge, MA 02139, USA*
[2] *HydroAnalysis, Inc., 33 Clark Road, No. 1, Brookline, MA 02445, USA*
E-mail: peteshan@mit.edu

Summary: Cities affect ground water through both hydrologic modification and pollution. The character of hydrologic modification changes with the growth and maturation of the city. Often, the initial growth of cities causes a rise in the water table and pollution level; transition to an industrial city leads to decline of the water table and continuing pollution; and maturation to a post-industrial city causes the water table to rise again with modest improvement in pollution. Each phase implies different effects on public health and urban infrastructure. While there are broad trends in these impacts, the effects in any one city and even at any location within a city are highly varied and site-specific, requiring detailed evaluation of each site and problem. Future research should therefore focus on case studies representative of the problems in multiple cities and, as a cross-cutting theme, methods for the integrated management of urban ground water and infrastructure.

INTRODUCTION

Even in simple environments, ground water is a complicated subject. The unaltered natural subsurface is difficult and expensive to sample and thus the hydrogeology is never known completely. Nevertheless, relatively minor features of the subsurface—a buried alluvial channel, for example—may significantly influence ground-water movement. Moreover, the main driver of the ground-water system—recharge—is difficult to measure and necessarily an estimate. Interaction with surface-water bodies is similarly difficult to quantify. These complications, already challenging in natural settings, are magnified in cities by the alteration of both the hydrology and the physical setting. The increased imperviousness of the surface reduces natural recharge, but the release of imported water from water-supply pipelines, sanitary sewers, and irrigation systems may more than replace the natural recharge. Subsurface utilities may profoundly alter the configuration of the water table and the rate and direction of ground-water flow. Withdrawal of large quantities of ground water for municipal and industrial water supply may greatly lower the water table, while release of polluted water from sewers, industries, and other sources may render urban water non-potable. The nature of urban ground-water hydrology is sufficiently distinct from natural systems that Vásquez-Suñé et al. (2005) have recently proposed that urban ground water be considered a distinct branch of hydrogeology. Recent conferences have also considered the unique problems of urban hydrology (Chilton et al., 1997; Wilkinson, 1994).

This paper provides an overview of issues associated with urban ground-water hydrology and provides illustrative case studies. Cities affect and interact with ground water in ways that are diverse and highly variable. This paper seeks to describe the overall patterns of urban effects on ground water, but to illustrate their diverse and site-specific nature through examples and a case study of Boston, Massachusetts, USA.

HYDROLOGIC MODIFICATION

Patterns of urban hydrologic modification

Figure 9.1 illustrates and contrasts the components of the ground-water balance for a natural system with that of an urban system (after Simpson, 1994). The natural system is driven by precipitation recharge, often augmented by inflow from adjacent aquifers and surface-water bodies. Eventually, the aquifer discharges to one or more surface-water bodies and, in some instances, adjacent aquifers.

In the urban setting, the natural fluxes illustrated in Figure 9.1 may be greatly diminished. Most or virtually the entire land surface is covered by buildings and pavement, reducing precipitation recharge and concentrating it in small areas where pavement is missing or cracked, or where roof and stormwater runoff is deliberately infiltrated into the ground. However, in place of natural recharge are a multitude of artificial sources. Chief among these are leakage from water-supply lines (often as high as 25% of the flow according to Barrett et al., 1999), leakage from sewer lines, and irrigation of parks and gardens.

Outflows from the urban aquifer may include the natural components of discharge to surface water or other aquifers, but these are often dwarfed by the withdrawal of water to wells for public and industrial water supply. In addition, in areas of substantial underground construction and high water table, ground water may be pumped or drained to prevent the intrusion of ground water into construction excavations, transportation tunnels, underground parking garages, basements, and utility conduits. Even if not deliberately pumped or drained, utility conduits and other constructed subsurface works may act as inadvertent drains.

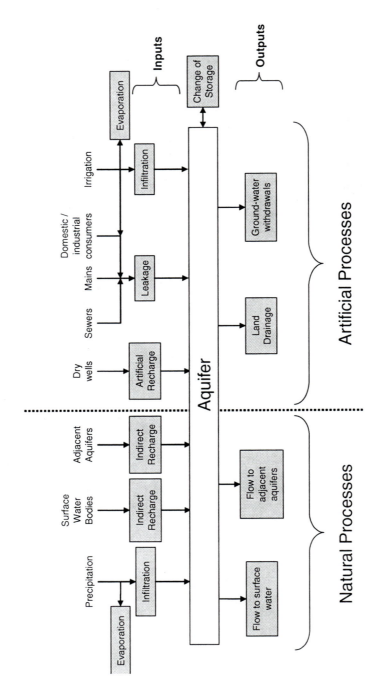

Figure 9.1.   The urban aquifer water balance (adapted from Simpson, 1994).

The growth and maturation of cities creates typical patterns of a changing ground-water balance over time. During the initial phases of growth, local ground water, if available, is tapped as a water supply. Typically, it is a shallow unconfined aquifer that is the main resource and the water table declines in response to the increasing extraction. Withdrawals from the shallow aquifer are often made by individual households providing their own water supply, and wastewater is also disposed on-site to cesspools or septic systems. Thus during this first phase of urban development there is widespread utilization of the shallow aquifer but also widespread recharge and contamination by on-site wastewater treatment.

Before long, growth of the city necessitates more complete water infrastructure. Water needs are greater and the distribution system becomes more-or-less complete. Sanitary sewers are developed to collect and remove wastewater. Leakage from both the water supply and wastewater collection systems becomes a significant sources of recharge, however the impact of this leakage may or may not be offset by ground-water withdrawals. In many cases, the unpressurized wastewater collection system may act as a ground-water sink if the water-surface elevation in the system conduits is less than the prevailing ground-water elevation. If the local aquifer is tapped, then leakage usually provides substantially less volume than is withdrawn by municipal and industrial wells. In this case, the decline of the water table accelerates, as does the pollution of the aquifer by industry. This period of lowered water table often coincides with substantial infrastructure development. Much of the infrastructure is underground, but constructed during a period of artificially lowered ground-water elevations. This second phase of urban development is therefore one of substantial infrastructure development, but under conditions of high ground-water utilization and a lowered water table.

Eventually, three forces can act to reduce dependence on local ground water. The first is the increasing pollution of the local aquifer due to urban and industrial pollution. Second is the senescence and departure of industry. These can cause the city to reduce local ground-water withdrawal, either by reducing need or forcing the city to alternative supplies. Thirdly, the availability of high quality surface-water sources may result in a reduced dependence on ground water, particularly in those cases where the ground-water yield is not adequate to satisfy the growing water demand. With decreased withdrawals of local ground water, the water table recovers, often rising dramatically. Thus, the mature phase of urban development often marks a return to ground-water levels similar to or even higher than pre-development conditions.

If a local aquifer is not present, then ground-water extraction does not factor nearly as significantly in the ground-water balance. In this case, leakage from water lines and sewers, as well as excess irrigation, creates a net increase in inflow to the ground-water system. Water tables rise, often to problematic levels, in both the second and third phases of development.

This phased pattern of development is idealized and local conditions may cause variations from this pattern, both in time and space. (See for example, Morris et al., 1997, for a variation on this analysis.) Despite these deviations, the idealized development model is a useful construct for categorizing experiences reported in the literature and for managing the city's ground-water resource, a topic to which we return later in this paper. Each phase of urban development is characterized by its own pattern of impacts on the populace, the environment, and the urban infrastructure.

Impacts of urban hydrologic modification

The phases of urban development may be marked by a water table that rises and falls over time. The potential impacts of rising or falling water tables are varied and many. Rising water tables may lead to the following adverse consequences, as listed by Johnson (1994), Abu-Rizaiza (1999) and Han (1998) among others:

1. Reduced bearing capacity of soils
2. Hydrostatic uplift pressures on floor slabs and loads on retaining walls
3. Swelling of clay soils
4. Expansion of compacted engineered fill
5. Compaction and settlement of loose fill
6. Leakage into basements and utilities
7. Surface ponding of and human exposure to sewage and contaminated ground water
8. Failure of existing septic absorption systems
9. Chemical attack on masonry and concrete (especially by polluted ground water)
10. Trapping and/or displacement of methane and other subsurface gases
11. Creation of a need for drainage
12. Creation of sinkholes in karstic areas

Many of these problems are geotechnical in nature, and can exact high costs in damage to buildings and structures. The proceedings edited by Wilkinson (1994) discuss a variety of geotechnical issues.

The problems due to falling water tables are somewhat fewer, but often very severe. They include (Braadbaart and Braadbaart, 1997; Downing, 1994):

1. Costs to deepen wells
2. Increased pumping costs for deeper wells
3. Salt-water intrusion in coastal areas
4. Induced flow from other aquifers or surface waters
5. Rotting of wooden structural piles
6. Subsidence

Either rising or falling water tables will alter the patterns of ground-water flow around the city, possibly causing the undesirable movement of polluted ground water. Moreover, alteration of the water balance may cause water tables to rise in some areas and fall in others, and also to experience altered seasonal patterns. For example, Nassau County, immediately east of New York City on Long Island, compensates for ground-water pumping by infiltrating stormwater in recharge basins (Ku et al., 1992). These basins recharge year-round, whereas evapotranspiration uptake eliminates most natural recharge during the growing season. The basins also tend to be located along the island's center spine, leading to a net increase in recharge in this inland area and a rise in water-table elevation. Meanwhile, the coastal areas are drained by storm sewers to the ocean but feature relatively few recharge basins. Thus, they not only lack the recharge augmentation of the inland areas, but are drained of water that would otherwise recharge during the dormant season. Consequently, the coastal areas see a net decrease in recharge and water-table elevations relative to predevelopment conditions.

Examples of urban hydrologic modification

While it is useful to outline the general patterns of hydrologic modification during urban development, and to list the potential effects of hydrologic modification, actual effects tend to be very variable and highly site-specific. The local hydrogeology is a major influence on the impact of hydrologic modification, but there are a myriad of other local factors. The following examples from the literature illustrate site-specific effects within the context of the three phases of urban development described above.

The initial phase of urban development—in which the population has increased but the infrastructure has yet to develop—can be seen today in the emerging megacities of the developing world. Typical cases are reported by Somasundaram et al. (1993) and Lawrence et al. (2000) for cities in India and Thailand respectively. Somasundaram et al. (1993) discuss the case of Madras, India, since renamed Chennai. The city is underlain by a shallow aquifer that is less than 20 meters thick. Although most of the city's water comes from outside sources and is provided via a municipal distribution system, the city's poorest residents still rely upon shallow wells within the city. Water-quality sampling of wells throughout the city illustrates the pollution of the local shallow aquifer: high concentrations of nitrate, metals, salts, and microbes were found as the result of highly polluted rivers, industrial pollution, and on-site wastewater disposal.

Lawrence et al. (2000) reported on Hat Yai, Thailand's third largest city with a population of 158,000 in 1995. The city has undergone rapid development, but has yet to develop a sewer system. Wastewater infiltration from surface canals introduces a substantial load of organic matter (and other pollutants) to the shallow aquifer. Microbial degradation of the organics has consumed all oxygen in the central portion of the aquifer, creating highly reducing chemical conditions that have caused the mobilization of natural, but previously insoluble, arsenic. Beneath the shallow aquifer is a semi-confined aquifer at a depth of 30 to 50 meters. This deeper aquifer is the main supply aquifer for the city and pumping has substantially lowered hydraulic head within the aquifer and induced recharge from the overlying shallow aquifer. Ground-water sampling by Lawrence et al. found that the pollutants of the shallow aquifer were beginning to intrude into the deep aquifer, threatening its viability as a supply.

In the case of both Madras and Hat Yai, the rapid development of the city combined with a lack of wastewater infrastructure has led to pollution of the shallow ground-water resource. While this is typical of the first stage of urban development, the different hydrogeology at the two cities means the difference between contaminating the major aquifer that supplies the entire city of Hat Yai versus only the local shallow aquifer that supplies a fraction of Madras' population.

The second phase of urban development marks the transition to the fully-developed infrastructure of the mature city. One might expect that such a city, with its increased impervious cover, would experience a net reduction in ground-water recharge compared to rural areas. While natural recharge is reduced, it is more than compensated by leakage from the water distribution and wastewater collection systems and relatively free use of supplied water for irrigation of public and private parks and gardens (Lerner, 2002).

In the absence of local withdrawals, the net enhancement of recharge has created such paradoxes as problematically high water tables in the very arid environments of Jeddah and Riyadh, Saudi Arabia (Al-Sefry and Sen, 2006 and Rushton and Al-Othman, 1994), Kuwait City, Kuwait (Al-Rashed and Sherif, 2001), and elsewhere in the Middle East (Abu-Rizaiza, 1999). These water-table rises are enhanced in arid areas by the presence of a shallow, low-hydraulic-conductivity layer of carbonate-cemented evaporite sediments. This layer impedes the deep infiltration of the excess recharge, accumulating water in the shallow soils. As a result, Kuwait City has seen water tables rise by as much as 5 meters, causing flooded basements and ponding in low areas (Al-Rashed and Sherif, 2001). Here again, local hydrogeologic factors make a significant difference. Absent the cemented layer typical of arid regions, there would be a lesser rise in the water table and fewer problems.

Where a local supply aquifer is utilized, the second phase of urban development is typically marked by dramatic declines in ground water elevations. The effect is nicely illustrated in a classic study of the confined Patuxent Formation aquifer beneath Baltimore, Maryland, USA, completed by Bennett and Meyer (1952). At the time of the study, Baltimore was highly industrialized and had been through a particularly intense period of production to support the US effort in World War II. Most of industry was self-supplied by on-site wells to the Patuxent Formation. As seen in Figure 9.2, the declines in potentiometric head were dramatic: several industrial centers showed head declines of over 50 feet (15 meters) below their natural near-sea-level elevations. This is a confined aquifer, so Figure 9.2 does not reflect the decline in the water table and many of the effects outlined above do not apply. Nonetheless, the drastic lowering of the potentiometric surface induced infiltration of salt water from the overlying Baltimore Harbor and eventually rendered the portions of the aquifer near the Patapsco River (to the north in Figure 9.2) unsuitable for use.

One of the major effects of significant ground-water-level drawdown is subsidence, the lowering of land-surface elevation due to consolidation of underlying compressible soils. The classic example of subsidence induced by ground-water withdrawals is Mexico City, which overlies a thick sequence of clayey lake deposits (Ortega-Guerrero et al., 1993). From 1930 to 1960, pumping below the city center led to as much as 7.5 meters of subsidence and consequent damage to buildings

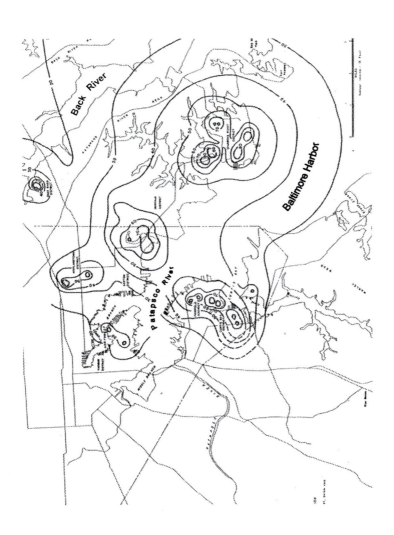

Figure 9.2.   Map of the potentiometric surface in the confined Patuxent Formation aquifer beneath Baltimore, Maryland, USA in 1945 (Bennett and Meyer, 1952). Potentiometric surface elevations are given in feet below mean sea level.

and structures. As a consequence of these problems, pumping beneath the city was reduced and new supply wells were tapped in the Valley of Mexico outside the city. This served only to extend the problems from the central city to the outlying areas (Ortega-Guerrero et al., 1993; Birkle et al., 1998). From 1960 to 1980, the water table in the Valley of Mexico fell by as much as 30 meters, ground-water flow patterns were altered, and springs disappeared. Subsidence in the city center had increased to as much 9 meters and was beginning to be problematic in the outlying areas. A water balance completed by Birkle et al. (1998) illustrates that withdrawals far exceed recharge and that active management of the ground-water resource is required. These problems are hardly unique to Mexico City: subsidence problems have also been experienced more recently around the major cities of China as they become major industrial centers (Zhang and Kennedy, 2006; Chen et al., 2003; Liu et al., 2001).

Overtaxing of ground water by growing cities is often facilitated by the absence of a regulatory or management framework (Braadbaart and Braadbaart, 1997). Frequently, ground-water extraction is unregulated and environmental regulations are applied casually during periods of rapid economic growth. This characterized industrial expansion in the U.S. and Europe historically, and is typical of expansion in developing countries today. The challenges in managing urban ground water are twofold. First, investigating and modeling ground-water systems is often technically demanding and expensive. Resources for this type of work typically do not become available until demanded by a crisis. Second, appropriate regulatory structures and institutions must be developed to manage ground water, often across political boundaries. This is a difficult and politically sensitive task that is, again, in most cases postponed until made unavoidable by crisis.

The final stage of urban development, in which the city transitions to a post-industrial city with diminished water demands, creates a new set of problems. In these cities, problems occur when pumping within the city is reduced and water tables recover to their former high levels. Unfortunately, the period of heavy ground-water use often coincided with rapid growth of the city and expansion of the infrastructure, including substantial subsurface infrastructure. Subway tunnels, underground parking garages, and building basements were built during a period of artificially lowered water tables. A return to former water-table levels implies the invasion of ground water into structures unprotected from ground-water intrusion. In London, for example, the water table has risen as much as 20 meters (Drangert and Cronin, 2004). Paris, New York and other major cities have experienced these problems (Drangert and Cronin, 2004; Cohen and Lenz, 2005). This has prompted some expensive remedies. In the Borough of Queens in New York City for example, ground-water pumping that has been suspended since the 1980s is being resumed, but with very expensive ground-water treatment required for the water and additional systems being constructed to control a plume of industrially contaminated ground water.

The problems described above are varied and depend upon site-specific characteristics: the presence of semi-confined aquifer conditions in Hat Yai and Baltimore; thick layers of clay soil beneath Mexico City; or the evaporite aquitard in the arid cities of the Middle East. These examples illustrate that each urban ground water problem will be unique in many ways, and will require site-specific investigations and solutions. Thus, while there are some broad patterns in the evolution of cities and the types of ground-water problems they experience, there will inevitably be a need for local solutions to local problems.

CITIES AND GROUND-WATER POLLUTION

Underlying the discussion of hydrologic modification is the issue of ground-water pollution, which is both caused by and a potential cause of hydrologic modification. The many potential sources of pollution from cities are well documented (see for example, OTA, 1984) and a complete list need not be repeated here. The major sources of urban contamination include pathogens and nitrates from on-site wastewater disposal and leaking sewers, synthetic organic chemicals from industrial and commercial facilities, petroleum hydrocarbons from underground storage tanks at gasoline stations and private fueling facilities, and, in cold climates, chloride and sodium from road salting during winter.

While heavy concentrations of on-site wastewater disposal are known to cause ground-water pollution, the importance of leaking sewers as a source of urban ground-water pollution is a matter of some debate. That sewers are a potential source is irrefutable: sanitary sewers are known to leak and sewage contains a multitude of pollutants. However, Lerner et al. (1994) point out that despite the widely held view that sewers are a pollution source, there is little actual documentation of such. Studies by, among others, Eisen and Anderson (1979), Apodaca et al. (2002) and Nazari et al. (1993) show clearly elevated concentrations of various pollutants in urban ground water, but without clear attribution to leaking sewers. Significantly, Barrett et al. (1999) sought specific chemical markers of sewer leakage in Nottingham, U.K. and concluded that the effects of sewers were "not high." Our conclusion is that leaking sewers are unlikely to be a universally important source. Unless sewers are located above the water table, they tend to act as drains rather than sources. Thus, with the exception of pressurized sewage force mains and in areas of low water table, sewers may be overemphasized as a source of contamination.

Despite the equivocation as to the significance of leaking sewers, there is ample evidence that there is commonplace pollution of urban aquifers. Somasundaram et al. (1993), for example, list thirty-five literature studies of urban ground-water pollution in addition to those mentioned above. The association is long established: Woodward (1961) for example documents ground-water contamination by on-site wastewater disposal in the Minneapolis, Minnesota suburbs in the 1950s.

Pollutants associated with urbanization, and particularly on-site wastewater disposal and/or leaking sanitary sewers, include nitrate and surfactants, lowered dissolved oxygen concentrations, and a variety of dissolved solids including sulfate, chloride, boron, sodium, and coliform bacteria (Trauth and Xanthopoulos, 1997; Eisen and Anderson, 1979; Nazari et al., 1993; Apodaca et al., 2002). We note that elevated concentrations of salts are not necessarily a cold-weather phenomenon alone: Abu-Rizaiza (1999) mentions high salt concentrations in ground water as one of the problems of the arid Middle East. Recent studies by the U.S. Geological Survey (Zogorski et al., 2006) have shown that volatile organic compounds are also common in urban ground water. The most commonly detected, in order, are trihalomethanes (drinking water disinfection byproduct), tetrachloroethylene (dry-cleaning fluid), methyl-tertiary butyl ether (gasoline additive), trichloroethylene (industrial solvent), toluene (gasoline component), dichlorodifluoromethane (refrigerant), and trichloroethene (industrial solvent).

In summary, although there is some ambiguity as to the significance of leaking sewers as a source of ground-water contamination, there is an unequivocal association between urbanization and ground-water pollution. The significance of leaking sewers is likely site specific but the general link between urbanization and ground-water pollution is universal. As discussed earlier, contamination of ground water often leads to hydrologic modification as the city seeks to replace contaminated local ground water with safer alternatives.

A CASE STUDY EXAMPLE: GROUND WATER IN BOSTON

The city of Boston lies on a one-time peninsula between Boston Harbor and the Charles River in eastern Massachusetts, USA. Many neighborhoods within Boston are constructed on fill over what were once either the Charles River mud flats or Boston Harbor (Figure 9.3). Buildings in these areas were almost exclusively

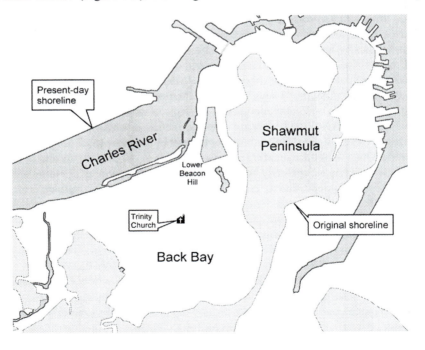

Figure 9.3.   Map of Boston showing present and original shorelines and areas of fill (based on Krieger and Cobb, 1999).

constructed on timber piles (Aldrich and Lambrechts, 1986). These include the three-story apartment buildings typical of the Back Bay and South End neighborhoods, as well as major buildings such as Trinity Church and the Boston Public Library. The construction of Trinity Church, for example, began in 1871 on newly filled ground. Its foundation lies on 4500 35-foot-long piles sunk to bedrock.

In settings where timber piles have been used to support building foundations, the depression of the water table can have significant consequences for building owners. Piles are columnar structures used to support building foundations in areas where surface soils are inadequate to support the weight of the building without substantial settling. Up to the 1900s, wood was the standard pile material. Standard practice in the use of timber piles was to excavate below basement level and drive the pile downward until the top of pile was below the minimum water-table elevation. The piles were capped by beams or large stones and the building constructed on this firm surface.

The anticipated life of timber piles is indefinite so long as the piles remain saturated in water. Wood piles of 800 years and more are routinely found beneath historic structures in Europe, still functioning as originally intended. Although bacteria constantly degrade the wood of these piles, the time scale of the decay process is very long if the piles remain saturated with water (Collin, 2002). Piles that become desaturated, either continuously or episodically, are subject to degradation by both fungus and insects as well as bacteria. The more aggressive decay under these conditions can compromise the structural integrity of the wood pile within roughly ten years or less. The decay time may be shorter or longer depending on the type of tree used for the pile, as well as local conditions including mineral composition of the soil, pH, and temperature.

As long as the water table remains above the top of the piles and keeps them saturated, they can be expected to retain their integrity and support the building without excessive settlement. However, if the water table is lowered, the exposed pile tops begin to deteriorate. As the piles decay, separation between the tops of individual piles and the overlying pile caps may occur, resulting in a shift of building loads from the decayed piles to other piles and to the underlying compressible soils. Differential settlement is then likely to occur due to localized settling at differing rates. This results in structural damage to the building often betrayed by a characteristic step-pattern of cracking to the brick façade.

Damage to timber piles, and consequent damage to building facades and structures, has occurred at many buildings in Boston, including the previously mentioned Trinity Church. Excavations beneath the church in 2001 revealed damage to piles on the eastern end of the building despite a system to artificially recharge ground water around the church since the 1930s. The St. James Street sewer, which runs alongside the Trinity Church, was found in the 1930s to be acting as a drain and lowering water-table elevations along its length (Figure 9.4). The sewer was partially dammed at that time to increase the water elevation within the sewer. This in turn increased the water-table elevation along the street to a sufficient level to saturate the wood piles of the church and other neighborhood buildings (Aldrich

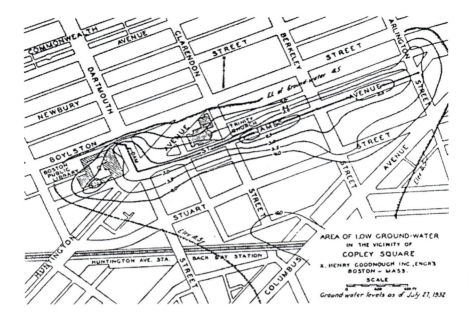

Figure 9.4.   Water table contours along St. James Street, Boston in 1932 showing impact of leaking St. James Aquifer. The location of the sewer dam is shown at the corner of Boylston Street and Dartmouth Street (Aldrich and Lambrechts, 1986).

and Lambrechts, 1986). Artificial recharge to ground-water wells was instituted at the same time. As the result of the damage discovered in 2001, church trustees installed monitoring wells to check on the elevation of the water table and improved the protocol for artificial recharge (Haynes, 2003).

The Lower Beacon Hill neighborhood built in Back Bay on filled land along the Charles River has been the locus of similar problems (Figure 9.5). Here, consistently low water-table elevations have been caused by the drainage of ground water to a combination of sewer and underground transportation structures (Lambrechts, 2000). Figure 9.5 shows the infrastructure. The Storrow Drive Tunnel was constructed in the 1950s but was soon found to be intercepting ground water. In order to maintain a dry road surface, a drain system was constructed that conveyed infiltrating ground water to wet wells that were pumped to the Charles River. The Boston Marginal Conduit (BMC) serves as a main interceptor for sanitary and combined sewer lines from Boston neighborhoods along the Charles River. Running through the Lower Beacon Hill neighborhood and connecting to the BMC are two combined sewer overflow pipelines. Water levels within the BMC had fluctuated with the tide up until the late 1970s, but then the pipe outlet was relocated and the BMC's water level maintained at a significantly lower elevation. When the operating level in the BMC was changed, ground-water monitoring wells at the Advent Church started to show extended periods of low water-table elevation. A

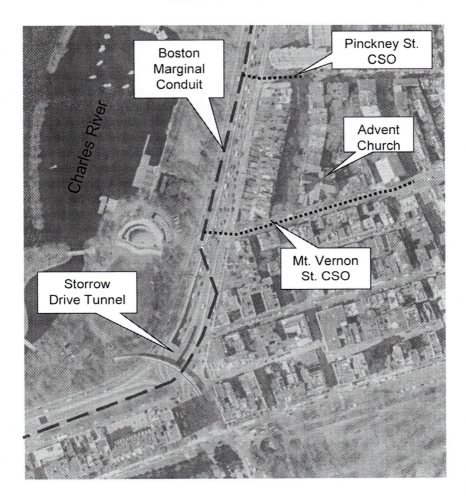

Figure 9.5.   Infrastructure affecting water-table elevations in Lower Beacon Hill, Boston.

detailed investigation revealed that the combined sewer overflows were acting as drains for ground water within the neighborhood and, with the outlet water level in the BMC lowered, the overflows carried more flow and were more effective in reducing water-table elevations in the Lower Beacon Hill neighborhood. Many homes have required expensive repairs as the result of damage to desaturated wood pile tops.

The interaction of lowered water-table elevations and wood pile foundations in Boston illustrates some of the difficulties facing the urban ground-water hydrologist. Firstly, the problems were caused by highly local circumstances: drainage into the sewer lines affected only the local neighborhood. Secondly, very detailed site-specific investigations were needed to uncover the relatively subtle triggers for lowering of the water table. Thirdly, individual actors in the urban scene made changes with dramatic effects on other parties without any regulatory oversight or integrated framework for managing ground water.

The absence of an integrated management framework is now being addressed by the City of Boston and other parties. In September 2005, the Boston Water Sewer Commission, the Massachusetts Bay Transportation Authority, the Massachusetts Water Resources Authority, the City of Boston and other state agencies signed a memorandum of understanding that states their joint commitment to participate in identifying areas of depressed ground water and finding appropriate resolutions (BGWT, 2005). Moreover, they have committed to "remedy any infrastructure under their agency's control that is reasonably demonstrated to contribute substantially to groundwater depletion ... where buildings are supported by wood pilings." Following this memorandum of understanding, the City implemented a Groundwater Conservation Overlay District that requires developers to evaluate the impacts of their plans on the ground-water elevation (Besser, 2006). New development projects in Boston now feature elaborate systems to capture roof runoff and recharge it to the aquifer (Palmer, 2005). A major force in the issue is the Boston Groundwater Trust, which was founded by the city in the mid-1980s to monitor ground-water levels. The trust has both collected and published ground-water data and drawn attention to the problems associated with falling water tables (BGWT, 2006).

RECOMMENDATIONS AND FUTURE RESEARCH

The problems of urban ground-water are diverse and varied: the problems of individual cities are often unique and require site-specific investigation and management. That said, there are broad trends at play. Over time, the typical city passes through three stages of development:

1. An initial phase in which growth precedes infrastructure development, and the city tends to rely on on-site wells for water supply as well as on-site wastewater disposal.
2. A phase of infrastructure and industrial development, but also of high demand for water supply (often from local aquifers), substantial hydrologic modification, and substantial ground-water pollution. This phase may be marked by a falling water table if water is supplied from a local aquifer or a rising water table if water is imported from outlying sources.
3. A post-industrial (or post-pollution) phase in which reduced demand for local ground water leads to rising water tables and intrusion of ground water into underground infrastructure.

Each of these stages is marked by modification of the hydrologic cycle. Natural inflow to the ground-water system from precipitation recharge is typically diminished over time as is natural outflow to surface water. In their place, artificial inflows from leaking water-supply and wastewater-removal infrastructure and irrigation of parks and gardens and artificial outflows to supply wells and urban infrastructure dominate. Often, the ground-water system responds with changes in the water level, and with a multitude of adverse consequences for constructed facilities, environmental quality, and human health. These changes usually occur in a regulatory and managerial vacuum: ground-water use is frequently unregulated and there is no managerial framework to monitor and control the urban ground-water resource. Management is itself a complicated issue and requires difficult political negotiations and demanding technical work.

This review suggests several areas for further research. A technical issue of some generality is the uncertainty regarding the magnitude of urban wastewater sewer leakage as a source of urban ground-water contamination. Other technical issues however tend to be site-specific: how do the problems of a particular city arise in light of its physical setting, state of infrastructure, and political framework? These technical issues are best addressed through case studies, but with particular attention to problems that typify multiple cities.

A more challenging area for research is the integrated management of urban ground water. As shown by the case study of Boston, problems are not only site-specific but often highly localized. This implies a need for information, and lots of it. Technological means to monitor and control urban ground water are clearly an area of potentially fruitful research. Studies of instrumentation and data retrieval technologies to enable real-time access to information on the ground-water system are recommended. Also needed is the development of robust management tools (i.e., computer models and information systems) to use that information to manage urban ground water in an integrated and comprehensive fashion. While theoretical research to develop these tools is needed, here also case studies are recommended as a logical way to develop, test, and refine management systems for urban ground water and infrastructure. A comprehensive urban hydrologic monitoring framework like the long-term urban environmental observatory in Baltimore (Welty et al., this monograph) would be an ideal setting for such a case study. Additionally, how new approaches to urban surface-water management (as discussed by Hill and Maimone et al., this monograph) affect ground water need to be evaluated in careful case studies.

The general failure to implement regulatory structures for urban ground water until necessitated by crisis suggests another area of research. Mechanisms for ground-water management and control are beginning to emerge; the ground-water recharge overlay district mandated in Boston's new rules is an example. A comparative analysis of different regulatory structures, funding mechanisms, and institutional mechanisms would be valuable to find which measures work in which political and cultural settings. Clearly, the widespread emergence of urban ground-water problems has created a need for these types of regulatory frameworks, but there is little information and almost no critical analysis on such frameworks in the existing literature.

Over all, the research needs for urban ground water would best focus on case studies to understand how technological, managerial, and regulatory approaches work in real systems with a diversity of problems.

REFERENCES

Abu-Rizaiza, O.S., 1999. Threats from groundwater table rise in urban areas in developing countries. *Water International*. Vol. 24, No. 1, Pg. 46–52. March 1999.

Aldrich, H.P., and J.R. Lambrechts, 1986. Back Bay Boston, Part II, Groundwater Levels, *Civil Engineering Practice*, Vol. 1, No. 2, Pg. 31–64. Fall 1986.

Al-Rashed, M.F., and M.M. Sherif, 2001. Hydrogeological aspects of groundwater drainage of the urban areas in Kuwait City. *Hydrological Processes*. Vol. 15, No. 5, Pg. 777–795. April 15, 2001.

Al-Sefry, S., and Z. Sen, 2006. Groundwater rise problem and risk evaluation in major cities of arid lands - Jeddah case in Kingdom of Saudi Arabia. *Water Resources Management*. Vol. 20, No. 1, Pg. 91–108. February 2006.

Apodaca, L.E., J.B. Bails, and C.M. Smith, 2002. Water Quality in Shallow Alluvial Aquifers, Upper Colorado River Basin, Colorado 1997. *Journal of the American Water Resources Association*. Vol. 38, No. 1, Pg. 133–149. February 2002.

Barrett, M.H., K.M. Hiscock, S. Pedley, D.N. Lerner, J. H. Tellam, and M. J. French, 1999. Marker species for identifying urban groundwater recharge sources: a review and case study in Nottingham, UK. *Water Research*. Vol. 33, No. 14, Pg. 3083–3097. October 1999.

Bennett, R.R., and R.R. Meyer, 1952. Geology and ground-water resources of the Baltimore area. Bulletin 4. Maryland Geological Survey, Baltimore, Maryland.

Besser, S., 2006. Overlay District Created to Prevent Further Groundwater Depletion. *Beacon Hill Times*. February 21, 2006. Boston, Massachusetts, USA. (http://www.bostongroundwater.org/ Article-75.html)

BGWT, 2005. Memorandum of Understanding. Boston Groundwater Trust, Boston, Massachusetts. September 15, 2005. (http://www.bostongroundwater.org/orgmou. html)

BGWT, 2006. Boston Groundwater Trust. http://www.bostongroundwater.org/. (accessed June 15, 2006.)

Birkle, P., V.T. Rodriguez, and E.G. Partida, 1998. The water balance for the Basin of the Valley of Mexico and implications for future water consumption. *Hydrogeology Journal*. Vol. 6, No. 4, Pg. 500–517. December 1998.

Braadbaart, O., and F. Braadbaart, 1997. Policing the urban pumping race: Industrial groundwater overexploitation in Indonesia. *World Development*. Vol. 25, No. 2, Pg. 199–210. February 1997.

Chen, C.X., S.P. Pei, and J.J. Jiao, 2003. Land subsidence caused by groundwater exploitation in Suzhou City, China. *Hydrogeology Journal*. Vol. 11, No. 2, Pg. 275–287. April 2003.

Chilton, J., K. Hiscock, P. Younger, B. Morris, S. Puri, H. Nach, P. Aldous, J. Tellam, R. Kimblin, and S. Hennings, Editors, 1997. *Groundwater in the Urban Environment: Proceedings of the XXVII IAH Congress on Groundwater in the Urban Environment, Nottingham, UK, 21–27 September 1997, Volume 1: Problems, Processes and Management*. Rotterdam, A.A. Balkema.

Cohen, D.K., and M. Lenz, 2005. Triple fix. *Civil Engineering*. Vol. 75, No. 6, Pg. 58–63. June 2005.

Collin, J.G., 2002. Timber Pile Design and Construction Manual. Timber Piling Council American Wood Preservers Institute, Fairfax, Virginia. (http://www.wwpinstitute. org/pdffiles/TimberPileManual.pdf)

Downing, R.A., 1994. Keynote paper: Falling groundwater levels—a cost-benefit analysis. In: W.B. Wilkinson, Editor. *Groundwater problems in urban areas*. Proceedings of the International Conference organised by the Institution of Civil Engineers and held in London, 2–3 June 1993. Pg. 213–236. Thomas Telford, London.

Drangert, J.O., and A.A. Cronin, 2004. Use and abuse of the urban groundwater resource: Implications for a new management strategy. *Hydrogeology Journal*. Vol. 12, No. 1, Pg. 94–102. February 2004.

Eisen, C., and M.P. Anderson, 1979. The Effects of Urbanization on Ground-Water Quality– A Case Study. *Ground Water*. Vol. 17, No. 5, Pg. 456–462. September-October 1979.

Han, Z.S., 1998. Groundwater for urban water supplies in northern China – An overview. *Hydrogeology Journal*. Vol. 6, No. 3, Pg. 416–420. October 1998.

Haynes, T., 2003. A Church on Stilts. *The Boston Globe*. August 26, 2003. Boston, Massachusetts, USA. (http://www. bostongroundwater.org/Article-26.html)

Hill, K., 2006. Urban ecological design and urban ecology: An assessment of the state of current knowledge and a suggested research agenda. *This monograph*.

Johnson, S.T., 1994. Keynote paper: Rising groundwater levels: engineering and environmental implications. In: W.B. Wilkinson, Editor. *Groundwater problems in urban areas*. Proceedings of the International Conference organised by the Institution of Civil Engineers and held in London, 2–3 June 1993. Pg. 285–298. Thomas Telford, London.

Krieger, A., and D. Cobb, 1999. *Mapping Boston*. MIT Press, Cambridge, Massachusetts, USA.

Ku, H.F.H., N.W. Hagelin, and H.T. Buxton, 1992. Effects of urban storm-runoff control on ground-water recharge in Nassau County, New York. *Ground Water*. Vol. 30, No. 4, Pg. 507–514. July–August 1992.

Lambrechts, J.R., 2000. Investigating the Cause of Rotted Wood Piles. In: K.L. Rens, O. Rendon-Herrero, and P.A. Bosela, Editors. *Forensic Engineering*. Proceedings of the Second Conference, San Juan, Puerto Rico, May 21–23, 2000. Pg. 590–599. American Society of Civil Engineers, Reston, Virginia, USA.

Lawrence, A.R., D.C. Gooddy, P. Kanatharana, W. Meesilp, and V. Ramnarong, 2000. Groundwater evolution beneath Hat Yai, a rapidly developing city in Thailand. *Hydrogeology Journal*. Vol. 8, No. 5, Pg. 564–575. October 2000.

Lerner, D.N., 2002. Identifying and quantifying urban recharge: a review. *Hydrogeology Journal*. Vol. 10, No. 1, Pg. 143–152. February 2002.

Lerner, D.N., D. Halliday, and M. Hoffman, 1994. The impact of sewers on groundwater quality. In: W.B. Wilkinson, Editor. *Groundwater problems in urban areas*. Proceedings of the International Conference organised by the Institution of Civil Engineers and held in London, 2–3 June 1993. Pg. 64–75. Thomas Telford, London.

Liu, C.M., J.J. Yu, and E. Kendy, 2001. Groundwater exploitation and its impact on the environment in the North China Plain. *Water International*. Vol. 26, No. 2, Pg. 265–272. June 2001.

Maimone, M., J. Smullen, B. Marengo, and C. Crosket, 2006. The role of low impact redevelopment/development in integrated watershed management planning: turning theory into practice. *This monopgraph*.

Morris, B.L., A.R. Lawrence, and S.S.D. Foster, 1997. Sustainable groundwater management for fast-growing cities: Mission achievable or mission impossible? In: J. Chilton et al., Editors. *Groundwater in the Urban Environment: Proceedings of the XXVII IAH Congress on Groundwater in the Urban Environment, Nottingham, UK, 21-27 September 1997, Volume 1: Problems, Processes and Management*. Pg. 55–66. A.A. Balkema, Rotterdam.

Nazari, M.M., M.W. Burston, P.K. Bishop, and D.N. Lerner, 1993. Urban Ground-Water Pollution – A Case Study from Coventry, United Kingdom. *Ground Water*. Vol. 31, No. 3, Pg. 417–424. May–June 1993.

Ortega-Guerrero, A., J.A. Cherry, and D.L. Rudolph, 1993. Large-Scale Aquifer Subsidence near Mexico City. *Ground Water*. Vol. 31, No. 5, Pg. 708–718. September–October 1993.

OTA, 1984. Protecting the Nation's Groundwater from Contamination. Report No. OTA-O-233. U.S. Congress, Office of Technology Assessment, Washington, D.C. October 1984.

Palmer, T.C., Jr., 2005. Columbus Center will make an impact on Boston's skyline . . . And under ground. *The Boston Globe*. February 21, 2005. Boston, Massachusetts. (http://www.bostongroundwater.org/ Article-48.html)

Rushton, K.R. and A.A.R. Al-Othman, 1994. Control of rising groundwater levels in Riyadh, Saudi Arabia. In: W.B. Wilkinson, Editor. *Groundwater problems in urban areas*. Proceedings of the International Conference organised by the Institution of Civil Engineers and held in London, 2–3 June 1993. Pg. 299–309. Thomas Telford, London.

Simpson, R.W., 1994. Keynote paper: Quantification of processes. In: W.B. Wilkinson, Editor. *Groundwater problems in urban areas*. Proceedings of the International Conference organised by the Institution of Civil Engineers and held in London, 2–3 June 1993. Pg. 105–120. Thomas Telford, London.

Somasundaram, M.V., G. Ravindran, and J.H. Tellam, 1993. Ground-Water Pollution of the Madras Urban Aquifer, India. *Ground Water*. Vol. 31, No. 1, Pg. 4–11. January–February 1993.

Trauth, R., and C. Xanthopoulos, 1997. Non-point Pollution of Groundwater in Urban Areas. *Water Research*. Vol. 31, No. 11, Pg. 2711. November 1997.

Vázquez-Suñé, E., X. Sánchez-Vila, and J. Carrera, 2005. Introductory review of specific factors influencing urban groundwater, an emerging branch of hydrogeology, with reference to Barcelona, Spain. *Hydrogeology Journal*. Vol. 13, No. 3, Pg. 522–533. June 2005.

Welty, C. et al., 2006. Designing an environmental field observatory for quantifying the urban water budget. *This Monograph*.

Wilkinson, W.B., 1994. Editor. *Groundwater problems in urban areas*. Proceedings of the International Conference organized by the Institution of Civil Engineers and held in London, 2–3 June 1993. Thomas Telford, London.

Woodward, L., 1961. Ground Water Contamination in the Minneapolis and St. Paul Suburbs. In: Ground Water Contamination, Proceedings of the 1961 Symposium. In: Ground Water Contamination, Proceedings of the 1961 Symposium, Cincinnati, Ohio, April 5–7, 1961. Pg. 66–71. Robert A. Taft Sanitary Engineering Center, Public Health Service, Cincinnati, Ohio, USA.

Zhang, L., and C. Kennedy, 2006. Determination of sustainable yield in urban groundwater systems: Beijing, China. *Journal of Hydrologic Engineering, ASCE*. Vol. 11, No. 1, Pg. 21–28. January-February 2006.

Zogorski, J.S., J.M. Carter, T. Ivahnenko, W.W. Lapham, M.J. Moran, B.L. Rowe, P.J. Squillace, and P.L. Toccalino, 2006. The quality of our nation's waters: volatile organic compounds in the nation's ground water and drinking-water supply wells. Circular 1292. U.S. Geological Survey, Reston, Virginia, USA. (http://pubs.usgs.gov/circ/circ1292/)

10

Framework for risk-based assessment of stream response to urbanization

B.P. Bledsoe

Colorado State University, Fort Collins, CO 80523-1320, USA
E-mail: bbledsoe@engr.colostate.edu

Summary: The standard practice of using simplified predictors of stream impact, such as watershed imperviousness, can lead to, among other things, poor allocation of resources with sensitive streams left underprotected and relatively resilient streams potentially overprotected against instability. This article argues against standardization of stormwater controls across stream types and proposes an alternative framework for risk-based modeling and scientific assessment of hydrologic-geomorphic-ecologic linkages in urbanizing streams for improved watershed management. The framework involves: 1) *a priori* stratification of a region's streams based on geomorphic context and susceptibility to changes in water, sediment, and wood regimes, 2) field monitoring of these strata across a gradient of urban influence, 3) coupling long term hydrologic simulation with geomorphic analysis to quantify key hydrogeomorphic metrics, and 4) using probabilistic modeling to identify links between hydrogeomorphic descriptors of urbanization effects with geomorphic and biotic endpoints of primary interest to stakeholders and decision-makers. The proposed framework is illustrated with an example of using logistic regression analysis to link hydrologic alteration, stream type, and erosion potential with the morphologic stability of streams in the arid southwest US.

INTRODUCTION

Watershed modifications typically accompanying urbanization have profound impacts on hydrologic and geomorphic processes in receiving streams. Urbanization frequently results in reduced infiltration and interception, conversion of subsurface flow to surface runoff, and more rapid conveyance of runoff via engineered drainage systems. By reducing natural watershed storage and vegetative cover, urbanization often intensifies the geomorphic processes of erosion and sedimentation through cumulative increases in flow energy, and causes an "urban stream syndrome" (Meyer et al., 2005; Walsh et al., 2005) with flashier hydrographs, altered channel morphology, and reduced biotic integrity (Jacobson et al., 2001; Konrad et al., 2005; Booth, 2005). Although best management practices (BMPs) intended to mitigate the effects of urbanization on receiving waterbodies have been widely implemented for decades, a lack of consideration of network scale hydrologic responses and geomorphic processes in standard design approaches can potentially exacerbate both flooding and geomorphic instability (Emerson *et al.*, 2005; MacRae, 1997).

Several efforts are underway to relate land-use changes to flow regimes with the intent of developing urban development practices that minimize hydrologic alteration of streams in urbanizing landscapes (e.g., King County Normative Flows Project, http://dnr.metrokc.gov/wlr/BASINS/flows/; State of New Jersey Ecological Flows Project, http://nj.usgs.gov/special/ecological_flow/). Exclusively focusing on the hydrologic effects of urbanization, however, is problematic because geomorphic responses to hydrologic change are mediated by channel boundary materials and context-specific geologic and human disturbance histories that may vary markedly within and among hydroclimatic regions (e.g., Knox, 1977; Trimble, 1974, 1983, 1997; Urban and Rhoads, 2003; Poff et al., 2006). As a result, stream geomorphic responses tend to be difficult to correlate with gross measures of imperviousness that do not reflect context-specific differences in stream sensitivity (Bledsoe, 2002). Unless management tools are designed to account for the differential sensitivity of stream types (*sensu* Downs and Gregory, 1995) within common management units (e.g., individual river basins, physiographic regions, or ecoregions), predicting the effects of urbanization on stream integrity is likely to be confounded by poor correlations between stream response, magnitude of developed area, and style of development and stormwater practices. Moreover, mitigation strategies may be confounded by one-size-fits all solutions that potentially underprotect the most vulnerable streams at the expense of overprotecting relatively resilient systems.

STATE OF THE ART – CURRENT KNOWLEDGE

Urbanization is a diverse collection of human influences as opposed to a single condition (Konrad and Booth, 2005). Accordingly, the effects of urbanization on fluxes of water, sediment, organic matter (including wood), nutrients, heat, and stream ecologic functioning vary significantly with watershed context and style of urbanization. Although all these fluxes affect stream integrity, the focus of this article is on the hydrologic and geomorphic processes controlling water and sediment regimes and the form of stream channels. The important influence of flow regime as mediated by geomorphic context on the structure, composition, and productivity of stream ecosystems is well-established (Konrad and Booth, 2006; Poff et al., 2006), and geomorphic stability within some range of variability is often a prerequisite for stream ecological integrity (Jacobson et al., 2001).

In a particular historical context, streams adjust over time to the flows of sediment and water delivered from their watershed (Schumm, 1969; Parker, 1991; Wilcock, 1997). As land uses change, spatial and temporal patterns in the transport capacity of stream channels relative to the type and amount of sediment supplied from the watershed are altered. Urban land uses tend to increase the frequency of high flows and daily variation in streamflows, and convert base flow to storm flow (Konrad and Booth, 2005; Poff et al., 2006). The effects of these changes in runoff and, consequently, sediment yield are often further exacerbated by direct channel disturbances that increase flow energy, decrease channel roughness, and reduce erosion resistance (Jacobson et al., 2001). In general, the response of streams to land use change fundamentally depends on cumulative excess specific stream power relative to the erodibility of channel boundary materials (MacRae, 1997; Rhoads, 1995; Bledsoe, 2002; Grant et al., 2003). For example, in an armored cobble bedded stream with sandy banks and little vegetative reinforcement, the dominant response to an increase in erosive power relative to sediment supply is likely to tend toward bank erosion and lateral adjustment (Downs, 1995). Conversely, in a sand bed stream with highly cohesive banks, the response will tend towards incision until bank failure results primarily from gravitational forces as opposed to direct hydraulic action (Schumm et al., 1984; Simon, 1989). Vegetation can play a critical role in affecting erosion resistance and channel response to land use change (Thorne, 1990; Dunaway et al., 1994; Anderson et al., 2004). Response potential also varies with the sequence of channel types distributed throughout a basin as segments transition between supply- and capacity-limitation and as floodplain connectivity varies with valley type (Montgomery and Buffington, 1998; Montgomery and MacDonald, 2002).

Scale also influences interpretation of geomorphic responses to hydrologic change. The time scales addressed here are intermediate (decadal) in that water and sediment discharge are both primary independent variables (Schumm and Lichty, 1965; Schumm, 1991). Stream responses to changes in these variables occur at spatial scales ranging from drainage networks, to reaches, to streambed patches. At larger scales, incision of a channel segment due to hydrologic change may exert widespread influence on entire tributary drainage networks through base level lowering and headcutting. In contrast, a stream reach that largely maintains its pattern and profile in response to land use change may be altered in terms of habitat complexity and patch scale substrate stability (Hashenburger and Wilcock, 2003; Booth and Henshaw, 2001; Konrad et al., 2005). Although some streams re-attain quasi-equilibrium in a coarse sense after land use change, this does not necessarily imply that the quality, quantity, and stability of habitats available to stream communities are comparable to pre-disturbance conditions.

Finally, geomorphic thresholds, temporal lags, non-linear behavior, and climatic variability further complicate stream responses to urbanization. Geomorphic thresholds relevant to the hydrologic changes frequently associated with urbanization include mass wasting of banks (Simon and Collison, 2002), planform change (Bledsoe and Watson, 2001), bedforms and flow resistance in sand bed channels (Simons and Richardson, 1966), and mobility of sediment mixtures (Jackson and Beschta, 1984; Wilcock, 1998). Yet, despite the recurring themes of non-linearity and uncertainty in the prediction of morphological change in hydrologically perturbed fluvial systems (Schumm, 1991; Richards and Lane, 1997), the following general conclusions may be drawn from previous research on the geomorphic effects of urbanization on streams:

- Different stream types have inherent system properties that create variable but predictable directional responses to urbanization.
- It is important to consider continuous flow regimes of both water and sediment as affected by the spatial and temporal aspects of land use change, drainage infrastructure, and BMPs.
- Effective stream assessment includes careful consideration of how time relates to responses observed in impacted streams. This includes response lag times, history, and the temporal sequence of geomorphically effective events. Historical influences, antecedent events, and infrastructure may "prime" or limit the system for a particular response trajectory.
- Restabilization of streams sometimes occurs in a few decades after land use changes but does not imply a return of comparable habitat quality and biological potential.

Modeling stream responses using hydrogeomorphic descriptors

Previous regional-scale studies of stream responses to changes in water and sediment regimes are often based on the assumption that the likelihood of channel instability can be assessed by identifying upper and lower boundary values of specific stream power for different stream types and boundary materials (Booth, 1990; Simon and Downs, 1995; Bledsoe and Watson, 2001a). This assertion is supported by several studies that have correlated bankfull specific stream power with channel stability. For example, surveys of channelized rivers in England, Wales, and Denmark indicated that for a limited range of channel gradients and bed material sizes, a threshold of specific stream power separating stable and unstable channels varies around a mean value of 35 Watts per square meter (W/m^2). Nanson and Croke (1992), in a classification of alluvial floodplains, suggested that floodplains of braided streams and rivers have specific stream power values ranging from 50–300 W/m^2 and that lateral migration, scrolled floodplains range from 10–60 W/m^2. Watson et al. (1998) observed a substantial increase in stability as disturbed channels evolve to a specific stream power less than about 35 W/m^2 in severely incised sand bed streams in northern Mississippi.

In a study of 270 streams and rivers, Bledsoe and Watson (2001b) demonstrated that logistic regression models (Menard, 1995) could accurately predict unstable channel forms with a "mobility index" based on slope, median annual flood, and median bed material size. The logistic regression analyses of stable and unstable channel forms suggested that simple indices describing the ratio of erosive energy to boundary material resistance can be robust predictors of channel planform and stability. The logistic models generally predicted the occurrence of unstable sand and gravel channel forms with more than 80% accuracy. In many cases, the predictive accuracy of logistic models utilizing the mobility index as the only independent variable exceeded 95%. A benefit of the logistic regression approach is that explicit probability statements may be attached to diagrams depicting channel stability and proximity to geomorphic thresholds. This provides users with a more useful and realistic assessment of risk when compared to the discrete thresholds of traditional approaches.

Given the ubiquitous degradation of streams occurring in urban areas despite the common use of structural stormwater management practices, there is a pressing need for a more process-based management framework for protecting the geomorphic stability and biotic integrity of streams. Such a framework should guide the evaluation of potential impacts and tailoring mitigation strategies to different stream types and regional contexts. However, it is clearly impossible to fully capture the multiplicity of factors influencing stream responses to urbanization in a mechanistic modeling approach. The following sections briefly outline a risk-based framework for predicting geomorphic and/or biotic responses to urbanization that couples the energy-based approaches described above with more detailed description of hydrologic alteration and boundary conditions influencing stream susceptibility.

PREDICTING STREAM RESPONSE TO URBANIZATION

The proposed framework for risk-based analysis of stream response to urbanization and selection of appropriate mitigation strategies involves four general steps: 1) *a priori* stratification of a region's streams based on geomorphic context and susceptibility to changes in water, sediment, and wood regimes, 2) field monitoring of these strata across a gradient of urban influence, 3) coupling long term hydrologic simulation with geomorphic analysis to quantify key hydrogeomorphic metrics, and 4) using probabilistic modeling to identify links between hydrogeomorphic descriptors of urbanization effects with geomorphic and biotic attributes of broad interest to stakeholders and decision-makers (Figure 10.1).

Because stream geomorphic types have inherent system properties that create variable responses to urbanization, the development of a region-specific, process-based classification of stream susceptibility to changes in water and sediment regimes provides a critical foundation for the other components of the approach. This is particularly true for regions that are relatively heterogeneous in terms of

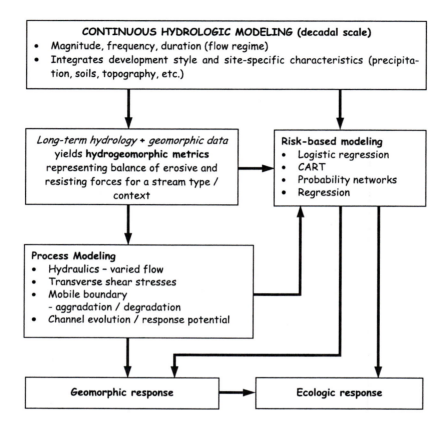

Figure 10.1.   Modeling framework for risk-based assessment of stream response to urbanization.

climate, geology, soils, topography, and geomorphic boundary conditions. Classification based on susceptibility can be accomplished through *a priori* stratification of the landscape into different geologic, valley, and stream segment types (Flores et al., 2006) using a geographic information system (GIS), and then conducting field reconnaissance and synoptic geomorphic surveys across a gradient of urbanization in each mapped geomorphic context. These synoptic surveys can be used to identify sites in each geomorphic context for more intensive field assessment. Sites selected for detailed geomorphic surveys will ideally span a range of channel evolution stages and sequences (Schumm et al., 1984; Simon, 1989; Downs, 1995) and variable degrees of hydromodification from least to heavily disturbed. Detailed geomorphic characteristics as well as development styles, drainage schemes (including connectivity of impervious areas and BMPs), and hydrologic regime should be documented and compared among study sites to the extent practicable.

    Although heterogeneity in history, boundary conditions, and processes will necessitate regionally-calibrated classifications, it is likely that the streams exhibiting

Table 10.1. Stream characteristics associated with risk of instability and loss of physical habitat (modified from Bledsoe (2002))

High-risk characteristics	Lower-risk characteristics
• High specific stream power relative to the most erodible channel boundary	• Low specific stream power relative to the most erodible channel boundary
• Capacity limited – fine bed material, esp. sand	• Supply limited – coarse bed material with potential for armoring
• Little or no grade control (geologic, wood, or artificial)	• Grade control sufficient to check incision (geologic, wood, or artificial)
• Low density of vegetation root volume in banks	• High vegetation root volume density in banks or cohesive / consolidated bank sediments (vegetation tends to override influence of cohesive bank material)
• Non-cohesive, fine grained, sparsely vegetated banks	• Instream form roughness and vegetation roughness on banks
• Entrenched channel – minimal floodplain energy dissipation at $Q > Q_2$, flows $> Q_2$ contained in channel	• Small ratio of woody debris size / channel width
• Near an energy threshold associated with abrupt changes in planform or initiation of incision	• Channel well-connected with rough riparian zone / floodplain that resists chutes cutoffs and avulsions / provides substantial overbank energy dissipation at $Q > Q_{1.5} - Q_2$
• Flashy flows result in pre-wetting / rapid wetting, drying, and drawdown	• Energy level not proximate to geomorphic threshold
• Low roughness – form and vegetative	• Flow regime results in gradual bank wetting and drawdown
• Floodplain susceptible to chutes cutoffs and avulsions	
Steep bank angles	
• Increased woody debris input may destabilize banks and/or enhance vertical stability	

the least geomorphic sensitivity to hydromodification will tend to be bedrock and alluvial threshold channels that have coarse beds with high armoring potential, geologic control, densely vegetated banks and well connected floodplains of high flow resistance. At the opposite end of the spectrum, the most susceptible channels frequently tend to have fine-grained beds, steep valleys, sandy banks composed of noncohesive material that is unprotected from high shear stresses by vegetation (Table 10.1). Several studies suggest that the ratio of stream power per unit channel area relative to the most erodible channel boundary is a robust indicator of channel adjustment potential (MacRae, 1997; Bledsoe, 2002; Grant et al., 2003).

Risk-based channel response analysis

Once the stream types of the region have been assessed in terms of their relative sensitivity to urbanization, potential changes in hydrology and sediment supply, the remaining steps in the proposed framework involve development, calibration, and integration of a suite of probabilistic modeling tools for assessing the anticipated effects of urbanization. The first critical tool is continuous simulation modeling of the hydrologic changes anticipated with different urbanization and land use scenarios. Indeed, there is a growing consensus among experts familiar with the hydrologic effects of urbanization and stormwater controls on stream physical processes that long-term continuous simulations of hydrologic change are essential for adequately assessing the magnitude, frequency, and duration characteristics of post-development flow regimes. This level of description is critical for subsequent geomorphic analyses and predictive assessment because it is the cumulative effect of all sediment transporting events that control geomorphic response.

Risk-based modeling as envisioned here is based on integrating hydrologic and geomorphic (hydrogeomorphic) data derived from the output of continuous hydrologic simulation models to generate metrics describing expected departures in fundamental geomorphic processes such as the cumulative distribution of specific stream power and sediment transport capacity (sometimes termed "erosion potential") across the entire range of relevant flows (Bledsoe et al., in press; Rohrer and Roesner, this volume). These physical metrics are provided as inputs to probabilistic models that estimate the risk of streams shifting to some undesirable state. Because the decision endpoint is often categorical (e.g. stable, good habitat, supporting aquatic life uses) the statistical tools of choice are often logistic regression, classification and regression trees (CART), and/or Bayesian probability networks. Figure 10.2 illustrates how logistic regression analysis can be used to estimate the likelihood of channel instability based on progressive degrees of erosion potential. This approach was recently used in the development of the Santa Clara Valley Urban Runoff Program Hydromodification Management Plan (www.SCVURPPP.org). This study demonstrated that a time-integrated index of erosion potential based on continuous hydrologic simulation and an assessment of stream power relative to the erodibility of channel boundary materials could be used to accurately predict which channels of a particular regional type are degraded by hydromodification in arid urban watersheds of southern California.

As suggested above, the variables included in risk-based models of stream response are not limited to the simple energy-based descriptors examined in previous research. Instead, additional multi-scale controls can be included. For example, simple categories of physical habitat condition and ecological integrity can be modeled by augmenting erosion potential metrics with descriptors of the condition of channel banks and riparian zones, geologic influences, floodplain connectedness, valley entrenchment, hydrologic metrics describing flashiness, proximity to known thresholds of planform change, and land use descriptors such as % connected imperviousness and BMP types.

The resulting probabilistic models can be used to conduct risk-based scientific assessments that account for geomorphic context and processes as opposed to simply defining "one-size-fits-all" threshold limits on imperviousness or some other surrogate. Risk-based modeling estimates the probability of stream states that are of interest to stakeholders. Decision-makers can then determine acceptable

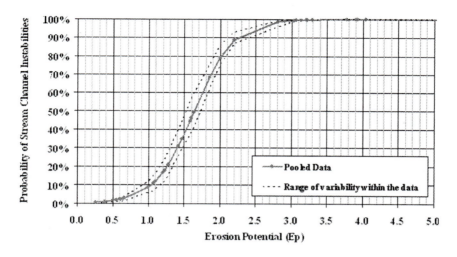

Figure 10.2. Logistic regression analysis showing the probability of channel instability given predicted changes in erosion potential (increase in time-integrated sediment transport capacity relative to pre-development condition).

risk levels based on an explicit estimate of prediction error. This type of risk-based approach is consistent with recommendations of Reckhow (1999a,b) and the National Research Council panel on TMDLs (NRC, 2001) that 1) the focus of scientific study in support of decision making should ultimately be on the decisions (or objectives) associated with the resource and not on the model or basic science, and 2) prediction error, not perception of mechanistic correctness, should be the most important criterion reflecting the usefulness of a model. The predictive models suggested here should be thought of as predictive scientific assessments, that is, a flexible, changeable mix of small mechanistic models, statistical analyses, and expert scientific judgment. A predictive scientific assessment should be evaluated in terms of its utility in addressing decisions and objectives of primary concern to stakeholders. Predictive model selection criteria must include factors such as prediction uncertainty, cost of calibration and testing, meaningful endpoints, appropriate spatial and temporal detail, and simplicity in application and understanding (NRC, 2001; Reckhow, 1999a,b).

The approach described above can be readily extended to the prediction of biological states in urban streams. Although the large number of potentially confounding influences makes prediction of biological responses to urbanization very challenging, the framework suggested here has the potential to provide a more rational and transparent basis for prediction and decision making by explicitly recognizing uncertainty in both the reasoning about stream response and the quality of information used to drive the models. Some critical limitations in our understanding of biotic responses to urbanization that currently inhibit such an approach are addressed in the following section.

RECOMMENDATIONS AND FUTURE RESEARCH

The framework proposed above is based on using metrics describing process linkages in regionally-calibrated, probabilistic models of stream responses to various
styles of urbanization. Successful implementation of this approach will undoubtedly necessitate communally developing strong conceptual models of the processes controlling stream response in urbanizing watersheds of different regions.
Moreover, conceptual models developed for predicting and mitigating the effects
of urbanization may also prove useful in the inverse problem of stream restoration. After identifying classes of urban streams that provide desirable benefits and
amenities despite altered regimes of water, sediment, wood, nutrients, and other
materials, a similar risk-based approach could be used to determine which physical
metrics best predict membership in the classes and thereby define potential restoration strategies. As high-resolution geospatial data become more widely available
(e.g., LIDAR), it may become increasingly feasible to map both putative stream
responses and restoration potential at the basin or regional scale by employing a
risk-based assessment approach similar to that described above.

Effective implementation of risk-based assessments will also depend on con-
tinual reevaluation of models through targeted monitoring and research. Much of
the existing body of research consists of one-time studies, which are only a static
snapshot of ongoing processes (Roesner and Bledsoe, 2002). To be effective and
defensible, strategies for protection and rehabilitation of streams impacted by urbanization must be underpinned with an understanding of fundamental geomorphic
processes. First and foremost, this necessitates comprehensive, long-term monitoring augmented with mathematical modeling of the linkages between development
style/drainage scheme, flow regime, and multi-scale changes in physical habitat
and biotic condition. Improved diagnosis and predictive understanding of future
change will require multifaceted, multiscale, and multidisciplinary studies based
on a firm understanding of the history and processes operating in a drainage basin
(Jacobson et al., 2001). Such multidisciplinary studies will necessarily combine
historical, associative, process-scale, and modeling approaches from hydrology,
geomorphology, sediment transport, water quality, and aquatic ecology.

Several studies over the last decade have underscored the importance of watershed land use and vegetative cover as well as valley context on local habitat and
biological condition. Despite this recognition, watershed managers currently lack a
conceptual framework for predicting the impact of large-scale watershed modifications and urbanization on ecological processes that influence stream communities.
Geomorphic disturbances resulting from urban land use have the capacity to alter ecological processes that operate over large spatial scales, while most studies
have been conducted at smaller spatial scales (Mathews and Heins, 1987; Roesner
and Bledsoe, 2002). A lack of understanding at this scale limits our ability to design effective restoration efforts in response to large-scale disturbances (Schlosser,
1995).

Although this article suggests that this general framework may be extended
to developing predictive models of biological response in urban watersheds, the

knowledge base for biota/habitat associations is not generally adequate to allow for prediction of how whole communities will change in response to environmental alterations associated with urbanization. Making such predictions requires a thorough knowledge of species-specific environmental responses, as well as an adequate (accurate) characterization of habitat structure and habitat dynamics (both of which are modified by urbanization). Viewing species in terms of their response potential to environmental factors is a common method in stream ecology, as seen for example in the use of functional feeding groups (Cummins, 1973) or pollution tolerance scores (Hilsenhoff, 1987; Lenat, 1993), which allow some expectation of how species may respond along environmental gradients of food resources or human pollution, respectively. However, it is difficult to predict biotic responses to environmental alteration involving multiple stressors. For example, species vary in their abilities to tolerate natural high rates of disturbance through colonization ability, resistance to flow disturbance, life cycle adaptation, fecundity, and other traits; however, characterizing species according to multiple traits that may be important in predicting local community structure under a specified environmental regime is generally lacking (Poff, 1997). Attempts to characterize community composition in terms of traits that are sensitive to multiple environmental factors (including disturbance) have shown some success both for fish (e.g., Poff and Allan, 1995) and for invertebrates (e.g., Richards et al., 1997). But, to date, a comprehensive, multi-trait characterization for invertebrates and fish that would incorporate sensitivity to disturbance, habitat conditions, and water quality has not been fully developed.

Finally, there is a need for better understanding of local biotic response in a landscape context. Aquatic organisms are highly mobile and generally excellent dispersers; therefore, it is not uncommon to find species in habitats not predicted from models (e.g., Poff, 1997). Further, this high mobility promotes rapid recolonization following disturbance, a phenomenon long appreciated in stream ecology literature (e.g., Larimore et al., 1959). Thus, the recovery potential for any particular stream segment that experiences a disturbance will be a function of the dispersal ability of the fauna and the availability of refugia from the local disturbance. Stream ecologists do not have a general framework for assessing this "recovery" potential, as the identification and quantification of refugia, and the scales at which they occur, remains a largely unsolved problem (e.g., Lancaster and Belyea, 1997; Lancaster, 2000). Further, the diversity of seemingly similar locations may be very different depending on neighboring habitat types. For example, small tributary streams to large rivers may have "inflated" species richness, because river fishes opportunistically move into them. Although there is growing appreciation of the importance of tributaries and other "anomalies" in influencing local diversity (Osborne and Wiley, 1992; Rice et al., 2001), there is no general theory in stream ecology that quantitatively incorporates the network structure of a drainage in a way that allows for formulation of expected diversity (or other) deviations from average conditions (Benda et al., 1998; Fisher, 1997). Thus, another heading of research needs would involve region-scale "neighboring" habitat issues.

ACKNOWLEDGEMENTS

I gratefully acknowledge support provided by the US Environmental Protection Agency and Water Environment Research Foundation. I thank Larry Roesner, Christine Rohrer, Derek Booth, Gary Palhegyi, and LeRoy Poff for many stimulating conversations about urban streams.

REFERENCES

Anderson, R.J., Bledsoe, B.P., and Hession W.C. (2004). Stream and river width response to vegetation, bank material, and other factors. Journal of the American Water Resources Association 40: 1159–1172.

Benda, L.E., Miller, D.J., Dunne, T., Reeves, G.H., and Agee, J.K., (1998). Dynamic landscape systems. In: Naiman, R.J. and Bilby, R.E. (Eds.), River Ecology and Management: Lessons from the Pacific Coastal Ecoregion: Springer, New York, New York, 261–288.

Bledsoe, B.P., Brown, M.C., and Raff, D.A. (in press). GeoTools: A toolkit for fluvial system analysis. Accepted by Journal of the American Water Resources Association.

Bledsoe, B.P. and Watson, C.C. (2001a). Effects of urbanization on channel instability. Journal of the American Water Resources Association 37(2):255–270.

Bledsoe, B.P. and Watson, C.C. (2001b). Logistic analysis of channel pattern thresholds: Meandering, braiding, and incising. Geomorphology 38: 281–300.

Bledsoe, B.P. (2002). Stream responses to hydrologic changes. In: Urbonas B R (Ed.), Linking Stormwater BMP Designs and Performance to Receiving Water Impacts Mitigation, American Society of Civil Engineers 127–144.

Booth, D.B. (1990). Stream-channel incision following drainage-basin urbanization. Water Resources Bulletin 26(3):407–417.

Booth, D.B. (2005). Challenges and prospects for restoring urban streams, a perspective from the Pacific Northwest of North America. Journal of the North American Benthological Society 24: 724–737.

Booth, D.B. and Henshaw, P.C. (2001). Rates of channel erosion in small urban streams. In: Wigmosta M S (Ed.), Influence of Urban and Forest Land Use on the Hydrologic-geomorphic Responses of Watersheds. AGU Monograph Series, Water Science and Applications 2: 17–38.

Breiman. L., Freidman, J.H., Olshen, R.A., and Stone, C.J. (1984). Classification and Regression Trees. Chapman and Hall/CRC, New York, New York, 358 pp.

Brice, J.C. (1975). Air Photo Interpretation of the Form and Behavior of Alluvial Rivers. Final report to the US Army Research Office.

Brookes, A. (1987a). River channel adjustments downstream from channelization works in England and Wales. Earth Surface Processes and Landforms 12: 337–351.

Brookes, A. (1987b). The distribution and management of channelized streams in Denmark. Regulated Rivers: Research & Management 1: 3–16.

Chin, A. and Gregory, K.J. (2001). Urbanization and adjustment of ephemeral stream channels. Annals of the Association of American Geographers 91(4):595–608.

Cummins, K.W. (1973). Trophic relations of aquatic insects. Annual Review of Entomology 18: 183–206.

Downs, P. (1995). Estimating the probability of river channel adjustment. Earth Surface Processes and Landforms 20: 687–705.

Downs, P.W. and Gregory, K.J. (1995). Approaches to river channel sensitivity. Professional Geographer 47(2):168–175.

Dunaway, D., Swanson, S.R., Wendel, J., and Clary, W. (1994). The effect of herbaceous plant communities and soil textures on particle erosion of alluvial streambanks. Geomorphology 9: 47–56.

Emerson, C.H., Welty, C., and Traver, R.G. (2005). Watershed-scale evaluation of a system of storm water detention basins. Journal of Hydrologic Engineering 10(3):237–242.

Fisher, S.G. (1997). Creativity, idea generation, and the functional morphology of streams. Journal of the North American Benthologic Society 16: 305–318.

Flores, A.N., Bledsoe, B.P., Cuhaciyan, C.O., and Wohl, E.E. (2006). Channel-reach morphology dependence on energy, scale, and hydroclimatic processes with implications for prediction using geospatial data. Water Resources Research 42: W06412 doi:10.1029/2005WR004226.

Graf, W.L. (1982). Spatial variation of fluvial processes in semiarid lands. In: Rhoads B L and Thorn C E (Eds.), The Scientific Nature of Geomorphology, Wiley, Chichester.

Grant, G.E., Schmidt, J.C., and Lewis, S.L. (2003). A geological framework for interpreting downstream effects of dams on rivers. In: A Unique River, Water Science and Application 7, American Geophysical Union, 209–225.

Haschenburger, J.K., and Wilcock, P.R. (2003). Partial transport in a natural gravel bed channel. Water Resources Research 38(1):1020 doi:10.1029/2002WR001532.

Hilsenhoff, W.L. (1987). An improved biotic index of organic stream pollution. Great Lakes Entomology 20, 31–39.

Jackson, W.L., and Beschta, R.L. (1984). Influences of increased sand delivery on the morphology of sand and gravel channels. Water Resources Bulletin 20(4):527.

Jacobson, R.B., Femmer, S.R., and McKenney, R.A. (2001). Land-use Changes and the Physical Habitat of Streams – A Review with Emphasis on Studies Within the U S Geological Survey Federal-State Cooperative Program. U S Geological Survey, Circular 1175, 63.

Knox, J.C. (1977). Human impacts on Wisconsin stream channels. Annals of the Association of American Geographers 67, 323–342.

Konrad, C.P., and D.B. Booth (2005). Hydrologic changes in urban streams and their ecological significance. American Fisheries Society Symposium 47, 157–177.

Konrad, C.P., Booth, D.B., and Burges, S.J. (2005). Effects of urban development in the Puget Lowland, Washington, on interannual streamflow patterns: Consequences for channel form and streambed disturbance. Water Resources Research 41(WO7009) dx.doi.org/10.1029/2005WR004097.

Lancaster, J. (2000). Geometric scaling of microhabitat patches and their efficacy as refugia during disturbance. Journal of Animal Ecology 69, 442–457.

Lancaster, J., and Belyea, L.R. (1997). Nested hierarchies and scale-dependence of mechanisms of flow refugium use. Journal of North American Benthological Society 16, 221–238.

Larimore, R.W., Childers, W.F., and Heckrotte C. (1959). Destruction and reestablishment of stream fish and invertebrates affected by drought. Transaction of the American Fisheries Society 88, 261–285.

Lenat, D.R. (1993). A biotic index for the southeastern United States – derivation and list of tolerance values with criteria for assigning water quality ratings. Journal of North American Benthological Society 12, 279–290.

MacRae, C.R. (1997). Experience from morphological research on Canadian streams: Is the control of the two-year frequency runoff event the best basis for stream channel protection? In: Roesner, L.A. (Ed.), Effects of Watershed Development and Management of Aquatic Ecosystems, ASCE, New York, New York, 144–162.

Mathews, W.J., and Heins, D.C. (1987). Community and Evolutionary Ecology of North American Stream Fishes. University of Oklahoma Press, Norman Oklahoma.

Montgomery, D.R. and Buffington, J.M. (1998). Channel processes, classification, and response. In: Naiman R and Bilby R (Eds.), River Ecology and Management, Springer-Verlag New York Inc., New York, New York, 13–42.

Montgomery, D.R. and MacDonald, L.H. (2002). Diagnostic approach to stream channel assessment and monitoring. Journal of the American Water Resources Association 38, 1–16.

Menard, S.W. (1995). Applied Logistic Regression Analysis. Sage Publications, Thousand Oaks, California, USA.

Meyer, J.L., Paul, M.J., and Taulbee, W.K. (2005). Stream ecosystem function in urbanizing landscapes. Journal of North American Benthological Society 24, 602–612.

Montgomery, D.R., and Buffington, J.M. (1998). Channel processes, classification, and response. In: Naiman, R. and Bilby, R. (Eds.), River Ecology and Management, Springer-Verlag New York Inc., New York, New York, 13–42.

Montgomery, D.R. and MacDonald, L.H. (2002). Diagnostic approach to stream channel assessment and monitoring. Journal of the American Water Resources Association 38, 1–16.

Nanson, G.C., and Croke, J.C. (1992). A genetic classification of floodplains. Geomorphology 4, 459–86.

National Research Council (2001). Assessing the TMDL Approach to Water Quality Management. National Academies Press, Washington, DC.

Osborne, L.L., and Wiley, M.J. (1992). Influence of tributary spatial position on the structure of warmwater fish communities. Canadian Journal of Fish and Aquatic Sciences 49, 671–681.

Parker, G. (1990). Surface-based bedload transport relation for gravel rivers. Journal of Hydraulic Research 28, 417–436.

Poff, N.L. (1997). Landscape filters and species traits: Towards mechanistic understanding and prediction in stream ecology. Journal of the North American Benthological Society 16(2):391–409.

Poff, N.L., and Allan, J.D. (1995). Functional organization of stream fish assemblages in relation to hydrological variability. Ecology 76(2):606–627.

Poff, N.L., Bledsoe, B.P., and Cuhaciyan, C.O. (2006). Hydrologic variation with land use across the contiguous United States: Geomorphic and ecological consequences for stream ecosystems. Geomorphology 79, 264–285.

Reckhow, K.H. (1999a). Lessons from risk assessment. Human and Ecological Risk Assessment 5, 245–253.

Reckhow, K.H. (1999b). NPSINFO Listserve comments by Kenneth Reckhow, Professor of Civil and Environmental Engineering, Nicholas School of the Environment, Duke University, Durham, North Carolina.

Rhoads, B.L. (1986). Flood hazard assessment for land-use planning near desert mountains. Environmental Management 10, 97–106.

Rhoads, B.L. (1988). Mutual adjustments between process and form in a desert mountain fluvial system. Annals of the Association of American Geographers 78, 271–287.

Rhoads, B.L. (1995). Stream power: A unifying theme for urban fluvial geomorphology. In: Herricks E E (Ed.), Stormwater Runoff and Receiving Systems: Impact, Monitoring, and Assessment, Chapter 5, CRC Press, Lewis Publishers, Boca Raton, Florida, 65–75.

Rice, S.P., Greenwood, M.T., and C.B. Joyce (2001). Tributaries, sediment sources, and the longitudinal organisation of macroinvertebrate fauna along river systems. Canadian Journal of Fish and Aquatic Sciences 58, 824–840.

Richards, C., Haro, R.J., Johnson, L.B., and Host, G.E. (1997). Catchment and reach-scale properties as indicators of macroinvertebrate species traits. Freshwater Biology 37, 219–230.

Richards, K.S., and Lane, S.N. (1997). Prediction of morphological changes in unstable channels. In: Thorne C R, Hey R D and Newsom M D (Eds.), Applied Fluvial Geomorphology for River Engineering and Management, Chapter 10, John Wiley and Sons Ltd., 269–292.

Roesner, L.A., and Bledsoe, B.P. (2002). Physical Effects of Wet Weather Flows on Aquatic Habitats – Present Knowledge and Research Needs. Final Report to Water Environment Research Foundation, WERF Project Number 00-WSM-4, 250 pp.

Rohrer, C.A., Roesner, L.A., and Bledsoe, B.P. (2004). The effect of stormwater controls on sediment transport in urban streams. In: Sehlke G, Hayes D F and Stevens D K (Eds.),

Critical Transitions in Water and Environmental Resources Management, American Society of Civil Engineers, World Water and Environmental Resources Conference – 2004, Salt Lake City, Utah, 1–13.

Schlosser, I.J. (1995). Critical landscape attributes that influence fish population dynamics in headwater streams. Hydrobiologia 303, 71–81.

Schumm, S.A. (1969). River metamorphosis. ASCE Journal of the Hydraulics Division 95(HY1):255–273.

Schumm, S.A. (1991). To Interpret the Earth, Ten Ways to be Wrong. Cambridge University Press, Great Britain.

Schumm, S.A., Harvey, M.D., and Watson, C.C. (1984). Incised Channels: Morphology, Dynamics, and Control. Water Resources Publications, Littleton, Colorado.

Schumm, S.A., and Lichty, R.W. (1965). Time, space, and causality in geomorphology. American Journal of Science 263, 110–119.

Simon, A. (1989). A model of channel response in disturbed alluvial channels. Earth Surface Processes and Landforms 14, 11–26.

Simon, A., and Collison, A.J.C. (2002). Quantifying the mechanical and hydrologic effects of riparian vegetation on streambank stability. Earth Surface Processes and Landforms 27, 527–546.

Simon, A., and Downs. P.W. (1995). An interdisciplinary approach to evaluation of potential instability in alluvial channels. Geomorphology 12, 215–232.

Simons, D.B., and Richardson, E.V. (1966). Resistance to Flow in Alluvial Channels. U S Geological Survey Professional Paper 422J.

Thorne, C.R. (1990). Effects of vegetation on riverbank erosion and stability. In: Thornes J B (Ed.), Vegetation and Erosion, John Wiley and Sons, New York, New York.

Trimble, S.W. (1974). Man-induced Soil Erosion on the Southern Piedmont, 1700–1970. Soil Conservation Society of America, Ankeny, Iowa.

Trimble, S.W. (1983). A sediment budget for Coon Creek basin in the Driftless Area, Wisconsin, 1853–1977. American Journal of Science 283, 454–474.

Trimble, S.W. (1995). Catchment sediment budgets and change. In: Gurnell A and Petts G (Eds.), Changing River Channels, Chapter 9, John Wiley and Sons Ltd., 201–215.

Trimble, S.W. (1997). Contribution of stream channel erosion to sediment yield from an urbanizing watershed. Science Magazine 278, 1442–1444.

Tung, Y. (1985). Channel scouring potential using logistic analysis. Journal of Hydraulic Engineering 111(2):194–205.

Urban, M.A., and Rhoads, B.L. (2003). Catastrophic human-induced change in stream-channel planform and geometry in an agricultural watershed, Illinois, USA. Annals of the Association of American Geographers 93, 783–796.

Waters, T.F. (1995). Sediment in streams – sources, biological effects, and control. American Fisheries Society Monograph 7, 251.

Watson, C.C., Thornton, C.I., Kozinski, P.B., and Bledsoe, B.P. (1998). Demonstration Erosion Control Monitoring Sites 1997 Evaluation. Submitted to the U S Army Corps of Engineers Waterways Experiment Station, Vicksburg, Mississippi, 129 pp.

Wilcock, P.R. (1997). Friction between science and practice: The case of river restoration. American Geophysical Union: EOS, Transactions 78(41):454.

Wilcock, P.R. (1998). Two-fraction model of initial sediment motion in gravel-bed rivers. Science 280, 410–412.

Walsh, C.J., Roy, A.H., Feminella, J.W., Cottingham, P.D., Groffman, P.M., and Morgan, R.P. (2005). The urban stream syndrome: Current knowledge and the search for a cure. Journal of the North American Benthological Society 24, 706–723.

Wilcock, P.R., and Kenworthy, S.T. (2002). A two-fraction model for the transport of sand/gravel mixtures. Water Resources Research 38 (10):1194, doi:10.1029/2001WR000684.

Williams, G.P., and M.G. Wolman (1984). Downstream Effects of Dams on Alluvial Rivers. U. S. Geological Survey Professional Paper 1286.

Wolman, M.G., and Gerson, R. (1978). Relative scales of time and effectiveness of climate in watershed geomorphology. Earth Surface Processes 3, 189–208.

11

Urban diffuse pollution and solutions in Japan

K. Yamada and V. S. Muhandiki

Ritsumeikan University, 1-1-1 Noji-Higashi, Kusatsu, Shiga, 525-8577 Japan
E-mail: yamada-k@se.ritsumei.ac.jp

Summary: Rapid urbanization and industrialization progressed in many parts of Japan over the past 50 years. This trend resulted in serious water pollution in many urban areas. To address the pollution problems, several measures such as construction of sewerage systems and enactment of laws and regulations have been undertaken. Because of these measures, the water quality has been improved to some extent. However, many water quality concerns still remain because the measures have been effective in controlling mainly point source pollution but not diffuse pollution. Thus, diffuse pollution remains a major problem in urban areas. In this chapter historical progress, present status and future aspects of diffuse pollution problems in urban areas in Japan are reviewed. It is noted that solving diffuse pollution problem will require combination of several measures including application of conventional treatment technologies and natural purification systems, implementation of appropriate policy measures and promotion of citizen participation.

INTRODUCTION

This chapter reviews the historical progress, current situation and future aspects of diffuse pollution problems in urban areas (hereafter referred to as UDP) in Japan. Particular emphasis is placed on Lake Biwa, the largest lake in Japan that provides drinking water for 14 million people. UDP in Japan has developed under unique conditions. Most of the land area in Japan (about 70%) is covered with forested mountains, the remainder being used for agriculture and urban development. Therefore, the population density in urban areas is very high. Furthermore, because industrial and commercial areas have expanded rapidly during the last 50 years, most cities and suburban areas are now faced with serious water pollution problems. To address these problems, a large number of sewerage systems were constructed mainly in urban areas. As a result, water quality has improved to some degree. However, in many areas water quality criteria have not been met because of UDP. The next sections describe the situation of UDPs in Japan and in the Lake Biwa Basin.

Urban diffuse pollution problems in japan

Sewerage systems in Japan were installed after urban areas including roads and housing had developed. Consequently, the cost of installation was very high. There were few sewerage systems in Japan 50 years ago. These systems were combined and existed mainly in central areas of large cities. In most urban areas there were no sewerage systems. Only night soil was collected and treated, while most other wastewaters were discharged into natural water systems without treatment. Because of industrial development and improvement in the standard of living, pollution loads to receiving water bodies increased. As a result the value of water bodies in urban areas as sources of drinking water decreased and alternative distant sources had to be sought. Therefore, it may be said that installation of sewerage systems in Japan lagged behind building other infrastructure because of the practice of using night soil as a fertilizer, and using existing small and medium-size rivers for drainage, among other reasons.

The government took a leading role in promoting construction of sewerage systems by providing funds. In Japan, sewerage systems are designed to cover river basins. Initially combined systems were constructed but later the practice shifted to building separate ones. Because of intensive use of land in urban areas, the area representing pavements and roofs increased, which resulted in decreasing open spaces and green tracts. Also, with construction of sewerage systems, rivers and canals that had earlier been used for drainage were filled up and converted into roads and construction sites (see also a chapter by Furumai in this book). These developments resulted in the following problems:

1) Increased rates of runoff from urban areas and increased flood risks,
2) Increased pollutant loads discharged from urban areas and increased combined sewer overflows (CSOs),
3) Even in separate systems, discharge of water with high pollutant concentration during rainfall increased, and
4) In semi-urban areas, water pollution increased due to poor management of agricultural activities.

Situation of the Lake Biwa Basin

The discussion of Japanese UDPs in this chapter draws heavily on the experience from the Lake Biwa Basin. Lake Biwa is the largest lake in Japan and is located

Figure 11.1.   Lake Biwa and Yodo River Basin (ILEC, 2005).

in Shiga Prefecture (Figure 11.1). It is an important water resource for 14 million people in the Kansai Metropolitan Region. In addition to water supply, the lake has multiple uses such as fishery, flood control, hydro-electric power generation and recreation. The population in the Lake Basin has grown over the past years because of the growth of industries in the basin and the existence of a convenient transportation network that connects the area to the neighbouring cities of Kyoto, Osaka and Kobe. The lake, believed to be more than 4 million years old, has become tremendously polluted over the past 50 years. Despite of implementation of pollution control measures, such as construction of sewerage systems and enforcement of effluent discharge standards, the lake water quality has not dramatically improved and still does not meet environmental standards (Shiga Prefectural Government, 2000, 2004). Figure 11.2 shows the trends in water quality of the lake, the environmental standards and water quality targets based on the Lake Law. As shown in Figure 11.2, even the North Lake that contains most of the lake water and receives relatively smaller pollution loads shows degradation of water quality.

Figure 11.3 shows estimated values of pollutant loads discharged to Lake Biwa from its basin. The steady decrease in pollutant load shown in Figure 11.3 contradicts the phenomenon shown in Figure 11.2. It is believed that this disagreement is caused by inaccurate estimation of pollutant loads for the following reasons:

1. Because it is not possible to directly measure the pollutant load associated with rainfall, its estimation is based on simulation of relational equations developed using few samples,

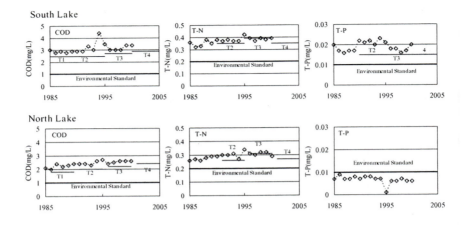

Figure 11.2.   Trends of water quality and targets (T1-T4) in Lake Biwa.

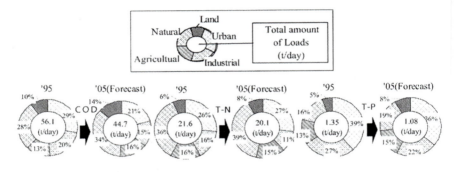

Figure 11.3.   Distribution of pollutant loads to Lake Biwa (Shiga Prefectural Government, 2004).

2. Inputs from industrial, agricultural, and forested areas, and groundwater are not accurately measured,
3. Continuous increases in recurrent and persistent pollutants, and
4. Decrease in in-lake self-purification capacity due to disappearance of wetlands and lakeshore vegetation.

Even though the above issues have not yet been clarified, they relate strongly to UDP.

NEED TO PROMOTE COUNTERMEASURES FOR UDPs

Mechanism of water pollution

Figure 11.4 shows the process of movement of pollutants from their source to Lake Biwa. The chemical constituents targeted for control include chemical oxygen

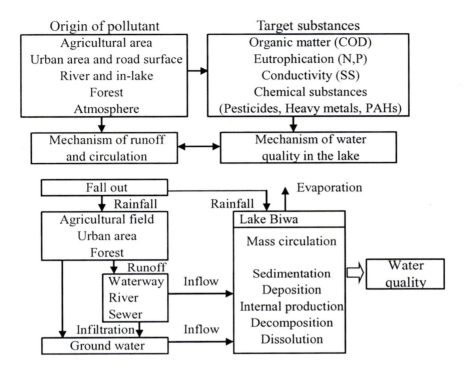

Figure 11.4.   Mechanism of diffuse pollution.

demand (COD), nitrogen (N) and phosphorus (P), suspended solids (SS), agricultural chemicals, heavy metals and polycyclic aromatic hydrocarbons (PAHs). It is difficult to implement control measures, because of a large number of sources, large quantity of water to be treated, difficulty to quantitatively evaluate the treatment of water with low concentrations of pollutants, and unclear mechanisms of generation of runoff and its pollution, among other reasons.

National regulations

The Lake Law has been recently revised by the national government. The revised law focuses on pollutant discharge including the activities related to UDP and also specifies the areas where diffuse pollution reduction should be implemented. This demonstrates the government's increased emphasis on diffuse pollution problems. Another important revision of national laws was the 1997 amendment of the River Law to incorporate environmental conservation. Prior to the amendment the objectives of river basin management were set primarily as water resources development and flood control. In the amended law, environmental conservation is included as an equally important objective in addition to water resources development and flood control. Additionally, the amended law requires that basin committees are established in all river basins and such committees incorporate opinions of local residents in the implementation of infrastructure development projects. With reference to this background, the need to promote pollution control measures for UDP is apparent.

Situation in the Lake Biwa Basin

Although Shiga Prefecture has put emphasis on remediation of diffuse pollution, the water quality of Lake Biwa has not improved as was expected (Figure 11.2), which makes the need for countermeasures even stronger. Moreover, further growth of population and industrial activities in the Lake Biwa Basin is forecasted. Such changes are expected to produce increased diffuse pollution loads and changes in the load quality.

EXISTING REMEDIAL MEASURES
AND THEIR EVALUATION
Existing remedial measures

Existing remedial measures may be classified into five categories according to the source of pollution, namely those for agricultural areas, urban areas and road surfaces, rivers and in-lake sources, forests and atmosphere. Among these categories urban areas and road surfaces are strongly related with UDP, while river and in-lake sources are slightly related to UDP. The existing measures for these categories are discussed below.

Agricultural areas

Agricultural areas in Japan are mostly covered with rice fields. A lot of water is withdrawn from rivers for rice field irrigation and the used water is discharged downstream causing a serious pollution of water bodies. Farmers are encouraged to use water-saving measures such as cascade irrigation and reduced application amounts of fertilizers and pesticides. Incentives such as direct payment to farmers who undertake good agricultural practices are also employed.

Urban areas and road surfaces

Many pollutants originate from urban areas, especially from paved surfaces such as roads. To remove these pollutants control measures such as sweeping, infiltration on road surfaces, solids deposition and runoff infiltration in gullies, and solids deposition and runoff purification at the end of the pipe are used. Solid pollutants are removed from major roads by sweeping at regular intervals, such as weekly or monthly. However, this removes mainly coarse particles, while fine grain particles are not removed. To reduce runoff, the use of porous pavements that allow runoff infiltration, has been encouraged recently. Such pavements also reduce road noise, but they are relatively expensive. To reduce peak flows and pollutant loads, porous gullies were found effective. Recently a device called FF cleaner was developed and installed in some gullies to intercept the first flush. Gullies installed with this device were more effective in controlling peak flows and loads. Pilot plants for treatment of runoff from road surfaces and full-scale plants for treatment of runoff from urban areas have been installed in the Lake Biwa Basin (as described in the next section).

Forests

The demand for domestic lumber has been declining in Japan over the past 50 years due to importation of lumber from other countries. Thus, large scale deforestation has not been a major problem in Japan in recent years. Appropriate management measures for forests have been proposed and regulations for forest restoration established.

Rivers and in-lake measures

Removal of bottom sediment has been carried out in Lake Biwa at different sites every year. It is very difficult to estimate the effect of the removal of sediments in the short term, but beneficial effects over a long term are expected. On the other hand, some purification facilities, combining a storage tank and vegetation, have been installed at the river mouth or at the sewer pipe outfalls. Two such facilities, namely Moriyama River Purification Facility and Oba River Bio Park, are located in the South Basin. Both facilities are operational and their evaluation is being carried out (as discussed in the next section). Regarding UDP, some projects for conservation and restoration of reed communities are being carried out along the lakeshore and in wetlands (also discussed in the next section). In addition, harvesting of water grass is also being carried out. However, since concentrations of pollutants in lake and river water are relatively low, it is difficult to evaluate the effect of such measures quantitatively, even though it is known that these measures produce some positive effect.

Atmosphere

Atmospheric sources contribute significantly to water pollution, especially in urban areas. Air pollution was a major problem in Japan in many cities during the period of rapid economic growth, reaching a peak in the 1960s. However, starting from the 1970s the national government, local governments and industries undertook various measures that have led to a steady improvement of air quality in Japan. In fact, air pollution control in Japan is often cited as one of the success stories (Fujikura, 2005). The measures undertaken include establishment and enforcement of environmental regulations, investment in pollution control technology, and financial incentives for pollution control such as soft loan programs.

Examples of studies of remedial measures employing purification systems

Example-1: Pilot study of a soil purification facility (filter) for road runoff

Demonstration experiments using a soil purification facility (filter) for treatment of road runoff were carried out during a period of 3 years between September 2001 and November 2004 on a section of a road located on the shore of Lake Biwa in Kusatsu City (Yamada et al., 2004, 2006). The outline and details of the facility are shown in Figure 11.5. The facility consisted of two parallel soil tanks; Tank A and Tank B, each filled with soil to a depth of 45 cm. Tank A consisted of two types of soil, 5 cm deep Akadama soil in the top layer and 40 cm deep pit sand in the bottom layer. Tank B consisted of only pit sand. To make these Best Management Practices (BMPs) less expensive and simple, soil used in the filter facility is a common natural sand found on hill slopes. The ratio of the surface areas of the soil filter and the road surface was 1:15. The facility did not have any pre-treatment equipment, such as a sedimentation tank. Runoff from the road surface (750 m^2) flows directly into the facility and filters through the soil, but during heavy storms part of the runoff overflows through a weir installed on the surface of the sand. Flow meters were installed at appropriate points to measure inflow, outflow and overflow. Water and soil were sampled and their quality analyzed for selected storm events. The facility was continuously operated over the study period of 3 years.

The removal of pollutant loads in the facility was investigated by examining 14 selected rainfall events. Removal efficiency was evaluated in terms of Gross Removal Efficiency (GRE) and Net Removal Efficiency (NRE) defined as follows: GRE = (Pollutant load retained in the soil/Pollutant load in the influent),

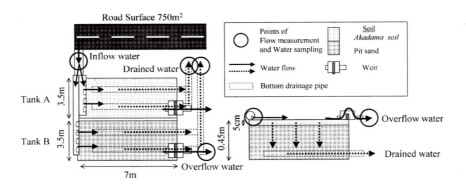

Figure 11.5.   Soil filter facility.

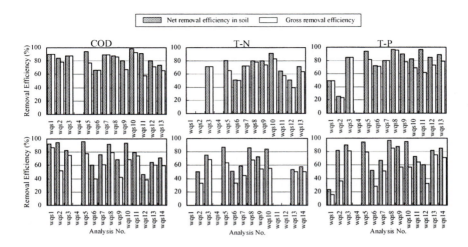

Figure 11.6.   Pollutant removal efficiency in Tank A (upper graphs) and Tank B (lower graphs).

NRE = [Pollutant load retained in the soil/(Pollutant load in Influent − Pollutant load in overflow)]. Results of GRE and NRE for the surveyed rainfall events are shown in Figure 11.6. Except for the first three initial observations, the average GRE in Tank A was about 75% for TCOD and TP, 65% for TN and 90% for SS. In tank B, though the ratio of surface overflow water to inflow water was comparatively high, the corresponding GRE values were about 60% for TCOD, 60% for TP, 50% for TN and 85% for SS. On the other hand, NRE was higher than GRE by about 10% in Tank A and 5–20% in tank B. Through simulation of a long-term mass balance of flow and pollutant load in the facility, it was found that the facility would remain in good condition in terms of infiltration capacity and pollutant removal efficiency over the duration of the experiments. It was also found that most of the pollutants removed accumulated in the top 5 cm layer of soil, thus the polluted soil could be easily replaced.

It was noted that generally much space is available on roadsides that could be potential sites for treatment of road runoff. These include both public and private spaces like parking lots and garden yards in housing areas and fallow fields in agricultural areas. Application of simple systems like the soil filter facility in such spaces to treat diffuse pollution is an important solution to diffuse pollution problem.

Example-2: Evaluation of a treatment facility for stormwater from an urban area

As one of the measures addressing the water quality concerns of the lake, Shiga Prefecture has constructed a multiple process treatment facility for urban stormwater

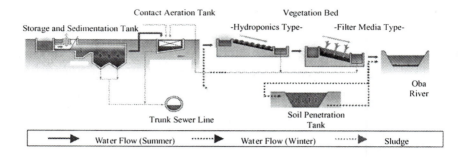

Figure 11.7.   Outline of the treatment facility.

runoff in Kusatsu City. The facility, which started operation in September 2003, is one of the largest pilot facilities for treatment of stormwater in Japan. Since its construction, the facility has been continuously monitored to investigate its effectiveness in removal of pollutants at the facility (Yamada et al., 2005a).

The outline of the facility is shown in Figure 11.7. The facility is designed to treat the first flush runoff from a catchment area of 80 ha (80% residential, 10% industrial and 10% non-urban). The facility treatment train is composed of a Storage and Sedimentation Tank (ST), Contact Aeration Tank (CA), Hydroponics Type Vegetation Bed (VB-H), Filter Media Type Vegetation Bed (VB-F) and Soil Penetration Tank (SP). The maximum treatment capacity of the facility, in one batch, is 7,200 m^3. Runoff water from the catchment area flows to ST during storm events and is stored for a minimum of 12 hours after which the supernatant flows to CA. Sludge settled in ST is drained to a trunk sewer pipe for treatment at a wastewater treatment plant. The effluent from CA flows to VB-H and VB-F in series, in that order. In the winter season, the effluent from CA also flows to SP. When there is no inflow of stormwater to the vegetation beds, the water needed to maintain the plants is taken from the nearby Oba River.

The removals of pollutant loads in each unit process of the treatment facility were investigated by examining 12 selected rainfall events; the results are summarized in Table 11.1. The ST was the most effective in removing pollutants,

Table 11.1.   Removal of pollutant load in the treatment facility (5)

	COD			T-N			T-P			SS		
	Max	Min	Ave	Max	Min	Ave	Max	Min	Ave	Max	Min	Ave
ST	87.8	30.1	64.1	88.2	29.4	59.7	91.5	13.1	69.4	99.2	64.0	88.1
CA	55.8	−13.4	15.2	66.3	−25.9	10.1	72.3	−25.5	13.9	96.8	−23.3	43.3
VB-H	66.7	−16.7	12.2	82.6	−1.2	23.7	49.8	−29.0	23.8	91.0	−18.9	45.6
VB-F	33.3	−41.4	−3.5	48.0	−30.7	5.8	55.2	−26.3	6.2	75.5	−81.1	−4.3
SP	57.1	23.6	45.1	47.7	16.5	33.1	78.2	69.4	74.1	100.0	50.0	78.8
Total	87.2	40.6	69.9	87.8	43.2	69.3	94.5	65.8	82.6	99.5	86.5	95.2

especially particulate substances that comprised the greater portion of the pollutant load entering the facility. In the CA, relatively high removal of COD load
was obtained compared to removals of N and P. On the other hand, relatively high
removals of N and P loads were obtained in the VB-H and VB-F. In spite of low
concentration, high removal of pollutant loads was obtained in the SP. In overall,
the ranges of pollutant removals were 41–93% for COD, 26–88% for N, 39–96%
for P, and 79–99% for SS. The results demonstrate the potential for removing
pollutant loads by optimizing the combination of mechanical (ST and CA) and
vegetation (natural)(VB-H, VB-F and SP) treatment systems.

Example-3: Study of restoration technique for reed communities on the Lake Biwa shore

The original area covered by reeds on the shore of Lake Biwa was reduced to less
than half due to lakeshore developments and land reclamation works. A restoration
program for reed communities was started in 1987, and reeds were planted to
preserve the natural environment. However, the growth of reeds has not been
successful in many places. Seven types of investigations have been carried out in
an experimental field (Biyo Center) and on the Lake Biwa shore since 1996 to
find out proper planting methods and functions of reed communities. In one of
these experiments factors affecting reed growth were investigated. Different reed
plantation methods were applied in the experimental field (Tanaka et al., 2002,
2004, 2006).

The effect of water depth on growth of reeds is shown in Figures 11.8 and 11.9.
Water depths greater than 30 cm in the first year and greater than 40 cm in the
second year tended to inhibit reed growth, but submerged conditions with depths
less than 30 cm promoted growth compared with dry condition (Figure 11.8). The
"mat" method was found to be the best method for planting reeds on grounds

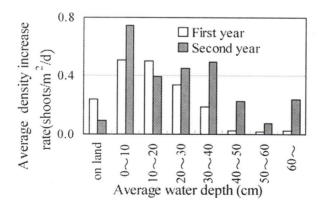

Figure 11.8. Relationship between water depths and reed growth (March–August).

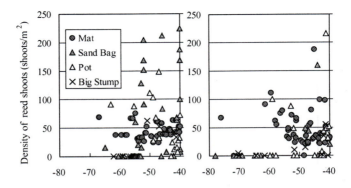

Figure 11.9.   Average density of reed shoots as a function of ground elevation.

lower than B.S.L.-40 cm (Figure 11.9). It was also shown that stabilization of ground (protection from soil erosion or deposition) after planting was important for effective reed plantation.

Invasion of naturalized plants and the effects of wave energy absorbing facilities were also investigated. A lot of reeds were planted with installation of wave absorbing facilities serving to reduce the effects of strong waves. However, some parts of reed communities were succeeded by other plant communities. One of the reasons might be an excessive inhibition of water movement in the plantation fields by wave absorbing facilities, which seem to allow other plants to invade reed communities. Research on reed communities around Lake Biwa is still ongoing. More focus on the function of micro-organisms in reed communities is required.

Social evaluation of remedial measures

Several measures have been put in place to conserve Lake Biwa and its environment, as described above. These measures contribute to solving UDP problems. Addressing many of the issues related to UDP requires cooperation of the citizens. Moreover, there is an intimate relationship between a comfortable human life and clean water environment. Therefore, promotion of citizen involvement in environmental activities has been of great interest in Shiga Prefecture. Environmental education centers and environmental museums have been established where both public and school education are emphasized. Research to evaluate the effect of these activities on citizens behaviour has been undertaken as described in the example below.

Example-4: Study on consensus building between citizens and the administration regarding public works

Consensus building between citizens and the administration is needed to ensure effective implementation of policies regarding public works. However, due to the lack of understanding of policies, it often takes time to reach consensus and some-times conflicts occur between citizens and the administration. In this study, a social experiment concerning the involvement of citizens in the decision making process on measures being implemented by Shiga Prefecture for conservation of the water environment of Lake Biwa was undertaken with a Non-Profit Organization (NPO) as a mediator (Yamada et al., 2005b). The study aimed to develop an understanding of the consensus building process between citizens and the administration. A ques-tionnaire survey was carried out three times targeting the same respondents. Prior to the each survey background information about Lake Biwa and sewerage systems around the lake was disseminated to deepen citizen's understanding of the lake.

A change in the awareness of citizens of Lake Biwa issues, as a result of informa-tion dissemination, has been recognized (Figure 11.10). Comparing the responses to the same questions in consecutive questionnaire surveys, interest in diffuse source pollution and toxic substances, such as trihalomethanes, was increased. Citizen's acceptance of implementation of advanced tertiary treatment system and measures to address the non-domestic sources of pollution has increased.

Willingness to pay (WTP) for conservation of Lake Biwa increased from 572 to 653 yen/household/per month between the first and second questionnaire surveys, prior to disseminating specific information about the monthly fees actually required per household to achieve the conservation objective. However, WTP decreased to 329 yen/household/per month in the third survey after specific information about the required payments was disseminated. It was therefore recognized that WTP was affected by dissemination of specific information about the actual cost of

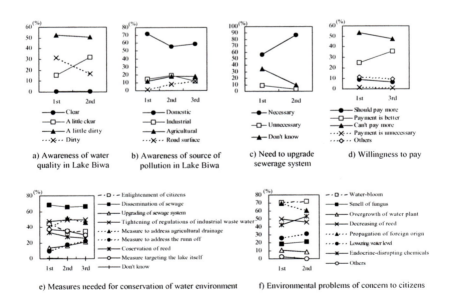

a) Awareness of water quality in Lake Biwa

b) Awareness of source of pollution in Lake Biwa

c) Need to upgrade sewerage system

d) Willingness to pay

e) Measures needed for conservation of water environment

f) Environmental problems of concern to citizens

Figure 11.10.   Change of the of citizen awareness between questionnaire surveys.

conservation. Using the AHP method, the prioritization of policy for conservation of water environment of Lake Biwa was shown. Among the three objectives of Lake Biwa Comprehensive Conservation Plan (LBCCP, also referred to as the Mother Lake 21 Plan), citizens considered protecting water quality more important than the other two objectives of recharging capacity of aquifers, and preserving the natural environment and scenic landscape. The results demonstrate the important role that the information dissemination plays in building consensus between citizens and the administration.

BASIC CONCEPTS OF REMEDIAL MEASURES FOR UDP

Technical aspects

Development of purification systems

Technological developments are essential to ensure environmentally safe water. Purification systems for diffuse pollution include both conventional treatment technologies and natural purification systems. Cooperation with local residents is essential to reduce pollutants at the source and control runoff pollution. Participation of citizens in running the facilities (e.g. harvesting of vegetation used in systems that employ phytotechnologies) is essential not only for maintenance of these facilities, but also for increasing environmental awareness of citizens.

Pollution mechanisms

To design appropriate remedial measures for UDP, it is essential to understand the mechanisms of pollution generation. This involves estimation and verification of unit pollutant loads and the understanding of the water cycle. Several types of data are required and are usually obtained through field surveys.

Model development and evaluation of the effectiveness of measures

Modeling is a policy support tool that facilitates selection of effective measures. Pollution mechanisms and material circulation are built into the model. Data collection, updating and sharing are essential components of model development.

Establishment of monitoring and measuring methods

Dynamic and integrated monitoring with the monitored parameters selected on the basis of diverse opinions is essential for UDP. The data collected are used in models described above.

Economic aspects

Economic methods and instruments

Economic methods and instruments are often applied in to remedial measures for UDP. Examples include application of economic methods in apportioning the responsibility for pollution and in valuing natural ecosystems, and using economic incentives.

Efficiency

Because of limited resources, remedial measures for UDP should be implemented with due consideration of the efficiency of these measures. One method commonly used to assess the efficiency is cost-benefit analysis. Evaluation of resource-energy using life cycle assessment (LCA) methods is also important. Based on such an evaluation, it is necessary to apply the "scrap and build" approach to political measures.

Social aspects

Integrated basin management

To adequately address water pollution problems, it is essential to consider them at the basin level. The concept of integrated basin management is generally applied. Involvement of all stakeholders including ordinary citizens is essential.

Appropriate land use

Planning and control of land use is needed to accelerate circulation of water of appropriate quality and to restore natural purification.

Strengthening of compliance

Compliance with laws and regulations, such as those related to proper land use and water use is essential for dealing with UDP.

FUTURE DIRECTIONS

Understanding of pollution mechanisms and water circulation

To effectively address UDP, it is necessary to advance the understanding of pollution mechanisms, water balance and pollution load mass balance. To achieve this, unit pollutant loads (pollutant potential) should be accurately estimated using such data as pollutant deposition and runoff for each land use. In particular, time series and their changes, taking into account seasonal variations and rainfall characteristics need to be described accurately. Also, pollutant loads discharged from urban areas should be characterized by actual field surveys. In particular, there is a need to collect data during rainfall events. Such data are generally deficient because of the difficulties associated with their collection.

Undertaking integrated diffuse pollution measures

By installing storage or infiltration facilities, water quality and circulation in urban areas can be improved. Restoration of wetlands in urban areas is important not only for improving water quality through natural purification but also for aesthetic reasons. Conserving water, or creating a society discharging low pollutant loads is also important for reducing the burden of sewerage facilities. Collaboration between the administration and citizens is essential to achieve this. Furthermore, socio-economic aspects should be appropriately considered. For example, policies providing economic incentives, appropriate regulations, and ecological planning taking into consideration the value of ecosystems should be promoted.

Undertaking measures to reduce pollutant loads for each land use

Construction of facilities (structural measures) to reduce pollutant loads and facilitation of institutional arrangements (non-structural measures) through which citizens can cooperate in running the facilities should be undertaken for each type of land use. For urban areas and roads, small scale purification facilities, constructed wetlands or more effective CSO controls should be developed. Financial assistance to private land owners to construct purification facilities should be provided and regular sweeping of roads and cleaning of gullies should be carried out. For rivers and lakes, associated lakes or wetlands should be restored and these areas should be maintained with cooperation of citizens.

Developing monitoring methods to evaluate BMPs

To monitor the progress and effectiveness of the various measures for UDP, it is necessary to develop management models capable of multiple evaluations because of the diversity of the measures involved. Appropriate indices for water quality should be developed and input parameters to the model should be determined by field work. The progress achieved by these measures should be managed using such concepts as Plan-Do-Check – Act (PDCA) management cycle.

Undertaking pilot studies to confirm effectiveness

It is necessary to undertake pilot studies in demonstration areas to confirm the effectiveness of the facilities and systems proposed above. Such testing will facilitate the selection of the most appropriate remedial measures for a given problem area.

By undertaking the above measures, it is hoped that the vision of "Cities of the Future-Blue Water in Green Cities" will be realized. It is emphasized that to realize such an integrated city, it is necessary to develop a system for collaboration between the public administration and citizens. Such collaboration can be effectively facilitated by third parties such as NPOs.

ACKNOWLEDGEMENTS

This chapter is based on a report entitled "The Future Aspects of Control Measures for Diffuse Pollution" prepared for the Shiga Prefecture Government by a

Committee chaired by the lead author and furthermore it incorporates the authors' research experiences. All those who provided resource materials are gratefully acknowledged.

REFERENCES

Fujikura, R. (2005). Successful air pollution control in Japan: History and implications. *In* Local Approaches to Environmental Compliance: Japanese Case Studies and Lessons for Developing Countries (WBI Learning Resources Series), Bianchi, A., Cruz, W., and Nakamura, M., eds., The World Bank, Washington DC, USA, 19–51.

International Lake Environment Committee Foundation, ILEC (2005). Managing Lakes and their Basins for Sustainable Use: A Report for Lake Basin Managers and Stakeholders. International Lake Environment Committee Foundation, Kusatsu, Japan. (available online at: http://www.ilec.or.jp/lbmi/index.html)

Shiga Prefectural Government (2000). Mother Lake 21 Plan: Lake Biwa Comprehensive Conservation Plan. Shiga Prefectural Government, Otsu, Japan.

Shiga Prefectural Government (2004). http://www.pref.shiga.jp/index.html

Tanaka, S., Fujii, S., Yamada, K., and Bito, T. (2002). Evaluation of influence factors on reed growth by survey on reed communities around Lake Biwa. Journal of Environmental engineering, 39, 459–465.

Tanaka, S., Fujii, S., Yamada, K., and Bito, T. (2004). Effects of soil properties on reed growth around Lake Biwa. Proceedings, The 8th International Conference on Diffuse/Nonpoint Pollution, Kyoto, Japan, 24–29 October, 567–574.

Tanaka, S., Fujii, S., and Yamada, K. (2006). A restoration technique for reed communities on Lake Biwa shore. Proceedings, International Symposium on Wetland Restoration, Kusatsu, Japan, 28–29 January.

Yamada, K., Shiota, T., Nagaoka, Y., Shiono, M., and Nishikawa, K. (2004). Empirical study on efficiency and sustainability of soil purification facility for discharged pollutants from road surface. Proceedings, The 8th International Conference on Diffuse/Nonpoint Pollution, Kyoto, Japan, 24–29 October, 149–156.

Yamada, K., Muhandiki, V.S., Nagaoka, Y., and Morino, K. (2005a). Evaluation of performance of a treatment facility for storm water from an urban area. CD Proceedings, The 9th International Conference on Diffuse/Nonpoint Pollution, Johannesburg, South Africa, 9–12 August, 8 pages.

Yamada, K., Ono, A., Hirai, S., and Muhandiki, V.S. (2005b). A social experiment by NPO on consensus building between citizens and the administration regarding public works. Abstracts volume, International Conference on the Conservation and Management of Lakes (11th World Lakes Conference), Nairobi, Kenya, 31 October–4 November, 67.

Yamada, K., Ujiie, D., and Nishikawa, K. (2006). Study on purification mechanism in soil penetration facility for effluents from urban area and control strategies. Water Science and Technology, 53(2): 858–866.

12

Tools for the evaluation of stormwater management practices that provide ecological stability in urban streams

C.A. Pomeroy and L.A. Roesner

Colorado State University, Fort Collins, CO 80523-1374, USA
E-mail: christine.rohrer@colostate.edu

Summary: Protocols and diagnostic measures are currently being developed at Colorado State University to help standardize data generation for identifying the linkages between urban land use policies and practices, stormwater runoff characteristics, geomorphic parameters, and effects on aquatic habitat and biota. Diagnostic measures include the evaluation of peak flow frequency exceedance curves, flow and shear stress duration curves, and hydrologic and geomorphic metrics. Interrelationships between biologic monitoring data and these descriptors of hydrologic and geomorphic regime can be explored to identify stormwater management practices that allow stream protection and restoration goals to be met. In the absence of monitoring data, continuous simulation hydrologic and hydraulic models can be used to generate long-term flow records that allow for evaluation of the geomorphic and ecologic implications of alternative land use and stormwater management practices.

INTRODUCTION

Land use changes, especially those related to urbanization, can have profound impacts on the characteristics of stormwater runoff, resulting in accelerated geomorphic changes that alter the quality of aquatic habitats and native biota of streams. Significant knowledge gaps exist with respect to developing cause-effect relationships between urban stormwater management (such as land cover and drainage system modifications) and observed alterations of physical habitats in receiving waters (Roesner and Bledsoe, 2003). Protocols and diagnostic measures are currently being developed at Colorado State University (CSU) to help standardize data generation for identifying the linkages between urban land use policies and practices, stormwater runoff characteristics, geomorphic parameters, and effects on aquatic habitat and biota. Identification of these linkages is needed when evaluating the effectiveness of urban stormwater runoff management, including the management of urban development and limiting percent of impervious surface cover to achieve the fewest ecological impacts and increase sustainable physical habitats and ecological conditions in urban streams. These linkages will also permit effective multi-scale functional stream restoration and rehabilitation activities.

This paper summarizes research and findings from studies conducted in the Urban Water Center at CSU regarding linkages between urban development patterns and urban drainage practices on urban stream hydrology, geomorphology and aquatic ecology. The goal of the program is to produce pragmatic design guidance for land use development and urban drainage infrastructure that will result in sustainable urban stream systems. The work presented here is in three parts: first, we examine the effects of development and stormwater controls on the hydrologic regime of urban streams; second, we look at the geomorphic implications of urban development and controls; and third, we outline a protocol for determining viable land development patterns and stormwater runoff controls that promote ecologic integrity in urban streams as development occurs.

Approach

Fifty-year continuous hydrologic and hydraulic model simulations were used to examine the entire spectrum of the runoff regime affected by urbanization and alternative stormwater control facilities in three separate watersheds. Hydrologic and geomorphic analyses were conducted for each of these watersheds. Hypothetical development was applied to two of the watersheds examined, both approximately 10 ha, referred to hereafter as Test Study Area 1 and Test Study Area 2. In Test Study Area 1 the development consisted of single family residences, an office/retail complex, and a central green space, as shown in Figure 12.1. The overall percent impervious of the development was 47 percent. In Test Study Area 2, the developed watershed was compared to the pre-development watershed by converting pasture land to a medium-density residential subdivision of representative lot sizes typical of existing residential subdivisions (0.15 ha, with 27 m of frontage). Average imperviousness of the developed area in Test Study Area 2 was 25 percent.

Hydrologic and geomorphic impacts of urbanization and stormwater controls were also evaluated in the Cedar Creek Watershed located in Lenexa, Kansas. This watershed is part of the rapidly growing Kansas City metropolitan suburbs. A 93 ha study area in the Cedar Creek watershed was selected for hydrologic and geomor-

Figure 12.1.   Watershed for hydrologic analysis.

phic analyses. The Cedar Creek Study Area is currently undeveloped, consisting mostly of pasture land and natural prairie. Development plans for the watershed include the addition of a business park and medium density residential neighborhoods that will increase imperviousness (approximately two percent, currently) to an average of 57 percent.

Runoff from all three Study Areas was simulated using the USEPA SWMM 4.4 h model. Test Study Area 1 was evaluated using typical watershed and hydraulic routing parameters; i.e., slopes of overland flow, channel and pipe elements, Manning n for pervious and impervious surfaces and conveyance elements, depression storage on pervious and impervious surfaces, Horton infiltration parameters, and evaporation rates (Nehrke, 2002). Test Study Area 2 and the Cedar Creek Study Area were evaluated using watershed and hydraulic routing parameters obtained from existing GIS data and site surveys (Rohrer, 2004; Rohrer et al., 2005), Under developed conditions, drainage systems were typical curb and gutter, with pipes connecting the inlets to the flow control facilities at the bottom of each catchment (indicated by the white oval on Figure 12.1). Figure 12.2 shows a cross section schematic of the peak flow attenuation basin, the best management practice (BMP), and the outlet positions. Stormwater detention ponds and outlets were sized in SWMM Extran, using simulations of the 100-, 10-, and 2-year design storms. Trial and error was used to design appropriate detention volumes as well as orifice outlet sizes and inverts.

In order to examine the entire spectrum of the runoff regime as affected by alternative runoff control facilities, the models were run continuously for a 50-year

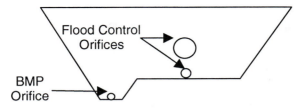

Figure 12.2.   Detention Basin outlet configuration.

period using hourly rainfall data. In the evaluations performed for Test Study Area 1 and Test Study Area 2, runoff for precipitation from Fort Collins, Colorado (National Climatic Data Center (NCDC) station 090451) and Atlanta, Georgia (NCDC station 053005) was evaluated from 1948 to 1998. These two cities were selected in order to evaluate the hydrologic response of both a semiarid region, and arid region with more substantial annual rainfall. The rainfall time series was the only SWMM variable changed between the two locales. For the Cedar Creek study area, a fifty-three year continuous simulation of runoff was performed using precipitation data from the Kansas City Downtown Airport (NCDC Station 234359) from 1948 to 2002. Resulting runoff time series were statistically analyzed using Excel spreadsheets to compute peak flow frequency curves; SAS, a statistical analysis/programming package, to compute flow-duration curves, erosion rates and erosion potential indices; and Matlab to calculate stream metrics.

EFFECTS OF URBANIZATION AND RUNOFF CONTROLS ON THE HYDROLOGIC REGIME OF URBAN STREAMS

It is well known that urbanization significantly alters watershed hydrology in urban areas (WEF and ASCE, 1998). Nehrke and Roesner (2004) demonstrated the severity of the hydrologic impact in Test Area 1 for pre- and post-development conditions using the Fort Collins, Colorado and Atlanta, Georgia precipitation data described in the previous section. Although the climatology of the two areas is very different, the results of urbanization, summarized in Table 12.1, reveal that the *number* of runoff events per year increases by a factor of 2 over the pre-development case in both areas, and the *volume* of runoff increases by an order of magnitude.

Table 12.1. Precipitation summary.

Location	Annual Precipitation Millimeters/Year	Mean Storm Depth* Millimeters	Runoff Events per Year Undeveloped	Developed	Annual Runoff (mm) Undeveloped	Developed
Fort Collins, CO	335	11	27	47	12	124
Atlanta, GA	1262	18	48	78	36	500

*Values obtained from Fig. 5.3 ASCE and WEF (1998).

Effects of uncontrolled development and runoff controls on the peak flow frequency curve

To further explore the hydrologic impacts of urbanization and the effectiveness of commonly used drainage facilities toward mitigating those impacts, Nehrke and Roesner (2004) examined the peak flow frequency curve as it is affected by uncontrolled development and by commonly used drainage facilities for flow control and pollutant removal. To examine the effectiveness of current stormwater management practices, computer simulations were made for the following two typical stormwater control scenarios: (1) control of the peak flow rates for the 100-yr, 10-yr and 2-yr storms to their predevelopment rates (peak shaving) and (2) 100-yr, 10-yr and 2-yr peak shaving with an extended detention BMP with 24-hr drawdown for removal of suspended sediments. The BMP was sized using the design criteria in the WEF/ASCE Manual of Practice on Urban Stormwater Quality Management (1998).

Figure 12.3 shows the peak flow frequency curve for Fort Collins that results when development occurs using typical curb and gutter drainage with no flow control. Similar behavior was observed for Atlanta. The abscissa of Figure two is exceedance frequency, i.e. the number of storms per year for which the indicated peak discharge is equaled or exceeded. The return period for any given peak flows is the reciprocal of the exceedance frequency so that the 100-yr storm has an

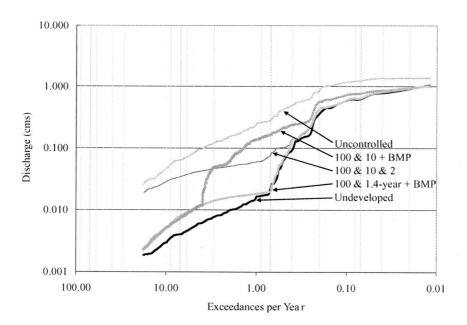

Figure 12.3.   Peak flow frequency exceedence curve for Study Area 1.

exceedance frequency of 0.01/yr and the 10-yr storm has an exceedance frequency of 0.1/yr. The curve shows that the peak runoff rate from a developed watershed for the 100-yr storm increases by a factor of about 1.5 times over the peak runoff rate from the undeveloped watershed; however, the peak rate of runoff is nearly one order of magnitude greater for storm smaller than the 2-yr storm (exceedance frequency of 0.5) revealing that the uncontrolled runoff from typical developments have a much greater effect on the peak flows from small storms that from large storms. The curves in Figure 12.3 also demonstrate how development changes the frequency of occurrence of pre-development peak flows. The peak runoff rate that occurred once in 100 years, now occurs about once every seven years when development is uncontrolled, the peak runoff rate that occurred once in 10 years, now occurs every three years, and the predevelopment 1-yr storm now occurs more than 10 times per year.

Figure 12.3 shows that controlling the 100-, 10-, and 2-yr peak flows results in a flow frequency curve very close to the pre-development curve for storms equal to and greater than the 2-yr storm, but this peak shaving does little for reducing the peak flow from smaller storms. Figure 12.3 also shows that adding the BMP results in much improved control of the small storms. Examination of the peak flow frequency curves revealed that traditional controls did not allow for a good match in the middle region of the curve, which was identified as the point at which the soil becomes saturated, and only minimal infiltration is occurring on the pervious areas for larger storms. This discovery led to the examination of a runoff control scenario that includes matching the pre-development peak flow for the 1.4-year storm for Fort Collins, and the 3-month storm for Atlanta. The result for Fort Collins, illustrated in Figure 12.3, shows that the resulting peak flow frequency curve is very close to the pre-development peak flow frequency curve. A similar result was obtained for Atlanta.

Flow duration analysis

In addition to examining the frequency of peak flows, analysis of the duration of flows is important from a geomorphic and ecologic perspective. This descriptor of hydrologic regime has been used to establish targets for ecological rehabilitation (Wiley et al., 1998) and can also be used to evaluate the effects of urbanization and measure how various stormwater management practices contribute to flow restoration. Although post-development peak flow frequencies can be controlled to predevelopment levels using traditional stormwater management practices such as extended detention, without reproducing the pre-development hydrologic regime by matching historic infiltration and evapotranspiration rates a post-development flow duration curve cannot be the same as the pre-development flow duration curve in its entirety because the total volume of stormwater runoff increases when development takes place.

Rohrer (2004) examined durations of flows in Study Area 2 for undeveloped and developed uncontrolled conditions, as well as three types of stormwater detention practices. In addition to the peak shaving and extended detention BMPs described

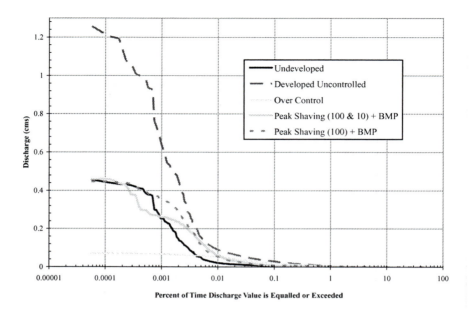

Figure 12.4.   Flow duration curve for 50-year simulation at 15-minute time steps

in the previous section, overcontrol detention practices were also examined. In this study, overcontrol detention requires that stormwater runoff be released from developments at a rate not greater than the 2-year historic runoff. The amount of runoff to be detained on-site is the difference between the 100-year runoff under developed conditions and the 2-year historic runoff. Figure 12.4 displays the percent of time discharges are equalled or exceeded over the 50-year simulation for undeveloped conditions and each post-development level of stormwater control examined. In this figure, one percent represents one percent of the 50-year simulation period, or approximately 183 days; 10 percent represents 1826 days (five years); 0.1 percent, 18.3 days, etc. Figure 12.4 shows that the Peak Shaving (100 & 10) + BMP scenario creates a flow duration curve most like undeveloped conditions; however, without further information about the types of flows that are necessary to minimize changes in channel erosion or meet ecological targets, it is unknown if this match in the flow duration curve is adequate to meet stream protection goals.

Hydrologic metrics and their sensitivity of to urban runoff control strategies

In addition to the analysis of the duration of flows, hydrologic metrics that indicate altered stream flows can also be used to evaluate the effects of development and stormwater controls. Hydrologic metrics that demonstrate altered stream flow regimes can, in some cases, also provide a direct mechanistic link between aspects of urban development and degraded stream ecosystems (Booth et al. 2004). Numerous methods exist for estimating desirable flow regimes in the form of hydrologic metrics. Some examples include the Instream Flow Incremental Methodology (IFIM) that correlates flow levels to habitat availability (Bovee et al., 1998); a flashiness index that describes how quickly flows change in a watershed and whether increases or decreases in flashiness occur over time (Baker et al., 2004); and the Indicators of Hydrologic Alteration (IHA) that measure the magnitude of both high and low flows, timing indexed by monthly statistics, frequency, and rate of flow change (Richter et al., 1996).

Hydrologic metrics have traditionally been computed using daily mean values of flow. While this method of computation is valid for large watersheds, the rainfall-runoff response of small watersheds, such as the one in Study Area 2, is so rapid that trends in hydrologic metrics are not easily detected using a mean daily flow time series. To make the statistics meaningful for small watersheds, it was necessary to modify the computation of the stream metric to work on runoff events, rather than mean daily flows. Edgerly (2006) examined nine hydrologic metrics proposed by various researchers to determine their sensitivity to urban runoff control strategies in Study Area 2. These include: $T_{0.5\,yr}$, which is the percent of time that the stream flow exceeds the peak flow of the 0.5-yr storm; *high pulse duration rise rate* and *fall rate* metrics from Richter et al. (1996); *event-mean rise and fall rate;* and the *ratio of event-peak-duration to total event duration*. Of the nine stream metrics examined, a *modified* $T_{0.5\,yr}$ metric proved to be the most promising metric with respect to its potential utility as a design parameter for stormwater management.

Figure 12.5 shows the response of the *modified* $T_{0.5\,yr}$ hydrologic metric to urban development with varying levels of controls for Study Area 2 modeled using precipitation data for both Atlanta, GA and Fort Collins, CO. As expected, the controls were most effective at mitigating the effects of urban development on *modified* $T_{0.5\,yr}$ since they were designed to prevent post-development flow rates from exceeding target flow rates. In both Atlanta and Fort Collins, the peak shaving + BMP controls also appeared to bring this metric closer to pre-development levels than the other designs. This makes sense because BMPs limit the range of possible flows by controlling smaller storms which are closer in size to the half-year storm. Also, by reducing the 100-year peak to the pre-development 2-year peak discharge, the over-control designs increase the durations of intermediate flows by releasing the excess volumes at lower rates for longer durations than the peak shaving designs.

EFFECTS OF URBANIZATION AND RUNOFF CONTROLS ON IN-STREAM EROSION POTENTIAL

When stormwater is left uncontrolled or is controlled inadequately, increases of the magnitude and duration of runoff can accelerate channel erosion; these increases allow a stream to carry more sediment than it could prior to watershed development.

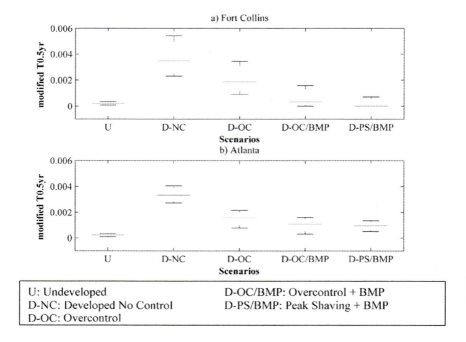

U: Undeveloped	D-OC/BMP: Overcontrol + BMP
D-NC: Developed No Control	D-PS/BMP: Peak Shaving + BMP
D-OC: Overcontrol	

Figure 12.5.   Median and interquartile range values of *modified* $T_{0.5\,\mathrm{yr}}$ for multiple scenarios of development in Fort Collins and Atlanta.

When a watershed cannot supply the stream with the volume of sediment it has the capacity to carry, channel degradation may occur in the form of incision, lateral migration or a combination of both (Bledsoe, 2002). Although many municipalities and agencies require some level of stormwater control when development or redevelopment takes place within their jurisdiction the intent of these practices is typically focused on flood prevention and/or protection of water quality. Currently, few municipalities require design of stormwater controls specifically to minimize stream channel degradation. This section describes an alternate method for designing stormwater controls to minimize the potential for stream channel degradation by matching the critical portion of shear stress duration curves and identifying the stormwater management scenarios that have an erosion potential index closest to one.

Erosion potential analysis

Incipient motion of a particle on a stream bottom or bank occurs when the fluid flow around a sediment particle exerts a force that is greater than the resisting force of the particle weight. This condition can be evaluated by comparing a calculated shear stress in the channel to a critical shear stress value for the bed material examined. Boundary shear stress, τ_o is calculated as:

$$\tau_o = \gamma R_h S_f$$

where, γ is the unit weight of water, R_h is the hydraulic radius of channel, and S_f is the energy gradient in the channel. Channel erosion is a function of the erodibility of the bed and bank materials and excess shear stress. Erosion rates can be estimated using an excess shear equation described by Foster et. al. (1977):

$$\varepsilon = k(\tau_o - \tau_c)^A$$

where ε is the erosion rate, k is an erodibility or detachment coefficient, τ_c is the critical shear stress, and A is an exponent with an average value of 3/2. For granular (non-cohesive) materials, τ_c can be estimated using the Shields criteria:

$$\tau^* = \tau_o/(\gamma_s - \gamma)d$$

where τ^* is the critical dimensionless shear stress, γ_s is the unit weight of sediment, γ is the unit weight of water, and d is the representative particle diameter. It is more difficult to estimate the erosion resistance of cohesive materials. Various studies on the erodibility of cohesive materials have shown that numerous soil properties influence erosion resistance, including antecedent moisture, clay mineralology and proportion, density, soil structure, organic content, and pore and water chemistry (Grissinger, 1982). Erodibility coefficients and critical shear stress values for cohesive sediments can be estimated by in-situ jet-testing using the methods described by Hanson and Cook (2004).

As suggested above, the potential for erosion in a channel can be quantitatively evaluated as the difference between *calculated* shear stress and *critical* shear stress. Bledsoe (2002) described an index which estimates an increase or decrease in erosion potential by calculating a ratio of post-development to pre-development "work" on a channel. This "work" on a channel can be estimated using erosion rates. For this study an overall erosion potential index was calculated as a ratio of the sum of excess shear values raised to the 3/2 power, ε, for each time step modelled using continuous simulation. This index uses a simple finite-difference approximation to estimate the time-integrated sediment transport capacity over the duration of the continuous flow record:

$$E = \frac{\sum_{t=0}^{T} \varepsilon_{post}}{\sum_{t=0}^{T} \varepsilon_{pre}}$$

where E is the instream erosion potential, ε is the erosion rate at time t, T is the length of the continuous simulation (50 years), and *pre* and *post* represent pre- and post-development conditions, respectively.

Sensitivity analysis of erosion potential

Using the methods described above, erosion potential for the channel cross-section draining Study Area 2 was evaluated for a variety of critical shear stress values, ranging from 0.2 Pa to 10 Pa. These erosion potential values are presented in Figure 12.6. In Figure 12.6, erosion potential values less than one indicate a decrease in erosion rates pre- to post-development, and erosion potential values greater than one indicate an increase in erosion rates. Erosion potential values equal to one indicate no change in erosion rates. Figure 12.6 demonstrates that the use of stormwater detention increases durations of the discharges that entrain sediment with shear stress values less than 0.7 Pa. When τ_c is between 0.2 and 0.7 Pa, the calculated erosion potential is larger when stormwater controls are in place than when stormwater is left uncontrolled. These results indicate that detention may exacerbate erosion in a stream channel above what might occur if the stormwater were left uncontrolled. This increase in erosion potential suggests that a reduction in the total volume of runoff, through the use of infiltration for example, may provide a better method for stormwater management for watersheds that drain to fine grained non-cohesive or erodible clay streams. Figure 12.6 also shows that as the τ_c values increase, stormwater management practices create erosion potentials closer to the ideal value of 1.0. The erosion potential values of zero for the Overcontrol scenario demonstrate that no erosion occurs in this scenario if the τ_c

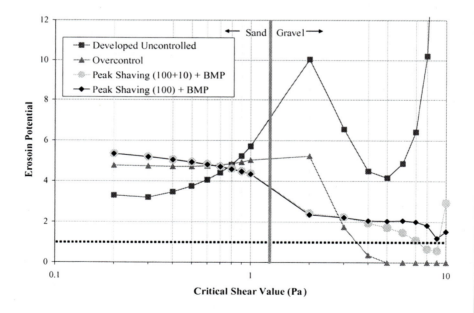

Figure 12.6.   Erosion potential at varied levels of critical shear stress. When $E = 1$ there is no change in erosion potential. For the developed, uncontrolled scenario when $\tau_c = 9$, $E = 21.7$ and when $\tau_c = 10$, $E = 1460$.

is greater than 4. Using the example of τ_c equal to seven Pa, Figure 12.6 shows that the "Peak Shaving (100+10) + BMP" scenario results in an erosion potential value closest to one, indicating that this type of stormwater control may be most appropriate to minimize changes in stream erosion potential when urbanization of the watershed takes place.

Erosion potential analysis for Lenexa, Kansas

The geomorphic impacts of urbanization and stormwater controls were evaluated in the Cedar Creek Watershed Study Area. Results from these analyses were used by the City to develop a strategy for the design of a system of multi-use stormwater facilities to protect the natural resources within the watershed, while providing flood control benefits and recreational amenities. Figure 12.7 displays the percent of time in-channel shear stresses are equalled or exceeded over the 53-year simulation for undeveloped conditions; developed, uncontrolled conditions; and the recommended developed, controlled conditions. The recommended stormwater controls include peak shaving of the one- and ten-year storm and control of the WQCV using extended detention with a 40 hour draw-down. Figure 12.7 shows that when development takes place without any type of stormwater control, the developed, uncontrolled shear stress values are greater in magnitude than undeveloped conditions for approximately 10.2 percent of the time, or approximately 5.5 years of the overall time period examined. The magnitude and duration of shear

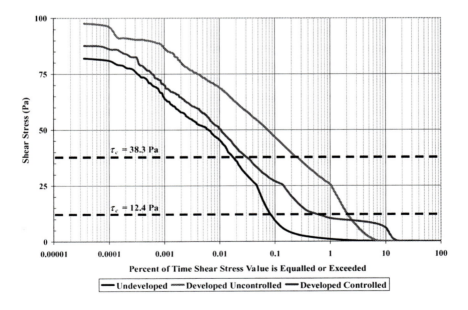

Figure 12.7.   Shear Stress Duration Curve.

Table 12.2. Erosion potential under developed conditions.

Scenario	Erosion Potential	
	$\tau_c = 12.4$ Pa	$\tau_c = 38.3$ Pa
Developed, Uncontrolled	19	13
Developed, Controlled	2.7	1.8

stresses in the developed, controlled scenario are greater in magnitude than undeveloped conditions for approximately 18.5 percent of the time, or approximately 10 years of the overall time period examined. Increases in overall durations of shear stress are not telling in and of themselves, however, unless stream channel critical shear stress values are identified and results are examined for the critical portion of the shear stress duration curve.

The critical portion of the shear stress duration curve varies depending on the critical shear stress (τ_c) of the stream bed and bank material to be protected. For the natural streams in this basin, the critical shear stress value for the upper tributaries is approximately 12.4 Pa, which represents colloidal alluvial silts, and approximately 38.3 Pa for the main stem, which consists of graded silts to cobbles, based on soil types and channel conditions observed in the field. Figure 12.7 shows that the duration of time the τ_c value for the main stem (38.3 Pa) is equalled or exceeded increases by an order of magnitude between undeveloped and developed, uncontrolled conditions. The developed, controlled condition allows a closer duration match to undeveloped conditions. Figure 12.7 also shows that if the τ_c value for the reach were a lower value, such as the value determined for the upper tributaries, 12.4 Pa, the critical portion of the shear stress duration curve would not be as well matched. The duration of time τ_c is greater than this lower critical shear stress value increases by an order of magnitude, regardless of the stormwater control, demonstrating the importance of identifying the critical shear stress threshold when designing stormwater management practices to minimize erosion potential.

Table 12.2 quantitatively describes differences in excess shear using the erosion potential index discussed in the previous section. Table 12.2 shows that for both τ_c values, erosion potential is decreased with the application of the developed, controlled scenario that includes peak shaving of the one- and ten-year storm and control of the WQCV using extended detention with a 40 hour draw-down.

LINKING URBAN DEVELOPMENT AND RUNOFF CONTROLS TO STREAM ECOLOGIC HEALTH – CURRENT RESEARCH

Review of recent literature reveals several notable advancements in research regarding the linkages between urbanization, hydrology, hydraulics, geomorphology, physical habitat, and stream ecology. Some promising work has been done

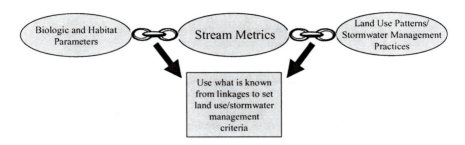

Figure 12.8.   Establishing the link between urban development and stream ecologic health.

exploring ways of moving beyond using gross measures of imperviousness as predictors of biologic integrity by developing more meaningful land use/land cover metrics, establishing clearer relationships between land cover and hydrologic response, and identifying mechanisms through which altered flow regimes affect stream ecosystems. Studies examining the factors important for the success of specific organisms or groups of organisms provide insight regarding critical physical habitat and flow requirements necessary to support healthy lotic ecosystems.

Regression of hydrologic metrics with biological data has become increasingly common in recent years. A number of researchers have used hydrologic metrics in urban gradient studies to establish statistical relationships between urbanization, hydrology, hydraulics, and biota (Scoggins, 2000; Kirby, 2003; Booth et al., 2004; Cassin et al., 2005). The results of these studies provide clues about types of metrics likely to be meaningful in urban applications; metrics that measures of flow flashiness, variability, and timing have been identified as useful predictors of macroinvertebrate community status. When combined with relationships between stormwater management practices and hydrologic metrics, relationships between biological data and hydrologic metrics can be used to identify land use and stormwater management criteria that will meet resource protection goals as is illustrated in Figure 12.8.

A protocol for linking land use patterns and urban runoff controls to ecologic health in urban streams

Linkages between land use patterns, urban runoff controls, and the ecologic health of urban streams can be developed using the protocol outlined in Figure 12.9. This protocol uses data from a continuous flow record to generate hydrologic (i.e., $T_{0.5\,yr}$) and geomorphic (i.e., erosion potential, E) metrics which can then be related to in-stream biologic data. The continuous flow record can be obtained from stream flow gage data or generated by a continuous simulation hydrologic and hydraulic model such as SWMM. Relationships between the hydrologic and geomorphic metrics and the biologic and habitat parameters are established across a gradient of urbanization; methods for developing these relationships include

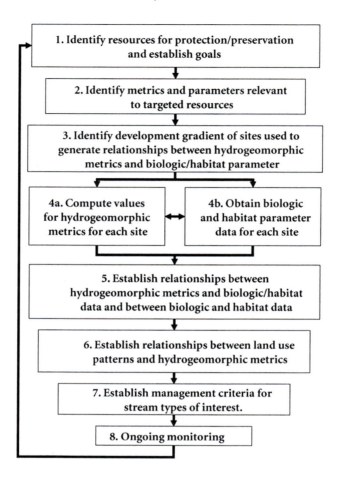

Figure 12.9.   The Protocol.

regression, logistic regression, a classification and regression tree (CART) analysis, or Bayesian networks.

In addition to linkage development, the protocol recommends that stream goals are set in order to determine the level of protection and/or restoration desired. Once community goals are related to designated use, biological integrity, stable channel, or protection of a specific species or group of species have been identified, a list of which biologic and habitat parameters to measure is created so that linkages can be developed between these parameters and the hydrogeomorphic metrics. This is accomplished by a process to 1) identifying sensitive and appropriate biological indicators, and 2) generating a list of possible or potential stressors in the watershed. If available, existing biological and habitat data collected by state and local agencies can provide the starting point for this process because they are often tied to existing goals for a region, and in some areas have already derived or explored the stressor gradients that may be important.

CONCLUSIONS

Numerous tools exist for evaluating the effects of urbanization and stormwater management practices on urban stream hydrology, geomorphology and stream ecology, including peak flow frequency exceedance curves, flow and shear stress duration curves, and hydrologic and geomorphic metrics. Interrelationships between biologic monitoring data and these descriptors of hydrologic and geomorphic regime can be explored to identify stormwater management practices that allow stream protection and restoration goals to be met. In the absence of monitoring data, continuous simulation hydrologic and hydraulic models can be used to generate long-term flow records that allow for evaluation of the geomorphic and ecologic implications of alternative land use and stormwater management practices.

ACKNOWLEDGEMENTS

This work was supported by Harold H. Short, CDM, the City of Lenexa, Kansas, and the Water Environment Research Foundation. The authors are grateful for the collaborative efforts of Seth Nehrke, John Edgerly, Jennifer Davis, Natalie Postel, Pat O'Neill, Jamie Coleman, Boris Kondratieff, Ed Rankin, and Brian Bledsoe.

REFERENCES

Baker, D.B., Richards, R.P., Loftus, T.T., and Kramer, J.W. (2004). A New Flashiness Index: Characteristics and applications to midwestern rivers and streams. American Water Resources Association 40:503–522.

Bledsoe, B.P. (2002). Stream Erosion Potential and Stormwater Management Strategies. Journal of Water Resources Planning and Management 128:451–455.

Booth, D.B., Karr, J.R., Schauman, S., Konrad, C.P., Morley, S.A., Larson, M.G., and Burger, S.J. (2004). Reviving urban streams: land use, hydrology, biology, and human behavior. Journal of the American Water Resources Association 40:1351–1364.

Bovee, K.D., Lamb, B.L., Bartholow, J.M., Stalnaker, C.B., Taylor, J., and Henriksen, J. (1998). Stream habitat analysis using the Instream Flow Incremental Methodology. Information and Technology Report USGS/BRD-1998-0004. US Geological Survey, Biological Research Division, Washington, DC.

Cassin, J., Fuerstenberg, R., Tear, L., Whiting, K., StJohn, D., Murray, B., Burkey J. (2005). Development of Hydrological and Biological Indicators of Flow Alteration in Puget Sound Lowland Streams. King County Water and Land Resources Division. Seattle, Washington.

Edgerly, J.L. (2006). Quantifying urban-induced flow regime alteration using mathematical models and hydrologic metrics. Master's Thesis. Colorado State University, Fort Collins, CO.

Foster, G.R., Meyer, L.D., and Onstad, C.A. (1977). An erosion equation derived from basic erosion principles. Transactions of the American Society of Agricultural Engineers 20: 678–682.

Grissinger, E.H. (1982). Bank erosion of cohesive materials. In: R.D. Hey, J.C. Bathurst, and C.R. Thorne (eds.), Gravel-bed Rivers. Wiley: Chichester, pp. 273–287.

Hanson, G.J. and Cook, K.R. (2004). Apparatus, test procedures, and analytical methods to measure soil erodibility in situ. Applied Engineering in Agriculture 20:455–462.

Kirby, C.W. (2003). Benthic macroinvertebrate response to post-development stream hydrology and hydraulics. PhD Dissertation. George Mason Univ., Fairfax, VA UMI Microform 3079343.

Nehrke, S.M. (2002). Influence of extended detention BMPs and traditional flood controls on the flow frequency curve of urban runoff. Master's Thesis. Colorado State University, Fort Collins, CO.

Nehrke, S.M,. and Roesner, L.A. (2004). Effect of Drainage Controls and BMPs on the Flow Frequency Curve of Urban Runoff. Journal of Water Resources Planning and Management 130:131–139.

Richter, B.D., Baumgartner, J.V., Powell, J., and Braun, D.P. (1996). A method for assessing hydrologic alteration within ecosystems. Conservation Biology 10:1163–1174.

Roesner, L.A. and Bledsoe, B.P. (2003). Physical Effects of Wet Weather Flows on Aquatic Habitats – Present Knowledge and Research Needs. Final Project Report No. 000-WSM-4, WERF, Alexandria Virginia.

Rohrer, C.A. (2004). Modeling the effect of stormwater controls on sediment transport in an urban stream. Master's Thesis. Colorado State University, Fort Collins, CO.

Rohrer, C.A, Postel, N.A., O'Neill, P.A., and Roesner, L.A. (2005). Development of Design Criteria for Regional Stormwater Management Facilities to Maintain Geomorphic Stability in Cedar Creek. World Water & Environmental Resources Congress: Impacts of Global Climate Change on Water Resources and the Environment Anchorage, Alaska, 15–19 May 2005.

Scoggins, M. (2000). Effects of Hydrologic Variability on Biological Assessments in Streams in Austin, TX. National Water Quality Monitoring Council (NWQMC) Monitoring Conference 2000: Monitoring for the Millennium. Austin, Texas April 5–27, 2000.

Water Environment Federation (WEF) and American Society of Civil Engineers (ASCE) (1998) Urban Runoff Quality Management WEF manual of practice; no. 23. ASCE manual and report on engineering practice; 87.

Wiley, M., Seelbach, P., and Bowler, S. (1998). Ecological targets for rehabilitation of the Rouge River. Rouge River National Wet Weather Demonstration Project. RPO-PI-SR21.00.

13

Effluent dominated water bodies, their reclamation and reuse to achieve sustainability

V. Novotny

Northeastern University, Boston, MA 02115, USA
E-mail: Novotny@coe.neu.edu

Summary: Effluent dominated/dependent streams are the most stressed receiving water bodies. Yet, the Clean Water Act does not allow downgrading the designated uses of the aquatic life protection and propagation or providing conditions for primary contact recreation. The water quality regulations outlining the Use Attainability Analysis reinforce the need for restoration and make a downgrade of the use difficult. In addition to overwhelming effluent discharges effluent domination/dependence is also caused by large water withdrawals and long distance water and sewage transfers. This chapter presents an historic outlook, possibilities to achieve the ecologic potential, and outlines solutions and social and legal barriers.

INTRODUCTION

Many urban waters have not met the goals of the Clean Water Act (Section 101) of attaining and preserving the physical, chemical, biological integrity of the nation's waters and ability of the receiving waters to provide conditions for contact and noncontact recreation. In the 2000 report to Congress, the US Environmental Protection Agency (2002) reported that 30% of assessed streams, 45%of lakes, and 51% of estuaries were found to be impaired. Urban sources of pollution and modifications of urban streams are the leading cause of impairment for coastal waters and second for urban streams and impoundments.

Since the passage of the Clean Water Act in 1972, billions of US dollars have been spent on building sewers, treatment plants, and reducing untreated overflows from sewerage systems. The situation is complicated by multipurpose uses of the urban waters (Novotny et al., 2005) and the conflicts between the economical (water supply, navigation, urban and suburban irrigation, power production, flood conveyance, wastewater disposal, cooling) and ecological (fish and wild life propagation, green space ecology, contact recreation) uses. Many economical uses are irreversible. Consequently, uses of urban water resources must be reconciled and optimized. However, while the economic uses can be optimized, attaining and maintaining the goals of ecologic integrity and protecting public health cannot be compromised by economic overuse.

The traditional goals of urban water management have been to provide a safe and adequate water supply, environmentally acceptable disposal of treated wastewater, and flood control. Such focused management goals have helped to protect human health and welfare in cities, but often at the expense of dramatic alterations in the ecosystem and, ultimately, endangered the human health and lives to be protected. Most pollution in cities today is caused by the urban landscape, which has been shaped by the preference for impervious over porous surfaces; fast "hard" conveyance infrastructure rather than "softer" approaches like ponds and vegetation; rigid stream channelization instead of natural stream courses, buffers and wetlands, and development in floodplains. The fast-conveyance drainage infrastructure conceived in Roman times to eliminate unwanted, highly-polluted runoff and sewage is now an impediment to solving the pollution problem – in spite of billions of dollars spent on costly "hard" solutions like sewers and treatment plants.

HISTORICAL PERSPECTIVES

Urbanization results in large and often long distance transfers of flows. With few notable exceptions (e.g., water transfers by Romans two thousand years ago) water supply and wastewater disposal were local until the first half of the last century. Water was withdrawn by the community from the river upstream or from nearby wells and discharged back into the river with or without treatment in the city or short distance downstream. It is interesting to note that early effluent discharge criteria in Great Britain in the first half of the 20th century considered dilution; for example, an effluent receiving only primary treatment required one part of effluent vs. 150 to 300 parts of river dilution flow during low flow conditions. Effluents receiving secondary treatment could have dilution ratio of less (but not by much) than 1:150 (Imhoff and Imhoff, 1993; Novotny et al., 1989). Effluent standards based on toxicity and expressed as maximum concentrations were introduced in the second half of the 20th century.

However, rivers are not used only for wastewater disposal. Large withdrawals for water supply, irrigation, and other economic uses, many of which resulted in high consumptive losses, were diminishing the flow. Because of the rapid urbanization in the last two centuries many rivers very quickly ran out of dilution flow. Essentially, dilution criteria were primitive and useless.

Water – sewage – water cycle

At the end of the nineteenth century, industrialization and urbanization reached levels that overwhelmed the receiving water bodies with untreated effluents. During the first seventy years of the twentieth century, some urban rivers were smelly because of hypoxia in summer, unsuitable for recreation, overloaded with sludge deposits and often dense mats of sludge worms developed on their bottoms (e.g., the Lower Des Plaines River in Illinois (Butts et al., 1975; AquaNova International/Hey and Associates, 2004)). In many places, rivers were covered and converted to sewers. For example, at the end of the twentieth century, the effluent dominated river transecting the seat of the European Community, Brussels, was still covered because of pollution by untreated discharges.

In the industrialized Ruhr district of Germany, several small rivers provided all water supplies and served as recipients of sewage from about eight million people and one third of the German heavy industry and coal mining. In 1908, Ruhr water management district was established and Karl Imhoff, a pioneer of environmental engineering (see Novotny, 2003) was put in charge. Imhoff implemented primary and later secondary treatment of waste to protect the rivers. But he realized that the loading capacity of the rivers was insufficient and implemented designated uses. For example, the Ruhr River provided most of the water supply while the smaller Emscher River (Figure 13.1) was designated a depressing use of receiving untreated or only partially treated sewage, essentially becoming an open sewer and an ultimate effluent dominated stream. The population equivalent of wastewater discharged from Dortmund, Essen, and several other communities and numerous industries into the Emscher River was over 2 million. The river was lined, fenced off in many places, and before entering the Rhine River the entire flow was treated in a primary treatment plant (Novotny and Olem, 1993). Because of insufficient water availability during dry seasons, reversible hydro power plants (turbines in the power plant could reverse the flow and serve as pumps) were installed on impoundments on the Ruhr that could back pump the Ruhr and Rhine Rivers flows and provide flow augmentation in upstream water deficient urban sections of the Ruhr.

Karl Imhoff realized that once an effluent is discharged into a receiving water body it will be reused, in a more or less diluted form, as a source of water somewhere downstream. This was called "water-sewage-water cycle" (WSW) (Imhoff and Imhoff, 1993) that occurs in almost every municipal water management with an exception of waste water disposal into the ocean. Municipalities relying on water supply from watercourses and environmental protection agencies must always consider that raw water contains residuals of upstream wastewater discharges and nonpoint pollution. Even groundwater can not be safe from wastewater contamination and, in general, the groundwater was once surface water subjected to pollution. Almost every waste water discharge into continental waters may later be

Figure 13.1.   The Emscher River in Germany before rehabilitation around 1950. Cascades and other aeration were implemented to keep the raw or poorly treated wastewater fresh. In spite of natural and man-made aeration, the bad smell during summer was strong.

uncontrollably reused by the population downstream (Imhoff, 1931). Imhoff also stated that the only thing engineers can do is to maximize the distance and time between the discharge and reuse.

The problems with the water-sewage-water cycle came to light many times during the last century. In the Ruhr area, the dry periods when back-pumping was activated occurred in 1929 and 1959. Water shortages in 1959 lasted four months and 109 million m^3 of Rhine-Ruhr mixture were pumped upstream towards the water intakes of Essen and other Ruhr cities (Koenig et al., 1970). As a result, the quality of potable water greatly worsened, concentrations of nitrate (NO_3^-) reached 35 mg/L, surfactants 1.5 mg/L, and chloride 490 mg/L, respectively. At the end of 1959, the water demand could only be met by back pumping the Rhine and Ruhr water. This resulted in a sewage reuse factor of 0.86 (Koenig et al., 1970) and about 10% of the people in Essen suffered gastrointestinal problems. In the succeeding years, the Ruhr River Management Association (Ruhrverband) increased the capacities of storage reservoirs and an occurrence of a similar back-pumping and effluent reuse has not happened and is unlikely to happen. Nevertheless, water-sewage-water cycles occur everywhere. Downstream users of the Mississippi River water reuse wastewater effluents from about one half of US population.

Today, the Emscher River, after a strong grass-root pressure, is being "renaturalized" by a massive project that includes a high degree of treatment in several new Emschergenossenschaft (Emscher association) treatment plants, augmenting flows with stored and treated urban runoff, daylighting (bringing buried stream back to the surface) and redesigning new and restoring the old more natural channels, floodplain restoration, clean-up and rehabilitation of brownfields of old abandoned industries and mines and by developing an urban landscape Emscher Park surrounding the river. The goals of this unique project are (Petruck et al., 2003):

- Safe waste water transport and treatment
- Ecological improvement of the water courses ("renaturalization")
- Prevention of flooding
- Integration of the restructured Emscher into the area that surrounds it
- Providing space for recreation

This approximately US$ 5.6 billion restoration and rehabilitation project for the river and its surroundings will take about twenty years (Petruck, 2003; USEPA, 2006).

Large transfers of water and sewage

In the last century, regional sewerage agencies used the argument of the economy of scale and of the possible adverse effects of the uncontrolled water-sewage-water cycle to promote and develop large regional sewer systems that transfer water and wastewater over long distances, often to another watershed. Long distance water transfers are nothing new, two thousand years ago the Romans built aqueducts that transferred water to Roman cities from as far as 50 kilometres that fed fountains, baths, and flushed excreta. The relatively small Tiber River in Rome was (and still is) a heavily polluted effluent dominated stream that receives, among other discharges, stormwater and wastewater flows brought into the river by the ancient Roman sewer Cloaka Maxima. In 1920s, the City of Los Angeles began buying land and water rights on Owens Valley in the Sierra Mountains and with the aqueduct drained the river. The water then appeared as a sewage effluent elsewhere and created, with other transfers from Northern California and the Colorado River, the effluent dominated Los Angeles and Santa Ana Rivers.

The long distance transfers dramatically changed the hydrology of the impacted surface waters, which became flow deficient after withdrawal and the water body receiving the effluent then became effluent dominated. However, even today with long distance water and sewage transfers and sewer separation, the problems with combined and sanitary sewer overflows (CSOs and SSOs) have not been and most likely will not be fully mitigated in the near future. Several cases will be discussed in the subsequent section of this chapter.

LEGAL DILEMMA AND MANDATE OF EFFLUENT DOMINATED OR DEPENDENT RECEIVING WATERS

The terms *effluent dominated* and *effluent dependent* water bodies have been used interchangeably (e.g., Pickus et al., 2002). To distinguish between the two categories the following definitions are proposed:

1. *Effluent dominated water body is a water body that predominantly contains waste water effluents during all or a part of a year.* For water quality analyses

and assessment purposes, this definition implies that the design low flow is predominantly made of upstream effluent. Effluent dominated streams do not have to be naturally ephemeral.

2. *Effluent dependent water body is generally an ephemeral (either natural or due to excessive upstream withdrawals) stream whose aquatic biota and other uses of the water body are or can be sustained by treated effluents creating perennial flows.*

The CWA and ensuing Water Quality Standards Regulations (40 CFR 131) do not give agencies or dischargers a "carte blanche" to exclude effluent dominated water bodies from attaining the integrity goals. A Use Attainability Analysis (UAA) is needed if a downgrade or any change of the designated use is contemplated (40 CFR 131.10; USEPA, 1994). In a large majority of cases, UAAs have not been prepared and these water bodies remain downgraded. Typically, these water bodies are water quality limited and have been put on the CWA Section 303(d) TMDL action list; however, the TMDL process as formulated in CWA Section 303(d) is only capable of dealing with the pollutant discharges and not with adverse stream morphology and hydrology modifications caused by urbanization that falls under the definition of "pollution" in the CWA Section 5 but it is not a discharge of pollutants. Furthermore, the TMDL process is not designed to change the designated use nor associated standards.

The Water Quality Standards Regulation (40 CFR 131.10) provides six reasons that can be used in petitioning (through a UAA) for a downgrade of the statutory water uses defined in the CWA, i.e., maintaining integrity of aquatic life and providing for primary recreation, considering also uses for water supply and navigation (USEPA, 1994; Novotny et al., 1997). However, §131.10(g)(2) allows the removal of an aquatic life use, that is not an existing use (i.e., the water body has been impaired) where "natural, ephemeral, intermittent or low flow conditions prevent the attainment of the use, *unless these conditions may be compensated for by the discharge of a sufficient volume of effluent discharge without violating State water conservation requirements to enable uses to be met* (emphasis added). This rule has been interpreted to imply (USEPA Region 8, 2003) that "**where an effluent discharge creates a perennial flow, the resulting aquatic community is to be fully protected**" which means the integrity goals specified in Section 101 of CWA should be attained. Integrity has been defined as a condition in the stream supporting balanced aquatic life (Karr et al., 1986) and supporting contact and noncontact recreation. Effluent dominated streams present a challenge but also a potential for restoration.

Hundreds of effluent dominated/dependent streams can be found in the US and thousands abroad. The categories may include:

1. Originally ephemeral streams that were made perennial by effluent discharges (Santa Ana River in Orange County, Los Angeles River in California, Gila River in Arizona).

2. Perennial rivers where flow was diminished by upstream withdrawals for water supply and irrigation and made effluent dominated by large effluent

discharges from municipal or regional treatment plants (e.g., the South Platte River in Colorado, the Trinity River in Texas)

3. Smaller perennial rivers that were overwhelmed by effluent discharges from sewer outfalls and regional treatment plants receiving wastewater from outside of the original watershed (the Des Plaines River in Chicago Metropolitan Area, the Emscher River in Germany).

As a result of effluent domination and/or excessive wastewater discharges, ecological and hydrological sustainability of urban receiving waters is compromised. The underlining causes are:

1. *Population increases and migration.* In the next 50 years the world population is expected to increase by 50% (NRC, 1999). Most of the increase will occur in developing countries; nevertheless, migration of people from the Northern Midwest and Northeast US to South, from cities to suburbs and immigration from other countries will continue, posing new demands on water supply and increasing waste water volumes which will magnify the problems.

2. *High imperviousness and sewers* built in the last hundred fifty years in cities have dramatically changed the hydrology of urban areas. Imperiousness increased peak high flows in urban streams by a factor of 4 to 10 and diminished the base flow. Increased variability and higher frequency of bank overtopping resulted in unstable eroding channels and a loss of habitat.

3. *Long distance transfer of water and sewage.* In the last thirty years, local water supply and sewerage systems were replaced by long distance transfers of water and sewage in regional systems (Breckenridge, 2004). The long transfers of raw water and sewage made water bodies in the source area deprived of flow and in the discharge zones overwhelmed by the effluents.

4. *Increasing living standards* of the urban and suburban population have increased the water demand and use as well as volumes of sewage. This development will continue (NRC, 1999).

5. *Multiple and often conflicting uses of urban waters.* The conflicts between uses must be reconciled and in most cases urban waters must be managed and the uses must be optimized.

6. *Adverse, ecosystem and habitat damaging water body modification* to preferentially accommodate competing uses (e.g., lining for flood conveyance, impoundments for navigation and hydropower).

In addition to effluent domination, many complex and interconnected issues contribute to the difficulty in achieving desired improvements in water quality and urban watershed ecosystem integrity and their sustainability and must be considered in an integrated fashion. They include the following:

Stormwater. In addition to wastewater effluent, urban watersheds are impacted by stormwater runoff and combined sewer overflows that contain pollutants washed from the city's many impervious surfaces. Stormwater drainage infrastructure itself can also contribute to non-attainment of aquatic life and recreation uses in rivers and streams, as conditions in large pipe and culvert systems can contribute to toxic "first flush" discharges. This runoff can also result in intermittent flows that erode the bank habitat and threaten the well-being of aquatic organisms (Novotny, 2003). However, after treatment, stored rain water and urban runoff can be a resource for providing flows and blending with the treated effluents that could be used for rehabilitation of streams. Rainwater has been collected for potable and nonpotable domestic and municipal use for more than two thousand years.

Dropping Groundwater Table and Loss of Base Flow. Impervious surfaces and the diversion of rainfall into sewers frequently prevents groundwater recharge, leads to lower groundwater tables, and deprives urban streams of valuable base flow. Much of pumped groundwater for water supply, rather than being returned to the river, is diverted to a regional treatment facility in another basin. As a result, many rivers experience little or no flow during the summer and do not support a healthy aquatic community, i.e., they could become effluent dependent such as some rivers in the Northeastern US and Pacific Coast region. In some older urban communities, low groundwater tables resulting from insufficient recharge have caused significant structural problems. Many historic buildings in Boston (see the chapter by Shanahan and Jacobs (2007) in this book) and other cities (Mexico City, Venice, Italy) are experiencing subsidence and structural damage because the wooden piles they were built on begin to rot when not submerged in groundwater (Aldrich and Lambrechts, 1986).

Contaminated sediments. Contaminated sediments are a consequence of historic effluent domination which presents a serious environmental challenge to water body restoration. In the US alone, the EPA estimates that 1.2 billion cubic meters in the upper 50 mm of sub aqueous sediments are sufficiently contaminated to pose health risks to the aquatic food chain (USEPA, 1997). Ninety-six watersheds have been identified by the EPA as suffering adverse effects from sediment contamination (USEPA, 1997). PCBs and metals are the two most prevalent classes of chemical pollutants.

Arguments about reversibility or irreversibility and calls for a downgrade of designated uses of effluent dominated and other impaired urban waters become less relevant if the integrity goals of currently impaired effluent dominated/dependent water bodies are considered in the long term, looking 20 to 50 years into the future instead of 10 to 15 years planning horizon typical for TMDL studies. If history is any proof of dramatic regional and national changes then in the past fifty years most heavy industries have abandoned many cities, public transportation by trolleys and trains was replaced by automobiles, almost all freeways were built, population moved from the cities to the suburbs, heavily polluted urban streams became much cleaner today than they were thirty to fifty years ago, chemical fertilizers and pesticides were introduced and became widely used for application on urban lawns and golf courses, and diffuse pollution was not known nor recognized as a problem fifty years ago.

PATH TO RESTORATION AND SUSTAINABILITY
Sustainable development

Sustainable development has been defined as "development that meets the needs of the present without compromising the ability of future generations to meet their own needs" (Brundtland, 1987). The chapters by Speers (2007) and Heaney (2007) in this book contain similar references of sustainability; The National Research Council (1999) outlined global pathways for fostering sustainable development. In elaborating concepts of sustainable development, the literature has emphasized that people – including city dwellers – are participants in ecosystems, and that they are ultimately dependent upon the renewability of ecosystem resources and services. Communities must therefore find ways to live adaptively within the loading capacity (waste assimilative capacity, carrying capacity) afforded to them by the ecosystems of which they are a part (Rees, 1992). The linkages between socioeconomic and ecological systems mean that people must pay attention to the protection, and if necessary, the re-creation of resilient, self-organizing ecosystems that have the capacity for self-renewal in the wake of disruptions.

It should be noted that if the definition of ecological sustainability is extended to urban ecosystems the understanding of "sustainability" does not necessarily imply a return to pre-development ecological conditions. The emphasis is on restoration of viable and resilient aquatic biota and letting the present and future generations enjoy and live in harmony with the urban water resources and their surroundings.

Level of treatment of effluents and stormwater and emerging technologies

Unlike fifty to eighty years ago, the level of treatment of urban wastewater already achieved in many communities and establishments (e.g., Sacramento, CA; Brookfield, WI, Patriots stadium waste water reuse and disposal facility in Foxborough, MA) has reached such efficiency that the effluent quality is not much different or is even better than the upstream water in the receiving stream. Some wastewater utilities, justifiably, have changed their name to "water reclamation district" (Chicago, Denver) rather than waste disposal and sewerage agency. Some communities now also effectively control combined and sanitary sewer overflows and store and treat urban runoff (e.g., Chicago (IL), Milwaukee (WI), Denver (CO), Malmö and Göteborg (SW)). Urban rainwater has been collected and used for millennia in some areas of the world. Technologies are available and within reach to rediscover the vast potential of these sources and redesign the urban drainage system to maximize the benefits of reuse. Planned and managed reuse of the highly treated/reclaimed effluent could be one important component of the rehabilitation of the effluent dominated/dependent streams.

Considering the treated effluent as a resource may require new ways of management. Today, some utilities argue that because the effluent dominated river flow is used for irrigation some distance downstream, nutrient removal is not needed in spite of the fact that the water quality downstream of the discharge point is degraded by dense algal growths and other symptoms of riverine eutrophication (low DO in the morning hours, turbidity, large quantities of biodegrading biomass). Again, the CWA and water quality regulation do not allow such water quality degradation and may require an advanced nutrient removal. However, nutrients in the effluent are a resource that, if advanced treatment is applied, would have to be replaced by industrial fertilizers if the river water is used for irrigation. A better solution would be a dual effluent scheme. A part of the effluent flow needed to sustain the balanced aquatic life would receive advanced treatment for nutrient removal and be released into the stream and the second part would receive secondary treatment and be transported by a man made conveyance channel/pipeline and marketed for irrigation. The mandatory integrity goals apply only to natural water bodies.

Total hydrologic (water cycle) and water quality balance

Making urban effluent dominated/dependent streams sustainable will require total urban hydrologic management with ecological integrity as a focus. Water supply, stormwater management, wastewater disposal, groundwater levels, and stream flow are components of the same system and should be managed in an integrated fashion with the integrity goals in focus. Figure 13.1 in the Preface to this monograph (Novotny and Brown, 2007) shows the concept of the total urban hydrological cycle and closing the loop. The Total (Integrated) Urban Water Management, its applications and ramifications are discussed in this book in a chapter by Heaney (2007). The integrated plans and management may use the following tools:

1. *Water conservation*
   Each m^3 of flow conserved and saved in the water distribution system represents more than 1 m^3 of flow (because of system losses) in the stream available for aquatic life and other in-stream uses. Typical conservation methods include low water flow toilet flushing, grey water reuse, use of stored rainwater for garden and lawn irrigation, change from water demanding lawns to xeriscape. Slowing or even reducing withdrawals is a global goal (NRC, 1999). Also minimizing water losses from the distribution systems (leaks) and storage reservoirs (evaporation) can bring significant savings as documented in this book by Breckenridge (2007). However, it is important that the saved water is used for ecologic restoration and not for further development. This rule of ecological reuse is in place in Reno, Nevada (Denis, 2006) and should be considered everywhere.

2. *Capture, store and reuse of rainwater* (surface runoff) and use it for
   - groundwater recharge
   - local irrigation (rain gardens)
   - low flow augmentation

3. *Effluent reclamation and reuse*
   - for irrigation (e.g., golf courses) to decrease withdrawals
   - flow augmentation to sustain aquatic life

– aesthetic enhancement of effluent dependent streams
– groundwater recharge and storage for future nonpotable reuse

The idea of the total water cycle balance is being promoted by urban hydrologists (Heaney et al., 2000; Marsalek et al., 2001) and is well advanced in Australia (Mitchell et al., 1996; Clark et al., 1997) or Sweden (Swedish Water Institute). Sustainable urban water management leads to (Grotte and Otterpohl, 1996; Heaney et al., 2000; Breckenridge, 2004):

1. Minimization of the distance of water and wastewater transportation
2. Use of stormwater from roofs and collected in cisterns, preferably for water supply (supplementing and blending with the other sources) instead of discharging it without use
3. Not mixing water food cycle with the water cycle. Not mixing wastewater of different origins, including mixing rainwater and waste water in combined sewers.
4. Decentralization of the urban water system and not allowing human activities with water if local integration into water cycle is not possible
5. Increasing the responsibility of individual homeowners for their impact on local water systems

This concept may lead to treated effluent and stored storm water *blending* to enhance flows and *water flow trading* being promoted and tested on a pilot scale by the Charles River Watershed Association in a USEPA sponsored project (Bowditch, 2007). Environmental water flow trading in effluent dominated/dependent and other impaired basins is based on the idea that stored and/or recharged stormwater with subsequent base flow enhancement may partially reduce the requirement for a high degree of treatment of effluents into impaired waters, especially during low or no natural flow situations when aquatic life would be in danger. Utilities managing their waters on the total hydrologic cycle concept should get credit or even market to other communities the flow enhancement achieved by water conservation and stormwater management.

Urban aquatic habitat

In spite of the challenges described above, effluent dominated/dependent waters can and should provide suitable habitat and water quality for the biological communities living in and around them. To rehabilitate and restore damaged ecosystems, new watershed and water body engineering and management technologies may need to be developed. Progress has been made in developing approaches that define the *ecological potential* of ecosystems, which can serve as a basis for establishing realistic goals in ecosystem restoration projects (Mitsch and Jørgensen, 1989; Pedroli and Postma, 1998). Essentially, the ecological potential identifies reasonable goals for rehabilitation and restoration projects in terms of reducing pollutant loads and restoring habitat quality and hydrology. Obviously, the quality of a stream reach running through a densely-populated urban area may never return to the condition of a pristine reach in a forested area. The ecological potential considers the reversible and irreversible modifications as an integral part of management. Ideally, achieving the ecological potential for an aquatic ecosystem should allow the ecosystem to conform to the standards of the CWA.

Methodologies for determining the ecological potential of surface waters and restoring urban water bodies to their ecological potential are concepts still in their infancy (Novotny et al., 2005). Several major restoration projects of urban streams have been conducted worldwide in the last ten years (Yoshikawa, 2004; Novotny et al., 2001). Some communities have focused on reclamation and beautification of the riparian zones of partially ephemeral effluent dominated/dependent streams (*e.g.*, the South Platte River in Denver, the San Antonio River in Texas or the Trinity River in Dallas) without attempting to fully restore the river flow and habitat to attain the ecologic potential of the water body.

Recent studies that relate Indices of Biotic Integrity (IBI) and their metrics (Karr et al., 1986; Barbour et al., 1999) to multiple stressors (Yuan and Norton, 2004; Virani et al., 2005) revealed that habitat and other nonpollutant stresses (e.g., embeddedness and bottom substrate quality, riparian zone quality) and morphological parameters (e.g., slope, flow and velocity) may have as much or more impact on the biotic integrity of the water body as concentrations of traditional pollutants, provided that mandated control of point source discharges are in place.

Defining the ecologic potential in quantitative terms is not straightforward and is evolving with time. For example, the goal of water quality abatement forty years ago was to prevent fish kills and major toxic impacts on aquatic biota and human health. Later the goal was elevated to prevention of chronic toxicity. Subsequent scientific and administrative efforts translated these narrative goals into numeric ambient water quality standards and, then, these standards were used to develop loading capacities of water bodies that limited the loads of pollutants. Today the goal is attaining and maintaining a "balanced aquatic biota" that would be resilient and protected, and providing primary sustainable recreation; however, corresponding numeric standards have not been developed for the biotic goals. Furthermore, it is becoming evident that the traditional focus on pollutant discharges may not be sufficient because impairment of integrity in many cases is caused by pollution, such as habitat impairment, that is not caused by pollutant discharges (Novotny et al., 2005). In an adaptive management approach the solutions focus on known causes of impairment and rank them according to their impact on impairment of integrity and develop integrated shorter term solutions for the most obvious stressors and monitor the impact of abatement. Concurrently, new research and modelling efforts will quantify the effects of less dominant stressors and, after the assessment of the phased abatement is completed, the ecologic goals and the plans are updated to achieve the new goals.

Once the ecologic potential is determined, restoration of damaged streams is a key step towards sustainability. It does not make much sense to require a high degree of treatment if the effluents being discharged into effluent dominated waters are discharged into a water body converted into a concrete lined flood conveyance channel. Unfortunately, the TMDL process is not designed to initiate restoration.

On the other hand, restoration can only take place if the pollution by pollutants in the contributing watershed is fully controlled, which can be accomplished in the TMDL planning process.

Green cities – smart growth initiatives

Mayors of many major cities, county executives of urban counties, USEPA, environmental activists (e.g., Sierra Club), and other community interests have been promoting the Green City – Smart Growth ideas and programs. In many respects, it is a grass roots effort to incorporate ecological principles into urban planning and development. The goals are generally broad, and ecologically balanced aquatic systems or addressing the problem of the effluent dominated streams may not be in the forefront of the public dialogue. For example, the criteria for the title of the "most Green City" includes a percentage of people participating in public transportation or walking to work, changing public transportation (buses) to more environmentally benign vehicles, building parks and greenbelts, reducing or removing freeways, opportunities for recreation, energy efficient buildings, or even an access to locally and organically grown fresh food (http://www.thegreenguide.com). Also, biotic integrity of urban waters can be linked to the good riparian habitat and conditions (Yuan and Norton, 2004; Ohio EPA, 1999) that mainly include parks and greenbelts desired by green city advocates and the population. In reverse, poor water quality and habitat make the riparian zones undesirable for green concepts. Replacing imperviousness or installing green zones and strips in streets and parking lots, an important component of green development, could be hydrologically functional and filter out pollutants and have significant pollution reduction benefits. Such measures are important Best Management Practices for reducing urban diffuse pollution. Therefore, there is an urgent need to find a common ground, goals and communication between urban architects, officials, urban diffuse pollution control managers and other specialists and stakeholders. Ecologically oriented ideas of urban landscape that would help to and dilute flows by green design aimed at groundwater recharge by excess runoff and improving base flow are discussed in chapters by Hill (2007) and Ahern (2007). Restoration, brownfield clean–up and greening of the urban landscape surrounding the rivers is an integral part of the restoration effort, which then may progress in the following phases:

- Mitigating, as best as possible, the cause of water quality and habitat deterioration, using TMDL planning that
  - estimates loading capacity of the water bodies and develops corresponding BMPs (storage, treatment, infiltration, etc.) for diffuse pollution and point source controls (storage and treatment) to meet the limits of the Loading Capacity with an adequate Margin of Safety and considering also background sources,
  - considers treated effluents and stormwater as a resource and develops necessary storage and reuse components.
- Restoring floodplain and riparian habitats and reconnecting the stream to them. This will also reduce flushness of the flows.

Figure 13.2. San Antonio River is a beautifully landscaped impounded semi-artificial channel attracting tourists but suffering from water quality degradation and impairment of integrity.

- Restoring and recreating riparian wetlands and green zones. This will improve water quality and wild life habitat and will have a positive effect on hydrology of the stream (e.g., improving base flow) and also on water quality.
- Restoring and recreating in stream habitat and eliminate habitat fragmentation.
- Daylighting the buried streams.
- Bringing the water body back to the community by providing pedestrian and bike paths, water boating and riparian zone parks

None of the above restoration components can be over-looked. An example of the San Antonio (TX) River shows the potential of urban rivers may have for revitalizing the city and tremendously improving the economy by developing the river surroundings (Figure 13.2). However, because of the river development into a semi-artificial beautifully land-scaped channel, the river is not a functioning ecosystem. The San Antonio River is still a degraded water body and essentially an artificial lined channel (personal observation by the author in June 2006) suffering from high turbidity, high nutrient content and hypoxia spots. Nevertheless, the city has become a prime tourist attraction.

EXAMPLES OF EFFLUENT DOMINATED WATERS AND THEIR POTENTIAL

The Lower Des Plaines River (Illinois). This river is the largest effluent dominated river in the US, carrying the flows and effluents from the Chicago Metropolitan Area. It becomes the Illinois River (still effluent dominated) after joining the Kankakee River and, after crisscrossing the state, the rivers joins the Mississippi River. The Lower Des Plaines River is impounded and is a major inland shipping artery connecting Lake Michigan with the Mississippi River. It carries almost 60 m^3/s of treated effluent from the Chicago Metropolitan Area, while the natural low flow in the Des Plaines River upstream from the confluence with the Chicago Sanitary and Ship Canal (CSSC) is less than 4.5 m^3/s (AquaNova International/Hey and Associates, 2004; Novotny et al., 2007).

Until the beginning of the 20th century, wastewater from Chicago, including CSOs, was discharged into the Chicago and Calumet Rivers and was conveyed into Lake Michigan or directly into the lake. The polluted discharges into the lake, which is the main source of drinking water for the metropolis, created WSW cycle with severe public health consequences. In the 1870s and 1880s, Chicago had the highest per capita municipal typhoid rate in the United States (Macaitis et al., 1977). In 1889, the Illinois State Legislature created the Chicago Sanitary District to solve this acute health problem. The District is the predecessor of the Metropolitan Water Reclamation District of Greater Chicago (MWRDGC). As a solution to problems with the unhealthy water quality of the Chicago and Calumet Rivers, ground was broken in 1892 by the District (MWRDGC) for the CSSC. The 45-km- (28-mile-) long canal, which is wider and as deep as the Suez Canal, was completed at the beginning of the 20th century. The canal reversed the flow direction of the Chicago River into the CSSC and, subsequently, into the Lower Des Plaines River. The Calumet–Sag Channel, also reversing the flow of the Calumet River into the CSSC, opened in 1922. The result of these massive projects was the diversion of all wastewater effluent flows, stormwater, and CSOs into CSSC as well as a virtual elimination of any overflows into the lake. By 1917, typhoid deaths (per capita) dropped to the lowest level for major cities in the nation (Chicago Public Library, 2005). Subsequent legal agreements between the Great Lakes states and Canadian provincial governments authorized a total diversion of 90 m^3/s (3200 cfs) of Lake Michigan water into the CSSC and the Lower Des Plaines River. This diversion allowance also included all wastewater and runoff discharges, which otherwise would be flowing into the lake. Figure 13.3 shows the channelized river in Joliet, IL.

The Lower Des Plaines River consists of two impoundments, an upper smaller and highly constricted by embankments Brandon Road Dam Pool, and a lower, more natural impoundment of the Dresden Island Pool. Significant progress has been made in improving water quality at the Stickney, Calumet, North Shore, Joliet, and other water reclamation (wastewater treatment) plants discharging into the Lower Des Plaines River system (Figure 13.4). Approximately 85% of the CSO discharges from the Chicago metropolitan area are now conveyed into the Tunnel

Figure 13.3.   The Brandon Pool of the Lower Des Plaines River in Joliet (IL). Most of the river flow is treated effluents from the Chicago metropolitan area.

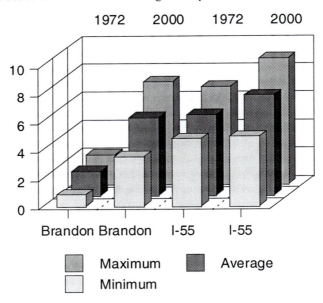

Figure 13.4.   Historic Dissolved Oxygen concentrations in the Lower Des Plaines River in the Brandon (upper) and Dresden (Interstate I-55 bridge) Pools. Dramatic improvement of DO in the Brandon Pool were achieved by treatment in MWRDGC plants. The Dresden Pool DO is also improved by aeration over the Brandon Rd. Dam spillway.

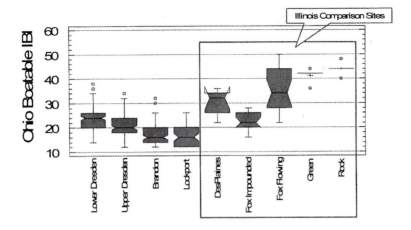

Figure 13.5.   Index of Biotic Integrity of the Lower Des Plaines Rives (Brandon and Dresden sites) and impounded reference streams. Before the confluence with the Kankakee River (Lower Dresden Pool) the river is reaching the ecologic potential of impounded Fox River. Lockport site is on CSSC. Green and Rock Rivers are medium size impounded references and represent the upper limit of integrity (Novotny et al., 2007).

and Reservoir Project (TARP) system and receive treatment in the Stickney plant, the largest secondary treatment plant in the world.

An extensive and successful Use Attainability Analysis was prepared in which an upgrade from an inferior use (Secondary Contact and Indigenous Aquatic Life) to a Modified General Use was proposed and is now being implemented by the Illinois Environmental Protection Agency (AquaNova/Hey and Associates, 2004; Novotny et al., 2007).The General Use in Illinois is the use complying with the CWA Section 101a integrity goals. Currently, the Lower Des Plaines River is moderately impaired due to limited sediment contamination and high temperature and will recover after further actions proposed by the UAA (that apparently substituted also for a TMDL) will be carried out following the implementation of the new Modified General Use standards. After removal of the last major pollution problem (the heated discharges from power plants and implementing disinfection and, possibly, supplemental aeration), the Lower Des Plaines River may achieve the ecologic potential for stressed impounded rivers. It is expected that the lower part of the river (the Dresden Pool) will be capable of supporting a balanced biota and be safe for limited primary recreation.

The South Platte River (Colorado). The South Platte River has a drainage area of 62,830 km^2 (24,300 sqmi). It originates in the Rocky Mountains. The river has been heavily hydrologically modified by changes in land use and diversions and today the river is highly regulated. Over 65% of the Colorado population lives along a 30 mile long segment of the river in Denver and other cities. Most of the river flow upstream of the densely populated areas is diverted for irrigation. By the time the South Platte River reaches downtown Denver from its pristine headwaters

Figure 13.6.   The flow deprived Trinity River near Dallas (TX). Flows have been withdrawn upstream to provide water for the Dallas – Ft. Worth Metropolitan Area. Downstream, the river becomes effluent dominated.

in the Rocky Mountains, the low flow rate is reduced to 1 m^3/s or less, a fraction of its original rate. Downstream from the Denver regional treatment plant, the treated wastewater effluent increases the river flow to 10 m^3/s, hence, the river water is 90 percent effluent dominated. During storms, the river also receives large quantities of urban runoff, causing flooding to occur. This river has a potential for flow augmentation, effluent and stormwater blending and reuse for irrigation or flow augmentation, and water conservation effects.

The Trinity River (Texas). This river (Figure 13.6) is a lifeline for the Dallas-Fort Worth Metropolitan Area and several smaller communities (Trinity River Authority, 2001). The waters in the Trinity River Basin are divided into the following groups: Effluent Dominated Streams, Small Urban Streams, Streams with Natural Base Flow or Reservoir Based Flow, Ephemeral Streams, and Lakes. Because of the withdrawals the river has very little flow near Dallas-Ft. Worth. The flows in the streams in the river basin are highly variable, most of the flow is surface runoff and between rains the flow is low or in summer time dry. Wherever there is a wastewater treatment plant discharging to these hydrologically ephemeral streams, the effluent flow constitutes the majority, often all, of the flow. There are 31 reservoirs in the Trinity River Basin and seven outside the Trinity River Basin supplying or are planned to supply water to Trinity Basin users. The benefit of storage and providing water by reservoirs may be partially offset in the warm and dry Texas climate by surface evaporation losses. Water quality of the basin and in the effluent dominated streams has been steadily improving. Several clean perennial streams can be used as reference.

Flow and effluent load trading are possible as well as stormwater storage and effluent reuse for restoration. Plans have been prepared for rehabilitation of the river near Dallas and in Forth Worth (Trinity River Authority, 2001).

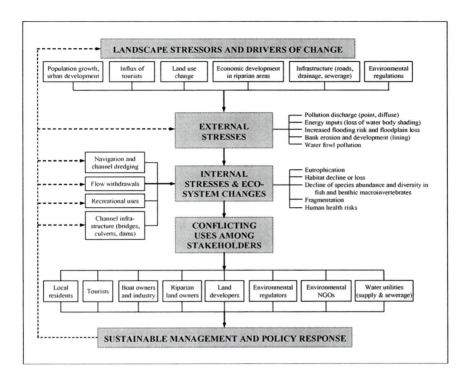

Figure 13.7.   Concept of urban ecosystems with technological and socioeconomic stressors and impacts (based on Turner et al. 2004).

BARRIERS ON THE PATH TOWARDS SUSTAINABILITY

Aquatic ecosystems are governed by a wide range of institutional arrangements at the federal, state, and local level. The policies of the federal Clean Water Act encompass far-reaching concerns with the sustainability of aquatic ecosystems, but federal regulatory programs such as the NPDES permit system are only one component of the multi-layered institutional frameworks that shape societal efforts in pursuing such policies. Municipal zoning laws, private and public land ownership, state water rights laws and state environmental regulatory systems all have significant effects in defining the interests of stakeholders and in establishing the scope of their authority to use, manage, and protect aquatic resources. The laws, regulations and agencies implementing them are fragmented and often do not co-operate. The complexity and multitude of players in the path towards sustainability is shown on Figure 13.7.

The federal Clean Water Act recognizes, through its broad definition of "pollution," and through its statement of policies and purposes in Section 101 of the act, that many types of man-made alterations in water bodies may impair the integrity of aquatic ecosystems. Designations of water uses that need protection under federally mandated water quality standards have focused attention on the many different ways that human activities affect ecological functions in water bodies. Recent developments in the implementation of the Clean Water Act have served to expand the effective reach of federal regulatory oversight in important ways. For example, the focus on pollution loadings from nonpoint sources (see court decision of *Pronsolino v. Nastri*, 291 F.3d 1123 (9th Cir. 2002), Northern California) has helped to broaden the scope of scientific studies and policy planning efforts conducted under the aegis of the Clean Water Act's TMDL program.

The Clean Water Act's regulatory programs have spurred state and local agencies to launch a variety of important initiatives for protecting or restoring healthy ecological conditions in urban waters. Such initiatives have included efforts to control diffuse pollution sources, remediate contaminated sediments, reduce urban erosion, control combined sewer overflows, reduce pollution from separate storm sewer systems, recharge aquifers, reuse stormwater, and protect and restore habitat. However, there are many barriers that should be overcome on the path towards sustainability.

Technical barriers

- Urban development has made urban areas highly impervious and hydrologically and ecologically unsustainable. Imperviousness would have to be reduced which may be resisted.
- Satisfying multiple demands on uses of urban water bodies while protecting ecosystem functions requires complex scientific modeling and data collection.
- Incorporating integrated total water balance concepts into basin planning and management requires significant conceptual and technical advances.
- Historical modifications to water bodies and watersheds may pose significant constraints on realistically-available options for changes in regional resource management.

Socio-economic barriers

- Stakeholders with significantly differing interests and preferences make multiple and often conflicting demands on uses of urban water bodies.
- Cooperation among planners is limited, and integrated approaches to urban management are rare, even where Green City and Ecosystem Restoration efforts are underway.
- Public financing for environmental restoration projects is limited, and public support for additional fees and taxes may be low.
- Consideration of all relevant ecological implications in construction and operation of urban water supply and sewerage projects may require significant changes in approaches to benefit/cost analyses
- Urban populations may be reluctant to give up wasteful water use practices in the absence of significant incentives or penalties.

- The financing and construction of major landscape and infrastructure modifications can sometimes raise important issues of equitable distribution of costs and benefits among different members of the urban population.

Legal barriers

- Jurisdictional divisions among federal, state, and local governments and among different administrative agencies at each level of government may hinder coordinated regulation and restoration of water resources.
- Boundary lines among property owners, both private and public, in land and water resources, may result in fragmented decision-making, and integrated water resource management in the public interest may be difficult.
- Programmatic distinctions among water supply services, stormwater management, and wastewater disposal make cooperation and integrated action by relevant authorities difficult.
- Zoning bylaws and other land use requirements may prevent construction of building projects in accordance with principles of low impact development.
- Subdivision ordinances and other requirements may encourage the creation of impervious surfaces and rapid conveyance of stormwater, with detrimental ecological consequences.
- Limited legal mechanisms are in place to regulate some aspects of ecological impairment in aquatic ecosystems, such as nonpoint source pollution.

CONCLUSIONS AND RECOMMENDATION FOR FUTURE RESEARCH

In the terms of the water resources system optimization, the total water cycle management conceived by urban water engineers (e.g., Mitchell et al., 1996) has the objective of maximizing water availability for urban uses and may have ecological quality limits as a constraint. Such limits are typically set by numeric chemical and bacteriological standards. The Unified Management for sustainable urban water system may reverse the focus. Achieving the ecologic potential of the receiving waters and aquifers is the moving goal, i.e., it will be approached by adaptive management. Water availability is then continuously or step by step optimized to achieve or approach the ecologic potential. Green development parameters (reducing imperviousness, natural drainage, infiltration, surface storage by vegetation, reclamation of riparian zones and their conversion to "green" buffers and storage) should be added to the decision variables. Flow optimization is then achieved by the total hydrologic balance model of the watershed.

It is quite likely the currently impaired effluent dominated urban waters will not return to their pre-development status because of

- Irreversible modifications and withdrawals
- Conflict between increased water scarcity and demand
- Irreparable legacy pollution in sediments

However, it is possible to achieve a resilient ecological status that would have a balanced indigenous biota and attain most of the water quality goals expressed by the standards. The ecologic potential is a moving target that is approached by adaptive management which is a stepwise long term process towards the ecological goal under the condition of uncertainty. Sustainability and resilience of the new ecologic potential can be maintained by management of the total hydrologic cycle.

Future research

The following areas of research can be identified:

- Unifying the concept of urban landscape ecology and water body integrity.
- Identifying the most efficient landscape surface, green areas, and drainage that would enhance and maintain the water body integrity.
- Developing regional total water balance models that would have attainment and maintaining (resilience/sustainability) of integrity of urban waters as a goal and optimizing the beneficial water uses for the population, reduce diffuse pollution and protect the receiving waters.
- Continuing and enhancing development of wastewater treatment technologies. The treatment should be tailored also to the safe and beneficial reclamation and reuse that will be increasing in future cities.
- Researching potential for using blended and treated urban stormwater and treated effluents for landscape irrigation and flow augmentation.
- Elimination of socio-economic and legal barriers must be researched, tested on pilot watersheds and potentially successful remedies included into the legal instruments and public education by schools and mass media.
- To achieve sustainability, most urban water bodies must be managed. This will require real time monitoring and control (RTMC). While RTCM has been researched on a plant by plant basis, models for basin wide real time control are still being researched and developed. RTCM of highly effluent stressed dominated streams considers variable effluent loads, flow augmentation, artificial supplemental aeration, storage on land (retention), in tunnels, and riverine reservoirs. While RTCM models have been developed, their pilot testing and implementation in the US has been limited to the control of combined sewer overflows and treatment plant bypasses (see chapter by Labadie, 2007).

ACKNOWLEDGEMENTS

The author acknowledges the assistance of Professor Lee Breckenridge of Northeastern University who provided the legal insight on the issues related to the effluent domination. The author's work on the use attainability issues was a part of the Lower Des Plaines River UAA study sponsored by the Illinois Environmental Protection Agency. The views expressed in this paper are solely those by the author and not of any funding agency.

REFERENCES

Ahern, J. (2007). Green infrastructure for cities: The spatial dimension, in *Cities of the Future: Towards integrated sustainable water and landscape management* (V. Novotny and P. Brown, eds.), IWA Publishing, London, Proc. of the 2006 Wingspread Conference.

Aldrich, H. P. Jr, and J.R. Lambrechts (1986). Back Bay Boston, Part II: Groundwater Levels. *Civil Engineering Practice: Journal of the Boston Society of Civil Engineers*/ASCE; 1986 Fall 1:2 ISSN: 0886–9685.

AquaNova International/Hey Associates (2004). *Lower Des Plaines River Use Attainability Analysis.* A report submitted to the Illinois Environmental Protection Agency, Springfield, IL.

Barbour, M.T., J. Gerritsen, B.D. Snyder, and J.B. Stribling (1999). *Rapid Bioassessment Protocols for Use in Streams and Wadeable Rivers: Periphyton, Benthic Macroinvertebrates, and Fish, Second edition,* EPS−841-B-−99/002, US Environmental Protection Agency, Washington, DC.

Bowditch, K. (2007). Restoring the Charles River watershed using flow trading, in *Cities of the Future. Towards integrated sustainable water and landscape management* (V. Novotny and P. Brown, eds.), IWA Publishing, London, Proc. of the 2006 Wingspread Conference.

Breckenridge, L.P. (2004). Maintaining instream flow and protecting aquatic habitat: Promise and perils on the path to regulated riparianism, *West Virginia Law Review* 106:595–628.

Breckenridge, L.P (2007). Ecosystem resilience and institutional change, in *Cities of the Future: Towards integrated sustainable water and landscape management* (V. Novotny and P. Brown, eds), IWA Publishing, London, Proc. of the 2006 Wingspread Conference.

Brundtland, G. (ed.). (1987). *Our Common Future: The World Commission on Environment and Development,* Oxford University Press, Oxford.

Butts, T.A., R.L. Evans, and S. Lin (1975). *Water Quality Features of the Upper Illinois Waterway,* Illinois State Water Survey, Report of Investigations No 79, Urbana, Illinois.

Chicago Public Library (2005). 4: The Big Ditch - http://www.chipublib.org/digital/sewers/history4.html.

Clark, R., A. Perkins, and S.E. Wood (1997). *Water sustainability in urban area – An Adelaide and Region Case Study.* Report One – An Exploration of the Concept. Dept. of Environment and Natural Resource, Adelaide, Australia.

Denis, G. (2006). The financial implications of supporting fast growth, Paper presented at the ACE06, *AWWA Annual Conf. and Exposition. Session TUE22*, San Antonio, June 11–15.

Grimmond, C.S.B., T.R. Oke, and D.G. Steyn (1986). Urban water balance 1. A model for daily totals, *Water Resources Research* 22(10):1397–1403.

Grotte, M., and R. Otterpohl (1996). Integrated Urban Water Concept, *Proc. 7th International Conference on Urban Storm Drainage*, Hanover, Germany, p. 1801–1806.

Heaney, J. (2007). Centralized and decentralized urban water, wastewater, and storm water systems, in *Cities of the Future: Towards integrated sustainable water and landscape management* (V. Novotny and P. Brown, eds), IWA Publishing, London, Proc. of the 2006 Wingspread Conference.

Heaney, J.P., L. Wright, and D. Sample (2000). Sustainable Urban Water Management, Chapter 3 in Field, R., J.P. Heaney, and R. Pitt (Eds.) *Innovative Urban Wet-Weather Flow Management Systems*. Technomic Publishing Co., Lancaster, PA, p. 75–120.

Hill, K. (2007). Urban ecological design and urban ecology: an assessment of the state of current knowledge and a suggested research agenda, in *Cities of the Future: Towards integrated sustainable water and landscape management* (V. Novotny and P. Brown, eds), IWA Publishing, London, Proc. of the 2006 Wingspread Conference.

Imhoff, K. (1931). Possibilities and limits of the water-sewage-water-cycle, *Eng. News Report,* May 28.

Imhoff, K. and KR. Imhoff (1993). *Taschenbuch der Stadtentwässerung (Pocket book of Urban Drainage – in German),* 28th edition, R. Oldenburg Verlag, Munich, Germany.

Karr, J.R., K.D. Fausch, P.L. Angermeier, P.R. Yant, and I.J. Schlosser (1986). (Assessing biological integrity of running waters. A method and its rationale, *Illinois Natural Hist. Survey, Spec. Publ. #5,* Champaign, Il.

Koenig, H.W., G. Rincke, and K.R. Imhoff (1970). Water re-use in the Ruhr Valley with particular reference to 1959 drought period, Pap I-4, Proc. of the IAWQ Water Pollution Research Conf, San Francisco, CA, *Advances In Water Pollution Research,* Pergamon Press, 1971.

Labadie, J.W. (2007). Automation and real-time control in urban stormwater management, in *Cities of the Future: Towards integrated sustainable water and landscape management* (V. Novotny and P. Brown, eds), IWA Publishing, London, Proc. of the 2006 Wingspread Conference.

Macaitis, B., S. J. Povilaitis, and E. B. Cameron (1977). "Lake Michigan diversion - stream quality planning," *Water Resources Bull.* 13(4):795–805.

Mitchell, V.G., R.G. Mein and T.A. McMahon (1996). Evaluating the resource potential of stormwater and wastewater: An Australian perspective; *Proc. 7th International Conference on Urban Storm Drainage,* Hannover, Germany, p. 1293–1298.

Marsalek, J., Q. Rochfort and D. Savic. 2001. Urban water as a part of integrated catchment management, chapter 2, p. 37–83. In C. Maksimovic and J.S. Tejada-Guibert (ed.), *Frontiers in urban water management: deadlock or hope?* IWA Publishing, London, UK.

Mitsch, W.J. and Jørgensen, S.E. (1989). *Ecological Engineering: An Introduction to Ecotechnology.* John Wiley & Sons, New York, NY.

National Research Council (1999). *Our Common Journey – A Transition toward Sustainability,* National Academy Press, Washington, DC.

Novotny, V. (2003). *Water Quality: Diffuse Pollution and Watershed Management.* 2nd., J. Wiley, Hoboken, NJ.

Novotny, V. (2005). The Next Step - Incorporating Diffuse Pollution Abatement into Watershed Management, Invited Key Note Opening Address, *Water Sci. & Technol.(2005)* 51(3–4):1–9.

Novotny, V. (2007). Preface – cities of the future: the fifth paradigm of urbanization, *This monograph.*

Novotny, V. and H. Olem (1993). *WATER QUALITY: Prevention, Identification and management of Diffuse Pollution,* Van Nostrand Reinhold, New York, distributed by J. Wiley, Hoboken, NJ.

Novotny, V., K.R. Imhoff, P.A. Krenkel, and M. Olthof (1989). *Karl Imhoff Handbook of Urban Drainage and Wastewater Disposal,* Wile y, New York

Novotny, V. et al. (1997). *Use Attainability Analysis: A Comprehensive UAA Technical Reference,* Water Environment Research Foundation, Alexandria, VA, 1997

Novotny, V., D. Clark, R. Griffin, A. Bartošová and D. Booth (2001). *Risk Based Urban Watershed Management - Integration of Water Quality and Flood Control Objectives, Final Report.* Institute for Urban Environmental Risk Management, Marquette University, Milwaukee, WI.

Novotny, V., A. Bartošová, N. O'Reilly, and T. Ehlinger (2005). Unlocking the relationship of biotic integrity of impaired waters to anthropogenic stresses. *Water Research* 39:184–198.

Novotny, V., N. O'Reilly, T. Ehlinger, T. Frevert, and S. Twait (2007). A River is Reborn: The UAA for the Lower Des Plaines River, Illinois, *Water Environment Research* 79(1):68–80.

Ohio EPA (1999). *Association between Nutrients Habitat, and the Aquatic Biota in Ohio Rivers and Streams.* Ohio EPA Technical Bulletin MAS/1999-1-1, Columbus, OH.

Pedroli, G.B.M. and Postma, R. (1998). Nature rehabilitation in European river ecosystems: Three cases. In *New Concepts for Sustainable Management of River Basins*, (P.H. Nienhuis, et al., Eds.). 67–84. Backhuys Publishers, Leiden, The Netherlands.

Petruck, A., M. Beckereit, and R. Hunrck (2003). Restoration of the River Emscher – From an open sewer to an urban water body, *Proc. World Water and Environmental Resources Congress 2003* (P. Bizier and P. DeBarry, eds.), ASCE/EWRI, Reston, VA, CD-ROM.

Pickus, J.M., W.B. Samulels, and D.E. Amstutz (2002). Applying GIS to identify effluent dominated waters in California, *ESRI Conference*, gis.esri.com/library/userconf/proc02/pap0393/p0393.htm.

Rees, W.E. (1992). Ecological footprints and appropriate carrying capacity: What urban economist leaves out, *Environment and Urbanization* **4**:121–130.

Shanahan, P. and B.L. Jacobs (2007). Ground water and cities, in *Cities of the Future: Towards integrated sustainable water and landscape management* (V. Novotny and P. Brown, eds), IWA Publishing, London, Proc. of the 2006 Wingspread Conference.

Speers, A. (2007). Water and cities - Overcoming inertia and achieving sustainable future, in *Cities of the Future: Towards integrated sustainable water and landscape management* (V. Novotny and P. Brown, eds), IWA Publishing, London, Proc. of the 2006 Wingspread Conference.

Trinity River Authority (2001). *Trinity River Basin Master Plan,* Arlington, TX.

Turner, R.K., Bateman, I.J., Georgiou, S., Jones, A., Langford, I.H., Matias, N.G.N. and Subramanian, L. (2004). An ecological economics approach to the management of a multi-purpose coastal wetland. *Regional Environmental Change.* 4:86–99.

US Environmental Protection Agency - Region 8 (2003). EPA Region 8 - Effluent-dependent streams and the Net Environmental Benefit Concept, Discussion with the Colorado 309 Workgroup, April 28, 2003, San Francisco.

US Environmental Protection Agency (1994). *Water Quality Standards Handbook,* 2nd ed., EPA-823-b-94-005A, Office of Water, Washington, DC.

US Environmental Protection Agency (1997). *The Incidence and Severity of Contamination in Surface Waters of the United States, Vol. 1.* EPA 823-R-97-006. U.S. Environmental Protection Agency, Office of Water, Washington, DC.

U.S. Environmental Protection Agency (2002). *National Water Quality Inventory - 2000 Report,* Washington, DC.

US Environmental Protection Agency (2006). International Brownfield Study: Emscher Park, Germany, http://www.epa.gov/brownfieds/partners/emscher.html.

Virani, H., E. Manolakos and V. Novotny (2005). *Self Organizing Feature Maps Combined with Ecological Ordination Techniques for Effective Watershed Management,* Tech. Report No.3, Center for Urban Environmental Studies, Northeastern University, Boston, MA.

Yoshikawa, K. (2004). On the progress of river restoration and the future view in Japan and Asia, *Proc.* 3rd *European Conf. On River restoration,* Zagreb, Croatia, May 17021, 2004.

Yuan, L.L. and S.B. Norton (2004). Assessing the relative severity of stressors at a watershed scale, *Environmental Monitoring and Assessment,* 98(1–3):323—349.

PART FIVE

Integrated Solutions – Water and Landscape

14

Reclaimed stormwater and wastewater and factors affecting their reuse

H. Furumai

The University of Tokyo, 7-3-1, Hongo, Bunkyo-ku, Tokyo 113-8656, Japan
E-mail: furumai@env.t.u-tokyo.ac.jp

Summary: Concern about the sustainability of urban water use is the strong motivation to understand the potential of rainwater and reclaimed wastewater use. The history of water works in Tokyo and their experience may provide useful information to develop sustainable urban water use and find future possible tasks. Besides, various innovative strategies to meet the current and future water demand may help us to consider new approaches adjusting to the developing mega cities. In this paper, the past and current practices on utilization of latent water resources such as rainwater, reclaimed wastewater and seepage water in Tokyo are summarized from the viewpoint of sustainable water use. From the aspect of human health risk, contamination by new micropollutants such as estrogens, endocrine disrupters and surfactants should be considered besides the conventional ones. Groundwater recharge through the infiltration facilities provide a potential storage of water resource which can be withdrawn in the future if necessary.

INTRODUCTION

As the demand for natural resources has been increased by growing population and industrial development, it is said that freshwater will be the first resource to

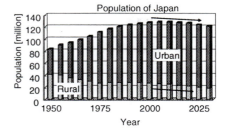

Figure 14.1. Urban and rural populations in Southeast Asia and Japan. The Southeast Asia includes Cambodia, Indonesia, Lao People's Dem. Rep., Malaysia, Myanmar, Philippines, Thailand, and Viet Nam. Data from United Nation Population Division (2003).

run short (Wagner et al., 2002). Due to the rapid progress of urbanization with economic development in Japan, natural water cycle has partially been damaged, and various problems have been occurring, such as instability of river flow, drying up of spring water and deterioration of the ecosystem. Our struggling experience should be transferred to other countries which are going to be faced to rapid growth of urban population for their developing urban water management. Especially, the stable water supply and efficient water use have become concerning in mega cities with high urban activities. Concern about the sustainability of urban water use is the strong motivation to understand the potential of rainwater use and water recycling.

Figure 14.1 shows the past and predicted growth of urban and rural populations in Southeast Asia and Japan. Economic developments in the last few decades in Southeast Asia have imposed urban areas to experience rapid demographic change. The predicted rapid growth in urban population in Southeast Asia shows similar trend in Japan's early urbanization stage from 1950s to 1970s. As the urbanization progresses, water abstraction from surface water and groundwater inevitably increases to support human activities. Thus urban water consumption has a major factor to impact on natural water cycle and brings quite a lot of changes in the aquatic environment.

The achievement of the sustainability is required ensuring a long-term supply of water with adequate quality for all designated purposes minimizing adverse economic, social and ecological impacts. Implementation of sustainable water management will likely require combination of policy/regulation and technology and participation of citizens. Thus, we believe that our past struggling experiences and current practices of water use in Tokyo are useful information to consider possible countermeasures towards sustainable urban water use in mega cities in Southeast Asian countries. In this paper, water supply and water balance in Tokyo are briefly summarized. From the viewpoint of sustainable water use, current practices on utilization of latent water resources such as rainwater and reclaimed wastewater are summarized.

HISTORY OF WATER SUPPY AND WATER BALANCE IN TOKYO

Escalated water demand and the following problems

Figure 14.2 shows the representative locations with ground subsidence problems and their cumulative subsidence changes over time. In Tokyo, the large scale groundwater pumping started in 1914. In the following years the number of deep wells with large diameter increased rapidly. Ground subsidence was first observed in Tokyo in the 1910's, and in Osaka in the 1920's. The groundwater level continued to fall due to the extensive pumping to support increased production activities. Ground subsidence caused the destruction of buildings and damages from floods and high tides, and raising public concern. By around the mid-1940s, the damage to industry in World War II reduced the industrial use of groundwater, thereby halting ground subsidence. With the rapid progress of urbanization with economic development after the World War II, water supply had to catch up with remarkably increased water demand for domestic and industrial purposes. There was a sudden surge in demand for groundwater, and striking subsidence was again seen, especially in large urban areas. Countermeasures against ground subsidence, e.g. control of groundwater pumping rates, were started in the 1960s and thereafter the rate of subsidence in metropolitan areas has slowed.

In some regions, however, large amounts of groundwater are still pumped for tap water, agriculture and melting snow, as well as for industry. At present, ground subsidence still continues to occur in such places as the suburbs of metropoli-

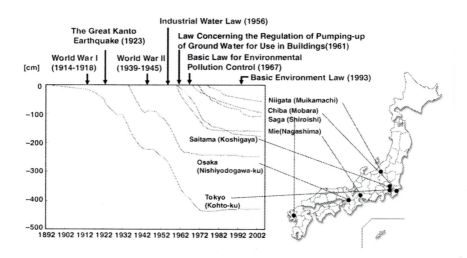

Figure 14.2.   Cumulative change in ground subsidence in typical area over time Source: http://www.env.go.jp/en/water/wq/pamph/page35-36.html.

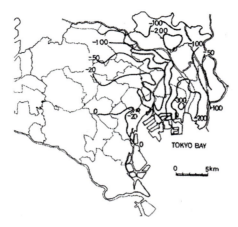

Figure 14.3.   Ground subsidence profile in Tokyo from 1935 to 1975 (depth unit: cm).

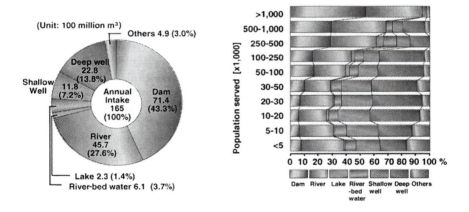

Figure 14.4.   Composition of water sources in Japan (FY2001) and their distribution by size of water supply works (Source: http://www.jwwa.or.jp/water-e04.html).

tan Tokyo (e.g. Saitama prefecture), rural regions (e.g. Saga Plain) and in snowy regions facing to the Japan Sea (e.g. Niigata prefecture). Ground subsidence problems remarkably appeared in wide region within Tokyo as shown in Figure 14.3. After the government imposed the withdrawal restriction to control ground subsidence, the limited availability of groundwater resulted in shifting water resource from groundwater to surface water. As shown in Figure 14.4, water supply in Japan depends on mainly river and impounding reservoirs to meet entire portion of water demand. As a result, the total volume of the use of impounding reservoirs as water resource in fiscal year (FY) 2001 was brought up to 43.3%, an increase of 12% compared with FY 1965. The ratio of reliance on impounding reservoirs is increasing year by year.

    The increase of water consumption in big cities made the water balance partially
distorted and various problems occurred such as decline of groundwater level, poor
river water flow and deterioration of the aquatic ecosystem by dam construction.
The water resource originally had depended on Tama River in Tokyo until the rapid
industrial development and economic growth took place. The escalating water
demand led to extensive water withdrawal at the upstream of Tama River and the
construction of Ogouchi-dam. In addition, new water resources were developed
in Tone and Sagami rivers flowing in outside of Tokyo. Currently, Tone river
system is the main source and contributes 78% of water supply in Tokyo, while
ground water occupies only 0.2%. Therefore, we have made efforts to economize
and make efficient use of water from watershed out of Tokyo, which leads to
reduce water intake from natural water system and to secure the sound water
cycle in and out of Tokyo. This indicates that the water demand in mega cities
is to be fulfilled with several alternatives of latent water resources under several
constraints.

Water balance in Tokyo

Figure 14.5 shows the water balance in Tokyo based on the average metrological
data during the period from 1988 to 1997 (Tokyo Metropolitan Government, 1999).
The average annual rainfall is 1405 mm in Tokyo. The precipitation is distributed
to runoff (634 mm: 45%), infiltration (359 mm: 26%), and evapotranspiration (412
mm: 29%). The runoff fraction is remarkably higher in Tokyo than the nationwide
average, which is almost one third of the precipitation. The imperviousness is
estimated to be greater than 80% within the special districts of the 23 wards. In
spite of the poor infiltration and limited water availability within Tokyo, the human
consumption is more than the 1100 mm equivalent rainfall including recycle water
use (199 mm) and reclaimed wastewater use (5 mm).

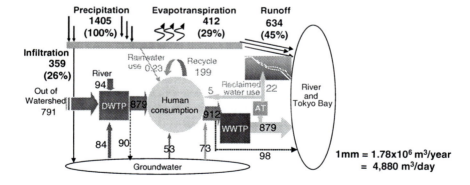

Figure 14.5.   Water balance and recycle in Tokyo (unit: mm/year) DWTP: Drinking water
treatment plant, WWTP: Wastewater treatment plant, AT: Advanced treatment, Data source
from Tokyo Metropolitan Government (1999).

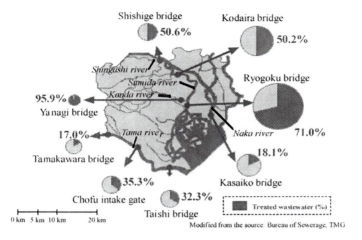

Figure 14.6. Proportion of treated wastewater in urban river flow at the ordinary water-level.

The Tokyo metropolitan area relies on surface waters for almost 90% of its water supply. Rapid increase of water consumption proportionally increases the wastewater discharge, which further worsen the urban aquatic environment until sewage treatment system is completed. As shown in Figure 14.6, proportion of treated wastewater has become larger in urban river flow at the ordinary water-level. There are significant impacts of effluent on river environment. A large amount of treated wastewater should be applied for advanced treatment and regarded as new water resources in urban area.

ALTERNATIVE WATER RESOURCES FOR SUNTAINABLE WATER USE IN TOKYO

Rainwater harvest and its use

Tokyo has been adopting various innovative approaches to supply water beyond its natural endowment, so as to elaborate a water resource management to meet the current and future demands. Rainwater harvesting for miscellaneous use such as toilet flushing and water cooling is employed in an individual scale as well as in a large scale. The Sumida ward is a frontier region, where rainwater harvesting systems are actively introduced in small to large scales and the subsidy system was early provided in 1995. Two sports stadiums (Kokugikan for Sumo wrestling and Tokyo Dome mainly for baseball) and the Sumida Ward office building have large-scale rainwater harvesting systems as pioneering approaches. Table 14.1 lists their detailed configuration. Such rainwater harvesting practices have met 20–60% of their water demand on a building basis.

Table 14.1. Large scale rainwater storage facilities for water use in Tokyo.

Place (year)	Effective capacity [m^3]	Purposes
Kokugi-kan (1985)	750	toilet flushing, cooling water
Sumida-ward office (1988)	1,000	toilet flushing
Tokyo Dome (1988)	1,000	toilet flushing

Rainwater use was a small amount (0.23 mm in 1998) in the whole water balance of Tokyo as shown in Figure 14.5. However, storage of rainwater is also a useful measure for water demand in emergency cases. In addition the rainwater use can work as a kind of environmental education to make citizens aware of sustainable urban water use. There are 850 facilities (566 public buildings and 284 private buildings in FY 2001) for rainwater use in Tokyo, which are exemplified by the above-mentioned rainwater storage tanks.

Reclaimed wastewater for miscellaneous use

Reclaimed wastewater use has the following benefits: it increases water supplies by reducing demand for higher-quality water; it reduces wastewater discharge, thus reducing water pollution; and it is economically efficient as it means lower water costs compared to transporting water from distant sources. Increased wastewater could become an abundant and stable resource to satisfy most water demands, as long as it is adequately treated to ensure the water quality appropriate for the use. It should be recognized that the reclaimed wastewater is obtainable within urban area.

In water-short cities and "green buildings", reclaimed wastewater is used for flushing toilets. Similarly, it is used for cooling, cleaning, and dilution in industrial plants, but separate pipeline systems are required for both uses (Ogoshi et al., 2001). An innovation is the processing of wastewater into ultra-clean water using membrane filtration technology. There is now such plants producing the so-called "NEWater" in Singapore, and the water has been used for indirect portable water and special industrial purposes that require water with a high degree of purity.

There are many practices in reclaimed wastewater use in Tokyo, in which a huge water volume has been utilized for various purposes such as washing, water-cooling, toilet flushing, waterway restoration and creation of recreational water-front. The reclaimed wastewater use has been carried out in both open-loop and closed-loop. Table 14.2 summarizes the recent state of water recycling for miscellaneous purposes in Tokyo. There are 560 facilities, which are categorized into individual building, block-size, and large-scale recycling systems. The large scale recycle is supported by the supply of treated wastewater from municipal reclamation plants. The total supply of the treated wastewater has reached around 500,000 m^3/day in FY2003, including reuse as cooling and cleaning water for inside of water reclamation plants.

Table 14.2. Facilities of water recycling for miscellaneous purposes in Tokyo (FY 2001).

Recycle type	Number of facilities	Recycled water use [m³/day]	Recycle percentage* [%]
Individual building	293	43,809	22
Block-wide	170 (50 blocks)	20,167	22
Large-scale	97 (4 regions)	17,062	27
Total	560	81,038	23

*Recycle percentage = (Recycled water use/Total water use) × 100.

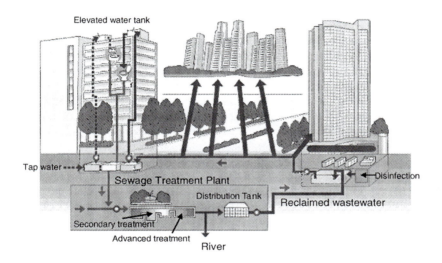

Figure 14.7. Reclaimed wastewater use in skyscrapers in Shinjuku, Tokyo.

A model example of closed-loop use for toilet flushing applied in skyscrapers in Shinjuku sub-center covering 80 ha in 1984 is shown in Figure 14.7. The large-scale recycling system was followed in other districts of Tokyo such as Ariake (1996), Shinagawa (1997), Osaki (1998), and Shiodeme (2002). In the Shiodome area, high quality reclaimed wastewater has been supplied by introducing the new wastewater reclamation system with ozone resistant membrane at the Shibaura reclamation plant since April 2004. It treats 4,300 m³/day of secondary efflu-ent. The new reclaimed wastewater was used for sprinkling over water retaining type pavements to alleviate a heat island phenomenon in the urban area as well as for toilet flushing and landscape irrigation. This would be the most innova-tive recycling use of treated water. The sprinkling reuse for streets might also contribute to reduction of urban non-point pollution loads from road surface in Tokyo.

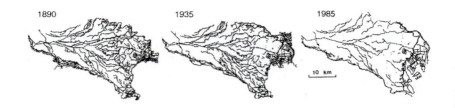

Figure 14.8.   Disappearance of small rivers and waterways over the past century in Tokyo.

Reclaimed wastewater for environmental use

The rapid urbanization caused small rivers and waterways to be covered by concrete and pavement in order to create new space and to keep citizens away from offensive odor and dirty view. Urban rivers and waterways were heavily polluted by discharged domestic wastewater, until the sewerage system was constructed. In addition, the excess water intake at the upstream of the Tama river could not provide distribution of water flow to the water channels flowing through the residential region in Tokyo. Figure 14.8 shows that a rapid dwindling of many waterways and streams with the urbanization. One of the recent challenges in Tokyo is to restore the disappeared waterways by introducing reclaimed wastewater.

Addition to close-loop use, the application of reclaimed wastewater to open-loop use has been also promoted in Tokyo, because citizens have high demands for regeneration and restoration of the dried-up river and waterways. Especially sewage treatment plants produce stable and abundant flow, which can supplement their demand to increase the water flow to restore the water front and aquatic ecosystem. There are several experiences of reclaimed wastewater use for landscape irrigation, recreational and environmental purposes in Tokyo. They are summarized in Table 14.3 with recent restoration practices by reclaimed wastewater.

Extensive built-up of subsurface railways and underground structures have occupied a large subsurface volume in Tokyo. Such constructions have seemed to

Table 14.3.   Use of reclaimed water and seepage water for landscape and recreational purposes.

Type	Source	Place	Flow rate [m³/day]
Reclaimed water	Tamagawa-jyoryu	Nobidome-channel	15,000
from sewage	Treatment plant	Tamagawa-channel	13,200
treatment plant		Senkawa-channel	10,000
	Ochiai Treatment plant	Shibuya river	
		Meguro river	86,400
		Nomi river	
Seepage water	JR East tunnel	Tachiai river	1,600
		Shugatami pond	1,400
		Shinomazu pond	270
	Tokyo Metro tunnel	Shibuya river	456

raise the groundwater level, being coupled with stringent control of groundwater withdrawal. Railway companies used to collect the seepage water in tunnels and drained into the sewerage system with payment. This was considered as the waste of good quality water, additional burden of charge for the companies and the extrawater to treatment system. Recently, collected seepage water began to be introduced into neighbouring recreational ponds and waterways for their restoration as listed in Table 14.3.

Standards and guidelines for reclaimed wastewater

It is very important to consider use of reclaimed wastewater from the viewpoint of not only quantity stability but also quality risk that could rise under different conditions of water use. Therefore, the water quality should be characterized to evaluate their potential for reuse and recycling. The Ministry of Land, Infrastructure and Transportation has established and revised water quality criteria and guidelines on the reclaimed wastewater use for miscellaneous purposes since 1981. In April 2005, the water quality standard and facility standard were established for 4 respective reuse of treated wastewater (Tajima, 2005). The standard values are summarized in Table 14.4. Primarily, the reclaimed wastewater should be utilized securing hygienic safety to protect the public health. At the same time, aesthetic and operational aspects are considered to apply the water quality standards. In the case of human health-related uses, pathogens including virus should be considered as quality guideline parameter. In fact, the introduction of virus parameter was discussed in the Exploratory Committee for the standard establishment, but it was not included because of absence of commonly available detection method of viruses.

Occurrence of water soluble organic micro-pollutants has been well documented (Alder et al., 1997). Secondary effluents contain various man-made chemicals, some of which are collectively termed endocrine disrupting chemical, and natural estrogens (i.e. estradiol: E2, estrone: E1 and estriol: E3) at the significant level (Desbrow et al., 1998). The presence of the micro-pollutants limits the use of reclaimed wastewater for some purposes. We performed soil column experiments in consideration of soil aquifer treatment (Nakada et al., 2007). Secondary effluent was continuously fed into the soil column in a down-flow mode for 80 days. Water quality of the influent and effluent was evaluated by instrumental analysis and yeast estrogen screen assay. Removal efficiency is shown in Figure 14.9.

The removal efficiency of nonylphenol (NP), which has low biodegradability, showed a relatively low, whereas that of estrogens and estrogenic activity was consistently high throughout the experimental period. While phosphorus was effectively removed, unstable nitrogen removal was observed. It suggested that groundwater pollution by nitrate without proper denitrification.

Table 14.4. Standards and guidelines for reclaimed wastewater quality for miscellaneous use.

Use parameters	Flushing water	Sprinkling water	Water for landscape use	Water for recreational use
Escherichia Coli (CFU/100ml)	Not detectable	Not detectable	<1000 (Tentative values)	Not detectable
pH	5.8–8.6	5.8–8.6	5.8–8.6	5.8–8.6
Appearance	Shall not be distasteful	Shall not be distasteful	Shall not be distasteful	Shall not be distasteful
Turbidity (degree)	<2 (Control target)	<2 (Control target)	<2 (Control target)	<2
Chromaticity (degree)	–*	–*	<40**	<10**
Odor	Shall not be distasteful*	Shall not be distasteful*	Shall not be distasteful*	Shall not be distasteful*
Residual chlorine (mg/l)	>0.1(free) >0.4(combined)	>0.1(free) >0.4(combined)	Not be stipulated	>0.1(free) >0.4(combined)
Facility standards	Sand filtration or its equivalent	Sand filtration or its equivalent	Sand filtration or its equivalent	Precipitation + Sand filtration or its equivalent

*Based on the intent of the users, standard value shall be set as required.
**Based on the intent of the users, more stringent value shall be set as required.

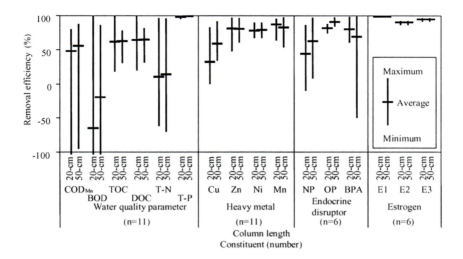

Figure 14.9. Averaged removal efficiency of each constituent with minimum and maximum value on the column experiment.

Infiltration facilities and runoff simulation

The sewerage system in Tokyo has been basically designed to cope with heavy rainfall of 50 mm/hr. The large impervious surface areas and the limited flow capacity of urban rivers have made the stormwater flow management extremely important. Being water conservation conscious city, the Tokyo Metropolitan Sewerage Bureau constructed a new type of sewerage system (Experimental Sewer System; ESS) in 1980's that incorporates infiltration facilities over a 1400 hectares area (Fujita, 1984). The infiltration facilities include infiltration inlets, trenches, LU curbs and permeable pavement. Figure 14.10 illustrates the basic scheme of infiltration facilities in the ESS.

The construction of infiltration facilities mainly aimed at inundation control in rapidly urbanized areas by reduced stormwater peak flows and at reduction in improvement works of streams and rivers. These infiltration facilities provide secondary benefits such as recharge of groundwater and reduction of non-point pollutant loads from urban surfaces. The groundwater recharge is considered essential to secure the sound water cycle in Tokyo. Groundwater aquifer is generally important for the potable water resource. Although Tokyo has restricted and controlled groundwater use these days, it is a potential storage of water resource and can be withdrawn in the future if necessary.

In order to promote introduction of infiltration facilities, it is necessary to evaluate their functions quantitatively. Runoff simulation is useful for evaluating reduction of non-point pollution load as well as effect of groundwater recharge. The positive evaluation of non-point pollutants trapping by infiltration facilities would promote the introduction of infiltration facilities in urbanized area. In that

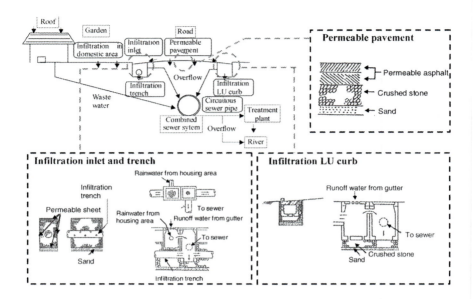

Figure 14.10.   Configuration of infiltration facilities in the experimental sewer system

case, it is especially important to express runoff behaviour quantitatively in frequently occurring rainfall events. The utilization of the distributed runoff model is
advantageous for separate handling of roof and road runoff pathways. We have focused urban non-point pollution and their runoff behaviour (Furumai et al., 2002;
Pengchai et al., 2004; Murkami et al., 2005a) and carried out runoff analysis using
a distributed model (Murakami et al., 2004; Fuurmai et al., 2005; Murakami et al.,
2005b). In addition, field surveys and investigation of heavy metal deposition have
been conducted at long-term working infiltration facilities in a highly urbanized
drainage area (Aryal et al., 2005; Aryal, 2006).

Recent study on field measurement and runoff simulation (Furumai et al., 2005;
Murakami et al., 2005b) showed that the infiltration facilities still work well and
contribute the urban runoff reduction by 80 percents for weak and medium rainfalls.
Continuous runoff monitoring was conducted at the outlets of a drainage district
which have 64 manholes, 161 infiltration inlets, 2329 m infiltration LU curbs
and 0.57 ha permeable pavements in the catchment area of 7.15 ha, by the Tokyo
Metropolitan Sewerage Bureau from June to December in 2002. The 66 monitored
events were classified into five groups using cluster analysis with the Ward method,
in which z-scores (standardized scores) of total rainfall, average rainfall intensity
(ARI) and maximum rainfall intensity (MRI) were variables. The events are plotted
within the domain of total rainfall and MRI in Figure 14.11.

Figure 14.12 shows the accumulative rainfall and calculated flow volumes of
runoff and infiltration for seven months. The main fractions of the total rainfall
volume of 67,400 m^3 (equivalent to 942 mm in this drainage area) are accounted

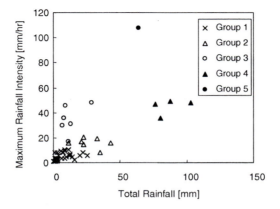

Figure 14.11.   Total rainfall and maximum rainfall intensity of 5 classified rainfall groups.

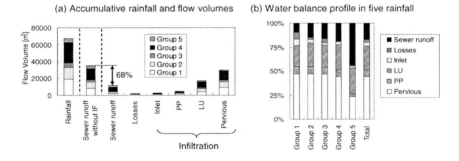

Figure 14.12.   Water balance in the runoff simulation for the serial 66 rainfall events. (a) Cumulative rainfall, outflow and infiltration volume, (b) Water balance profiles in 5 rainfall groups (SR: sewer runoff which means the excess runoff from infiltration facilities to sewer pipes, IF: infiltration facilities, Losses: rainfall losses on impervious area, Inlet: infiltration inlets, PP: permeable pavements, LU: infiltration LU curbs, Pervious: pervious areas such as park and field).

by Group 1, 2, and 4 rainfalls. The result shows that infiltration facilities functions to reduce 68% of the excess runoff volume to sewer pipes during the period of the seven months. This graph also shows the infiltration volume through different types of infiltration facilities. This comparative illustration is helpful to understand the contribution of each infiltration facility to groundwater recharge. The infiltration LU curbs contributed to groundwater recharge among the facilities, although the flow volume through pervious surface is the greatest. Except the Group 5, there is no significant difference in the water balance profiles among the rainfall groups.

The Group 5 rainfall caused greater sewer runoff than other rainfall groups due to more surface runoff from pervious areas at strong rainfall intensity.

Non-point pollutants in infiltration facilities

The performance of the infiltration facilities in pollutants control has not been investigated in detail. The aim of our research work (Aryal et al., 2005; Aryal, 2006) is to understand the accumulation of heavy metals in the infiltration facilities in Tokyo with depth and possible release of heavy metals from the infiltration facilities. Mason et al. (1999) and Pitt et al. (1999) suggested that stormwater pollutant possibly causes the groundwater contamination through infiltration facilities. On the contrary, it is reported that road runoff pollutants such as heavy metals and polycyclic aromatic hydrocarbons are trapped at the infiltration facilities and their deposition occurs slowly with time (Mikkelsen et al., 1996). However, most of the heavy metals in sediment are readily mobilized in low pH (Tyler and McBride, 1982; Lee et al., 1996). Prolonged deposition of heavy metals in the infiltration inlets might have caused to release the heavy metals to the infiltrated water.

Figure 14.13 shows the heavy metal contents in the top 1 cm sediment samples of infiltration inlets, road dust and soil samples in pervious fields. When we compare the sediment and the soil, Zn and Pb content are significantly higher in the infiltration inlet. This shows a potential accumulation of Zn and Pb (compared to other metals) in the area. The heavy metals content in the sediment were similar to the road dust and much higher than the soil in most of the cases. This indicated that the origin of the heavy metals is probably the road dusts. The higher heavy metals content in the infiltration inlets shows their accumulation and possible threat to the groundwater contamination by leaching. To the contrary, the low heavy metal con-

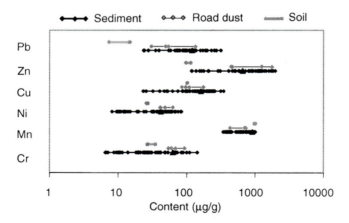

Figure 14.13.   Heavy metal content range in sediment, road dust and soil Triangle plot (▲) in the bar of sediment indicates the average content.

tent in the sediment than the soil or road dusts indicates the possibility of leaching of heavy metals from infiltration inlets to the ground. The heavy metal contents in the sediment were high compared to the background soil (except for Mn). This indicates that heavy metals accumulated in the infiltration inlets are mainly from anthropogenic sources. A similar range of heavy metal contents in sediments and road dusts indicate that road dust is one of the major probable sources to the sediment although roof and other impervious surface runoffs may also contribute.

The infiltration facilities possibly serve as place for both "sink" and "source" of urban non-point pollutants. Therefore, it is important to clarify the mechanism of the wash-off and resuspension of sediment within infiltration inlets for runoff quality behaviour in order to discuss their function of pollutant reduction and pollutant sources. It is also desirable to investigate the role of physical (pH, temperature, particle sizes) and chemical (organic content, metal oxides) parameters in heavy metals release/desorption from the infiltration inlet sediments.

CONCLUSION

Tokyo has taken several measures to meet the increased water demand and to find alternative ways towards sustainable water use, while urban development has altered the nature of water cycle. As a result, rainwater and reclaimed wastewater have been applied for miscellaneous purposes. As well as use of reclaimed wastewater in closed-loop system, restoration projects of small rivers and waterways by reclaimed wastewater has also been promoted, which is recognized as the forerunner of water recycling measures from the viewpoint of the sustainable urban water use. Similarly, the rainwater harvesting in a small scale provides a model example for the sustainable approach in mega cities. The application of infiltration facilities for the flood control is an example to promote groundwater recharge for securing the sound water cycle. Besides, it would be a farsighted approach to provide potential water storage for the future. These past and current practices and concept of rainwater and reclaimed wastewater use in Tokyo will be worthy of consideration in the urban water management of future mega cities.

The use of reclaimed wastewater and recharged groundwater is highly expected especially in urban areas where great amount of water resource is dependent to outside of their watershed. Considering that wastewater can contain various pollutants including emerging micropollutants such as estrogenic compounds, pharmaceuticals, and personal care products, we almost reach to the limit to rigorous and minute water quality monitoring by chemical analysis. Perfect determination of the margin of safety is a really difficult task due to limited information on monitoring quality data and toxicity/infectiveness for enormous chemicals and pathogens. Therefore, a multiple evaluation method is required based on chemical analysis and bioassay to characterize water quality and label their safety/risk for various beneficial water uses.

Regarding to reclamation of road runoff water by infiltration facilities, the heavy metal release/desorption from deposited sediment particles should be investigated under possible environmental conditions in the infiltration inlets. In addition, we have to develop better structure and design of facilities considering the function of non-point pollutant trapping and maintenance method. In other words, infiltration facilities should be defined as a urban infrastructure not only for runoff control but also non-point pollutant control.

ACKNOWLEDGEMENTS

The study was conducted as a part of a Core Research for Evolutional Science and Technology (CREST) entitled "Risk-based Management of Self-regulated Urban Water Recycle and Reuse System" in "Hydrological Modeling and Water Resources System" supported by Japan Science and Technology Agency.

REFERENCES

Alder, A.C., Siegrist, H., Fent, K., Egli, T., Molnar, E., Poiger, T., Schaffner, C., and Giger, W. (1997). The fate of organic pollutants in wastewater and sludge treatment. Significant processes and impact of compound properties. CHIMIA 51(12):922–928.

Aryal, R.K., Furumai, H., Nakajima, F., and Murakami, M. (2005). Sediment deposition and heavy metals accumulation in infiltration inlets installed for two decades in Tokyo, 1st IWA-ASPIRE conference, an Singapore.

Aryal, R.K., Furumai, H., Nakajima, F., and Murakami, M. (2006). Prolonged deposition of heavy metals in infiltration facilities and its possible threat to groundwater contamination, *Water Science & Technology*, 54(6–7):205–212.

Desbrow, C., Routledge, E.J., Brighty, G.C., Sumpter, J.P., and Waldock, M. (1998). Identification of estrogenic chemicals in STW effluent. 1. Chemical fractionation and in vitro biological screening. Environ. Sci. Technol. 32(11):1549–1558.

Fujita, S. (1984). Experimental sewer system for reduction of urban storm runoff. Proc. of 3rd Int. Conf. on Urban Storm Drainage, Gutenberg, Sweden.

Furumai, H., Balmer, H., and Boller, M. (2002). Dynamic behavior of suspended pollutants and particle size distribution in highway runoff, Water Science & Technology 46(11–12):413–418

Furumai, H., Jinadasa, H.K.P.K., Murakami, M., Nakajima, F., and Aryal, R.K. (2005). Model description of storage and infiltration functions of infiltration facilities for urban runoff analysis by a distributed model, Water Science & Technology 52(5): 53–60

Lee, S-Z., Allen, H.E., Huang, C.P., Sparks, D.L., Sanders P.F., and Peijnenburg, W.J.G.M. (1996). Predicting soil-water partition coefficients for cadmium. Environmental Science and Technology 30: 3418–3424.

Mason, Y, Ammann, A.A., Ulrich, A., and Sigg, L. (1999). Behavior of heavy metals, nutrients, and major components during roof runoff infiltration, Environ. Sci. Technol., 33(10):1588–1597.

Mikkelsen, P.S., Hafliger, M., Ochs, M., Tjell, J.C., Jacobsen, P., and Boller, M. (1996). Experimental assessment of soil and groundwater contamination from two old infitratio systems for road run-off in Switzerland. Science of the total Environment 189/190 (8–9):341–347.

Murakami, M., Nakajima, F., and Furumai, H. (2004). Modelling of Runoff Behaviour of Particle-Bound Polycyclic Aromatic Hydrocarbons (PAHs) from Roads and Roofs Water Research, 38(20):4475–4483.

Murakami, M., Nakajima, F., and Furumai, H. (2005a). Size- and density-distributions and sources of polycyclic aromatic hydrocarbons in urban road dust, Chemoshpere, 61(6):783–791.

Murakami, M., Furumai, H., Nakajima, F., Jinadasa, H.K.P.K., and Aryal, R.K. (2005b). Comparison of infiltration flows at three types of infiltration facilities in serial rainfall events with different characteristics, 10th International Conference on Urban Drainage (10ICUD), Copenhagen.

Nakada, N., Yaymashita, N., Miyajima, K., Suzuki, Y., Tanaka, H., Shinohara, H., Takada, H., Sato, N., Suzuki, M., Ito, M., Nakajima, F., and Furumai, H. (2007). *Multiple Evaluation of Soil Aquifer Treatment for Water Reclamation using Instrumental Analysis and Bioassay*, in H. Furumai, F. Kurisu, H. Katayama, H. Satoh, S. Ohgaki, N.C. Thanh (eds.) *Southeast Water Environment 2* London: IWA Publishing, pp. 303–310.

Ogoshi, M., Suzuki, Y., and Asano, T. (2001). Water reuse in Japan. Water Science & Technology 43(10): 17–23.

Pengchai, P., Furumai, H., and Nakajima, F. (2004). Source apportionment of polycyclic aromatic hydrocarbons in road dust in Tokyo, Polycyclic Aromatic Compounds, 24(4–5):713–789

Pitt, R., Clark, S., and Field, R. (1999). Groundwater contamination potential from stormwater infiltration practices. Urban Water 1(3):217–236.

Tajima, A. (2005). The establishment of the new technical standard for the treated wastewater reuse, Proc. Of the International Workshop on Rainwater and Reclaimed Water for Urban Sustainable Water Use, Tokyo, Japan. (http://env.t.u-tokyo.ac.jp/furumailab/crest/workshop05/june10pm_2.pdf).

Tokyo Metropolitan Government (1999). "Master plan of water recycle" (in Japanese). http://www.metro.tokyo.jp/INET/KEIKAKU/SHOUSAI/7094G100.HTM#B

Tyler, L.D. and McBride, M.B. (1982). Mobility and extractability of cadmium, copper, nickel and zinc in organic and mineral soil columns. Soil Science 134:198–204.

United Nation Population Division (2003). "Urban and Rural Areas 2003" http://www.un.org/esa/population/publications/wup2003/2003urban_rural.htm

Wagner, W., Gawel, J., Furumai, H., De Souza, M.P., Teixeira, D., Rios, L., Ohgaki, S., Zehnder, A.J.B., and Hemond, H.F. (2002). Sustainable Watershed management: An International Multi-watershed Case Study. Ambio 31(1):2–13.

15

Centralized and decentralized urban water, wastewater & storm water systems

James P. Heaney

University of Florida, Gainesville, FL 32611, USA
E-mail: heaney@ufl.edu

Summary: General principles of urban water management are presented that might permit the development of more sustainable systems by integrating the traditionally separate functions of providing water supply, collecting, treating, and disposing of or reusing wastewater, and managing urban wet-weather flows for flood control, drainage and water quality control. Improved sustainability is defined in terms of the extent to which treated wastewater and stormwater can be reused as sources of supply for nonpotable water supply needs such as irrigation and toilet flushing. Several years ago, we were asked to explore innovative wet-weather control systems for the 21st century. A catalyst for this effort was earlier findings from the literature on sustainable urban water systems that decentralized control systems are preferable to maximize reuse. This chapter compares the findings from seven years ago with research results and trends that have developed during the interim period. Finally, projected future directions are described.

INTRODUCTION

With sponsorship from US EPA, the University of Colorado research team under the author's leadership and a team of stormwater consultants were asked to take a holistic view of options for innovative wet-weather control systems for the 21st century. With a primary focus of urban stormwater quality management, EPA's specific interest in the problem, we were also charged to evaluate newer ideas. This evaluation included urban water supply and wastewater systems because of the linkage among the three water infrastructure components via reuse to reduce the need for water supply for nonpotable purposes (Heaney et al., 1999). This chapter presents historical perspectives of the last forty years and an update to incorporate the more recent research findings.

PRINCIPLES OF INTEGRATED URBAN WATER MANAGEMENT

As early as the 1960's, urban water experts were advocating the need to take an integrated systems view of urban water supply, wastewater, and stormwater systems (Heaney et al., 1999). While these investigators were able to scope out the problem, they lacked the essential computational tools to conduct these evaluations. The unprecedented federal commitment to upgrade wastewater systems and correct combined sewer overflows along with areawide wastewater master planning (208 planning) moved the wastewater and stormwater fields forward during the 1970s. Simulation models such as EPA SWMM were used routinely to evaluate these complex wastewater and stormwater systems. Concurrently, research funding was available to stimulate new thinking in the field. The results were impressive:

- Virtually every major city in the United States upgraded their wastewater collection and treatment systems with the bulk of the funding coming from federal construction grants. These systems were typically large, centralized, facilities that were felt to be more cost-effective and easier to manage than dispersed, decentralized, facilities.
- Research and the results of numerous 208 studies showed that initiating wet-weather quality controls could be more cost-effective than tertiary treatment of wastewater.
- The results of infiltration/inflow studies typically indicated that it was more cost-effective to accept the infiltration/inflow, expand the capacity of the wastewater treatment plant, and add storage facilities to accommodate these larger flows.
- Major improvements in controlling CSOs were initiated that continue to this date, e.g., Seattle, Washington's 45 year history of CSO controls.
- Stormwater detention systems became popular as required control devices to reduce off-site flooding and drainage impacts and improve water quality.

Water supply planning during the latter part of the 20th century proceeded along traditional lines of upgrading and expanding facilities as needed. Growing concerns about water supply led some communities to initiate water reuse programs, e.g., in southern California and Florida.

During the mid 1990's, sustainability became a popular issue across the entire infrastructure field including the urban water field. A variety of definitions of sustainability for urban water systems have been advanced. A report by Heaney et al. (1999) listed the following principles for sustainable water infrastructure systems for the 21st century:

- Ideally, individual urban activities should minimize the external inputs to support their activities at the parcel level. For water supply, import only essential water for high valued uses such as drinking water, cooking, showers and baths. Reuse wastewater and stormwater for less important uses such as lawn watering and toilet flushing. Minimize the demand for water by utilizing less water intensive technologies where possible. For transportation, minimize the generation of impervious areas, especially directly connected impervious areas, for providing traffic flow and parking in low use areas.
- Minimize the external export of residuals from individual parcels and local neighborhoods. For wastewater, export only highly concentrated wastes that need to be treated off-site. Reuse less contaminated wastes such as shower water for lawn watering. For storm water, minimize off-site discharge by encouraging infiltration of less contaminated stormwater and use cisterns or other collection devices to capture and reuse stormwater for lawn watering and toilet flushing.
- Structure the economic evaluation of infrastructure options to maximize the incentive to manage demand by using commodity use charges instead of fixed charges. For water supply, assess charges based on the cost of service with emphasis on commodity charges. Charges should be a combination of a level of service that specifies flow, quality, and pressure. For wastewater, assess charges based on the cost of service with emphasis on commodity charges. Charges should be a combination of a level of service that specifies flow and quality. For stormwater, assess charges based on the cost of providing stormwater quality control for smaller storms and flood control for larger storms. Charges should be based on the imperviousness with higher charges for directly connected imperviousness and the nature of the use of the impervious areas and their pollutant potential. Some charge should be assessed for pervious areas. Credit should be given for on-site storage and infiltration.
- Assess new development for the full cost of providing the infrastructure that it demands, not only within the development, but also external support services.
- Implement policies to make drivers pay the full cost of using personal automobiles.

Sustainability continues to be a major goal of urban development. However, it is viewed in many different ways in terms of how it is actually implemented. The next four sections present the author's views on key advances and challenges in urban water supply, wastewater, stormwater, and the integration of these systems via reuse. The findings are drawn from the research during the past decade.

WATER SUPPLY

Supply

Water and energy shortages in the United States and around the world have emerged as major issues for the 21st century. In rapidly growing areas like Florida, cities are now expected to evaluate a variety of alternative water supplies as part of their application for new consumptive use permits or renewal of existing consumptive use permits. Traditionally, more than 90% of water use in Florida has been satisfied by pumping from local groundwater sources. These sources have been overdeveloped in many areas. Now cities must also evaluate the following alternative water supplies:

- Surface water sources.
- Desalination.
- Water reuse.
- Aquifer storage recovery.

These evaluations are done for local sources and nearby regional sources that are serviced by regional water supply authorities. Interregional water transfers can be considered but only after a strong argument has been made that cost-effective local options have been exhausted. There is strong political opposition to inter-regional water transfers.

Demand management

Major gains have been made in demand management during the past decade because the results of large-scale studies of water conservation studies have shown that these conservation practices are cost-effective and reliable. DeOreo et al. (1996) described a new method to measure water use for individual water use events at the household level. A continuous flow data or "flow trace" is recorded at 10 second intervals from a standard water meter using a data logger. This flow trace is disaggregated into discrete water use events using signal processing software called Trace Wizard. A pattern recognition program is used to assign fixture designations, e.g., toilet flush, clothes washer cycle, irrigation. A sample trace is shown in Figure 15.1 for a shower in a combination bathtub/shower. During the initial half minute, the user adjusts the flow rate and temperature of the tub faucet. Then the user switches to the shower option for the balance of the shower. This concept was originally developed for 16 homes in Boulder, Colorado. This success led to a national AWWARF sponsored study of 1,200 homes in 12 cities across North America (Mayer et al., 1999).

The results of the national water use study provided a firm database that showed conclusively that indoor water conservation devices such as low flow toilets work well and provide a high level of reliability that they will reduce indoor water demand. The distribution of residential water use for the 12 cities is shown in Table 15.1. Indoor water use averages about 226 liters per capita per

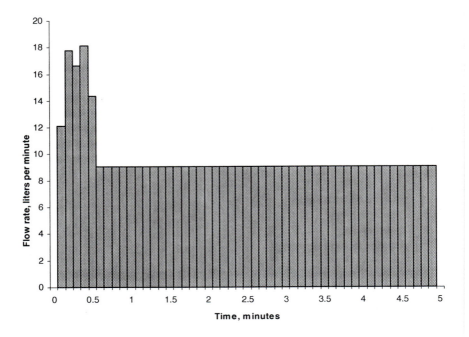

Figure 15.1.   Flow trace for a shower in a combination bathtub/shower.

day (liters/capita-day). Toilet flushing is the largest indoor water use followed by clothes washers and showers. Water efficient fixtures are required for most new developments. For example, a national regulation mandates the use of low flush toilets with a maximum flush rate of 6.0 liters per flush. Studies during the past decade indicate that per capita indoor water use can be reduced to about 115 to 150 liters/capita-day with an aggressive water conservation program. Indoor per capita use and its distribution are very consistent across the United States. Thus, it is appropriate to extrapolate findings from one city to other cities. Policy issues regarding water conservation relate to the how aggressive the utility should be in promoting or requiring the use of water conserving fixtures. More aggressive programs mandate changing to more efficient fixtures, e.g., toilet retrofits that are paid for by the utility.

Whereas indoor water conservation can be implemented for well defined costs with a high degree of reliability, outdoor water use patterns are much less predictable. For the 12 city study shown in Table 15.1, average outdoor water use is 382 liters/capita-day, much larger than the 226 liters/capita-day for indoor water use. However, outdoor water use is highly variable across the United States. For a given city, in-ground sprinkler systems use more water than manual systems. Often, the timers are set for maximum water use throughout the summer. Irrigation water use also heavily influences peak water use.

Table 15.1. Per capita water use statistics for the 12
city AWWARF study (Mayer et al., 1999).

Fixture/End Use	Average liters per capita per day	% of total
Toilet	70.0	10.8%
Clothes washer	56.8	8.7%
Shower	43.9	6.8%
Faucet	41.3	6.3%
Other domestic	6.1	0.9%
Bath	4.5	0.7%
Dishwasher	3.8	0.6%
Indoor Total	226.3	34.8%
Outdoor	381.5	58.7%
Leak	36.0	5.5%
Unknown	6.4	1.0%
Total	650.3	100.0%

http://www.aquacraft.com/Publications/resident.htm

Economists have advocated conservation pricing as a means to manage demand. Whitcomb (2005) divided single-family residential water customers across the state of Florida into four profiles based on assessed property values of homes. The demand curve for the average residential water user in Florida is shown in Figure 15.2. At relatively low water prices of $0.50/1,000 liters, per capita demand would be about 650 liters/capita-day. Indoor water use is about 225 liters/capita-day so about 2/3 of the water use is outdoor. If the water price is $1.00/1,000 liters, then the demand decreases sharply to about 450 liters/capita-day. Now the mix of indoor and outdoor water use is about equal. If the water price is $1.50/1,000 liters, then total water use drops to about 350 liters/capita-day and outdoor water use is only about 33% of total use. Finally, at $2.00/1,000 liters, water demand is about 250 liters/capita-day with outdoor water use constituting an even smaller percentage of total water use. As price increases, indoor water use can be expected to decline. Best estimates at present are that indoor water use will decrease from 225 to about 150 liters/capita-day due to the installation of low-flush toilets and other water saving devices. Thus, given expected increasing scarcity of water, prices should increase and people will use water more efficiently

Conclusions on water supply

The following conclusions can be reached regarding trends in water supply:

- Outdoor water use exceeds indoor water use in the more arid and warmer parts of the United States. Outdoor water use is relatively less important in the northeastern United States but it does have a strong effect on peak water demand.
- Accurate high frequency measurements and process models provide convincing evidence of the cost-effectiveness and reliability of conservation.

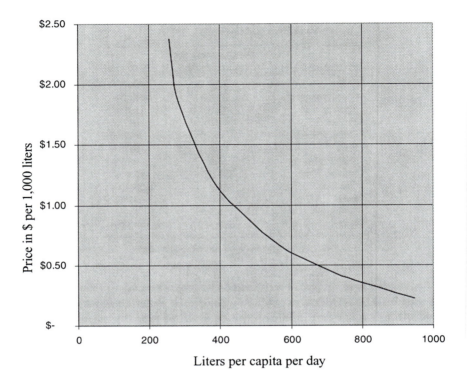

Figure 15.2.   Demand for water as a function of price in Florida (Whitcomb 2005).

- Indoor water demand can be reduced by up to 50% by retrofitting toilets, clothes washers, faucets, and fixing leaks.
- Outdoor water use is increasing in many areas due to more lawn area per capita and higher application rates.
- Conservation rates that charge a higher rate for outdoor water use can be an effective way to reduce demand especially for outdoor water use.
- Some of the water supply needs could be met by treated wastewater and recycled stormwater.

WASTEWATER

The issue of centralization vs. decentralization was an important topic in the late 1960's and 1970's as US EPA embarked on an unprecedented program to upgrade wastewater systems throughout the United States. Many urbanized areas had a large number of wastewater collection and treatment systems dispersed over the entire service area. The question arose as to whether an optimal size service area existed. Adams et al. (1972) evaluated how population density and size of service area affected unit cost per customer served. Total costs are the sum of collection plus treatment costs. Their results indicated that the optimal size of service area was less than 1,600 hectares if the population densities were less than 75 persons per hectare. Even for high population densities, per capita costs did not decrease significantly beyond service area sizes of 4,000 acres. It is difficult to generalize about the optimal size of service area for wastewater systems alone or integrated with water supply and stormwater systems since the solution is so dependent on the availability of receiving waters or land for the treated effluent.

For residential areas, wastewater flows should closely approximate indoor water use. However, this is rarely the case due to infiltration and inflow (I/I) into sewers. I/I consist of dry-weather and wet-weather sources. Wet-weather I/I can overload the sanitary or combined sewer system and cause backups into basements and onto the surface within the service area and/or treatment plant bypasses into receiving waters. I/I can be reduced considerably by improved sewer design, installation, and operation and maintenance practices. Special techniques are needed to locate I/I in order to optimize the maintenance and replacement schedules (Wright et al., 2006).

Control of combined (CSO) and sanitary sewer (SSO) overflows requires a mix of actions. Seattle, Washington has implemented numerous changes to ameliorate its combined sewer overflow problems during the past 40 years. They pioneered the use of inline controls to maximize wet-weather storage within the combined sewer system. They also initiated a sewer separation program. The net result has been a 90 percent reduction in annual overflow volumes (King County, 2005).

Many of the sanitary sewer systems in the United States are nearing the end of their service lives. Thus, strategies for handling CSO and SSO issues will impact the system for most, if not all, of the 21st century. Extensive I/I studies were done during the 1970's as part of the major EPA supported upgrade of wastewater treatment plants (WWTP). Cities had to evaluate whether it was more cost-effective to control I/I and thereby reduce the peak flows at the WWTP or to accept the I/I and provide centralized storage and additional treatment capacity to handle these larger flows. Generous federal subsidies for WWTPs induced cities to invest in WWTP rather than I/I control within the collection system.

STORMWATER

Urban stormwater varies in relative importance because of climatic variability. Water budgets indicated that annual urban runoff was about 20% of water use for Denver while it was about 100% of water use for New York City (Heaney et al., 1999). The quantity of stormwater per capita increased dramatically during the 20th century due to:

- more automobiles which require more streets and parking and create additional pollutant sources.
- larger houses on larger lots that increase runoff quantity and pollutant loads
- Growth of suburbia and much lower population densities.
- More contemporary urban area is devoted to parking than to human habitat and commercial activities.

- Low density urbanization generates over three times as much stormwater runoff per family than did pre-automobile land use patterns.

Major trends since 1999 related to urban stormwater systems have been:

- Significant growth in interest in "green developments" that are more energy and water efficient.
- Significant increase in use of low impact development concepts.
- Trend towards higher density mixed use developments that reduce the dependency on automobiles.
- Auto use continues to increase.

Wet-weather pollutant loads depend on several factors, including the characteristics of the source area and the rain energy and volume. Directly connected impervious areas (DCIA) contribute a disproportionate amount of the annual runoff and associated as shown in Figure 15.3 (Lee and Heaney, 2003). Over 70% of the annual runoff from the high density residential site is from DCIA that only comprises 44% of the area. Similarly, over 80% of the runoff from the highway site is from DCIA even though DCIA is only 18% of the study area.

Much of our research during the past five years has been directed to the question of how well stormwater BMPs, including decentralized on-site controls, work based on the available measurements of control effectiveness. These results are published in reports to the National Cooperative Highway Research Board (Low Impact Development Center et al., 2006; Geosyntec et al., 2006; Oregon State U. et al., 2006) and the Water Environment Research Foundation (Strecker et al.,

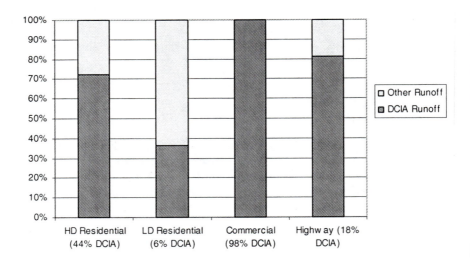

Figure 15.3. Comparison of the total annual runoff from directly connected impervious areas to the percent DCIA for high and low density residential, commercial, and highway sites in southeast Florida (Lee and Heaney, 2003).

2005). Key findings from these research efforts are listed below:

- There is an almost total reliance on measuring BMP performance based on comparing event mean concentrations for inflows and outflows without accounting for intra-event behavior or flow and storage effects.
- Suspended solids are measured using total suspended solids instead of the more complete suspended sediment concentration that includes larger particle sizes that can contain a significant portion of the pollutant load.
- Stormwater BMPs should be viewed as being ubiquitous within the urban area. Thus, control can be achieved by manipulating the rainfall-runoff-transport relationship as well as installing a downstream control.
- Low impact development (LID) onsite controls are popular because of the perception that they reflect a green engineering approach.
- LID controls are much more complex to analyze because there may be several of them just on a single parcel and they tend to rely on more complex infiltration processes.
- BMPs can be evaluated using common principles of physical, chemical, and biological process engineering.
- It is more productive to evaluate wet-weather loadings based on functional units such as roofs and parking lots as opposed to more aggregate categories such as medium density residential. GIS tools make it realistic to use this more accurate spatial disaggregation.
- EPA SWMM 5.0 can be used to evaluate a wide variety of BMPs either as separate controls or by simply altering parameters such as infiltration rates to represent the expected impact of a BMP. SWMM can be used to route water from subcatchment to subcatchment without restriction as to the nature of the subcatchment, e.g., pervious to DCIA to pervious.
- A spreadsheet version of SWMM has been developed that is linked to an optimizer to find least cost combinations of on-site and off-site controls (Lee et al., 2005; Heaney and Lee, 2006). The next step is to devise efficient ways to iterate between complex process models like SWMM 5.0 and simpler simulation/optimization models that can select cost-effective control systems efficiently.
- Little progress has been made in linking stormwater loads with receiving water impacts in a rigorous manner. Total maximum daily load (TMDL) studies should stimulate better data collection in this area since TMDL relies on measuring the health of the receiving water using an indicator constituent such as nitrogen concentration. If BMPs are installed and the water quality does not improve, then additional data will be collected to better define this linkage.

INTEGRATED SYSTEMS

Reuse of treated wastewater and/or stormwater can have a major impact on the demand for additional water supply since irrigation and toilet flushing represent a major part of urban water demand. Wastewater reuse is more popular than stormwater reuse at present.

The greatest use of reclaimed water has been in regions suffering water scarcity or severe restrictions on disposal of treated effluents (Koopman et al., 2006). Water utilities facing these difficulties have devised and implemented a variety of innovative approaches that utilize reclaimed water to help meet their communities' needs. Case studies of these projects have been summarized in several recent reviews, e.g., Crook (2004), Law (2003), and Radcliffe (2004). The water reuse industry in the U.S. has experienced rapid growth in recent years. Over the 18 year period from 1986 to 2004, both reuse flow and capacity in the state of Florida have increased by more than 300% reaching 2,380 ML/day and 4,810 ML/day, respectively, in 2004 (Florida DEP, 2005).

A major catalyst for reuse in Florida is that discharge of treated wastewaters to receiving waters is not generally allowed. Thus, land application is a very popular alternative. While Florida relies on groundwater as its main source of water, many Coloradoans rely on surface water. Boulder, Colorado uses surface water from the Rocky Mountains. All of its treated wastewater is discharged to Boulder Creek where it is reused multiple times by downstream cities and agricultural areas. Thus, we need to distinguish between the case of a city reusing its own treated wastewater and a downstream city using a combination of treated wastewater and "virgin" water that comprises its source water. A key contemporary water supply issue is to determine flow needs for the natural system. In south Florida, the water needs of the Everglades National Park as part of the $10 billion Everglades Restoration project have heightened competition for this water. In Boulder, Colorado, the City of Boulder purchased water rights to maintain base flows in Boulder Creek during the summer low flow periods.

The most significant option to integrate water supply, wastewater, and stormwater systems is to use reclaimed water and stormwater to satisfy outdoor water use demands and perhaps nonpotable indoor demands for toilet flushing. The technical feasibility of this integration depends upon the supply-demand balance for water supply, wastewater, and stormwater. Heaney et al. (1999) developed monthly water budgets for cities across the United States. Mitchell et al. (2003) describe Aquacycle, an Australian urban water budget model, that is used to evaluate integration of water supply, wastewater and stormwater. Sample and Heaney (2006) describe a methodology for evaluating the effectiveness of an integrated irrigation-stormwater control system using a continuous simulation model. Heaney et al. (1999) demonstrated that sufficient stormwater is available in many parts of the United States to satisfy average irrigation demands. A key question is the size of storage that is necessary to have stormwater available when it is needed for irrigation. Given that stormwater detention ponds exist in many urban areas, it can be cost-effective to retrofit them to provide irrigation water. The quality of the stormwater is a potential public health issue for irrigation reuse. Treatment requirements for reclaimed water exist and it is straightforward to meet them since the wastewater is delivered on a continuous basis. However, stormwater treatment is more problematic.

Satellite treatment

Regional wastewater management systems may be more expensive if reuse is included because the reclaimed water needs to be returned to the original source areas over longer distances. Reclaimed water distribution costs can be reduced by integrating satellite facilities for water reclamation into regional systems (Butler and MacCormick, 1996). Satellite facilities withdraw wastewater from a sewer near an area that needs the water, reclaim the liquid portion, and return the solids to the sewer (Okun, 2000). They maintain economies of scale for biosolids management, since the biosolids are still processed in a regional facility (Koopman et al., 2006). Satellite facilities lessen the hydraulic load on the regional treatment plant, thus delaying or ameliorating the need for capacity upgrades. They can also achieve higher qualities of reclaimed water. Satellite water reclamation facilities have been integrated into regional wastewater management systems since 1962. Most of the facilities use conventional process trains that include preliminary treatment to remove screenable materials and grit, primary settling, activated sludge, filtration, and disinfection with chlorine or UV. A few facilities with membrane bioreactors substituting for activated sludge and filtration are in operation, ranging in size from a 9 MGD plant in Tempe, Arizona to 0.01 MGD units demonstrated in Melbourne, Australia. Satellite water reclamation facilities greatly expand the potential for supplying reclaimed water to users throughout the sewer collection system at reasonable distribution costs.

RECOMMENDATIONS AND FUTURE RESEARCH

A variety of catalysts are stimulating new thinking and attitudes towards urban water infrastructure systems including:

- The utter devastation of Hurricane Katrina and the failure of the flood control and public health infrastructure associated with this disaster.
- Intensified competition for water supplies including provision of water for natural systems.
- Decaying infrastructure that needs to be rehabilitated or replaced.
- More aggressive regulation of stormwater quality.
- Desire for more sustainable infrastructure that is not as resource dependent.

This chapter examined the concepts of sustainable urban development. Then, the separate topics of water supply, wastewater, and stormwater were discussed. Finally, opportunities to better integrate these systems to promote water conservation and reuse were presented.

Research initiatives that would advance our ability to address these issues are listed below:

- Establish long-term experimental watersheds in the United States that can provide the foundation for understanding and better managing and these complex urban water systems. Include the essential cyberinfrastructure as part of this program to allow researchers to collaborate. The National

Science Foundation is currently linking its research programs in the hydro-logic sciences and environmental engineering. This linkage may provide an opportunity to develop such systems.

- Establish and implement rigorous protocols for evaluating the effectiveness of urban stormwater BMPS. These protocols need to outline in detail the required experimental design to achieve meaningful results. The current approach of simply evaluating performance based on event mean concentrations is inadequate and is leading to wasteful investments of large sums of money on ad hoc trial and error procedures.
- Reestablish the equivalent of the Office of Water Resources Research at the federal level to provide funding for research that looks at integrated urban water systems. At present, a wide variety of federal agencies deal with components of urban water systems but in a very fragmentary way, e.g., US EPA deals with stormwater quality while the Corps of Engineers deals with flood control.
- Evaluate the efficacy of multi-purpose urban storage-treatment systems for reclaimed wastewater and stormwater as integral components of reuse systems. These satellite systems can be integrated with existing centralized control systems.

ACKNOWLEDGEMENTS

Key professional collaborators include William DeOreo and Peter Mayer of Aquacraft, Inc., Wayne Huber of Oregon State University, Ben Koopman and John Sansalone of the University of Florida, Eric Strecker and Marcus Quigley of Geosyntec, Inc., and Neil Weinstein of the Low Impact Development Center. Numerous graduate students at the University of Colorado (Lynn Buhlig, Peter Mayer, Jeff Harpring, Laurel Stadjuhar, Beorn Courtney, Istvan Lippai, Joong Lee, David Sample, Leonard Wright, Donald Alexander, Derek Rapp, and Chelisa Pack) and the University of Florida (Ruben Kertesz, Scott Knight, Dan Reisinger, and Matthew Rembold) conducted this research and generated many new ideas.

Research during the past several years on urban water systems decentralized systems has been funded by: the Amercian Water Works Association Research Foundation for the national study of residential water use; US EPA Edison to evaluate innovative practices and develop spreadsheet methods to simulate and optimize BMPs; the National Cooperative Highway Research Program to evaluate highway BMPs including LID; the Water Environment Research Foundation to develop principles of BMP evaluations including LID; and the Florida Dept. of Environmental Protection on water reuse as an option for ocean outfalls in southeast Florida.

REFERENCES

Adams, B.J., Dajani, J.S., and Gemmell, R.S. (1972). On the centralization of wastewater treatment facilities. Water Resources Bulletin 8(4): 669–678.

Butler, R. and MacCormick, T. (1996). Opportunities for decentralized treatment, sewer mining and effluent re-use. Desalination 106:273–283.

Crook, J. (2004). Innovative applications in water reuse: ten case studies. WateReuse Association, Alexandria, Virginia.

DeOreo, W.B., Heaney, J.P., and Mayer, P. (1996). Flow trace analysis to assess water use. Journal of the American Water Works Association 88(1): 79–90.

Florida Department of Environmental Protection (2005). 2004 reuse inventory. Florida Department of Environmental Protection, Division of Water Resource Management. June. http://www.dep.state.fl.us/water/reuse/inventory.htm.

GeoSyntec Consultants, Oregon State University, University of Florida, and The Low Impact Development Center (2006). Evaluation of best management practices and low impact development for highway runoff control- user's guide for BMP/LID selection. Final Report and Appendices to the National Cooperative Highway Research Program, Transportation Research Board, National Research Council, Washington, D.C.

Heaney, J.P. and Lee, J.G. (2006). Methods for optimizing urban wet-weather control systems. Final Report to US EPA, Edison, NJ.

Heaney, J.P., Pitt, R., and Field, R. (eds.) (2000). Innovative urban wet-weather flow management systems. EPA-600-R-99-029, Cincinnati, OH. http://www.epa.gov/ORD/NRMRL/pubs/600r99029/600R99029prelim.pdf.

King County (2005). http://dnr.metrokc.gov/WTD/cso/page02graph.htm.

Koopman, B., Heaney, J.P., et al. (2006). Reuse options for ocean outfalls in southeast Florida. Final Report to Florida Department of Environmental Protection, Tallahassee, FL

Law, I.B. (2003). Advanced reuse – from Windhoek to Singapore and beyond. Water 30(5): 31–36

Lee, J.G. and Heaney, J.P. (2003). Urban imperviousness and its impacts on stormwater systems. Journal of Water Resources Planning and Management 129(5): 419–426.

Lee, J.G., Heaney, J.P., and Lai, D. (2005). Optimization of integrated urban wet-weather control strategies. Journal of Water Resources Planning and Management 131(4):307–315.

Low Impact Development Center, GeoSyntec Consultants, University of Florida, and Oregon State University (2006). *Low impact development design manual for highway runoff control-design manual. Final report to the National Cooperative Highway Research Program, Transportation Research Board.* National Research Council, Washington, D.C.

Mayer, P. et al. (1999). Residential end uses of water. Final report to the American Water Works Association Research Foundation, Denver, CO.

Mitchell, V.G., McMahon, T.A., and Mein, R.G. (2003). Components of the total water balance of an urban catchment. Environmental Management 32: 736–746.

Okun, D.A. (2000). Water reclamation and unrestricted nonpotable reuse: a new tool in urban water management. Annual Reviews Public Health 21: 223–245.

Oregon State University, GeoSyntec Consultants, University of Florida, and The Low Impact Development Center (2006). Evaluation of best management practices and low impact development for highway runoff control-research report. Final report to the National Cooperative Highway Research Program, Transportation Research Board, National Research Council, Washington, D.C.

Radcliffe, J.C. (2004). Water recycling in Australia. Australian Academy of Technological Sciences and Engineering, Parkville, Victoria. May.

Sample, D.J. and Heaney, J.P. (2006). Integrated management of irrigation and urban stormwater infiltration. Journal of Water Resources Planning and Management 132(5).

Strecker, E., Huber, W., Heaney, J., Bodine, D., Sansalone, J., Quigley, M., Leisenring, M., Pankini, D., and Thayumanavan, A. (2005). Critical assessment of stormwater treatment and control selection issues. Water Environment Research Foundation Report 02-SW-1, Alexandria, VA, 290 p. plus appendices

Whitcomb, J.B. (2005). Florida water rates evaluation of single-family homes. Prepared for Southwest Florida, St. Johns River, South Florida, and Northwest Florida Water Management Districts. July.

Wright, L.T., Heaney, J.P., and Dent, S. (2006). Prioritizing sanitary sewers for rehabilitation using least-cost classifiers. Journal of Infrastructure Management 12(3).

16

Urban ecological design and urban ecology: An assessment of the state of current knowledge and a suggested research agenda

K. Hill

University of Virginia, Charlottesville, VA 22904, USA
E-mail: kzhill@virginia.edu

Summary: A new type of city emerged in the 19th century. Urban designers and urban ecologists now seek to synthesize observations about this new type of city. They lack a common conceptual framework that recognizes history, the political economy of practice, and the need for an experimental approach that can refine predictive ability.

INTRODUCTION

My purpose in this paper is to use a particular lens on the current state of the field, in order to frame the potential for urban design to affect the hydrological and ecological performance of cities. The study and practice of urban design involves a diverse, multi-disciplinary field that draws on physical planning, building architecture, landscape architecture, and civil engineering. Thus there are many possible perspectives from which to assess the current state of knowledge in this broad field. My emphasis is intended to provide a consistent focus on the areas of urban design that have been most influenced by research in scientific ecology, and which have sought to produce tangible ecological benefits. I will also focus on cities and movements in the United States, since they share a common history and often also share a common set of influences. Some background is included on the history of ideas in this sub-field, which might be called "urban ecological design," to provide context for the reader.

Like engineering, urban design is an applied practice as well as a scholarly discipline. The practice of urban design is very heavily influenced by local and global socioeconomic trends, by changing philosophies of governance, and by what can only be called the idiosyncracies of personalities in the public and private sectors who influence large civic investments. Therefore its theoretical base tends to be dominated by normative theories that articulate what is seen as "good" during a particular historical era, with a lesser emphasis on predictive theories that are tested using the scientific method.

STATE OF THE ART – CURRENT KNOWLEDGE
Urban design with ecological principles

Urban designers have historically been interested in understanding the effect of urban forms and structures on human experience, including parks, residential districts, marketplaces, transportation infrastructure, waterfront districts, and all the other elements of cities. Urban design emerged as a specific academic discipline only during the past century, although cities have been designed by humans for a much longer period, approximately 7,000 years. Individuals and organizations with significant economic and political power were the original "urban designers." Architects, engineers and the emerging profession of landscape architecture began to play a significant role in the 19th century. These professionals pursued various social and aesthetic goals primarily by studying and proposing changes in the morphology of urban spaces, and in the surfaces of urban structures. Engineers and landscape architects exerted a significant influence on American cities of the mid- to late-nineteenth century, as road systems, park systems and water systems were developed (Sutcliffe, 1981; Melosi, 2000; Spirn, 2002).

One of the most influential sets of ideas that shaped American urban form in the 19th century was articulated by Frederick Law Olmsted, who named the field of landscape architecture and defined it as constituting both an art and a science (Howett, 1998). Olmsted's basic idea was that parks should be organized into a system of spaces that served functional public infrastructure needs as well as democratic social goals. His vision that cities should be designed with and around park systems evolved during an era when civil engineers and other professionals were developing America's urban water, sewer and drainage systems (Melosi, 2000). But in part because of the growing separation among the disciplines during that century, it was not obvious at the time that park systems should be used as a multi-purpose infrastructure to protect drinking water quality, provide surface conduits and biological filters for polluted drainage water, establish human recreational networks, and simultaneously serve as the defining element of new residential districts.

Olmsted's parks were often fully constructed landscapes, not remnants of earlier, pre-urban woodlands – as in the case of New York City's famous Central Park, which was designed to present a pastoral aesthetic that Olmsted and others believed would promote the biological, psychological and social health of urban people (Zaitzevsky, 1982). Although the concept is sometimes attributed to Olmsted, the idea that urban parks were analogous to organs of the body, and specifically that parks served as "lungs" by providing access to fresh air in cities where the burning of coal and other carbon-based fuels often polluted it, originated in the previous century. Sir William Pitt was quoted in 1798 as frequently remarking that parks are the lungs of London (Oxford Dictionary of Quotations, 2006). This biological analogy had come into common use in the 19th century.

In the late 19th century, Ebenezer Howard developed an idea for what he called the "Garden City," a concept he felt was suited to the expansion of British cities, after observing Olmsted's proposals for suburban residential areas outside Chicago (Howard, 1985; Parsons and Schuyler, 2002). Howard was an inventor who was interested in the future relationship between countryside areas and expanding cities, and in finding ways to preserve a high quality of residential environment for industrial workers. Several prototypes of garden cities were built in England that embodied Howard's utopian ideals, creating new models of suburbs that were based on his concentric scheme for urban form in satellite cities that would surround large, older cities and allow the population to grow in an orderly way. The center of the Garden City was to be a park, with the next concentric ring containing the bulk of the town's housing. Radiating boulevards would provide access in and out of the "hub" of the central city, with ring roads providing local access. Outside the residential ring would be a more rural zone, acting as a kind of matrix that supported access to light, air, and food for urban residents. The outermost zone would contain everything from managed forests and crop fields to convalescent homes and an agricultural college. In a sense, Howard's attempt to create a holistic diagram of what constituted a good small city or town was an attempt to represent the town as an organism, complete in its elements and – although workers might well travel to the larger central city for jobs – even in its ability to support itself with food and some local employment.

Patrick Geddes, a Scottish botanist by training who also engaged in formal studies in engineering and economics, addressed similar issues in his proposals for urban design during the late 19th and early 20th centuries (Geddes, 1915; Meller, 1990). Geddes was one of the first, if not the very first, to propose combining the knowledge of economics, biology, and sociology to study flows of energy in cities. In his teaching at the University of Edinburgh, Geddes used an actual tower to represent layers of transects through that city's region. He is sometimes credited as inventing the contemporary notion of an urban region as a functioning whole, that could be described using gradients or transects that revealed the interactions among its concentric zones as well as what might be called the suitability of various sectors for particular uses. Geddes, like Howard, was interested in how cities could be designed as more humane places that would contain gardens woven into their residential fabric in order to provide food as well as access to light, air, and recreation.

The primary driver for these inventive explorations of city form was the drastic growth that occurred in the urban populations of particular cities such as London and Berlin from the 1840's to the 1930's. New industries needed workers, and their promoters encouraged rural people to move to the cities for a better wage and what was described as a better life (Richie, 1998). This centralization of a working population occurred at a rate that typically outpaced the provision of drinking water and drainage infrastructure, so that overcrowding was accompanied by cholera outbreaks and other diseases (Warner, 1972; Sutcliffe, 1981; Cuthbert, 2006). Rail lines were developed in Berlin and London as a means of bringing new industrial workers in from the countryside for whom beds simply could not be found in the city (Richie, 1998). This period of dramatic growth is significant because it precipitated changes in cities that were both quantitative and qualitative. In a period of fewer than 100 years, cities changed from being made up of at most a few square miles of dense, walkable space – which had been their condition since the first cities of the Indus Valley – to encompassing whole landscapes and regions. Olmsted, Howard, Geddes and others were dealing with a qualitatively different "thing" that emerged in the mid- to late-19th century, an entity that required a very different approach.

This period of urban change corresponds to the period in which ecology first developed as a scientific sub-discipline, along with a kaleidoscopic array of many new specialized professions and fields of knowledge, including sociology and economics. Large-scale urban industrialization, the extremely rapid growth of the world's first megacities, and the chaos of early modern cities themselves all appeared at essentially the same time that biologists began their study of evolution, engineers began building systematized urban infrastructure systems, ecology began its attempt to understand the relationship of organisms to their environment, and our fundamental approach to the design of cities came into question. Concerns about human health and water supply drove a wide range of reforms in Europe and America, as evidenced in Ellen Swallow's pioneering work with water quality and urban health at MIT where she founded the first laboratory in "sanitary chemistry" in 1884 (Weingardt, 2005). Most of the same basic elements of today's attempt to understand urban sustainability were in place and were already of broad public concern more than 100 years ago.

Building architects developed a more influential voice in the design of American cities as the 20th century began, joining a movement of European modernists who proposed various utopian forms for future cities (Gosling, 2003). These proposals included road systems and park patterns as well as building types. After Daniel Burnham's architectural design for the World's Columbian Exposition of 1893 in Chicago, in collaboration with landscape architect Frederick Law Olmsted (Wilson, 1989), the architect Frank Lloyd Wright's proposal for "Broadacre City" replaced Olmsted's earlier proposal for suburban villages by articulating a very different comprehensive form for the expansion of American cities that was designed specifically for an American context. This form was radically different from Ebenezer Howard's centralized proposals of the late 19th century, in which cities were treated as contained by a matrix of countryside. Wright's proposal was to transform the countryside into a low-density city, essentially making this low-density city itself the matrix land-use type by granting each family 1 to 5 acres within a grid of roads, that was to be traversed by private automobiles and airborne, propeller-driven commuter vehicles (Herberger, 1994–1995).

Political and economic dynamics, however, have continued to be the primary drivers of actual change, and have most heavily influenced the implementation of particular urban forms. Cedric Price, a British architect, is quoted as having joked that cities have evolved through three stages: the "boiled egg" stage of the walled medieval city, the "fried egg" stage of the industrial city divided by railroad tracks and centered around a core high-rent district, and the "scrambled egg" stage of the post-WWII city in which access to personal automobiles allowed an unprecedented reduction of density and a dispersion of commercial, residential, and industrial uses in strips and clusters (Shane, 2003). The economic philosophy of Fordism (i.e., systematic production that reduces the cost of goods, allowing mass consumption) has often been cited in architectural essays as the driver of the current stage of urban form in the United States in particular, along with the political advantage gained by wealthy suburban communities whose growth was fueled in the 1950's and later by federal subsidies, institutionalized racism, and the widespread belief that families needed private yards to support the healthy development of children.

Between the 1970's and the end of the 20th century, new medium-density satellite towns were constructed on former agricultural and forest lands to respond to market demands for new single-family housing in walkable, mixed-use neighborhoods. Urban designers responded to these drivers by developing proposals for a design philosophy known as New Urbanism (Katz, 1994) that sought to produce an experience of community cohesion by re-introducing building forms and street patterns that often dated from the colonial period in American history, or in some cases, utilized the geometric symmetry of urban district and circulation designs of the Beaux Arts period in Europe.

More recently, in the 1990's and early 21st century, re-development has become a more typical pattern of growth, often resulting in increased urban residential density in newer cities like Seattle, Washington or Portland, Oregon. New medium- to high-density neighborhoods have been constructed in the former warehouse districts of existing cities. These re-developed districts are often intended for young childless couples with high personal incomes and for older couples whose single-family homes have generated wealth by increasing in value. Often the children of these couples have matured and left home, with the parents deciding to relocate or add second residences in areas with greater proximity to a higher density and quality of social and cultural amenities. American political and economic institutions facilitated new investments of wealth in cities that support public transit, sanitation, water quality improvements, communications infrastructure upgrades, road repair, and park development.

These most recent social, political and economic changes created a demand for re-designed urban mixed-use districts that include restaurants, specialty food stores, art galleries and boutique shops that provide a high quality of life without requiring the use of an automobile. Architects, engineers, landscape architects and planners have responded to this trend with a renewed interest in designs that promote the concept and experience of urban sustainability (see for example Nassauer, 1997; Singh, 2001; Marchettini, 2004). In the case of urban redevelopment, the location of the road systems and other underground infrastructure is typically already well established, and the enormous capital cost of relocation prevents designers and planners from pursuing utopian formal proposals that would require changes in the spatial patterns of these systems. Instead, a movement known variously as Green Urbanism, Landscape Urbanism, Green Infrastructure, or Green Design has evolved that, at its best, proposes ways of adapting existing infrastructure to new functional goals.

The emphasis of this movement is currently on the inclusion of design elements such as green roofs or recycled materials, the purpose of which is to allow urban structures to manage water, material and energy flows more efficiently, rather than on developing a particular aesthetic model. The structures and dynamics of urban landscapes figure prominently in this evolving school of thought, including park systems, street rights-of-way, car parking areas, waterfront districts, rainwater collection systems, stream corridors, and water bodies.

ASSESSMENT

Adaptive urbanism: Limits to the state of the art

The current approach to re-development of urban districts is evolving rapidly, but is so thoroughly driven by specific projects rather than systematic goals that it falls short of constituting a comprehensive proposal for cities. On this question, a comparison of US and European approaches is useful.

The most systematic element of current practice in the US has been the development of the US Green Building Council's LEED standards for site development (USGBC, 2006), which have begun to be adopted as standards by government bodies and private corporations. These standards include a comprehensive set of issues, but the intent to apply them at the site and building scale means that these standards do not provide a district- or landscape-scale approach. Interestingly, two US Federal laws (the Endangered Species Act and the Clean Water Act) have spurred district and landscape-scale approaches, particularly where both laws are triggered simultaneously, as in the Pacific Northwest cities of Portland, Oregon and Seattle, Washington, or where the declining health of a major waterbody has prompted widespread public concern, as in the Chesapeake Bay region of the east coast. However, the typical US development pattern of parcel-by-parcel decision-making by private landowners has rarely lent itself to experiments that transform an entire urban district and that could provide new models of social organization, transportation and water infrastructure, energy use and energy generation. The general lack of leadership on these issues at the federal, regional, and state levels, however, has left American cities to find local financial and organizational tools to deal with problems that are genuinely regional and national, such as limiting energy use, creating opportunities for decentralized alternative energy generation, providing mobility options other than private automobiles, creating decentralized water collection options, and building an infrastructure of open space that allows cities to maintain regional biodiversity.

In contrast, the European Commission and its constituent nations have developed comprehensive sets of goals for energy use and generation, pollution reduction, waste reduction, new housing and transit systems, water conservation, and other essential steps towards making cities more socially and economically vibrant while reducing negative impacts on their regions and their inhabitants (European Commission, 2006). Through the Agenda 21 planning process, these goals have commonly been adopted at regional and local scales of government and are being implemented in a growing number of innovative urban district and new town experiments that use far less water and energy, and even generate energy locally from waste products such as biogas (see for example the Hammarby Sjöstad urban district in Stockholm, Sweden; City of Stockholm, 2006). The implementation of these goals is boldly challenging, if not transforming, the economy of Europe. However, the developmental history of the Old World has left it absent some of the diverse animal species that were present before widespread agricultural and urban development occurred. Fewer urban design proposals address biodiversity, and specific plans for particular species, than can be found in North America.

Of course, the greater number of opportunities to develop urban designs that might help to conserve biodiversity in American cities does not indicate that those designs will succeed. It indicates only that the more recent industrialization of the North American continent has not yet eliminated all of the ecosystems that preceded it. The European model is clearly far more advanced in creating prototypes for new cities and urban regions that consume far less energy, promote healthier inhabitants who can walk or bike to their destinations, and where personal transportation is less culpable for reductions in water quality.

To the extent that the characteristic terrestrial and aquatic biodiversity of our regions is a compelling concern in North American cities, mainstream urban design has not yet built sufficient bridges with ecological science to achieve the goal of sustaining that biodiversity. The sub-field of urban ecological design has begun to base its proposals in scientific knowledge, but that effort began only in the late 1960's with the work of the firm Wallace, McHarg, Roberts and Todd, and in the urban planning and landscape architecture departments of the University of Pennsylvania under the intellectual leadership of Ian McHarg. The ecological sciences have undergone a major paradigm shift in the intervening decades, and many individual designers and design firms that wish to make science-based proposals for cities have not caught up with that paradigm shift. There are several reasons for this disjunction and manifestations of it that limit the current practice of urban design.

The first limit generated by this disjunction between the fields of ecology and urban design is a limit on the imagination of designers as they conceive of the possible goals of design in and around cities. In 1973, urban designer and architect Kevin Lynch wrote that the purpose of urban design was to provide for the healthy and successful development of individual humans and human culture. "We could go further and consider the effect of the man-made environment on the growth and development of other living species. . . . The ideal is enticing, but unfortunately its meaning is uncertain. For the time being, we must speak for ourselves," (quoted in Banerjee and Southworth, 1990). Lynch became famous for his use of ideas from a sub-field of psychology that was known at the time as ecological psychology (Gibson, 1979). This field considered interactions with the environment to be a major influence on the development of human cognitive abilities. Lynch used its basic ideas to develop an approach to urban space and form that emphasized the ability of urban residents to move through that space and recognize its organization. His contention was that improving the legibility of otherwise chaotically-changing American urban spaces was essential to producing productive citizens who could participate in the social, economic and political life of cities (Lynch, 1960).

Lynch's terminology for legible urban spaces (words like edge, landmark, district, path, etc.) is still central to the way most urban designers analyze and propose form, although they may not recognize Lynch as the originator of the various offshoots of his generic typology which they employ. This powerful language of urban forms can be applied to the movement and perceptual abilities of other things besides people: salmon, crabs, great blue herons, beetles, coyotes, and other non-human organisms also move according to perceptual cues. Other materials and energy flows, along with the larval life stage of many marine organisms, interact through physical and biological processes with the non-perceptual elements of Lynch's urban structures, such as edges, nodes, districts, and paths. If urban designers choose to conceive of their work as including influences on flows of organisms (including humans), materials and energy as well as human perception over time, they can overcome this first limit to their knowledge and practice. Urban designers could then do more to help envision and articulate goals for biodiversity in urban regions, in both aquatic and terrestrial systems.

The second limit is also related to Lynch's quote regarding the ability of designers to consider the needs of non-human species, and in addition, draws on the influential urban design observations of Jane Jacobs. Both writers point to difficulties generated by the complexity of taking a multiple-perspectives approach to understanding the dynamics of urban environments. In 1964, Jacobs published a controversial book based on her observations of New York City and the large-scale designs that were being proposed for that City as part of the agenda of "urban renewal" (Jacobs, 1964). Her argument included the idea that cities are complex systems, using a biological analogy to the complexity of interactions within the human body that were being explored in medicine at the time. Jacobs argued that the complexity wasn't "bad" and in need of clarification, but rather that it was the nature of the beast; that good cities develop from complex local interactions that reflect intimate social dynamics, and produce livable urban spaces incrementally, over many decades. Jacobs argued that complexity is a defining aspect of the kind of problem a city is, and that the livability of cities (unlike their traffic patterns) could not be improved by attempts to reduce their complexity.

While this insight was critical to stopping the destruction of historical urban neighbourhoods that were the prototypes of livability within megacities like New York, it has not lead in the short term to greater predictive clarity about how complex urban structures and interactions can be changed to adapt to new functional needs. Jacob's recognition of complexity reduced some of the hubris of the post-World-War-II planners and architects (and their contemporary descendants) who believed the past stood for chaos, and that the future should involve simpler urban blocks to achieve "better cities" for commuters who drive in from the suburbs. But in my view, this same awareness of complexity has sometimes led urban designers to fear and avoid bold structural proposals that might transform the environmental performance of cities by insisting that large urban districts can't be built with one shared vision, or by working against new transit system proposals in existing cities.

Lynch wrote that it was impossible to know what the needs of non-human species are, and Jacobs' observations of social complexity left some design readers uncertain of whether self-organizing human activities can change or be altered in helpful ways over time. Both writers used their awareness of complexity as a reason not to engage in unnecessary changes or suspect design ventures. But if cities now need to change, in terms of their performance as habitat, as "neighbors" to other species, and as filters and conduits for water flow, their complexity must be engaged more directly. With caution, to be sure; and incrementally, whenever possible. But change must be proposed.

Third, there is a limit in urban design that is generated by fragmented education of its theorists and practitioners. Most urban designers are still trained in building architecture, rather than landscape architecture, physical planning, or geography. In spite of the increasing recognition within urban design that building systems must become more efficient in their management of energy and material flows such as water, architecturally-trained urban designers are often silent on the need for cities to set goals for biodiversity conservation and to increase resilience in the face of increasingly extreme hydrologic or climatic events. The differences in training that are required to analyze and consider events and dynamics that occur at the scale of landscapes (i.e., tens of square miles and larger) rather than sites (typically measured in square feet) may explain this tendency for most architecturally-trained urban designers to use morphological strategies at the landscape scale that reflect a lack of awareness of dynamic changes and interactive effects in the non-human world.

For example, in a recent design competition held to re-design the Seattle waterfront (Seattle Channel, 2004), only one of many teams used the opportunity to re-design the city's seawall as an opportunity to create shallow sub-tidal and intertidal habitat for marine species. If corridors of suitable eelgrass and kelp beds can be re-established, either on natural substrates or on floating modular substrates manufactured by humans, urban marine waterfronts can act as important stepping stones in the dispersal of marine organisms that have commercial, ecological and cultural significance. So why did almost no one among the design teams recognize that Seattle has a 9–13 foot tidal change daily that provides a dynamic environment for visually-compelling shoreline changes? The answer may be that few if any of them had an education that included examples of design with marine processes, or even landscape-scale terrestrial processes. Most urban designers simply don't anticipate the opportunity to use those processes in their work.

For a different scale of example, in addition to the obvious need for non-human species to be able to disperse along elevational and latitudinal corridors as a response to changing regional climates, there is also an important potential human health impact of sharing space with species whose health has become compromised. Urban concerns for regional biodiversity are not necessarily altruistic; instead, they can and perhaps should be self-serving. Historical examples of human health being affected by negative impacts on species in the immediate area are primarily related to the expansion of agriculture, but in many ways this phenomenon can be considered analogous to urban expansion.

The best-known examples include the development of the human immunedeficiency virus, which is now thought to have originated in a population of West African monkeys whose ability to find food and other resources was severely compromised by agricultural expansion, and the spread of the deer tick population that carries the bacillus which causes Lyme disease in humans, after the Northeast's deer population was confined to a very small land area by deforestation of that region during the 19th century (Grifo and Rosenthal, 1997). Similarly, fish that host a tiny snail parasite known as Nanophyetus salmincola may be infected with a bacteria called Neorickettsia. This bacterial agent can be transferred, along with the parasite, to dogs that eat raw fish along creeks and beaches, resulting in death (WSU, 2006). It is not inconceivable that other aquatic bacteria could be transferred to human hosts via dogs or other domestic animals, as well as by swimming in water that contains these organisms. Interdisciplinary education that includes physical geography, landscape ecology, and public health issues ties urban design at the site scale to the lived experience of urban regions. Without it, urban designers are very limited indeed in their ability to understand the interactions among landscape, biodiversity, water, and human health.

In summary, three significant limits to the current knowledge and practice of urban design are: (1) limits on the goals set for urban design due to an exclusive emphasis on human social and aesthetic experience, (2) limits on the willingness of urban designers to partner with engineers and ecologists to grapple with biological and social complexity in experimental ways that seek better predictive understanding of design outcomes, and (3) limits on the educational exposure of urban designers to landscape-scale dynamics that involve non-human species, flows of materials and energy, and public health.

RECOMMENDATIONS FOR FUTURE RESEARCH

Outline of a research agenda
Brief discussion of the proposed research agenda

The outline of the proposed research agenda is summarized on Figure 16.1.

A1. Eco-mimicry

The goal of this scholarship would be to help urban designers learn to design cities that mimic the ecosystems we build them in – a design goal sometimes referred to as

A. Overcoming limits to imagination requires setting more ambitious ecosystem-related goals for urban design

    1. Eco-mimicry / characteristic regional biodiversity and ecosystem services;

    2. "faktor 10" / reducing net energy usage by a factor of 10 in urban districts;

    3. transit and walkability/bikability / reducing transportation impacts, improving human health and quality of life;

    4. re-development / infill / brownfields : achieving urban growth by increasing density inside existing urban boundaries.

B. Confronting complexity in practice and scholarship

    1. Representation in design: representing complexity for insight and rhetoric (scales, interactions, multiple perspectives – predictive insights via new tools for evaluating design that go beyond the tradition of plan, perspective and section);

    2. decentralized vs. centralized urban systems and designs (water, energy, wastes, transit).

C. Education across scales – landscapes, public health, sites, bodies

    1. A comprehensive, clear, nested framework that links ecology to design (and economic incentives, small and large-scale) to produce outcomes that increase the adaptability/resilience of urban regions;

    2. curriculum reform – identifying critical regional issues, teaching design options as experimental approaches;

    3. re-evaluation of the incentives for professionals to update their skills and knowledge throughout their careers, so that it will occur more often.

Figure 16.1. Outline for a research agenda in urban ecological design.

"eco-mimicry." Eco-mimicry, in this sense, does not include any prescription that cities should be designed using naturalistic forms. The functions of these designs are the most critical characteristic to predict, monitor, and adjust to achieve success. Forms, materials, and surface treatments can and should vary as an expression of culture as long as they don't impair the ecological functions of a designed structure. It is also of primary importance that all designs be a locus of cultural meaning, particularly in urban areas where they may play multiple functional roles – and if not, they may represent areas that have become inaccessible to humans and require some kind of interpretation of that status.

For example, hydrologists have estimated that the former conifer forest that once covered the landscape we now call "Seattle" was able to hold back about $4\frac{1}{2}$

inches of rain water over every square inch of the land's surface. The enormous surface area of conifer trees, with all their small needles or scales, contributed to this sponge effect – as did the thick layer of springy plant debris on the forest floor that ecologists call the "duff layer." When Seattle designers take on the task of mimicking that big ecological sponge on developed sites, they're talking about finding ways to hold back 13,400 gallons of water for a typical residential lot, measuring 5000 square feet.

There's more to it than volume, of course. Eco-mimicry also involves trying to understand and recreate the appropriate seasonal rate and timing of rain water releases to creeks, and figuring out how to get the right proportion of water to soak down into the soil instead of flowing directly to surface streams and other water bodies. When the water comes into important habitat areas, like shallow lakes or bays and streams, it's also important to consider the temperature of the water and whether it is not only clean, but also full of beneficial components that support fish and other animals. That could mean making sure it's carrying insects and detritus from decaying plant matter; because as we learn more about the food needs of the fish that are suffering declines in their population size, we find out that these can either feed them directly or support the food needs of the smaller animals that these fish feed upon. This is, in a sense, another way of discussing the need for urban areas to conserve the ecosystem services that were once provided by the pre-existing landscape.

There are limits to what can be mimicked in an urban environment as well. When pathogens, competitive species or predators enter into the picture the solution is often not a design solution, but rather a management solution. A quasi-experimental approach to research in this area of work that avoids irreversible impacts may be the best way to identify whether physical design changes can make a difference, or whether other less obvious variables limit the ability of the particular biological system to respond to physical changes.

A2. Factor 10

Factor 10 is a short-hand name for an efficiency goal that proposes reducing energy use or energy impacts (along with other resource uses) by a factor of 10. It is widely discussed among designers in Europe, and some countries and regions there have adopted it as an explicit goal. This is a critical goal for American urban design, because energy usage in North America is extreme in comparison to most of the world's older urban areas. What role can and should urban systems and multi-parcel design opportunities play in reducing the energy use of urban districts? Much of the energy use question may need to be resolved in individual buildings, but there are also steam energy redistribution districts in many cities that allow for efficiencies to be gained at larger scales. Certainly transit is also one of these opportunities, but only one of many possible angles on energy conservation and reduced consumption. How can urban design contribute to this in a North American context, given the structure of our energy markets and the ability of some municipalities and large private clients to control their energy purchases?

A3. Human circulation in city regions: Walking, biking, and taking the train

The Robert Wood Johnson Foundation, among others, has recently begun funding efforts to support research into the qualities of urban neighborhoods where people walk more frequently. There are many variables that may underlie this behavior, from the morphology of sidewalks and streetscapes to the number of available retail or transit destinations available to a pedestrian.

North American urban designers have also not taken the bicycle connections as seriously as they should, given the potential for this form of transit to allow people to avoid the use of their private automobiles, and the number of actual physical barriers to bicycling that often exist in urban street systems. Documentation of the elements of urban design that support bicycling in Europe and Asia would help bring this aspect of urban design into the American mainstream.

Rail transit, meanwhile, is the ultimate logical goal of dense urban neighborhoods that need alternatives to private cars – not because it will reduce congestion, but rather because it provides alternatives to being trapped in congested traffic in a car or on a bus. More research needs to be done on what actually works in cities that have invested in transit, so that American cities can make smart future investments. Are buses most effective at a particular scale and/or type of route, for instance, and is fixed guideway transit easier to justify in other situations? What technologies have been tested in revenue service and proven effective, and at what costs? What are the implications for the physical design of the street-level pedestrian environment of choosing particular technologies, such as bus rapid transit? These are critical questions for urban design, since they involve such large investments of public funds; and yet very little empirical work exists on these subjects in the design fields.

A4. Urban brownfield re-development

This area of urban in-fill development involves many significant social and political questions. How can the health risks of living on sites that have some level of soil contamination be addressed through design decisions that are made on a multi-parcel basis? Are there effective technologies for in situ remediation of these contaminated urban sites, and what are their design implications? Are there opportunities to phase in different land uses over time, so that the development can be economically successful while contaminants are treated or isolated? The importance of these questions is evident when one considers how much vacant urban land is unavailable for development because of low-level soil contamination, at a time when it is more important than ever that cities grow by increasing their internal density rather than increasing their geographic extent.

B1. Representation – beyond plan, section, and perspective

The traditions of two and three-dimensional representation in design have allowed designers and critics to evaluate buildings, infrastructure and urban spaces in particular ways. But perhaps it's time to reconsider these traditions. Are there ways of representing flows in space, or change over time, which would help decision-makers determine whether a design is good from an ecological perspective? Should we develop and standardize these tools for use in public and professional critique of designed proposed for urban spaces?

B2. Decentralized vs. centralized urban systems and designs (water, energy, wastes, transit)

One solution to the many demands on urban infrastructure that has begun to emerge from our contemporary "scrambled-egg" urban landscapes is the strategy of decentralizing formerly centralized services. This choice is often driven by today's higher levels of expectation for a clean and healthy environment, environmental regulation, and the political nature of the oversight of public budgets. Decentralization has one very interesting implication for design, which is that it often involves providing more functions above ground that used to be provided below ground. This can create opportunities for urban design, architecture, and landscape architecture to reconsider some of the fundamental relationships between form and function for urban sites and systems that have in the past been disguised by the role of centralized, underground systems. Equally important, it suggests that urban residents may have new opportunities to participate in shaping the performance of urban ecosystems at the scale of a neighborhood and an individual household.

As soon as it involves the daily patterns of life, this change becomes much more than functional. It is a cultural change that could engage people in "performance" in the sense of the arts, as well as in the sense of designed structures, and could be the basis for a significant re-engagement by diverse kinds of people in shaping the future of cities. But to date we know very little about urban residents' perception of decentralized infrastructure elements, or their willingness to participate in supporting them financially or with other contributions of time and labor.

C1. Developing a shared conceptual framework across the urban design disciplines, as well as ecology and engineering

What frameworks exist that can bridge – if not hybridize – the different epistemologies of ecological science, engineering and urban design? This question is a classic for scholars, and as yet remains unresolved. If the common questions revolve around the structure and function of resilient, adaptive cities, there must be some set of categories for investigation and action that all of these fields can agree upon. Even if the questions are different, at least some sort of joint investigative agenda could draw out the differences of method or emphasis that would help practitioners and scholars in each field understand their divergences.

C2. Reforming urban design curricula

There is currently no single standard curriculum in urban design, and perhaps this is as it should be. But guidance could be provided on the most significant biophysical issues, such as determining what types of knowledge of biophysical processes or analytical methods should be taught to designers. Perhaps the shared framework called for in C2 would provide the basis for comparing curricula, and suggest appropriate reforms. Alternatively, the representation challenges identified in B1 might be the impetus to change. Most obvious is the need for theories and methods that shape the thinking of urban designers that allow them to question or test the implications of designing according to the normative aesthetic theories of any particular cultural era.

C3. Continuing professional education and creating incentives to update skills/knowledge

While this one may seem the least interesting of the possible research topics suggested here, it may be the most critical to the practice of the field for the next 30 to 40 years. Given the frequent disjuncture between theory and practice in professional fields such as design and engineering, practitioners often take the energy-conserving path of substituting experience for the effort to stay current. What will persuade our professional colleagues that strategies which have succeeded in the past are actually counter-productive to the performance requirements of today's cities, if that is indeed the case? Unless professional societies commit to restructuring the incentives for learning the value of new approaches, we will miss our current opportunity to start practicing urban design with ecological knowledge when, as this paper has argued, we are already about 100 years behind our own best judgment.

REFERENCES

Banerjee, T. and M. Southworth, eds. (1990). *City Sense and City Design: Writings and Projects of Kevin Lynch,* Cambridge, Mass., MIT Press.

City of Stockholm (2006). Website accessed June 20, 2006. http://www. hammarbysjostad.se/

Cuthbert, A. (2006). *The Form of Cities: Political Economy and Urban Design,* Oxford, Blackwell Publishers.

European Commission (2006). "Agenda" 21: The first five years," website accessed June 10, 2006. http://ec.europa.eu/environment/agend21/implem.htm

Geddes, P. (1915). *Cities in Evolution; An Introduction to the Town Planning Movement and to the Study of Civics,* London, Williams & Norgate.

Gibson, J. (1979). *The Ecological Approach to Visual Perception,* Boston: Houghton Mifflin.

Gosling, D. (2003). *The Evolution of American Urban Design: A Chronological Anthology,* Chichester, West Sussex, England; Hoboken, NJ: Wiley-Academy.

Grifo, F. and J. Rosenthal, eds. (1997). *Biodiversity and Human Health,* Washington DC, Island Press.

Herberger Center for Design Excellence (1994–1995). *Frank Lloyd Wright: the Phoenix papers,* Tempe, AZ: College of Architecture and Environmental Design, Arizona State University.

Howard, E. (1985). *Garden Cities of Tomorrow,* (reprint). Eastbourne, Attic Books.

Howett, C. (1998). "Ecological values in twentieth-century landscape design: a history and hermeneutics," pp. 80–98 in *Nature Constructed, Nature Revealed: Eco-Revelatory Design,* Special Issue, *Landscape Journal.*

Jacobs, J. (1964). *The Death and Life of Great American Cities,* New York, Random House.

Katz, P., ed. (1994). The new urbanism: toward an architecture of community, New York, McGraw-Hill.

Lynch, K. (1960). *The Image of the City,* Cambridge [Mass.] Technology Press.

Marchettini, N., ed. (2004). *The Sustainable City III: Urban Regeneration and Sustainability,* International Conference on Urban Regeneration and Sustainability (Siena, Italy)., Southampton, WIT.

Meller, H. (1990). *Patrick Geddes: Social Evolutionist and City Planner,* London, Routledge.

Melosi, M. (2000). *The Sanitary City: Urban Infrastructure in America from Colonial Times to the Present.* Baltimore, Johns Hopkins University Press.

Nassauer, J., ed. (1997). *Placing Nature: Culture and Landscape Ecology,* Washington DC, Island Press.

Oxford Dictionary of Quotations (2006). ed. Elizabeth Knowles. "Lungs of London," Oxford University Press, *Oxford Reference Online.* Accessed on June 10, 2006. http://www.oxfordreference.com.offcampus.lib.washington.edu/views/ BOOK_SEARCH.html?book=t115

Parsons, K. and D. Schuyler, eds., (2002). *From Garden City to Green City: The Legacy of Ebenezer Howard,* Baltimore, Johns Hopkins University Press.

Richie, A. (1998). *Faust's Metropolis: A History of Berlin,* New York, Carroll & Graf.

Seattle Channel (2004). *Waterfront Charrette Presentations.* Website accessed June 12, 2006. http://www.seattlechannel.org/videos/watchVideos.asp?program=viaduct

Shane, G. (2003). "The Emergence of Landscape Urbanism," pp. 1–8, *Harvard Design Magazine,* number 19, Fall.

Singh, R.B., ed. (2001). *Urban Sustainability in the Context of Global Change: Towards Promoting Healthy and Green Cities,* Enfield, N.H.: Science Publishers.

Spirn, A. (2002). "The Authority of Nature: Conflict, Confusion and Renewal in Design, Planning, Ecology," pp. 29–50 in *Ecology and Design: Frameworks for Learning,* ed. by B. Johnson and K. Hill, Washington, DC, Island Press.

Sutcliffe, A. (1981). *Towards the Planned City: Germany, Britain, the United States, and France, 1780–1914.* New York, St. Martin's Press.

United Nations Department for Social and Economic Affairs, Division for Sustainable Development (2006). Agenda 21 summary. Website accessed June 10, 2006. http://www.un.org/esa/sustdev/documents/agenda21/index.htm

United States Green Building Council (USGBC). (2006). website accessed June 10, 2006. http://www.usgbc.org/DisplayPage.aspx?CategoryID=19

Warner, S.B. (1972). *The Urban Wilderness; A History of the American City,* New York, Harper & Row.

Weingardt, R. (2005). *Engineering Legends: Great American Civil Engineers: 32 Profiles of Inspiration and Achievement,* Reston, Va., American Society of Civil Engineers.

Wilson, W. (1989). *The City Beautiful Movement,* Baltimore: Johns Hopkins University Press.

WSU, College of Veterinary Medicine (2006). "Pet Health Topics." Washington State University. Online resource, accessed June 20, 2006. http://www.vetmed.wsu.edu/ ClientED/salmon.asp

Zaitzevsky, C. (1982). *Frederick Law Olmsted and the Boston Park System,* Cambridge, Mass., Belknap Press.

17

Green infrastructure for cities: The spatial dimension

J. Ahern

University of Massachusetts, Amherst MA 01003, USA
E-mail: jfa@larp.umass.edu

Summary: Planning for sustainable cities is a complex process addressing the fundamental areas of economic, environmental and socially-equitable sustainability. This chapter focuses on the environmental area, with theories, models, and applications illustrating possible spatial configurations of a green infrastructure to support ecological and physical processes in the built environment including: hydrology, biodiversity, and cultural/human activities. Green infrastructure is an emerging planning and design concept that is principally structured by a hybrid hydrological/drainage network, complementing and linking relict green areas with built infrastructure that provides ecological functions. Green infrastructure plans apply key principles of landscape ecology to urban environments, specifically: a multi-scale approach with explicit attention to pattern:process relationships, and an emphasis on connectivity. The chapter provides theoretical models and guidelines for understanding and comparing green infrastructure approaches. International examples at multiple scales are discussed to illustrate the concepts and principles introduced.

INTRODUCTION

The aim of this chapter is to introduce and explore the concept of urban green infrastructure as a means of spatially organizing urban environments to support a suite of ecological and cultural functions. In contemporary urban planning and design literature there is a convergence of research and case applications addressing sustainable cities and sustainable urbanism (Low et al., 2005; Moughtin and Shirley,

2005; Wooley, 2003; Steiner, 2002; Beatley, 2000; Van der Ryn and Cowan, 1996; Hough, 1995). This emerging focus reflects a broader international awareness of sustainability across its basic tripartite dimensions: economy, environment and (social) equity – often known as the "three E's" of sustainability (Wheeler and Beatley, 2000). As the sustainable development concept has matured and gained greater acceptance over the past two decades, it has directly and increasingly influenced regional and municipal policy and plans (Benedict and McMahon, 2006). As with the tripartite principles of sustainability, the policies and plans developed to advance urban sustainability through policy and practice address the economic, social, and environmental dimensions of sustainability. This chapter focuses primarily on the environmental dimension of urban sustainability, and more specifically, the role of spatial configuration of the urban environment in supporting key ecological functions through a "green infrastructure" in a sustainable manner. In addition, green infrastructure is also presented as a strategy to achieve abiotic, biotic and cultural goals. The chapter starts with a review of key ecological processes and principles of landscape ecology, with respect to sustainable planning, and of the particular importance of spatial configuration of urban environments.

KEY ECOLOGICAL PROCESSES AND FUNCTIONS

Ecological processes are the mechanisms by which landscapes function – over time, and across space – and are therefore appropriate to use as the goals for – and the indicators of – sustainability. Landscape ecology provides a theoretical perspective and the analytical tools to understand how complex and diverse landscapes, including urban environments function with respect to specific ecological processes (Pickett et al., 2004).

The Ecological Society of America defines ecological functions as those that provide "services" that moderate climatic extremes, cycle nutrients, detoxify wastes, control pests, maintain biodiversity and purify air and water (among other services) (ESA, 2006). The ecosystem services concept helps to place value on ecological functions, often to the direct benefit of human populations in physical health, economic or social terms.

A widely accepted resource model for landscape planning is the Abiotic, Biotic and Cultural (ABC) resource model (Ndubisi, 2002; Ahern, 1995). This comprehensive and inclusive model is consistent with the landscape ecology perspective that explicitly recognizes the needs and reciprocal impacts of humans on biotic and abiotic systems and processes. The ABC resource model is applied here to articulate the key ecological functions of a green urban infrastructure (Table 17.1).

The ABC functions described in Table 17.1 are intended to be illustrative, but not comprehensive. It is important to note how this broad, multipurpose, and multifunctional suite of ecological and cultural functions supports the broad principles of sustainability, in contrast with single-purpose policies or plans that address more focused goals (e.g. managing water quality, endangered species protection or pollution remediation). And because this suite of functions spans an abiotic-biotic-cultural continuum – it is inherently more likely to enjoy a broad base of

Table 17.1. Key abiotic, biotic and cultural functions of a green urban infrastructure

Abiotic	Biotic	Cultural
Surface:groundwater interactions	Habitat for generalist species	Direct experience of natural ecosystems
Soil development process	Habitat for specialist species	Physical recreation
Maintenance of hydrological regime(s)	Species movement routes and corridors	Experience and interpretation of cultural history
Accommodation of disturbance regime(s)	Maintenance of disturbance and successional regimes	Provide a sense of solitude and inspiration
Buffering of nutrient cycling	Biomass production	Opportunities for healthy social interactions
Sequestration of carbon and (greenhouse gasses)	Provision of genetic reserves	Stimulus of artistic/abstract expression(s)
Modification and buffering of climatic extremes	Support of flora:fauna interactions	Environmental education

This figure articulates what a green urban infrastructure can explicitly do to contribute to sustainability.

public support – an essential characteristic for a successful urban sustainability program.

LANDSCAPE ECOLOGY PRINCIPLES FOR GREEN URBAN INFRASTRUCTURE

Key ideas from landscape ecology that are relevant to green urban infrastructure for sustainable cities include: a multi-scale approach with an explicit recognition of pattern:process relationships and an emphasis on physical and functional connectivity.

A multi-scaled approach is based on hierarchy theory that addresses the structure and behavior of systems that function simultaneously at multiple scales. For example, hierarchy theory is widely used in transportation planning, for to understand the dynamics and capacity of local road traffic, one must understand the larger highway system with which local roads are connected. The same applies to landscapes which are also hierarchical systems. While landscapes are, by definition, broad heterogeneous areas of land, they are also by definition "nested" within larger areas of land that often constrain, or control the ecological processes – particularly those associated with species movement or hydrology. In applied landscape ecology, a multi-scaled approach – addressing spatial patterns and ecological processes – is the accepted norm (Leitão and Ahern, 2002; Ndubisi, 2002). The multi-scaled approach involves assessment and planning of spatial configuration of landscape patterns and ecological processes at multiple scales, and how these patterns and processes interact. This analysis typically indicates key points for physical linkages, where important connections exist, or where connections should be made. In urban environments the appropriate scales are: the metropolitan region or city, the districts or neighborhoods, and individual sites.

The pattern:process dynamic is arguably the fundamental axiom of landscape ecology because the spatial composition and configuration of landscape elements directly determines how landscapes function, particularly in terms of species movement, nutrient and water flows (Turner, 1989). Because landscape pattern and process are highly interrelated and interdependent, both must be understood to plan for sustainability. Landscape architects, and applied landscape ecologists have advanced theories, guidelines and models for landscape patterns that support a desired, or maximum level of ecological functions in a sustainable manner (Dramstad et al., 1996). The ecological network concept, in particular, has been implemented worldwide to address the intriguing promise of an optimal spatial strategy at broad scales including continents, nations and regions (Jongman and Pungetti, 2004). The ecological network concept, however, has aimed primarily at maintaining biodiversity and has been rarely applied in urban contexts. This trend is changing with a focus on urban environments through the green infrastructure movement.

Connectivity is a property of landscapes that illustrates the relationship between landscape structure and function. In general, connectivity refers to the degree to which a landscape facilitates or impedes the flow of energy, materials, nutrients, species, and people across a landscape. Connectivity is an emergent property of landscapes that results from the interaction of landscape structure and function, for example: water flow, nutrient cycling and the maintenance of biological diversity (Leitão et al., 2006). In highly modified landscapes, and especially in urban environments, connectivity is greatly reduced, often resulting in fragmentation – the separation and isolation of landscape elements with significant impacts on the ecological processes that require connectivity. The concept of connectivity applies directly to water flow, arguably the most important flow in any landscape, particularly in human-dominated and urban environments. Disruption of hydrologic connectivity is a major concern when planning for sustainability. Because human culture relies on water in many respects, maintaining a connected and healthy hydrological system supports multiple ABC functions. In urban, or built environments, roads represent the greatest barrier to connectivity and are the primary contributor to fragmentation (Forman et al., 2003).

Spatial configuration

With an understanding of ecological processes, the pattern:process dynamic and the importance of connectivity, spatial configuration is the point of integration. In applied landscape ecology, the mosaic model for describing and understanding the spatial configuration of landscapes is almost universally accepted. The model uses three fundamental landscape elements to define landscape structure: patches,

Table 17.2. Examples of Urban Landscape Elements Classified in the Patch-Corridor-Matrix Model

Urban Patches	Urban Corridors	Urban Matrix
• Parks	• Rivers	• Residential Neighborhoods
• Sportsfields	• Canals	• Industrial Districts
• Wetlands	• Drainageways	• Waste Disposal Areas
• Community Gardens	• Riverways	• Commercial Areas
• Cemeteries	• Roads	• Mixed Use Districts
• Campuses	• Powerlines	
• Vacant Lots		

Table 17.3. A typology of planning strategies, classifying representative actions that planners and designers routinely practice (Ahern 1995)

Protective	Defensive
Taking preventative actions to preserve well functioning, intact landscape elements before they are threatened by change or development:	Implementing actions to defend landscape elements that are suffering from development pressure:
• World Heritage Areas	• Regional, local parks
• National Parks	• Buffer zones
• "Big" patches of native vegetation	• Environmental impact mitigation
• Nature preserves	• Corridors that are pressured from adjacent land use(s)
Offensive	**Opportunistic**
Taking remedial or restorative actions to reintroduce abiotic, biotic or cultural functions where they do not currently exist:	Recognizing the potential for non-contributing landscape elements to be managed or structured differently to provide specific functions.
• Ecological restoration	• Many greenways
• Brownfields	• Most urban/green infrastructure
• Daylighted streams	• Transportation and utility infrastructure
• Bioremediation	

corridors, and the matrix. A patch is a relatively homogeneous nonlinear area that differs from its surroundings. Patches provide multiple functions including wildlife habitat, aquifer recharge areas, or sources and sinks for species or nutrients. A corridor is a linear area of a particular land cover type that is different in content and physical structure from its context (Forman, 1995). Corridors serve many functions within the landscape including habitat for wildlife, pathways or conduits for the movement of plants, animals, nutrients, and wind, or as barriers to such movement. The matrix is the dominant land cover type in terms of area, degree of connectivity and continuity, and control that is exerted over the dynamics of the landscape (Forman, 1995; Forman and Godron, 1986). Table 17.2 provides examples of urban landscape elements classified in the Patch-Corridor-Matrix Model.

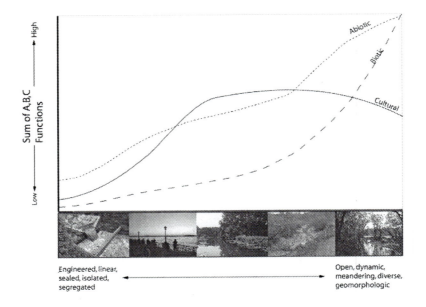

Figure 17.1.   A continuum of hydrological/stream types and associated abiotic, biotic and cultural functions

Figure 17.1 presents a continuum of urban water courses from highly engineered, linear sewers to diverse meandering river channels. Note how the associated ABC functions respond differently across the continuum. For example, streams with lower biological value may have relatively high cultural value, and high biological functions may have lower cultural functional value. The implications being that planning and management needs to consider, and employ a mixed range of hydrological types to provide a complete suite of ABC functions as part of a sustainable urban landscape. And important to accept reduced or minimal values on one category if valued functions are provided in other areas.

Forman's "indispensable patterns" are perhaps the most succinct, compelling and memorable of the landscape ecology-based guidelines (Forman, 1995) as shown in Figure 17.2. These indispensable patterns are equally relevant in urban environments as they are in landscapes that are less dominated by human development and built infrastructure. Forman argues that these patterns are fundamental, for without them specific ecological functions will not be supported.

GUIDELINES FOR PLANNING AND DESIGNING A GREEN URBAN INFRASTRUCTURE

As discussed above, landscape ecology provides scientifically-based principles for landscape planning including a multi-scaled perspective, recognition of

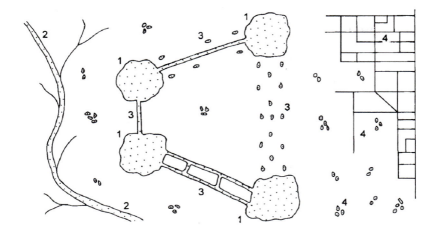

Figure 17.2.   Forman's "indispensable" patterns for planning a landscape: (1) large patches of natural vegetation, (2) stream/river corridor, (3) connectivity between patches and stepping stones, and (4) small "bits of nature" (Forman, 1995, p. 452)

pattern:process relationships, the fundamental importance of connectivity and specific guidelines for planning the spatial configuration of landscapes. To successfully apply these principles in landscape or urban planning, they must be associated with, and related to planning guidelines which enable the "good science" of landscape ecology to be effectively applied in the service of sustainability. Following are five proposed guidelines for planning and designing a green urban infrastructure based on landscape ecology principles.

1. Articulate a spatial concept

Spatial concepts guide, inspire and communicate the essence of a plan or planning strategy to provide for specific ABC functions. Spatial concepts are often articulated as metaphors that are highly imaginable and understandable by the public, but which also can support and inspire the planning process (Zonneveld, 1991). Examples include: "green heart", "ring city", and "edge city". The green heart spatial concept, for example, has guided national and regional planning in the Netherlands for decades by protecting a mostly agricultural "green heart" surrounded by a ring of urbanization including the cities of Amsterdam, Rotterdam and Utrecht. Spatial concepts are well understood in planning, but less so in science. They represent an important interface of empirical and intuitive knowledge through which rational knowledge is complemented with creative insights. Spatial concepts are essential tools for proactive, or innovative planning, and can structure and inspire the planning process, particularly with respect to achieving genuine and effective public participation.

2. Strategic thinking

Employ a strategic approach, appropriate to the spatial context and planning goals, potentially including: protective, defensive, offensive or opportunistic strategies (Ahern, 1995). Defining these strategies also helps to place the planning activity within a broader context that is particularly relevant when planning methods are transferred or adopted for use in different locations, contexts or for different applications. A planner should be aware of the macro drivers of change in a given landscape with respect to the goals of a particular plan. This awareness is the basis for informing a planner's choice of, or combination of, methods, and for engaging the appropriate participants in the planning process.

When the existing landscape supports sustainable processes and patterns, a protective strategy may be employed. Essentially, this strategy defines an eventual, or optimal landscape pattern that is proactively protected from change while the landscape around it may be allowed to change. Benton MacKaye's (1928) vision of a metropolitan open space system structured by a system of protected "dams" and "levees" is a classic example from North America. It can be effective to prevent landscape fragmentation in urbanizing landscapes by pre-defining a patch and corridor network for protection. This strategy employs planning knowledge, regulation, and land acquisition to achieve the desired spatial configuration (goal).

When the existing landscape is already fragmented, and core areas already limited in area and isolated, a defensive strategy is often applied. This strategy seeks to arrest /control the negative processes of fragmentation or urbanization. As a last resort, the defensive strategy is often necessary, but it can also be seen as a reactive strategy which attempts to "catch up with" or "put on the brakes", against the inevitable process of landscape change, in defense of an ever-decreasing nature (Sijmons, 1990).

An offensive strategy is based on a vision, or a possible landscape configuration that is articulated, understood and accepted as a goal. The offensive strategy differs from protective and defensive strategies in that it employs restoration, or reconstruction, to re-build landscape elements in previously disturbed or fragmented landscapes. The offensive strategy relies on planning knowledge, knowledge of ecological restoration, and significant public support/ funding. It requires, by definition, the displacement, or replacement of intensive land uses (e.g. urbanization, agriculture) with extensive land uses, green corridors or new open spaces in urban areas.

A landscape often contains unique elements or configurations that represent special opportunities for sustainable landscape planning. These unique elements may or may not be optimally located, but represent the potential to provide particular desired functions. The opportunistic strategy is conceptually aligned with the concept of green infrastructure by seeking new or innovative "opportunities" to provide ABC functions in association with urban infrastructure.

3. The greening of infrastructure

To achieve sustainability in urban landscapes, infrastructure must be conceived of, and understood as a bona fide means to improve, and contribute to sustainability. If one only thinks about avoiding or minimizing impact related to infrastructure development, the possibility to innovate is greatly diminished. Stormwater management provides a good example. Until recently, stormwater management aimed at controlling development-related stormwater management at pre-development levels. This "damage control" mentality produced the familiar sterile, unvegetated, inaccessible stormwater retention and detention ponds that are common throughout the USA. While this stormwater infrastructure accomplished the primary goal of controlling runoff, it failed to provide other ABC functions (water quality, ecological integrity). In contrast, consider a "green infrastructure" stormwater system that incorporates green roofs, infiltration wells, vegetated bioswales, small ponds and created wetlands. This infrastructure adds a wealth of ABC functions to the stormwater system and improves liveability for humans (van Bohemen, 2002).

4. Plan for multiple use

As discussed above under planning strategies, planning and implementing urban infrastructure presents a fundamental spatial challenge: how can new functions be added when the built environment has already displaced or replaced "natural" areas and functions? It is naive and impractical to believe that stakeholders and decision-makers will make sweeping substitutions of built forms with green areas, regardless of how committed to sustainability they are. The political, economic and social costs of such wholesale replacements are too great. Rather, it is incumbent on planners and designers to think strategically to find new ways to reconceive "grey" infrastructure to provide for sustainable ABC functions. This can be accomplished by intertwining/combining functions (Tjalllingii, 2000), as described above for stormwater management. Another design strategy is vertical integration, where multiple functions can be "stacked" in one location, as with wildlife crossings under/over roads, infiltration systems beneath building or parking lots, or green roofs on buildings (van Boheman, 2002). Innovative scheduling can also be employed to take integrate and coordinate the time dimension of ABC functions. Examples of infrastructure scheduling include limited human use of hydrological systems during periods of high flows, restrictions of recreational use of habitat areas during sensitive breeding periods, or the closing of roads at night when nocturnal species movement is concentrated. Planning for multiple use of green infrastructure can also be a useful strategy for cost effectiveness and for building a broad constituency of public support.

5. Learn by doing

A fundamental challenge and impediment to applying landscape ecology-based principles is the common lack of empirical evidence of the effectiveness of a given intervention in a specific location. Wildlife corridors provide an example. While corridors have been implemented across the world to move species across agricultural and suburban locations (Bennett, 1999), the recommendations for corridor width, length or structure are specific to the particular species and the landscape context involved. Thus, a corridor system for Koalas in Australia, has questionable transferability for planning a moose corridor in the northeastern USA. The dilemma faced by planners is that the specific recommendations needed to implement a corridor system cannot be proven by applications elsewhere for different species. Unfortunately, the result is too often inaction. Adaptive planning provides an alternative strategy. Under an adaptive approach, plans and policies are based on the best available knowledge, structured as experiments and monitored to learn how the actions result in specific goals for ABC functions. For example, to monitor cultural functions, surveys and observations of green corridor users can be kept systematically over time to track not only numbers of users but their motivations, their expectations and their impressions of the resource. Implicit in the adaptive approach is the potential to fail, but also the possibility to succeed. An adaptive approach requires a transdisciplinary effort involving, scientists, stakeholders, decision makers and planning and design professionals.

The adaptive approach is promising for green infrastructure because the knowledge to plan and implement these systems is evolving. If experimental applications can be practiced routinely, the potential to build empirical knowledge, while exploring sustainability is quite profound.

EXAMPLES OF GREEN URBAN INFRASTRUCTURE

Landscape ecology holds great potential to guide and inform the application of green urban infrastructure at a range of scales and in diverse contexts. Following are examples that illustrate green urban infrastructure across a range of scales: metropolitan/city, neighborhood/district, and site scale. The examples have been selected to also explore a broad geographical range including Asia, Europe and North America.

Taizhou City China: Metropolitan green infrastructure

Taizhou is a metropolis located on the southeast coast of China that occupies about 1000 square kilometers and has a current population of 5.5 million people. The metropolitan region is expecting a 115% population increase in the next 25 years. In response to routine flooding, the city historically developed an extensive water network integrating natural water courses, wetlands and human-made ditches. The water system significantly defined the cultural landscape character of this region, but is now suffering disturbance and destruction from rapid and extensive infrastructure construction to serve the booming economy.

An ecological infrastructure plan was designed by landscape architect Kongian Yu of Turenscape and Bejing University to support important abiotic, biotic and cultural resources, while structuring future urban development and to avoid

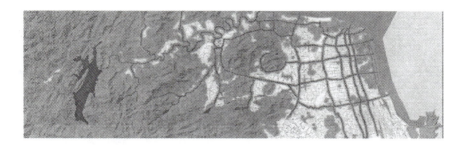

Figure 17.3.   Alternatives for the Taizhou regional ecological infrastructure at three "security levels", dark grey minimal security, medium grey, medium security and light grey high security. The medium security alternative was adopted (Turenscape, 2006)

sprawl. The ecological infrastructure is conceived to support abiotic, biotic and cultural functions, here defined as security patterns to provide sustainable ecosystem services. Security patterns represent the areas that provide important ecosystem services (flood protection) and therefore provide "security" against disturbance. Each security pattern was separately assessed, then synthesized into three alternatives (Figure 17.3).

The Taizhou plan demonstrates the application of landscape ecology guidelines including a multi-scaled approach with plan alternatives developed at regional and district scales, a linkage of pattern and process through the security patterns, and an emphasis on connectivity, particularly with respect to the multipurpose water systems. The plan also illustrates the green infrastructure guidelines proposed earlier in this chapter. The plan's security patterns present a clear spatial concept that communicates the essence of the plan effectively. The plan applies multiple planning strategies, including the defensive security patterns, and the opportunistic integration of water throughout the urban area. The ecological infrastructure is conceived as beneficial, and even essential to the city's future. Multiple use is demonstrated in many of the plan's components including the water system and green fingers that penetrate neighborhoods (Figure 17.4). There was not an adaptive component identified in the research on this example.

The Taizhou ecological infrastructure plan is an innovative and proactive response for a metropolitan region that is experiencing extreme pressure for urbanization. Although the metropolitan region includes 5.5 million people, important decisions remain to made about the future regional and urban form, influencing future sustainability. The concept is potentially transferable in urban areas where the impacts of expected population growth can motivate decision makers to explore and implement innovative ideas.

The staten Island bluebelt: neighborhood/district green infrastructure

Staten Island is the least populated borough of New York City and has a relatively intact mosaic of undisturbed wetlands. In the 1980's New York City started

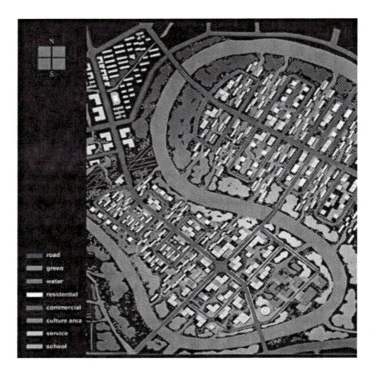

Figure 17.4. In the Taizhou "water town" alternative plan, the river is split and diverted through the city, managing the flood hazard and distributing the ecological functions provided by the river to residential neighborhoods (Turenscape, 2005)

planning to address flooding and water quality problems, including a major combined sewer overflow (CSO) problem. Unlike most cities addressing the CSO problem, New York City integrated the extensive existing wetlands into their water management plans for a 4000 hectare section of southwestern Staten Island involving some 16 small urban watersheds (Figure 17.5). The resulting Bluebelt plan was a direct result of Ian McHarg's Staten Island Study (McHarg, 1969). The Bluebelt plan has proven successful from a water quality and economic perspective, with over $80 million in savings to date (New York City DEP, 2003).

The plan had two principle components, construct a separate sanitary sewer system, and build a separate stormwater system using existing wetlands and best management practices. The stormwater system was conceived as an early example of green urban infrastructure by integrating multipurpose stormwater and wetland systems thoroughly into the fabric of the city. The Bluebelt has been successful in reducing the quantity and velocity of runoff, and removing contaminants from the runoff by introducing aquatic plants for bioremediation.

Figure 17.5.   The Staten Island Bluebelt includes 16 sub watersheds and 2500 hectares on Staten Island, New York. (http://statenislandusa.com/2004/bluebelt.htm)

The Staten Island Bluebelt anticipated many of the principles of landscape ecology. It employed a multiscale approach addressing watersheds, subwatersheds and isolated wetlands. With its focus on water management, it successfully applied a pattern:process understanding to mitigate problems and advance beneficial opportunities. The bluebelt is a model of understanding the importance of connectivity in a complex and hybrid hydrological system. It employed a logical spatial concept based on the district's hydrological patterns, which were revealed and interpreted by pioneering ecological designer Ian McHarg. Although initially motivated and focused on water quality issues, it recognized the potential to provide multiple functions, including wildlife habitat, recreational trails, and the protection of wetlands within the city. It combined a protective strategy for existing wetlands with offensive and opportunistic strategies to integrate the system with stormwater management infrastructure (Figure 17.6). The plan demonstrates the potential of "beneficial infrastructure" and has learned by doing, through water quality monitoring and the application of emerging and evolving best management practices.

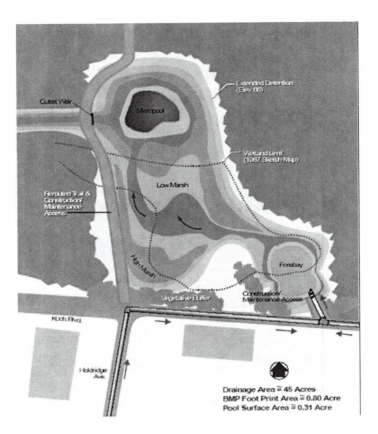

Figure 17.6.   The neighborhood scale of the Staten Island Bluebelt showing the
integrated stormwater collection system, stormwater best management practices, and a
pre-existing wetland. http://www.ci.nyc.ny.us/html/dep/html/news/bluebelt.html

The Berlin biotope/green area factor, site scale green infrastructure

The Biotope/Green Area Factor program of Berlin, Germany is an innovative example of green urban infrastructure implemented at the parcel or building scale.
From 1945 until 1990, West Berlin was an urban island within the German Democratic Republic and this unique isolation motivated research and public interest in
urban ecology. Since the 1980's West Berlin has had an active green movement,
reflecting national policies such as the National Environmental Protection Law that
empowered local authorities to develop landscape plans for urban areas, including
the Biotope/Green Area Factor program.

The Biotope/Green Area Factor program is based on the principle that modest, incremental and decentralized green infrastructure can have a significant cumulative effect to improve the urban ecology. Under the program, each parcel

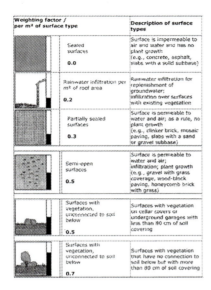

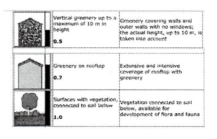

Figure 17.7.   The weighting system of Berlin's Biotope/Green Area Factor program is based on the percentage of imperviousness and the amount of vegetation present per square meter at the building or site level. http://www.stadtentwicklung.berlin.de/umwelt/landschaftsplanung/bff/en/bff_berechnung.shtml

must mitigate its impacts on-site. A primary goal of the program is to counteract "creeping impermeability" by mandating that new or renovated buildings achieve a prescribed green factor rating. The "greening" is intended to provide several functions: evapotranspiration of water, retain and infiltrate stormwater, remove airborne particulates, support natural soil functions and provide plant and animal habitat. The program is implemented at the neighborhood level, where priorities are decided, technologies selected and performance data collected and evaluated to measure progress towards goals.

The program sets green area targets based on land use: residential 60%, mixed use 40% and commercial/city center at 30% – recognizing that the targets must differ in response to land use intensity. When the policy is activated by a property sale or renovation, the owner is required to meet these targets by implementing greening techniques selected from a menu. Each technique is assigned a weight based on its contribution to the program goals and calculated as a poercentage of site area to determine the green factor. Techniques include: green roofs, bioswales, façade greening, pervious paving and plantings (Figure 17.7).

The Biotope/Green Area Factor demonstrates a "bottom-up" decentralized approach to green infrastructure planning (Keeley, 2004). While it aims at multiple goals and emphasizes the beneficial aspects of infrastructure, it does not have an explicit spatial concept. The program employs a fully opportunistic strategy and includes an adaptive component realized through monitoring of the cumulative effectiveness of the greening techniques (urban climate recording, urban species diversity, and water quality and total runoff) .

CONCLUSIONS

Green urban infrastructure is an evolving concept to provide abiotic, biotic and cultural functions in support of sustainability. Examples cited in this chapter illustrate how the green infrastructure planning and design benefit from landscape ecology principles, and how they tend to follow and support the five guidelines proposed. For green infrastructure to advance and to make legitimate contributions to urban sustainability, it must be practiced in a transdisciplinary manner – for it must meet the needs of stakeholders, benefit from the support of decision makers, engage scientists and engineers and challenge planners and designers to innovate. The proof of its success depends on the extent to which monitoring and systematic evaluations of long and short term results are made. To those who understand the green infrastructure concept, and its promise, the needs and opportunity to apply it in the pursuit of sustainability are quite profound.

ACKNOWLEDGMENTS

Support for this research was provided by the Massachusetts Agricultural Expreiment Station, Project #868. Important contributions were provided by University of Massachusetts graduate students in landscape architecture and regional planning from the Spring 2006 Green Urbanism Seminar: Taizou City, Sada Kato and Rumika Chaudry; Staten Island Greenbelt, Mark O'Rourke; and Berlin Green Factor, Susan Fitzgerald.

REFERENCES

Ahern, J. (1995). Greenways as a Planning Strategy. *Landscape and Urban Planning*, Special Greenways Issue. 33(1–3): 131–155.
Beatley, T. (2000). Green Urbanism: Learning from European Cities. Island Press, Washington.
Benedict, M.A. and McMahon, E.T. (2002). Green Infrastructure: Smart Conservation for the 21st Century. Sprawlwatch Clearinghouse Monograph Series. The Conservation Fund, Washington DC.
Bennett, A. (1999). Linkages in the Landscape: the Role of Corridors and Connectivity in Wildlife Conservation. The World Conservation Union, Gland.
Dramstad, W.E., Olson, J.D., and Forman, R.T.T. (1996). Landscape Ecology Principles in Landscape Architecture and Land-Use Planning. Island Press, Washington.
Ecological Society of America (2006). http://www.actionbioscience.org/environment/ esa.html (accessed June 30, 2006).
Forman, R.T.T., Sperling, D., Bissonette, J., Clevenger, A.P., Cutshall, C.D., Dale, V.H., Fahrig, L., France, R., Goldman, C.R., Heanue, K., Jones, J.A., Swanson, F.J., Turrentine, T., and Winter, T.C. (2003). Road Ecology: Science and Solutions, Island Press, Washington.
Forman, R.T.T. (1995). Land Mosaics. Cambridge University Press, Cambridge.
Forman, R.T.T. and Godron, M. (1986). Landscape Ecology. John Wiley, York.

Hough, M. (1995). Cities and Natural Process: A Basis for Sustainability. Routledge, New York.

Jongman, R. and Pungetti, G., Editors (2003). Ecological Networks and Greenways: Concept, Design, Implementation. Cambridge University Press. Cambridge.

Keeley, M. (2004). Green Roof Incentives: Tried and True Techniques from Europe. Proceedings of the Second Annual Green Roof for Healthy Cities Conference.

Lazaro, T.R. (1990). Urban Hydrology. Technomic, Lancaster, PA.

Leitão, A.B., Miller, J., Ahern, J., and McGarigal, K. (2006). Measuring Landscapes: A Planner's Handbook. Island Press, Washington.

Leitão, A.B. and Ahern, J. (2002). Applying landscape ecological concepts and metrics in sustainable landscape planning. Landscape and Urban Planning, 59(2): 65–93.

Low, N., Gleeson, B., Green, R., and Radovic, D. (2005). The Green City: Sustainable Homes Sustainable Suburbs. Taylor and Francis, New York.

MacKaye, B. (1928). The New Exploration University of Illinios Press, Urbana.

McHarg, I.L. (1969) Design with Nature. Natural History Press, Garden City.

Moughtin, C. and Shirley, P. (2005). Urban Design: Green Dimensions, Second Edition. Architectural Press, Amsterdam.

New York City DEP (2003). The staten island bluebelt: A natural solution to storm water management. accessed on-line (April 20, 2006). <http://www.ci.nyc.ny.us/html/dep/html/news/bluebelt.html>.

Ndubisi, F. (2002). Ecological Planning: A Historical and Comparative Synthesis. Johns Hopkins University Press, Baltimore.

Pickett, S.T.A., Cadenassso, M.L., and Grove, J.M. (2004). Resilent cities: meaning, models, and metaphor for integrating the ecological, socio-economic, and planning realms. Landscape and Urban Planning 69(4): 369–384.

Sijmons, D. (1990). Regional Planning as a Strategy. Landscape and Urban Planning. 18(3–4):265–273.

Steiner, F. (2002). Human Ecology: Following Nature's Lead. Island Press, Washington.

Tjallingii, S.P. (2000). Ecology on the edge: Landscape and ecology between town and country. Landscape and Urban Planning. 48(3–4): 103–119.

Turenscape (2006). Accessed at: http://www.turenscape.com/english/index.asp

Turner, M.G. (1989). Landscape Ecology: the Effect of pattern on process. Annual Review of Ecological Systematics. 20:171–197.

Van Bohemen, H. (2002). Infrastructure, ecology and art, Landscape and Urban Planning, 59: 189–201.

Van der Ryn, S. and Cowan, S. (1996). Ecological Design. Island Press, Washington.

Wheeler, S.M. and Beatley, T. (2002). The Sustainable Urban Development Reader: Second Edition. Routledge, New York.

Woolley, H. (2003). Urban Open Spaces, Spon Press, London.

Zonneveld, W. (1991). Conceptvorming in de Ruimtelijke Planning. Universetiet van Amsterdam.

18

Strategic planning of the sustainable future wastewater and biowaste system in Göteborg, Sweden

Per-Arne Malmqvist

Urban Water AB. Chalmers Industriteknik, SE 41288 Göteborg, Sweden
E-mail: pam@urbanwater.chalmers.se

Gerald Heinicke

DHI Water & Environment AB, Kyrkogatan 3, SE 222 22 Lund
E-mail: gerald.heinicke@dhi.se

Summary: The research program Urban Water has developed methodologies and tools for strategic planning of water and wastewater systems. The basis for the systems analysis is the Urban Water Framework, describing the system as composed of the sub-systems organization, technology and users. Criteria and indicators have been developed. The Urban Water methodology was used in the project "System Study Wastewater", aiming for developing a basis for decisions on the future sustainable wastewater system in the city of Göteborg, Sweden. Among the criteria was recycling of the nutrients in wastewater back to agriculture. The time horizon was until the year 2050. The study included wastewater and biowaste in the Göteborg region. Out of eight investigated system alternatives, four were found to be equally sustainable: Source control with household waste disposers; Source control with digestion of the biowaste; Extraction of pure nutrients at the treatment plant; and Incineration with use of the ashes.

INTRODUCTION

The water and wastewater systems in Sweden have during the last decade been questioned from the viewpoint of sustainability. New systems, often comprising local initiatives and separating technology, have been developed and are running in a small scale as alternatives or complements to the central systems. It is essential to identify the future possibilities and the limitations of different systems. Will the sustainable water and wastewater systems of the future be improved versions of those existing today, or will there be some radical changes?

In Sweden, abundant of water and nature, there are also severe environmental problems, such as high nutrient levels in the Baltic Sea, acidification of lakes and land, and continuing exposure to a number of toxic substances. Since 1999, the Swedish Parliament adopted sixteen objectives relating to the quality of Sweden's environment, and most of them are to be achieved by the year 2020 (Environmental Objectives Council, 2006). The aim is to pass on to the next generation a society in which all major environmental problems have been solved.

Among these sixteen goals, three are particularly relevant for the urban wastewater sector: Zero Eutrophication, A Non-Toxic Environment and A Good Built Environment. Other goals of concern for the urban wastewater sector are Reduced Climate Impact, Natural Acidification Only and Good-Quality Groundwater.

Zero eutrophication

The main goal is: "*Nutrient levels in soil and water must not be such that they adversely affect human health, the conditions for biological diversity or the possibility of varied use of land and water.*"

Discharges of phosphorus and nitrogen to the receiving water is a main task for all Swedish wastewater treatment plants (WWTPs). Most of them are designed for phosphorus and nitrogen removal (nitrogen removal is required only if more than 10 000 persons are connected).

A non-toxic environment

The main goal is: "*The environment must be free from manmade or extracted compounds and metals that represent a threat to human health or biological diversity.*"

Of concern to the wastewater sector is the discharge of heavy metals and organic hazardous substances to the receiving water and to soil in case the sludge is used in agriculture.

A good built environment

The main goal is: "*Cities, towns and other built-up areas must provide a good, healthy living environment and contribute to a good regional and global environment. Natural and cultural assets must be protected and developed. Buildings and amenities must be located and designed in accordance with sound environmental*

*principles and in such a way as to promote sustainable management of land, water
and other resources."*
Among the short-term goals are:

- *"To promote more efficient energy use, use of renewable energy resources and
  development of production plants for district heating, solar energy, biofuels and
  wind power".*
- *"The quantity of household waste landfilled, excluding mining waste, will be
  reduced by at least 50% by 2005 compared with 1994, at the same time as the
  total quantity of waste generated does not increase." The first part of the goal
  was achieved, the second not (Environmental Objectives Council 2006).*
- *"By 2010 at least 35% of food waste from households, restaurants, caterers
  and retail premises will be recovered by means of biological treatment. This
  target relates to food waste separated at source for both home composting and
  centralized treatment."*
- *"By 2015 at least 60% of the contents of phosphorus in wastewater will be
  brought back to productive land, which of at least half will be brought back to
  agricultural soils."*

THE GÖTEBORG WASTEWATER SYSTEM

The regional wastewater company "Gryaab" is owned by and treats wastewater
from seven municipalities. Each municipality owns and operates their local sewer
systems. Most of the sewer systems are duplicated, except in the city centre of
Göteborg where there is a combined sewer system. Gryaab owns and operates
a 124 km long tunnel system that leads to the treatment plant. Around 770 000
person equivalents were connected in 2005.

The treatment plant has a modern chemical-biological process for phosphorus
and nitrogen removal. Annually about 50 000 tons of sludge is produced (15 000
tons of dry solids, DS), and anaerobically digested for the production of biogas.
Heat pumps and turbines make use of the energy contents in the wastewater. The
sludge is composted and mixed with peat and bark, and used as construction soil
on golf courses, noise barriers alongside roads etc.

In the city of Göteborg, the Department of Sustainable Water and Waste is responsible for all water, wastewater and solid waste management. The operation and
maintenance of the water and wastewater system is the responsibility of Göteborg
Water.

The City of Göteborg, Department of Sustainable Water and Waste, has developed a "Recycling Plan" in which goals for the city are specified. Among the eight
challenges defined, three are especially relevant in this context:

- Recycling of nutrients
- Management and decrease of hazardous substances
- Increased knowledge of material flows

THE URBAN WATER TOOLBOX

The Urban Water interdisciplinary research program has been developing tools for strategic planning of urban water and wastewater systems (Malmqvist et al., 2006). The program ended in 2006. A small researcher-owned company has been founded with the purpose of making the tools available nationally and internationally, and to further develop the tools. 15 PhD exams have been another major outcome of the program. Many of the research projects in the program have been developed in co-operation with other universities in Sweden and abroad, with consulting companies and with the participating municipalities.

The toolbox contains the following tools:

- The computer model URWARE® for substance flow analysis of wastewater systems
- The computer model SEWSYS® for substance flow analysis of stormwater systems
- A computer model for assessment of microbial risks
- A framework and model for assessment of chemical risks
- A computer model for cost estimations
- Criteria for a successful organization
- Criteria for designing systems that are understood and accepted by the users
- A framework and model for multicriteria decisions

The Urban Water conceptual framework forms the basis for all systems analysis. It is a comprehensive and pragmatic approach that makes it possible to compare water and wastewater systems from a sustainability point of view.

STRATEGIC PLANNING OF THE FUTURE SUSTAINABLE WASTEWATER AND BIOWASTE SYSTEM IN GÖTEBORG

The project system study wastewater

As a means of making the Recycling Plan more strategically operative, the project System Study Wastewater was started in 2005. The project is a joint commitment between the three main actors in Göteborg: the Department of Sustainable Water and Waste, Goteborg Water and Gryaab. The project has a board with representatives from the three organizations, and a working group that also includes researchers from Urban Water.

A project plan was adopted with the following key contents:

1. Setting of goals,
2. Future scenarios for Göteborg,
3. Criteria and indicators,
4. Development of system alternatives,
5. Assessments,

6. Sensitivity analyses,
7. Multi-criteria analysis, and
8. Conclusions.

Goals

The project goal was *"to develop a basis for strategic decisions for the management of wastewater, based on which future systems that are most sustainable, i.e. cause the least environmental impact, lead to a good recycling of nutrients, and that are economically and socially acceptable."*
Key questions were:

- Which criteria should be used to declare that a system is sustainable?
- Which system or combination of systems is most sustainable – is it possible to reject any system entirely?
- Are there measures that are unambiguously justified, regardless of the design of the system?
- Should we continue the attempts to use sludge as fertilizers on arable soils, or are there better ways?
- Should we continue to trace down and decrease sources of pollutants, or is this approach futile, since we cannot use the sludge in agriculture anyway?
- Should we continue to successively separate the combined sewer system?
- Should the basic principles of the sewage system be changed and replaced by, e.g., a principle of separating the flows at the source?

Future scenarios for Göteborg

The time horizon of the study was set to the year 2050. This long time span allowed the project team to consider technologies that are not fully developed today, and not to overrate the existing infrastructure and the investments already made. By including the results from earlier studies of the future development in Sweden and the Göteborg region, and by intensive discussions among the partners in the project, two possible future scenarios were developed. In both scenarios the number of people in the region will increase by 50%. In the "goal-scenario" there is a positive development from an environmental point-of-view concerning e.g. the use of construction materials (heavy metals, organic hazardous substances), household chemicals, pharmaceuticals and other substances. In a "threat-scenario" the situation is similar to or worse than the situation today. In both scenarios there will be a scarcity of fossil fuel and of phosphorus, which will result in significantly higher prices for these resources, regulations regarding their usage, or both.

Criteria and indicators

It was decided to use the criteria defined by the Urban Water framework, see Figure 18.1.

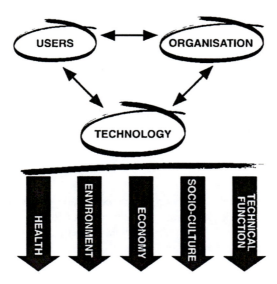

Figure 18.1.   The Urban Water conceptual framework for sustainable water and wastewater systems.

Environmental indicators

The most relevant environmental indicators were considered to be

- Eutrophication (phosphorus and nitrogen to the recipients)
- Cadmium to water
- Copper to water
- Cadmium to soils
- Copper to soils
- Pharmaceutical residues to water
- Pharmaceutical residues to soils
- Other hazardous organic substances to water
- Other organic hazardous substances to soils
- Acidification
- Climate change
- Recycling of phosphorus to arable soil
- Recycling of other in principle limited nutrients (K and S) to arable soil
- Use and production of energy (calculated as exergy)

Cadmium and copper were chosen as representatives for the heavy metal content in sludge. Compared to the rate of accumulation of metals in agricultural soils, also silver, gold, lead and mercury should be included, probably also zinc and arsenic. The Swedish Environmental Protection Agency (SEPA) has adopted limit values for the use of wastewater sludge in agriculture, see Table 18.1. The only metal for which tightened requirements are suggested is cadmium. However, seen to the risk of accumulating metals in the arable soils, the ratio between the concentrations in wastewater sludge and in average arable soils in Sweden are highest for silver, gold, copper and mercury (>10 times).

Little is known of the contents of pharmaceutical residues in sludge, and the hazards of bringing them to agricultural soils. Nitrogen is not a limited resource, but the energy required to produce it from the air was included in the calculations.

Economic indicators

The total annual costs until 2050 were estimated for all systems alternatives. The total annual costs include capital costs and operation and maintenance costs, but neither income from the sale of rest products or energy, nor external costs that occur outside the system boundaries.

Indicators for household aspects

The research within the Urban Water program has developed some key criteria for a successful involvement and acceptance of the households when new or controversial technologies are introduced (such as urine separation):

- The technology must be practical to operate, not take considerable extra time, nor imply significantly higher costs for the user than the existing system.
- The users need to have trust in the administrations responsible, or other key actors.
- The users need to have trust in the environmental benefits of the new technology.

Indicators for organizational aspects

The research within the Urban Water program has developed some key criteria for success of strategic planning that involves new or controversial technologies or management schemes:

- Shared values between key stakeholders (shared ideas, shared goals, or at least a common understanding of existing disagreement)
- Freedom of action (laws and political support)
- Available resources (financing and knowledge)
- Communication with the users of the system
- Arenas for solving of possible conflicts
- Stakeholders who act responsibly and who actively promote the new system.

Table 18.1. Current limit values for the transport of metals to soils and for concentrations in the soil.

	Limit concentrations in wastewater (mg me/kg DS)	Limit concentrations in wastewater (mg me/kg P)	Max amounts brought to agricultural soils (g/ha,year)	Limit concentrations in soil (mg me/kg DS)
Lead	**100**	3600	**25**	**40**
Cadmium	**2** (1.7)	71 (61)	**0.75** 0.55 (2010) 0.45 (2015) 0.35 (2020)	**0.4**
Copper	**600**	21000	**300**	**40**
Chromium	**100**	3600	**40**	**60**
Mercury	**2.5** (1.8)	89 (64)	**1.5** (1)	**0.3**
Nickel	**50**	1800	**25**	**30**
Silver	15	540	8	–
Tin	35	1200	–	–
Zinc	**800**	29000	**600**	**100**

Enforced limit values are marked in bold.
Note the suggested lowered limit values for cadmium and mercury from 2010 on (SEPA, 2002)

Indicators for hygiene

The hygienic aspects were assessed by Urban Water's Microbial Risk Assessment (MRA) tool. This tool comprises risk identification, an estimation of exposure, dose-response correlations and finally risk assessment. The model is based on identification of the points at which human beings are exposed to pathogens from wastewater. For each exposure point, exposure and risks are then quantified by simulation. The indicator pathogens used in the simulations represent the following groups: bacteria, virus, protozoa and worms. The method accords well with methodology developed by the World Health Organization.

Development of system alternatives

The geographical system boundary has been the Göteborg region: the seven municipalities that deliver their wastewater to a common wastewater treatment plant. The technical system boundaries have been the wastewater system and the system for collecting and treating the biowaste from households and industries.

The following eight systems were compared in this study:

1. The reference system with use of the sludge for construction soils
2. Source control with composting
3. Source control with household waste disposers
4. Source control with digestion
5. Blackwater separation with food waste disposers
6. Extraction of pure nutrients at the treatment plant
7. Incineration with use of the ashes
8. Incineration with depositing of the ashes

1. The **reference** system is similar to today's system, but under the conditions prevailing in 2050. This means a central sewer system, partly combined, and a mechanical, chemical and biological treatment process at the central treatment plant. The sludge is used for production of construction soil by composting it and mixing it with peat and bark. The household biowaste is composted and also used as construction soil.
2. The **source control systems** (systems 2–4) imply far-reaching measures towards a cleaner society, in order to produce sludge with a low content of hazardous substances. Source control with **composting** means that the sludge is composted and thereafter used in agriculture. The biowaste is composted and used for construction soils.
3. In the source control system with **food waste disposers**, biowaste is ground and transported in the wastewater pipe system to the WWTP. The sludge is used in agriculture.
4. The Source control system with **digestion** is similar to the second system, but the biowaste is collected by truck and digested separately from the wastewater sludge. The rest products are used in agriculture.

5. **Blackwater** separation means that a second pipe is installed in each household, collecting only the water from the toilets. Household waste disposers are introduced. The blackwater flows are transported in pipes to four local plants in the outskirts of Göteborg, in which the water is concentrated about ten times and sanitized before transport to farms for direct use in agriculture. This system is introduced in the "New Göteborg", that is the parts of Göteborg where new buildings are expected to be built until 2050, and in districts where far-reaching renovations to old buildings will be carried out. Altogether, about 70% of the inhabitants in Göteborg 2050 will live in such buildings. The flows from the "Old Göteborg" are treated as in the reference alternative.
6. In the **extraction** system a pure phosphorus product is extracted at the treatment plant by a chemical process. Also some 20% of the nitrogen is extracted from the internal reject flow from the digest chamber. The biowaste is digested and used in agriculture.
7. The **incineration** system with use of the ashes is similar to the reference system, except that all sludge is incinerated and the ashes are used in agriculture. The biowaste is digested and used in agriculture.
8. In the incineration system with **deposit** of the ashes, the sludge and biowaste are incinerated and the ashes deposited.

In addition to these systems alternatives, separate measures for reducing pollution from stormwater were suggested. Also specific measures at the treatment plant were considered, such as separate treatment of storm water and sludge treatment during wet weather periods.

Assessment of environmental impact

The substance flow model URWARE was used for simulation of the environmental aspects. URWARE simulates flows of matter and energy in the *core system* (the technical water system) and the *extended system* (including upstream and downstream effects. An example of upstream effects is the environmental impacts of production and transport of chemicals used for treatment. An example of downstream effects is leakage of nutrients from farmland to water bodies). The model also considers *compensatory effects* to allow a just comparison between differing system alternatives. The compensatory system in URWARE makes sure that all system alternatives are assumed to produce the same amount of district heating, fertilizer products etc. In practice, the compensatory system adds virtual contributions of energy consumption and pollution load to those system alternatives that produce less benefits than others. The model is able to simulate the sources and pathways for 83 components in the wastewater and solid waste system, Figures 18.2–18.7 show *examples* of simulation results for some relevant indicators. The simulations in the enlarged system are in URWARE made according to the principles of LCA (Life Cycle Assessment) –"from cradle to grave".

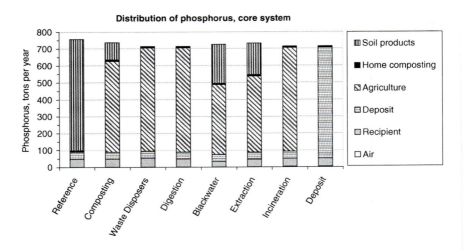

Figure 18.2.   Distribution of phosphorus.

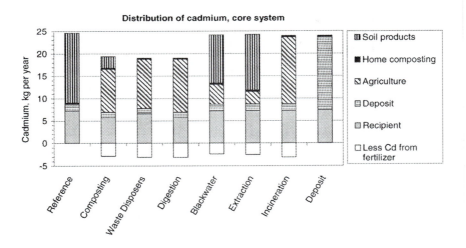

Figure 18.3.   Distribution of cadmium.

Multi-criteria analysis

Several methods for multi-criteria decision aid have been developed internationally, some of which involve support from computer models (Ashley et al., 2004). An example of a simplified method without computer support has been demonstrated by Lindholm (2002). In this study we have used a simplified method involving scoring and weighting.

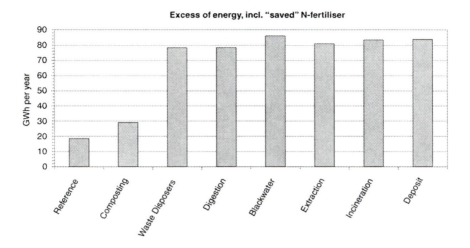

Figure 18.4.   Excess of energy.

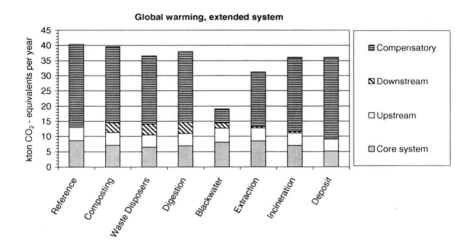

Figure 18.5.   Global warming.

In order to compare both quantitative and qualitative data, the results were normalized using a scale 0–4, where the score 2 was given when the system was considered to meet future demands. To account for the relative importance, each indicator and group of indicators were assigned a weight. An average score was then calculated by multiplying the scores with their corresponding weights.

As a basis for the weighting, the relative importance of the Swedish environmental goals in the Göteborg region were used, together with evidence or qualified

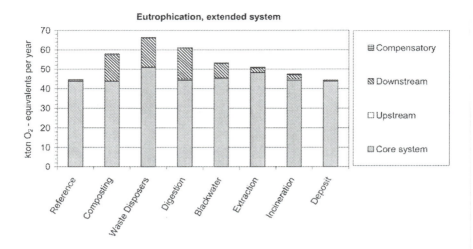

Figure 18.6.   Eutrophication.

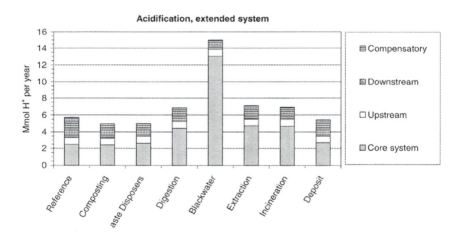

Figure 18.7.   Acidification

guesses of how important the wastewater sector is for the environmental aspects studied in relation to the total anthropogenic contributions (Hellström et al., 2004). When setting the scores for the indicators, future requirements from the food industry, the farmers and the public were also considered. The setting of weights is a highly political matter; the project group developed a set of weights that was assumed to be in accordance with the current main-stream political visions.

As a result four systems alternatives turned out to be more sustainable than the others:

Table 18.2. Some quantitative data for the most sustainable systems alternatives.

Extended system	Source control with waste disposers	Source control with digestion	Incineration with use of ashes	Extraction
Eutrophication (kton O_2-equiavalents/year)	66	61	47	51
Cadmium to water (kg Cd/year)	7	6	7	7
Copper to water (kg Cu/year)	500	430	600	600
Cadmium to agriculture (kg Cd/year)	11	12	15	3
Copper to agriculture (kg Cu/year)	1900	1900	2500	300
Global warming (kton CO_2-eqv./year)	37	38	36	31
Acidification (kmol H^+/year)	5000	6800	6900	7100
Recycling of phosphorus (ton P/year)	610	620	610	450
Recycling of nitrogen (ton N/year)	350	720	440	1210
Exergy excess (GWh/year)	78	78	83	82
Quality of sludge or ashes (mg Cd/kg P)	18	19	24	
Quality of sludge or ashes (mg Cu/kg P)	3000	3000	4100	
Annual costs (Euro/person,year)	140	120	110	110

no. 3: Source control with household waste disposers
no. 4: Source control with digestion
no. 6: Extraction of pure nutrients at the treatment plant
no. 7: Incineration with use of the ashes

Some quantified data for these four alternatives are given in Table 18.2. The qualitative data for e.g. organizational and the user aspects is not presented here.

Sensitivity analyses

Sensitivity analyses have been made regarding: (1) if the "goal scenario" is replaced by the "threat scenario" and (2) with alternative weights for the indicators used. The analyses performed so far do not change the main conclusions.

PRELIMINARY CONCLUSIONS

The methodology for strategic planning developed by the Urban Water research program was proven to be suitable for the task of planning a future sustainable wastewater system. The project described above has been a learning process, in which the partners have gained considerably more knowledge, both of the existing system and possible alternative systems. The preliminary general findings from the study are:

- Source-separating sewer systems should not be chosen as the main principle.
- Anaerobic digestion of biowaste is more sustainable than composting.
- Separation of combined wastewater pipes is no cost-effective measure to reduce the pollution of sewage sludge.
- Source-control measures are advantageous regardless the choice of system.

ACKNOWLEDGEMENTS

We acknowledge the support of the Swedish Foundation for Strategic Environmental Research (MISTRA), the main financier of the Urban Water research program. We also acknowledge the support of the City of Göteborg, Department of Sustainable Water and Waste, for making this project possible.

REFERENCES

Ashley, R., Blackwood, D., Butler, D., and Jowitt, P. (2004). Sustainable Water Services – A Procedural Guide. IWA Publishing, London. ISBN 1-84339-065-5
Hellström, D., Hjerpe, M., Jönsson, H., Malmqvist, P.-A., and Palmquist, H. (2004). Implications of ecological sustainability for urban water management. *Vatten* no. 60-1, pages 33–41. ISSN 0042–2886.
Lindholm, O. (2002). *Analyse av kriterier og vektemetoder for bærekraftighet av avløpssystemer* (Analysis of criteria and weighting methods for sustainable wastewater

systems, In Norwegian). Report 4553–2002, Norwegian Institute for Water Research, Oslo. ISBN 82-577-4208-2

Malmqvist, P.-A., Heinicke, G., Kärrman, E., Stenström, T.A., and Svensson, G. (2006). *Strategic Planning of Sustainable Urban Water Management*. IWA Publishing, London. ISBN: 1843391058.

SEPA (2002). *Aktionsplan för återföring av fosfor ur avlopp*. (Action plan for recycling of phosphorus from wastewater. In Swedish with English summary). Report no. 5214. Swedish Environmental Protection Agency (Naturvårdsverket), Stockholm. ISBN 91-620-5214-4, pdf: www.naturvardsverket.se

Environmental Objectives Council (2006). *Sweden's environmental objectives – Buying into a better future, de Facto 2006*. Swedish Environmental Objectives Council, Stockholm. ISBN 91-620-1240-1, pdf: www.miljomalen.nu.

19

The role of low impact redevelopment/development in integrated watershed management planning: Turning theory into practice

M. Maimone[1], J. Smullen[2], B. Marengo[3] and C. Crockett[3]

[1]*CDM, Woodbury, NY 11797, USA*
[2]*CDM, Edison, NJ 08818, USA*
[3]*Philadelphia Water Department, Philadelphia, PA 19107, USA*
E-mail: MaimoneM@cdm.com

Summary: The Philadelphia Water Department is meeting its commitment to reduce combined sewer overflow with an innovative program. Key elements of the program include the establishment of more realistic implementation targets coupled with land based stormwater management practices based on principles of low impact development. Combining their water management program with City Greening initiatives has broadened the base of support for the program, and expanded its benefits to include all the citizens of Philadelphia.

INTRODUCTION

The Philadelphia Water Department (PWD) has provided integrated water, wastewater, and storm water services for almost 200 years to Philadelphia and outlying communities. In addition to providing the more traditional utility services, its mission includes sustaining and enhancing the region's watersheds and quality of life by managing wastewater and storm water effectively. To this end, PWD has embarked on an ambitious program of watershed-based, water management that includes integration of structural projects, regulatory changes, and non-structural measures. PWD is integrating all aspects of its water management program into integrated watershed management plans for each of its five watersheds.

PWD's mission is to meet its obligations under the Clean Water Act and the Pennsylvania Clean Streams Law to protect the integrity of its receiving waters. PWD's approach to attainment and maintenance of the designated and beneficial uses of these waters is through planning, developing, and implementing technically viable, cost-effective improvements and operational changes. Some of the watersheds within the city are served by combined sewers, others by separate sewers (see Figure 19.1). Each presents unique challenges to improving and protecting water quality. PWD is investing in necessary capital projects to increase the sewer system's ability to store and treat combined sewer and stormwater flows. Of equal importance, however, is PWD's innovative planning approach to promote control of stormwater at the source through low impact development and redevelopment retrofits, supported by new stormwater regulations and other progressive practices such as street tree planting and riparian buffer restoration. These practices are designed to improve conditions in both the combined and separate sewered areas and better manage water, protect and ensure beneficial uses, restore stream habitat, meet the priorities of residents for cleaner and safer streams, while optimizing the use of infrastructure planned and constructed in the past for larger populations than those realized today.

IMPLEMENTATION CHALLENGES

Planning must constantly consider feasibility of implementation or it is doomed to failure (failure being defined here not as a poor plan, but a plan not implemented). Without the ability to implement actions, a plan is useless. This does not imply that a plan has failed if it is not implemented in its entirety, or even as recommended. The implementation phase is truly where an incremental approach is most realistic and applicable. In developing an implementation plan, a number of practical considerations can make the difference between a plan that is implemented, and one that is not.

- Range of Implementation Tools: It is important to look at the range of implementation tools available. These can be regulatory tools (e.g. the aggressive use of discharge permits, the application of water quality standards, enforcement actions against violators), structural tools (e.g. the construction of water control or

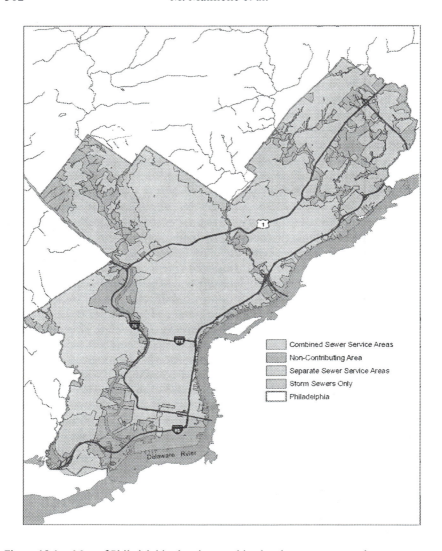

Figure 19.1. Map of Philadelphia showing combined and separate sewered areas.

treatment structures), changes in process or procedure (e.g. hazardous material handling), or motivational/voluntary tools (such as public education campaigns, school curricula, volunteer stream cleanup days).

- Incremental or Coordinated Implementation: The implementation plan should be very clear about which tools can be implemented individually, and which must be considered in a more coordinated fashion. For example, stream restoration often occurs piecemeal, as sections of a stream show signs of erosion or siltation. Yet restoring a stretch of stream without considering solutions for the stormwater flows that are causing it, or consideration of changes occurring upstream that will influence the restoration, usually results in failure of the restoration. In contrast, the reduction in pollutant loading from a point source such as a wastewater treatment plant through a new treatment process rarely needs to be coordinated with other actions. Its value is easily calculated and not influenced much by other actions.

- Identify the Implementer: An implementation plan that does not specify the agency or group that needs to carry out action is also likely to fail. If the implementing agency is not identified, action is invariably left to "someone else".
- Set a schedule: Implementation schedules are rarely met in water resource planning. This does not mean, however, that a schedule is not required. Schedules motivate implementers to action, and provide a means to check on progress (or lack thereof). Without a schedule, stakeholders cannot point out that implementation is "behind schedule".
- Determine the appropriate level of detail: Water resource planning spans such a variety of plan types and scales of implementation that specific guidelines on level of detail cannot be given. Usually, plans are made on a watershed level, and budget and time constraints do not allow for details of implementation sites to be included in the recommendations. Because many engineers consider implementation plans to be synonymous with detailed facility plans (e.g. the conceptual design of a treatment plant or stream restoration), this is considered as a reason to end the planning at the recommendation step. Implementation plans can be presented in a nested level of detail, however, starting with general planning guidelines, followed by specific tasks, leading to detailed plans and specifications or task descriptions for individual actions. The implementation plan itself can be structured to provide the guidelines and a general list of tasks and schedule, with the details left as the first step in implementation of recommended actions.

In developing an implementation approach, PWD identified a critical impediment to achieving a balance in meeting overall program objectives: a strict adherence to water quality standards under all conditions is impossible to achieve for an urban stream, and might even be counter-productive to achieving the broader goals articulated by the stakeholders in each watershed. Therefore, a new approach to classifying and attaining the identified water body improvement goals was developed consisting of three general planning and implementation targets.

Target A – Meeting water quality goals in dry weather: Improving dry-weather water quality and the stream's aesthetic quality through such measures as trash removal, litter prevention, and elimination of wastewater discharges during dry weather is critical. These measures should enable swimable and human health water quality standards to be met nearly 70% of the time on an average annual basis. Improvements to the aesthetic appearance and health of the riparian habitat will make the stream an asset to the community during the times most appropriate for engaging in the intended beneficial uses.

Target B – Healthy living resources: improve conditions for fish propagation and population stability, benthic organisms abundance and diversity, and the pollutant reduction functions of riparian and wetland plants. To achieve this target of fishable conditions at all times, management measures address the fluvial geomorphology of the stream system to provide improvements to riparian habitat, eroding and undercut banks, scoured streambeds, excessive silt deposits and burial of organisms, channelized and armoured stream sections, trash build-up, anthropogenic encroachment and invasive species.

Target C – Meeting swimable and human health water quality criteria during wet weather and addressing flooding issues: this target is the most difficult to achieve in an urban environment and must be met in phases, with interim targets for reducing wet-weather pollutant loads and stormwater flows, and with ongoing monitoring to determine the efficacy of control measures. This target also includes the review, over time, of water quality standards and designated uses in the City's CSO-receiving waters.

These targets provide the underlying basis for PWD's comprehensive watershed management planning approach and for the development of plan implementation strategies. Although structural measures to detain and treat stormwater are an important element in PWD's approach to Target C, it is only one aspect of their comprehensive approach. Reducing stormwater runoff through hydrologic source controls is an equally critical component of both the CSO control program and the stormwater management program. PWD has developed a multi-faceted program to demonstrate the feasibility of these practices in reducing CSO and stormwater discharges and to ensure that the technology is broadly applied as the City develops and redevelops. The program includes the following components.

- New Stormwater Regulations that require management of rate and volume of stormwater for water quality, channel protection, and flood protection.
- Low impact development and redevelopment of the City's vacant or abandoned land as part of a Green City Program
- Encouraging the disconnection of impervious cover and implementation of greening practices on City School lands through the Campus Parks initiative.
- Implementing a PWD Facilities LID Program to demonstrate the technology to others and to verify its effectiveness in mitigating the impacts of CSO and Stormwater discharges.
- Implementing a low-impact development grants program to provide control incentives for small properties to voluntarily implement LID practices on private property.
- Implementing a comprehensive tree planting and greenway development program to increase tree canopy throughout the City.

Water resource planning usually has a very long implementation horizon (decades), and the likelihood that the implementation will proceed according to plan is very small. PWD believes that the only viable approach to implementation is through adaptive management. Adaptive management puts forth the concept that water resource management must be adaptive and flexible, treating management policies and actions as experiments, not fixed policies. This implies that management must continually improve by learning from the ecosystems being affected. Adaptive management links science, values, and the experience of stakeholders and managers to the art of making management decisions as implementation proceeds. The concept of adaptive management appears to be the most viable approach in implementing the planned options of a water resource plan because of the uncertainties in knowledge and the variability of societal attitudes toward the resource. Unfortunately, an adaptive management approach is often in conflict with the need to meet specific regulatory requirements, and PWD's target approach to implementation is still under discussion with regulators at the State and Federal level.

In the following sections, elements of PWD's implementation approach to meeting Target C, or wet weather water quality, are briefly described.

STORMWATER REGULATIONS

In 2006, PWD implemented a new set of stormwater regulations to require development and redevelopment projects to control the rate and volume of stormwater runoff to meet water quality, channel protection, and flood protection criteria. These regulations require that controls be applied for all new impervious surfaces within the limits of earth disturbance for any earth disturbance greater than 15,000 square feet. In addition, PWD developed a new Urban Stormwater Management Guidance Document to help developers to develop and redevelop Philadelphia's vacant land in accordance with the new stormwater regulations. PWD will also investigate a revision of the stormwater rate system to encourage stormwater controls by linking stormwater rates to impervious cover and project the effectiveness of this program in reducing CSO discharges.

There are three major elements to PWD's new stormwater regulations: Water Quality, Channel Protection, and Flood Control Requirements.

Water quality requirement

The Water Quality Requirement manages the first one inch of precipitation over directly connected impervious areas (DCIA) from each storm and is established to recharge the groundwater table and increase stream base flows, and to reduce contaminated runoff from sites as well as decrease combined sewer overflow (CSO) discharges in the City's combined sewer systems. The management technique required is infiltration unless infiltration is determined to be physically impossible (due to contamination, high groundwater table, shallow bed rock, poor infiltration rates) or where it can be shown that doing so would cause property or environmental damage. Where infiltration is not feasible for the entire inch, any remaining portion of the initial inch of precipitation from a storm that cannot be infiltrated must be treated for water quality. The treatment is different for separate sewer areas than for combined sewer areas:

- Separate sewer areas: stormwater management practice for water quality
- Combined sewer areas: detain and release at an average rate of 0.12 cfs per acre and maximum rate of 0.24 cfs per acre in no less than 24 hours and no more than 72 hours.

Channel protection requirement

The Channel Protection Requirement is a slow release of the 1-year, 24-hour storm event, a 2.7 inch event for Philadelphia, detained from DCIA. The Channel Protection Requirement is established to protect quality of stream channels and banks, fish habitat, and man-made infrastructure from the influences of high stream velocity erosive forces. The Channel Protection Requirement requires that all new development or redevelopment detain and release runoff from DCIA at an average rate of 0.12 cfs per acre and maximum rate of 0.24 cfs per acre in no less than 24 hours and no more than 72 hours. To stimulate low impact development, a reduction in DCIA by 20% between the post-development and pre-development condition exempts redevelopment projects from the Channel Protection Requirement.

Flood control requirement

The Flood Control Requirement is established to reduce or prevent the occurrence of flooding in areas downstream of the development site, as may be caused by in-adequate sewer capacity or stream bank overflow. The Flood Control Requirement is based upon ongoing watershed wide stormwater planning under Pennsylvania Act 167 that results in determining flood management districts for controlling peak rates of runoff. In general, a development project is required to make peak rates of runoff post-development equal to pre-development conditions. Reducing DCIA by 20% between the post-development and pre-development condition also exempts redevelopment projects from the Flood Control Requirement.

As an example of the expected efficacy of the new stormwater regulations, projected reductions in CSO's to city waters attributable to redevelopment under the new stormwater regulations were modeled using the EPA SWMM model. Model results suggest that redevelopment under the new stormwater regulations of about 100 acres City-wide in the combined sewered areas would increase capture of combined stormwater and wastewater from approximately 60% to 65% in the coming 20-years, a substantial reduction in combined sewer overflow and pollutant loads during wet weather.

LID PRINCIPLES

The new stormwater regulations are designed to change the face of development in the city from traditional ways of handling stormwater to development maximizing the use of Low Impact Development (LID) principles. The primary Goal of LID is to design each development site to protect, or restore, the natural hydrology of the site so that the overall integrity of the watershed is protected. This is done by creating a "hydrologically" functional landscape. Some basic principles of LID include:

- Reduction of roads and other infrastructure necessary to support the development;
- Conservation of significant natural resources and habitat;
- Minimization of the environmental impact resulting from the change in land use (minimum disturbance, minimum maintenance);
- Maintenance of a balanced water budget by avoiding direct connection of impervious surfaces to drainage-ways, making use of available site characteristics and natural infiltration;
- Incorporation of unique site features (natural, scenic and historic) into the configuration of the development to increase property owners' enjoyment of and access to those features;
- Reduction of property owner / homeowner association / community maintenance responsibility for the number and size of structural or engineered stormwater management practices.

As described in the Philadelphia BMP manual, the integrated site design procedure can be summarized in three steps:

1. Protect and utilize existing site features
2. Reduce impervious cover to be managed
3. Manage the remaining stormwater using a systems approach to stormwater management practice (SMP) design

These steps are implemented initially in sequence and then in an iterative approach leading to formulation of a site stormwater management plan. The intent of the planning process is to promote development of stormwater management solutions that protect receiving waters in a cost effective manner. This procedure replaces conventional stormwater management which has existed in Philadelphia for many years. Perhaps most importantly, the procedure involves the total site design process. Conventional stormwater management was usually relegated to the final stages of the site design and overall land development process, after most other site issues have been determined and accommodated (and thus the frequent relegation of stormwater management practices to what appears to be the "leftover" areas of the site). PWD's new recommended site design procedure places stormwater management in the initial stages of the site planning process, when the building program is being fitted and tested against the site conditions. In this way, comprehensive stormwater management can be integrated effectively into the site design process.

Protecting and utilizing existing site features is primarily encouraged through protection of sensitive areas, through clustering and concentrating site development, and by making every attempt to minimize site disturbance. A reduction to Directly Connected Impervious Area (DCIA) is acknowledged for a site when:

- downspouts are disconnected and then directed to a pervious area which allow for infiltration, filtration, and increased time of concentration.
- pavement runoff is directed to a pervious area which allows for infiltration, filtration, and increase the time of concentration.
- new or existing tree canopy from an approved species list extends over or is in close proximity to the impervious cover.
- a green roof is installed on a proposed building.
- a porous pavement system is installed on the site such that it does not create any areas of concentrated infiltration. Porous pavement systems, including porous asphalt; porous concrete; porous pavers with at least 50% void space; and other approved porous structural surfaces can be considered to be disconnected if they receive direct rainfall only if underlain by a crushed stone infiltration bed that is at least 8 inches deep.

GREEN CITY PROGRAM

PWD is not only thinking of LID as an approach to stormwater management and CSO control. It is also linking the program to other city initiatives. For example, PWD is emphasizing the use of sustainable, locally based greening projects, such as street tree plantings, small lot conversions to green space, community gardens and rain barrels. The City's Neighbourhood Transformation Initiative (NTI) has focused on razing abandoned and structurally deficient housing throughout the combined sewer service area. As a result of this practice, impervious cover has been reduced significantly. PWD has developed specifications for grading that will allow for enhanced infiltration to occur on these sites, and therefore contribute to CSO reductions. In addition, PWD is supporting projects to improve runoff controls for city streets as part of a Street Greening Program to facilitate the planting of trees, the greening of medians and traffic islands and the creation of sidewalk planters. In order to demonstrate the effectiveness of the LID and the Philadelphia Green program, a number of demonstration projects have been completed and others will be completed in the next five years.

PROGRESS TO DATE

The program is fully outlined in planning documents, but PWD has already completed several projects. A few examples are provided below.

Mill creek public housing redevelopment project

As part of the complete redevelopment of the Mill Creek Public Housing development in West Philadelphia, the Philadelphia Housing Authority teamed up with the Philadelphia Water Department to integrate stormwater management features that exceed City Code requirements and are designed to mitigate the development's contribution to combined sewer overflows. By incorporating a large underground detention/infiltration facility as part of the development's open space plan, the design separates a sizeable portion of site runoff from the existing combined sewer system.

Mill creek – Fairmount avenue

The Fairmount avenue project was designed to provide a prototype for basic stormwater management, via infiltration when appropriate, for programs such as the Neighbourhood Transformation Initiative (NTI). The site preparation involved regrading to the center, planting trees for soil detention and erosion control, and blanketing the site with woodchips.

Mill creek – Reno street

The Reno Street project involved the creation of a community "block" park that invites residents to gather and share tips on planting or life. A vegetated bioswale in the southwestern corner of the park detains and cleans stormwater runoff from the site (the storage portion is underground). The bioswale has an overflow pipe to the city sewer should the swale become saturated.

Courtesy stables project

The Courtesy Stables Runoff Treatment Project was aimed at correcting a suite of problems contributing to nutrient-laden stormwater that flows from the barnyard through an adjacent wetland and into a tributary of the Wissahickon Creek. The intent of this project was to route stormwater from the barnyard and surrounding area into a grassed waterway/filter strip where nutrients and sediment are removed and a portion of the water infiltrates before reaching the wetland. Flow from a springhouse is routed directly to the wetland, serving as a continuous source of clean water, rather than through the riding ring, where it adsorbs nutrients and creates muddy conditions. Invasive plant species onsite were removed and replaced with Philadelphia-native trees and shrubs and educational signage was erected, linking the nutrient runoff reduction to the improvement of the Delaware Estuary.

PWD plans additional projects during the next five years to further its LID program. Some examples include:

Mill creek recreation center

The Mill Creek Recreation Center (approximately 95,000 ft^2) is used heavily by the community for sports, activities, and meetings. The site contains two basketball courts, a playground, a recreation center, a baseball field and a swimming pool, which were all built directly above the streambed of the buried Mill Creek. Over the years, differential subsidence of the poor quality fill material has led to significant cracking and low spots throughout the paved areas. To ameliorate the drainage and subsidence problems, both basketball court areas will be replaced, one with porous pavement and the other one with standard asphalt. An underground stormwater detention system with an overflow pipe will be installed under the porous paved court, while the other court and surrounding paved areas will be graded to drain runoff to the porous pavement. In addition, a storm inlet that drains a large portion of the site will also be connected to the underground storage area. Rain barrels and a raised planter box will be connected to the rain leaders of both the recreation center and the pool maintenance building in order to collect and manage rooftop runoff.

Traffic triangle & street median stormwater demonstration

This project will redesign a traffic triangle in West Philadelphia and 3200 linear feet of median in Southwest Philadelphia to accommodate stormwater inflow and infiltration and serve as a demonstration of appropriate techniques for naturalized stormwater management when streets are rebuilt. The vegetated areas will serve as bioretention gardens and contribute to reducing runoff flows to the combined sewer system and improving runoff water quality in separate storm sewer areas.

Norris square – El Mercado

This project proposes to build a green roof and rain garden on the grounds of the Mercado at Front and Palmer Street.

Woodmere museum porous pavement parking lot

A new porous pavement parking lot with subsurface gravel storage will be constructed as part of a planned facility expansion. In addition, roof runoff from the building addition will be piped into the subsurface storage beds. Water will be allowed to infiltrate, with any excess water released at a controlled rate.

Hawthorne community park (12th & Catharine sts.)

This project has been designed and includes landscape features within the park to manage all site runoff and specialized tree trenches to collect, store, and infiltrate runoff from two existing street inlets.

Lansdale borough park improvements

Concept planning for this project is complete and includes creating stormwater treatment/infiltration gardens for five stormwater outfalls and improved stream-bank buffer vegetation along 900 feet at the headwaters of Wissahickon Creek.

Northern liberties community center

Concept planning for this project is complete and includes a vegetated roof, stormwater collection system for site irrigation, and rain gardens to manage rooftop and site runoff.

PWD is also concentrating on education to improve citizen awareness of environmental issues. PWD's Campus Park Initiative aims to create outdoor spaces that are educationally, environmentally, and socially vibrant at every school throughout the District over the next five years. The program is rooted in the belief that students must have access to outdoor facilities that are stimulating, safe, and conducive to an array of appropriate uses. As part of this initiative, PWD will continue to develop a program element that encourages schools to add stormwater management aspects to the campus park and greening program. The Philadelphia Water Department's role within the initiative is to supply technical assistance to landscape architects and designers to incorporate stormwater management best management practices (BMPs) and outdoor watershed education areas into the site designs. Already there have been two completed projects at the Penn-Alexander School and Sulzberger Middle School. Projects in design and planned for the coming years include:

Philadelphia school district green roof program

The Delaware Valley Green Building Council, in association with PWD has been working with the School District to implement the U.S. Green Building Council's Leadership in Energy and Environmental Design (LEED) standards for developing high-performance, sustainable buildings. However, building to LEED standards often involves higher up-front costs. Because low-impact stormwater management systems such as vegetated roofs (commonly referred to as green roofs), satisfy LEED guidelines, PWD intends to offset the marginal cost difference bourn by the School District when installing green roofs.

Saul high school

This project will combine urban stormwater and agricultural BMPs to reduce the harmful impact of the school's runoff on the water quality of the Wissahickon Creek, a tributary of the Schuylkill River. Cattle crossings, cattle fencing, and riparian buffers will be used to limit the impact of livestock on the runoff quality and prevent harmful pathogens and nutrients from entering the watershed. Prior to discharging into the sewer, which flows to the Wissahickon, agricultural runoff from the livestock and farming practices, as well as stormwater runoff from the school's roofs and parking lots, will be captured and treated though a series of long pools connected by wetland swales.

Wissahickon charter school

The final design is nearing completion for an outdoor learning lab incorporating stormwater BMP's on the grounds of the Wissahickon Charter School. Site runoff will be routed through a vegetated swale into a large rain garden. As a school focused on environmental education, this project will fully integrate several educational opportunities for the school children.

Part of the overall program includes rethinking how PWD manages its own facilities. Facilities managed by PWD account for more than 400 acres of impervious surfaces within the City of Philadelphia. Implementation of low-impact development techniques on PWD properties will provide invaluable demonstrations of many stormwater best management practices (BMP), reduce CSOs, and offer opportunities for ongoing monitoring of BMP effectiveness. As the major watershed steward for the Southeastern Pennsylvania region, PWD is in a unique position to lead by example. But leading by example will not be enough. With stormwater regulations already in place to stimulate low impact redevelopment across the city, PWD will initiate a low impact development grants program to provide incentives for controls on private property. The program will include the following components:

- Redevelopment Enhancement: Grants to offset stormwater management costs at sites that increase the level of management beyond the proposed Regulation limit of 1 inch of infiltration to enhance the level of CSO control.
- Maximize Controls on Additions: incentives to developers who increase capacity in their management facility to manage additional site runoff when they complete an addition.
- Parking Lot Program: The City's proposed Stormwater Regulations do not require stormwater management controls when top grinding and repaving parking lots. PWD plans to develop a program element to subsidize the construction of bio-retention islands and medians in parking facilities that are being repaved.
- Green Roofs: PWD is also considering developing a funding program element to encourage the installation of green roofs by subsidizing conversion on buildings that are undertaking a roof replacement or overhaul. This also could be incorporated into a program that would require any facility, building, or structure built using public money to have at least a 50% green roof cover.
- Property-Level Stormwater Management - Roof Drainage Disconnection and Rain Barrels: This program element is already being carried out to encourage homeowners, in areas where it is feasible, to disconnect roof leader downspouts by providing a small rebate to compensate them for the time and materials necessary to undertake downspout disconnection. This element also subsidizes barrel installation and requires proper operation.

HOW EFFECTIVE IS LID?

PWD's program is extensive and varied, which makes it difficult to gauge its effectiveness, or even to set realistic goals for improved stormwater management. As part of one of PWD's comprehensive watershed management plans (Cobbs Creek), modelling was used to provide estimates of expected benefits of LID. Analysis of

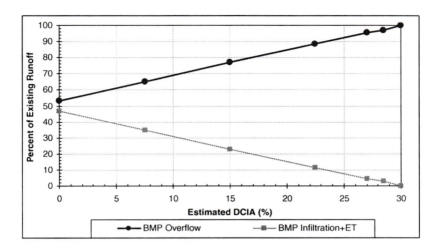

Figure 19.2.   RUNOFF example: Impervious area routed to pervious.

available GIS data indicates that approximately 50% of the Cobbs Creek drainage area is covered by impervious surfaces, including roads, parking lots, and rooftops. About 20% of the total area is served by a combined sewer system, 75% is served by separate storm sewer systems, and 5% is unsewered. In a highly urbanized environment where a large percentage of the land surface consists of roads, parking lots, rooftops, driveways, and other impervious surfaces, complete restoration of natural flow patterns in the urban streams is usually impractical. LID principles applied to new development, when combined with structural best management practices incorporated in redevelopment offer the primary opportunity to reduce runoff volume and pollutant loads while slowing the rate of stormwater releases to the sewers and streams in existing urban areas.

To assess the potential for PWD's program to succeed in managing stormwater, The US EPA's Storm Water Management Model (SWMM) was used to evaluate the operational characteristics and benefits of structural BMPs because it is suitable to the hydrologic and hydraulic complexity of the system (Roesner et al., 1988) and provides the capability to simulate the operation of most structural options under consideration. PWD developed a comprehensive hydrologic and hydraulic model of the Cobbs Creek watershed including a sub-model of the combined sewer system (EXTRAN) and its tributary area (RUNOFF), and a RUNOFF representation of the drainage area tributary to the areas served by separate storm sewer systems. The stream system itself was modeled using EXTRAN, with open channels representing the natural stream cross sections.

Figure 19.2 shows the decrease in untreated runoff as BMP elements or LID principles are applied to sites slated for redevelopment. On the right side of the figure, the current condition of the watershed (30% DCIA) is represented by 100% runoff and 0% infiltration. As LID and BMPs are added, DCIA is successively reduced. The dark blue line shows the potential reduction in runoff, while the pink

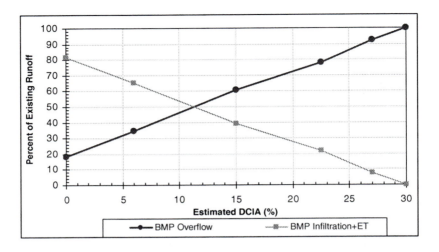

Figure 19.3.   RUNOFF example: Impervious area converted to pervious.

line tracks the increased infiltration expected as redevelopment progresses. The modelling results for the watershed as a whole show a decrease in untreated runoff of almost 50% when the entire directly connected impervious area in the watershed is served by BMPs (DCIA is zero). The decrease in runoff is nearly linear given the soil hydraulic conductivity and size of the BMP elements.

The effects of LID and adding BMP elements may be contrasted with the effects of removing impervious area entirely. In Figure 19.3, the points corresponding to 0 DCIA (all impervious cover removed but urban soils remaining) yield an 80% reduction in untreated runoff, compared to the 50% reduction when LID/BMPs are added.

CONCLUSIONS

CSO/stormwater management in urban areas is often driven by regulatory pressure towards the construction of large tanks to capture a greater percentage of the combined sewer overflow. In Philadelphia, a more holistic approach has been developed, aimed at improving streams and coupling stormwater to Green City and urban renewal initiatives. Perhaps the key obstacle to overcome is acceptance of a set of practical targets for CSO and stormwater to avoid just building tanks.

The principal advantage in applying a systems approach to stormwater management through hydrologically functional landscape design is the reduction of effective impervious cover. That is, implementing low impact redevelopment and development concepts has the effect of altering the hydrologic response of cityscapes to that of watersheds with less development. This ameliorates stormwater hydraulic impacts on receiving waters through decreasing the peak discharge rate, the total runoff volume, the stream velocities, and the frequency and severity of flooding during storms, while increasing the time for stormwater to reach a stream and the stream baseflow rate between storms. These aspects have direct benefit for preservation or restoration of stream habitat for biological resources. In addition to the benefits of the altered hydrologic and hydraulic response, receiving water quality benefits directly from implementing LID concepts through reduced discharge of stormwater-borne pollutants. Today, for communities served by combined sewer systems, the adoption of LID approaches translates directly into reduced public-sector costs of regulatory compliance with national and state water quality protection policies, laws, and regulations. In the near future, as water quality regulation of communities served by separate storm sewer systems evolves, the public-sector costs of compliance will be greatly reduced for communities that have adopted stormwater regulations largely reliant on low impact development approaches.

ACKNOWLEDGEMENTS

We acknowledge the support of the staff of PWD and CDM, who are too numerous to name here. The development of PWD's overall approach, and the planning involved in the Cobbs Creek Integrated Watershed Management Plan were truly a team effort.

REFERENCES

Maimone, M. (2006). Water Resource Planning, Turning Theory into Practice, Geo Press, Groningen, The Netherlands

Myers, R.D., Maimone, M., Smullen, J.T., and Marengo, B. (2004). Simulation of Urban Wet Weather Best Management Practices at the Watershed Scale, in Innovative Modeling of Urban Water Systems, Monograph 12, CHI, Guelph, Ontario: 237–256.

PWD (2004). Cobbs Creek Integrated Watershed Management Plan, Philadelphia Water Department Office of Watersheds, Philadelphia, PA.

Roesner, L.A., Aldrich, J.A., and Dickinson, R.E. (1988). Storm Water Management Model User's Manual, Version 4: Addendum I. EXTRAN; Cooperative Agreement CR-811607, U.S. EPA, Cincinnati, Ohio.

20

Automation and real-time control in urban stormwater management

J.W. Labadie

*Department of Civil and Environmental Engineering, Colorado State University,
Fort Collins, Colorado 80523-1372, USA*
E-mail: labadie@engr.colostate.edu

Summary: Two case studies are presented demonstrating that current advances in hardware and software technologies have overcome many past obstacles in implementation of automation and real-time control in urban stormwater management. The first case study addresses the optimal real-time regulation of in-system storage in combined sewer systems for minimizing untreated overflows, which is challenging due to the need for integrating optimization with urban stormwater runoff prediction and fully dynamic routing of sewer flows. A neural-optimal control algorithm provides fully integrated real-time control with consideration of unsteady sewer routing for the King County combined sewer system, Seattle, Washington. For the second case study, many coastal ecosystems have been adversely impacted by increased stormwater drainage due to expanding urbanization. Multipurpose stormwater control facilities provide for restoration of these ecosystems by regulating long-term frequency distributions of stormwater discharges to estuaries to coincide with desired distributions. A genetic algorithm is coupled with a daily stormwater drainage simulation model to optimize the sizing and fuzzy operating rules of reservoirs for real-time control of stormwater discharges to the St. Lucie Estuary on the southeast coast of Florida. The success of these case studies should provide encouragement to urban water managers to consider implementation of automation and real-time control in urban stormwater management.

INTRODUCTION

Computer software and hardware technologies have attained a level of maturity that is conducive to successful application of automation and real-time control in urban stormwater management. Two case studies are presented that demonstrate the applicability of advanced technologies to real-time control of urban stormwater for pollution control and ecological restoration. These case studies involve application of classical optimization algorithms based on optimal control theory, in concert with modern methods of artificial intelligence including neural networks, genetic algorithms, and fuzzy rule-based systems. These applications are designed to allow integration with realistic models describing the hydrologic and hydraulic behavior of stormwater drainage systems and their ecological impacts, while allowing implementation under the severe constraints of an on-line, real-time computing environment.

The first case study focuses on real-time control of combined sewer systems (CSS's) as a cost-effective means of reducing pollution from untreated overflows. The aggregate storage capacity in combined sewer networks may be sizable in many cases, and can help reduce pollution from untreated combined sewer overflows (CSO's) to adjacent water bodies if properly managed over time and space, along with any available detention storage in the sewer network. Real-time control (RTC) is most effective if integrated over the entire sewer network, resulting in a large-scale, spatially distributed optimal control problem. Unfortunately, most current RTC implementations are limited to reactive or local supervisory control (Pleau et al., 2005). The fully integrated, system-wide or global optimal control problem is large-scale, dynamic, and nonlinear, requiring integration with stormwater runoff prediction models and fully dynamic unsteady flow models of the sewer network. In a real-time environment, the computational requirements of realistic global optimal control may exceed the on-line computer processing capabilities since iterative solution of the optimal control model may require successful completion within 5 to 15 min. time intervals as rainfall forecasts and measured levels and flows are updated in real-time.

This case study explores the potential usefulness of dynamic artificial neural networks (ANN's) as a means of overcoming the computational challenges of the complex models necessary for global real-time control of combined sewer systems. A *machine learning* approach is utilized whereby a highly accurate, but computationally time consuming, optimal control model is utilized to provide the training data set for a recurrent ANN under a wide range of sewer inflow conditions. Performance of the ANN is compared with the optimal control module using validation data sets not included in the ANN training. The neural-optimal control algorithm is applied to the combined sewer system of the King County Wastewater Treatment Division, Washington USA as a demonstration of its viability.

The second case study concentrates on coastal ecosystems adversely impacted by increased stormwater drainage due to expanding urbanization. The ecosystem of the St. Lucie Estuary (SLE), located on the east coast of south Florida, has been significantly impacted by development of a complex network of stormwater drainage canals in the tributary watershed. A suite of models dealing with watershed hydrology, reservoir optimization, estuary salinity and ecology are applied for optimal sizing and real-time operation of stormwater retention reservoirs. The multipurpose stormwater control facilities provide for hydrologic restoration to pre-drained or natural hydrologic conditions for recovery of salinity sensitive biota in the SLE, as well as supplemental irrigation water and pollution control through connected stormwater treatment areas (STA's). The optimization is challenging since the ecological goal is for mean monthly stormwater discharges to the SLE to coincide with the desired natural frequency distribution, rather than simply attempting to control individual extreme events. The OPTI6 optimization model applies a genetic algorithm, coupled with a daily simulation model of the stormwater drainage network, to optimize the sizing and real-time operation of the stormwater control facilities.

CASE STUDY 1: REAL-TIME CONTROL OF COMBINED SEWER SYSTEMS

Background

Although real-time regulation of in-system storage in combined sewer systems has been successfully demonstrated in several cities in the U.S., Canada, and Europe since the mid-1970's, there are few ongoing implementations in existence today. This is due to past difficulties in the robustness and reliability of required computer control equipment, sensor and communication devices, as well as inadequate software and modeling capabilities. According to Schutze et al. (2002), however, current advances in hardware and software technologies have overcome many of these past obstacles at reasonable cost, thereby providing a renewed impetus for implementation of real-time control of CSS's. Maximum utilization of spatially distributed in-system storage in a combined sewer system is an inexpensive method (relative to capital construction) of reducing the polluting effects of untreated spills to receiving waters (Labadie, 1993). The goal is to provide optimal regulation of control structures in the sewer network such that CSO's are minimized or even eliminated. Reducing the occurrence and magnitude of CSOs, and thereby reducing pollution impacts on receiving waters, is the primary goal of the real-time control system. Therefore, minimization of overflows and maximization of through-flows to the wastewater treatment plant are the primary objectives.

Optimal control module

Darsono and Labadie (2003) describe the optimal control module OPTCON based on Pontryagin's maximum principle (Pontryagin et al., 1962) for minimizing untreated spills at regulator stations throughout a combined sewer or stormwater drainage network. The objective function in OPTCON minimizes total weighted overflows (squared) from the CSS, with associated weighting coefficients that can vary both spatially and temporally. That is, receiving water impacts may be more sensitive to overflows at certain locations than others. Tidal and other temporal influences may also necessitate changes in the weighting coefficients over time. Spills are *squared* in the objective function to produce a *smoothing* effect on the solution that provides more stable discharges, thereby avoiding oscillations and surges in the sewer system. In addition, adjacent water bodies can better absorb the polluting effects of spills if they are smoothed over time, rather than a first-flush shock of untreated spills that can be damaging to aquatic habitat and species. A final term in the objective function minimizes residual storage in the system at the end of the operational time horizon, which indirectly maximizes through-flow to the wastewater treatment plant.

The primary decision variables are dynamic controls at the regulator and outfall gates. It is assumed that check structures are not directly located within interceptor sewers, but rather regulators control discharges from each lateral into the interceptor as well as bypass overflows to the receiving waters if the interceptor sewer is surcharged at that location. Lateral sewers may also include check structures and pump stations in the drainage network model. It is assumed that the state vector in the formulation represents temporary storage accumulated behind these structures located in the lateral or trunk sewers.

Routing matrices are included in the system dynamics of the optimal control model that provide current estimates of routing coefficients calculated from the fully dynamic sewer hydraulics model UNSTDY (Chen and Chai, 1991). Sewer discharges and heads are calculated over each sewer section and discrete time by the hydraulic model. Since maximum capacities on temporary storage and node discharge are dynamic functions of hydraulic flows and heads throughout the sewer network, an iterative strategy adjusts restrictions on temporary in-line storage and discharges to prevent surcharge conditions occurring upstream in the trunk sewers. System dynamics constraints in the optimal control model essentially maintain mass balance in each sewer reach, assuming distributed stormwater inflows to the CSS are predicted from an urban stormwater runoff model such as the RUNOFF module of the U.S. EPA SWMM model (Huber and Dickinson, 1988). These predictions are assumed to be updated in real-time as the storm event progresses and new rainfall forecasts are generated. A flow chart of the solution process is given in Figure 20.1.

Neural network module

The OPTCON optimal control module for minimizing CSO's is easily solved off-line for large sewer networks, but integration with the RUNOFF model for watershed inflow predictions and successive execution of the sewer hydraulics model UNSTDY exceeds the clock time limitations required for real-time implementation. Several studies have shown that artificial neural networks (ANN) are an effective tool for controlling complex, nonlinear systems (Parisini and Zoppoly, 1994). Adaptive control of combined sewer systems in real-time requires a recurrent ANN to model dynamic operational trajectories.

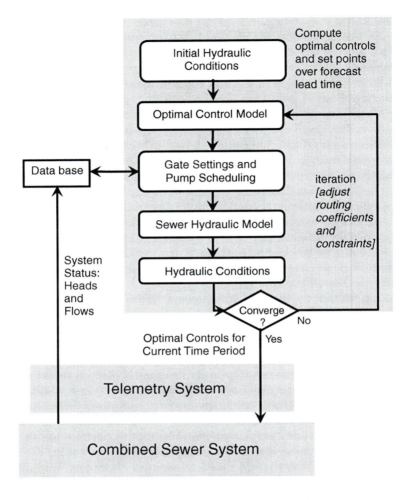

Figure 20.1.   Flow chart of successive, iterative solution of optimal control model and hydraulic sewer model.

The Jordan architecture (Figure 20.2) selected for the ANN structure adapts well with time-varying systems, since the main purpose of the dynamic neural control module is to compute optimal real-time gate controls as output based on current and previous rainfall data and previous gate control decisions. This dynamic or recurrent architecture is characterized by portions of the ANN output (optimal gate controls) returning as inputs to the ANN in the next time step. External inputs to the ANN are current, lag-1, and lag-2 spatially distributed rainfall data, recognizing the likely correlation of successive rainfall inputs over time and space.

Training of the dynamic neural control module is a supervised learning process for determining the optimal connection weights and bias weights from the

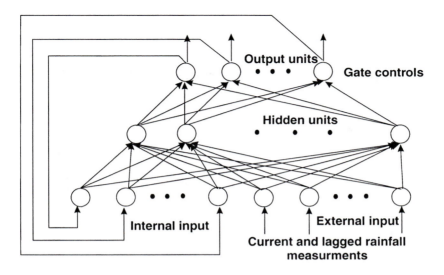

Figure 20.2.   Schematic of Jordan architecture recurrent artificial neural network (Parisini and Zoppoli, 1994).

input/output training data set. The input data are rain gauge measurements for various historical storm events and the output data are the optimal gate controls calculated off-line by the optimal control module. In effect, this *machine-learning* process results in a trained ANN that effectively *mimics* the performance of the complex optimal control module, but without the requisite computer processing time requirements.

Application to west point combined sewer system, King County, Washington USA

Darsono and Labadie (2003) applied the neural-optimal control algorithm to the West Point Treatment Plant collection system of the King County Wastewater Treatment Division, Seattle, Washington USA as a demonstration of its capabilities. The portion of the service area of the West Point Treatment Plant included in this study covers over 26,000 ha which includes 160 km of gravity sewers with diameters up to 3.66 m, 11 pumping stations and 17 regulator stations (Figure 20.3). Although the average capacity of the West Point Treatment Plant is 5.83 m^3/s, the plant has been recently expanded to handle wet weather peak flow rates up to 19.3 m^3/s. Although this configuration does not reflect recent expansions, upgrades and improvements in the King County wastewater system (King County, 2004), it was deemed acceptable for demonstrating the viability of the neural-optimal control algorithm.

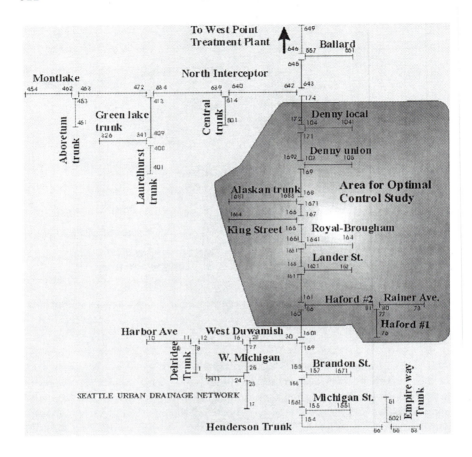

Figure 20.3. Schematic diagram of the drainage network of the West Point Combined Sewer System of King County WA.

Labadie (1993) describes calibration of the RUNOFF model for predicting stormwater inflows from Seattle area urban catchments into the combined sewer system as conducted by staff of the King County Department of Natural Resources, Wastewater Treatment Division (formerly Metro Seattle). The sewer network modeled in UNSTDY is confined to only those portions of the sewer network with 1.22 m diameter pipe sizes and higher, with smaller pipe sections modeled using kinematic wave approximations. For the numerical modeling, the sewer network was divided into 260 sections, with distances between sections ranging from 18.3 m to 91.4 m based on desired numerical accuracy, changes in slope or existence of a weir or other control structure. Pipe and gate sizes, slopes, roughness coefficients and junction data were provided to the hydraulic model, along with rating tables for downstream boundary conditions based on the normal flow approximation. To facilitate linkage of the hydraulic model with the optimal control module, UNSTDY

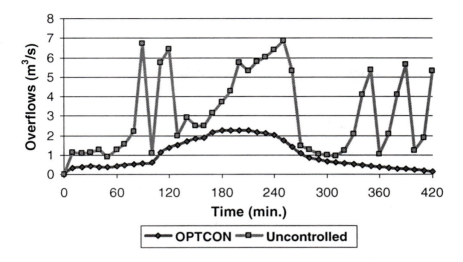

Figure 20.4. Comparison of total overflows from OPTCON solution versus uncontrolled operation for Storm #11.

allows specification of optimal regulator station releases calculated from OPTCON as interior boundary conditions, and then calculates the required gate openings to produce those releases.

Rain gauge data in the Seattle area were used to define 11 diverse, spatially distributed storm events over the study area. Rainfall data in 10 min. time increments were obtained from National Weather Service stations in the area: SeaTac airport south of the study area and Sand Point located north of the study area. Although other rain gauges are maintained by the City of Seattle, the NWS gauges provide the most complete data set and at the desired time increment of 10 min. The first ten events were utilized as training data sets for the neural-optimal control model, with Storm #11 used for testing and validation. Although future work can consider a larger number and variety of storm events, these were considered to be a reasonably diverse representation of storm events suitable for this demonstration. Each of the 11 storm events produced from the rainfall database provide input to OPTCON to compute optimal gate controls for each event in 10 min. time increments. For the routing links, routing coefficients are set to 1.0 as initial approximations in the iterative process. Convergence of the routing coefficients typically requires eight iterations.

Figure 20.4 demonstrates the results of application of OPTCON in reducing combined sewer overflows under fully integrated dynamic optimal control for Storm #11. These results are contrasted with the uncontrolled solution where diversions from the trunk sewers are allowed to enter the interceptor sewer until surcharging occurs at that location, resulting in spills or overflows at that regulator station. Storm #11 is a large event which results in overflows in spite of

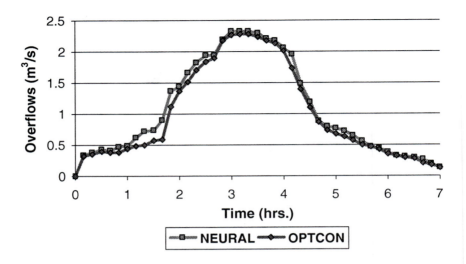

Figure 20.5.   Comparison of gate operations under neural control and OPTCON at
Connecticut Street regulator station

the application of the optimal control strategy, but with total overflows and peak
discharge rates substantially reduced under optimal control. However, since these
optimal controls are based on perfect foreknowledge of the storm event, they cannot be implemented for actual real-time control. Therefore, OPTCON solutions
for Storms #1 through #10 are utilized as training data sets for the recurrent neural
network. The trained ANN can then be validated using results from Storm #11,
which allows demonstration of optimal real-time control without the presumption
of perfect foreknowledge of the storm event.

Figure 20.5 provides a comparison of the gate controls produced by the recurrent ANN with those produced under perfect foreknowledge by OPTCON for the
regulator gates at the Connecticut Street regulator station. It should be emphasized
that the neural controls are generated with only current rainfall, past rainfall, and
previous gate controls as inputs. The ANN is provided no forecast information
or foreknowledge of the storm event, and yet the gate controls are quite close to
those produced by OPTCON under perfect foreknowledge. This is in spite of the
fact that the validation event Storm #11 is dissimilar to any of the storms used
during training of the ANN. Similar results are found in comparing the neural and
OPTCON gate controls for the other regulator stations in the study area. This indicates that the neural network displays a learning capability that adapts the control
strategy to the ongoing storm event as rainfall data are being collected.

It was originally hypothesized that providing direct rainfall measurements as
inputs to the ANN would not be successful since rainfall data can be noisy and
sporadic, and that it would be necessary to process the rainfall data through the
RUNOFF model and provide the resulting sewer inflow predictions as the input

data sets for the ANN. These results indicate, however, that rainfall data can indeed be provided as direct inputs to the ANN, although it is believed that the dynamic nature of the recurrent ANN and use of time-lagged inputs facilitates the direct use of rainfall data. In spite of the high variability of the rainfall inputs, the gate controls produced by the ANN are smooth and stable. The neural-optimal control model can be easily implemented for adaptive, real-time control of combined sewer systems since, for this application, execution of the recurrent ANN at each time step required only 0.02 sec. of CPU time on a 2 GHz Pentium 4 workstation.

CASE STUDY 2: REAL-TIME CONTROL OF STORMWATER DISCHARGES FOR COASTAL ECOSYSTEM RESTORATION

Background

Along many coastal areas, shoreline and estuarine ecosystems have been adversely impacted by increased stormwater discharges and pollutant loadings from expanding urbanization and development. The ecosystem of the St. Lucie Estuary (SLE) on Florida's east coast has been greatly influenced by construction since the early 1900's of an elaborate network of stormwater drainage canals in the watershed (Figure 20.6). These canals drained many historic wetlands, and promoted extensive agricultural and urban development. The current SLE watershed covers over 200,000 ha (500,000 acres) with about 50% irrigated agricultural land (primarily citrus), 17% rangeland, pasture and forest; 17% urban and only 16% wetlands. The quantity, quality, timing and distribution of stormwater runoff into the SLE has been significantly altered by these developments, further exacerbated by emergency freshwater releases from Lake Okeechobee. Increased drainage from these man-made alterations has lowered groundwater tables and degraded the quality of urban stormwater runoff. These changes in basin hydrology and water quality have significantly affected the sensitive salinity regime supporting the SLE ecosystem to the extent that seagrasses and oysters, once abundant in the Estuary, have completely disappeared (Haunert and Startzman, 1985).

Restoration of the SLE ecosystem is a component of the Comprehensive Everglades Restoration Plan (CERP) jointly undertaken by the South Florida Water Management District (SFWMD) and the U.S. Army Corps of Engineers (USCOE). The proposed restoration plan focuses on recapturing the important hydrologic characteristics of the pre-drained or natural watershed using stormwater retention reservoirs for recovery and protection of salinity sensitive biota (USCOE and SFWMD, 2004). During the restoration plan formulation, a suite of models for watershed hydrology, reservoir optimization, and estuarine salinity and ecology were employed for establishing hydrologic restoration targets and refining the alternatives (Wan et al., 2002). Aquatic biologists have determined that biota in the SLE are more sensitive to the long term frequency distribution of mean monthly inflows, rather than to individual extreme hydrologic events (Haunert and Konyha, 2001).

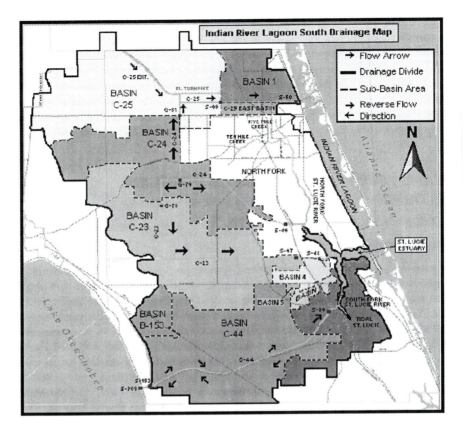

Figure 20.6.   Location map of St. Lucie Estuary and major drainage canals in the watershed

Reservoir sizing and rule optimization model

The optimization model OPTI6 was developed to determine the optimal sizing and operating rules for retention reservoirs in the SLE watershed that: (1) achieve the target long-term frequency distribution of stormwater discharges to the SLE, (2) provide supplemental water supply at a specified reliability, and (3) minimize required capacities of retention reservoirs. The optimization requires daily simulation of the drainage network for calculation of the mean monthly probabilities for all frequency classes of stormwater releases to the SLE under alternative reservoir sizing and operational decisions.

The drainage network simulation assumes construction of offstream reservoirs requiring pumping facilities for both diversion into and release from the basins, with a multi-cell STA connected to each retention reservoir for reducing nutrient, pesticide, and other pollutant loadings from stormwater runoff. The drainage network simulation model calculates mass balance for the reservoirs with connected

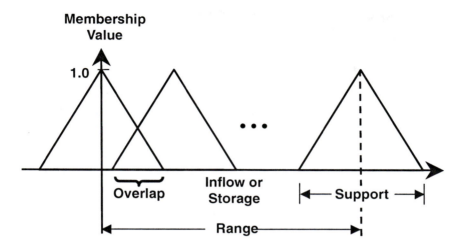

Figure 20.7.   Triangular fuzzy numbers for premises (Inflow or Storage) for fuzzy rule-based system

STA's, along with additional constraints designed to maintain nonnegative flows. The primary decision variables in the optimization are the scheduling of diversions pumped into each reservoir/STA and discharges either pumped out or released by gravity. Transbasin diversions are given a high priority in the simulation and are assumed to occur up to the pumping capacity if flow is available.

Fuzzy operating rules

The optimal reservoir operating rules $q_i^*(I_{it}, s_{it})$ represent flexible and adaptive feedback policies whereby operators measure current day inflows I_{it} and reservoir storage s_{it}, and then obtain reservoir operation guidelines from the rules based on those measurements and the time of year. Rather than presuming an *apriori* mathematical structure for the operating rules, a fuzzy rule-based system is developed (Bárdossy and Duckstein, 1995). As shown in Figure 20.7, triangular fuzzy sets are utilized to condition fuzzy set membership values on current day reservoir storage and inflows. The membership values represents *degrees-of-truth* that an inflow or storage measurement belongs to that fuzzy set. Each fuzzy rule represents only a portion of the data set, but a degree of overlap is established to assure a continuous rule-based system. This means that certain inflow or storage measurements can belong to two fuzzy sets with a nonzero membership value. *Product inference* is utilized to calculate degree of fulfillment of each fuzzy rule, with the *normed weighted sum* combination method used to combine the fuzzy responses from all the rules. The *mean defuzzification* method is utilized to *defuzzify* the fuzzy combinations of the rule responses. The advantage of these methods is that characterization of the structure of the membership function of the fuzzy consequence is not explicitly required. Only the *means* of the fuzzy consequences for each basin and rule are needed, so they can be regarded as decision variables in the optimization.

Rainfall in South Florida varies seasonally, with distinct wet and dry seasons. Wet season or summer rainfall results primarily from convective and tropical storms, whereas frontal systems govern dry season or winter rainfall. In order to relate the operating rules to seasonal influences, distinct rules $q_i^w(I_{it}, s_{it})$ and $q_i^s(I_{it}, s_{it})$ are developed for each season by optimizing the means of the fuzzy consequences $\bar{q}_{\text{in}}^w$ and $\bar{q}_{\text{in}}^s$ for the winter and summer seasons, respectively.

Genetic algorithm

Solution of the optimization model requires execution of an imbedded daily simulation model over a 31 year historical period for producing the mean monthly frequency distribution for stormwater discharges to the SLE, as well as probabilities of failing to satisfy irrigation demands for each basin. Traditional stochastic optimization algorithms are not well suited to solution of this problem since the goal is to directly optimize a probability distribution. A genetic algorithm (GA) was therefore selected since it requires no explicit analytical representation of the objective function and constraint sets in the optimization. Gradient information is not required and discontinuities in the objective function have little effect since GA's are resistant to becoming trapped in local optima.

Rooted in the mechanisms of natural selection in biology, genetic algorithms were first proposed by Holland (1975) whose goal was to design computing systems for modeling natural systems. With publication of Goldberg (1989), researchers in a wide variety of fields have attempted to apply GA's. GA's have proven to be particularly attractive for solving complex combinatorial problems, as well as providing easy interfacing to existing simulation models. As a heuristic search procedure, convergence of a GA to optimal solutions cannot be guaranteed. However, practical experience with GA's finds they are often able to locate global optimal solutions to nonconvex, multimodal optimization problems when other methods fail to do so (Michalewicz, 1996). GA's attempt to maximize the genetic *fitness* of the individuals (i.e., variables) in the population (i.e., solution set), which corresponds to maximizing (or minimizing) the objective function.

Figure 20.8 shows the connection between the GA and drainage network simulation model, where the GA selects populations of fuzzy consequences of the fuzzy rule-based system which are then evaluated through the daily simulation model. The daily simulation evaluates mean monthly frequency distributions of stormwater discharges, water supply reliability, and required storage capacity, which are returned to the GA for improvement in the fitness.

Optimal restoration of the St. Lucie Estuary

Application of OPT16 requires daily watershed stormwater discharges into the SLE as well as irrigation demand on pumpage from the Floridian Aquifer as input data. Hydrologic simulation of current watershed conditions was conducted using the

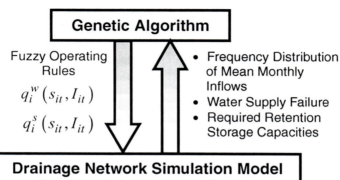

Figure 20.8.   OPTI6: Interaction of genetic algorithm for with drainage network simulation for optimizing fuzzy operating rules.

Hydrologic Simulation Program-FORTRAN (HSPF), Version 12 (Bicknell et al., 2001). Enhancements to HSPF for simulating the high water table and wetland conditions prevalent in South Florida were incorporated prior to model calibration and application (Aqua Terra Consultants, 1996). The model was calibrated for drainage basins where long-term flow monitoring data are available at the discharge control structures. The model was applied to six major drainage basins within the watershed (Figure 20.6), which were further divided into subbasins, with each segmented into six landuse types. These landuse categories, including irrigated agriculture (primarily citrus), nonirrigated pasture, forest, wetland, and urban lands, are considered to be the most important factors determining hydrologic response in the watershed. The 1995 landuse coverage was used to represent current development, with the resulting simulation under these conditions referred to as the *1995 base*. Projected landuse in 2050 constitutes the future development condition, with the associated simulation designated as the *2050 base*. Substantial expansion of urban development is expected in 2050, but with irrigated agriculture remaining at current levels along with decreases in forest and pasture areas. Wetlands are not anticipated to decline in total acreage. HSPF was applied to predicting how these changes in land use could impact the hydrology of the basin.

An important aspect of the modeling process is to simulate the allocation of available water supplies to meet irrigation demands. The SLE watershed includes substantial irrigated acreage, but limited surface storage capacity. The major source of water supply has been the Floridian Aquifer, an artesian aquifer with salinity concentrations generally exceeding acceptable levels for direct use in agriculture. During extended periods of drought, these salt levels can significantly reduce citrus yields. HSPF attempts to satisfy the irrigation demands by drawing water from specified canals. When the canals are dry, however, irrigation demands are unmet since groundwater cannot be mixed with better quality surface water. The Agricultural Field Scale Irrigation Requirements Simulation (AFSIRS) model (Smajstrla, 1990) was applied to determining irrigation demands, water availability, and Floridian Aquifer withdrawals within the SLE watershed. Daily stormwater discharges into the SLE and supplemental irrigation from the Floridian Aquifer were processed as input data for OPTI6 based on results of HSPF and AFSIRS.

Restoration target flow distribution

To establish the target flow distribution, a favorable range of watershed stormwater discharges for salinity sensitive biota in SLE, called the *salinity envelope*, was first established. Salinity modeling in the SLE indicated that once watershed inflows exceed 56.6 m^3/s (2000 ft^3/s) to 85 m^3/s (3000 ft^3/s), salinity in the upper SLE can be close to zero even during high tide conditions (Hu, 1999). The combination of biological understanding of the ecosystem and salinity modeling in SLE led to the establishment a salinity envelope ranging from 10 m^3/s (350 ft^3/s) to 56.6 m^3/s (2000 ft^3/s) for juvenile marine fish and shellfish, oysters, and submerged aquatic vegetation. After establishing the restoration target, Haunert and Konyha (2001) determined acceptable violations of this range that can occur and still sustain the Estuary ecosystem. It is assumed that acceptable violations are confined by the temporal and spatial hydrologic variability of the pre-drained watershed, a concept supported by other efforts undertaken to restore freshwater riverine ecosystems (Richter et al., 1997).

The *Target* monthly flow frequency distributions were developed using the Natural System Model (NSM) (Van Zee, 1999) which attempts to simulate pre-drainage conditions in the SLE. This distribution is compared with *Current* (1995 base) conditions in the SLE in Figure 20.9. Compared with the Current condition (1995 base), the pre-drained watershed had a substantially lower probability of high flows. The NSM model also showed that lower flow rates entered the Estuary during dry periods, suggesting that the Estuary may not require flow augmentation during dry periods and, consequently, that irrigation demands do not directly compete with environmental demands from the Estuary. This observation also supports the contention that hydrologic restoration should focus on reducing the frequency of high flows to the SLE.

Reservoir optimization results

To achieve the *Target* monthly flow frequency distribution, capacities and fuzzy operating rules of the reservoir/STA were optimized using the OPTI6 model. A

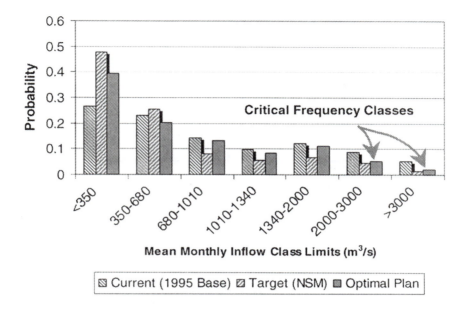

Figure 20.9.   Comparison of mean monthly frequency distributions of SLE inflows between Current Distribution (1995 Base), Target (NSM Model) and Optimal Plan

multiobjective analysis was conducted by varying the weighting factors in the objective function until a suitable compromise solution was obtained between the three criteria: (i) matching the frequency distribution of stormwater discharges to the SLE; (ii) satisfying the minimum irrigation water supply reliability requirements; and (iii) and minimizing required storage capacities of the reservoirs. The highest penalties on deviation from the target frequencies were assigned to flows outside the favorable range 10–56.6 m^3/s (350–2000 ft^3/s), particularly high flow frequencies above 56.6 m^3/s (2000 ft^3/s). Flows in the range <10 m^3/s (<350 ft^3/s) were assigned the next highest penalty, with less important intermediate ranges given low weights. The year 2050 was used to represent future watershed conditions, with projected landuse in 2050 used for modeling watershed hydrology in HSPF and irrigation demands with AFSIRS. The selected alternative consists of four off-line water storage reservoirs in the C-23, C-24, North Fork (NF), and C-44 basins. Construction of these features includes water control structures, pumps, levees, canals and acquisition of approximately 4,937 ha (12,200 acres) of land. These reservoirs, along with 36,422 ha (90,000 acres) of re-hydrated wetlands and flow diversion from C-23/C-24 to the North and South Forks, create a hydrologic regime similar to the natural flow frequency distribution.

Figure 20.9 shows that the optimal restoration plan provides an excellent match to the NSM targets for the important high flow frequency classes, with lower flow

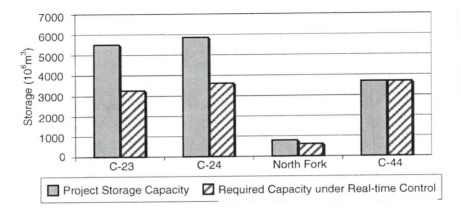

Figure 20.10.   Reduction in required storage capacity under optimal plan with real-time control.

frequency classes of less importance displaying acceptable agreement. Significant improvement over the current frequency distributions without stormwater control is evident in Figure 20.9, particularly for the critical frequency classes. Results indicate that 20% of stormwater inflows would be treated in the STA's for all basins, resulting in reduced pollutant loadings to the SLE. The multipurpose benefits of the retention reservoirs is also evident with risks of failing to satisfy irrigation water supply requirements falling below the 10% risk target, providing significant improvement over the current uncontrolled conditions. In addition, as seen in Figure 20.10, significant cost savings are achieved by OPTI6 in reducing sizing requirements of the retention reservoirs/STA's by 45.4 10^6m^3 (36,800 acre-ft), or almost 30% of the original capacity estimates with inclusion of real-time optimal control.

SUMMARY AND CONCLUSIONS

In the first case study, a neural-optimal control model is presented that integrates the realistic, fully dynamic hydraulic model UNSTDY with a dynamic optimization model OPTCON for optimal regulation of control structures to minimize untreated overflows from a combined sewer system. Incorporating hydraulic realism in the optimization model requires a computationally expensive iterative process which violates the clock time constraints of real-time implementation. A strategy is adopted whereby OPTCON/UNSTDY computations are conducted off-line for a wide range of spatially distributed, historical storm events. The resulting optimal control outputs, along with the rainfall data sets as inputs, provide training data sets for a recurrent artificial neural network (ANN). The Jordan architecture adopted for the ANN effectively captures the dynamic, rapid response characteristics of combined sewer systems. Inputs to the ANN include current and past spatially distributed rainfall data, as well as previous controls, with synaptic weights and other parameters in the ANN optimized to reproduce the optimal gate controls from OPTCON/UNSTDY as closely as possible. Once trained, the recurrent ANN can be implemented for real-time control of combined sewer systems with full consideration of complex system hydraulics and fully integrated, spatially distributed system-wide control. The neural-optimal control algorithm was demonstrated on the West Point Treatment Plant collection system of the King County Wastewater Treatment Division, Seattle, Washington USA. Validation results indicate that the neural-optimal control algorithm closely tracks the optimal gate controls from OPTCON/UNSTDY for a storm event *not included* in the training data sets for the recurrent ANN, while requiring only inputs of current and past rainfall measurements and previous gate controls. The ability of the ANN to adapt the optimal controls to changes in rainfall intensity and distribution in the ongoing event is clearly evident in this demonstration.

The second case study focuses on the St. Lucie Estuary (SLE) ecosystem restoration plan that has been developed based on the integration of a suite of models to simulate watershed hydrology, reservoir optimization, estuary salinity and ecology. The Natural System Model (NSM) is applied for establishing the hydrologic restoration targets and justifying flow transfers between basins. The OPTI6 model couples a genetic algorithm with a daily drainage network simulation model using HSPF inflow calculations for optimal sizing and operation of the reservoir/STA's in the SLE watershed. Robust operating rules obtained through application of a fuzzy rule-based system is the key to providing optimal solutions that achieve target mean monthly frequency distributions for stormwater inflows to the SLE for restoration of the estuarine ecosystem. In addition, the multipurpose benefits of the retention reservoirs are clearly evident by maximizing use of the attached storage-treatment areas (STA's) for pollutant load reductions, as well as maintaining desirable risk targets for supplemental irrigation water supply from stormwater. Significant cost reductions were also achieved under the optimal plan through reduction of total sizing requirements for the retentions basins by over 30% from the initial estimates. Results indicate that the optimal restoration plan has the potential to restore and protect the mesohaline ecosystem in the SLE.

These case studies demonstrate the viability of automation and real-time control in providing a cost-effective means of reducing the polluting impacts of urban stormwater on aquatic ecosystems, while maintaining essential flood control capabilities. The first case study showed that substantial reductions in combined sewer overflows could be achieved by optimum utilization of existing in-line storage capacity. In the second case study, incorporation of automation and real-time control in the planning of stormwater control facilities resulted in substantial reductions in sizing requirements and associated costs. The low cost and high reliability of modern data acquisition and control technology and the powerful control software such as demonstrated herein should help remove the obstacles that have heretofore impeded the implementation of automation and real-time control in urban stormwater management.

REFERENCES

Aqua Terra Consultants. (1996). Modifications to HSPF for High Water Table and Wetlands Conditions in South Florida. Report submitted to South Florida Water Management District, West Palm Beach, Fla.

Bárdossy, A. and Duckstein, L. (1995). Fuzzy Rule-based Modeling with Applications to Geophysical, Biological, and Engineering Systems. CRC Press, Boca Raton, Fla.

Bicknell, B., Imhoff, J., Kittle, J., Jobes, T., and Donigan, A. (2001). Hydrologic Simulation Program-FORTRAN, Vs. 12, User's Manual. Office of Research and Development, U.S. Environmental Protection Agency, Athens, Ga.

Chen, Y.-H. and Chai, S.-Y. (1991). UNSTDY: Combined Sewer Model User's Manual, Chen Engineering Technology, Inc., Ft. Collins, CO.

Darsono, S. and Labadie, J. (2003). Neural optimal control algorithm for real time regulation of in-system storage in combined sewer systems. Proceedings of the World Water and Environmental Resources Congress. EWRI and ASCE, Philadelphia, PA, June 22–26.

Goldberg, D. (1989). Genetic Algorithms in Search, Optimization and Machine Learning. Addison-Wesley Publishing Company Inc., Reading, Mass.

Haunert, D., and Startzman, J. (1985). Short Term Effects of a Freshwater Discharge on the Biota of St. Lucie Estuary, Florida, Technical Publication 85-1, South Florida Water Management District, West Palm Beach, Fla.

Haunert, D. and Konyha, K. (2001). Establishing St. Lucie Estuary Watershed Inflow Targets to Enhance Mesohaline Biota, Appendix E., Indian River Lagoon—South Feasibility Study, South Florida Water Management District, West Palm Beach, Fla.

Holland, J. (1975). Adaptation in Natural and Artificial Systems, The University of Michigan Press, Ann Arbor, Mich.

Hu, G. (1999). Two-dimensional hydrodynamic model of St. Lucie Estuary. Proceedings of ASCE-CSCE National Conference on Environmental Engineering, ASCE, Reston, Va.

Huber, W. and Dickinson, R. (1988). Stormwater Management Model, Version 4: User's Manual, Report No. EPA-600-3-88-001a, U.S. Environmental Protection Agency, Washington D.C.

King County, Washington USA (2004). 2003–2004 Annual Combined Sewer Overflow Report, Department of Natural Resources and Parks, Wastewater Treatment Division, Seattle, WA.

Labadie, J. (1993). Optimal use of in-system storage for real-time urban stormwater control. In: Cao, C., Yen, B. C., Benedini, M., (Eds.), Urban Storm Drainage. Water Resources Publications, Inc., Highlands Ranch, Colorado.

Michalewicz, Z. (1996). Genetic Algorithms + Data Structures = Evolution Programs, Springer-Verlag, Berlin, 1996.

Parisini, T. and Zoppoli, R. (1994). Neural networks for feedback feed-forward nonlinear control systems. IEEE Transaction on Neural Networks 5(3): 436–449.

Pleau, M., Colas, H., Lavallée, P., Pelletier, G., and Bonin, R. (2005). Global optimal real-time control of the Quebec urban drainage system. Environmental Modeling & Software 20: 401–413.

Pontryagin, L., Boltyanskii, V., Gamkrelidze, R., and Mishchenko, E. (1962). The Mathematical Theory of Optimal Processes. Wiley-Interscience Publishing Co., New York.

Richter, B., Baumgartner, J., Wigington, R., and Braun, D. (1997). "How much water does a river need?" Freshwater Biology, 37: 231–249.

Schutze, M., Campisano, A., Colas, H., Schilling, W., and Vanrolleghem, P. (2002). Real-time control of urban wastewater systems–where do we stand today? Proceedings of the Ninth International Conference on Urban Drainage, Portland, OR, September 8–13.

Smajstrla, A.G. (1990). Agricultural Field Scale Irrigation Requirements Simulation (AF-SIRS) Model, Version 5.5., Technical Manual, Univ. of Florida, Gainesville, Fla.

U.S. Army Corps of Engineers (USCOE). and South Florida Water Management District (SFWMD) (2004) Central and Southern Florida Project: Indian River Lagoon – South: Final Integrated Project Implementation Report and Environmental Impact Statement, U.S. Army Corps of Engineers, Jacksonville, Florida.

Van Zee, R.J. (1999). Natural System Model Documentation, Version 4.5. South Florida Water Management District, West Palm Beach, Fla.

Wan, Y., Konyha, K., and Sculley, S. (2002). "An integrated modeling approach for coastal ecosystems restoration." Proceedings of the Second Inter-Agency Hydrologic Modeling Conference, July 28– August 1, 2002, Las Vegas, Nev., p. 13.

PART SIX

Implementing Future Urban Hydrological and Ecological Systems

21

Urban drainage at cross-roads: Four future scenarios ranging from business-as-usual to sustainability

J. Marsalek[1], R. Ashley[2], B. Chocat[3], M.R. Matos[4], W. Rauch[5], W. Schilling[6], and B. Urbonas[7]

[1] National Water Research Institute, Environment Canada, 867 Lakeshore Rd, Burlington, Ontario, Canada, L7R 4A6
[2] Pennine Water Group, University of Sheffield, Mappin Street, Sheffield S13JD, UK
[3] URGC, I.N.S.A. Lyon, 69621 Villeurbanne Cedex, France
[4] LNEC, Avenida do Brasil – 101, 1700-066 Lisboa Cedex, Portugal
[5] University of Innsbruck, Unit of Environmental Engineering, Technikerstrasse 13, A-6020 Innsbruck, Austria
[6] Urban Water Management Consultant
E-mail: schilling.wolfgang@gmail.com
[7] Urban Drainage & Flood Control District, 2480 West 26th Ave, 80211 Denver, CO, USA
E-mail: Jiri.Marsalek@ec.gc.ca

Summary: An international group of researchers affiliated with the International Water Association (IWA)/International Association of Hydraulic Engineering and Research (IAHR) Joint Committee on Urban Drainage prepared a position paper on urban drainage, in which they set out to analyze the current status and predict future developments in this field. While the assessment of the current status was relatively straightforward, the future developments were considered uncertain and therefore were described by four possible scenarios: (a) Business-as-usual, (b) Privatization, (c) Technocratic, and (d) Green scenarios. Individual scenarios were further characterized with respect to their main features, risks and problems, and reasons why they could prevail. The likelihood of prevalence of individual scenarios would depend on local conditions; the authors believe that in the immediate future, the business-as-usual scenario, enhanced by elements of the green scenario, seems to have the best chance of actually coming about in many parts of the world. Ongoing research on sustainable urban drainage, described here as the green scenario, offers good reasons for guarded "water optimism".

INTRODUCTION

An international group of researchers affiliated with the International Water Association (IWA)/International Association of Hydraulic Engineering and Research (IAHR) Joint Committee on Urban Drainage recently updated their 2004 position paper on urban drainage (Chocat et al., 2004) by addressing the current status of urban drainage practice and attempting to predict future trends and developments. While there was a fair agreement among the group members on the current status of urban drainage, somewhat uncertain predictions of future developments were described by four scenarios: (a) Business-as-usual scenario, (b) Privatization scenario, (c) Technocratic scenario, and (d) Green scenario. The group then characterized the individual scenarios with respect to their main features, risks and potential problems, and reasons why a particular scenario could prevail. Thus, the findings of this analysis reflect informed opinions and a broad experience from a fair number of countries, and as such, should be of interest when discussing "cities of the future".

CURRENT STATUS OF URBAN DRAINAGE AND NEEDS FOR IMPROVEMENT

Currently, the level of development of urban drainage systems varies greatly between countries and regions, within individual countries, within cities, and sometimes even locally. Thus, when assessing the current status, a reference is made to the "typical" developed country conditions representing a "middle" sector of the spectrum of drainage systems, as done, for example, in the earlier national reports on urban drainage commissioned by IWA (Marsalek and Chocat, 2002).

The major objectives of urban drainage are the protection of urban areas against recurrent flooding and excessive water ponding, provision of public health, and more recently, environmental/ecosystem protection and provision of amenities. In developed countries, the first two objectives have been to a large extent achieved, with the current emphasis on pollution control for protecting the environment and creation of such amenities as recreational opportunities and aquatic habitat (Chocat et al., 2004). In some countries (e.g., Canada and USA), there are also large differences between the drainage practices of the last 20–30 years, and the older drainage systems implemented earlier. The situation is much different in developing countries, where urban infrastructures are inadequate (or even missing) and fail to provide the minimal urban services required (Niemczynowicz, 1999). Some persisting problems with the existing urban drainage systems are further described below.

Continuing expansion of urban areas

The expansion of urban areas strongly affects the landscape and hydrological cycle, including fluxes of water, sediments and solids, chemicals, and waste heat. Of the current world population of 6.5 billion, about 54% live in urban areas, and in some parts of the world the urban population represents more than 90% of the total population (UN Secretariat, 2005). The growing urban population and continuing migration of rural population into cities contribute to large increases in impervious surfaces and expansion of drainage systems. Changes in urban landscape and water use patterns then contribute to large changes in the hydrological cycle, which is transformed into an urban water cycle encompassing municipal infrastructures (Marsalek et al., 2006). Basic urbanization effects on the water cycle include reduced soil infiltration, higher surface runoff and reduced recharge of groundwater (causing soil subsidence) (e.g., Leopold, 1968). There are also other additional effects caused directly by urban water infrastructures. For example, leakage from water supply networks can contribute to recharging groundwater at rates equivalent to annual precipitation in hundreds of millimeters, and both exfiltration from sewers or stormwater management facilities and infiltration of groundwater into sewers also affect the groundwater regime (Marsalek et al., 2006).

Increasing exports of sediments and chemicals from urban areas

Urban drainage contributes to the export of sediments and chemicals from urban areas, as reported in numerous references (e.g., U.S. EPA, 1983; Duncan, 1999). Particularly important water quality constituents conveyed by urban drainage effluents include suspended solids, biodegradable organics, trace organics (particularly polycyclic aromatic hydrocarbons, PAHs), trace metals (particularly Cu, Pb and Zn), chloride and pathogens. As the drainage systems expand, the constituent loads conveyed by drainage systems usually also increase. Furthermore, new chemicals of concern have been recently identified and can be found in combined sewer overflows (CSOs) and storm runoff, including POPs (persistent organic pollutants), flame retardants, endocrine disruptors, pharmaceuticals and personal care products (PPCPs) (e.g., Snyder et al., 2003). The continuing emergence of new chemicals of concern points to the need of assigning even more importance to source control in pollution protection.

Ageing drainage systems

Many existing drainage systems are now ageing to the extent that this is leading to problems (older components may be more than 150 years old), with some parts becoming obsolete as the current performance standards for pollution control continue to increase, and the required sewer maintenance or rehabilitation often does not keep pace with the system requirements. To a lesser degree, these concerns also apply to stormwater management facilities built in the late 20th century that are approaching the age when they require major repairs and or upgrades to meet the current expectations (Ashley et al., 2006a). While some urban drainage service providers may be equipped for corrective action, in other cases, the financing of rehabilitation and upgrading of drainage systems must be planned in competition with other priorities and drainage is often rated as of lesser importance (Gaudreault and Lemire, 2006). There is a lesson to be learned from this situation by developing countries without extensive centralized infrastructures: distributed systems may offer better services.

Potential impacts of climate change/variation

Frequent occurrences of extreme weather events in recent years (e.g., hurricanes) contribute to the ongoing debate as to the effects of climate change/variation and the associated effects on drainage systems. The predicted effects include rising sea levels, higher rainfall intensities in some regions, less rainfall in others, and higher frequency of occurrence of extreme events. While all drainage systems have some flexibility in coping with different hydraulic and pollution loads, in some regions, this capacity is likely to be exceeded by the expected changes during the life expectancy of existing drainage systems, often in the range of less than 100 years (Dlugolecki and Loster, 2003; Evans et al., 2004). Whatever the reality of the future, increases in the frequency of extreme events and expected climate changes make the prediction of future needs and performance of urban drainage systems less and less certain. Hence, uncertainty is increasing (e.g., Ashley et al., 2006a; Evans et al., 2004) and needs to be accounted for in the design.

Progressing deterioration of receiving waters

The arguments presented so far point to greater stresses being exerted on receiving waters with limited capacity to cope with these growing stresses (U.S. EPA, 1994), with some 2 million tones of waste being discharged daily, polluting some 12,000 m^3 of receiving waters (UN, 2003). Thus, in many locations, the long-term cumulative effects of drainage effluents on receiving waters are growing, particularly with respect to deposition of polluted sediments. The cumulative effects then impact on landscape aesthetics, aquatic ecology, and beneficial water uses, including potable water supply, bathing, fishing, aesthetics, and the recreational potential of the aquatic and surrounding urban landscape. In the case of aquatic biota, degradation is observed with respect to the reduced abundance and biodiversity.

While examining these challenges and concluding that, on the whole, the current drainage systems are generally not sustainable, Chocat et al. (2001, 2004) have also recognized the tremendous progress in this field, particularly since urban runoff is now widely regarded as a resource, rather than nuisance (Marsalek and

Chocat, 2002). This progress is particularly evident in very recent urban developments, or in pioneering small projects in highly valued parts of urban areas, where adequate resources and financing are available (Andersen et al., 2001). However, the principle of sustainability and particularly its component that no "environmental deficits" are transferred to future generations brings in very high expectations on system performance (Ashley et al., 2006b).

FUTURE SCENARIOS

Authors' opinions on future developments in the planning and operation of urban drainage systems were synthesized into four scenarios: the business as usual, privatization, technocratic and green scenarios. In the following discussion, each scenario is presented briefly and characterized with respect to the main features, pros and cons, and the feasibility.

Business as usual scenario

The business-as-usual (BAU) scenario is relatively poorly defined, depending on local conditions, and often represents a mixture of elements of the other three scenarios. The majority of the authors argued that the BAU scenario is unsustainable and that ageing infrastructure and increasing frequency of problems/failures and increasingly high costs of managing these will increase the pressure on moving away from this scenario, at least in some areas. On the other hand, so far in most areas that continue to develop using the planning processes of the past the incentives for change are small or missing, and the scenario may persevere more or less by the inertia of those responsible (Ashley et al., 2006b).

BAU scenario main characteristics

The main operating principle of the BAU scenario is the provision of a publicly acceptable level of service with respect to flooding and water ponding, but the environmental goals and objectives are either missing or vaguely formulated and their attainment is often delayed by low levels of funding and the slow implementation of drainage master plans. The prevailing drainage system architecture is a centralized system, with some experimental use of decentralized facilities (particularly in suburbs or satellite developments). The system ownership can be either public or private. The weaknesses of this scenario perceived by the authors included the lack of any explicit consideration of the risks, opportunities and needs for action; inadequate and insufficient funding and financing; low-level involvement of stakeholders; small investment into R&D and hesitation to apply innovative approaches; and, in spite of good progress in new developments, insufficient attention being paid to older areas that were originally designed for much lower environmental performance standards (e.g., without modern stormwater controls). Thus, the BAU approach is responsible for many of the current problems.

From a more positive perspective under BAU the most pressing drainage problems are addressed, certainly those causing system failure. The master plan implementation is under way in most areas, but with frequent political interventions after incidents and accidents, and progress is limited by low funding distributed over long periods (e.g., 25 years). When drainage problems are encountered, conventional simple solutions are often implemented based on expediency rather than detailed study and as consequence of political pressure. Many but the largest utilities experience a brain drain, and the associated loss of local expertise may result in propagation of conventional or simplistic solutions. Overall, this approach contributes to a gradually decreasing quality of systems and somewhat variable quality of services, and poorly planned operation in a reactive mode.

This scenario does, however, offer opportunities for experimenting with and testing new ideas that could contribute to the selection of the best options for future urban water management. However, this eventuality would require a number of prerequisites, including the collection of proper information and data, open-minded discussions, sufficient public interest and the availability of enough time for such a process (i.e., slow urbanization) (Chocat et al., 2004). This condition could potentially be fulfilled in certain developed countries, but not in those still under development.

Risks and problems of this scenario

Hesitating to make a choice is obviously a dangerous idea, resulting in an accumulation of "deficits" with respect to the infrastructure state of repair and environmental protection (a loss of serviceability). In the developed world this situation is further disguised by a rather low risk of acute (catastrophic) failures, because water systems usually work reasonably well and service deterioration is so slow that it may go unnoticed (e.g., NAO, 2004). However, large problems are already being experienced due to catastrophic events (e.g., hurricanes) that have been so far considered as "unusual".

Other problems may emerge from poorly defined performance objectives and the lack of clear-cut priorities. Practical solutions require frequent changes between approaches that are neither consistent nor continuing, and therefore unsustainable. Operations are under-funded and have to rely on general budget allocations and subsidies from senior levels of government, while the costs are high due to the necessity to manage different systems that are tried out in parallel. Finding money is more and more difficult, particularly when social and environmental concerns about water services decrease and professional marketing is lacking. Public urban water utilities find themselves squeezed between quarrelling pressure groups, that do not even acknowledge basic achievements, and the private industry that is eagerly waiting to take over. Risks and acute problems of this scenario might not be so severe in the developed world, but they are unbearable for developing countries, where neither inherited infrastructure, nor money, nor implementation capacity is available.

Reasons why this scenario could prevail

The most important reason is the resistance against change, visible as the inertia of technical, administrative, economic and political systems. Stated positively, in the absence of acute problems (catastrophes), the incentive for change is low. In fact, in most areas urban water services operate without dramatic failures and thus attract limited attention of both the public and politicians, except in those periods immediately following catastrophic events, such as a flood or fish kill.

The technocratic scenario

In the technocratic scenario, engineers are fully in charge and strive for technical excellence, with a minimum involvement of the public or politicians. In some way, this scenario resembles the way public utilities used to be operated in wealthy countries 30 or more years ago, but with a greater emphasis on technical excellence and performance standards. From the capacity and technological point of view this scenario would be achievable, but paying the associated costs would be challenging. Another weakness is the fact that it is very difficult to maintain economic efficiency while dealing with the whole spectrum of storm events of widely varying magnitude and frequency of occurrence. Analogous experience from flood protection shows that frequent storm events can be fully controlled, but extreme events remain essentially uncontrolled and can cause catastrophic impacts in such systems, since these are unexpected.

Main characteristics of the technocratic scenario

The main operating principle of this scenario is technological excellence, based on application of well-proven technology coupled with redundancy and adequate safety factors for a chosen design event return frequency. The resulting solutions are robust and conservative in terms of system performance and safety. The system is protected by fail-safe devices and fall-back alternatives to keep operational risks small. Advanced technologies applied include, for example, automation, robotics, operation in real-time, the use of third generation communication – computers and mobile phones, advanced levels of treatment including microfiltration and the use of biotechnology for water quality control, and new microprocessor-controlled field measuring equipment. Drainage systems rely both on advanced source controls (e.g., rainwater harvesting from roofs and reuse for landscape irrigation) and end-of-pipe solutions. Water saving technologies are used widely in households and industry, and water reuse and recycling is used in water-scarce regions, with strict supervision by the central authority to minimize any inherent risks.

Whilst the traditional cost-benefit analysis is undertaken, the emphasis is placed on benefits and system performance, rather than on balancing such factors against costs. Long-term planning (development of master drainage plans) is emphasized and such plans are frequently updated. Furthermore, retrofitting and renewal of the central drainage systems is a top priority. Responsibility for system operation and maintenance is centralized and mostly public; the provision of water services remains a monopoly.

In essence, this scenario would eliminate socio-economic factors from planning, design and operation of drainage systems. It assumes that: (i) it is unrealistic to expect essential changes in individual and corporate behavior with respect to environmental protection and consequently such changes are not needed; (ii) most urban citizens are not interested in urban drainage issues; and, (iii) politicians are satisfied with a low level of control, as long as there is no trouble. Thus, the drainage system could be operated largely independently of the economic, social and political context.

Risks and problems of the technocratic scenario

In terms of expertise and technology, there is capacity to solve urban water management problems in a "technocratic" way and protect cities against flooding, with some acceptable level of risk, and even to rehabilitate receiving waters to provide such valuable amenities as ecology, recreation and attractive aesthetics. There are growing numbers of examples where such solutions have been implemented on a small scale on land with high value and public interest. However, a broad application of this approach would encounter a large number of barriers. The first one is financial. Purely technocratic solutions are expensive and may not gather sufficient public and political support for their introduction and maintenance. Furthermore, these systems would require appreciable maintenance and capitalization costs, which would be difficult to provide extensively, and this scenario could eventually slide toward the "privatization" scenario, in which private companies may find it easier to raise the capital needed for sustained operation. The technocratic scenario is feasible (affordable) in many developed countries, but is essentially irrelevant for developing countries, where the lack of funds and engineering expertise or operation and maintenance capacities prevents adoption of such solutions. Thus, the scenario is not applicable to areas inhabited by about 4/5 of the world population. The second weakness of this scenario follows from the fact that designs for specific return frequencies of runoff (typically 2–10 years for minor drainage, 25 to 100 years for major drainage; ASCE, 1992) remain vulnerable when such design conditions are exceeded. In such cases, consequences can be very dramatic. In other words, technological perfection will avoid all but the most extreme and rare problems, leaving people unprepared for such terrible catastrophes.

Reasons why this scenario could prevail

The technocratic scenario offers many benefits in the form of "robust" solutions (e.g., reliable and robust sewage conveyance systems using storage tunnels in downtown areas, also resembling the solution proposed for London; Thames Water, 2005) guaranteeing performance under design conditions and meeting direct environmental protection goals. The scenario philosophy fits well the mentality in developed countries and furthermore, the scenario would benefit from progress made in other fields of science (biotechnology, genetic manipulation, computers, nano-technologies, etc.). Many of these aspects are also appealing to politicians, government and decision makers by being in line with the latest technological progress. It may appear that this scenario could prevail in view of the continuing advances in our knowledge and expertise, and support for the scenario by municipal engineers, but the assumed detachment of the public and politicians from urban drainage issues (given by the system complexity and relatively problem-free operation) is probably unlikely, particularly if the associated costs increase too quickly.

The privatization scenario

The privatization scenario as described here reflects some early experiences with privatization of water services in Europe. It assumes a systematic involvement of the private sector in buying (or assuming a license for) the entire deteriorating water infrastructure and providing water services for contracted fees. This process usually encompasses all water services, rather than just urban drainage emphasized in this paper. Even though the privatization scenario refers just to the ownership of the infrastructure and/or the operating company, this ownership implies many other characteristics of water service delivery and asset management. It should be acknowledged that many variations of this scenario currently exist with different types of involvement of the private sector, often in combination with public ownership of assets (e.g., Juuti and Katko, 2005).

The governing principle of this scenario is making profit by providing the urban water services described in the operating contract. The main benefits of this scenario are often stated as the economic efficiency and effectiveness, and operations based on the "true" value of water. The main risks are rapidly increasing prices of water services and the perception of water becoming just another trading commodity. The service provider/customer relationship may be further complicated by the deficiencies of contracts signed between the municipality and the private utility, which may not sufficiently protect the customers or perhaps not provide adequate revenue to properly invest in the assets (Ashley and Cashman, 2006).

In some cases, the privatization scenario has resulted from a particular situation in which the reliability of water services delivery was undermined by repeated failures, reported waste of public money, or chronic under-funding as happened in Britain in the 1980s (e.g., House of Lords, 2006). Under such circumstances, the privatization of the urban water system is very tempting for politicians – they eliminate these problems by privatizing the system, selling off licensing use of, or giving away, the assets, and using the revenue from the sale for other purposes. This idea appears particularly attractive where ageing infrastructure requires large investments, which would have to be generated by raising taxes.

Main characteristics of the privatization scenario

Two operating principles govern this scenario – economic efficiency and the 'true' value of water as an economic good. Economic efficiency is achieved by control of labor costs and limited investments into infrastructure (particularly where the operating contract is limited to durations shorter than the life expectancy of the assets). When funding shortfalls and economic losses loom, the private operator often requests higher service fees (using the clauses put into the contract with this eventuality in mind), and/or renegotiation of the original contract. In certain cases (e.g., in Britain), this cannot happen due to the functioning of an economic regulator for the water industry (House of Lords, 2006). Environmental issues are often addressed at the level which was in effect when signing the contract; upgrading/renegotiation of such standards may be difficult without large fee increases.

In terms of the system architecture, the technological solutions will comprise "adequately maintained" centralized systems with some "decentralized pockets". Such systems are easier to manage, and the service charges are easier to invoice. In addition there is very limited technological risk using tried and tested systems which the financial backers are more comfortable with. Competition only happens once, i.e., when the initial contract is at stake, although surrogate systems are increasingly popular, attempting to represent competition artificially (Ashley and Cashman, 2006). The success of this scenario (from the public point of view) depends on the formulation of the initial contract and also the quality of the services delivered.

Risks and problems of the privatization scenario

The main risk of the privatization scenario is that the price for water service will become unreasonable, especially for developing countries (this has already happened in South America, in the case of drinking water, Ashley and Cashman (2006)). Market concentration could be a significant problem with only a few companies managing water resources worldwide. Water could assume the same role as energy today: being no longer a natural resource, but a tradable commodity such as crude oil (Bryce, 2001). While the main privatization drive often focuses on the supply of drinking water, other services, including drainage, are likely to be included in integrated packages.

The dilemma of water service companies is that regulators want to keep the fees down while investors require revenues and profits, which may be reinvested in other fields entirely unrelated to water. This may also happen in areas where these services are municipally owned, but operated by specialist 'quasi-private' organizations as in Australia, where the States take the profits generated annually by the water wholesalers. A solution to the dilemma could be to offer "new" but lower level services to cut costs (e.g., bottled water for potable purposes in exchange for lower quality tap water adequate for non-potable purposes). However, there are major problems in doing this as water transport costs, the use of plastics for bottles, waste generated and the generally poorer quality of bottled water compared with tap water make this risky (e.g., Gleick, 2004).

Private companies might renege on operating contracts accompanied by legal disputes when facing significant financial difficulties. They even might become bankrupt and stop operating overnight. Another risk is that companies could develop short-term strategies guaranteeing returns on investment, but disregard long-term sustainability issues. Other activities typical for the private sector, including mergers and take-overs, accompanied by re-organization and asset stripping, will frequently lead to changes of staff and their duties. In this process, experienced engineers may be lost to the industry. In addition, where private sector service providers are strongly regulated, the regulator, usually with government interference, may prevent the service provider from operating efficiently (e.g., House of Lords, 2006).

Reasons why this scenario could prevail

The authors identified four major driving forces that can make this scenario happen (Chocat et al., 2004): (i) Selling public water infrastructure assets generates large one-time income that can be used for other high priority purposes; (ii) All urban dwellers need urban water services; given a "well-formulated" contract, water service can be a profitable business and an attractive investment for private interests; (iii) Actual (or perceived) failures of the conventional technocratic approach will support the opinion that "private is better than public"; and, (iv) Where privatization is one of several options (e.g., in neighboring cities), private industry will attract the best engineers and offer them opportunities and resources to apply their technical talents. Thus, the public utilities will encounter a steady brain drain.

Furthermore, this scenario is somehow auto-engaging. Once it has reached a certain level of development and acceptance, it is difficult to change the strategy. For private companies it is tempting to gain a business monopoly. Experience of recent years indicates strong promotion of privatization as the (only) strategy to "make the urban water system work". When trying to penetrate the market, the private water companies show some generosity towards researchers and international agencies by awarding them financial grants. On the political side, the privatization benefits from the argument that customer costs will become lower because of the greater economic efficiency of the private sector. For politicians, it offers an opportunity to shed responsibility for under-funded, ageing water infrastructure and the need to raise capital, deal with municipal employee unions, public complaints about poor service, increases in water fees, etc. Thus, this scenario may prevail, because of its political attractiveness of generating income by selling off an ageing infrastructure (i.e., a liability) and promoting economic efficiency. However, storm drainage systems which are of less direct interest to the public are currently of less interest to private companies than water supply and sanitary systems, since it is difficult and unusual to install flow monitors on storm drainage systems. However, in parts of UK, partial privatization has been implemented whereby the private company operates the sewer system based on an initial measured flow rate at the start of a 25–30 year concession (Brown et al., 2005).

The green scenario

The final potential scenario is termed the "green" scenario. Some aspects (particularly those dealing with drainage) of the green scenario are currently promoted in many areas as low impact development (LID), smart growth, sustainable drainage systems (SUDS), water sensitive urban design (WSUD), and others. The main characteristics of this scenario are the replacement of conventional central water service systems by distributed systems, with more account for sustainability, attention to environmental concerns, restoration and re-naturalization of receiving waters, etc. In spite of crediting this approach to recent activities described by the new terminology, essential principles of this approach to urban drainage were formulated in the mid 1970s and implemented (with some success) in the Woodlands Planned Community (Texas), where natural drainage was preserved as much possible with respect to drainage patterns and infiltration areas, source controls included the use of porous pavement, additional drainage capacity was created in the form of vegetated open channels, and runoff flow rates were balanced in aesthetically attractive impoundments (U.S. EPA, 1979).

The above examples cover just one aspect of urban water, surface drainage and its impacts on receiving waters. Experience from regions with water scarcity indicates that for the green scenario to be really viable, it should address all aspects of the management of urban waters in the form of the "total urban water cycle based management", which encompasses: (a) reuse of reclaimed wastewater (for pollution prevention, sub-potable water supply); (b) integrated stormwater, groundwater, water supply and wastewater based management (water supply, flow management, water and landscape provision, substituting sub-potable water sources where feasible, and protection of downstream areas against urban impacts); and, (c) water conservation based approaches (efficient water use, reduced water demand for landscape irrigation, and substitution of industrial processes with reduced water demands). This type of management can be achieved only with decentralized solutions applied everywhere (Lawrence et al., 1999), but currently it represents a planning goal rather than a routinely implemented task.

Generally speaking, the objectives and performance criteria of the green scenario are not well defined, as reflected e.g. by the existence of more than 200 definitions of "sustainable development". The scenario is based on appealing principles, ideas and visions, but its sustainability when used at a large scale is unclear (Ashley et al., 2006b).

Main characteristics of the green scenario

The theoretical governing principle of this scenario is sustainability. But sustainability is difficult to define at the scale of a city and promoters often use such slogans as "back-to-nature", "green cities", "cleaning the future", or applying engineering "married" with social sciences. The technical measures applied include source control techniques and infiltration systems (rainwater management rather than stormwater management); water supply with waters of different qualities; separation of blue, grey, black and yellow wastewater; recycling of both wastewater and solid waste in small local cycles (house, lot, neighborhood); and, rainwater harvesting. Ecologists promote these systems and have imposed a "balanced integrated approach" attempting to combine both "soft" and "hard" technologies for achieving sustainability of urban drainage in decentralized systems. The responsibility for such systems becomes decentralized, requiring participation of property owners for their operation and management (U.S. EPA, 2002). Long-term master plans for large-scale infrastructure gradually disappear. Many innovative ideas, devices and systems are available, and numerous small companies offer support in the form of a wide range of technical and operational services.

Risks and problems of the green scenario

Many of the perceived or real risks of the green scenario arise from the fact that it is a new concept that has not been truly tested in the field, certainly not on sufficiently large scales and for sufficiently long periods for larger urban areas. With this limitation in mind, the authors identified three risks or problems of the green scenario.

Firstly, the notion of "no impact development" is unrealistic (Strecker, 2001). Urban development does produce impacts on the land cover, soils, hydrologic cycle including receiving waters and local microclimate (Marsalek et al., 2006). These impacts can be low if the infrastructure presence is small, and the quantity of water used or effluent released, waste heat and pollutants generated are limited, but nevertheless they do exist. Furthermore, low impact development needs low urban density, which means larger space consumption and longer travel distances; costly in energy and pollutant emissions unless all services and facilities follow a similar decentralized pattern, like in Swedish eco-villages (Hedberg, 1999). Under typical circumstances, there is a lack of certainty that the green solutions promoted as described above are really harmless. Note that their potential effects may be of a cumulative and long-term nature, thus taking a long time to manifest. There is a risk that nature alone cannot cope with the deleterious effects, particularly in the case of persistent chemicals (e.g., heavy metals or some trace organic toxicants). Once released into the local environment, these chemicals are likely to accumulate in benthic sediments in ponds and streams, and may enter the food chain or be released into the water column under certain ambient conditions. Other effects are likely to increase in severity with respect to the sediment regime and aquatic habitat modified by urbanization. Where the green scenario includes water reclamation and reuse, there will be public health risks potentially caused by unqualified personnel (property owners) operating such systems. Thus, while the emerging "green" developments appear more ecological than the conventional developments, their sustainability needs to be thoroughly studied and assessed, particularly with respect to the long-term operation and maintenance of decentralized facilities (ASCE, 1998). For example, the lack of master planning can result in huge cumulative impacts on the water regime at the watershed scale, even when no impact can be seen at local scales.

Secondly, other risks arise from the fact that green solutions transfer much of the maintenance responsibility to the property owners, on whose land the on-site stormwater management (and other) facilities would be built and operated. This transfer of responsibility to the end-users might result in no or poor maintenance and, consequently, numerous partial system failures, which due to the drainage system connectivity, would affect not only the particular owner, but others as well. Such incidents could lead to litigations among the property owners. Quite likely, the need for maintenance services would lead to the development of new businesses providing such services in urban areas. This was recognized by the U.S. EPA in the early 21st century when promoting the use of on-site systems (U.S. EPA, 2002) and since the early trials, specialist contractors are now recommended to service groups of on-site systems (Ashley et al., 2004).

Finally, effecting the transition from the existing centralized systems to future decentralized systems is not clear. The existing urban water systems in developed countries took a long time to build and represent high value assets. Decentralized systems would also require large investments, with wide-scale implementation stretched over many decades. During the transition, both systems would have to be operated in parallel and financed. When finally the centralized systems would no longer be needed and could be turned off (in an unfeasibly long time, since pipes would be difficult to replace by on-site systems in the densest parts of cities), the resulting write-off of (financial) capital would be unprecedented. This, however, is not the situation in developing countries where such assets do not already exist. Here the use of decentralized systems would appear feasible and cost-effective. Unfortunately, individual expertise and financial ability would mean that these systems would typically need to be provided externally by donor agencies (Ashley and Cashman, 2006).

Reasons why this scenario could prevail

The main driving forces supporting the green scenario are its positive political appeal and economic appeal (at least in a short run) to those public utilities that struggle financially. The scenario's objectives are undisputed, both internationally and locally (e.g., the Brundtland Commission, local Agenda 21) and it receives "green political support", particularly in relatively affluent countries where stakeholders are concerned about the over-exploitation of nature and want "to do something good". Furthermore, it is defended by a lot of enthusiastic supporters. From the sociological point of view, it appeals to well-educated, well-to-do part of the society, often living in up-scale developments or eco-villages. Furthermore, the green scenario may be even more feasible in developing countries, where large central infrastructures are almost non-existent and their construction is hardly feasible (Stephenson, 2001; Matsui et al., 2001). Thus, it is attractive to seek alternative green approaches to urban water management, while there are doubts whether the present system can be regarded as sustainable. Thus, although radical, the green scenario is not impossible.

SUMMARY AND CONCLUSIONS ABOUT FUTURE DEVELOPMENTS

Sustainability is a relatively new concept which is becoming a major driving force in the evolution of urban water management (Harremoes, 1996; Otterpohl et al.,

1997; Ellis et al., 2004). Increasing water scarcity due to an increasing world population and the concept of equitable provision for all is raising the importance of water as a fundamental human right and a cornerstone for economic, health and food security. In addition concerns about real or perceived climate change have focused thinking about the future uncertainty and demand for water (New Scientist, 2006). Three areas that need to be addressed to respond to current shortages and future needs are: Institutional systems, technological advances, and stakeholder responsibility. Water could become a catalyst for greater world-wide co-operation, rather than a source of conflict (Ashley and Cashman, 2006; Asmal, 2001). However, at present, the vision of a coherent world-wide water policy is utopian and will be so for decades. Industrialized nations do have fully functional urban water systems, whereas developing and emerging countries do not (Stephenson, 2001), prompting the prominence of water in the Millennium Development Goals (UN, 2000).

The optimum scenario in the industrialized world is probably some combination of the beneficial properties of the scenarios discussed earlier. The existing systems have to be adapted by introducing "green" solutions where appropriate with strong incentives to adopt novel solutions, but also with a global vision and strong and competent government regulatory control. To achieve greater water security, the Stockholm Water Symposium (SIWI, 2001) called for: (i) More flexible institutional arrangements; and, (ii) Increased water awareness among all stakeholder groups. The importance of ensuring that water professionals were competent, multidisciplinary and up-to-date was emphasized. The globalization of economic systems was seen as a potential impediment with consequential institutional arrangements constraining innovation and the need to introduce more appropriate "local" solutions, as also echoed in the EU Daywater project (http://www.daywater.enpc.fr).

There is agreement that both storm and sanitary "wastewaters" provide a resource opportunity within the Integrated Water Resource Management framework (Ellis, 1995). Where feasible, all opportunities for resource recovery should be taken, with water re-use and recycling, and utilization of nutrients from human wastes (Malmquist, 1999). This approach would: (i) create a greater awareness for water in the urban environment, which is largely unknown today, except in certain eco-villages (Hedberg, 1999), and (ii) ease a major economic and infrastructure problem of current water supply provision in many regions of the world.

The cornerstone of a realistic future vision is therefore decentralized wastewater treatment and localized urban drainage networks comprising mostly surface rather than underground systems that may then be utilized as resources. This requires both technological development and greater individual and community responsibility. It may be achieved by a combination of source controls (provided by the industry) and technological development (particularly biotechnology research). By industrial and technological re-engineering of substances and materials used in the household, the liquid waste stream from this source will become of comparable quality to rain runoff (HHWF, 1999) and a major portion of the pollution problem in urban drainage would be solved. By pollution minimization and also applying infiltration systems to the discharge of household effluent, a sustainable solution, at least for the dry weather situation, is feasible, provided that groundwater quality is not compromised and soils and groundwater aquifers can accept these extra inputs. The need for integration of these efforts with other measures, i.e., total urban water cycle management, is obvious.

What remains is the discharge of excess rain runoff. In densely urbanized areas there is usually no way of avoiding the use of a traditional pipe system, at least for the foreseeable future and such an approach may be the one which is the most sustainable in the context in which it is applied (DTI, 2006). However, large improvements over conventional systems can be attained by applying source controls as much as possible (water harvesting, infiltration, etc.) and new generation technology for water routing and runoff treatment. Special attention needs to be paid to road runoff by focusing on minimizing toxic and harmful chemicals.

The key feature of any optimum drainage solution in industrialized areas is the fusion of the existing traditional technology with recent ecological and technological achievements, taking due account of the increased responsibility of the individual and local community. Applying higher technology to decentralized solutions would shift a significant part of the cost (and responsibility) from the public to the individual. It is unlikely that this new system would function without an effective inspection and enforcement system and clearly defined performance criteria and indicators.

For such an approach to be viable, it would be necessary to change the current institutional systems, in which the water utility (i.e., the asset owner or operator) is valued according to the infrastructure assets it owns, and the revenue income is based on volumes and pollutants handled. Both "hard" infrastructure and handled wastewater volumes would clearly be reduced under such a scenario. Other barriers include professional reluctance to change from the "way-it-has-always-been-done" (Geldof, 1999; NAO, 2004). Thus, new approaches to educating professionals will be needed, and new ways of providing real knowledge about water/wastewater systems will be required for new system "operators" (consumers) at the local level.

Optimum approaches to future urban drainage systems for developing countries will be different from what is envisaged for developed countries, inasmuch as the fusion between existing and new technologies is not likely to be as big an issue. Megacities in developing countries face essentially the same problem with respect to urban water management as densely populated urban areas in the first world: land use. Here the key principle must be the utilization of all "wastes" wherever feasible. It is likely that in many cases this may require high-technology solutions and the fusion of traditional (pipe systems where necessary) with "green and ecological technologies" (infiltration, ponds, natural waterways, water harvesting) and modern technology (treatment facilities where affordable). The principal consideration will be economic, and it is difficult to see how adequate resources can be made available to achieve what is required within the current globalized economic systems (Ashley and Cashman, 2006).

Where possible, the approaches adopted must take into account the need for easily repairable technological implementation, increased employment opportunities for local inhabitants, and the provision of basic education for installers and operators. Inherent flexibility and adaptability (key criteria for sustainability) must be built into water/wastewater systems due to the rate of change of circumstances in developing country communities. Greater "consumer" involvement and responsibility for water and wastewater services has been shown to work effectively at the community level in developing countries (Stephenson, 2001), but where sophisticated technologies have to be used, this may be problematic.

With respect to urban drainage, large progress and innovation is occurring in two classes of projects: (i) new suburban developments (particularly in upscale developments with large plots, where innovative on-site water management is sometimes a prerequisite for receiving building approval), and (ii) retrofits in attractive locations with a high land value. In the former case, advanced planning, land availability for siting nature-mimicking management measures/facilities, and the availability of funds are important; in the latter case, the availability of funds makes it feasible to employ high-technology solutions on a small footprint (including membrane technology in water reuse) and design self-reliant water systems. Both types of developments can serve well for studying modern concepts included in the green scenario and preparing for their wider applications, perhaps during various stages of urban redevelopment.

REFERENCES

American Society of Civil Engineers (ASCE) and Water Environment Federation (WEF) (1992). Design and Construction of urban stormwater management systems. ASCE, New York, NY.

American Society of Civil Engineers (ASCE) (1998). Sustainability criteria for water resource systems, WRPM Division and UNESCO International Hydrological Programme IV, Project M-4.3, Task Committee on Sustainability Criteria. ASCE, New York, NY.

Andersen, T., Eklund, J., Tollefsen, T., Lindheim, T., and Schilling, W. (2001). The role of stormwater source control in the conversion of the National Hospital of Norway into an ecological residential complex. Advances in Urban Stormwater and Agricultural Runoff Source Controls. Ed. Marsalek, J., Watt, E., Zeman, E., Sieker, H., NATO Science Series, IV. Earth and Environmental Sciences, Vol. 6, Kluwer Academic Publishers, Dordrecht, The Netherlands, 159–168.

Ashley, R.M. and Cashman, A. (2006). Assessing the Likely Impacts of Socio-Economic, Technological, Environmental and Political Change on the Long-Term Future Demand for Water Sector Infrastructure. OECD report, Pennine Water Group, Sheffield, UK.

Ashley, R.M., Clemens, F., and Veldkamp, R. (2004). The Environmental engineer – a step too far? Proc. NOVATECH 2004, 5th Int. Conf. on Sustainable Techniques and Strategies in Urban Water Management, Lyon (France), June 6–10, Vol. 1, 79–86.

Ashley, R.M., Tait, S.J., Styan, E., and Cashman, A. (2006a). Sewer system design moving into the 21st century – a UK perspective. Proc. 7th Int Conf on Urban Drainage Modelling/4th Int Conf on Water sensitive urban design, Melbourne, Apr. 4–7. Ed. Deletic A, Fletcher T, ISBN 0-646-45903-1. Vol. 2, 559–566.

Ashley, R.M., Tait, S.J., Cashman, A., Blanksby, J.R., Hurley, A.L., Sandlands, L., and Saul, A.J. (2006b). 21ST CENTURY SEWERAGE DESIGN. Part 1: Summary report. UK Water Industry Research Ltd Report WM07.

Asmal, K. (2001). Water is a catalyst for peace. Water Science and Technology 43(4): 24–30.

Brown, R.R., Sharp, L., and Ashley, R.M. (2005). Implementation impediments to insti-tutionalising the practice of sustainable urban water management (Keynote paper). Proc. 10th Int. Conference on Urban Drainage, Copenhagen (Denmark), Aug. 21–26. CD-ROM, 8 p.

Bryce, S. (2001). The Privatisation of Water: The trend towards privatising the world's water supplies and applying full-cost pricing policies means that millions of people are losing access to an already scarce resource. Nexus Magazine, 8(3), available at http://www.nexusmagazine.com/articles/waterprivat.html

Chocat, B., Krebs, P., Marsalek, J., Rauch, W., and Schilling, W. (2001). Urban drainage redefined: from stormwater removal to integrated management. Water Science and Technology 43(5): 61–68.

Chocat, B., Ashley, R., Marsalek, J., Matos, M.R., Rauch, W., Schilling, W., and Urbonas, B. (2004). Urban drainage: Out-Of-Sight-Out-of-Mind. Proc. NOVATECH 2004, 5th Int. Conf. on Sustainable Techniques and Strategies in Urban Water Management. Lyon (France), June 6–10, 2004, Vol. 2, 1659–1690.

Dlugolecki, A. and Loster, T. (2003). Climate change and the financial services sector: an appreciation of the UNEPFI study. Geneva Papers on Risk and Insurance 28: 382–393.

DTI (2006) GLOBAL WATCH MISSION REPORT - Sustainable drainage systems: a mission to the USA. March. UK Department of Trade and Industry website (http://www.globalwatchservice.com/pages/Three Columns.aspx?PageID=102).

Duncan, H.P. (1999). Urban Stormwater Quality: a Statistical Overview. Cooperative Re-search Centre for Catchment Hydrology, Melbourne, Australia, Report 99/3.

Ellis, J.B. (1995). Integrated approaches for achieving sustainable development of urban storm drainage. Water Science and Technology 32(1): 1–6.

Ellis, J.B., Deutch, J.-C., Mouchel, J.-M., Scholes, L., and Revitt, M.D. (2004). Multicri-teria decision approaches to support sustainable drainage options for the treatment of highway and urban runoff. Science of the Total Environment 334–335: 251–260.

Evans, E.P., Ashley, R.M., Hall, J., Penning-Rowsell, E., Saul, A., Sayers, P., Thorne, C., and Watkinson, A. (2004). Foresight. Future Flooding Vol 1–Future risks and their drivers. UK Office of Science and Technology. April. Crown Copyright. DTI/Pub 7183/2k/04/04/NP.URN 04/939.

Gaudreault, V. and Lemire, P. (2006). The Age of Public Infrastructure in Canada. Analysis in brief, No. 11-612-MIE2006035, Statistics Canada, Ottawa, Ont.

Geldof, G. (1999). QWERTIES in integrated urban water management. Proc. 8th Int. Conf. on Urban Drainage, I B Joliffe and J E Ball Eds, Sydney (Australia), Aug. 30–Sept. 2. Vol. 2, ISBN 0 85825 718 1, 809–816.

Gleick, P.H. (2004). The World's Water 2004–2005. Island Press. Washington, D.C., USA, ISBN 1-55963-812-5.

Harremoes, P. (1996). Dilemmas in Ethics: Towards a Sustainable Society. Ambio – A Journal of the Human Environment 25(6): 390–395.

Hedberg, T. (1999). Attitudes to traditional and alternative sustainable sanitary systems. Water Science and Technology 39(5): 9–16.

House of Lords (2006). Water Management. UK House of Lords Select Committee on Science and Technology report. London UK.

Household Hazardous Waste Forum (HHWF) (1999). Good practice guide: The defini-tive Guide to Best Practice in Household Hazardous Waste Management. Household Hazardous Waste Forum, UK.

Juuti, P.S. and Katko, T.S. (Eds.) (2005). Water, Time and European Cities - History matters for the Futures. WaterTime project. EU Contract No: EVK4–2002–0095. http://www.watertime.net/DOCS/WP3/WTEC.pdf

Lawrence, A.I., Ellis, J.B., Marsalek, J., Urbonas, B., and Phillips, B.C. (1999). Total urban water cycle based management. Proc. of the 8th Int. Conf. on Urban Storm Drainage, Sydney (Australia), Eds Joliffe, I.B., and Ball, J.E., Eds. Aug. 30–Sept. 3, 1999, vol 3, pp. 1142–1149.

Leopold, L. (1968). Hydrology for Urban Land Planning – a Guidebook on the Hydrologic Effects of Urban Land Use. U.S. Geol. Survey Circular 554, USGS, Reston, VA.

Malmqvist, P.A. (1999). Sustainable Urban Water Management. Vatten 55(1): 7–17.

Marsalek, J. and Chocat, B. (2002). International report: stormwater management (SWM). Water Science and Technology 46(6–7): 1–17.

Marsalek, J., Jimenez-Cisneros, B.E., Malmquist, P.-A., Karamouz, M., Goldenfum, J., and Chocat, B. (2006). Urban Water Processes and Interactions. UNESCO Press, Paris, December.

Matsui, S., Henze, M., Ho, G., and Otterpohl, R. (2001). Emerging paradigms in water supply and sanitation. In: Frontiers in Urban Water Management – Deadlock or Hope. Ed. Maksimovic, C., Tejada-Guibert, J.A., IWA Publishing, London, UK, ISBN 1 900222 76 0. pp 229–263.

NAO (National Audit Office) (2004). 'Out of sight - not out of mind'. Ofwat and the public sewer network in England and Wales. HC 161 Session 2003–2004: 16 January 2004. Report by The UK Comptroller And Auditor General.

New Scientist (2006). The Parched Planet. 25th February. p. 32–36.

Niemczynovicz, J. (1999). Urban hydrology and water management - present and future challenges. Journal of Urban Water 1(1): 1–14.

Otterpohl, R., Grottker, M., and Lange, J. (1997). Sustainable water and waste management in urban areas. Water Science and Technology 35(9): 121–133.

Snyder, S.A., Westerhoff, P., Yoon, Y., and Sedlak, D.L. (2003). Pharmaceuticals, personal care products, and endocrine disruptors in water: Implications for the water industry. Environmental Engineering Science 20(5): 449–469.

Stephenson, D. (2001). Problems of developing countries. In: C. Maksimovic and J.A. Tejada-Guibert Eds. Frontiers in Urban Water Management: Deadlock or hope? IWA Publishing, London, UK, ISBN 1 900222 76 0, 264–312.

Stockholm International Water Institute (SIWI) (2001). Water Security for the 21st Century – Innovative Approaches. Water Science and Technology 43(4): 7–9.

Strecker, E. (2001). Low-Impact Development (LID) – Is It Really Low or Just Lower? In: Linking Stormwater BMP designs and performance to receiving water impacts mitigation. B R Urbonas Ed., Proc. of UEF conference. Snowmass, CO (USA), published by ASCE, Reston, VA. ISBN 0-7844-0602-2, 210–222.

Thames Water (2005). Thames Tideway Strategic Study. Steering Group Report. February. (available from the Thames Water website: wwwthamestidewaystrategic-study.co.uk/pdfs/TTSS_Steering_Group_Report.pdf)

United Nations (2000). Resolution adopted by the General Assembly [without reference to a Main Committee (A/55/L.2)] 55/2. United Nations Millennium Declaration. September. http://www.un.org/millennium/. http://www.undp.org/mdg/basics.shtml

UN (2003). Water for People, Water for Life – UN World development Report (WWDRI). UNESCO and Berghan books, UK.

UN Secretariat (2005). World population prospects: The 2005 revision. UN, New York, NY.

U.S. EPA (1979). Maximum Utilization of Water Resources in a Planned Community, Report EPA-600/2-79-050a, U.S. Environmental Protection Agency, Cincinnati, Ohio, July.

U.S. EPA (1983). Results of the Nationwide Urban Runoff Program Vol. I - Final Report. U.S. Environmental Protection Agency, Water Planning Division, U.S.EPA, Washington, DC.

U.S. EPA (1994). National water quality inventory. Report EPA 841-F-94-002, U.S. Environmental Protection Agency, Washington, DC, USA.

U.S. EPA (2002). Onsite Wastewater Treatment Systems Manual. Report EPA/625/R-00/008, U.S. Environmental Protection Agency, Washington, DC, USA.

22

Overcoming legal barriers to hydrological sustainability of urban systems

R. Adler

University of Utah, S.J. Quinney College of Law, Salt Lake City, UT 84112, USA
E-mail: Adlerr@law.utah.edu

Summary: Conflicts between private property and public resources remains the major legal barrier to restoring and maintaining hydrological sustainability in urban areas. Past laws and practices allowed or encouraged extensive development "at the water's edge," leaving inadequate natural buffers in transition zones between land and water. Solutions to this problem that involve modifications to existing development face constitutional and other legal protections provided to private property. One solution that would be consistent with existing property rights would involve a competitive bidding process, modeled after a program designed to reduce salinity in the Colorado River, to provide economic incentives to reduce hydrological impacts of development. Alternative solutions would involve reduced private property rights in transition zones between land and water areas in recognition of the public character of those areas. That solution, while probably a more effective long-term approach, would involve considerably more political controversy and potential economic disruption.

INTRODUCTION

Many legal doctrines influence implementation of measures to achieve hydrological sustainability in urban systems. The most important of these is the protection of private property rights under the Fifth and Fourteenth Amendments to the U.S.

Constitution. I suggest two possible solutions to this issue, one economic and one legal in nature.

CURRENT KNOWLEDGE AND UNDERSTANDING OF THE PROBLEM

Private property law and waterfront development

Virtually from the beginning of recorded history, people have settled at the water's edge (Worster, 1985). Waterfront areas have been among the first to be developed, because proximity to water was useful for navigation, irrigation, industry, defense, and culinary uses. Relatively flat, fertile flood plain lands were easier to build on and more productive to farm. Likewise, from the earliest days of the American Republic, waterways were critical for transportation and trade, settlement, fishing, and defense (Wilkinson, 1985). As the demand for waterfront land increases, those who "own" property at the water's edge find their lands increasing in value, whether due to traditional commercial values or pure aesthetics (Adler et al., 1993; EPA, 2000; Archer and Stone, 1995). At the same time, however, waterfront land uses cause serious harm to aquatic ecosystems.

Under traditional theories of property, private ownership promotes both individual and societal welfare. Therefore, waterfront landowners have at least presumptive rights to develop right up to the water's edge. Other societies, and some Anglo-American scholars, believe that land contains a hybrid of private and public and ecological values, not all of which are necessarily maximized by private ownership (Rose, 1994; Rieser, 1991; Sax, 1989). Aquatic and other ecosystems provide a range of "ecosystem services" that are not captured by private economic markets (Daily, 1997). While many of these services benefit human economies, their ecological "value" is not limited to humanity alone. Thus, the assumption that an individual landowner will maximize the overall value of the land, as opposed to the value to that owner alone, is open to question.

Under traditional notions of property law, landowners have an incentive to intensify waterfront land uses wherever increased profits outweigh losses in value to that landowner. Property law doctrines designed to balance the rights of that landowner against the rights of other landowners or society as a whole— for example, to prevent downstream pollution or flooding—can be invoked to limit or alter those detrimental land uses. But to the extent that waterfront development impairs ecosystem uses that do not have understood or accepted economic values, and therefore that are not considered by those doctrines, current notions of property law provide ineffective relief (Karp, 1989; Freyfogle, 1993).

Anglo-American property law also relies heavily on fixed boundaries and stability of title, without which landowners would lack the requisite certainty of return necessary to invest labor and capital (Freyfogle, 2002). Fixed land usually can be characterized by fixed boundaries. Aquatic ecosystems, however, frequently are not so amenable to such rigid delineation. Large lakes and coastal waters are subject to the ebb and flow of the tides. In their natural state, rivers and streams expand and contract seasonally and annually through a shifting mosaic of flood plains and side channels. Other aquatic ecosystems, from wetlands to ephemeral streams, can vary along both temporal and geographic spectra. Thus, while it is desirable for purposes of the law of property and the human economy to delineate fixed boundaries between land and water, from an ecological perspective such boundaries simply do not exist. The Supreme Court has recognized this problem in the context of regulating discharges to aquatic ecosystems under the federal Clean Water Act:

> In determining the limits of its power to regulate discharges under the Act, the Corps must necessarily choose some point at which water ends and land begins. Our common experience tells us that this is often no easy task: the transition from water to solid ground is not necessarily or even typically an abrupt one. Rather, between open waters and dry land may lie shallows, marshes, mudflats, swamps, bogs–in short, a huge array of areas that are not wholly aquatic but nevertheless fall far short of being dry land. Where on this continuum to find the limit of 'waters' is far from obvious.

(Riverside Bayview Homes, 1985).

For purposes of determining "ownership" in areas of fluctuating or otherwise variable shorelines, the law has developed various methods to fix property boundaries (Sax, 2000). To the extent that aquatic ecosystems include ecological and other values that transcend the commodity value of private ownership, however, the law has found no adequate solutions. For most purposes, the water is considered "public", although we dole out usufructuary rights to water if the exercise of those rights is consistent with the public welfare. Land, by contrast, is considered "private" in the sense that it is "ownable" by private parties or by a government. That distinction, however, assumes a sharp dividing line between land and water. On one side of this ecologically- and hydrologically-artificial dividing line presumptive development rights apply; on the other, restrictions to protect shared aquatic resources are relatively stronger. This existing legal regime typically allows extensive development in the transition zone between land and water.

As addressed in other chapters in this book, waterfront development has profound effects on hydrologic sustainability (National Research Council, 2002). Aquatic ecosystems in the United States are in serious peril, in large part due to the massive loss and alteration of aquatic habitats both within the bodies of rivers, lakes, and coastal waters and in their ecologically- and hydrologically-important affiliated wetlands, floodplains, and other transition zones between land and water. Millions of miles of America's aquatic ecosystems have been dammed, diked, levied, channelized, filled, drained, and otherwise altered. Most experts believe that this habitat loss and degradation is the single leading cause of aquatic ecosystem decline (Abell et al., 2000; Adler et al., 1993).

Inadequacy of current legal solutions

Existing legal approaches have not succeeded in balancing private, public, and ecosystem interests at the water's edge. Two major bodies of law attempt to strike a balance between private property rights and ecological health. Public law solutions involve statutes and regulations designed to prevent, minimize, or mitigate the adverse effects of development through federal, state, or local regulation (including planning and zoning laws). While those efforts provide some solutions, and could be even more beneficial if implemented more consistently and effectively, they are hampered by jurisdictional, political, practical, and perceived constitutional problems. Existing property law solutions seek to protect common "public" resources through governmental ownership interests in some aquatic lands and resources. Those doctrines, however, are also limited by issues of funding, jurisdiction and scope, and by judicial reluctance to expand or redefine traditional doctrines designed to protect common economic resources rather than newly-recognized ecological values.

Public law solutions

All levels of government regulate development in aquatic or semi-aquatic zones to some degree, through planning and zoning, pollution control, fish and wildlife protection, flood plain protection, and other programs. These efforts have not succeeded in preventing extensive damage to aquatic ecosystem values and services, although they have had some mitigating effects. Two examples explored below are the federal Clean Water Act (CWA) and federal floodplain protection programs.

The overriding objective of the CWA is to "restore and maintain the chemical, physical, and biological integrity of the Nation's waters," a potentially far-reaching aim which could address not only the types of chemical impairment most often associated with "water pollution" control but a much wider range of issues related to hydrologic sustainability. However, both EPA and the states have focused almost exclusively on the "chemical" to the exclusion of the "physical" and "biological" integrity components of the CWA (Adler, 2003). The scope and jurisdiction of the CWA, moreover, is limited in ways that prevent it from addressing all impairments to hydrologic sustainability in urban and other areas.

First, the CWA is limited in the *types of activities* that can be regulated. The statute prohibits the discharge of any pollutant into the "waters of the United States" without a permit issued by EPA or a delegated state under section 402, or by the U.S. Army Corps of Engineers (ACE) or a delegated state under section 404, and unless in compliance with various controls and limitations imposed on those discharges. By and large, that provision limits the discharge of pollutants into water bodies, and thus serves as a constraint, but not a prohibition, on building or other development at the water's edge. Channelized erosion constitutes the "discharge of a pollutant" which requires a stormwater permit under section 402. Those permits, however, typically require only that builders implement "best management practices" or other controls designed to minimize the effects of erosion and sedimentation. To date, broader municipal stormwater programs have not succeeded in resolving either chemical or hydrological impairments to urban waterways.

In theory, more significant limitations could be placed on waterfront development that adversely affects aquatic resources via the "nonpoint source" control provisions in sections 208 and 319 of the Act, and through the provisions designed to implement ambient water quality standards under section 303. State-adopted programs under those provisions could include such limitations as minimum building setbacks from water bodies or limitations or prohibitions against construction in floodplains or other aquatic transition zones (ELI, 1997). Due to political and other factors, however, most states have not elected to use their nonpoint source pollution control authority under the CWA to impose those types of restrictions, and nothing in the statute affirmatively requires them to do so (Zaring, 1996). When and if they do so, or where similar controls are imposed under local zoning laws, constitutional takings claims are likely to arise (discussed below).

Filling and impairment of wetlands has particularly significant impact on hydrological sustainability, because wetlands filter pollutants and slow the runoff and infiltration of storm flows into surface waters (National Academy of Sciences, 2002; Keller, 2005). Section 404 of the CWA authorizes the Secretary of the Army (through the Army Corps of Engineers) to issue permits for the discharge of dredged or fill material into waters of the United States. Section 404 allows permits for the discharge of "clean" fill into wetlands and other waters for purposes of converting aquatic areas into "useable" land, absent a finding by EPA that such filling will have "unacceptable adverse effects" on aquatic resources. EPA has made that finding only a handful of times in more than thirty years (Adler, 2003). The annual loss of wetlands has slowed considerably in the United States over the past several decades, but wetlands and other aquatic ecosystem components continue to be lost at an alarming rate even under this regulatory system (Dahl, 2000).

Second, the effectiveness of the CWA in protecting the land-water transition zone is limited in *geographic scope*. While the U.S. Supreme Court confirmed in 1985 that the CWA applies to wetlands immediately adjacent to navigable surface waters, another decision issued in 2001 cast a shadow of doubt about the degree to which the Act applies to so-called "isolated" wetlands and other water bodies (Riverside Bayview Homes, 1995; Solid Waste Ass'n of Northern Cook County, 2001). More recently, with ramifications which will not be known for several years, a plurality opinion joined by four justices on the court suggested that the Act applies only to "relatively permanent, standing, or continuously flowing bodies of water," and not to channels through which water flows intermittently or ephemerally, or channels that provide drainage for rainfall, and that only wetlands with direct surface water connections to navigable waters or their tributaries are "adjacent" waters under Riverside Bayview Homes. Four other justices would have adopted a much more liberal test, however, and the tie-breaking justice (Justice Kennedy) would require only an adequate showing that the water or wetland to be regulated has a "'significant nexus' to waters that are or were navigable in fact or that could reasonably be so made.'" (Rapanos, 2006). From an ecological and hydrological perspective, efforts to draw clear lines between "isolated" waters and "navigable" waters are largely arbitrary, and fail to reflect the nature of the hydrological cycle, critical hydrological linkages between surface and ground water, and ecological linkages between various components of broader aquatic ecosystems (U.S. Fish and Wildlife Service, 2002; National Research Council, 1995). There are likely to be both hydrological and ecological linkages between all water bodies within a given watershed (National Research Council, 1999).

Finally, for the most part the CWA, however broadly or narrowly its jurisdiction is defined, protects only those components of riparian ecosystems which can be defined in some way as "aquatic". Riparian, lacustrine, and coastal ecosystems, however, often are comprised of mosaics of upland and wetland habitats, and both water quality and hydrological components of aquatic ecosystem health rely on these connections (Karr and Schlosser, 1978). The existing legal boundary between land and water does not account fully for the far more complex "boundary" between aquatic and terrestrial ecosystems, one which is more properly characterized as either a continuum or a mosaic hybrid.

In its efforts to protect water's edge ecosystems under floodplain management programs, the federal government has taken an even less direct approach than in the CWA, despite the facial focus of the program on development itself as opposed to the water polluting-impacts of that development. Indeed, federal flood plain management efforts are based largely on incentives as opposed to direct controls, and those approaches arguably have done more to promote development than to protect floodplains via structural controls, subsidized federal insurance, and locally-adopted and enforced development restrictions which appear to have been largely ineffective over time. Historically, the federal government's approach to floodplain management was to promote and protect human intrusion into floodplains by building dams and levees designed in part to hold back spring floods; and by channelizing, straightening, and dredging waterways in an effort to speed water through the system rather than spreading across natural flood plains (Adler, 1994). The very structural flood controls engineered to protect human development, of course, wrought some of the most serious damage to the integrity of aquatic ecosystems and their adjacent hybrid habitats.

Because of these problems, in 1968 Congress adopted the National Flood Insurance Program (NFIP), which provided federally-subsidized flood insurance not available from the private insurance industry, in return for the adoption and enforcement of local land use controls designed to restrict further development in coastal and riparian flood risk zones. Those subsidies promote waterside development that many would not otherwise risk absent private insurance. While program participation has been high, accompanying local control efforts have not curbed floodplain development significantly (Adler, 1994). In 1974, Congress also amended the federal Flood Control Act to promote the consideration of nonstructural approaches to flood control focused on preventing further development, relocation of existing structures, and acquisition of flood plain areas for open space, ecological, and recreational resources. Those reforms only required nonstructural controls to be "considered", however, and actually included financial preferences for the existing structural approach.

National flood control legislation and policy is built on the same foundation of private property rights as are regulatory controls. Moreover, the reluctance of federal, state, and local governments to prohibit further floodplain development is a continued testament to the prevailing notion that landowners have a presumptive right to build in those areas, and that government interference with those rights should be disfavored.

Property law solutions

To the extent that unbridled private ownership causes continued degradation of aquatic ecosystems and their associated transition zones, presumably public ownership can provide some degree of protection. In many aquatic ecosystems, however, a substantial percentage of the waterfront is privately owned, and public purchase, by eminent domain or otherwise, is expensive. Moreover, even a moderate amount of private ownership can cause significant harm, or pose barriers to ecosystem-based restoration efforts.

Two related common law doctrines, however, constrain the extent to which the government can cede to private interests commonly-owned resources and values within the affected areas. These doctrines are the public trust doctrine and the navigation servitude with respect to state- and federally-owned shore lands and submerged lands. These doctrines, though, have their own limitations for purposes of protecting the common and ecological resources associated with those lands. While the federal navigation servitude serves largely to protect commercial and other interests in navigability, the public trust doctrine protects a broader range of common interests and values (Phillips Petroleum, 1988).

The public trust doctrine has its origins in Roman law, and was embedded in modified form in Anglo-American common law. Under the original doctrine, resources such as air, water, the ocean, and the shores of the sea cannot be owned by individuals, but instead are held for the common benefit of the public at large (Wilkinson, 1989; Lazarus, 1986; Johnson, 1989). At least some commentators, however, read the original Roman law as suggesting that some common resources cannot be owned *at all* (Bean and Rowland, 1997; Rieser, 1991). As adopted in England, the trust doctrine entailed ownership by the sovereign on behalf of the people in common, and restricted the ability of the Crown to grant those trust resources to private individuals. The trust concept imposed on the government a duty to manage and protect those resources for the common purposes of commerce, navigation, and fishing.

In the seminal U.S. public trust doctrine case, the U.S. Supreme Court held that the Illinois legislature and other government trustees have only limited discretion to dispose of public trust resources, and may not make a disposition that is fundamentally inconsistent with the purposes of the trust (Illinois Central R.R. Co., 1892).

However, the public trust doctrine is constrained in its ability to prevent the impairment of aquatic ecosystems and transition zones. The geographic scope of the doctrine does not always reach into the critical transition zone in which most damaging private development occurs, and varies from state to state (Shively, 1894; Philips Petroleum, 1988; Archer and Stone, 1995). The doctrine could be useful when private owners attempt to fill large areas that were formerly below the high water mark, but not necessarily with respect to riparian wetlands, floodplains, and similar transition zones. And the original doctrine was limited to the common law triad of commerce, navigation, and fisheries. Additionally, the public trust doctrine is not self-enforcing, but rather depends on discretionary decisions by state agencies, subject to limited judicial review (National Audubon Society, 1983; Arnold, 2004).

Beginning in the early 1970s, however, the public trust doctrine was revitalized in an effort to provide a common law basis for broader environmental protection. Professor Joseph Sax argued for the renovation and expansion of the doctrine to provide a legal right to vindicate commonly-held expectations in environmental values (Sax, 1970; Sax, 1980). Others argue for extension of the doctrine to include environmental issues such as over-appropriation of water, wildlife, biodiversity, wetlands and coastal habitats, water pollution, and public lands (Blumm and Schwartz, 1995; Archer and Stone, 1995; Johnson, 1989; Rieser, 1991; Meyers, 1989; Wilkinson, 1980). Some state courts expanded the purposes of the trust to include a wide range of other public and environmental values (Marks, 1971; Just, 1973; National Audubon Society, 1983). Some courts also expanded the doctrine geographically, to include land-water transition zones such as non-navigable tributaries, all waters usable for public recreation, and dry sand beaches (Phillips Petroleum, 1988; Van Naess, 1978; Paepke, 1970; Gould, 1966). As of 1991, at least 39 states had adopted and applied the doctrine in some way (Adler, 2005).

Despite these conceptual advances, however, the public trust doctrine has not been an adequate solution to the ongoing problem of waterfront building and other activities that continue to impair aquatic ecosystems. Each state has the authority to define the scope of the trust and its attendant duties as it sees fit (Phillips Petroleum, 1988). On a national scale the doctrine remains limited to issues of public access and navigability, and some courts have applied the doctrine to *sanction* economic development activities at the water's edge, such as transportation systems, public utilities, oil production, and urban and commercial expansion (Lazarus, 1986). Moreover, as with the statutory applications and solutions discussed above, some argue that expanding the doctrine beyond its traditional reach violates private property rights and other constitutional limitations (Huffman, 1987).

Constitutional barriers

The underlying problem with both regulatory programs (such as the CWA) and property-based doctrines (such as the public trust doctrine), is that they presume a background of private ownership of even the aquatic component of land parcels. Statutory jurisdiction is defined by reference to traditional notions of navigability rather than ecology, as confirmed recently by the Supreme Court (Rapanos, 2006). From both a legal and political perspective, agencies implementing those laws will tread lightly on perceived property rights and in so doing, will not fully address the ecological harms of development at the water's edge. And when agencies seek to exert their authority more broadly to protect aquatic ecosystem resources, courts will be faced with takings claims based on the same perception of private property rights, although judicial reaction to such challenges has been mixed (Loveladies Harbor, 1994; Tahoe-Sierra Preservation Council, 2002).

Several U.S Supreme Court cases in particular highlight the potential barriers to hydrological sustainability posed by constitutionally-based principles of takings law. In one case, the Court limited government ability to impose public easements through private beachfront property (Nollan, 1987). In another, the Court required the state to pay compensation when its state coastal zone program restricted development on coastal property beaches to prevent erosion and other adverse impacts (Lucas, 1992). Most relevant to urban hydrologic sustainability, in another case the Court restricted government's ability to reduce stormwater impacts on an adjacent stream by limiting the degree to which a private landowner could increase impervious surface on her property, absent a clear showing that the government regulation was proportionate to the impacts of the proposed development, a difficult technical analysis (Dolan, 1994).

Jurisprudence in this area is extremely context-specific, however, and in other recent decisions courts have supported efforts to limit waterfront development for purposes of environmental protection, including elements of hydrological sustainability (Tahoe-Sierra Preservation Council, Inc. v. Tahoe Regional Planning Comm'n, 2002; Pallazzolo, 2001). At best, however, the current state of takings and related legal doctrines is such that the federal, state, and local governments act under a cloud of constitutional uncertainty when planning and implementing programs to achieve hydrological sustainability.

RECOMMENDATIONS AND FUTURE RESEARCH

I propose two very different possible ways to improve governmental ability to achieve hydrological sustainability given the legal barriers discussed above. One would embrace the private property model as a way to avoid public-private conflicts by providing economic incentives to landowners to reduce the hydrological impacts of activities on their property. The second would change the background law of property in a way that more properly recognizes public rights within the critical transition zone between land and water.

A potential economic solution

The proposed economic solution could be patterned on a competitive bidding process used in the Colorado River Basin Salinity Control Program (CRBSCP) to reduce anthropogenic sources of salinity (measured as total suspended solids, or TSS) in the Colorado River. The program is designed primarily to meet dual regulatory objectives: (1) international treaty obligations with Mexico regarding the quality and quantity of Colorado River water at the international boundary; and (2) interstate water quality criteria for salinity adopted by the seven basin states and approved by EPA pursuant to the CWA (U.S. Department of the Interior, 2005; U.S Department of the Interior, 2001; Adler and Straube, 2000).

The salinity program evolved considerably over time. From roughly 1974 to 1984, the program consisted mainly of large program of public works projects identified by the U.S. Bureau of Reclamation (BOR) and approved by Congress on a project-specific basis, combined with limited changes in on-farm management practices and other land use changes. These programs made some progress in implementing salinity control programs. However, they were expensive and ignored other, potentially more cost-effective means of controlling salinity through changes in land and water use and management practices. The original salinity control program focused on large, capital-intensive off-farm construction programs rather than changes in on-farm water use and management practices that would reduce irrigation of saline soils (Environmental Defense Fund, 1981).

By the end of federal fiscal year 1994, the federal government had spent $362 million on salinity control, and plans were in place to spend an additional $430 million. The results of this massive investment were mixed. The salinity standards reportedly were met consistently since 1974, and continued compliance was projected for another decade and a half assuming implementation of then-planned efforts. Monitoring of selected salinity control projects suggested that the types of controls that were being implemented were succeeding in reducing salinity inputs in the basin. However, BOR also predicted that the standards would be violated by 2000 without funding for additional control measures.

Moreover, studies questioned whether the program provided sufficient flexibility to provide the most salinity control at the lowest cost (U.S. General Accounting Office, 1995; U.S Department of the Interior, 1993; U.S. Department of the Interior, 1994). The estimated cost-effectiveness of BOR salinity projects at the time ranged from a low of $5 per ton of salt removed per year (tpy) to $138/ton, with most projects ranging from $35 to $102/ton. By that time, BOR controls had reduced salinity inputs by an estimated 341,000 tpy. USDA costs were generally lower and more consistent, ranging from $29/ton to $70/ton, with total estimated salt savings of 191,000 tpy.

Given estimates of salinity damages of between $750 million and $1.5 billion per year (or $340/ton at the $750 million estimate), these expenditures clearly seemed justified from a cost-benefit perspective (Lohman and Milliken, 1988). The wide range of cost-effectiveness, however, suggested that the same benefits might be achieved at lower cost. BOR concluded that the program should consider alternatives to purely government-planned projects; allow non-federal project construction; consider proposals anywhere in the basin rather than within specific, congressionally-approved "units"; consider non-traditional proposed methods of salinity control; be competitive after considering both estimated cost-effectiveness and performance risk; and continue to be voluntary (non-regulatory) (Bureau of Reclamation, 1994).

In 1995, Congress amended the statute to provide for a basinwide salinity control program under which BOR could invite any party, public, private or mixed, to bid for salinity control funding (Pub. L. 104-20, 1995). Congress also established cost-effectiveness as a program criterion by directing BOR and USDA to give preference to projects with the "least cost per unit of salinity reductions." This mandate recognized the fundamental reality that the salinity "problem" derived from a huge number of individual sources, not all of which could be addressed with limited available resources. Therefore, absent mandatory regulation of all salinity inputs, "targeting" those control opportunities where the most salt could be controlled for the fewest dollars was essential to long-range program success.

BOR and the basin states implement this authority through an open competitive bidding process under which the most cost-effective salinity control projects are selected for funding on an annual basis. Every year, BOR issues a request for proposals for projects to reduce salinity from any source in the basin above the Hoover Dam, point or nonpoint, public or private. Controls are negotiated by the parties on a flexible, case-by-case basis (grants, contracts, memoranda of agreement, cooperative agreements, etc.). Each proposal must identify the specific salinity control methods to be used; a project management plan and schedule (how, when and by whom); annual projected salt load reductions supported either using BOR's approved analysis or other documented methods; the expected project life; the proposed payment method and schedule; an evaluation of environmental impacts and proposed mitigation measures; proposed maximum project costs; and a risk analysis characterizing the likelihood of project success. A ranking committee composed of representatives of both BOR and the forum states evaluate and rank the projects, and forward successful projects to BOR contract personnel to negotiate final project terms. The committee selects projects based on a combination of cost effectiveness and risk, including an assessment of uncertainties in cost and salt reductions.

The new program has been phenomenally successful. When the program was initially designed, BOR expected the average cost-effectiveness of controls to be $50/ton, and indicated that proposals above $100/ton were "not likely to be competitive." (Bureau of Reclamation, 1996). In fact, the costs of the projects selected to date have been far lower than expected. Four years into the new program, selected projects ranged between $11 and $36 per ton of salinity control, and averaged $26 per ton, less than half of the average cost-effectiveness of projects under the old program ($70/ton). Moreover, while program officials expected that bids for new projects probably would begin to increase as the most cost-effective available projects were "skimmed" off the top, project bids remained low or even declined for several years (Bureau of Reclamation, 2001). Costs have increased somewhat since then, at an average of $42/ton in 2005, but remain below those in the prior program (U.S. Department of the Interior, 2005).

The competitive bidding process may provide a model for improving the cost-effectiveness of other watershed protection efforts, including those designed to achieve urban hydrological sustainability. Like the salinity program, many and diverse sources contribute to urban hydrological sustainability problems, and many of those occur on private land. I do not suggest that national, state or local regulatory efforts to reduce or mitigate those impacts should be abandoned; indeed, they should be strengthened in many ways that are beyond the scope of this analysis.

However, large public investments are being made to implement stormwater and other programs around the country, with questionable success. It is possible that stormwater management agencies and other public entities could employ competitive bidding approaches to obtain more cost-effective solutions. Private landowners, for example, could bid to sell conservation easements (providing for reduced impervious surfaces, for example), buffer strips, or structural controls on their properties. This might improve the location and hence effectiveness of stormwater controls currently installed on public lands.

Using a competitive bidding process for urban hydrological sustainability will be more complicated than the CRBSCP in several ways. The salinity program focuses on a single metric (TSS measured in tons of salt removed per year), while stormwater and other urban hydrological sustainability efforts must address multiple chemical and hydrological factors. While it is possible to estimate within reasonable bounds the amount of salt reductions likely to be achieved through irrigation improvements and other strategies, those estimates might be more difficult for urban hydrological sustainability. Those kinds of issues are the subject of additional research before the competitive bidding model could be used in this area. If those questions can be resolved adequately, however, competitive bidding in conjunction with existing regulatory controls might provide another tool to achieve urban hydrological sustainability without generating even more severe conflicts between private property rights and public resources and values, and more constitutional battles over the validity of governmental actions in this area.

A potential legal solution

Another possible solution to the existing conflict between private property rights and public values in aquatic transition zones would reject rather than embrace the supremacy of private property rights in the land-water transition zone which is so important to achieving urban hydrological sustainability. It would establish that private parties cannot "own" and use those components of property at the water's edge in ways that fundamentally impair hydrological sustainability and related public resources and values.

In 1979, in overturning a state statute which prohibited the commercial export of minnows taken from Oklahoma waters, the Supreme Court ruled that states cannot "own" wildlife, even for purposes of legitimate regulation (Hughes, 1979). While states have authority to regulate wildlife for legitimate purposes, that power is based on inherent sovereign authority to protect common resources and the common welfare. It is not based on ownership.

The non-ownership principle conforms with modern notions of environmental ethics (Sax, 1993). Nonhuman components of the natural world are not merely resources for human use and consumption, but have their own intrinsic value (Leopold, 1949; Arnold, 2002; Freyfogle, 1996). At least since the early 1970s, commentators began to propose legal rights for nonhuman species (Stone, 1972). If a component of nature enjoys its own legal rights, it is axiomatic that it cannot be "owned" by anyone or anything else. The idea that wildlife cannot be owned also makes sense in light of our growing realization that species provide "ecosystem services" that transcend the confines of our existing market economy (Daily, 1997). Ecosystem services typically are optimized when the ecosystem is left alone, not when it is altered through human activity for purposes of generating wealth as traditionally defined in modern human economies.

The "non-ownership" doctrine developed in the context of wildlife may be useful to the issues of hydrological sustainability and the health of aquatic ecosystems. Application of the non-ownership doctrine fits the aquatic transition zone because it can be molded to the transitory nature of the ecosystem in question. Rather than deciding that ownership is an all-or-nothing proposition, with the private party owning all on one side of the artificially-defined boundary (the "dry" side), and the government owning all on the other ("wet") side, the "non-ownership" doctrine is amenable to a sliding scale in which the relative degree of ownership rights shifts along with the temporal and geographic transition from water to land.

The realization that various values, services, and components of aquatic ecosystems cannot be "owned" does not render private title in the land-water transition zone without any value. It does, however, limit the degree to which owners have a property right to develop in ways that would infringe on those aquatic ecosystem values. Geographically, the zone of private title and the area within which aquatic ecosystem services exist which cannot be owned will necessarily overlap. The landowner maintains the right to use property for the traditional purposes for which land is put, so long as the aquatic ecosystem values that cannot be owned are not substantially impaired. Government will continue to have a role to play in making those determinations, either through legislative and regulatory or judicial actions.

Likewise, the non-ownership concept will affect the manner in which government agencies and the courts interpret and administer existing regulatory statutes designed to protect aquatic resources (like the CWA and the ESA), and in which legislatures enact such statutes in the future. In particular, the doctrine will alter the interaction between the regulatory regime and takings provisions of the federal and state constitutions. Traditionally, when governments impose restrictions on the use and development of private land in order to protect aquatic environmental resources, landowners will assert constitutional takings claims. To the extent that resources are not subject to private ownership, by definition they cannot be "taken" by a government acting in its capacity as legal guardian of those resources. The landowner has a legal right to use the resources properly included in private title, but only so long as substantial impairment of the resources that are *not* subject to private ownership does not result. In short, one cannot assert a "taking" with respect to property "rights" that are inherently incapable of "ownership" (Stevens, 1992)

There are several ways in which the non-ownership doctrine as applied to aquatic ecosystem resources and values differs from the existing public trust doctrine, and is likely to be a superior tool to protect those resources and values. First, the public trust doctrine is constrained by artificial geographic boundaries. The non-ownership doctrine is defined not by artificial geographic limits, but by actual determinations of the degree to which aquatic ecosystem values and services exist. Second, rather than requiring the government to prove that it owns or otherwise controls a resource under the public trust doctrine in order to justify protection, a landowner presumptively has no rights to impair ecosystem components, values, or services in a significant way.

Application of the public trust doctrine also relies heavily on a subjective set of judgments about what advances public trust values and how those values should be balanced against other resources and values, both public and private. The "non-ownership" doctrine will also require sometimes difficult case-by-case judgments. The core governing principle of "non-ownership", however, is amenable to a far greater degree of uniformity. Once it is recognized that private property rights do not include the right to destroy or degrade aquatic ecosystem resources, the role of government as guardian of those resources is less open to the type of discretion that characterizes the public trust doctrine.

Finally, as I have suggested, adoption of the non-ownership principle in the context of the constitutional law of takings could free regulatory programs under statutes such as the CWA and state and local zoning laws from the lurking shadow of the takings threat, and provide much more latitude to protect aquatic transition zone resources from the continuing onslaught of development at the water's edge.

REFERENCES

Abell, R.A., et al. (2000). *Freshwater Ecoregions of North America, A Conservation Assessment*. Island Press, Covelo, CA.

Adler, R.W. (2005). The Law at the Water's Edge: Limits to "Ownership" of Aquatic Ecosystems. In Craig, A.A., ed. *Wet Growth: Should Water Law Control Land Use?* Washington, D.C.: Environmental Law Inst.

Adler, R.W., et al. (1993). *The Clean Water Act: 20 Years Later 93*. Island Press, Covelo, CA.

Adler, R.W. (1994) Addressing Barriers to Watershed Protection, *25 Envtl. L.* 973.

Adler, R.W. (2003). The Two Lost Books in the Water Quality Trilogy: The Elusive Objectives of Physical and Biological Integrity, *33 Envtl. L.* 29.

Adler, R.W. and Straube, M. (2000). Lessons from Large Watershed Programs, Learning from Innovations in Environmental Protection, Research Paper Number 10, National Academy of Public Administration, Washington, D.C.

Arnold, C.A. (2004). Working Out and Environmental Ethic: Anniversary Lessons from Mono Lake, 4 *Wyo. L. Rev.* 1.

Arnold, C.A. (2002). The Reconstitution of Property: Property as a Web of Interests, *26 Harv. Envtl. L.* Rev. 281.

Archer, J.H. and Stone, T.W. (1995). The Interaction of the Public Trust and the "Takings" Doctrines: Protecting Wetlands and Critical Coastal Areas, *20 Vt. L. Rev.* 81 (1995).

Bean, M.J. and Rowland, M.J. (1997). *The Evolution of National Wildlife Law* (3d ed.). Praeger, Westport, CT.

Blumm, M.C. and Schwartz, T. (1995). Mono Lake and the Evolving Public Trust in Western Water, *37 Ariz. L. Rev.* 701.

Clean Water Act, 33 U.S.C. §1251 *et seq.* (2000).

Dahl, T.A. (2000). *Status and Trends of Wetlands in the Coterminous United States,* 1986 to 1997. U.S. Fish & Wildlife Service. Washington, DC.

Daily, G.C., ed. (1997). *Nature's Services: Societal Dependence on Natural Ecosystems.* Island Press, Covelo, CA.

Dolan v. City of Tigard, 512 U.S. 374 (1994).

Environmental Defense Fund v. Costle, 657 F.2d 275 (D.C. Cir. 1981).

Environmental Law Institute (1997). Enforceable State Mechanisms for the Control of Nonpoint Source Water Pollution., Washington, DC.

Florida Rock Industries, Inc. v. U.S., 45 Fed. Cl. 21 (1999).

Freyfogle, E.T. (1993). Ownership and Ecology, 43 Case W. L. Res. Rev. 1269.

Freyfogle, E.T. Freyfogle (2002). Community and the Market in Modern American Property Law, in John, F. Richards ed. Land, Property, and the Environment. ICS Press, Oakland CA.

Gould v. Greylock Reservation Comm'n, 215 N.E.2d 114 (1966).

Hughes v. Oklahoma, 441 U.S. 322 (1979).

Illinois Central R.R. Co. v. Illinois, 146 U.S. 387 (1892).

Johnson, R.W. (1989). Water Pollution and the Public Trust Doctrine, 19 Envtl. L. 485, 490.

Just v. Marinette County, 201 N.W.2d 761 (Wis. 1972).

Karp, J.P. (1989). Aldo Leopold's Land Ethic: Is an Ecological Conscience Evolving in Land Development Law?, 19 Envtl. L. 737, 742 (1989).

James R. Karr and Isaac J. Schlosser (1978). Water Resources and the Land-Water Interface, 201 Science 201:229.

Keller, B. (2005). What We Always Knew: Wetlands Win Hands Down at Pollution Mitigation. National Wetlands Newsletter, September-October 2005:12–14.

Lazarur, R.J. (1986). Changing Conceptions of Property and Sovereignty in Natural Resources: Questioning the Public Trust Doctrine, 71 Iowa L. Rev.

Leopold, Aldo (1949). A Sand County Almanac. Oxford University Press, New York.

Lohman, L.C. and J.G. Milliken, et al. (1988). Estimating Economic Impacts of Salinity of the Colorado River. Milliken Chapman Research Group for the U.S. Department of the Interior, Bureau of Reclamation, Denver, CO.

Loveladies Harbor, Inc. v. United States, 28 F.3d 1171 (Fed. Cir. 1994).

Lucas v. South Carolina Coastal Council, 505 U.S. 1003 (1992).

Marks v. Whitney, 491 P.2d 374 (Cal. 1971).

Meyers, G.D. (1989). Variation on a Theme: Expanding the Public Trust Doctrine to Include Protection of Wildlife, 19 Envtl. L. 723.

National Audubon Society v. Superior Court, 658 P.2d 709 (Cal. 1983).

National Flood Insurance Act, Pub. L. No. 90-448, tit. XIII, §1302, 82 Stat. 572 (1968), codified as amended at 42 U.S.C. §§4001-28.

National Research Council (2002). Riparian Areas: Functions and Strategies for Management. Washington, D.C.: National Academy Press, Washington, DC.

National Research Council (1999). New Strategies for America's Watersheds. Washington, D.C.: National Academy Press, Washington, DC.

National Research Council (1995). Wetlands: Characteristics and Boundaries. Washington, D.C.: National Academy Press, Washington, DC.

Nollan v. California Coastal Comm'n., 483 U.S. 825 (1987).

Paepke v. Public Building Comm'n., 263 N.E.2d 11 (Ill. 1970).

Palazzolo v. Rhode Island, 553 U.S. 606 (2001).

Phillips Petroleum Co. v. Mississippi, 484 U.S. 469 (1988).

Pub. L. 104-20, 109 Stat. 255, July 28, 1995.

Rapanos v. United States, 126 S. Ct. 2208 (2006).

Reiser, A. (1991). Ecological Preservation as a Public Property Right: An Emerging Doctrine in Search of a Theory, 15 Harv. Envtl. L. Rev. 393 (1991).

Rose, C.M. (1994). Given-ness and Gift: Property and the Quest for Environmental Ethics, 24 Envtl. L.

Sax, J.L. (1993). Property Rights and the Economy of Nature: Understanding Lucas v. South Carolina Coastal Council, 45 Stan. L. Rev. 1433.

Sax, J.L. Sax (1989). The Limits of Private Rights in Public Waters, 19 Envtl. L. 473 (1989).

Sax, J.L. et al. (2000). *Legal Control of Water Resources*, 3d ed. 473–479. West Group, St. Paul. MN.

Sax, J.L. (1970). The Public Trust Doctrine in Natural Resources Law: Effective Judicial Intervention, *68 Mich. L. Rev.* 471 (1970).

Sax, J.K. (1980). Liberating the Public Trust Doctrine from Its Historical Shackles, 14 U.C. *Davis L. Rev.* 185.

Shively v. Bowlby, 152 U.S. 1 (1894).

Solid Waste Ass'n of Northern Cook County v. U.S. Army Corps of Engineers, 531 U.S. 159 (2001).

Stevens v. Cannon Beach, 835 P.2d 940, 942 (Ore. App. 1992).

Stone, C. (1972). Should Trees Have Standing? Toward Legal Rights for Natural Objects, 45 S. Cal. L. Rev. 450 (1972).

Tahoe-Sierra Preservation Council, Inc. v. Tahoe Regional Planning Comm'n, 122 S.Ct. 1465 (2002).

U.S. Bureau of Reclamation (1996). *Colorado River Basin Salinity Control Program*, Report to Congress on the Bureau of Reclamation Basinwide Program. Salt Lake City, UT.

U.S. Bureau of Reclamation (1994). Colorado River Basin Salinity Control Act, Report on Public and Agency Review of the Program and Suggested Revisions to the Program. Salt Lake City, UT.

U.S. Department of the Interior (2005). Quality of Water, Colorado River Basin, Progress Report No. 22., Salt Lake City, UT.

U.S. Department of the Interior (2001). Quality of Water, Colorado River Basin, Progress Report No. 20, Salt Lake City, UT.

U.S. Department of the Interior (1993). Audit Report 93-I-810, Implementation of the Colorado River Basin Salinity Control Program, Washington, D.C.

U.S. Department of the Interior (1992). Audit Report 93-I-258, Operation and Maintenance Contracts, Colorado River Basin Salinity Control Program, Bureau of Reclamation, Washington, D.C.

U.S. Environmental Protection Agency (2000). Liquid Assets 2000: America's Water Resources at a Turning Point, EPA-840-B-00-001, Washington, D.C.

U.S. Fish and Wildlife Service (2002). Geographically Isolated Wetlands: A Preliminary Assessment of their Characteristics and Status in Selected Areas of the United States, Washington, D.C.

U.S. General Accounting Office (1995). Water Quality, Information on Salinity Control Projects in the Colorado River Basin. GAO/RCED-95-98, Washington, DC.

United States v. Riverside Bayview Homes, 474 U.S. 121 (1985).

Van Ness v. Borough of Deal, 393 A.2d 571 (1978).

Wilkinson, C.F. (1989). The Headwaters of the Public Trust: Some Thoughts on the Sources and Scope of the Traditional Doctrine, 19 Envtl. L. 425 (1989).

Wilkinson, C.F. (1980). The Public Trust Doctrine in Public Land Law, *14 U.C. Davis L. Rev.* 269.

Worster, D. (1985). Rivers of Empire 19-21, Oxford University Press, New York, Oxford.

Zaring, D. (1996). Agriculture, Nonpoint Source Pollution, and Regulatory Control: The Clean Water Act's Bleak and Present Future, *20 Harv. Envtl. L. Rev.* 515.

23

Ecosystem resilience and institutional change: the evolving role of public water suppliers

L.P. Breckenridge

Northeastern University School of Law, 400 Huntington Ave., Boston, MA 02115, USA
E-mail: l.breckenridge@neu.edu

Summary: Water and sewer organizations serving large cities have long played key roles in managing water resources to protect public health and support economic development. Increasingly, they also act to protect broader concepts of public welfare, minimizing environmental repercussions and seeking to maintain the long-term viability of resilient aquatic ecosystems. The Boston metropolitan water supply system offers an informative history. This paper focuses on the evolving roles of the agencies responsible for the water supply, highlighting issues that affect the sustainable management of water resources in urban settings. The paper describes changes in the legal framework that have shaped the increasing attention to ecological goals, and it considers the policy challenges that lie ahead as the Massachusetts Water Resources Authority seeks to capitalize on successful conservation measures by allocating water resources to new communities in an expanded service area.

INTRODUCTION

The Massachusetts Water Resources Authority (MWRA) is an independent public authority created by the Massachusetts legislature in 1984. It provides wholesale water and sewer services to dozens of communities in eastern Massachusetts, primarily in the greater Boston metropolitan area. Its most important responsibilities include interbasin transfers of water from one of the great water supply projects of the United States, the Quabbin Reservoir, located many miles to the west of Boston in the Connecticut River watershed.

From the early days of Boston's water system, the responsible government authorities sought to ensure a reliable and plentiful supply of clean water to promote the health of city residents and foster economic growth. The creation and expansion of the modern-day system involved a series of far-sighted legislative actions and engineering accomplishments that extended the health and economic benefits of a bountiful drinking water supply to growing numbers of urban residents. The MWRA now provides drinking water for approximately 2.2 million people.

While the engineering achievements of the past are widely recognized, recent events have highlighted emerging controversies over policies to govern future operations. The MWRA has announced that it is developing plans to expand its service area to include new communities, given the availability of an "excess" water supply capacity. The perceived water surplus has resulted from highly successful measures to conserve water and enhance the efficiency of MWRA operations, implemented pursuant to laws and policies that encouraged a multi-faceted approach to water management, using a long planning horizon.

The MWRA's water savings have given rise to questions about how the agency ought to manage available water supplies in the future, in light of increasingly complex notions of the public interest. Demands for more economic growth conflict with modern scientific understandings of the needs of aquatic ecosystems for instream flows. The potential uses for water currently within the MWRA's purview include the protection and restoration of aquatic ecosystems, not simply the protection of human health and the short-term enhancement of human welfare.

This paper takes a look at the policy developments that have shaped the role of the MWRA and its predecessor agencies in managing water to serve public goals. It traces some key changes in the legal framework that have provided the institutional backdrop for the evolving managerial role, using the example of the greater Boston water supply system to highlight broader issues affecting the sustainable management of water resources in urban settings. It proceeds to consider how new understandings about the importance of integrating human and ecological needs are affecting the potential roles of public water organizations serving cities.

EVOLVING GOALS AND RESPONSIBILITIES OF URBAN WATER SYSTEMS: THE METROPOLITAN BOSTON EXAMPLE

The history of the metropolitan Boston water supply system illustrates three types of important advances in sustainable resource management. First, the acquisition and protection of land and water rights on a large scale reflected legislative attention to water resource development as a major public health and economic issue that needed to be addressed on a regional basis through centralized governmental authority. Second, increasingly sophisticated efforts to integrate land and water management through multi-factor planning and regulatory controls have reflected a growing attention to synergistic interactions between human uses of water resources and ecosystem functions. Third, successful efforts to identify new means for conserving water and eliminating wasteful uses have manifested a commitment to efficient operation and an ever more sophisticated understanding of economic and ecological tradeoffs.

As discussed in more detail below, these advances have been shaped by legal frameworks governing water rights and establishing agency authority and duties to serve the public interest. New challenges lie ahead. Current debates focus on the future role of the water authority in providing water to meet ecological as well as socioeconomic needs, and in determining how to integrate the operation of the urban water system with modern understandings of the instream flow regimes necessary to support resilient aquatic ecosystems.

Acquiring water and land to protect public health and support economic development

As Boston grew, the Massachusetts government met increasing demands for urban water supplies by acquiring watershed lands and water rights at ever greater distances from the city. By the end of the nineteenth century, the state legislature had authorized the construction of the Wachusett Reservoir, a river impoundment which was, at the time it was completed, the largest water supply reservoir in the world. (Nesson, 1983). Despite the size of the project, the urban demand for water again outgrew the available supplies. By the 1920's, the Massachusetts legislature had passed new laws authorizing diversions of waters even further west, in the Connecticut River watershed. The westward expansion of the system culminated in the construction of the Quabbin Reservoir, a project that entirely inundated four towns in western Massachusetts.

The water impoundment and diversion projects of the early twentieth century reflected a resource development policy focused on sustainable deliveries of a single valuable product – clean drinking water (Adler, 2002). Recognizing human dependence on the environment, these projects were based on long-range planning and expert technical investigations. They sought to conserve and use renewable natural resources to provide for the health and welfare of a large urban population well into the future. In this regard they were a remarkable success. Reservoirs located away from urban development had successfully protected Boston from epidemic water-borne diseases in the nineteenth century, and the Quabbin Reservoir system replicated that basic approach on a grander scale (Nesson, 1983).

While there have been subsequent technological enhancements, and several legislative changes in the jurisdictional authority of the responsible government agencies, the major projects and basic configuration of the water supply system remain in place today. The Massachusetts Water Resources Authority continues to import unfiltered water from the relatively undeveloped lands to the west, serving an expanded number of communities in the greater Boston area.

The legal framework enabling water resource development rested on the acquisition and protection of property rights for public use. It relied on sovereign powers of ownership to exclude incompatible private land and water uses, and to manage and transfer the resources to serve well-defined purposes of public health and economic development.

At the time that the Quabbin Reservoir and related projects were constructed, Article 49 of the Amendments to the Massachusetts Constitution authorized the legislature to acquire private rights to land and water by eminent domain in the following terms:

> The conservation, development and utilization of the agricultural, mineral, forest, water and other natural resources of the commonwealth are public uses, and the general court shall have power to provide for the taking, upon payment of just compensation therefor, of lands and easements or interests therein, including water and mineral rights, for the purpose of securing and promoting the proper conservation, development, utilization and control thereof and to enact legislation necessary or expedient therefor.

This constitutional language, alluding to the "conservation, development and utilization" of resources as recognized public uses, emphasized the role of government in exploiting renewable natural resources to serve public interests. The concept of public control of natural resources expressed here implicitly encourages rather than discourages economic development.

The Quabbin water supply system is premised on a rigorous separation of human uses of the environment – effectively, a specialization of zones in the landscape defined by property boundaries. The proponents of the Quabbin Reservoir explicitly sought to avoid filtration of the water supply, as well as any need to constrain the growth of industry in urban areas. Thus, on the one hand, the government authorized the exercise of eminent domain to take land and water rights in the donor watersheds, effectively putting those resources off limits for other economic development. At the same time, the project effectively supported economic development elsewhere in an unquestioning manner by making plentiful water supplies available.

This segregation of uses through exercise of proprietary powers was an approach that did not involve integration of watershed protection goals with industrial or other urban activities through refined regulation or ongoing adjustment of competing uses in a shared location. On the contrary, these land use arrangements envisioned autonomous action by both public and private actors, separated by well-defined geographical boundaries.

In the watershed lands, the Quabbin project displaced private rights to water as well as rights to ownership of land. In Massachusetts, as under the common law of other eastern states in the United States, the common law system of riparian rights defined private rights to water use and also provided a means for adjusting activities of landowners along watercourses to enhance communal interests

(Breckenridge, 2005). U.S. courts, in case-by-case decision-making, developed legal principles governing the "reasonable use" of water. The courts' decisions allowed some pollution and consumptive uses of water, but limited the total use of waterways and encouraged sequential reuses of water by designated users (the riparians or riverbank landowners), thus ensuring that water bodies would continue to sustain multiple enterprises over time (Rose, 1990).

In short, the common-law system of riparian water rights aimed to adjust multiple private uses of shared resources in order to enhance the overall welfare of a community. But it did not provide for large-scale management of water or interbasin transfers to serve regional public interests, nor did it in practice ensure that water quality would remain sufficiently unimpaired for drinking water purposes (Adler, 2002). When Massachusetts exercised eminent domain powers to take riparian rights in the Connecticut River watershed, its actions thus displaced a smaller-scale form of communal resource management in favour of more centralized expert administration.

The decisions to centralize ownership in the hands of state agencies represented an implicit judgment about the appropriate scale of resource management. Today, we might see these actions as ecosystem management decisions, adjusting the scale of control to fit the geographic scope of the primary management purpose at that time – providing safe water to an urban population.

The water impoundments and diversions had repercussions not just for private landowners but also for other sovereign authorities. The diversion of water reduced the amount of water flowing into the Connecticut River, thus potentially affecting the United States' interests in the free navigability of the river, as well as Connecticut's interests in uses of the waterway.

The possible impairment of navigation in a nationally significant waterway meant that Massachusetts had to obtain permission for the diversions from the federal government under the Rivers and Harbors Act of 1899. The federal government ultimately approved the diversions, subject to certain seasonal conditions and requirements for water releases. Despite the federal approval, the State of Connecticut strongly opposed the export of waters from the Connecticut River watershed. In a suit asserting its interests in the Connecticut River, it asked the U.S. Supreme Court to stop Massachusetts' new water supply projects, arguing that the projects would cause injury in Connecticut by altering the amount of water in the Connecticut River and affecting the water quality. The U.S. Supreme Court, implicitly endorsing the importance of Massachusetts' goals and the wisdom of obtaining abundant and unpolluted water supplies to support Boston's urban growth, found that Connecticut had not proven enough harm to justify judicial intervention in the confrontation between the states (Connecticut v. Massachusetts). The resolution of these issues effectively reinforced Massachusetts' sovereign authority, as well as the legitimacy of its proprietary control over the water supply.

Aside from the debates over competing sovereign claims to water flows in the Connecticut River, the impacts of Massachusetts' water supply project on the environment did not receive much independent assessment. Despite the lack of attention to environmental consequences and the absence of modern-day regulatory controls, however, the project has had many significant environmental benefits quite apart from the resulting purity and quantity of the water supplies. The watershed lands have been called an "accidental" wilderness (Conuel, 1990). With human habitations and other forms of economic development eliminated, the protected landscape supports a much altered but nonetheless rich and diverse habitat for wildlife. This outcome had little to do with the original purposes of the project. It was at that time a fortuitous side-effect, rather than the product of an integrated planning process incorporating attention to wide-ranging socioeconomic and ecological concerns. Nevertheless, as explored next, the management of the system over the years has required increasing attention to the interactions between ecosystem functions and human activities.

Protecting ecosystem processes with an integrated approach to land and water

The metropolitan Boston water supply system has always involved certain efforts to integrate management of land and water resources. From the early days of planning, there has been an understanding that the public health of the urban population was dependent on environmental quality, and that maintaining an abundant, pure, drinking water supply required close attention to activities on land nearby. The system designers knew that large-scale acquisition and protection of undeveloped watershed lands might obviate the need to filter polluted water supplies, and that watershed lands, in this sense, could substitute for filtration facilities (Nesson, 1983).

Recent decisions regarding the water supply have reflected an increasingly sophisticated appreciation for integrated planning that sees human resource demands within a larger ecological context. This perspective recognizes the importance of "nature's services" – contributions of organisms and ecosystems to human welfare that may be highly beneficial even though they are not necessarily easy to value in economic terms (Daily, 1997).

The renewed attention to the interdependencies of land and water use, and the recognition of complex ecosystem dynamics in the watershed lands, have stemmed in part from federal and state laws establishing new environmental requirements. Water suppliers must now account for their actions to other administrative agencies charged with regulating environmental harms. Citizens, too, have gained expanded rights to scrutinize agency actions by pursuing or intervening in administrative and judicial proceedings.

Among the many new forms of accountability motivating new efforts at integrated land and water management are the requirements of the federal Safe Drinking Water Act, a law protecting the purity and safety of public drinking water supplies. The U.S. Environmental Protection Agency, exercising its regulatory authority, has issued a Surface Water Treatment Rule requiring all public water systems to install filtration systems unless they meet certain criteria for avoiding compliance. One such criterion for avoiding the filtration requirements is that the public water supplier must have an effective watershed control program to prevent microbiological contamination of the water supply.

Faced with EPA's filtration rule, the Massachusetts Water Resources Authority sought to avoid constructing a filtration plant by embarking on extensive new efforts to protect the water supply from contamination. Its Integrated Water Supply Improvement Program involved extensive new watershed protection measures, along with construction of ozonation facilities, a massive new water supply tunnel, and covered storage facilities.

Notably, the MWRA's program statements endorsing watershed protection measures as part of the program incorporated themes of "adaptive management" typical of the ecosystem management literature (Doremus, 2001; Karkkainen, 2003). The MWRA advocated a "flexible and adaptable" approach to managing an integrated and dynamic complex system consisting of both natural and built components (MWRA, 1998). It sought to engage in an ongoing process, using experimentation, monitoring, and adjustment to achieve incremental improvements in water quality. Although the MWRA remained singleminded in its primary focus – delivering abundant and safe urban drinking water – the means chosen reflected a newfound attention to interactions between ecosystem dynamics and human activities.

The MWRA's watershed protection plans entailed use of several different types of legal tools, as well as collaboration with other state agencies and nonprofit organizations. The plans involved new land acquisition, as well as new conservation restrictions on land remaining in private hands. They also involved regulation of private activities. Although most land use controls in Massachusetts are implemented by municipalities, the Massachusetts Watershed Protection Act created a state-run regulatory program that imposed new controls on a regional scale, extending across municipal boundaries. Unlike the direct takings of property that dominated the earlier era of water supply development, the regulatory approach to land use control under the Watershed Protection Act relies on the state's police powers to channel private activities while leaving private ownership patterns intact. The regulatory program thus envisions continuing human habitation, occurring within buffer zones that serve as natural filters and contaminant barriers.

Given the coordinated implementation of these different measures, MWRA took the position that the natural "services" provided by the landscape could be just as effective as an industrial-scale filtration plant in protecting the safety of the drinking water supply. In adopting this view, it had the support of the Massachusetts Audubon Society, watershed organizations, and other citizens groups. These groups saw the watershed protection program as an important means for preserving natural areas and maintaining fish and wildlife habitat, not simply as a guarantee of the safety of water supplies. They were concerned that if a filtration plant were built, the public's perception of ecological interconnections and societal dependence on ecosystem functions would erode, and the state's vigilance in protecting natural areas would diminish (Colburn and Hubley, 1990).

In this fashion, the MWRA's response to the strict health-based requirements of the Safe Drinking Water Act became the springboard for an ecologically-minded planning process and a new collaborative relationship with nongovernmental organizations. Even though the federal law aimed only to address the safety of the drinking water supply, the EPA's filtration avoidance criteria led the MWRA to engage in far-reaching efforts to integrate the activities of human society with the healthy functioning of ecosystems on a regional scale.

When the U.S. Environmental Protection Agency sued Massachusetts to force construction of a filtration plant, the Massachusetts Audubon Society and other environmental advocates joined Massachusetts in defending the watershed protection plans. Ultimately, Massachusetts prevailed in its arguments that natural filtration, coupled with subsequent treatment of the water by ozonation, could reliably achieve the requisite health-based standards of purity in the water supply (United States v. Massachusetts Water Resources Authority). Boston consequently remains one of the few major metropolitan areas in the United States that does not have water filtration facilities, and that relies instead on an array of ownership, regulation, and management activities to insure that watershed lands continue to provide an adequate alternative to filtration while supporting other human uses and wildlife habitat.

The decision to forego filtration and spend available funds on other aspects of system management and rehabilitation was controversial. Some health specialists advocated a more precautionary approach that might have sacrificed ecological protections in favour of more technological controls to limit risks to public health. Nevertheless, by opting to monitor and protect the natural capacities of the land, the MWRA has achieved significant success in pursuing an ecosystem-based management approach that integrates goals of human protection with other environmental objectives.

Conserving water and eliminating waste

A third major accomplishment spearheaded by the MWRA has been a remarkable reduction in the use of water through implementation of conservation and efficiency measures.

After the creation of the Quabbin Reservoir, water use grew. New member communities were added to the service area, in part because of the increasing pollution of local water supplies lying between the state's watershed lands to the west and Boston to the east. Although the population did not grow as dramatically as some had predicted, expansion of the service area, and higher per capita usage, led to an increase in total water use from 143 mgd in 1930 to 306 mgd in 1970 (Wallace, Floyd, Associates Inc., 1984). A drought in the 1960's highlighted the possibility that the system would again be unable to meet growing urban demands.

The MWRA estimates the safe yield of the water supply system to be about 300 mgd, but actual usage regularly exceeded this amount in the 1980's. (Figure 23.1). The growing demands for water, and the longstanding assumption that plentiful supplies were necessary for public health protection and economic development, led to a renewed search for additional water to augment the Quabbin system. Studies conducted in the 1980's explored alternatives for expansion that included diverting

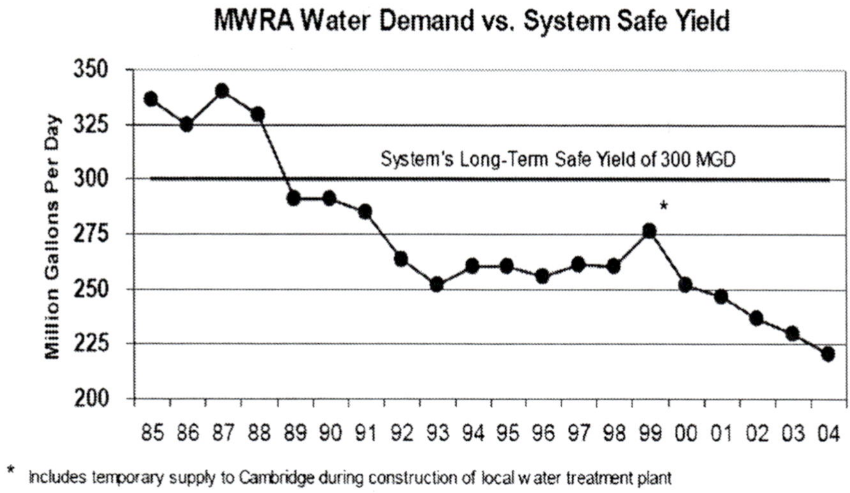

MWRA Water Demand vs. System Safe Yield

* Includes temporary supply to Cambridge during construction of local water treatment plant

Figure 23.1.   Water demand in the MWRA water supply system, 1985–2004, compared with MWRA's estimated safe yield. (Source: MWRA, 2006).

additional water from Connecticut River tributaries (Wallace, Floyd, Associates Inc., 1986).

By the 1980's much had changed in the legal frameworks governing water allocation and management. The understanding of environmental repercussions and the political landscape had changed as well. Laws aimed at ensuring attention to environmental concerns influenced governmental decision-making.

Today, a proposal to expand a water supply system triggers a variety of environmental requirements. For example, the Massachusetts Water Management Act imposes permit requirements and other statutory controls on large new consumptive water withdrawals. This law creates a "regulated riparian" statutory system of water rights similar to that of other eastern states (ASCE, 2004; Breckenridge, 2004). The state Department of Environmental Protection must consider permit applications in light of numerous public interest criteria, including environmental and economic impacts, effects on other water users, and availability of conservation measures. The law requires denial of all new permits if the safe yield of a water source will be exceeded.

For large new transfers of water, the Massachusetts Interbasin Transfer Act establishes important presumptions against diversions out of a watershed, imposing a variety of conservation and efficiency measures. The law allows exports out-of-basin only when all practical measures including metering, leak detection, and conservation of the receiving basin's water supplies have been implemented. It also requires exploration of alternatives to the project, and protection of reasonable instream flows (Breckenridge, 2005).

The enabling legislation that created the MWRA refers explicitly to goals of water conservation and efficiency. The Enabling Act directs the MWRA, wherever reasonably practicable, to implement water conservation solutions in preference to increased water withdrawals. More generally, the Massachusetts Environmental Policy Act (MEPA), a state law that resembles the National Environmental Policy Act (NEPA), also requires environmental impact assessments of state activities and evaluation of alternatives to proposed projects in environmental impact reports.

The Massachusetts Constitution, too, has been amended to express broader ideas about the scope of public interests in the environment, and the public trust responsibilities of the state government. Article 97 of the Amendments replaced terms that implicitly emphasized the government's responsibilities in fostering economic development with new provisions referring to the public's rights to the "natural qualities" of the environment:

> The people shall have the right to clean air and water, freedom from excessive and unnecessary noise, and the natural, scenic, historic, and esthetic qualities of their environment; and the protection of the people in their right to the conservation, development and utilization of the agricultural, mineral, forest, water, air and other natural resources is hereby declared to be a public purpose.

> The general court shall have the power to enact legislation necessary or expedient to protect such rights.

The combined effect of modern environmental laws has been to add mechanisms for ensuring consideration of a wide array of public interests in the allocation of water, and to highlight state agencies' stewardship responsibilities. Collectively, the procedural frameworks created through legislation have enabled closer scrutiny of environmental impacts, both by other state agencies and by members of the public. Looking beyond the safety of the water supply and economic impacts, they encourage decision-makers to consider indirect ecological repercussions.

The MWRA ultimately rejected proposals to construct new water diversion projects to meet demand. Responding to recommendations of a citizens' advisory group, and considering studies suggesting that the projected shortfall in water supply could be recouped through feasible repairs and water-saving technologies, the agency chose instead to recover needed water through extensive conservation and efficiency measures (Platt, 1995).

Water usage fell dramatically pursuant to the MWRA's demand management program (Figure 23.1). Working with its member communities, MWRA reduced average demand from a peak of 340 mgd in the 1980's to 285 mgd by 1990 and 220 mgd in 2004 (MWRA, 2000; MWRA, 2004). These reductions were achieved, first, through extensive efforts to detect and repair leaks in the MWRA's distribution system, and in the systems of member communities. In a broad outreach program, MWRA also installed more than a million water saving devices in residential locations. Rising water and sewer rates, based on MWRA's costs of constructing expensive new sewage and water treatment works, have also provided significant incentives for residents to conserve water.

Consequently, the MWRA has not pursued additional diversions and acquisitions that would have added a new westward expansionary chapter to the history of the water supply. Instead, it sought to restrict the "footprint" of urban activities on the state's water resources, in an effort to derive larger public health and economic benefits from a smaller quantity of water within the existing system. Overall, the remarkable accomplishments of the MWRA in achieving its water conservation goals reflect important changes in the perceived purposes of metropolitan water supply management, a broadened perspective on goals of managing urban infrastructure, and a successful incorporation of environmental concerns into the legal framework governing new development projects.

At the same time, as explored next, the water savings accomplished through conservation measures have given rise to new controversies, highlighting gaps and uncertainties in the legal framework, and triggering expression of differing views about the ultimate scope of the MWRA's authority and duties. The MWRA sees itself as exercising control over the "surplus" water that it has worked so hard to save. What factors govern disposition of water conserved in this fashion? Who is ultimately in charge of the decision-making, and how will the wide-ranging and conflicting views of modern public interests be resolved? The situation provides a window into future policy decisions and directions.

Allocating water to serve instream ecological needs in addition to meeting direct human demands

The success of conservation efforts over the past two decades has led MWRA to propose an expansion of its service district to include additional communities in the metropolitan area. The expansion proposal targets communities that have remained dependent on local water supplies but that face problems with water shortages and water contamination.

The MWRA has offered several justifications for its proposal (MWRA, 2006; Pioneer Institute, 2006). First, it has cited the "excess capacity" or "surplus" water in the system that has resulted from conservation measures. Estimating the long-term "safe yield" of the system at 300 mgd, and accounting for projected growth and additional needs of communities already participating in the system until the year 2030, the MWRA concludes that at least 36 mgd could be made available to additional communities without affecting current operations. Second, the MWRA has pointed to water demands in surrounding watersheds, noting that some of these watersheds are "highly stressed" as a result of depletion of groundwater resources and instream flows. Third, the MWRA has noted that as water and sewer rates have gone up dramatically in its service area, the need for new sources of revenue to offset financial burdens on residents has risen.

The proposal is motivated by several quite different concerns. On the one hand, the desire to market water to new customers reflects in part a wish to spread the costs of the service among a larger number of participants. An expansion would increase the number of ratepayers, and more importantly, it would require new communities to reimburse existing participants for a portion of their past investments in the system.

On the other hand, the proposal reflects an increasing attention to environmental benefits, and to the MWRA's potential affirmative role in restoring impaired aquatic ecosystems by delivering water services to "stressed" basins or to communities now drawing water from over-allocated water sources. The Ipswich River to the north of Boston, in particular, appears to be a candidate for such restoration efforts (Breckenridge, 2004).

The Ipswich River is a picturesque waterway, relatively unaffected by nineteenth century industrial development, known for its beauty, wildlife habitat, and canoeing and other recreational opportunities. The river has been severely depleted in recent years, however, by insufficiently controlled municipal water withdrawals. In times of drought, consumptive uses of water (including extensive use of water supplies for suburban lawn watering and irrigation of golf courses) leave the riverbed dry, killing fish and other organisms, and profoundly altering the types of aquatic life that the free-flowing stretches of the river once supported (Glennon, 2002; Breckenridge, 2004).

The impairment of the river has been so dramatic that the situation has garnered national attention. It has become a leading example of how the "water-rich" eastern states in the United States have come to face water shortages akin to those of the arid West (Glennon, 2002). The ecological degradation of the Ipswich is a manifestation of legal failures. Although water quality standards have been developed to protect the biological, chemical and physical integrity of waterways, existing federal and state laws have not in practice been interpreted or implemented to provide the instream flows that are needed for native aquatic life to survive and reproduce (Adler, 2003).

In the specific context of Massachusetts law, the Ipswich situation has focused attention on inadequacies in the Massachusetts water management system (Breckenridge, 2004). A more fully developed regulatory approach would integrate consideration of socioeconomic and ecological needs in planning for uses of water in aquatic ecosystems. It would limit calculations of water available for withdrawal so that instream flows, including the dynamics of the "flow regime" were sufficient to maintain the resilient, self-organizing patterns characteristic of the aquatic habitat. (Instream Flow Council, 2004; Holling and Gunderson, 2002). It would make allocations of water to instream uses equivalent to allocations of water for consumptive uses, putting protection of ecological needs and human consumptive needs on an equal footing (Instream Flow Council, 2004; Postel and Richter, 2003). This is the approach generally taken by the American Society of Civil Engineers' Model Regulated Riparian Water Code, which explicitly ties determinations of "safe yield" to concepts of ecological integrity and sustainable use (ASCE, 2004).

Given the uncertainties and inadequacies of current Massachusetts laws and regulations, however, the role of the MWRA, and the outcome of legal controversies, are not clear, and difficult questions are now on the table. How far must, or should, the MWRA go in assessing and balancing the importance of ecological needs? For example, should it provide communities with water that will be used for watering lawns, or is it in a position to prevent such uses of potable water, imported from another basin, as inherently wasteful? Residents of watersheds that currently export water to Boston and surrounding communities seek a return of water supplies to the stretches of river below the dams, where aquatic habitat has been impaired by water depletion and changes in water quality. Should ecological needs in the source watersheds trump the potential ecological uses of water out-of-basin? Should water that has been saved through conservation measures presumptively return to the river where the water has been withdrawn?

Such questions about who should benefit from conservation measures raise issues about stewardship and control of water resources. The concern, as repeatedly expressed in public hearings and written comments on the MWRA expansion proposal, is that the MWRA will leave the more difficult aspects of ecological planning to other regulatory agencies, each exercising fragmented jurisdiction, without affirmatively integrating ecological considerations into its own decision-making, or imposing its own conditions on water allocations to meet ecological goals. If the MWRA expands its system on a first-come first-served basis, or fails to insist on conservation measures by recipient communities, the growth of its service area might simply foster suburban sprawl, while continuing to deprive impaired rivers and streams of needed instream flows (Arnold, 2005).

A different scenario will occur if the MWRA sees itself as serving broader concepts of public welfare, rather than acting simply as a supplier of urban drinking water. The MWRA's decisions about how to reallocate water saved through conservation measures could potentially fill in some of the regulatory and policy gaps and inadequacies in the state's water management system. By adhering to aims of long-term sustainable use of renewable resources, and giving attention to ecological needs, the MWRA could bring about some important advances in integrated water management to meet the needs of both urban communities and aquatic ecosystems.

RECOMMENDATIONS AND CONCLUSIONS

The Massachusetts Water Resources Authority, asserting general control over the management of the Boston metropolitan water supply, has claimed authority to evaluate competing claims and to allocate water to new uses that include restoration of impaired aquatic ecosystems. Despite uncertainties in the scope of its duties and responsibilities, the MWRA is in a powerful position to take a leadership role in defining and reorienting water management policies to encompass a newly expanded ecological focus, attuned to coordinating human socioeconomic activities with the dynamics of ecosystem functions.

Modern scientific understandings of aquatic ecosystems include ever more precise insights into the dynamic characteristics of flow regimes. The developments in Boston's water system point the way toward a more widespread and robust notion of water infrastructure to serve the public welfare – a concept that includes water supply services to serve ecological goals as well as human health needs. This approach would place the management and allocation of water for ecological purposes on an equal footing with the allocation of off-stream consumptive water rights directly serving the needs of human populations. If the MWRA does in fact take seriously its responsibility to evaluate and serve a growing array of potential water uses, including instream uses of water to meet the needs of aquatic ecosystems, then it may bring about an important new ecologically-minded chapter in its role as the manager of a great urban water supply.

REFERENCES

Adler, R.W. (2003). The two lost books in the water quality trilogy: the elusive objectives of physical and biological integrity. Environmental Law 33: 29–77.

Adler, R.W. (2002). Fresh water. In Dernbach J. C., (ed.) Stumbling Toward Sustainability 197–225. Environmental Law Institute, Washington, DC.

Arnold, C.A. (2005). Introduction: Integrating water controls and land use controls: New ideas and old obstacles. In Wet Growth: Should Water Law Control Land Use? 1–55. Environmental Law Institute, Washington, DC.

American Society of Civil Engineers (ASCE) (2004). Regulated Riparian Model Water Code, ASCE/EWRI 40-03. American Society of Civil Engineers, Reston, VA.

Breckenridge, L.P. (2004). Maintaining instream flow and protecting aquatic habitat: promise and perils on the path to regulated riparianism. West Virginia Law Review 106: 595–628.

Breckenridge, L.P. (2005). State survey: Massachusetts. In Beck, R. E. (ed.) Waters and Water Rights, 6: 657–59.

Colburn, E.A. and Hubley, R. (eds.) (1990). Watershed Decisions: The Case for Watershed Protection in Massachusetts. Massachusetts Audubon Society, Lincoln, MA.

Connecticut v. Massachusetts, 282 U.S. 660 (1931).

Conuel, T. (1990). Quabbin: The Accidental Wilderness. University of Massachusetts Press, Amherst, MA.

Daily, G.C. (1997). Nature's Services: Societal Dependence on Natural Ecosystems. Island Press, Washington, DC.

Doremus, H. (2001). Adaptive management, the endangered species act, and the institutional challenges of "new age" environmental protection. Washburn Law Journal 41: 50–89.

Glennon, R. (2002). Water Follies: Groundwater Pumping and the Fate of America's Fresh Waters. Island Press, Washington, DC.

Holling, C.S. and Gunderson, L.H. (2002). Resilience and adaptive cycles. In Gunderson, L.H. and Holling, C.S., (eds.) Panarchy: Understanding Transformations in Human and Natural Systems 25–62.

Instream Flow Council (2004). Instream Flows for Riverine Resource Stewardship, revised edition. Instream Flow Council.

Karkkainen, B.C. (2003). Adaptive ecosystem management and regulatory penalty defaults: Toward a bounded pragmatism. Minnesota Law Review 87: 943–998.

Massachusetts Constitution, Amendments, Articles 49 and 97.

Massachusetts Interbasin Transfer Act, Mass. Gen. Laws ch. 21, §§8B–8D.

Massachusetts Water Management Act, Mass. Gen. Laws ch. 21G.

Massachusetts Water Resources Authority Enabling Act, Ch. 372, Acts of 1984.

Massachusetts Water Resources Authority (2006). MWRA Long-Range Water Supply Planning: Presentation. MWRA Board of Directors Public Forum, June 28, 2006. http://www.mwra.state.ma.us/04water/2006/062806testimony/062806testimony.htm

Massachusetts Water Resources Authority (2004). Five-Year Progress Report, 2000–2004. Massachusetts Water Resources Authority, Boston, MA. Available at http://www. mwra.state.ma.us/publications/5yearreport0004/5yearreport0004.pdf

Massachusetts Water Resources Authority (2000). Summary Report of MWRA Demand Management Program, Fiscal Year 2000. Available at http://www.mwra.state. ma.us/harbor/enquad/pdf/ms-061.pdf

Massachusetts Water Resources Authority (1998). MWRA's New Walnut Hill Treatment Plant: Strategies to Best Protect and Strengthen MWRA Water Quality.

Nesson, F.L. (1983). Great Waters: A History of Boston's Water Supply. University Press of New England, Hanover, NH.

Massachusetts Environmental Protection Act, Mass. Gen. Laws ch. 30 §§61-62H.

Massachusetts Watershed Protection Act, Mass. Gen. Laws ch. 92½.

Platt, R.H. (1995). The 2020 water supply study for metropolitan Boston: The demise of diversion. Journal of the American Planning Association 61: 185–199.

Pioneer Institute for Public Policy Research and Massachusetts Clean Water Council (2006). Water Management and the MWRA: Would MWRA Expansion Benefit Massachusetts Communities and Our Environment? Available at http://www.pioneerinstitute.org/ pdf/06_mwra_transcript.pdf

Postel, S. and Richter, B. (2003). Rivers for Life: Managing Water for People and Nature. Island Press, Washington, DC.

Rast, W. and Holland, M.M. (2003). Sustainable freshwater resources: Achieving secure water supplies. In Holland, M.M., Blood, E.R., and Shaffer, L.R., (eds.) Achieving Sustainable Freshwater Systems: A Web of Connections 283–315.

Rivers and Harbors Act of 1899. Codified as amended, 33 U.S.C. §§401–418.

Rose, C.M. (1990). Energy and efficiency in the realignment of common-law water rights, Journal of Legal Studies 19: 261–296.

Safe Drinking Water Act, 42 U.S.C. §§300f et seq.

Surface Water Treatment Rule, 40 C.F.R. Part 141.

United States v. Massachusetts Water Resources Authority, 97 F. Supp.2d 155 (D. Mass. 2000), aff'd, 256 F.3d 36 (1st Cir. 2001).

Wallace, Floyd, Associates, Inc. (1986). Summary Report: Impact Assessment and Criteria for Evaluation of Alternatives, Water Supply Study and Environmental Impact Report-2020 (EOEA #03518).

Wallace, Floyd, Associates, Inc. (1984). A History of the Development of the Metropolitan District Commission Water Supply System, Metropolitan District Commission Water Supply Study and Environmental Impact Report-2020, Task 18.20. Available at http://www.mwra.state.ma.us/04water/pdf/ws1984book.pdf

24

Financial, economic, and institutional barriers to "green" urban development: The case of stormwater

W.H. Clune and J.B. Braden

University of Illinois at Urbana-Champaign, Urbana, IL, 61801 USA
E-mail: jbb@uiuc.edu

Summary: "Green" development refers to urban development practices and architecture intended to reduce environmental impacts, including water quality problems from stormwater and runoff. However, there are substantial financial, economic, and institutional barriers to the acceptance and application of these techniques. Problems with information, financing, development and building codes, and project costs are among the most challenging of these barriers. Good policy should focus on removing the impediments that prevent market actors from employing green design strategies.

INTRODUCTION

Urban water quality problems from stormwater and runoff, as well as increasing treatment costs, have prompted close examination of methods to better manage stormwater. "Green" stormwater designs that attempt to retain, replicate, or restore natural processes of infiltration and recharge are being used in many different commercial and residential projects.

"Green development" generally encompasses many different strategies employed to reduce the environmental impacts of urbanization. Successful green development may integrate multiple, complementary components designed not only to moderate stormwater flows, but also to reduce energy use and transportation requirements while incorporating natural aesthetics. Examining a single component of green development, stormwater in this paper, achieves some focus in the course of identifying financial, economic, and institutional barriers that may slow the adoption of low-impact strategies more broadly.

Many of the costs of poor stormwater management take the form of environmental externalities, whereby one community imposes the costs of its environmental activities on other, perhaps distant, groups of people. However, there is substantial evidence showing that a large share of the costs of environmentally unsound stormwater development is born locally by owners, neighbors, and their immediate municipal organizations. In fact, the financial and other motivating benefits of green development are well-known at the municipal level, and many communities already try in various ways to encourage on-site management of stormwater.

The range of pre-existing incentives for better stormwater management practices makes the inquiry into barriers all the more important: not only to discover why present incentives have been slow to induce solutions, but also to formulate better solutions. In the presence of positive incentives, policy should focus on removing the impediments that prevent market actors from employing green design strategies.

URBAN STORMWATER PROBLEMS AND SOLUTIONS

Stormwater problems present a number of challenges to builders and communities trying to develop in an environmentally sustainable manner. This section discusses some of the major environmental concerns with stormwater and runoff, and introduces some of the green development solutions.

The underlying urban design challenge for stormwater is the impermeability of much of the landscape. Roads, rooftops, and even grassed lawns are far less able to soak up and store water than are landscapes with trees and other native vegetation. More stormwater runs quickly across impermeable surfaces into creeks, streams, and lakes, meaning less rainwater seeps through the soil profile into shallow aquifers or continues percolating through soil layers to deeper aquifers that may be sources of potable water. Water table levels often drop and streams become much more "flashy," with high flows after storms and mere trickles, or less, in drier periods. The accelerated runoff can worsen flooding locally and downstream. The loss of vegetated surfaces also decreases annual evapotranspiration, which further hinders successful recharge to groundwater (Whitely, 2006). Accelerated runoff is especially problematic in cities where stormwater and sanitary wastes commingle in the same system of pipes. Large storms can overwhelm the collection and treatment systems, causing untreated releases into receiving waters.

Stormwater and runoff in urban areas also tend to be polluted by a variety of chemical compounds and microbial organisms (Whitely, 2006; USEPA, 1992; Bauer et al., 2004). Much of the contamination would be naturally cleansed if allowed to filter through soils. However, stormwater that flows directly to streams does not benefit from these natural processes.

Apart from improvements to the stormwater collection and treatment system, solutions designed to prevent hyper-runoff episodes and encourage infiltration include restorative landscaping, roof and rain gardens, smaller streets and parking lots, permeable paving, increased use of margins and buffer strips as infiltration zones, detention basins, and bioretention basins. The latter are vegetated shallow basins that both slow and treat stormwater as it infiltrates back into the soil (DuluthStreams.org, 2006). In Long Island, NY, for instance, the use of these stormwater recharge measures has increased water table levels compared to the significant declines observed in areas where stormwater discharges directly to rivers (Ku et al., 1992).

Methods for dealing with stormwater work best when applied in a coordinated fashion. This is the guiding principal behind green design, which aims to reduce runoff volume and, therefore, also reduce the movement of waste products into sewer systems. Green design tactics manage stormwater by facilitating infiltration to groundwater, evaporation back into the atmosphere, or by finding some other beneficial use for it (Holz, 2001).

BARRIERS TO GREEN DESIGN AND SOME PROPOSED SOLUTIONS

A number of factors cause urban-dwellers to prefer (and choose) something other than green strategies for stormwater management. Other factors limit the ability of developers and builders to supply crucial components of these strategies.

Key barriers to "green" stormwater management practices

Lack of Information. A lack of information is one of the most significant barriers to green development practices. To begin with, consumers are generally unfamiliar with green design and construction, let alone the potential benefits and opportunities they offer (Frej, 2003). Many view green construction as being new and risky (Lucuik, 2005). As a result, many builders find that buyers want more information and evidence of successful projects before they will consider green building options (Natural Strategies, 2001).

These information problems are not limited to buyers. There is still widespread uncertainty among builders and designers. Only recently have authoritative guidance documents begun to emerge for sustainable urban technologies (e.g., Ferguson, 2005). The available information continues to be piecemeal, focusing on individual technologies. A failure to understand the integrated nature of the various elements of sustainable design has sometimes led to poor results (Lucuik, 2005). Developers and builders lack information as to where to acquire the plans, materials, and expertise to complete projects successfully (Wisconsin Environmental Initiative, 2005). Many of the important real estate intermediaries, such as project financers and real estate brokers, are not well informed about the value or features of green buildings; as a result, these key project facilitators are not always able to assess the market value of those design elements or communicate those advantages to buyers (Frej, 2003).

Continuing education and information clearinghouses are addressing some of these barriers. In fact, several public and private organizations have emerged to promote green building practices and better stormwater design. These include:

- The Greater Vancouver Sewerage and Drainage District has published on-line stormwater source control tools, including information regarding the design and installation of green roofs (http://www.gvrd.bc.ca/buildsmart/).
- DuluthStreams.org has an on-line Conservation Tools section for stormwater management; guidance, design suggestions, and plans are all available (http://duluthstreams.org/).

In short, many of the information problems are being solved through outreach and education, along with the emergence of a green construction industry. It helps that green development has typically enjoyed free and favorable publicity in the media (Wilson et al., 1998).

Another aspect of the information problem may be a lack of coursework and formal training regarding green design in education and training programs for engineers, contractors, and architects. Even this situation is changing, however, with green construction and design becoming part of mainstream curriculums at many universities as well as in continuing education programs offered by professional associations.

While awareness remains a major barrier to green stormwater management, with initiatives such as those noted above the information gaps are closing. With increasing resources devoted to innovative urban design, and more frequent use of green building techniques, the quantity and quality of information should continue to improve.

Zoning, Regulation, and Building Codes. Zoning, regulation, and building codes have also been identified as major problem areas holding back green development and better stormwater management (Eisenberg et al., 2002). The process of establishing codes can be inflexible, often times relying on over-designed standards to protect health and safety to the detriment of legitimate conservation efforts (Zeigler, 2005). Perceived conflicts with code intent can be as likely as technical code specifications to lead to denials (Eisenberg et al., 2002).

Code approval and administration can involve many different government agencies, so every proposed variance can involve complicated sequences of negotiation with multiple parties (1000 Friends, 2004). Many different suggestions have been made for improving this process: fewer codes, clearer requirements, a more integrated process, the insertion of green building exceptions, or the creation of governmental green building champions to mediate code issues in favor of good green designs (West Coast, 2002). The following are examples of progressive building codes that encourage green stormwater design:

- The City of Lacey, WA, adopted a Zero Impact Development Ordinance in August, 1999. This new law aims to preserve hydrologic function after development so that there is "near zero" impact of impervious surfaces. The ordinance specifically encourages green construction and design (Holz, 2001).
- The United States Green Building Council (USGBC) has developed national building certification standards (called LEED) intended to promote high-quality green buildings; certification points are granted for multiple criteria such as the use of integrated green design elements, as well as resource efficiency (US Green Building Council, 2005).

Developers, of course, prefer an approval process that is clear and certain. It can be costly to have plans rejected, particularly if the rejection occurs far along into the approval process (Green Roundtable, 2003). However, an overly prescriptive or proscriptive certification process can discourage innovation by both agencies and developers. Green projects can generate a great deal of public support, which can lead to greater willingness on the part of public officials to work cooperatively with developers (Wilson et al., 1998). On the other hand, code requirements must balance different interests. For example, in Wisconsin, narrowing streets in the interest of reducing impervious surfaces was resisted because of the desire for rapid delivery of public safety services such as fire fighting (Nelson et al., 2003).

Zoning requirements may restrict the minimum building size, minimum frontage widths, exclude multifamily dwellings, restrict the number of bedrooms, and impose deed restrictions requiring specified types of building size and design (Wisconsin Environmental Initiative, 2005). All of these restrictions may stymie innovative and effective stormwater design. For example, Maryland green builders have complained about minimum lot size requirements and prohibitions against narrowing street widths; the former code provision doesn't allow for a redistribution of housing and open spaces, even in cases that leave overall housing density unchanged (1000 Friends, 2004).

In fact, high density urban developments or multi-family dwellings have been shown to have significantly less impact on water quality than similarly populated areas dominated by detached single family homes spread across larger areas (Goonetilleke et al., 2005). Clustering is a commonly used conservation zoning measure for better stormwater management (Yaggi, 2001). This technique requires flexibility with respect to lot sizes in order to build homes closer together while preserving open spaces where stormwater can be managed.

While competing public goals and concerns with liability are reasons to be cautious about green design, some critics argue that zoning and building codes have not gone far enough with respect to green standards or their implementation. Most current building codes require only minimum environmental standards, while most green construction designs exceed those minimums (Lucuik, 2005). By comparison, many countries in the European Union have incorporated sustainability requirements into their building codes and ordinances (King and King, 2005). Then again, many well-intended green code requirements are simply evaded: developers faced with impervious area limitations sometimes purchase more land, which enables code compliance but may do nothing to alter stormwater management or improve receiving water quality (Jones et al., 2005).

One potential solution to some of the barriers caused by building codes and zoning – and an idea that also addresses the tension between more or less regulation – is for municipalities to adopt performance-based standards rather than design standards (Zeigler, 2005). Many performance-based codes, like those published in the International Building Code guidelines, include provisions for alternative materials, designs, and methods of construction. Performance-based standards require the active involvement of architects and engineers capable of certifying specific practices and materials to achieve the required levels of performance (Zeigler, 2005).

In addition, regulatory and reporting requirements pertinent to stormwater are burdensome and expensive, particularly for municipalities. The National Pollution Discharge Elimination System (NPDES) requires medium and large sized cities to obtain a federal permit for their stormwater systems. The clear trend in recent years has been to increase municipal responsibility, costs, and potential liability. Cities have been asked to become more like pollution control and enforcement agencies with respect to stormwater.

Public managers have watched these developments warily, not only with respect to their budgetary and decision-making impacts, but also to their growing exposure to lawsuits, the erosion of government immunity, and increasing exposure of government employees to criminal prosecutions for environmental damages (O'Leary, 1993). As such, green stormwater design offers some interesting cost-saving opportunities by limiting the need for new sewer systems and infrastructure, and reducing the water pollution created in runoff. Given the strong municipal interest in reducing costs and liabilities, it may be in the interest of local governments to reward developers able to provide projects that minimize community environmental costs.

Financial Markets and Project Financing. Obtaining financing is a major challenge for development projects, and it can present barriers to green design and development strategies. Banks and most conventional sources of real estate financing are conservative and risk averse with respect to building projects. A conservative financing culture is reluctant to assume that innovative or untested projects will be accepted by consumers. Moreover, financiers may lack confidence in the cost and performance specifications of unfamiliar designs (Canadian Housing, 2000). The following examples illustrate problems that can arise in financing unusual development projects:

- The Second Street Studios in Santa Fe, New Mexico, designed to combine living and working spaces and thereby reduce the need for commuting, were viewed by banks as a risky undertaking with no proven market. The project didn't fit any traditional lending criteria that would allow it to be sold on the secondary market with lending agency credit enhancements; it was simultaneously commercial and residential, whereas most projects have only one such designation (Wilson et al., 1998).

- The Spring Island development in South Carolina faced reluctance from banks unused to the project's various green components, including a trust created to permanently exclude a large portion of the island's acreage from future development. The banks eventually required the possibility of mortgage foreclosure on the preserve until the project debt was reduced to specified levels (Wilson et al., 1998).

There are also numerous examples showing that banks often badly underestimate the financial value and marketability of green projects:

- The Village Homes community in Davis, California, found it difficult to find funding and achieve approval. The developers discovered that emphasizing green features like energy conservation and natural drainage made bankers more skeptical about the financial prospects. However, marketing the environmentally-friendly aspects has helped the project achieve annual returns of around 30 percent for its investors (Wilson et al., 1998).
- At the Spring Island development in South Carolina, noted above, the completed project reported a very high absorption rate and market prices (Wilson et al., 1998).

With respect to lender conservatism and risk aversion, the burden is on green developers to quantify as many environmental features as possible in financial terms that relate to potential return on investment. The economics literature contains a growing volume of studies that link the value of housing to environmental amenities such as open space (e.g., Irwin, 2003; Earnhart, 2006) and wetlands (e.g., Earnhart, 2001). Johnston, Braden, and Price (2005) illustrated the use of valuation techniques to assign benefits to stormwater mitigation. However, these insights do not appear to have systematically penetrated the development finance community.

The fact that so many banks and lenders do not yet understand the value potential of green buildings is addressed by some green developers with presales that demonstrate interest and demand (Wilson et al., 1998). But financing green projects is likely to remain a challenge until the number of successful projects increases financier familiarity, expertise, and comfort level.

Once green construction becomes more mainstream, banks should have an easier time calculating expected costs and returns. Until that time, additional funding incentives for green projects may be required to accelerate the process of familiarizing the financial community (e.g., Wisconsin Environmental Initiative, 2005).

Green Development Costs and Price Competitiveness. We have seen that inadequate performance documentation, inflexible public codes, and lack of builder and consumer information are potential barriers to the adoption of "green" development methods for stormwater. In addition, real or perceived financial costs associated with employing green designs have the potential to thwart project adoption. Builders, in particular, typically assume that the up-front costs of green projects are greater and, thus, that the finished product may have to be priced above the competition.

Green construction projects, whether because they tend to be highly customized or include premium features, or because they are still relatively novel, can take more time in their initial planning and design. Customization can mean dependence on specialized consultants, working with unfamiliar or unproven materials and techniques, and increasing exposure to call-backs (Natural Strategies et al., 2001). There is also considerable evidence that materials and expertise for many green projects are more costly because they are still relatively uncommon. The supply sector has not achieved the economies of scale required to reduce costs to levels that compare favorably to conventional materials (Wisconsin Environmental Initiative, 2005).

Each of these factors can add to the cost of a green project. Nevertheless, a review of several hundred existing cases showed that sustainable buildings are cost-competitive when both initial and operating costs are considered (Kats, 2003). In some cases, operating cost savings can be significant: for example, heating, cooling, and energy costs can be reduced in green buildings. One approach, therefore, to handling the perceived cost barrier is to emphasize the long-run cost efficiency in consumer marketing efforts.

Furthermore, not all green construction is more expensive. In fact, cost-efficiency is a selling point for some projects. In the specific case of stormwater, better design may reduce construction, maintenance, regulatory, and operation costs, while employing natural features for drainage and the control of water flow may eliminate the need for sewer infrastructure (Holz, 2001). For example:

- The Ford Motor Company's Rouge complex in Dearborn, Michigan, has the world's largest green roof as part of an extensive and integrated system for managing stormwater. Undertaking these projects was based in part on the fact that Ford kept a $48 million funding source to deal with stormwater and runoff. The green roof allowed them to spend that money on the roof's construction rather than for ongoing remediation efforts (Frej, 2003).
- The Prairie Crossing development near Chicago saved in sewer construction costs by designing narrower streets, reducing the overall amount of impervious concrete surfaces, and using vegetated detention basins for stormwater control (Wilson et al., 1998).
- The Dewees Island development in South Carolina included extensive flood management in the construction planning, and as a result received a 5 percent discount on flood insurance from FEMA on all buildings (Wilson et al., 1998).

Although green designs may be cost-neutral or better in the long-run, consumers tend to focus on the purchase price and to discount the potential for long-term savings in operating costs (Natural Strategies et al., 2001). Similarly, consumers tend to discount environmental hazards, such as the potential to incur flood damages, when purchasing homes (Chivers and Flores, 2002). This suggests a need for on-going efforts to reduce construction costs through standardization and technology development.

While the development, construction, and financial communities express concern about the cost and price implications of green design, there is evidence to suggest that some consumers consider green building to be a premium product worth the extra price (Wilson et al., 1998). In fact, green construction is creating its own successful marketing niche, and many green projects enjoy above-average sales rates (Wilson et al., 1998).

Funding Stormwater Management. The methods employed by urban communities to fund and finance their stormwater management systems play a major role in providing price and incentive signals to market actors. Traditionally, stormwater was financed through general tax revenues (Morandi, 1992) or bond issues repaid from special service fees (Keller, 2001). However, as local level responsibilities for water quality and stormwater management have grown, budget shortfalls have forced the selection of alternative funding methods. A variety of methods have been used to fund stormwater systems, including: inspection and permit fees; higher taxes on new developments; and the formation of large regional stormwater management districts with distinctive service fees (NRDC, 1999). Today, development fees and, to an even greater extent, stormwater user fees are two of the most important funding methods in the United States.

Development fees (or impact fees, or capital recovery fees) are typically charged to project developers, and are meant to cover the costs of expanded services made necessary by new residents. In fact, development fees are now widely used to finance many municipal services, such as stormwater, fire, police, and libraries (impactfees.com, 2006).

There are two main approaches to setting development fees: an average cost flat fee; and a marginal cost fee that divides the charges into costs for additional facilities, costs for the connections necessary to deliver the service, and costs of producing the service itself (Libby et al., 2003). In terms of taxation incidence, these costs are likely to be shared between buyers paying higher prices and the landowners of undeveloped land receiving lower prices from developers (impactfees.com, 2006).

A contentious issue has arisen from the use of development fees about whether the costs of infrastructure are distributed fairly between new and ongoing residents. In fact, development fees have faced numerous legal and constitutional challenges on due process grounds (whether the fee is really a tax), equal protection grounds (whether the fee discriminates between groups of people unduly or unfairly), and on takings grounds (whether the assessment is an uncompensated taking not closely related to the stated objective). Most U.S. jurisdictions have upheld the use of development fees. In fact, many states legislate the control and regulation of development fees, with applicable statutes ranging from complex and restrictive, as in Illinois and Texas, to general and permissive, as in New Jersey and Indiana (Libby et al., 2003).

User fees, on the other hand, charge for the operating costs of stormwater systems, with rates set to produce revenues equal to total operational and infrastructure costs. These user fee arrangements are primarily operated in the U.S.

through special waste water utilities (Cameron et al., 1999). These utilities have the authority to generate funding to be used specifically for stormwater management and expenses. The number of waste water utilities is growing quickly: there are several hundred in the U.S. today with projections for over 2,500 within ten years (Kaspersen, 2000).

Two of the potential benefits of user fees include requiring beneficiaries to pay for the services they receive, and providing a disincentive for activities that cause more pollution or runoff. Of course, these benefits will depend upon the type of fee employed: there is a marked difference between flat fees, fees based upon property type or size, and fees linked to the actual quantity of stormwater discharge. In fact, a survey of several hundred U.S. stormwater utilities showed that about one-half charge fees based on the amount of impervious area, one-quarter have fees based on both impervious area and total area, and the rest reporting rates based upon other factors (Hoag, 2004). There is a growing trend of setting fees based upon pollutant concentration, while only the most cutting edge fee schedules also reward activities that improve the system's performance or reduce its costs (Reese, 1996). The latter type of fee does most to encourage property owners to mitigate runoff. However, since many of the mitigation methods must be built in at the development state, development fees structured to reward on-site management have a very significant role to play (Kobza, 1995).

To relate stormwater funding to the discussion of barriers and incentives, it can be problematic when developers and consumers are not rewarded in the market for the community benefits of green design, such as savings in stormwater infrastructure and treatment costs. In these cases, policy makers may want to consider measures that level the playing field in order to give builders and buyers more accurate incentives with respect to true costs. Of course, caution is needed when tinkering with price signals. There can be unintended consequences. However, as development and stormwater charges indicate, there are numerous ways to encourage green strategies through fiscal policies that are well supported by real resource costs to society.

In addition to accurate pricing of stormwater services, a number of U.S. states and localities offer general tax incentives like deductions or credits for employing green building features (King, 2005). These measures can improve the initial costs of a project. Coupled with pricing that recognizes performance, they can reward both developers and consumers for measures that reduce offsite environmental costs.

To the extent that the costs or benefits of development decisions are hidden from the decision makers, they will be led towards unbalanced choices. With stormwater, for example, if developments that increase runoff rates and downstream flooding had to pay for the resulting damages, they might choose designs that allow for higher rates of infiltration. In short, the stormwater problem is also complicated by jurisdictional boundaries that do not respect watershed processes. Proper resolution might require cooperation between jurisdictions to ensure that cause and effect are coupled (Eisen, 1995).

Other barriers to better stormwater management

Consumer Interest. Many current barriers to green development practices would become irrelevant with a strong, sustained consumer market for green building reinforced by price signals that accurately convey the differential environmental costs of property ownership. Once consumers become educated about green building options, many respond enthusiastically (Wilson et al., 1998). However, others consider green practices to be strange, unattractive, or even a cause for concern (Wisconsin Environmental Initiative, 2005). Many consumers prefer comfort and ease of maintenance to environmental features that might produce wet basements (Cairncross, 1994). For example, rain gardens that retain standing water near homes and require maintenance, are unattractive to many homeowners (Nelson, 2003).

Green development strategies not only apply to residential areas, but also affect many other aspects of daily life such as transportation and parking. Reducing the amount of car habitat has great potential for reducing the amount of impervious cover (Yaggi, 2001), with corresponding benefits for stormwater management. Addressing the impacts of transportation infrastructure, however, raises complex issues of urban form, congestion, pricing, service delivery, and consumer preferences.

Legal Liability. Another claim in favor of green development is that it might help prevent fires, floods, or other disasters, thereby reducing legal liability for owners (Wilson et al., 1998). Flooding is the most likely category of liability associated with stormwater. In most cases, construction must meet all applicable building code requirements, including those for stormwater. Therefore, it is most often the municipality that is sued if a flood occurs after permits were granted. A fear of lawsuits may make city planners cautious about development strategies that appear to create potential hazards, such as wetlands, in an effort to better manage stormwater (West Coast, 2002). This suggests the need for safety-testing of green development or building codes. Expedited reviews or relaxed requirements for green building projects (King, 2005) may overlook important issues of public safety and potential liability.

Potential liability applies equally to conventional stormwater management as to green techniques. What is important here is that any method used (and certified) must be sufficient to control runoff. As mentioned above, this liability issue may be driving some of the caution with which city planners approach the granting of variances to their generic requirements (West Coast, 2002). With respect to actual liability for flooding, there does not appear to be an appreciable (or recorded) difference for conventional planning compared to green stormwater management.

In fact, the current liability regime actually enhances stormwater management in the sense that it provides a powerful incentive for community vigilance. It would be mistake to grant legal immunity to municipalities that give code waivers for innovative but untested project designs. If unsustainable management of sprawl and stormwater has led to increasing incidents of flooding and legal liability, the costs of resulting fines and judgments reinforce the incentives for municipalities and taxpayers to encourage more effective practices, which should favor proven green designs (Lachman, 2001).

CONCLUSIONS

As the market for development based on green design principles expands, the nature and severity of the barriers to adoption will change. The barriers intertwine. A useful way to think about them is in terms of the ways they interact to influence the supply of green development, buyer preferences, and market prices.

Correcting information problems seems to be a major issue with respect both to consumer perceptions of green development and the willingness of developers, builders, and financiers to engage in it. Information problems will be resolved through research, development, and, especially, certification programs for on-site management methods. Demonstration projects can help practitioners gain first-hand experience and increase confidence in cost estimation. These measures should provide a wider range of methods while also resulting in standardization that can reduce risks and costs.

Regulatory barriers are important to recognize and address. Inflexible zoning and building codes protect municipalities and their employees while deterring innovations that might reduce overall costs to the community. Technical certification of onsite technologies could help to alleviate the liability concerns. The adoption of performance-based standards in lieu of design standards adds flexibility, but reliable technical guidance for developers and regulators is an essential ingredient, and communities must commit to ongoing monitoring of system performance. The development of model zoning and building codes may help communities to adapt their current regulations. Finally, development fees and stormwater management fees that reward onsite management measures engage the economic interest of both developers and property owners in the cause of stormwater management following principles of green development.

ACKNOWLEDGEMENTS

The authors acknowledge financial support received from the International Joint Commission and project ACE 0305 of the Office of Research, College of Agricultural, Consumer, and Environmental Sciences, University of Illinois at Urbana-Champaign. Any opinions, findings, and conclusions or recommendations expressed are those of the authors and do not necessarily reflect the views of these sponsors.

REFERENCES

1000 Friends of Maryland (2004). Barriers to Environmental Design in Maryland.
Bauer, S., Bayer-Raich, M., Holder, T., Kolesan, C., Muller, D., and Ptak, T. Quantification of groundwater contamination in an urban area using integral pumping tests. Journal of Contaminant Hydrology 75 (3–4):183–233.

Brisman, A. (2002). Considerations in establishing a stormwater utility. Southern Illinois University Law Journal, 26 (Spring):505–528.

Clarke, R.A., Stavins, R.N., Greeno, J.L., Bavaria, J.L., Cairncross, F., Esty, D.C., Smart, B., Piet, J., Wells, R.P., Gray, R., Fischer, K., and Schot, J. (1994). Harvard Business Review 72 (4):37–50.

Cameron, J., Cincar, C., Trudeau, M., Marsalek, J., and Schaefer, K. (1999). User pay financing of stormwater management: A case-study in Ottawa-Carleton, Ontario. Journal of Environmental Management 57 (4):253–65.

Canadian Housing and Mortgage Corporation (2000). Implementing Sustainable Community Development: Charting a Federal Role for the 21st Century.

Chivers, J. and Flores, N. (2002). Market failure in information: The National Flood Insurance Program. Land Economics 78 (4):515–521.

DuluthStreams.org (2006). Conservation Tools, Stormwater Management. http://www.duluthstreams.org/general/aboutus.html.

Earnhart, D. (2001). Combining revealed and stated preference methods to value environmental amenities at residential locations. Land Economics 77 (1):12–29.

Earnhart, D. (2006). Using contingent-pricing analysis to value open space and its duration at residential locations. Land Economics 82 (1):17–35.

Eisen, J. (1995). Toward a sustainable urbanism: Lessons from federal regulation of urban stormwater runoff. Washington University Journal of Urban and Contemporary Law 48 (summer):1–86.

Eisenberg, D., Done, R., and Ishida, L. (2002). Breaking Down the Barriers: Challenges and Solutions to Code Approval of Green Building. The Development Center for Appropriate Technology.

Ferguson, B.F. (2005). Porous Pavements. Boca Raton: CRC Press.

Frej, A. (2003). Report: Green Buildings and Sustainable Development: Making the Business Case. Urban Land Institute.

Goonetilleke, A., Thomas, E., Ginn, S., and Gilbert, D. (2005). Understanding the Role of Land Use in Urban Stormwater Quality Management. Journal of Environmental Management 74 (1):31–42.

Greater Vancouver Sewerage and Drainage District (2005). Stormwater Source Control Design Guidelines 2005 – Green Roofs.

Green Roundtable (2003). An Analysis of Barriers and Opportunities to Green Development in Boston.

Hoag, G. (2004). Developing equitable stormwater fees. Stormwater Magazine 5 (1).

Holz, T. (2001). Stormwater Strategies. SCA Consulting Group.

impactfees.com. FAQ's from on online impact fee resource site, provided by Duncan Associates. http://www.impactfees.com/.

Irwin, E.G. (2002). The effects of open space on residential property values. Land Economics 78 (4):465–481.

Johnston, D.M., Braden, J.B., and Price, T.H. (2006). The downstream economic benefits of stormwater management: A comparative analysis. Journal of Water Resources Planning and Management 132 (1):35–43.

Jones, J., Earles, A., Fassman, E., Herricks, E., Urbonas, B., and Clary, J. (2005). Urban stormwater regulations – Are impervious area limits a good idea? Journal of Environmental Engineering 131 (2):176–79.

Kaspersen, J. (2000). The stormwater utility: Will it work in your community? Stormwater Magazine 1(1).

Kats, G. (2003). Report to California's Sustainable Building Task Force: The Costs and Financial Benefits of Green Buildings. Capital E Consultants.

Keller, B. (2001). Buddy can you spare a dime? What's stormwater funding. Stormwater Magazine 2(2).

King, N.J. and King, B.J. (2004–2005). Creating incentives for sustainable buildings: A comparative law approach featuring the United States and the European Union. Virginia Environmental Law Journal 23:397–460.

Kobza, L. (1995). Controlling stormwater in Wisconsin: Municipal considerations and strategies. Wisconsin Environmental Law Journal 2:1–34.

Ku, H.F.H, Hagelin, N.W., and Buxton, H.T. (1992). Effects of urban storm-runoff control on ground-water recharge in Naussau County, New York. Ground Water 30 (4):507–513.

Lachman, S. (2001). Should municipalities be liable for development-related flooding? Natural Resources Journal 41 (4):945–980.

Libby, L. and Carrion, C. (2003). Development Impact Fees. Ohio State University Extension, Columbus.

Lucuik, M. (2005). A Business Case for Green Buildings in Canada. Morrison Hirschfield, Report No. 2052223.00, Ottawa, Canada, May.

Morandi, L. (1992). Wastewater Permitting and Finance: New Issues in Water Quality Protection. National Conference of State Legislatures (NCSL) State Legislative Report, May.

Natural Strategies Inc. and VITETTA Public Management Consulting (2001). Barriers to Building Green. Report submitted to the California Integrated Waste Management Board.

Nelson, A. (1995). Calculating System Development Charges for Stormwater Facilities. Boca Raton: CRC Press.

Nowacek, D. (2003). Social and Institutional Barriers to Stormwater Infiltration. Report pursuant to EPA/NSF/USDA 1999 Water and Watershed Research Grant No 99-STAR-L1, Department of Sociology, University of Wisconsin, July.

Lehner, P., Aponte Clark, G.P., Cameron, D.M. and Frank, A.G. (1999). Stormwater Strategies, Community Responses to Runoff Pollution. New York: Natural Resources Defense Council.

O'Leary, R. (1993). Five trends in government liability under environmental laws: Implications for public administration. Public Administration Review 53 (6):542–549.

Reese, A. (1996). Storm-water utility user fee credits. Journal of Water Resources Planning and Management 122 (1):49–56.

US Green Building Council (2005) LEED-NC Application Guide for Multiple Buildings and On-Campus Buildings Projects. https://www.usgbc.org/ShowFile.aspx?DocumentID=1097

U.S. Environmental Protection Agency (1992). Environmental Impacts of Stormwater Discharges: A National Profile. Report No. 841R92001, Washington, DC, June.

West Coast Environmental Law (2002). Cutting Green Tape, An Action Plan for Removing Regulatory Barriers to Green Innovations. April.

Whiteley, H. (2006). Effects of Urban Areas on Groundwater in the Great Lakes. Report prepared for the International Joint Commission, Great Lakes Science Advisory Board, Workgroup on Parties' Implementation.

Wilson, A., Uncapher, J., McManigal, L., Lovins, L., Cureton, M., and Browning, W. (1998). Rocky Mountain Institute. Green Development: Integrating Ecology and Real Estate. New York: John Wiley & Sons.

Wisconsin Environmental Initiative (2005). A Series of Goals and Strategies for the Effective Greening of Affordable Housing. Madison.

Yaggi, M. (2001). Impervious surfaces in the New York City watershed. Fordham Environmental Law Journal 12:489–522.

Zeigler, P.M. (2005). Building Codes and Regulations: Removing Barriers to Sustainable Development and Use of Green Building Technologies. Governor's Green Government Council, Commonwealth of Pennsylvania.

25

Restoring the Charles River watershed using flow trading

Kate Bowditch

Charles River Watershed Association, 190 Park Road, Weston, Massachusetts 02493
E-mail: kbowditch@crwa.org

Summary: The impacts of urban development on wetland and water resources are widespread and well documented. The impacts vary both spatially and temporally, and are often the result of numerous interrelated activities in the urban environment. Restoration of urban watersheds requires an interdisciplinary approach, working to reduce impacts across a broad spectrum of land use and resource management practices. Efficient and effective restoration programs are difficult to develop, however, in part because of uneven regulatory authority. The complexity of watershed dynamics, and the existing regulatory framework, present an opportunity to use trading techniques to improve urban water resource management. The Charles River Watershed Association has developed a framework for a model trading program to restore urban and suburban watersheds using "base flow credits" as a basis for trading.

INTRODUCTION

In spite of the significant progress in river restoration efforts across the country since the passage of the federal Clean Water Act in 1972, nearly 40% of the rivers in the United States remain impaired, and there are some indications that this percentage is actually on the rise after many years of decline (Otto et al., 2004). These impairments have numerous sources: point and nonpoint source pollution; hydrologic alterations; withdrawals and diversions; dams; loss of basin storage, especially wetlands; channelization and dredging; alterations to groundwater flow paths; loss of riparian cover and bank alterations; and excessive thermal loading.

Impairments in urban rivers are particularly severe, and difficult to address, because of the numerous activities that contribute to those impairments. Restoration in urban watersheds can be extremely expensive, requiring repair and replacement of existing infrastructure and retrofitting of existing development. In many cases, land use patterns and channel alterations significantly limit the restoration opportunities in an urban environment. Across the country, urban river restoration proponents have initiated programs under a "watershed approach," which seeks "an integrated perspective in water resources planning that provides a framework for integrating economic, natural and social considerations that share the same geographic space" (Deason, 2001). Integrated programs, based on the collection of data and the development of predictive models, and undertaken with the partnership of broad stakeholder groups that include government agencies, business, environmental groups and local residents, are prevalent, especially in river systems where human demands are already outstripping available supplies (Poff et al., 2003).

Even within the framework of a broadly accepted watershed plan, however, current regulatory, financing and policy programs can make efficient implementation of urban watershed restoration difficult. Most regulatory controls, implemented through a variety of permits, policies and incentive programs, are not suited for dispersed, diverse and interacting impacts. Two major institutional impediments to urban watershed restoration are the uneven regulation of the causes of watershed impairment; and the isolation of regulatory programs.

To deal with these impediments, numerous programs have begun to be implemented across the country that attempt to integrate regulatory programs and to diversify the opportunities for meeting regulatory requirements. These programs typically include the establishment of watershed-based restoration goals; a system of environmental trading that allows cost-effective strategies to be used to meet goals; and an administrative framework to manage the integration of activities that are geographically dispersed and unevenly regulated.

The Charles River Watershed Association (CRWA) is working to develop an integrated, permit-based program that will reduce pollution loads and restore a more natural hydrologic regime to the Charles River. The program relies on the development of "base flow credits," for trading among both permitted and unpermitted water resource management activities.

EXISTING REGULATORY FRAMEWORK FOR WATER RESOURCE MANAGEMENT

Despite the Clean Water Act's established goal "to restore and maintain the chemical, physical, and biological integrity of the Nation's waters," its primary regulatory authority is focused on the discharge of pollutants (US EPA, 2002). The Act provides no regulatory mechanism to support, for example, adequate base flows, or to prevent excessive stormwater peak flows, although addressing these issues is critical to achieving the Act's goals.

Even within the context of regulated pollutants, some sources of pollution, primarily point source discharges, are heavily regulated, while others, especially nonpoint source discharges, are not. One of the most significant sources of pollution to the nation's water resources, and the most significant in urban areas, is stormwater. Some stormwater discharges, such as municipal separate storm sewer systems (MS4s), are generically regulated, but current federal regulations have no numeric limits for pollutants and rely instead on general narrative goals, standard practices and education programs (Kosco et al., 2003). Other sources of stormwater, such as runoff from large parking lots, commercial sites, residential areas, agricultural uses, and many industrial properties, are completely unregulated. Furthermore, sources of pollution that discharge into MS4s are generally unregulated, leaving a regulatory gap even in regulated systems.

Paradoxically, regulation of stormwater is especially weak in urban areas, where stormwater impacts are often the most severe. In the states and municipalities that have implemented stormwater policies to regulate stormwater discharge volume and/or control stormwater quality prior to its discharge into an MS4, developed areas are generally exempt or allowed to meet much less stringent stormwater discharge standards.

Combined sewer systems, those that carry both sanitary sewage and stormwater runoff, are generally regulated through the wastewater treatment plant permitting program. Regulatory programs tend to focus on the combined sewer overflows (CSOs) in these systems, with permit emphasis placed on the frequency and volume of CSO activations into receiving water bodies. But few of these programs set limits on stormwater discharges into the combined system, which are the cause of the CSO activations.

This regulatory framework creates an uneven burden of control for a wide variety of pollutants, with little if any regulation of some of the most significant sources of urban stream degradation. Some dischargers, such as wastewater treatment plants or certain industrial dischargers, have significant technical and financial obligations to remove pollutants. In contrast, other dischargers, whose pollutant loads may be significant at either the storm event scale or on an annual basis, are weakly regulated or left completely unregulated.

Current regulation of water resources frequently occurs in isolation, with little recognition of the interactions and impacts among the various elements of the water cycle. In urban and suburban watersheds, this problem is acute, with infrastructure for water supply, wastewater and stormwater all having interactive impacts throughout the watershed. At the same time, the impacts of development, including the removal of vegetation, filling of wetlands and increased imperviousness, alter local water resources. Further hydrologic alterations such as damming, channelization, and diversions radically alter watershed dynamics.

These alterations, impacting one another as well as the watershed and river, have historically been regulated in isolation from one another. Although regulating the same resource, water supply permitting programs have not been integrated with wastewater permitting programs. Land use planning, zoning and development rarely take into account natural resources, and tend not to consider a watershed's carrying capacity, or the impacts land uses will have on long term aquifer recharge. Decisions about dams and other hydrologic alterations have only recently begun to consider multiple resource impacts, and tend to focus primarily on balancing the needs of irrigation, hydropower, flood control, public water supplies and endangered species, with little focus on wetlands, ecosystem health, or sustainability (FitzHugh and Richter, 2004).

Stormwater infrastructure has tremendous impacts on wastewater systems, water supplies and river health. CSO discharges, the cause of many major storm-related impairments to urban rivers across the country, occur because of stormwater flows into wastewater systems. But stormwater regulatory programs typically do not regulate stormwater in combined sewer areas because the stormwater flows do not discharge into a water body. The impacts of centralized stormwater conveyance structures on water supplies include loss of recharge, alterations in reservoir and dam management, and potential supply contamination. Yet stormwater management and water supply protection are generally permitted in isolation. Stormwater runoff from urban areas has been identified as one of the most significant sources of ecosystem degradation, water quality impairments and aesthetic degradation (US EPA, 2000). Yet stormwater regulation is rarely integrated with other programs such as endangered species protection programs or historic preservation.

The regulatory structures that exist for managing water resources are limited in their potential for urban river restoration. Even if significant resources are dedicated to individual programs, such as CSO control or wastewater treatment plant upgrades, they may not achieve any significant river restoration goals if they occur in isolation from overall watershed program management. New regulatory, financing and program management structures are needed to overcome the institutional barriers that exist today.

ENVIRONMENTAL TRADING

Environmental trading began to be considered as an alternative way to achieve environmental goals as early as the 1960's (Keiser and Fang, 2004). In theory, a market-based trading program can achieve environmental standards at lower cost than a standard "command and control" approach. Successful trading programs rely on strong, uniform regulatory authority in a given sector, and heterogeneity between methods for achieving an environmental goal (and meeting a regulatory requirement).

In the United States, air emissions trading began in the 1970's, followed in the 1980's by trades involving lead gasoline and chlorofluorocarbons, and the 1990's by acid rain emissions trading. The success of the acid rain emission trading program, which resulted in sulfur dioxide (SO_2) reductions of 30% more than the initial target at significant cost savings, led to the expansion of environmental trading programs.

Today, numerous environmental trading programs exist in the United States. Two models for trading dominate: so-called "cap and trade" programs; and "mitigation" (sometimes called "offset") programs. The cap and trade model is based on pollution limits (caps) that are established through a regulatory permitting program. Permit holders are allowed to discharge only a certain volume of a pollutant and must offset any discharge above their cap by purchasing "credits." Credits are typically purchased from someone who is discharging less than their permit allows and thus has unused pollution "credits." In this system, while a certain volume of pollution is "free of charge" for all polluters (just as it is in traditional permitting programs), pollution above the cap has a cost, and "unused pollution" below the cap has value.

Mitigation/offset trading programs are based on regulations that require mitigation for "unavoidable impacts" created by development or another activity. If a project will create unavoidable impacts to a wetland, rare species habitat, or other regulated resource, the developer is required to pay for measures that will mitigate (or offset) those impacts. Markets are created when regulators allow this mitigation to occur off-site, and allow a developer to purchase mitigation "credits" from another party.

TRADING AND OTHER REGULATORY PROGRAMS IN WATER RESOURCE MANAGEMENT

Regulators, the regulated community, environmental groups, and planners recognize the limitations of existing water resource regulatory programs, and new regulatory structures are emerging across the country. Many of these programs are attempting to include trading and other market-based approaches to maximizing environmental outcomes at the least cost. To date, the largest water trading programs involve water pollution trading programs, where a specific pollution load is managed through a market of tradable credits.

Total Maximum Daily Load (TMDL) analyses are usually used to set a "cap" in a water pollution trading program. The cap is based on the total load that a water body can receive without impairing its designated uses. TMDL studies can be used to allocate pollutant loads for both regulated and unregulated discharges in a watershed and can, in some instances, be used to integrate planning and permitting programs. "Watershed permitting" regulatory programs typically regulate a type of discharge (such as wastewater) based on the discharge's impacts on an entire watershed or subwatershed, and have been implemented in some states.

In the past several years, a number of programs have been initiated to integrate regulation of multiple sources of urban river impacts. The Tualatin River Watershed Permitting Project in Oregon, established to reduce thermal loads to the river, allows trading between wastewater treatment plants and projects that will increase riparian shading, augment streamflows, or modify streambed geomorphology to increase hyporheic exchange flow (the movement of water between the stream and the subsurface – the hyporheic zone) (Oregon Department of Environmental Quality, 2005). While this program uses thermal heat load as the regulatory basis for permitted trades, the nature of the activities that can be used to generate thermal credits create numerous benefits in addition to cooling.

In Cincinnati, Ohio, there is a proposal to establish a trading program in the Shepherd Creek watershed using "stormwater runoff credits" (U.S. E.P.A. 2006 b) The program depends on establishing a target stormwater runoff volume from developed parcels, based on the site's soils and topography. A Watershed Authority would be established to manage the market in credits, allowing property owners to determine, on a site by site basis, whether it is in their economic interest to build Best Management Practices (BMPs) on their site to meet their runoff volume requirement, to purchase credits, or to control more runoff than is required on their site, thereby creating credits which could be sold to others. A similar stormwater control offset program has been established in Vermont, in which new development, or redevelopment, is required to implement BMPs to control both water quality and peak flows, or to offset their impacts by paying an in lieu fee used to fund watershed restoration programs (Vermont Department of Environmental Conservation, 2004).

The Vermillion River Joint Powers Organization in Apple Valley, Minnesota is developing a new regulatory framework based on infiltration and riparian shading requirements (Vermillion River Joint Powers Organization, 2005). The program envisions creating a watershed-based set of zoning and planning requirements, and establishing National Pollutant Discharge Elimination System (NPDES) Phase II stormwater permit limits based on design-storm discharge rate and volume standards. NPDES permitees would have the option of meeting infiltration requirements set for their own site, or trading with other projects that would result in infiltration equal to that required on their own site.

THE CHARLES RIVER

The Charles River, in eastern Massachusetts, is typical of many urbanized watersheds across the country (see Figure 25.1). Its 308-square mile watershed includes 35 urban, suburban and exurban communities, as well as small areas of semi-rural and agricultural land use. Water supplies and wastewater management in the upper half of the watershed are dominated by local systems, with five wastewater treatment plants discharging treated effluent into the mainstem or tributaries of the Charles River. The lower half of the watershed is served by a large regional water supply and wastewater system, with supplies brought from the Quabbin and Wachusett reservoirs, and wastewater collected at the Deer Island Treatment Works and discharged to Massachusetts Bay. These water supply and wastewater systems significantly alter natural hydrologic regimes and impact water quality.

In the lower basin, in spite of a 10-year CSO control plan, there are still seven active CSOs that discharge directly into the river during large storm events, and six of them discharge untreated combined sanitary and stormwater flows. There are 20 dams along the river, including one at the river's mouth that eliminated tidal flows and created a lake in the estuary that once reached over ten miles up the river from Boston harbor. Saltwater intrusion into the lower river basin has created an

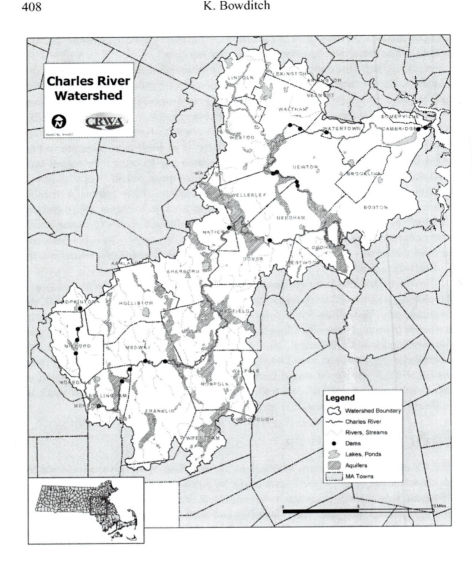

Figure 25.1.   The Charles River Watershed.

anoxic "dead zone" in the bottom of the water column in some areas, and many areas of the river have significant levels of accumulated toxic sediment.

In spite of these challenging circumstances, significant improvements in water quality have been made in the past decades. Once one of the nation's most infamous dirty rivers, the Charles River is now among the nation's cleanest urban rivers (US EPA Region 1, 2006 a). Most of the water quality improvements have been the result of permit-based regulatory control programs. Point source discharges have pollutant-based discharge requirements. The Long Term CSO Control Plan for the Charles River has been implemented through a federal court order as part of the Boston Harbor clean-up. In 1995, EPA initiated a rigorous program to eliminate cross connections between sanitary sewers and storm drains in nine cities and towns in the lower Charles River watershed. EPA's NPDES Phase II stormwater permitting program has also required improved stormwater management in municipal drainage systems.

Nevertheless, these programs target only some sources of pollution, and do not address many other activities that contribute to the significant impairments in the river. The Charles River still violates water quality standards in some locations after even moderate rainfall events, and nearly all of the river violates water quality standards following large rainfall events, even areas with no CSO discharge. Low base flows, especially in summer, severely impair aquatic habitat, and contribute to eutrophic conditions in many impounded segments of the river. In some areas of the Charles River, wastewater effluent comprises over 90% of the instream flow during average August flow conditions.

Watershed studies and restoration efforts increasingly point to the need to reduce the impacts of urbanization in order to restore the river. Rather than trying to address one problem area, such as nutrient loads, or CSOs, some studies support using natural hydrologic regimes as an overall watershed restoration target (Poff et al., 1997; Coffman, 2000; Postel and Richter, 2005). By managing water resources so that they closely mimic the natural flow regime of the region, numerous other water resource impairments, including water quality degradation, erosion and scouring, flooding and ecosystem impairment, can be reduced.

TOWARD A CHARLES RIVER FLOW TRADING PROGRAM

The natural flow regime of the Charles River is significantly altered. Of major concern, declining base flows contribute significantly to water quality impairments, and are a primary cause of habitat degradation and non-attainment of designated uses. One source of base flow decline is water supply withdrawals. In some areas of the river, these withdrawals are returned to the river as wastewater effluent downstream of the point of withdrawal, creating river segments that are significantly dewatered. In five towns in the mid-watershed, water supplies are withdrawn from local sources, but discharged into the Metropolitan Water Resources Authority (MWRA) regional wastewater system and into the Massachusetts Bay, creating a net loss to the Charles River of over 10 million gallons per day. Inflow and infiltration (I/I) into sewer pipes results in additional water losses from the basin.

Restoring Charles River base flows can be accomplished in part by limiting withdrawals, through strict conservation measures, and under certain circumstances, banning non-essential outdoor water uses. However, withdrawal limits may ultimately prove insufficient in a growing community if current infrastructure and management practices continue. In order to meet the goal of increasing base flows and to sustain long-term public water supply needs, it is necessary to undertake activities beyond restricting water supply pumping. If successful, these measures may actually reduce the need for limits on pumping, and will provide for both water supply and ecosystem health.

Given the flexibility and larger context that trading offers, CRWA has developed a conceptual regulatory model that uses trading as a tool to help restore declining base flows and to support long term viability of water resources. Base flow restoration projects typically have multiple benefits: restoring natural hydrologic regimes; reducing pollution loads; and protecting groundwater resources. Thus, other regulated programs that target water quality, can be used to generate base flow credits.

The Charles River Flow Trading Program (CRFTP), as currently framed, uses water withdrawal limits to drive trading. The program is based on the establishment of a total cap on withdrawals from surface or groundwater supplies in any subbasin. This cap would be established based upon the estimated pre-development natural base flow in each subwatershed. The cap would be implemented in the water supply permit.

In Massachusetts, all water supply withdrawals over 100,000 gallons per day are permitted through the Water Management Act (WMA) regulatory program. The WMA requires that the Massachusetts Department of Environmental Protection (DEP), who regulates the program, balance competing uses in their permitting process, creating an obligation that the department issue permits that limit pumping when a river is under stress due to reduced flow levels. In addition, pursuant to the WMA, DEP is prohibited from issuing permits that would result in an exceedence of the basin's safe yield. Massachusetts Water Management Policy has defined safe yield as the volume of water that can be removed from a basin without unreasonable damage to the water resource. These regulatory requirements provide the jurisdiction for DEP to impose a cap on withdrawals.

Under CRFTP, a cap on the amount of water that could be pumped from each well in a sub-basin in a given month would be established. This cap would be based on a sub-basin Water Balance analysis, and would be designed to reflect natural variations in flow. The cap would require a reduction of impacts to base flow, dependent on the level of stress in the sub-basin, phased in over a several year period. In order to pump more than the allotted cap for a given well in a given month, a water supplier would have to obtain base flow credits from another source in the same sub-basin, or from an upstream sub-basin. Potential sources of base flow credits include another well with excess capacity under its own cap; projects that recharge stormwater or treated wastewater; projects that reduce effective imperviousness in developed areas; and water releases from reservoir storage or other projects to augment base flows. The pumping cap would need to be phased in over time to allow the development of a market for groundwater credits.

In this model, the trading unit would be gallons of groundwater pumped, or gallons recharged in a sub-basin in a given month. Such a trade would not be perfectly equal. Recharge to groundwater generally does not increase base flows in the same location or the same month that well withdrawal impacts deplete base flows. A trading program should thus weigh, or credit, recharge differently depending on its location and the estimated season of impact. In a trade involving a recharge program where the impacts were significantly lagged, a three- or five-to one ratio might be needed. In a case where trading occurs between two pumping wells, reducing pumping from a well in a highly sensitive area might generate credits at a larger ratio.

This type of cap and trade approach would allow for direct trades, for example involving two wells in the same sub-basin; or between two different water utilities in a sub-basin, for example between a wastewater treatment plant and a well. It could also allow for third-party trades, for example, private developers, by building recharge of stormwater or wastewater into their projects, could sell the credits to the town.

The programs most likely to generate the necessary credits to support such trades are regulated under the NPDES program, and include wastewater treatment plant (WWTP) and stormwater discharge permits. While these programs do not regulate water quantity, the water quality requirements could drive water resource management projects that would increase groundwater recharge and/or reduce imperviousness.

Under current regulatory programs, low effluent limits typically drive direct trades between dischargers of that pollutant, and do not have significant impacts on streamflow or hydrology. However, if a market existed for base flow credits, projects that recharge water would have an additional advantage.

The WWTP permitting program would work as a potential driver if effluent limits, especially for nutrients, were significantly reduced from current permit limits. Existing limits on wastewater plants in the Charles River tend not to support trading programs because effluent limits are fairly high, and current technologies to achieve those limits are inexpensive. The Charles River is impaired by nutrients, however, especially phosphorus, and several TMDL analyses that are underway are likely to drive lower effluent limits in future permits.

NPDES stormwater permits, which are required for all small municipal storm drain systems, could also work as a catalyst for groundwater recharge and increased base flows. Like wastewater discharge permits, stormwater discharge permits are water quality permits. To meet these limits, a municipality seeks the most cost-effective programs to control pollutants, which may or may not include programs that would increase base flows and restore natural hydrologic function. However, if a trading program were in effect, the municipality might undertake programs to control stormwater runoff that generate base flow credits as well as meet permit requirements, such as projects to reduce impervious surfaces, recharge stormwater, daylight streams, construct wetlands or other approaches to improve recharge.

CHALLENGES

The CRFTP is still a conceptual model, and has not yet been implemented. Two primary objectives to move towards implementation of a trading program are to finalize the credit structure for base flow credit; and to identify and implement the regulatory structure for implementing and overseeing the program. CRWA continues to work closely with state and federal regulators on both of these efforts.

In addition to the challenges of implementing the trading program as currently structured, the program will also need to be modified in order to achieve its potential in the most urbanized sections of the Charles River watershed. Because the oldest, most urbanized municipalities in the Charles River watershed do not use local water supplies, the CRFTP will need to be modified to reflect the permit controls applicable in those communities. In these communities, the NPDES stormwater permit program, with recharge requirements similar to the program being developed in the Vermillion River watershed, is the most likely regulatory driver.

There is already some precedent for this approach. In Boston, where groundwater depletion has led to significant structural problems for buildings constructed in areas of fill, groundwater recharge has been established as a permit requirement for new and redevelopment programs in several zoning overlay districts. The Vermont stormwater offset program is recognized as a successful regional model for generating flexible, effective stormwater control options. Many state and municipal stormwater management programs require some type of stormwater controls for new and redevelopment projects.

CRWA is continuing to develop the CRFTP model for those towns with local water supplies, with a goal of expanding it to include regulated activities in municipalities without local water supplies. CRWA's assessment indicates that DEP has the regulatory authority it needs to implement a successful trading program based on base flow credits, driven by a cap placed on WMA withdrawal volumes, and supported by credits generated through multiple programs. Implementing a cross-regulatory program may require the development of a new regulatory body, similar to the Vermillion River Joint Powers Organization, or more aggressive promotion of the use of financing and regulatory mechanisms such as stormwater utilities.

Environmental trading is still relatively new in environmental management, and there are issues that an effective trading program must address. Trading programs must avoid creating "hot spots" where too much pollution occurs or localized flow is significantly altered. Lag times and local geology must be carefully considered as credit ratios are determined. The availability of trading programs should not short circuit alternatives analyses, encourage development in inappropriate locations, or reduce efforts to avoid and minimize the impacts of a project.

Because of uncertainties about the long term success of trading programs, the difficulties in monitoring and compliance review, and the legal and liability concerns, most trading programs still require significant regulatory oversight, and often have an expiration or reauthorization date. Higher ratios are often used to ensure a "measure of safety" for the impact that is being allowed through the trade.

For all trading partners, the risks of entering a trading program are real. Water resource trading programs require integrated planning, modeling and targets, and there are numerous uncertainties about the long term success of specific projects. Financial planning is difficult and all parties struggle somewhat to adapt to new regulatory programs. Most important, trading programs require willing players: regulators, the regulated community, the unregulated community, and environmentalists, all of whom must be willing to take risks and work together.

FLOW TRADING BENEFITS

Like most successful trading programs, trading to increase base flows will be most effective if it encourages flexible ways to accomplish a goal, allowing local hydrologic, geographic and economic circumstances to drive solutions. Strict water conservation, bans on outdoor watering, and high water fees will decrease per capita water demand and will help to reduce groundwater impacts from wells, but these unpopular measures may not be able to increase base flows as much as recharging treated wastewater or reducing infiltration into leaky sewer pipes. In addition, many of the trades that could be developed would have multiple benefits, improving water quality, reducing stormwater runoff, and even increasing open space.

In urban environments, where there are numerous causes of impacts, regulation of those activities that cause impacts is very uneven, and restoration is expensive, trading may provide a real opportunity to make another significant step forward in river restoration.

ACKNOWLEDGEMENTS

Charles River Watershed Association is grateful for the support of the United States Environmental Protection Agency (EPA), who awarded a three year Targeted Watershed Initiative Grant in support of CRWA's flow trading program. We also thank Chris Kilian at the Conservation Law Foundation for program and conceptual support. Original analysis and program development was assisted by Kathleen Baskin of the Massachusetts Executive Office of Environmental Affairs. CRWA received additional grant support for this project from an anonymous source.

REFERENCES

Coffman, L.S. (2000). Low Impact Development Design: A New Paradigm for Stormwater Management Mimicking and Restoring the Natural Hydrologic Regime: An Alternative Stormwater Management Technology. Proceedings, National Conference on Tools for Urban Water Resource Management and Protection. US EPA. February 7–10, 2000, Chicago, IL p. 158–167.

Deason, J.P. (2001). The Passaic River Restoration Provides a Nationwide Model for Addressing Polluted Urban Rivers Pollution Engineering, available from http://www.gwu.edu/ eemnews/fall2001-PollutionEngineeringArticle.htm.

FitzHugh, T.W. and B.D. Richter (2004). Quenching Urban Thirst: Growing Cities and Their Impacts on Freshwater Ecosystems. *BioScience* 54(8):741–54.

Kieser, M.S. and A.F. Fang (2004). Economic and Environmental Benefits of Water Quality Trading – An Overview of U.S. Trading Programs. Paper presented at the Workshop on Urban Renaissance and Watershed Management, Japan, 2004. Available from http://www.envtn.org/docs/Japan_paper.pdf

Kosco, J., W. Ganter, J. Collins, L. Gentile, and J. Tinger (2003). Lessons Learned From In-Field Evaluations of Phase I Municipal Stormwater Programs. National Conference on Urban Stormwater: Enhancing Programs at the Local Level. US EPA. February 17–20, 2003. Chicago, IL p. 191–197.

Oregon Department of Environmental Quality (2005). Issues First in the Nation Watershed-Based Permit to Clean Water Services, 2005. Available from http://www.deq.state.or.us/news/prDisplay.asp?docID=1494.

Otto, B., K. McCormick, and M. Leccese (2004). "Ecological Riverfront Design: Restoring Rivers, Connecting Communities." American Planning Association, Planning Advisory Service Report number 518–519, Chicago, IL.

Poff, N.L., Allan, J.D., M.B. Bain, J.R. Karr, K.L. Prestegaard, B.D. Richter, R.E. Sparks, and J.C. Stromberg (1997).The Natural Flow Regime: A Paradigm for River Conservation and Restoration. *BioScience* 47:769–784.

Poff, N.L., J.D. Allan, M.A. Palmer, D.D. Hart, B.D. Richter, A.H. Arthington, K.H. Rogers, J.L. Meyer, and J.A. Stanford (2003). River Flows and Water Wars: Emerging Science for Environmental Decision Making. *Frontiers in Ecology* 1(6): 298–306.

Postel, S. and B.D. Richter (2005). Rivers for Life: Managing Water for People and Nature: Island Press, Washington, DC.

US Environmental protection Agency (2000). The National Water Quality Inventory: 2000 Report to Congress, 2000. Available from http://www.epa.gov/305b/2000report/.

US Environmental Protection Agency (2002). Federal Water Pollution Control Act, (33 U.S.C. 1251 *Et Seq.*), As Amended Through P.l. 107–303, Available from http://www.epa.gov/r5water/cwa.htm.

US Environmental Protection Agency (2006). Massachusetts Water Resources Authority to Implement Sharp Reductions in Sewage Contamination of Charles River Press Release, 2006 . Available from http://yosemite.epa.gov/opa/admpress.nsf/1367c41957 02d16a85257018004c771c/36af973cd6fcf4728525715700532418!OpenDocument.

US Environmental Protection Agency (2006). Using Tradable Credits to Control Excess Stormwater Runoff, Available from http://www.envtn.org/docs/tradeflows.pdf

Vermillion River Joint Powers Organization (2005). Creating the Optimal Regulatory and Market Framework to Preserve Stream Flow and Temperature Stability in an Urbanizing Trout Stream in the Midwest. Available from http://www.epa.gov/owow/watershed/initiative/2005proposals/05vermillion.pdf

Vermont Department of Environmental Conservation (2004). VTDEC Procedure for Evaluation of Stormwater Discharges and Offsets in Stormwater Impaired Watersheds. Available from www.anr.state.vt.us/dec/waterq/ stormwater/docs/sw_rule_impaired.pdf

Appendix

WINGSPREAD WORKHOP SPEAKERS LIST

NAME	ADDRESS
Adler, Robert	S.J. Quinney College of Law, The University of Utah, 322 South 1400 East, Salt Lake City, UT 84112–0730
Ahern, Jack	Department of Landscape Architecture and Regional Planning, University of Massachusetts 109 Hills Rd., Amherst, MA 01003
Baker, Lawrence A.	University of Minnesota, 173 McNeal Hall, 1985 Buford Ave. St.Paul, MN 55108
Betrand-Krajewski, Jean Luc	URGC Hydrologie Urbaine INSA de Lyon, 34 Avenue des Arts 69621 VILLEURBANE Cedex, FRANCE
Bledsoe, Brian	Engineering Research Center, Colorado State University, Fort Collins, CO 80523
Bowditch, Kate	Charles River Watershed Association, 190 Park Road, Weston, MA, 02493
Braden, John B.	Department of Agricultural and Consumer Economics, University of Illinois, 1301 W, Gregory Drive, Rm. 304A Urbana, IL 61801
Breckenridge, Lee	Northeastern University, 58 Cargil Hall, 400 Huntington Avenue, Boston, MA 02115
Brezonik, Patrick	Division of Bioengineering and Environmental Systems, National Science Foundation, 4201 Wilson Blvd, Arlington, VA 22230
Brown, Paul	CDM, 1925 Palomar Oaks Way, Suite 300, Carlsbad, CA 92008

Englande, Andrew J., Jr.	Environmental Health Sciences, Tulane University 1440 Canal Street, Suite 2133, New Orleans, LA 70112
Furumai, Hiroaki	Department of Urban Engineering, University of Tokyo, 7-3-1 Hongo, Bunkyo-Ku, Tokyo, 113-8656, Japan
Goodman, Iris	U.S. EPA, National Center for Environmental Research, 1200 Pennsylvania Ave., NW (8723F), Washington, DC 20460
Heaney, Jim	Environmental Engineering Sciences, P.O. Box 116450, University of Florida, Gainesville, FL 32611-6450
Hill, Kristina	Program in Landscape Architecture, 230 Cambell Hall, The University of Virginia, P.O. Box 400122, Charlottesville, VA 22904-4122
Labadie, John	Colorado State University, Department of Civil Engineering, Fort Collins, CO 80523-1372
Lanyon, Richard	Metropolitan Water Reclamation District of Greater Chicago, 100 East Erie St. Chicago, IL 606112803
Maimone, Mark	CDM, 100 Crossways Park Drive W. Woodbury, NY 11797
Malmqvist, Per-Arne	Chalmers University of Technology Department of Civil and Environmental Engineering SE-412-96 Göteborg, Sweden
Marsalek, Jiri	Urban Water Management Program, National Water Research Institute, Canada Centre for Inland Waters, 867 Lakeshore Rd., PO Box 5050, Burlington, Ontario L7R 4A6, Canada
Novotny, Vladimir	Department of Civil and Environmental Engineering, Snell 429, Northeastern University, Boston 02115
Rankin, Ed	ILGARD/Midwest Biodiversity Institute, The Ridges, Bldg 22, Athens OH 45701
Reiter, Paul	Exec. Director, International Water Association, Alliance House, 12 Caxton Street, London, SW1H 0QS, United Kingdom
Roesner, Larry	Department of Civil Engineering, Colorado State University, Fort Collins, CO 80523-1372
Shanahan, Peter	Department of Civil and Environmental Engineering, MIT, 77 Mass. Ave 1-280, Cambridge, MA-02139
Soyster, Al	Director, Division of Engineering Education and Research Centers, National Science Foundation, 4201 Wilson Blvd., Arlington, VA 22230
van Heerden, Ivor	Hurricane Center Suite 3513 CEBA Building, Louisiana State University – Baton Rouge, LA 70803
Welty, Claire	Center for Env. Research and Education UMBC, Technology Research Center 102, 1000 Hilltop Circle, Baltimore, Maryland 21250
Yamada, Kiyoshi	Department of Environmental Systems Engineering, Ritsumeikan University, Noji, Kuisatsu City, Shiga, 525 Japan

PANEL EXPERTS

NAME	ADDRESS
Haas, Charles	Drexel University, College of Engineering, Department of Civil, Architectural and Environmental Engineering 3141 Chestnut Street, Philadelphia, PA 19104
Hao, X.D.	Center for Sustainable Environmental Biotechnology, Beijing Institute of Civil Eng, and Architecture, 1 Zhanlanguan Road, Xicheng District, Beijing 100044, P.R. China
Hranova, Roumiana	Civil Engineering Department, University of Botswana P. Bag 0061, Gaborone, Botswana
Lucey, Patrick	Aqua-Tex Scientific Consulting Ltd., 390 7th Avenue, Kimberley, B.C. Canada VIA2Z7
Lue-Hing, Cecil	6815 County Line Lane, Burr Ridge, IL, 60527
Magruder, Chris	Milwaukee Metropolitan Sewerage District, 260 West Seeboth Street, WI 53204
Morea, Susan	CDM Colorado, 1331 17th Street, Suite 1200, Denver, CO 80202
Nelson, Valerie	Coalition for Alternative Wastewater Treatment, P.O. box 7041, Gloucester, MA 01930
O'Reilly, Neal	Hey and Associates, Inc., 240 Regency Court, Suite 301, Brookfield, WI 53045
Smullen, James T.	CDM, Raritan Plaza I, Suite 300 Raritan Center Edison, NJ 08818

Index

LaVergne, TN USA
12 December 2009
166829LV00001B/51/A

9 781843 391364